# Developing and Managing Embedded Systems and Products

# Developing and Managing Embedded Systems and Products

## Methods, Techniques, Tools, Processes, and Teamwork

Kim R. Fowler

Craig L. Silver

AMSTERDAM • BOSTON • HEIDELBERG • LONDON
NEW YORK • OXFORD • PARIS • SAN DIEGO
SAN FRANCISCO • SINGAPORE • SYDNEY • TOKYO
Newnes is an imprint of Elsevier

Newnes is an imprint of Elsevier
The Boulevard, Langford Lane, Kidlington, Oxford OX5 1GB, UK
225 Wyman Street, Waltham, MA 02451, USA

**Notice**
No responsibility is assumed by the publisher for any injury and/or damage to persons or property
as a matter of products liability, negligence or otherwise, or from any use or operation of any
methods, products, instructions or ideas contained in the material herein. Because of rapid
advances in the medical sciences, in particular, independent verification of diagnoses and drug
dosages should be made.

**British Library Cataloguing-in-Publication Data**
A catalogue record for this book is available from the British Library

**Library of Congress Cataloging-in-Publication Data**
A catalog record for this book is available from the Library of Congress

ISBN: 978-0-12-405879-8

For information on all Newnes publications
visit our website at http://store.elsevier.com/

Printed and bound in the United States of America

15  16  17  18  19    10 9 8 7 6 5 4 3 2 1

# Contents

# List of Contributors

**Kim R. Fowler**  IEEE Fellow, Consultant

**Allison Fritz**  Organization Development and Training Consultant, The Johns Hopkins HealthCare LLC

**Michael F. (Mike) Gard**  Senior Product Design Engineer The Charles Machine Works Perry, OK, USA

**Robert Oshana**  Director, Software Research and Development, Digital Networking, Freescale Semiconductor

**Geoff Patch**  CEA Technologies Pty. Ltd.

**Craig L. Silver**  Director—Strategic Initiatives/General Counsel, Amches, Inc.

**Eugene Vasserman**  Kansas State University

**Tim Wescott**  IEEE Senior Member, Owner, Wescott Design Services

**Steve Zeise**  Aerospace Electronics Industry

# *About the Editor*

**Kim R. Fowler** has spent over 30 years in the design, development, and project management of medical, military, and satellite equipment. His interest is the rigorous development of diverse, mission-critical, embedded systems. He co-founded Stimsoft, a medical products company, in 1998 and sold it in 2003. He also has worked for JHU/APL designing embedded systems, for a company now part of Curtiss-Wright Embedded Computing that built digital signal processing boards, and consulted for both commercial companies and government agencies. He is a Fellow of the IEEE and lectures internationally on systems engineering and developing real-time embedded products. He has been president of the IEEE Instrumentation & Measurement society and an adjunct professor for the Johns Hopkins University Engineering Professional Program. He has published widely and has written three textbooks—this book is his fourth. He has 18 patents—granted, pending, or disclosed. He is currently a graduate student in Electrical Engineering at Kansas State University to finally get his PhD.

# Co-Author Biography

**Craig L. Silver** has over 30 years of diverse legal experience for private, commercial, start-ups, and nonprofit entities, serving as litigation counsel for complex commercial disputes, constitutional law claims, aviation torts, and criminal defense. Having served as a general counsel, general manager, and president for defense electronic firms and telecommunications companies, he is experienced with technology companies dealing with software licensing issues, IP protection, and international transactions. He has previously published for the IEEE—"Silver Bullets" and has worn various hats in high technology companies that include business development and  technical liaison with field application engineers. He holds a BA degree from the Universityof Maryland and a JD degree from George Mason University Law School. He is a licensed pilot and ham radio operator. He is married and resides in Maryland, USA.

# Author's Biographies

## Chapter Authors

**Allison Fritz** is an Organization Development and Facilitation
professional with over 20 years' experience in a variety of industries.
Presently working as a Sr. Organization Development and Training
Consultant with the Johns Hopkins Health System, she has also
worked within higher education, the petroleum industry, and
independent consulting, serving both Fortune 100 and small
business, designing and facilitating processes. Allison's expertise is
in team development, change management, leader development,
strategic visioning, and coaching. With 14 years in management
roles, she applies her experience to her work. Allison has a doctoral

degree in Organization and Staff Development from the University
of Maryland College Park, a Master's degree in Counseling and Student Personnel, and a
Bachelor's degree in Communications and Psychology from the University of Delaware;
as well as holds several certifications including, *Emotional Intelligence (EQ2.0, 360),
Crucial Conversations*, Strong Interest Inventory, and MBTI. Allison focuses her work on
encouraging leaders, teams and organizations to realize positive change.

**Michael F. (Mike) Gard,** received his BSEE from Kansas State
University, MSEE (Interdepartmental Program in Biomedical
Engineering) from Washington University in St. Louis, and
PhDEE (Geophysics minor) from Southern Methodist University.
He has over 40 years of industrial experience in aircraft, medical
equipment, clinical engineering, petroleum, and construction
industries. He is presently Sr. Product Design Engineer at The Charles

Machine Works, Perry, OK. An adjunct professor, he occasionally
teaches at Oklahoma State University. He is a registered professional
engineer, patent agent, inventor (34 US patents), author, member of
the IEEE Instrumentation and Measurement Society's Administrative Committee, and
editor-in-chief of *IEEE Instrumentation and Measurement Magazine*. His technical interests
include real-time data acquisition and precision analog and analog/digital systems for low
power and hostile environments.

**Robert Oshana** has 30 years of experience in the software industry, primarily focused on embedded and real-time systems for the defense and semiconductor industries. He has BSEE, MSEE, MSCS, and MBA degrees and is a senior member of IEEE. He is a member of several Advisory Boards including the Embedded Systems group, where he is also an international speaker. He has over 200 presentations and publications in various technology fields and has written several books on Embedded software technology including "Software Engineering for Embedded Systems." He is an adjunct professor at Southern Methodist University where he teaches graduate software engineering courses. He is a distinguished member of Technical Staff and Director of Global Software R&D for Digital Networking at Freescale Semiconductor.

**Geoff Patch** has over 30 years experience as a software engineer. He has worked for the Australian government, in academia, and for a number of engineering companies. Since 1987, he has specialized in embedded systems, primarily in the areas of radar target tracking, radar signal processing, and command and control systems. He is also keenly interested in software process improvement, technical team leadership, and technical management. He has developed software for numerous commercially successful radar systems ranging from conventional maritime surveillance, through specialized applications such as submarine periscope detection and up to large air defense systems. He is currently the manager of a team of nearly 30 software engineers involved in the development of new radar systems at CEA Technologies in Canberra, Australia.

**Eugene Vasserman** received his PhD and master's degrees in Computer Science in 2010 and 2008, respectively, from the University of Minnesota. His BS, in Biochemistry and Neuroscience with a Computer Science minor, is also from the University of Minnesota (2003). In 2013, he received the NSF CAREER award for work on secure next generation medical systems.

**Tim Wescott** has 25 years of real-world experience in embedded systems design, with roles ranging from software designer to circuit designer to systems architect. Tim has worked on small, inexpensive hand-held instruments, on large airborne imaging systems, and on nearly everything in between. He has experience in all phases of system life cycles, ranging from designing new systems from a clean sheet of paper to extending the useful lives of systems that are on the verge of obsolescence. Tim is author of "Applied Control Theory for Embedded Systems", aimed at engineers who slept through control theory class in University, and who now need to design a system that must successfully implement a feedback control loop. Tim is the owner of Wescott Design Services, which provides analysis, design, and troubleshooting of embedded control systems, with a particular emphasis on control of dynamic systems, low-level communications systems, and metrology. Wescott Design Systems has helped customers of all sorts of problems ranging from drives for 1/2-inch diameter brushless motors to implementing communications systems for deep-well drilling platforms.

**Steve Zeise** is a mechanical engineer and designer with 30 years' experience in all things mechanical. He received a BS in Mechanical Engineering from Rose-Hulman Institute of Technology and immediately went to work for Westinghouse Defense and Electronics Systems Center designing mechanisms, structures, and cooling systems supporting embedded systems in night vision cameras. With positions at Northrop-Grumman and Lockheed Martin, he gained experience in structural analysis and environmental testing. He is currently with FLIR Systems where he helped to setup a small R&D facility in Orlando, FL and for the past 15 years has worked to help FLIR Systems solve complex vibration problems.

## Case Study Authors

**David von Oheimb** received his PhD in computer science in 2001 from the Munich University of Technology, where he focused on machine-assisted formal modeling and verification of the programming language Java. He joined Siemens Corporate Technology, where he became a senior researcher, developer, and key expert consultant on IT security. His specific areas of expertise are security architecture, formal analysis, and IT security certification according to the Common Criteria. He has been involved as participant and leader of various Siemens-internal and EU-funded R&D projects on security protocol and information flow analysis using model checkers and theorem provers and of various industrial projects dealing for instance with Infineon smart cards, software update mechanisms for Boeing and Continental Automotive, and German and Austrian smart metering systems.

**Kenneth W. Tobin** is the Director of the Electrical and Electronics Systems Research (EESR) Division at the Oak Ridge National Laboratory (ORNL), Oak Ridge, Tennessee, USA, where he has been working in various R&D and leadership capacities since 1987. The EESR Division is composed of 150 staff who perform R&D in electronics, sensors, communications, and controls for energy efficiency, resiliency, and security. His personal research areas encompass photonics, neutronics, x-ray, SEM, electronic imaging and microscopy coupled with signal processing and machine learning. Science and technology specialty in computational imaging, image metrology, object segmentation, and feature generation from multi-spectral, multi-source imagery for inverse imaging, robust human-level classifiers, image archival and retrieval applications, and image-based informatics. Dr. Tobin was named an ORNL Corporate Research Fellow in 2003 for his contributions to the field of applied computer vision research. He has authored and co-authored over 164 publications and he currently holds fourteen U.S. Patents in areas of computer vision, photonics, radiography, and microscopy. Dr. Tobin is a Fellow of the Institute of Electrical and Electronics Engineers (IEEE) and a Fellow of the International Society for Optics and Photonics (SPIE), where he is currently an Associate Editor for the Journal of Electronic Imaging. Dr. Tobin has a Ph.D. in Nuclear Engineering from the University of Virginia, an M.S. in Nuclear Engineering from Virginia Tech, and a B.S. in Physics also from Virginia Tech.

**Dwight A. Clayton** is the group leader of the Electronic and Embedded Systems group at the Oak Ridge National Laboratory (ORNL), Oak Ridge, TN. The mission of the Electronic and Embedded Systems (EESG) group is to apply modern electronic methods to provide solutions to challenges that are important to the ORNL, the Department of Energy, other federal agencies, and private industry. He joined ORNL in 1983 as a development staff member in the Instrumentation and Controls Division. In 1994, he was named leader of the Electronic and Embedded Systems Group. Since 2000, the innovative efforts of the Electronic and Embedded Systems group have resulted in the receipt of four R&D 100 awards. He has an MS and BS in electrical engineering from Tennessee Technological University.

**Bogdan Vacaliuc** is a research and development staff member in the Electronic and Embedded Systems Group of the Oak Ridge National Laboratory's Measurement Science and Systems Engineering Division. His entrepreneurial career has spanned several small and medium-size startup companies developing products in signal intelligence, telecommunications, visual image processing, and consumer electronics manufacturing. Prior to joining Oak Ridge National Laboratory in 2009, he served as Chief Technical Officer for Sundance DSP, Inc., a maker of modular hybrid signal processing computing hardware for portable and military applications. He emigrated to the United States in 1973 from Romania and earned bachelor and master's degrees in Electrical Engineering from Northwestern University in 1990 and 1992, respectively.

**Lee Barford** is master scientist at Keysight Laboratories and professor of Computer Science and Engineering (adjunct) at the University of Nevada, Reno, NV. He leads Keysight's efforts in applying parallel computing to speed electronic measurements. He also leads research to identify and apply emerging technologies in software and applied mathematics to enable new kinds of measurements and increase measurement accuracy and speed. His work has been used to improve R&D productivity and reduce manufacturing cost in the leading companies in the technology and transportation industries, including Apple, Boeing, Cisco, Ford, HP, Microsoft, and NASA. Previously, he managed a number of research projects at Agilent and Hewlett-Packard Laboratories, for example, in visible light and X-ray imaging systems, calibration methods for nonlinear and dynamical disturbances, and fault isolation from automatic test equipment results. He is an author of over 40 peer-reviewed publications and an inventor of approximately 60 patents.

**Hong-Liang Xu** is a senior research engineer at Keysight Laboratories. He joined Agilent Laboratories in 2007 after earning a master's degree from Beijing University of Posts and Telecommunications. In the field of parallel computing, he focuses on the methods to accelerate DSP and measurement algorithms on common platforms, like multicore CPUs and general-purpose GPUs. His recent work has included a demo of a purely software defined LTE base station with industry partners, including IBM and China Mobile Research Institute, demonstrating the DSP capabilities of multicore CPUs in the handling of wideband wireless communication protocols.

**Chun-Hong Zhang** is a scientific research staff member at Keysight Laboratories. He focuses on how to efficiently parallelize digital signal processing algorithms on heterogeneous computing platforms and optimize the parallel algorithms based on the advanced parallel computing features provided by different platforms. Previously, he did research projects on high-efficiency, nonreference digital voice and video quality assessment at Agilent Laboratories.

**Jake Brodsky** has been practicing the art of Control Systems and SCADA Engineering at the Washington Suburban Sanitary Commission for over 28 years. He intends to continue practicing until he gets it right. He is a registered professional engineer of control systems, a ham radio enthusiast, an instrument rated private pilot, a firearms instructor, and an amateur beer brewer—but not all at the same time. He co-founded and moderates the SCADASEC email list, he is co-author and co-editor of the Handbook of Control System/SCADA Security, published by CRC Press, and was recently re-elected Chair of the DNP Users Group.

**Daryl Beetner** is a professor of Electrical and Computer Engineering at the Missouri University of Science and Technology (formerly called the University of Missouri—Rolla). He received his BS degree in Electrical Engineering from Southern Illinois University at Edwardsville in 1990. He received an MS and DSc degree in Electrical Engineering from Washington University in St. Louis in 1994 and 1997, respectively. He conducts research with the Electromagnetic Compatibility Laboratory at Missouri S&T on a wide variety of topics including EMC of integrated circuits, EMC within embedded systems, and detection and neutralization of explosive devices. He is an associate editor for the IEEE Transactions on Instrumentation and Measurement.

**Natalia Bondarenko** received the BSc degree in Computer Science from Tbilisi State University, Georgia, Europe, in 2006, and received the MSc degree in Electrical and Electronics Engineering from the same university in 2009. Since 2009, she has been pursuing her PhD degree in Electrical Engineering in the EMC Laboratory at the Missouri University of Science and Technology, Rolla, MO. From 2005 to 2009, she was with EMCoS, Ltd., working on various research/consulting projects for automotive EMC. Her research interests include EM modeling and EMC/EMI measurements methods.

**Peng Shao** received the BS degree in Physics from Nanjing University, China, in 2006, received the MS degree in Physics from Missouri University of Science and Technology, USA in 2008, and received the MS degree in Electrical Engineering with the EMC Laboratory from the same university in 2011. Since 2011, he has been working with Cisco Systems as a hardware engineer in the signal integrity area.

**Tom Van Doren** has conducted research and education in electromagnetic compatibility for the past 31 years. More than 19,000 engineers and technicians from 108 companies and government agencies have attended his "Grounding and Shielding" and "Circuit Board Layout" courses. He has received two Outstanding Teacher Awards from Missouri S&T, the Richard R. Stoddard award from the IEEE EMC Society for contributions to EMC technology and education, and he is a life fellow of the IEEE and Honored Life member of the EMC Society. Much of his professional work has been devoted to helping engineers understand, diagnose, and reduce signal integrity and electrical interference problems. He can be contacted at vandoren@mst.edu or at 573-578-4193. The Van Doren Company website is www.emc-education.com.

# Developing and Managing Embedded Systems and Products: The Roadmap

Kim R. Fowler

*This book is for the entire project team!* The book's material contains best practices for developing embedded systems, which includes technical design, teamwork, collaboration, management attitudes, development processes, and legal liabilities. The book addresses these various topics because project development is not just a technical endeavor; it is a human endeavor.

The chapters present in a sequential fashion, but development of an embedded system is anything but sequential. Figure 1, below, is a repeat of Figure 1.2 from Chapter 1 and placed here for your convenience. The figure lists the chapters and when they might be most useful within a project. The multiple paths and connections indicate the parallel and concurrent nature of design and development efforts.

## Chapter 1: Introduction to Good Development

Chapter 1 describes the book's purpose to identify important issues in developing and managing embedded systems. The material outlines the technical aspects, the teamwork, the effort, and the cost you encounter in developing an embedded system. It reveals how technical issues impact business and schedules, how personal interactions are just as important as technical breakthroughs, and the interplay of various disciplines to realize the final product. Consequently, *this book is not solely for managers, it is written for every member of the project team.*

## Chapter 2: Drivers of Success in Engineering Teams

While technical work and knowledge are at the core of the engineering team, its success can rest equally on how well the team addresses the human aspects of daily activity. Organizational culture, team dynamics, and individual responses play significant roles in the

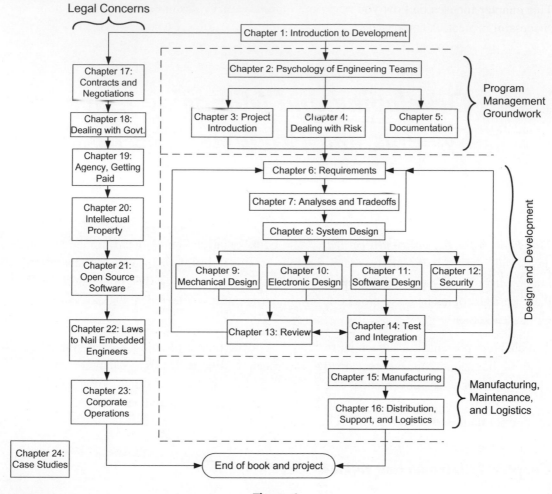

**Figure 1:**
Book organization and a suggested approach to reading it.

outcomes of the team's efforts. This chapter offers an overview of concepts and theories that relate to the functioning of the engineering team. It addresses the role of the team member, the role of the team leader, engagement, team development, dialogue, emotional awareness, and handling conflict to encourage leaders and members to pursue further study and skills training.

## Chapter 3: Project Introduction

What you do in the beginning of a project, sets precedence for the remainder of the project. Start on the right foot.

This chapter focuses on how you begin the long process of designing and building a successful project. It emphasizes the following:

- communicate the vision, mission, goals, and objectives;
- establish the "who, what, when, where, why, and how...";
- define the roles of the team and within the team;
- provide a clear communications plan;
- make the business case; and
- then establish the project plan, the administration, and the resources for the project.

## Chapter 4: Dealing with Risk

Risk is the potential to stray outside the defined cost, schedule, performance, or safety constraints. Every project is risky to some degree but risk can be managed. The chapter follows this format:

- first, identify the various risks and analyze them;
- then control risks by reducing, constraining, or transferring them;
- assess the state of the risks by analyzing them again; and
- repeat this cycle throughout the project's development.

## Chapter 5: Documentation

Documentation conveys the right information to the right person at the right time. Documentation is corporate communication and memory. Everyone working on a technical project spends 50–80% of their time preparing documentation.

This chapter has a number of suggested outlines and templates of documents. It also has examples of content within important documents.

## Chapter 6: System Requirements

This chapter will give readers a number of best practices to improve the quality of the requirements elicitation and development process in their organization. Formulation of high-quality requirements (complete, concise, accurate, modular, prioritized, analyzed, verified, and testable) reduce project risk, improve product quality, and allow for effective control of requirements volatility, which increases the likelihood of a successful project.

## Chapter 7: Analyses and Tradeoffs

Analyses and tradeoffs are absolutely necessary to the success of product design and development. Analyses form the feedback between requirements and design. Analyses gauge how well a design meets the requirements. If a design does not match the requirements, then either the design or the requirements need modification.

This chapter provides the motivation for analyses with the business case. It then discusses tradeoffs you might make to produce a design concept. Finally, it discusses a number of different analyses that you might use to refine a design.

## Chapter 8: The Discipline of System Design

This chapter shows you how to design the whole system by starting with goals in plain language that a business person might use. Then it proceeds through detailed system design by the team, a process that avoids either redundant effort or missing pieces.

Areas of particular focus in this chapter include:

- how to communicate effectively with nonengineering personnel,
- how to partition a large system into manageable subsystems,
- how to partition system requirements into requirements for individual disciplines, i.e., software, electronics, and mechanical engineering, and
- how to control costs for projects of various sizes and production volumes.

## Chapter 9: Mechanical Design

The chapter reviews basics of mechanical design to help nonmechanical staff communicate more effectively with their mechanical engineering colleagues. It addresses the fundamentals of packaging and thermal design so that the project manager may guide product development. The chapter goes on to discuss mechanisms; it focuses on robust design and methods for calculating loads and forces. The chapter looks at analysis and test, discusses how to use Finite Element Analysis effectively, and ends with a simplified approach to solving vibration problems.

## Chapter 10: Electronic Design

The chapter considers basics of electronic design allowing project leaders or nonelectrical engineers to communicate effectively with their colleagues in the electrical and electronic arena. Electronic circuit design involves the selection and interconnection of physical devices in a variety of topologies to meet performance specifications, environmental requirements, power and cost budgets, operating life requirements, and other design constraints in agreement with an overall schedule.

The chapter begins with basic components—resistors, capacitors, inductors, transistors, and display devices. Then it moves on to integrated circuits, in particular processors and controllers, and then moves on to more complex modules like solid-state relays. It covers circuit boards, connectors, cables, and conductors, topics often viewed with some disdain. It also discusses some issues with power supplies because so many folks get these wrong so often. It presents some analysis methods and test considerations and finishes with tradeoffs between various concerns.

## Chapter 11: Software Design and Development

Software development for embedded systems is not like software development for desktop systems. This chapter provides detailed descriptions of the key differences between the two domains. It examines operating system support, real-time requirements, resource constraints, and safety. It then provides advice about tools and techniques to develop embedded system software, with the primary emphasis on processes that help the embedded system developer to reliably and repeatedly produce high-quality embedded systems. The advice is based upon techniques that have developed over many years, and have been proven to work effectively in highly complex real-world military and industrial applications.

## Chapter 12: Security

Historically, system security has received little attention in embedded systems. Now embedded systems play a huge role in the operation of our modern world; many of these devices expose their presence on the Internet or through other means. Security in embedded devices is a daunting task—the challenging problem of security is compounded by resource limits which are far more restrictive in embedded systems than in desktop systems.

Security is defined according to the application. This chapter points to aspects of those applications and their security concerns.

## Chapter 13: Review

Review is a feedback path within system development. The act of review has two primary objectives: (1) to confirm correct design and development, and (2) to expose and identify problems with design, development, or processes.

This chapter outlines the processes and procedures for various types of reviews, provides the formats for review, and describes when to conduct reviews. It gives examples and templates for minutes, action items, and agendas. It has a checklists for preparing and debriefing a formal review and for basic topics to cover in formal design reviews.

## Chapter 14: Test and Integration

Test and integration, in this chapter, refer to verifying and validating the design and the development effort. Test and integration, coupled with design review, are the primary feedback activities to assure the project goals. The goals for test and integration are to analyze for design readiness, discover and characterize system behaviors, discover the product's limits, and prevent and constraint defects.

## Chapter 15: Manufacturing

Manufacturing is the transformation of design concept into physical realization. It converts energy, materials, labor, and thought into tangible products. Manufacturing is arguably the most obvious and necessary step for ideas to make money.

This chapter aims to make you aware of some the aspects of manufacturing that affect embedded systems. It will introduce manufacturing issues in the following areas:

- Circuit boards
- Wiring and cabling
- Enclosures
- Mechanisms

It will also give examples of how materials, fabrication of components, and assembly of systems interact within manufacturing. These examples should illuminate some of the time and effort involved in producing embedded systems.

## Chapter 16: Logistics, Distribution, and Support

Logistics manages the flow of product and information from manufacturing to the customer. Logistics integrates packaging, inventory, transportation, warehousing, delivery, and technical sales support. Everyone of these arenas fold into the ultimate management of the development of embedded systems; their operations and parameters directly affect the final cost of the system.

The chapter addresses distribution logistics, support, maintenance and repair, and disposal. Distribution logistics considers the delivery of the finished products to the customer and focuses on order processing, warehousing, and transportation. Support is the timely distribution of correct information to implement or fix the function of the embedded system. It also is the intangible perception that the supplier knows the situation and is there to help customer.

## Chapter 17: Agreements, Contracts, and Negotiations

This chapter addresses some salient ways that contracts are interpreted, gives particular attention to the conduct of the parties, and mentions the details associated with the signing of agreements. It mentions several types of contracts germane to the embedded engineer, such as NDAs and MOUs. Some consideration is given to various styles of negotiation and some suggested best ways to negotiate, such as the role of humility in negotiations.

## Chapter 18: Dealing with the Government

This chapter outlines the basics of government contracting and touches upon the important Christian Doctrine as it applies to the government. Additionally, it addresses changes to contracts, as well as termination procedures as they are uniquely applied in the government contracting context. It covers ethical concerns, as well as relevant statutes with a mention of the government contracting defense as it applies to possible tort litigation.

## Chapter 19: Agency and Getting Paid

This chapter outlines the concepts around creation of agency, agents operating outside the scope of their agency, and the duties of the principal once he learns of an agent exceeding the scope of the agency. The Getting Paid section highlights concerns about getting paid and gives practical advice regarding how larger companies manage their accounts payable. The chapter also outlines methods for getting paid on international accounts such as using Letters of Credit. Lastly, it has a section that discusses the esoteric way that imminent debtor bankruptcy can interfere with payment for services rendered.

## Chapter 20: Intellectual Property etc.

This chapter discusses source code licenses and rights of the bankruptcy trustee to interfere with the license. It addresses the scope of a software license in general is. The chapter also addresses the protection of intellectual property and copyrights; what constitutes a trade secret as well as the protection of thereof; what is a trademark and the enforcement of a trademark under the Lanham Act. Regarding patents, the chapter addresses what is patentable especially in regard to software patents and how the patent office is acting on software patents. There is a section on patent trolls and how to deal with them, as well.

## Chapter 21: Open Source Software

This chapter gives a history behind the open source initiative and describes the general licensing arrangement associated with open source software. It describes the various attributes of the GPL, as well as the various types of open source licenses that are in use. It describes considerations of the DMCA, as well as public domain and shareware licensing concerns and litigation that can possibly ensue.

## Chapter 22: Laws That Can Nail Embedded Engineers

This chapter outlines concerns of the DMCA and how it can be applied to the embedded engineer. It also discusses various criminal statutes that can be applied to the embedded engineer's activities. Moreover, the chapter addresses civil liability that can be charged to the engineer in the form of negligence or products liability and further addresses the way the engineer can protect himself from liability such as employing the use of indemnification agreements.

## Chapter 23: Corporate Operations, Export, and Compliance

This chapter addresses ways to maintain your corporate charter, the issuance of shares of stock, aspects of LLCs, and the hiring of out of state persons. The chapter also addresses some aspects of exporting particular to embedded devices, such as high-performance computing, as well as the antiboycott laws. It mentions the Foreign Corrupt Practices Act and various export controls on Cryptography. It addresses international arbitration clauses along with insurance and the international CB Scheme as it applies to compliance issues.

## Chapter 24: Case Studies

*Case Study 1:* Describes the development of a portable chemical and biological mass spectrometer for use on the battlefield.

*Case Study 2:* Describes a real-time radar simulation environment designed to test and validate radar systems for the U.S. Army. Staff at Oak Ridge National Laboratory integrated commercial off-the-shelf technologies with unique components, software, and control strategies to produce one-of-a-kind solutions to challenging measurement environments.

*Case Study 3:* Manufacturing test of digital television equipment and commissioning and maintenance of digital television systems requires instrumentation specialized for each transmission standard. This instrument measures both RF and the quality of the specific digital modulation. The instrument operates in a streaming mode, that is, the incoming video signal is processed without any temporal gaps.

*Case Study 4:* This is a tale of diagnosing a failed boiler control board. The board had erratic firmware behavior, the company had poor Internet presence, and the boiler manufacturer had a bad manual. This is the hazard of modern controls: unless you give a control narrative and define exactly what a system is supposed to do at every step of the way and what stimuli it expects, there is no way to know that anything you have is in specification or that it is even broken.

*Case Study 5:* Electromagnetic compatibility (EMC) problems must be solved before they hit the market. Debugging these issues can be challenging. This case study follows a problem on a display module used in transportation systems. Even very low emissions from the display module could prevent a GPS receiver in the vehicle from receiving a strong signal. The study showed how they found and eliminated the primary source of 1.2 GHz noise emissions.

# List of Acronyms

| | |
|---|---|
| AC | Alternating current |
| AFD | Arc fault detection |
| API | Application programming interface |
| APP | Adjusted peak performance |
| ASR | Alternative systems review |
| ASTD | American Society for Training and Development |
| ATE | Automatic test equipment |
| ATM | Automated teller machine |
| BIST | Built-in-self-test |
| BIT | Built-in-test |
| BITE | Built-in-test-equipment |
| BOM | Bill of materials |
| CAN | Controller area network |
| CB | Competent body |
| CBMS | Chemical and biological mass spectrometer |
| CBTL | Competent body test lab |
| CCB | Change control board |
| CDR | Critical design review |
| CE | Conformite Europeene |
| CEB | Corporate Executive Board |
| CMMI | Capability maturity model integration |
| COGS | Cost of goods sold |
| COTS | Commercial-off-the-shelf |
| CoDR | Concept of design review |
| CPU | Central processing unit |
| C-RES | Common radar environment simulator |
| DAU | Defense Acquisition University |
| DC | Direct current |

| | |
|---|---|
| DFM | Design for manufacturing (or maintenance) |
| DFx | Design for "x" |
| DMCA | Digital Millennium Copyright Act |
| DMM | Digital multimeter |
| EDR | Engineering design review |
| EMC | Electromagnetic compatibility |
| EMI | Electromagnetic interference |
| ESD | Electrostatic discharge |
| ESR | Equivalent series resistance |
| ETA | Event tree analysis |
| FCPA | Foreign Corrupt Practices Act |
| FDA | Food and Drug Administration |
| FEA | Finite element analysis |
| FFT | Fast Fourier transform |
| FMECA | Failure modes effects criticality analysis |
| FRACAS | Failure reporting, analysis and corrective action system |
| FRB | Failure review board |
| FRR | Flight readiness review |
| FTA | Fault tree analysis |
| FTCA | Federal Tort Claims Act |
| FTP | File Transfer Protocol |
| FUS | Follow-up service |
| GFCI | Ground-fault circuit interrupter |
| GeAs | Gallium arsenide |
| GNU | G'noo not Unix |
| GPL | General Public License |
| GPS | Global positioning system |
| GUI | Graphical user interface |
| HALT | Highly accelerated life test |
| HASS | Highly accelerated stress screen |
| HPC | High-performance computers |
| HR | Human resources |
| ICD | Interface control document |
| IEC | International Electrotechnical Commission |
| INCOSE | International Council on Systems Engineering |
| ISO | International Organization for Standards |
| IP | Intellectual property |

| | |
|---|---|
| ISR | In-service review |
| IT | Information technology |
| ITAR | International Traffic in Arms |
| ITR | Initial technical review |
| JMRI | Java model railroad interface |
| LED | Light emitting diode |
| LLC | Limited liability company |
| LOC | Letters of credit |
| MBTI | Myers—Briggs type indicator |
| MDB | Model-based design |
| MISRA | Motor Industry Software Reliability Association |
| MOU | Memorandum of understanding |
| MPUs | Multiple processing units |
| MTBF | Mean-time-between-failures |
| MTTF | Mean-time-to-failure |
| MTTR | Mean-time-to-repair |
| NASA | National Aeronautics and Space Administration |
| NDA | Nondisclosure agreement |
| NDIA | National Defense Industrial Association |
| NEMA | National Electrical Manufacturers Association |
| NPE | Nonpracticing entity |
| NRE | Nonrecurring engineering |
| ORNL | Oak Ridge National Laboratory |
| OS | Operating system |
| PACE | Password Authenticated Connection Establishment |
| PCB | Printed circuit board |
| PDR | Preliminary design review |
| PERRU | Plan, execute, review, report, and update |
| PM | Project manager |
| PMBOK | Project management book of knowledge |
| PMP | Project management plan |
| PM-RADARS | Project manager—Radars |
| PNA | Petri net analysis |
| PRA | Probabilistic risk assessment |
| PRR | Production readiness review |
| PSD | Power spectral density |
| PWB | Printed wiring board |

| | |
|---|---|
| QA | Quality assurance |
| QMS | Quality management system |
| RCA | Root cause analysis |
| RE | Recurring expense |
| RFI | Radio frequency interference |
| RHA | Risk and hazard analysis |
| RMS | Root mean square |
| RoHS | Restriction of hazardous substances |
| ROI | Return on investment |
| RTES | Radar test environment simulator |
| RTOS | Real time operating system |
| SDK | Software developer kit |
| SEI | Software Engineering Institute |
| SFR | System functional review |
| SRR | System requirements review |
| SSR | Solid-state relay |
| STK | Software tool kit |
| STPA | System theoretic process analysis |
| SVR | System verification review |
| SysDR | System design review |
| SWEBOK | Software engineering body of knowledge |
| TCB | Trusted computing base |
| TLC'ed | Think, learn, communicate, enforce discipline |
| TIR | Test and integration review |
| TPM | Trusted platform module |
| TRR | Test readiness review |
| UCC | Uniform Commercial Code |
| UML | Unified modeling language |
| UL | Underwriters laboratories |
| USPTO | United States Patent and Trademark Office |
| UTSA | Uniform Trade Secrets Act |
| V&V | Verification and validation |
| WEEE | Waste electrical and electronic equipment |

## CHAPTER 1

# Introduction to Good Development

**Kim R. Fowler**

*IEEE Fellow, Consultant*

**Chapter Outline**

Developing and Managing Embedded Systems and Products.
DOI: http://dx.doi.org/10.1016/B978-0-12-405879-8.00001-5

**1**

## *About this book*

### *Purpose*

The purpose of this book is to identify important issues in developing and managing an embedded system. It should aid your understanding of both the technical aspects of the project and the teamwork involved. The book provides guidelines and bounds on the effort and cost you might encounter in developing an embedded system. It illustrates principles with examples and case studies.

Much like a good project team, the authors selected to write chapters in this book have specific expertise and understand the interactions within a development team. They know that a good team will knit the various disciplines into a fine tapestry of an excellent project.

### *Audience*

This book is for all members of a project team: program managers, technical staff, administrative staff, and support staff. Every project team member should find insight from this book for understanding the issues—technical, managerial, business, and administrative—within a project. Generally, this book aims at project teams with 6–10 people working on a single, full-time project; it can also address many issues and help project teams up to 30 or more people.

This book examines the interactions and connections within a project and its team. It reveals how technical issues impact business and schedules, how personal interactions are just as important as technical breakthroughs, and the interplay of various disciplines to realize the final product. Consequently, ***this book is not solely for managers, it is written for every member of the project team*** (Figure 1.1).

**Figure 1.1:**
An abstract view that indicates many different disciplines are needed for project development.
*Illustration from iStockPhoto.*

*Road map*

The material in this book presents best practices for developing embedded systems. Best practices include technical design, teamwork, collaboration, management attitudes, and development processes. Consequently, this book is not restricted to technical topics because developing projects is not only a technical endeavor. As in project development, the technical topics in this book intertwine with managerial issues, teamwork concerns, and legal liabilities; they cannot be separated unless artificially done so. The book's most obvious departures from a traditional engineering text are in Chapter 2, which provides insight into the psychology of the team and personnel dynamics, and in Chapters 17 through 23, which discuss legal concerns.

Though the chapters present in a sequential fashion within this book, development of an embedded system is anything but sequential. Figure 1.2 is a diagram of the chapters and when they might be most useful within a project. The multiple paths and connections should clue you into the parallel and concurrent nature of most design and development. Furthermore, much of the design activity is iterative. Feedback is necessary for refinement of requirements and the design, it is also integral to verification and validation.

*What you can get from this book*

This book concentrates on the framework that you should use to manage, develop, produce, and support embedded products. It delves into the interactions between disciplines and the potential effort that you might expect to expend. Consequently, most chapters regularly refer to the resulting effects by particular activities on quality, schedule, and budget.

The focus of this book is embedded systems and products. Figure 1.3 illustrates a sampling of products and systems that contain embedded systems. The electric streetcar or tram has numerous embedded systems—motor controls, environmental controls, brakes, power conversion, communications, signboards—to name a few. The coffee maker has a microcontroller-based circuit board that receives button inputs, drives the display, and commands the power control of the heater. The radar system has high frequency signal conversion, digital filtering and tracking, communications, and displays. The custom motorcycle has a flat panel display, a biometric switch in place of a key, and engine control. The electric guitar has a microcontroller-based embedded system that detects the pitch of each string and automatically tunes the strings by small motor driven mechanisms, as shown in cutaway in the next photograph. Finally, the four-legged crawler and crane has embedded systems to control its movements as the crew replaces the store's heavy glass window.

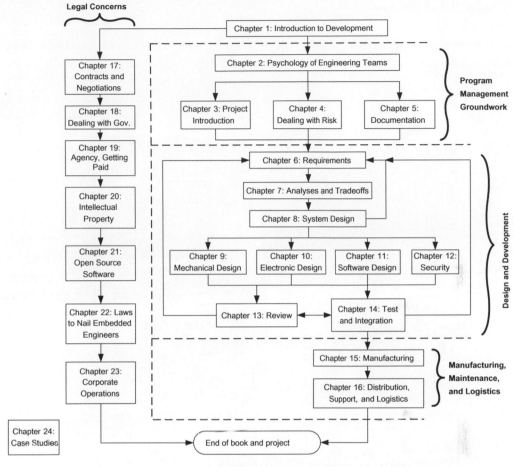

**Figure 1.2:**
Suggested approach to reading this book.

The book addresses disciplines for developing products and embedded systems. The disciplines include:

- Program management
- Systems engineering (SE)
- Operations research
- Design and development
  - Software
  - Electronic hardware
  - Mechanics
  - Human interface

- Testing and integration
- Manufacturing
- Distribution, logistics, and support
- Legal concerns.

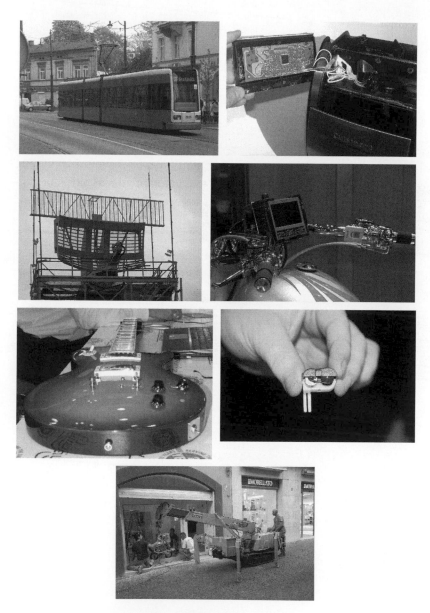

**Figure 1.3:**
A sample collection of photographs of embedded systems and equipment with embedded systems. © *2007–2014 by Kim R. Fowler. Used with permission. All rights reserved.*

## What you won't get from this book

Important areas not addressed in this book are large system applications, systems of systems, information technology (IT), marketing, business, and accounting. Furthermore, the book does not cover specific techniques or technical advances, except by way of example to address aspects of quality, schedule, and budget. In particular, it does not cover specific components for development tools, coding techniques, designs, manufacturing, or tooling. Finally, while the principles, guidelines, and metrics espoused in this book suffice for embedded systems, they are not sufficient for whole vehicles or large systems—they form a necessary subset but are not a complete set for such systems.

## Definitions and some basic concepts

To better understand this book, you will want to understand some definitions used throughout its pages. These definitions may be less than completely rigorous but they will be useful for conveying the necessary concepts:

*System*: A combination of elements or parts forming a complex or unitary whole composed of components, attributes, and relationships. Typically these elements within a system form definable inputs, processing, and outputs. The interrelated components work together toward a common objective [1].

*Systems engineering (SE)*: An interdisciplinary approach encompassing the entire technical effort to evolve and verify an integrated and life cycle balanced set of system, people, product, and process solutions that satisfy customer needs [2].

*Project*: Development work that has a clear beginning and end and is performed once to produce something unique [3]. This work is different than ongoing operations such as sales, support, and manufacturing—but it does include the insights and input from personnel in these activities to inform development of the product.

*Project management*: Verzuh claims that project management is "art informed by science." He lists five characteristics of a successful project management [3]:

- Agreement among the project team, customers, and management on the goals of the project.
- A plan that shows an overall path and clear responsibilities and will be used to measure progress during the project.
- Constant, effective communications among everyone involved in the project.
- A controlled scope.
- Management support.

*Analysis*: The examination of a concept or plan to better understand its constraints and limitations.

*Tradeoff*: A form of analysis that compares one design approach with another, different design approach or approaches.

*Synthesis*: The act of pulling elements together to create a new system, service, or operation; often new behaviors and functions arise.

*Design*: The iterative combination of analysis and synthesis to refine a concept into a realized product, which includes creation, modification, and analysis.

*Design process*: The methodology to implement design and design activities.

*Review*: A group activity that examines design, development, processes, and procedures for the purpose of comparing plans, operations, or accomplishments with the project's mission or goals.

*Test*: A set of physical challenges to a component, subsystem, or system to address a specific metric attached to a requirement.

*Integration*: The synthesis of subsystems into a larger system with testing to confirm, demonstrate, or clarify system behavior.

*Embedded system*: A combination of computer hardware and software, and perhaps additional mechanical or other parts, designed to perform a dedicated function [4]. Or, in other words, a system that depends on a computer as a critical element in its function.

*Real time*: Completing tasks within specified deadlines; this definition is not defined or limited by a specific execution speed.

*Quality*: The degree to which the sum total of product characteristics fulfills all of the requirements of customers.

*Process*: A group of interrelated activities and resources that transform inputs into outputs, often described by a block or flow diagram of events.

*Procedure*: Specific implementation of the process for a single, focused area of concern, typically step-by-step instructions.

*Validation*: The confirmation that the design, function, and operation of the final product satisfies the customer's intent.

*Verification*: The objective tests of metrics that shows that the final product meets the quantitative requirements.

*NRE*: The cost associated with "nonrecurring engineering" effort.

*COGS*: The cost of goods sold.

*Integrity*: The seamless whole, which requires a "big picture view" of how the parts fit into the whole (more on integrity later in the chapter plus the recommended book by Henry Cloud, *Integrity*.)

## Focus

The primary focus of this book is integrity, the seamless whole. Integrity sees that various topics fit together to function properly.

This book concentrates on the following topics: guiding principles, good technical design practices, good development practices, good business practices, life cycle perspective, and teamwork. If you understand and incorporate these concerns within a project, your probability of success will be good. Incorporating these concerns will not guarantee success, but they are very important. If you ignore any of these concerns, your probability of success will drop dramatically.

### Five guiding principles

Each of these principles has the overriding directive of integrity. These principles bound all the discussions within this book and should guide successful development of any embedded product:

- There are no silver bullets in development.
- Appropriate feedback stabilizes a defined system.
- All important actions occur at interfaces within systems.
- All problems have a human origin.
- Good development and engineering requires good relationships.

#### No silver bullets

Every interesting problem is multidimensional and requires the involvement of diverse disciplines. There are no "silver bullets."[1] Simply no tool or method or process fits all operations. Don't look for one. Any "silver bullet" will narrow your focus and you will overlook important issues and concerns. It would be worth your time to read Fred Brooks' paper "No Silver Bullet—Essence and Accident in Software Engineering" [5]. The 20-year retrospective, which is very short, by Mancl, Fraser, and Opdyke is a useful companion paper to Brooks' original paper [6].

#### Feedback stabilizes

Development needs closed-loop, feedback processes that regularly use various reviews to inform and adjust development. As in most closed-loop, feedback systems, you can over-control or under-control the feedback causing inappropriate results. Appropriate feedback requires study and experience. You will find more on feedback later in this chapter with the PERRU concept.

#### Interfaces are important

All, or nearly all, important actions occur at interfaces. If you think about it, you will see that most interesting things happen where two different domains meet. The physical environment certainly demonstrates this with wave propagation, which doesn't get interesting until it encounters an obstacle, then diffraction, refraction, and reflection take

---

[1] Silver bullet—supposedly a magical method to kill werewolves—implies that one solution solves all parts of a complex problem.

place. Extending the analogy further, semiconductor operations occur at the boundary of different materials. And finally, the interface between human operations and machine functions is interesting and very important.

Consequently, successful development requires you to understand the different types of interfaces (e.g., electrical, material, mechanical, software, signal flow, environmental, human–machine interactions). You use interfaces to define the boundaries and subsystems and then use these partitions to drive development. Good systems engineering (SE) does this well.

### All problems have a human origin

All problems have a human origin. Even the very best designs will fail or outlive their usefulness. The human capacity for design cannot account for all possibilities—circumstances that are unexpected or unknown, multiple and simultaneous interactions that lead to failure, and even human abuse from inappropriate, stupid, or malicious operations. Beyond these concerns, finite lifetime will limit normal use over an extended time. Failure, furthermore, can have other causes, such as business climate, socioeconomics, or politics that defy technical solution.

### Good development and engineering require good relationships

The best products are derived from fully functional teams that work together. I believe that *good interpersonal relationships are absolutely necessary for teams to work together and to develop successful products.* Chapter 2 will bear out my contention.

---

### Example: Bad relationships dissolve company

There are good historical examples of companies that thrived because of good relationships. Jim Collins in his books, *Good to Great* and *Built to Last*, discusses a number of examples of successful companies built on the right people relating well together [7,8].

Counterexamples to those successful companies are not always as obvious. I worked in a company where several interpersonal relationships were not just bad, but toxic. Gossip, backbiting, defamation of character, and nasty, untrue allegations of fraud ultimately destroyed the company. The disappointing thing was that sharp, capable people comprised the staff and the company's product designs were outstanding. *Bright ideas could not overcome bad relationships!* Consequently, we did not finish developing a product and the company foundered.

---

### Reliability, fault avoidance and tolerance, and error recovery

Closely related to problems, interfaces, and SE are the concepts of reliability, fault avoidance and tolerance, and error recovery. The interplay between these concepts is complex; hopefully, the remainder of the book will reinforce that interplay for you. One dimension of that interplay is the priority of consequences that a system displays when encountering a problem; systems can survive and demonstrate increasing robustness by incorporating structures and designs with the ascending order of complexity that follows:

- Reliable—Stands the test of time
- Available—Minimal downtime
- Gold-plated design—No catastrophic errors
- Redundancy with reliable and available attributes—Results are bounded and predictable
- Robust—Recovers gracefully from errors.

### The business case

***The end result of nearly all projects is to solve problems and make money from the solutions.*** This book is written with that thought in mind that we are all in business to make money. The technical concerns addressed in the majority of the book ultimately tie back to solutions that make for successful products. While this is not a book on program management *per se*, some business administration, particularly for program managers is included:

- Configuration management
- Risk management
- Scheduling
- Budgeting
- Contract(s)
- Legal liabilities.

### Life cycle

For a product to be successful, we must consider all facets of its life cycle. Problems and failures will inevitably sneak in if your team ignores or overlooks a phase. The major phases of the life cycle follow:

- Concept
- Requirements
- Analyses and tradeoffs
- Design
- Development
- Review
- Test
- Integration
- Manufacturing

- Delivery, commissioning, sales
- Support, maintenance, and repair
- Upgrades and modifications and reuse
- Disposal.

The book covers most of these bullet points in separate chapters that match closely to this list. Figure 1.2 shows the chapters in the book that align with these points.

### Types of markets and development

The book concentrates on embedded systems within certain general markets. Much of what is said here, however, is valid to any design effort for any embedded system. The markets that the book specifically addresses are as follows:

- Consumer appliances and products
- Industrial process controls
- Medical devices
- Aerospace instruments
- Military equipment.

The book also considers the range of production volumes for development within these markets. The range starts at very small volumes with custom, one-off projects that build anywhere from 1 to 50 systems. It continues through low volume (50−5000 units per year), high mix (many models) manufacturing. Finally, it covers high volume manufacturing, which usually goes from 5000 units to millions per year.

### Recent research

The authors of this book have also incorporated the results of recent research, along with their own experience, to give you a well-rounded approach to development and management. Some of the topics include systems thinking, complexity metrics, and psychological and legal concerns within development.

## Team attributes

### Working together

Working together requires team members to understand individual assignments and to relate well one with another. Relating well together without knowing who does what is just a party, it's not a project. Focusing only on individual assignments without relating to one another is drudgery that loses sight of the big picture and end goal.

*Individual assignments*

The first thing each of us must do is understand our individual assignments. Every team member usually wrestles, in some measure, with the following concerns within every project:

- Authority
- Responsibility
- Accountability
- Delegation
- Learning
- Rigor and discipline.

Often authority is mandated from upper management or the project manager. Interestingly and in spite of direction from above, a well-qualified individual can often carve out an area of authority by virtue of his or her capabilities.

It is important to draw distinctions among authority, responsibility, and accountability. *Authority* is the power and the right to accomplish, command, or influence something, often by directing other people or establishing a course of action or requisitioning resources. Tightly coupled with authority is *responsibility*, which encompasses the liability of articles assigned to or assumed by someone. More colloquially, authority and responsibility are the parts of the project that the person "owns." *Accountability* completes the circle with authority and responsibility; accountability means that an individual provides record, reasons, causes, grounds, or motives for using authority to accomplish or complete a responsibility.

---

### Example: Authority, responsibility, and accountability

If you have been given the task of designing a circuit board that is your responsibility, the objective is to complete a functional design that meets project requirements. You will be accountable for the circuit board design by providing record of your effort in the form of schematics, parts list, analysis results, test results, documentation, and review presentations. You have the authority to use company-supplied tools to prepare the design, obtain parts for bench testing of prototypes, conduct analyses and tests, and call for reviews. All of your efforts, of course, are part of the larger project and your authority, responsibility, and accountability integrate with the efforts of the other team members; you do not work in isolation.

Responsibility and accountability for a task without the requisite authority make your effort impossible to complete satisfactorily. Being responsible for the design of a circuit board but without the authority to get parts for bench testing limits thorough development. This may seem trivial but I have seen imbalances in assigning authority, responsibility, and accountability that prevented successful completion of projects.

---

*Delegation* is the parceling out of authority, responsibility, and accountability to another person to complete a subset of a project. As seen in the example, good delegation balances these components to avoid over constraining development. Delegation is a matter of authority. The person designing the circuit board, in the example, does not have the authority to delegate (i.e., push off all the design work onto someone else) and then go drink coffee all day in a cafe. In some more complex projects that same person might have the authority to delegate the development of a component or module to another team member.

*Learning* should be a lifelong pursuit for all of us. Within a project, learning might include developing new techniques and processes, applying new technology, or honing our skills. It may be investigating new business ventures. Or it may be refining our relationships.

Finally, all of us have the obligation to our company and other team members to exercise *rigor and discipline* in performing tasks. Applying rigor and discipline to a task means that you consider many aspects to completing the task; it does not mean that you mindlessly perform a single test repetitively in the hopes that the results average out well. Just spending a lot of time on a project does not indicate thoroughness, rigor, or discipline.

---

### Example: Rigor and discipline

Here is another very simple example. Jumping into a project to "crank out" a lot of code without thoughtful and careful planning, design, review, and test is not rigorous or disciplined effort. "Cranking out" code does expend a lot of effort but it is not necessarily efficient or productive.

Rigor means that you follow thorough guidelines for developing the code. It recognizes that good quality code development follows defined processes and procedures, which include many activities such as specifying requirements, planning, designing, reviewing, and testing.

Discipline means that you follow the guidelines carefully and consistently. Discipline means that you apply yourself to continuous effort for high quality results.

---

A manager or project lead who encourages rigor and discipline is valuable indeed!

### Relating together

Once we understand our individual assignments, the next step is to relate together in a productive manner. Chapter 2 is devoted to team dynamics and relating together well.

A productive team that relates well with one another is marked by good communications and active efforts to support one another. A team covers one another and picks up the slack whenever one team member needs extra help. Sports analogies abound here...

**Two examples of what to do and what not to do**

Too many times I have heard, "It's not my problem." Well, if you are part of the team, then yes, it is your problem. Regardless that it is not your direct responsibility, you are still accountable to the team. If it needs correction, you owe it to your team to speak up, to fix it, and to help cover the various aspects of concern.

I have seen both good and bad examples of a technician finding a design problem that the cognizant engineer really needed to understand. A good technician will discuss the problem with the engineer and point out the consequences and the possible fixes. I have appreciated the advice of a number of good technicians over the years. Other times I have heard, "It's not my problem," and the problem takes much longer to resolve.

Another problem is the attitude, voiced in various ways, that comes out as, "If it is not my field of expertise, it's trivial to fix." For example, a hardware engineer sees a problem that he thinks can be fixed in a couple lines of code by a software engineer, "Come on, it's only a couple lines of code. How hard can it be?" Clearly, the hardware engineer does not appreciate the effort to change the code, to perform regression testing, and to change the documentation. The reverse can happen, too. A software engineer can think that just changing out a component might fix the problem without appreciating the potential effort to revise the layout of the circuit board, update the analyses, and change the documentation.

In the final analysis, relating well is critical to the successful completion of the project. Another colloquialism is appropriate here, "None of us is as smart as all of us" (variously attributed to an ancient Japanese proverb, Abraham Lincoln, and the cartoon character Pogo). Of course, herd mentality can make the converse true, "None of us is as dumb as all of us," as pointed out to me by Tim Wescott, but let's not go there.

### Attributes of a good manager

With this background, let's look at several attributes of good managers. A good manager always has the end goal, or the vision, in mind. A good manager will maintain a big picture perspective with how activities and people fit together. A good manager will encourage, mentor, and enforce rigor and discipline. Finally, a good manager will publically praise a job well done but reserve correction for one-on-one encounters: ***public praise, private exhortation.***

### Attributes of good technical and support staff

A good team member implements the individual assignments discussed above, maintains a positive attitude, supports the team's effort in completing the project, and strives to communicate and to relate well. A good team member is a learner. Be proactive, reduce reactionary responses.

Chapter 2 speaks in detail about team development and function.

*TLC'ed*

Finally, with all that background, let me introduce an acronym that highlights some of the issues. You might recognize the acronym, TLC, as Tender, Loving Care. I suggest another acronym with some of the same letters but a different meaning—"TLC'ed," which stands for

- THINK
- LEARN
- COMMUNICATE
- Encourage Discipline.

The previous three pages explain the basics of TLC'ed. Nuff said.[2]

## Ethics

I think that most ethics can be encapsulated by *integrity*—the seamless whole, which requires a "big picture view" of how the parts fit into the whole. This requires vision.

I highly recommend the book by Henry Cloud, *Integrity: The Courage to Meet the Demands of Reality*. In that book, Henry Cloud states and supports six essential qualities of ethics, and particularly integrity, that support business success [9]:

1. Connect authentically to build trust.
2. Orient toward the truth by operating in reality.
3. Be effective to complete the mission.
4. Deal with the negative and work toward resolution.
5. Orient towards growth to see increase.
6. Be transcendent to understand and enlarge the picture/situation.

Each of these qualities support the attributes of team members from the previous section. Maintaining integrity, as defined and supported by Henry Cloud, will help us to be disciplined and rigorous, to commit to learning, to communicate well, and to relate well.

## Success and failure

"Success may be grand, but disappointment can often teach us more" [10]. All of us have experienced some measure and variety of failures. We have all learned from those failures. We just cannot learn enough from success to push the state of the art. We must push the boundaries of knowledge, experience failure, and learn to circumvent it.

---

[2] Nuff said is a slang American expression that means "enough said."

Any project team willing to risk and learn from the results has the poise and attitude to face failure and learn from it. In a way, bench tests and prototype field tests can push the boundaries and examine the onset of failure. Taking calculated risks that might result in failure but provides controls to allow efficient learning is key to a highly successful endeavor.

## Interview with Jack Ganssle

This interview distills thoughts from a conversation that I had with Jack Ganssle on August 18, 2011.

1. *Have you added to or changed anything about your 7-step plan? (Drawn from Ref. [11]).*
   - *Use a Version Control System*
   - *Have a Firmware/Software Standards Manual*
   - *Do code inspections*
   - *Create a quiet environment for thinking*
   - *Measure bug rates*
   - *Measure code production rates*
   - *Constantly study software engineering.*

   "Still sticking to it." Jack started in the embedded industry back in 1972 and founded an emulator business in 1983. Even today he says, "Same problems, very little change."

2. *What are the two or three most important things that engineering managers and project leads can do to be better at their jobs and more likely to develop successful products?*
   - Get management training. Most managers are engineers and got their jobs by default.
   - Impose discipline on the design/development team. An aside—we have created an environment to accept bugs. [The implication is that we need to encourage rigor and discipline and not accept the status quo that there will always be bugs.]
   - Leaders hold themselves accountable for the development and the team.

3. *What are the three or four most prevalent problems you see in companies when they develop embedded systems?*
   - Lack of complete requirements.
   - Jumping into building too early; teams do not spend enough time to design the product.
   - Lack of focus on quality. (Teams need to manage quality to achieve appropriate schedule and features.)
   - Focus on fixing bugs, which does not achieve quality.

4. *How would you compare the value of people versus the value of good engineering processes?*
   The Agile manifesto and method is brilliant but too often an excuse for glorified hacking. Test-driven development is often taken to the extreme. A real need for balance exists— "Great people don't scale" and you can't keep development teams constantly firefighting. An interesting twist of logic in this is that companies won't write the checks for good tools, even though the cost—benefit crossover points can be within weeks and even days.

5. *How much of a role do Quality Assurance (QA) standards really play in most companies? and should they?*
   Very few companies have an active QA program. Only military, aerospace, some manufacturing, and some medical companies have active QA programs. Surprisingly, the Nevada Gaming Commission requires source code for gambling machines placed in Nevada. Jack predicts that the "FDA will crack down" on the other medical companies

that have inadequate QA programs. Jack also predicts "Litigation will force changes [in companies by requiring active QA programs]."

6. *What are some examples of really wrong things that a manager or leader can do?*

   Other than lacking a focus on quality and lacking accountability and discipline, as already mentioned, it is becoming a question of this, "In a sinking boat, do you fix the leak or bail the water out?" Here are some specific examples of problems:

   - Bad science, e.g., not understanding signal-to-noise (SNR) and amplifier gain or correlating anything with something else.
   - Not fixing bugs and not understanding their sources. Being willing to allow outstanding bugs in products.
   - Managers believing that developers are artists and not engineers.
   - Managers allowing research to be mixed up with development.

7. *How have things changed during your 40 years in the industry?*

   Very little has changed. There are glimmers of hope that change may be coming. Jack sees that most change is occurring in the safety critical and security arenas.

## Systems engineering

SE, project management, and system architecting are separate and different activities. Figure 1.4 gives a graphical perspective for some of the responsibilities of these different offices. Often smaller teams combine these responsibilities into one person (as in Figure 1.5). I will treat them as combined in this book to be closely aligned to how many, if not most, projects run in practice. Figure 1.5 also illustrates three other dysfunctional configurations for how the systems engineer, project manager, and system architect might interact; these configurations impede communications and knowledge transfer.

SE is important; to some extent we all do it—or should. SE accounts for how the different disciplines relate during a project and how the different technical systems and components come together. Figure 1.6 provides another view and graphically outlines the disciplines for which a systems engineer might be responsible.

### Bob Rassa on SE

"The complexity of modern systems demands systems engineering," says IEEE Fellow Robert Rassa, Director of Engineering Programs for Raytheon SAS and past president of the IEEE Systems Council.

*. . ."First of all," says Rassa, "you need to have a broad engineering background." He recommends having both discipline experience—electrical engineering, software engineering, etc.—as well as domain experience—space, transportation, communications and the like. "Second, you need to understand a lot of the basics—mechanical engineering, software, logistics, reliability—and have an understanding of how those things play into design."*

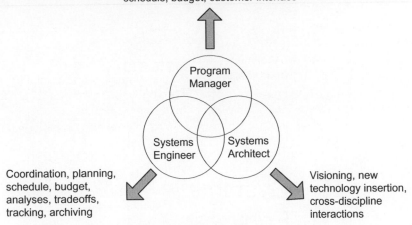

Management, company goals and positioning, planning, schedule, budget, customer interface

Program Manager

Systems Engineer

Systems Architect

Coordination, planning, schedule, budget, analyses, tradeoffs, tracking, archiving

Visioning, new technology insertion, cross-discipline interactions

Note:
1. QA should be an integrated component that spans all disciplines.
2. This "ideal" organization rarely, if ever, happens

**Figure 1.4:**
The division of responsibilities in a larger project with three separate people filling the different job descriptions. *© 2014 by Kim R. Fowler. Used with permission. All rights reserved.*

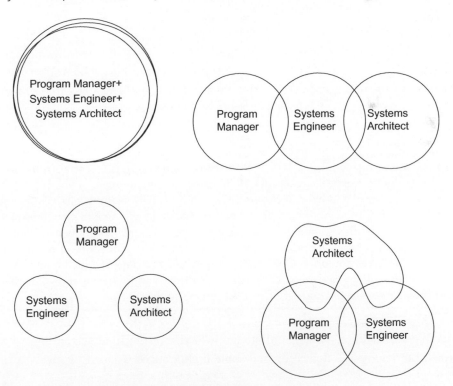

**Figure 1.5:**
The variety of different overlaps between the responsibilities in other companies; other than small projects where one person fulfills all three job descriptions (upper left in the diagram), the other three illustrate dysfunctional organizations that impede communication and knowledge transfer.
*© 2014 by Kim R. Fowler. Used with permission. All rights reserved.*

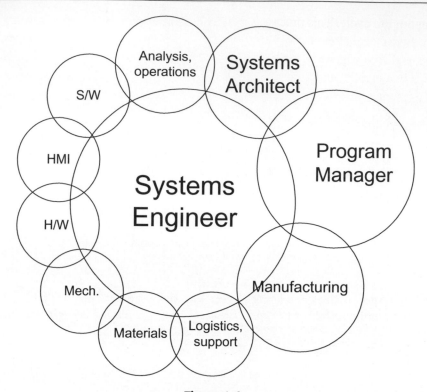

**Figure 1.6:**
Another view of the responsibilities of a systems engineer. © 2014 by Kim R. Fowler. Used with permission. All rights reserved.

The third important quality, Rassa says, is leadership. "Leadership is essential. You have to have good communication skills. Consider all inputs from the members of your team and be able to make both good and hard decisions." He says decision making is often influenced by what is called systems thinking, which he defined as the ability to think in system-level attributes. "What does this system we're creating have to do *in toto* and how can I get there, looking at the pieces I have, the pieces I have to design, and the pieces I have to change." Since this often involves components from multiple suppliers, some of which might not be changeable, a systems engineer must understand what they can influence and what they can't and then be able to integrate it all into a working system [12].

SE assures that the function, configuration, and behavior of the final system meets customer expectations. Let's look at a definition and some highlights of what SE does.

*Systems Engineering*: An "…engineering discipline whose responsibility is creating and executing an interdisciplinary process to ensure that the customer and stakeholder's needs are satisfied in a high quality, trustworthy, cost efficient and schedule compliant manner throughout a system's entire life cycle. This process is usually comprised of the following seven tasks: **S**tate the problem, **I**nvestigate alternatives, **M**odel the system, **I**ntegrate, **L**aunch the system, **A**ssess performance, and **R**e-evaluate. . . Systems Engineering Process is not sequential. The functions are performed in a parallel and iterative manner." Farther down the page, a commentary by Brian Mar states that systems engineers should follow these basic core concepts [13]:

- *Understand the whole problem before you try to solve it.*
- *Translate the problem into measurable requirements.*
- *Examine all feasible alternatives before selecting a solution.*
- *Make sure you consider the total system life cycle.*
- *Test the total system before delivering it.*
- *Document everything.*

SE encompasses [2]:

- the technical efforts related to the development, manufacturing, verification, deployment, operations, support, disposal of, and user training for, system products and processes;
- the definition and management of the system configuration;
- the translation of the system definition into work breakdown structures;
- development of information for management decision making.

The objective of the International Council on Systems Engineering (INCOSE) *Systems Engineering Handbook* is to provide a description of key process activities performed by systems engineers. The intended audience is the new systems engineer, an engineer in another discipline who needs to perform SE, or an experienced systems engineer who needs a convenient reference [14].

## INCOSE Systems Engineering Handbook

The INCOSE *Systems Engineering Handbook* [14] divides into a number of chapters: **Technical Processes** (see Chapter 4) include stakeholder requirements definition, requirements analysis, architectural design, implementation, integration, verification, transition, validation, operation, maintenance, and disposal. **Project Processes** (see Chapter 5) include project planning, project assessment and control, decision management, risk management, configuration management, information management, and measurement.

SE, when properly applied, contains cost and maintains schedule much more effectively than *ad hoc* efforts. The INCOSE *Systems Engineering Handbook* provides some good evidence:

> *Cost overrun lessens with increasing SE effort and appears to minimize at something greater than 10% SE effort. . . Variance in the cost overrun also lessens with increasing SE effort. At low SE effort, a project has difficulty predicting its overrun, which may be between 0% (actual = planned) and 200% (actual = 3 × planned). At 12% SE effort, the*

*project cost is more predictable, falling between minus 20% (actual = 0.80 × planned) and 41% (actual = 1.41 × planned) [14].*

*Schedule overrun lessens with increasing SE effort and appears to minimize at something greater than 10% SE effort, although few data points exist to support a reliable calculation. Variance in the schedule overrun also lessens with increasing SE effort. At low SE effort, a project has difficulty predicting its overrun, which may be between minus 35% (actual = 0.65 × planned) and 300% (actual = 4 × planned). At 12% SE effort, the project schedule is more predictable, falling between minus 22% (actual = 0.78 × planned) and 22% (actual = 1.22 × planned) [14].*

## NDIA and SEI report

A collaborative effort between the National Defense Industrial Association (NDIA) Systems Engineering Effectiveness Committee (SEEC) and the Software Engineering Institute (SEI) of Carnegie Mellon University produced the report *A Survey of Systems Engineering Effectiveness*, which found that Systems Engineering can contain costs and maintain schedule. "...improving Systems Engineering capabilities clearly can result in better Project Performance. However, more consideration also must be paid to ways of reducing Project Challenge. Doing so is a major challenge prior to the establishment of the development project, beginning during the pre-acquisition period. Earlier application of Systems Engineering practices and principles may go a long way towards reducing that challenge" [15].

## NASA report on cost escalation

The cost of changing the design in the concept stage is cheaper than changing the design in the preliminary stage, which is cheaper than changing the design in the final stage. Fixing problems is cheaper when done earlier than later. NASA performed a study on the cost to fix requirement errors at various stages of development. "If the cost of fixing a requirements error discovered during the requirements phase is defined to be 1 unit, the cost to fix that error if found during the design phase increases to 3−8 units; at the manufacturing/build phase, the cost to fix the error is 7−16 units; at the integration and test phase, the cost to fix the error becomes 21−78 units; and at the operations phase, the cost to fix the requirements error ranged from 29 units to more than 1500 units" [16]. Figure 1.7 graphs this cost escalation.

## NASA Systems Engineering Handbook

A good template for a large project is the NASA Systems Engineering Handbook. Obviously, it focuses on spacecraft but you can tailor it to a variety of projects [17].

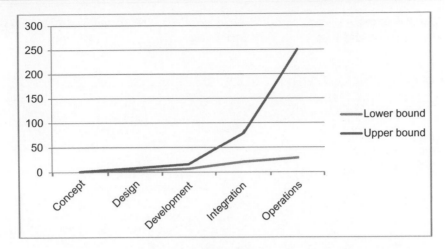

**Figure 1.7:**
Cost escalation from NASA report for finding and fixing errors in requirements.

## Various approaches to development processes

### Process models for development

A process model illustrates the high-level activities and their phasing during development. The activities include concept specification, design, development, test and integration, and iteration. The phases might include concept, design, evaluation, integration, acceptance sign off, production, and operation. There are three primary types of process models and hybrid combinations thereof: V-model, spiral, and prototyping.

So why are these process models important to understand? *Process models provide high-level perspective that helps team members understand what activities to do and what progress has been made on each of those development activities.* That perspective is foundational for the rest of this book to help you understand why particular subjects are covered. Like any model, they each have strengths and weaknesses—clarity and prominence for important features, lack of detail for the rest.

### V-Model

Figure 1.8 is a simple representation of the V-model of development. The figure marks five phases of the process. It illustrates how verification and validation connect between the various activities, across phases. According to the definitions earlier in this chapter, verification is an objective set of tests to confirm that the product meets the metrics of the requirements, while validation seeks to demonstrate that the product meets the original intent.

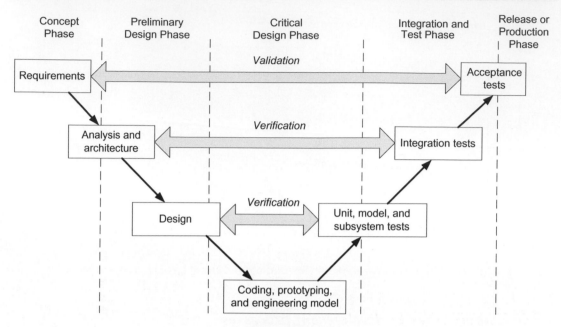

**Figure 1.8:**
Simplified V-model of development that shows how it fits the phases of development.
© 2008–2014 by Kim R. Fowler. Used with permission. All rights reserved.

A limitation of the V-model is that it implies that the requirements are complete in the conceptual or preliminary stage. This is certainly not reality in most, if not all, product developments. Requirements, design, and evaluation often iterate several times before final integration and acceptance. This iterative situation leads to spiral development as a defined discipline, which is next.

### Spiral model

Figure 1.9 shows spiral development. The goal here is to do the most important functions first. Spiral development adds components, modules, and subsystems, in planned succession, to a system. Spiral development thoroughly tests the system for functionality and ensures that development meets the appropriate set of requirements after each module addition; this is one form of verification and integration. Spiral development builds system functionality in stages. Each stage plans for a new set of features, tests and evaluates them, and then updates the system configuration to include them.

The spiral model fits very well Geoff Patch's advocacy in Chapter 11, "…once the initial up-front design process is completed, commence a series of iterative development cycles. The phrase 'design a little, code a little, test a little' best captures the essence of what is

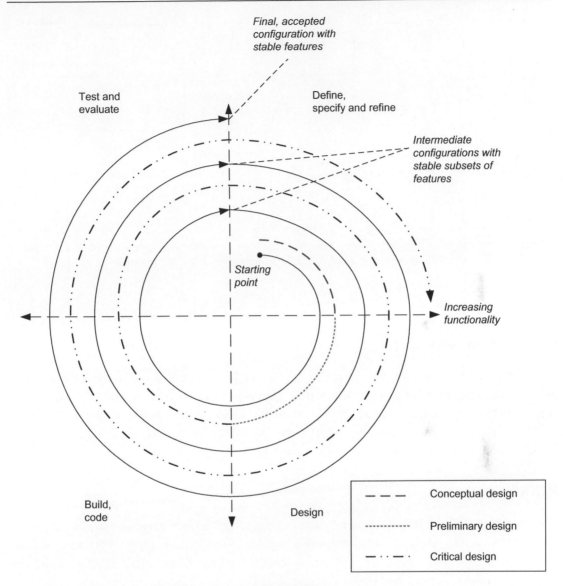

**Figure 1.9:**
Simplified Spiral model of development; a system develops in stages, each stage passes through four separate phases of development: design, build, test and evaluate, and refine. © *2008—2014 by Kim R. Fowler. Used with permission. All rights reserved.*

involved. The feedback process from the test phases is critical. You must recognize and correct design or implementation errors as early as possible, because such errors take more time and money to correct as the system gradually takes shape." Patch's contention also confirms the importance of early design and fixes to save money and time.

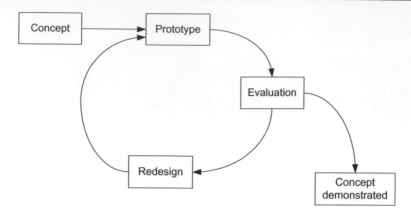

**Figure 1.10:**
Prototyping model of development; its primary purpose is to demonstrate a proof of concept.

### Prototyping model

Figure 1.10 shows development via prototype. Prototyping is best suited to demonstrate or prove a concept, often for a subsystem within the project. It is useful for showing that a component or subsystem can perform in a particular capacity. It is not adequate for large-scale development or commercial product development. It does not fit a phased scheme of development, unless you formalize it into spiral development.

### PERRU

Within any process model are a set of procedures, such as coding or design or analysis or documentation, that are consistent, thorough, and amenable to evaluation. Procedures generally should have a structure that can be summarized by the acronym PERRU: Plan, Execute, Review, Report, and Update. Figure 1.11 illustrates this structure of information flow; it also describes some documents generated during a procedure within a project.

### Quality Assurance (QA)

QA is a generic term for the effort to achieve the right product at the right price for the right people at the right time. QA should not be viewed as a separate function that "bolts onto" project activities. QA, rather, should permeate all facets of product development and be indistinguishable from good SE. Effective design, development, review, analysis, test, and documentation are good QA.

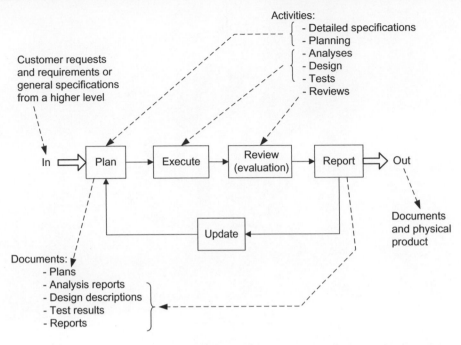

**Figure 1.11:**

PERRU—The basic feedback operation of a procedure that suits any level of detail within a project. © 2008–2014 by Kim R. Fowler. Used with permission. All rights reserved.

A Quality Management System (QMS) implements QA policy through company-specific procedures. Consistent application of the QMS across the enterprise is important. A good QMS will combine staff, tools of industry and business, and procedures in an effective and efficient manner (Figure 1.12). Three prominent standards for QA outline expectations for processes used in many industries are: ISO 9001, Six Sigma, and CMMI (Capability Maturity Model Integration). One of these can form the basis for the QMS in your particular company; choosing one depends on its fit to the industry and the understanding by both the employees and customers of the particular process.

## ISO 9001

ISO is the acronym for International Organization for Standardization, which is based in Geneva, Switzerland. ISO 9001 provides a QMS defined as, "a series of components logically linked together that provides measures and controls to manage and improve products" [1]. For ISO certification, a company needs to purchase the standard, implement it, and then undergo regular audits to confirm conformance to the standard.

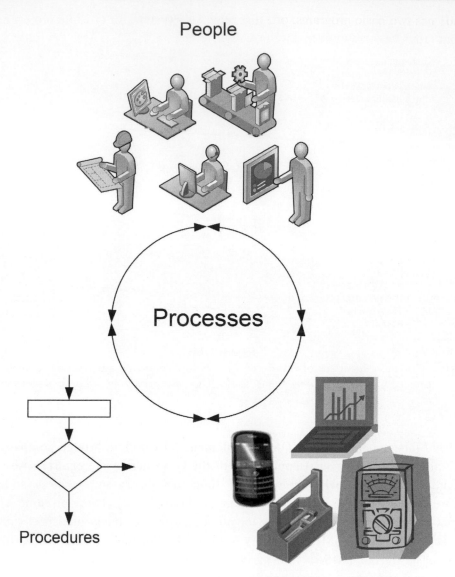

**Figure 1.12:**
Processes link people, tools, and procedures to support consistent application of effort towards quality across the enterprise. © *2008—2014 by Kim R. Fowler. Used with permission. All rights reserved.*

ISO 9001 has two basic programs, one that provides structure for QA to a project and the other that provides for improvement in QA. The structure program does the following [18]:

- Identifies processes and documents them
- Describes sequences and interactions
- Ensures resources to run the QMS.

The improvement program provides the following:

- Measure performance
- Judge effectiveness
- Improve the QMS.

ISO 9001 requires a QMS that has eight sections. The most important and most used sections are the last five. Section 4 discusses documentation, particularly the quality manual. Section 5 discusses management responsibility. Section 6 discusses resource management. Section 7 discusses product realization. Section 8 discusses measurement, analysis, and improvement. (*Note*: AS9100 is defined specifically by and for the aerospace industry. It adopted the 20 elements of ISO-9001-1994 and aligned them with the needs of the air transport industry. The AS9100 standard incorporates additional provisions for civil and military aviation and aerospace industry standard requirements. It also covers the needs of suppliers, regulatory bodies, and customers of the aerospace industry [18]).

### Six Sigma

The Six Sigma quality system originated in high volume production and manufacturing. Its main goal is to identify and measure variances. Six Sigma does not guarantee quality but provides expectations of program performance based on customer satisfaction [19].

### Capability Maturity Model Integration (CMMI)

CMMI grew out of guidelines for software development from the Software Engineering Institute at Carnegie Mellon University. The guidelines have unlimited distribution rights and can be downloaded from www.sei.cmu.edu.

CMMI has five levels of maturity that represent stages and capability of processes within a company. Table 1.1 lists the different levels of maturity and makes a brief descriptive comment [20].

CMMI has 22 Process Areas (PAs), and each PA is a cluster of related best practices to satisfy goals to improve processes. Table 1.2 lists the PAs along with the maturity level for a staged representation. Table 1.3 provides examples of project procedures paired to levels of PAs and maturity [20].

**Table 1.1: Maturity levels within CMMI for a staged representation [20]**

| Level | Name | Comments |
|---|---|---|
| 1 | Initial | Success depends on competence and heroics |
| 2 | Managed | Processes are planned and executed according to policy |
| 3 | Defined | Processes tailored to each project and described more rigorously |
| 4 | Quantitatively managed | Quantitative objectives measured, predictable performance |
| 5 | Optimizing | Continually improves processes based on quantitative understanding |

**Table 1.2: General areas and PAs for stage representation of CMMI [20]**

| Areas | Maturity Level |
|---|---|
| Requirements Management<br>Project Planning<br>Project Monitoring and Control<br>Supplier Agreement Management<br>Measurement and Analysis<br>Process and Product QA<br>Configuration Management | 2 |
| Requirements Development<br>Technical Solution<br>Product Integration<br>Verification<br>Validation<br>Organizational Training<br>Integrated Project Management<br>Risk Management<br>Integrated Teaming<br>Integrated Supplier Management<br>Decision Analysis and Resolution<br>Organizational Environment for Integration | 3 |
| Organizational Process Performance<br>Quantitative Project Management | 4 |
| Organizational Innovation and Deployment<br>Causal Analysis and Resolution | 5 |

*Comparison between ISO 9001 and CMMI*

ISO 9001 tends to be an "all or nothing pass the test" effort. Your company's QMS describes its processes and then shows how the company meets and performs these processes. ISO 9001 does not provide as many guidelines to incorporate into processes.

**Table 1.3: Examples of CMMI PAs established during each stage of implementation**

| General CMMI Level | Components Implemented | Process Area | Specific CMMI Level |
|---|---|---|---|
| 1 | Work orders, problem reporting and correction, requirements development | Project Planning | 2 |
| | | Requirements Development | 3 |
| | | Requirements Management | 2 |
| | | Measurement and Analysis | 2 |
| | | Process and Product QA | 2 |
| 2 | Engineering processes and checklists, configuration management, document templates | Project Monitoring and Control | 2 |
| | | Technical Solution | 3 |
| | | Supplier Agreement Management | 2 |
| | | Configuration Management | 2 |
| | | Organizational Training | 3 |
| | | Integrated Project Management | 3 |
| 3 | Engineering updates, manufacturing updates, vendor qualification and management | Product Integration | 3 |
| | | Verification | 3 |
| | | Validation | 3 |
| | | Risk Management | 3 |
| | | Decision Analysis and Resolution | 3 |
| | | Organizational Environment for Integration | 3 |
| 4 | R&D, project processes integrated | Integrated Teaming | 3 |
| | | Integrated Supplier Management | 3 |
| | | Organizational Process Performance | 4 |
| | | Quantitative Project Management | 4 |
| 5 | Business processes integrated | Organizational Innovation and Deployment | 5 |
| | | Causal Analysis and Resolution | 5 |

**Table 1.4: ISO 9001 versus CMMI**

| | ISO 9001 | CMMI |
|---|---|---|
| Characteristics | Stereotyped, "pass—fail" test | More flexible, focus on assessment and improvement |
| Advantages | • Easily adaptable to manufacturing<br>• More customers tend to understand | • Adapts well to software<br>• Easy to get standard |
| Disadvantages | • Less direction for improvement<br>• Cost to undergo audits | • Learning curve for the 22 PAs<br>• Must train customers to understand |

CMMI is more highly resolved than ISO 9001 and provides more guidance. The intent of CMMI is to focus on process improvement through assessment and maturity levels. Table 1.4 provides a comparison between ISO 9001 and CMMI.

## Life cycle phases

Describing phases within development is useful for highlighting specific activities particular to each time span. Setting phases can be a convenient way to select tasks and milestones within development. Phases also define the time frames within the process model for development. *Please note*: not all companies use these exact phases, though the first four—Concept, Preliminary, Critical, and Test and Integration—tend to be common terms often used by many companies.

### Concept

The Concept Phase is the first phase. It kicks off a project development. This is where you develop a concept and form the initial set of requirements. The project also puts in place the processes and procedures for risk management, configuration management, and hazard analysis, if needed. You might perform tradeoff analyses and in a few cases you might do some bench tests of new technology or applications. Successful completion of the Concept Design Review (CoDR), sometimes called the Engineering Design Review (EDR), marks the end of the phase.

### Preliminary

The Preliminary Design Phase follows the Concept Phase. The project team designs the basic architecture and completes the tradeoff analyses. The team also begins and then completes many, if not most, of the analyses for function and operation. The team will also order components and subsystems that have long lead times. Successful completion of the Preliminary Design Review (PDR) marks the end of the phase.

### Critical

The Critical Design Phase follows the Preliminary Design Phase. The project team executes detailed designs for the entire system. The team orders components and subsystems in preparation for test and integration and then look ahead to production. Component and module verification begins. Successful completion of the Critical Design Review (CDR) marks the end of the phase.

### Test and integration

The Test and Integration Phase follows the Critical Design Phase. The project completes verification of the system's components and modules. The project integrates the components and modules into a final system configuration and demonstrates function. Successful completion of the Test and Integration Review (TIR) marks the end of the phase. Chapter 14 addresses issues in test and integration.

### Compliance and system acceptance

To comply with standards in specific markets requires testing your product to those standards. Examples of compliance testing include electromagnetic compatibility, environmental extremes, and appliance safety. Compliance and system acceptance typically are important parts of the Test and Integration Phase.

### Production

The Production Phase follows the Test and Integration Phase. Products often cannot be sold before they are certified to comply with the requisite market standards. Production is ongoing and terminates in a fashion specific to the market and your company's decision. Chapter 15 addresses issues in manufacturing and production.

### Shipping and delivery

Shipping and delivery begins shortly after the Production Phase starts. Chapter 16 addresses issues in logistics and distribution.

### Operations and support

Operations and support usually begin concurrently with shipping and delivery. Chapter 16 addresses some issues in operations and support.

### Disposal

Disposal is defined by the company or when the product wears out or no longer has a purpose. Chapter 16 addresses some issues in disposal.

## Case Study: Disastrous engineering processes fixed

This story really happened. I was in the middle of it. The good news is that this story has a happy ending. I have changed names and obscured the product design to protect the innocent (and guilty).

Some years ago I was called into help with an aerospace subsystem. The subsystem collected high-bandwidth data, compressed it, and transmitted it to a central data center. In the data center, the collected data could be displayed in real time and then archived. Stored data could also be called up for display, as well.

Here were the requirements:

1. Collect data from five imaging instruments; the data had a video format.
2. Compress the data streams from the five imaging instruments.
3. Types of data and compression:
   * Lossless compression for radiometric observations.
   * Lossy compression for video streams.
   * Housekeeping data which was primarily temperatures from thermocouples.
   * GPS time and 1 pulse per second (pps) to synchronize the data.
4. Multiplex the data into a single serial data stream for telemetry from the aerospace vehicle.
5. Select and sequence instruments during a mission or have all five going at once.
6. Demultiplex the data, decompress images, store on disk, and display video images on a computer in the data center.
7. Survive a harsh environment onboard the aerospace vehicle until the end of the mission.

The electronic circuit hardware included a circuit board for compression, a circuit board for multiplexing, a circuit board for the power supply, and a circuit board for analog housekeeping. A Field Programmable Gate Array (FPGA) on the multiplexer board buffered the data streams. Digital Signal Processing (DSP) chips compressed the images and multiplexed the data. The software was custom C code running on the DSP chips. The software in the computer in the data center was to be custom C code, as well.

OK. That was the basic architecture of the system and its major requirements. What should be done, according to good practices in SE and QA? Here are the major activities that should have been in place (but none of these activities were in place):

1. Plan, execute, review, report iterate (remember PERRU from above?).
2. Define the development plans.
3. Define and operate a configuration management plan.
4. Use a style guide for software coding that included headers, comments, formats.
5. Maintain records of production metrics and bug rates—LOC per hour or per day, bugs found, severity, and status of fixes.
6. Hold regular code reviews.
7. Hold regular project reviews.
8. Provide incremental releases of the product with expanding functionality.

The background to the project was interesting but a bit convoluted. The system was represented as a catalog item that needed some nonrecurring engineering (NRE) to fit the customer's particular needs. The vendor also claimed to have done this sort of thing before and had a long series of successful missions to their credit.

But, the reality was a bit different. While the customer's system was to be nearly all digital, the vendor's previous systems were nearly all analog. The vendor had built a digital system once before for another customer but it was a much less ambitious system with only two imagers and they still ran over budget and over schedule on completing that system. On top of all this, the current customer executed a weak contract that did not specify staged deliverables or incremental releases of functionality—partly because the customer expected the system to be off-the-shelf in some measure. They should have insisted on a list of deliverables and a schedule! So, let's look at the good, the bad, and the ugly.

### The good

The vendor did have a remarkably good mechanical design and development for the enclosure that was to ride on the aerospace vehicle. The vendor also produced a good electronic hardware design and had quite reasonable development processes for the circuit boards.

### The bad

With about a month to go before delivery of the subsystem, the customer called the vendor and asked, "Is it working? Demonstrate its current operation to me." The vendor came back and said, "Ah, no, it's not working and we don't know when it will be working." At this point I and several others were called in by the customer to help assess the situation and fix it, if possible.

### The ugly

Within a few short days, we flew coast to coast and met with the vendor. Getting down to business, we discovered the following problems:

1. Software development was in chaos. A single developer—a "lone ranger" sort of character within the vendor's company—was writing the code. He had no accountability, no code reviews, no development plans, no design, no schedule, and no test plans. He could not demonstrate any operations.
2. He had used legacy code that did not fit this project and tried to shoe horn it into the new product. He had written spaghetti code with the following problems:
   * Nested interrupts
   * No common event handlers
   * Orphan code segments
   * Didn't follow a style guide
   * No documentation or comments.
3. The vendor had bitten off more than he could chew—planning to develop new code for the compressor, the multiplexor, and data center all in a very short span of six months. And with one guy doing the work, no less!

4. The vendor designed in an unfamiliar set of DSP processors and an equally unfamiliar development system. The vendor had to learn the "ins and outs" of using caches, serial ports, and internal memory.

### The turn around

After assessing the situation, we explained to the vendor just how bad the situation really was. There was no way the vendor was going to make the contracted delivery date. On the other hand, the customer had no alternative solutions, so we had to rewrite the contract for a new delivery date. Then we explained what we expected—that the vendor would deliver a working product within a redefined schedule for delivery. We outlined the following requirements to achieve a working product:

- staged releases of the software with demonstrated operations of the subsystem,
- monthly reviews of development,
- the data center software would be a purchased COTS package and not custom developed,
- new software processes that were defined, followed, code reviewed, with metrics for production and anomalies,
- a complete set of documents.

The customer supplied several staff members to prepare some DSP code modules for the vendor. These were folks with a long history of rigorous development ethics. The vendor then laid down the "new world order" to its staff, much to the credit of the project manager in the vendor company. The "new world order" included doing all that we requested, particularly new software processes and the vendor staff had to like it or lump it.

In the process of instituting the "new world order," I predicted that "lone ranger" would hang on but cause trouble by not following the new software processes. Unfortunately, that is what he did within two short weeks; he caused trouble and did not follow the new software processes. The vendor and the lone ranger developer agreed to a mutual separation.

The remainder of the vendor's team dived in and worked 70–80 hour weeks for six months. We all followed the new schedule of staged deliveries and gradually saw the system begin to function.

### Trials and tribulations

A number of things cropped up during this intense period of redevelopment. Almost all of them had to do with "features" of the selected DSP chip; these included:

- peculiarities with the serial ports and caches,
- floating reset lines that were not tied down correctly,

- queues versus buffers and differences in operations,
- delays in turning on transistor drives to open covers over the imagers.

### The final product

The final product delivered a bit later than the reworked contract specified but it worked as advertised. The vendor was proud (and rightly so) that they turned out a good product that worked and was robust in unusual and extreme environments.

## Conclusion

1. It is highly unusual for a company and group of people to admit failure and then regroup to push through to a successful end result. Far more typical is for human nature and tendency either to fall on their sword and quit or to deliver a half working product, which we would then have to work around and use its diminished capability.
2. Failure in this project could have severely damaged the customer's reputation and relationship with its sponsor (the final customer), but almost miraculously it did not because of hard work, discipline, and rigor.
3. Both the vendor and the customer agreed to tough measures—the customer put its foot down and then worked as a team member, the vendor agreed to changes in processes.
4. The customer gained a new industry partner in the vendor who proved itself and now is a trusted resource. I am guessing that they will get follow-on work and contracts from the customer.
5. This story has a happy ending, which is very unusual. It had complex interactions and a complicated structure. It came down to good people with good attitudes, doing good work with the right tools. If anything else had occurred, this story would have ended much differently and not nearly as satisfactorily.

## Acknowledgments

I thank Geoff Patch and Tim Wescott for their reviews of this chapter. I incorporated their insights and enjoyed the humor in their comments, which, in addition to pointing out that "none of us is as dumb as all of us," got into skinning raccoons.

## References

[1] Blanchard, Fabrycky, Systems Engineering and Analysis, fourth ed., Prentice-Hall, Upper Saddle River, NJ, 2005, pp. 1−2, 23−24
[2] MIL-STD-499B (Draft), Systems Engineering, May 6, 1994, p. 40.
[3] E. Verzuh, The Fast Forward MBA in Project Management, fourth ed., John Wiley & Sons, Inc., Hoboken, NJ, 2012, pp. 7−8, 16−17
[4] J. Ganssle, M. Barr, Embedded Systems Dictionary, CMP Books, San Francisco, CA, 2003, pp. 90−91

[5] F.P. Brooks, No silver bullet—essence and accident in software engineering, in: Proceedings of the IFIP Tenth World Computing Conference, 1986, pp. 1069–1076.

[6] D. Mancl, S. Fraser, W. Opdyke, No silver bullet: a retrospective on the essence and accidents of software engineering, Proceeding OOPSLA '07 on Companion to the 22nd ACM SIGPLAN Conference on Object-Oriented Programming Systems and Applications, ACM, New York, NY, 2007, pp. 758–759

[7] J. Collins, J. Porras, Good to Great, Why Some Companies Make the Leap...And Others Don't, HarperBusiness Essentials, New York, NY, 2001.

[8] J. Collins, J. Porras, Built to Last: Successful Habits of Visionary Companies, third ed., HarperBusiness Essentials, New York, NY, 2002.

[9] H. Cloud, Integrity: The Courage to Meet the Demands of Reality, Collins and imprint of HarperCollins Publishers, New York, NY, 2006.

[10] H. Petroski, To Engineer Is Human, The Role of Failure in Successful Design, Vintage Books, A Division of Random House, Inc., New York, NY, 1992. p. 9

[11] J.G. Ganssle, The Art of Designing Embedded Systems, Butterworth-Heinemann, Boston, Woburn, MA, 2000, pp. 13–34

[12] J.R. Platt, Career Focus: Systems Engineering. IEEE-USA Today's Engineer, <http://www.todaysengineer.org/2011/Nov/career-focus.asp> November 11, 2011 (accessed 23.11.11).

[13] <http://www.incose.org/practice/fellowsconsensus.aspx>.

[14] Systems Engineering Handbook, A Guide for System Life Cycle Processes and Activities, INCOSE-TP-2003-002-03.2.2, October 2011, Prepared by SE Handbook Working Group, International Council on Systems Engineering (INCOSE), San Diego, CA pp. iv, 2, 17, 18.

[15] A Survey of Systems Engineering Effectiveness, December 2008, SPECIAL REPORT CMU/SEI-2008-SR-034, p. 100. This report is the result of a collaborative effort between the National Defense Industrial Association (NDIA), Systems Engineering Effectiveness Committee (SEEC) and the Software Engineering Institute (SEI) of Carnegie Mellon University.

[16] J.M. Stecklein, et al., Error Cost Escalation Through the Project Life Cycle, Report Number: JSC-CN-8435, 14th Annual International Symposium, Toulouse, France, June 19–24, 2004. <http://ntrs.nasa.gov/search.jsp?R = 20100036670>, 2004 (accessed 18.01.14).

[17] Systems Engineering Handbook. NASA/SP-2007-6105 Rev1, <http://www.acq.osd.mil/se/docs/NASA-SP-2007-6105-Rev-1-Final-31Dec2007.pdf>, December 2007 (accessed 18.01.14).

[18] J.R. Sedlak, Quality Management System Aerospace Requirements, Differences between ISO 9001:2000 and AS9100B, Smithers Quality Assessments, Inc., 2006<http://www.smithersregistrar.com/as9100/page-deltas-between-iso9001-as9100.shtml> (accessed 19.01.14)

[19] L. Penn, CMMI and Six Sigma, in: M.B. Chrissis, M. Conrad, S. Shrum (Eds.), CMMI, Guidelines for Process Integration and Product Improvement, second ed., Addison-Wesley, Upper Saddle River, NJ, 2007, pp. 5–8.

[20] W. Humphrey, CMMI: history and direction, in: M.B. Chrissis, M. Conrad, S. Shrum (Eds.), CMMI, Guidelines for Process Integration and Product Improvement, second ed., Addison-Wesley, Upper Saddle River, NJ, 2007, pp. 5–8.

## Suggested reading

H. Cloud, Integrity: The Courage to Meet the Demands of Reality, Collins and imprint of HarperCollins Publishers, New York, NY, 2006. A fine business book that provides good support for why integrity is so important to all that we do as professionals. Dr. Cloud writes well and has interesting examples to illustrate the principles he espouses. New York, NY.

E. Verzuh, The Fast Forward MBA in Project Management, fourth ed., John Wiley & Sons, Inc., Hoboken, NJ, 2012. This book is very readable and composed very well. Mr. Verzuh writes succintly and describes concepts clearly. Hoboken, NJ.

# Drivers of Success in Engineering Teams

**Allison Fritz**

*Organization Development and Training Consultant, The Johns Hopkins HealthCare LLC*

## Chapter Outline

Developing and Managing Embedded Systems and Products.
DOI: http://dx.doi.org/10.1016/B978-0-12-405879-8.00002-7

## *Overview of organizational and psychological drivers*

In the workplace, much more activity impacts our individual and organizational success than simply applying our technical skills. While this may be easy to observe, still many highly technical professionals function as if the technical knowledge is all that matters. Although we may believe that other aspects of the working environment *should not matter* or impinge on us or our time, leaders have come to realize that they do.

A savvy engineering professional knows she or he needs to learn the basics of this world outside the hard science of the work. Increasingly, knowledge about functioning within teams is explicit in job descriptions. Organizational initiatives of talent and performance management demand your attention on nontechnical behaviors and competencies to be considered a "high potential" or "high performer." Ultimately understanding the psychology of the work environment, and your team, can improve recognition of your contributions.

### *Take a panoramic view of your workplace*

As we widen the lens on the working environment, the first view that comes into focus is that the workplace revolves around human dynamics, no matter how hard we try to make it all about the task. What does this mean? Let's say that you solve a challenging problem, derive a new approach, or develop a new model. How effective will you be if you cannot get others to see its value, be motivated to support or work on it, or translate it into practical application? Likewise, changes made on one level of the organization and then handed to another to implement are often met not only with criticism, but also with seemingly hostile resistance to the new approach or direction. The human factors and the processes that help us to manage effectively begin to come into focus.

### Step on the three-legged stool

One metaphor for the organizational workplace is a three-legged stool. Three critical components are essential to maintain and to manage, for the organization to stand with stability. These are:

1. The *tasks*, or the work we are attempting to accomplish.
2. The *processes*, or the means by which we are getting those tasks completed.
3. The *relationships*, or the people and groups with whom we work, such as project teams, coworkers, managers, and customers.

The task leg of your work, or that of your team, may be quite solid. Your processes, though somewhat imperfect (as most are), may also be sturdy enough to hold up. But, if the relationship leg of your effectiveness is fractured or undeveloped, your ultimate results will suffer in one way or the other. We will focus this chapter primarily on the relationship aspects of the workplace, as well as on some processes that can keep your stool standing firm.

## The role of the team member

Your first role is that of a team member even if you are the team leader or project manager. Many engineers prefer an introverted style of work, which may include a high degree of independence and even solitude. You work, however, in the context of a team (or group) and a workplace, even if virtually. How you relate in that context will impact your outcomes, as well as the perception of your value.

### Expectations of team members

#### Team player redefined

One expectation that most organizations have of their team members today is that they will be effective "team players." The metaphor of team work has been used for decades to illustrate effective means for working interdependently. Inherent in the usage of the term is the message that we cannot get to the goal without working together. It may never be more important than it is today. Collaboration, partnerships, alliances, affiliations, and shared services are examples of how we live in an age where relating well with others while working toward common goals, or at least interdependent outcomes, is essential. The term integration has made its way into many fields, magnifying that connectivity to others is a constant expectation, as is greater transparency and shared information. As discussed later in this chapter, you will see that you can rarely go it alone. The better you function as a team player, contributing to the multiplicity of "teams" you participate in, the more likely you are to be successful.

### Some common expectations

Expectations of team members vary from team to team. Rather than assuming that what has worked well for us in the past will work again, it is useful to clarify hidden expectations. Most professionals look only to basic job or technical expectations and miss the implicit messages about what is expected from leaders and coworkers that are embedded in the team's *culture*. Culture in this context is *the way we do things around here*—our norms, values, and beliefs that govern our behaviors and priorities.

One way to make these culture expectations explicit is to raise the discussion with the team. Start some one-on-one dialogue with team members, asking questions such as *What do you need or expect from me that will help make us both most effective?*" You can negotiate these expectations, so that what you agree to do is reasonable to complete. This sounds simple but is often avoided. As a result, you need to identify specific, implicit expectations and make them as explicit as possible. You can often frustrate or disappoint a team without being aware of it by following your own expectations or some set of assumed expectations, which seem reasonable to you.

Teams can clarify expectations as a group, expediting this understanding. Having worked with many different teams through the years to help them set team expectations, I have seen patterns. Here are a few of the general expectations that team members hold as ideal for their coworkers:

1. *Respect*: This tops the list. The topic of respect appears in virtually every group and team building I conduct. Respect is a bit like money, in that it means different things to people and is very emotionally charged. The key here is to clarify what respect looks like to your coworkers and deliver it. Typically it is a way of speaking and behaving that communicates the coworker has dignity, value, and perspectives worth considering.

    Consider your impact, as well as your intentions. Many agree that we have a natural tendency to assume the best in our own intentions but focus on the impact of others' behavior. So we excuse ourselves and are quick to judge others. If you look past behavior and consider others' intentions and then consider the impact your behavior has, *despite* your intentions, you stand the best chance of communicating respect.

2. *Open to ideas and perspectives*: Team members want teammates who will consider their ideas, even when they pull against the current paradigms.

3. *Honesty*: While this should be an assumption, it makes almost every list. If we look at statistics in the United States about lying it becomes clearer. But honesty addresses more than lying. They often mean they want team members to be forthright, offering their honest views and thoughts that would ultimately help the team become more successful, as well as not pretending to go along with things they do not agree with.

4. *Produce what is promised, when promised*: Nearly all teams want team members who do not overpromise or pass off work to others. It is not likely irresponsibility that produces the need to voice these expectations, but a combination of desired accountability (not passing the buck) and prioritizing what matters most to the team. If the team sets a goal, they want all team members to work diligently to achieve it.

5. *Willing to be flexible*: An adaptable team member is of high value, since work and life's demands change. Wasted energy or unpleasant negativism develops from those unwilling to move when the team needs to move. Most team members need some support because they cannot meet an expectation or have to change plans; they want team members who will understand. Adaptability also tops the list for many authors of books about leadership and teams, including John Maxwell, who sees it as an essential quality of a team player [1].

6. *Communication*: It is fairly unanimous that people do not like having to read others' minds; they don't do it very well, anyway. Team members want others to be willing to share their thoughts. But even more importantly, they want members to keep them informed, if for instance, the deadlines can't be met or they cannot make the meeting or the data gained would help them.

    Additionally, quiet, reserved individuals can be misread as withholding, or even focused on selfish gain. Communication builds trust. If communication is effective, it can build credibility, as well.

7. *Give the benefit of the doubt*: According to Patterson et al. [2], we observe something happening and build our own "story" around it. We, then, react emotionally to that story and act on it. Teammates prefer members who ask and assume the best before building a negative story. They want to be treated as reasonable, good people, with an assumption of their good intentions.

8. *Contribution*: Teams want the various contributions of their members to add quality, diversity, and to spread out responsibilities. Groups get concerned about uneven workload and expect members to offer their best to make the team successful and avoid inequity.

These expectations may seem to be common sense, but they are not always common practice.

## Characteristics of high performers

While job expectations differ and needed skill sets vary, there are characteristics associated with high performers that can be modeled in any engineering role. A key to being a high performer in today's workplace is the ability to **navigate through complex change**. Associated with this ability is being able to **collaborate with those differing from you in multiple ways**.

Managing change and diversity are key areas that will enhance your overall effectiveness regardless of the projects you tackle. Related characteristics that Corporate Executive Board (CEB) found associated with high performance in a study of organizations representing

20,000 employees include the following, which are also identified in much of the literature on emotional intelligence: "(1) *ability to prioritize,* (2) *team work,* (3) *awareness of the organization,* (4) *problem solving effectively,* (5) *self-awareness,* (6) *proactive approach,* (7) *ability to influence,* (8) *effective decision making,* (9) *open and agile to learning, and* (10) *technical knowledge* [3]." Technical knowledge is the only characteristic here that is not identified in the emotional intelligence models. These generalized competencies can serve as a focus for professional development.

### Managing priorities

With complex changes, restructuring, and rapid technological advancement, it is essential to be able to set priorities. Often, staff feels squeezed between competing voices of urgency. Stakeholders may have different priorities than your manager; a vice president may have different priorities than other leaders who depend on your work; even your coworkers may have priorities different from yours. Everything can seem urgent. You need to be able to read the political playing field and set your priorities accordingly. It helps to clarify what needs your attention, versus those activities that will eat up time without meaningful, progressive results. Additionally, trying to please everyone simultaneously may splinter efforts toward ineffectiveness. Those who prioritize well and align with the organization's strategic direction are likely to be most effective.

### Learning with agility

With the increasing emphasis on knowledge and information, remaining updated in your field is an assumed activity. Those who explore new approaches and are open to learning are also recognized as valued players in change initiatives. As a result, you cannot rest alone on your degrees or your theoretical underpinning. You need to seek out opportunities to stay current and explore new ways of doing what you do. That may mean allowing the younger or newer coworker to teach you; attending training classes or online webinars; obtaining new certifications; and finding opportunities to network with others.

*Informal learning* has become a key approach to staying relevant and up to date. An online study of informal learning done in 2008 by the American Society for Training and Development (ASTD) with the Institute for Corporate Productivity defined it this way: "a learning activity not easily recognized as a formal training and performance support [4]." They view it as an employee-driven process that happens without traditional style trainers. According to this research, learners look to technologies, such as e-mail, intranets, and online searches for information, as well as in-person conversation, such as coaching, mentoring, and communities of practice to gain information. Online informal learning communities are growing to keep conversations and learning active and relevant. Remaining agile and open to learning is a characteristic of today's effective professional.

## The role of the team leader

The role of the team leader has a significant impact of the outcomes and experience of the team. Anyone who has worked with a poor leader knows what research supports: that the direct supervisor has the greatest impact on the experience and ultimate retention of the team members.

Whether you are the functional manager, team lead, or project manager, your role should involve *leading*. We often think of leadership as top-down authority. Yet, leading is a reciprocal relationship in that others make you a leader by following or emulating you. If you march on, only to turn around and find no one there, you are marching, not leading.

*Leadership differs from management*. Management is most often focused on monitoring and tracking quality and results, while leadership is about moving ahead toward a desired vision or goal. Tracking and monitoring is important, but will fall short in this competitive, fast paced and changing technological world. Kotter [5] explains that there are fundamental differences between management and leadership, and companies need both. The stability of management is essential, but most companies have been "over-managed and under-led," according to Kotter, who sees change and increasing complexity demanding more leadership within US corporations. The higher up the level of leadership, the more future focused the role needs to be. Most employees wish their managers were more future-focused. More than likely, your role requires not just management of a team or project management, but leadership as well.

## The team leader's role in managing change

Leaders and team members are now residing in a professional world of constant change. How a leader manages those changes, especially the larger scale initiatives, makes a significant difference in the productivity and satisfaction of the team.

Your own responses to change can influence your management style, as well as effectiveness. A manager in the technical fields needs to be flexible and accommodate change well. An unwillingness to be adaptable can stifle a team and truncate a leader's career. Too much flexibility may seem insensitive to those who don't adjust well or are facing significant loss during a change initiative. It also may communicate lack of passion or commitment. Either way, if you are negative, your team is likely to be negative.

It's easy to see how your response to change is influenced by the nature of the change in the situation. Yet, you also carry within your personality an orientation toward change in general. If you are aware that stability and small incremental changes that respect the current system are your typical preferences, you will need to give yourself time and attention to deal with significant upheaval in your organization and avoid letting your preferences dominate the way you handle the change with your team. Being open and honest about your preferences also helps others to relate with you, as well as understand your reactions. If you tend to prefer transforming or redesigning as your orientation to change, you may need to ensure you are communicating a respect for the standard processes, as well as those deeply invested in maintaining them.

Your approach to change will certainly have an effect on your team. A strong manager will recognize the emotional aspects accompanying change and work with her team to address these, even when moving at lightning speed toward a deadline.

### Responses to change in the workplace

A team leader is challenged with navigating large and small organizational changes, straddling the practical realities of the present priorities, and dealing with the uncertainty of the next urgent wave of updates rushing toward the team. Bridges [6], in his book *Managing Transitions*, finds that it is not the change itself that is the issue for most people, it is the *transition*—or the process happening psychologically as one comes to terms with the change. Bridges finds that we move from a place of letting go to a new beginning. But in between we pass through a "neutral zone" where we feel suspended in the air, because the old is fading away but the new is not fully functional yet. It is an astute team leader who can recognize the process of transition and learn to manage people through it.

In the neutral zone, your team may feel disconnected and even insecure. Supporting your team members and helping them understand the impact of change will facilitate the process of accommodation. Encouraging team members to experiment with new methods in ways that reduce fear of failure of frustration serves as an investment in the future change. While changes may be transformational, implement them incrementally. Start small and where the energy is the highest and move from there when possible. Many organizations are experiencing a sense of overwhelm and overwork from too much change at once. This stress is an influential factor in causing disengagement which leads to lowered productivity. So keep whatever you can stable, while major changes are impacting the team. For example, if your group is going through a system change, keep smaller things in a routine, like your staff meetings, and save the physical move to a new floor or building for a later date.

What is important for you to grasp here as a manager who wants to see progress fast is that transition takes time and attention. You may install the new system or change the structure, and feel you have done your work, but the full acceptance of the new approaches will still happen over time, as the human aspects of the change process acts as the pace car to progress. Transitions are challenging. Some chaos and intense reactions are expected. But we can encourage a better, more efficient transition if we attend to the needs, responses, and challenges of those engaged in change. With awareness, a leader can ease the intensity on the team and avoid the sometimes unpleasant expressions of transition from stalling the change initiative.

Leaders are often surprised by the intensity of responses they get from their teams to organizational changes, especially those that they think may not be significant, or are obviously needed. I have observed the annoyance of managers who call reactions from their team as *push back* or *resistance*. A manager can even take the responses personally or judge the team members negatively. It is often the emotional response to change in these circumstances with which the leader is colliding. It is certainly the human factor that can and often does derail organizational change initiatives.

### Change and loss

A primary reason for the reactions we get is that most leaders want to plow ahead with changes without looking back. But all changes involve loss—even welcome changes. For example, when someone marries, he may experience some loss of independence. Losses related to organizational changes may include loss of status, relationships, or flexibility. A new leadership structure, for example, may pose a loss of an established reputation and relationships for a high performing employee that he now needs to regain. A seasoned employee may experience loss of his status as the "expert," as new systems and approaches are implemented. A new strategy or organizational vision may prompt an employee to feel she is losing her investment of time and commitment to projects or products she deemed valuable.

Depending on what we stand to lose and how much or significant that loss is, people will respond with varying degrees of emotional reaction. ***Loss evokes emotions.***

Many have used Kubler-Ross's [7] stages of grief as a template for how we relate to change that involves significant loss. These stages begin with shock or denial and move to anger, depression, bargaining, and finally to acceptance. Bridges writes that leaders need to accept the reality that team members are experiencing loss, even if they are intangible perceptions; acknowledge the losses openly, since discussing it may help speed recovery; and compensate in some way by giving something to balance the loss. Compensation may take the form of a career opportunity, positive recognition, or an action that may mitigate or counterbalance the loss.

As your team engages in new initiatives or projects, optimism in their expectations is a factor in the final results. Kelley and Conner's [8] emotional cycle of change overlays the grief cycle, identifying stages of emotions relating to two factors: optimism and level of information available. They found that people begin in a state of *"uninformed optimism,"* similar to what some may call the "honeymoon" stage. But that "honeymoon" may take the form of cynicism or assumptions that the change will simply not occur, if it is unwanted. They do not have enough information to prove otherwise. For instance, the employee who has stayed with a company through many changes may write off the latest initiative, assuming it will never happen or may be heard saying things like *"I'll be retired by the time that happens."*

As time passes, reality sets in that the change is not going away. Even if the change is perceived as positive, over time a more informed perspective begins to burst the idealistic image of what was originally expected. The individual may shift to *"pessimism"* since the unknown is now seen, as he wrestles with reality. For example, the new position to which you promoted your team member may involve duties she did not anticipate. She now feels pessimistic about her role and may even wonder, at this point, if she has made a mistake in taking the job. If the feelings are not challenged with exploration of how to cope and improve, she may experience disillusionment and then actually quit. If she begins to explore the opportunities to overcome her challenges, she will move into *"hopeful realism,"* tackling questions of how to conform her new tasks to her values and desires, as well as trying new approaches. Once she tests some new methods, she may experience what is tagged as *"informed optimism,"* and ultimately move to a satisfied state of *"rewarding completion."*

Change is so rapid now that this emotional process may recycle before we ever reach full satisfaction and completion. But we may recognize and move through the cycle more quickly once we can predict it and avoid overreacting at any single point in the cycle. Understanding emotions associated with change helps the team leader respond more effectively. He will more likely avoid the pitfalls of contagious negativity, as he identifies the cycle of emotions in himself.

We often can predict responses of our team members to change we are leading or promoting on behalf of the organization, according to Zagarmi [9] who offers six predictable concerns to consider in your management of the change process. The first two are pivotal. *Informational* concerns show up first, in that people generally want to feel "in" on the changes and will be most concerned with getting the information in a timely way: What is happening, when will it happen, to whom will it happen, what role do I play, etc. *Personal* concerns arise quickly, as well, when change is presented. People are concerned about how the changes will impact them. When a leader knows to expect these concerns, he can prepare to reduce them early in the process, as well as experience less frustration.

---

### Example: Personal Concerns During a Change of Business

I recall a change process in one organization that was moving to an inspiring new approach in their work. However, the initial conversations about the change defaulted to the details of whether the staff would be able to continue to telecommute. The people seemed resistant or to be ignoring the big picture and its potential, frustrating their leaders. They eventually made the shift to the new way of doing business, but their first point of focus was on how the change would impact them, personally.

*Individual patterns of response*

Some personality types are always future focused, while others pride themselves in maintaining stability in the present and slowly making incremental improvements. These personalities will respond very differently to the same change initiative or paradigm shift. Change is just more easily accepted by some than others; many times, it depends on the change itself—whether it is desired. A key to understanding resistance to change is recognizing that the more a person feels they will lose control, the less likely they are to move on [10]. Before starting any change process, it is essential to assess the readiness of the team and its members in two specific areas: capability and willingness. How capable are the team members to make the change? Do they have the knowledge, skills, experience to do what is asked? Secondly, how interested are they in making the change? If they do not buy in to the future picture, that requires a different focus for your energy.

Bunker from the Center for Creative Leadership [11] describes four patterns of responses that individuals may have to organizational change. These patterns are not types of people that are unchanging. They are behaviors that you may see people express based on their preparedness for the change:

- The first is the *entrenched*, who is not seeing himself as ready for the change in any way, is likely angry and anxious, while probably expressing some denial. A person who is entrenched may be openly critical of the company, the management, and sees the work environment as chaotic. This person can be encouraged to go beyond feeling stuck, if given the opportunity to explore the realities of the impending change, in a safe way.
- The *overwhelmed* person experiences fear, feeling that they are not ready despite possibly being able to make the transition. This person may isolate or retreat, and even become discouraged or depressed.

### Example: An Overwhelmed Employee

I recall a woman in a workplace who had been in her position for years, making it to a supervisor level without a lot of education. When the company began changing and her role was going away, she avoided a year or more of offers from me to help with her resume, retreated quietly to her office, and ultimately had an emotional breakdown when the company went out of business. The overwhelmed may not always seem this dramatic, but the signs of overwhelm are similar.

- The *poser* according to Bunker is one who seems very ready to make the change but doesn't have the capabilities in terms of knowledge or skill. While his eagerness may be attractive, this is not the person you want to see in charge of the process. This person can do damage, so a good deal of accountability is needed for him. With the right amount of learning opportunities and development, he can contribute well.

- Finally, there is the *learner*, who is willing to make the change and open to learning what is needed to go forward, with cautious optimism. This is a role model for others to encourage the change process, but be aware of overusing the person and burning him out.

To support a smooth transition to new paradigms, initiatives, and projects, each team member's concerns need to be considered and strengths should be used in the process. What leaders consider resistance often is a combination of the kinds of issues discussed in this section.

Aside from remaining alert to predicable responses of people to change, leaders can make their teams change-nimble, by keeping their eyes on the future and inspiring their teams to be thinking innovatively.

Essential to effectively managing change with your team is to avoid having the team members feeling like victims of change. One key way is to involve them in the process. The NTL Handbook of Organization Development and Change Principles reminds us that to reduce resistance, we need to create meaningful roles for our team members in the change process [12]. Harvard Business Professor and author, Moss-Kanter [13], encourages inclusion of the team members in decisions related to the change process, expressing that "smart leaders leave room for those affected by change to make choices."

## Change and future focus

Effective change initiatives have at their center a view of the future state. Whether you're the top executive or a team leader tucked in the middle of a large organization, a leader needs to have a sense of vision for the future. The higher the level, the more oriented to the future the leader likely needs to be. Leaders inspire others toward a vision. "Being forward-looking—envisioning exciting possibilities and enlisting others in a shared view of the future—is the attribute that most distinguishes leaders from non-leaders," according to Kouzes and Posner [14]. Vision and change are intricately intertwined, as are leadership and vision.

Despite their focus on daily managerial functions, managers are called on to be leaders. Ask yourself:

- Have I ever been inspired by a leader or a manager's vision at any time in my life? What vision and why? What did the leader do that influenced me?
- What has been my professional experience in trying to advance an organizational or project vision with my own team? What worked and what didn't?

Managers who are effective in leading the vision:

- Have vision for their personal life, as well as their work (they are not afraid to dream a little).

- Understand the mission and vision of their organizations and promote participation in the larger context to bring greater meaning to team members' work.
- Help their teams see the connection between their work and the larger vision.
- Communicate a positive future that inspires others and do it in an effective way through repetition and various forms of media.

Your connection to a higher purpose, one beyond your immediate self-gain, elevates the sense of spirit on your team. This can generate renewable energy toward your team's work and will positively impact your productivity and results. A sense of purpose leads to better performance. Many managers offer monetary incentives to their team as motivation. Pink [15], citing a study from MIT, notes that *bad results* come from unhitching the "purpose motive" from the monetary incentives offered employees.

If you survey great accomplishments through history, it becomes clear that vision helps ordinary people do extraordinary things. ***Vision is an essential part of change and leadership***.

## Aligning yourself with organizational goals and strategy

It's assumed that we need to identify and secure our focus on the goals and objectives of our program or project. We may be less driven to remain aware and committed to the overall organizational strategy. Yet, doing so will provide the context and the expectations for our work. How many professionals have settled into their task, put their heads down and plowed ahead in their work, only to discover that the project to which they have devoted long hours being slashed for another initiative? We may not be privy to the strategic decisions, but tuning into the larger picture of the organization can avert frustration, as well as wasted effort. Additionally, it will help us determine our genuine fit in the organization.

### Organizational mission and priorities

Do you know the mission and strategic priorities of your organization, at least on a conversational level? Or do you need to flip over a badge, a button, or a card to read what it's about, that quite honestly seems irrelevant to you? A first step in aligning is to observe if your organization is acting in concert with its professed mission statement. Some statements can become almost irrelevant. They hang on the wall or on badges, while leaders go about their business, adapting as necessary to real world pressures. Pay attention to what the leaders are saying and doing that communicates its true direction, values, and strategies. Once you are clear about the overall mission and current strategies, you will need to set a strategy and team vision that is aligned with the larger organizational strategy. This helps keep your expectations and those of team members realistic and relevant. It's wise to involve a facilitator/consultant to help your team develop a vision and plan that is not only well linked to the organization's business strategy but is also motivational, clear, and balances the real world with the ideal.

## Self-awareness and assessment

To lead, or even participate effectively in, an engineering team, it helps to have a good assessment of your own potential impact on the team. This requires some intentional self-assessment and self-awareness. What do you need to know?

Ideally, we want to have a good understanding of our own personality style. This increases our ability to adjust our style to communicate more effectively with our teammates. Why bother assessing our style?

---

### Case Study: Personal Style Affects Your Team

Jan was new to an internal consulting team. She had the technical knowledge needed for the job and some good ideas. But she lacked an awareness of how her style was impacting her teammates. In her eagerness to contribute and make improvement to the project, she appeared combative; and in her desire to be seen as competent, she came off arrogant and unyielding. Frustration, confusion, and misunderstandings swelled around her until she was forced to leave the job. Even in transitioning out, she continued to blame others for her inability to be successful. The professional who followed her in the role made the same suggestions that Jan had made, only to find the team very receptive. Had Jan assessed her own style and needs, and adjusted accordingly, she might have gained the trust of the team, rather than an unwelcome loss of her new job.

---

It is also useful to know how we tend to handle conflict, how engaged we are in elements of emotional intelligence, our orientation toward change, what roles we tend to play on a team, and our needs as they are expressed in the work environment. All of this may sound like overkill or a distraction from focus on our objectives but the greater understanding we have of ourselves, the more intentional we can be about influencing our environment, projects, and teams, while getting the results we want.

### Leader self-awareness and style

A leader's behavior is often experienced by his team as if amplified. Imagine the old time coaches with a megaphone in hand calling out to their teams. This may not seem remotely close to your intention. Yet, the effect of leader behavior and decisions are significant. You may feel like a regular teammate with no inordinate power, especially if you are sandwiched in the middle of the organizational structure. But you are intensely scrutinized if you carry any authority over others' careers. As a result, self-awareness can make a vast difference in how successful you can be, as well as in how much motivation and loyalty you can cultivate toward productive ends.

### The power of personality

A good beginning in assessing yourself is considering personality style. We may perceive activities very differently than others on our team, and act in ways that confuse those who lean on opposite preferences to ours. Additionally, personality style differences account for many avoidable conflicts on teams.

The well-known *Myers–Briggs Type Indicator* (MBTI), based on Jung's type theory, can help you identify your preferred tendencies in perception, decision making, and your relationship to the world around you. This particular tool, developed by Katherine C. Briggs and Isabel Briggs Myers, assesses four dichotomies and presents 16 personality types, based on these preferences, denoted by four letters (ISTJ, ESFP, etc.) [16].

The four dichotomies that the MBTI illuminate are introversion versus extroversion; intuition versus sensing; thinking versus feeling; and judging versus perception. The assessment can be taken and interpreted by a certified trainer or coach, who will explain, in detail, the particular meaning of these dichotomies on the tool. For example, those who prefer introversion get their energy from within, need time alone, and usually prefer to reflect on a problem, or think before speaking. They often do not initiate contact in social settings and may prefer to keep information more private. So, what impact might this have on a team? A leader who prefers introversion and expresses it in these ways might provoke insecurity in a new team member who wonders what the leader's perception of her is or whether his suggestion is being considered seriously. An extroverted team member may take offense to the leader's closed door or absence of a morning *"hello"* when entering the office.

There are a multitude of instruments that measure aspects of personality as well as behaviors. The key is to take time for some personal reflection and assessment to enhance your professional self-awareness.

### Leader style

There are many leadership style assessments to consider competencies and characteristics. You are in a better position to communicate with your team, particularly in helping them understand your meaning and intentions, as well as the areas you need them to support you, if you are more self-aware. You are also better equipped to hire and structure your teams to balance your strengths and weaknesses. Without intentional assessment, a leader's view of herself can be skewed by the reality that few want to confront their leader on his or her weaknesses, blind spots, or misunderstandings. Informal or formal assessments will give you a spring board for transparency and dialogue with your team.

*The 360° perspective—soliciting feedback*

Asking for and listening to feedback is an essential part of leadership, especially since negative feedback from staff is usually conveyed indirectly if at all, due to the perceived risk of sharing it. Choose four to seven people from your inner circle, personally and professionally, that you trust to be honest with you and ask them a few key questions about your style and behavior. For example, *"In the area of communicating, what do I do that works well? What is an area that you think I can improve?"* It can be challenging to avoid acting defensively when confronted with some answers that may differ from your self-perception. Resolve to record their responses and not react, until after reflection.

One way to gain a realistic view of your professional or leadership image is to conduct a 360° survey. This survey can be either formal or informal. Inviting those who work with you as well as your direct supervisor(s) and reports to offer you anonymous feedback about your behavior is invaluable information. The coworkers, whom you choose, respond to a customized survey about you; you then receive a report based on their results. There are 360° instruments based on a variety of areas relating to behavior and performance, as well as emotional intelligence. If the assessment tool is given to the appropriate people—those willing to be honest and without *an axe to grind*—the data can be useful in improving your overall effectiveness, as well as satisfaction.

### Blind spots

Everyone has areas of behavior or performance that we are not aware are affecting us in negative ways. Yet, the usual dynamic that accompanies blind spots is that we tend to deny the data that defies our perception. Even when confronted with survey responses and feedback from trusted sources, many of us continue to target the tool or the person as inaccurate, because the data does not agree with our perception. Being open to data that challenge our perceptions is helpful in individual and team effectiveness.

## Establishing essential relationships

Establishing relationships is a critical success factor for a manager. Managers need to assess their desired sphere of influence, as well as the interdependencies that will impact their work. Setting up channels of communication and a network of support is a priority. Key players in your network are your stakeholders.

### Identifying your stakeholders

To identify your stakeholders, you need to first define who a stakeholder is. In your work, stakeholders are those who, in one way or another, will be impacted or touched by the work of your team. They have some degree of investment in your work. Think broadly. One way to find them is to map out your business process. Note every touch point of interaction with other departments, customers, and collaborators. After doing so, make a list of every role or person with whom you interact and who will be affected by the impact of your work. Then, ask the questions, "Who else is affected by what we do? With whom do we interact indirectly? Who cares about what we do that may not be obvious?" Identify who best represents the roles on your combined lists. This is a good start. You may want to prioritize the list by level of investment and who you need to be satisfied with your work. Be careful not to make the mistake of assuming your primary stakeholder is your direct supervisor. While he or she is a key to your satisfaction and success, and it is essential to manage that relationship effectively, too much attention may detract you from serving stakeholders that will ultimately affect your results or provide additional security to you for your professional or project's success.

## Relationship building

Meeting with a representative of every stakeholder will open the door for future interaction. These meetings should be inquisitive and focused on learning, as much, or more so than sharing your perspectives. Treat the time as an informal interview.

Be mindful that the interview should be reciprocal, in terms of what each may gain. A meeting that seems contrived or self-centered may backfire in terms of establishing future collaboration. Additionally, trust can begin to grow with some initial sharing of personal data and questions that promote deeper understanding of the other's stake or perspectives.

### Functional and project manager tension

Depending on how the structure of the organization is set up, there can be varying degrees of tension between functional managers and project managers. The project manager (PM), focused on targeted tasks and deadlines, often feels the pressure of leadership expectations. The pressure of the project's advancement, success, and timeline may drive the project manager to frustrate inadvertently the functional manager who is focused on more long-term strategy, team engagement, and behaviors. The project manager may view the functional manager as impeding progress or resisting team assignments. Conflict may emerge and slow progress, or submerge and create a negative atmosphere of work.

It is wise to invest in team development and relationship building between the functional managers and project managers who will have access to the same staff. Natural tensions built in to the roles then can be mitigated by healthy dialogue and collaborative problem solving.

## Team development

### Building blocks for team development

Why is it that some teams or work groups seem to really click and produce terrific results while others seem to get stuck, or worse, become embroiled in petty, destructive conflict or infuriating inefficiency? Is it really just that one guy who annoys you that provokes this team's ineptitude? I have worked with many teams, from those that seem unable to be repaired to those on fire for the future. The difference is rarely singular or a person. The culture of the team develops from particular behaviors and messages that become tolerated and then ingrained. People just feel helpless to change what they know is not working. The leader can play a significant role in improving the team's effectiveness, but even the leader often lacks the awareness of how to turn the ship around.

The cause of the issues may vary from lack of trust to lack of focus. Gibb [17] presented four factors affecting any "social system's" progress, such as your team, in moving toward a strong and trusting organization. These factors can help you as a team leader or member point to team strengths, as well as identify where the team needs to improve. Gibb's theory implies that a team can grow in significant ways, and that these aspects may build upon one another toward maturation. These areas are:

1. *Trust* or getting to know each other, rather than staying in depersonalizing roles.
2. *Openness*, allowing for the free flow of data, contributions, and knowledge among the members and outside networks.
3. *Realization* and self-determination of team goals and direction, with focused productivity.
4. *Interdependence*, rather than overdependence on the leader or project manager, sharing the leadership as needed and relying on one another's abilities.

Once aware of where the team may be most challenged, you are able to focus the group on setting goals for improvement in these areas. It is useful to look to the prior stage to target improvement. For instance, if your team members seem reluctant to share information, resources, or data, look to the stage of trust as a reason this may be challenging and focus efforts on building trust. If a team is struggling with unclear goals, look to the stage of openness to identify if data is being easily shared and how you might improve. For example, I have worked with teams that have been at cross purposes, members having varying understanding of the team's goals. Sometimes, the high degree of structure of a group does not promote all members sharing their perspectives, talents, and knowledge. As a result, members may differ in terms of what they think the team ought to be focusing on. Individual agendas, preferences, and ideas about direction remain untapped, yet at work in the dynamics of the team. Expression of these promotes teams aligning on their goals as well as better accessing the resources available to them. In other words, encourage sharing to gain "realization" of unified direction—clearer goals.

### Critical team member needs

Leaning on Schutz's [18] theory of individual needs, we can assume that any staff or team member that comes into your group brings three significant needs: *inclusion*, *control*, and *affection/affiliation*. While the intensity of the need might vary by the individual, it aids the leader to remain aware that these needs are present. In this way you can be intentional about acknowledging and addressing them in your leadership, as well as in your meeting management, and aim at greater team effectiveness. For example, many managers, especially introverted ones, may reject the idea of icebreakers at meetings or team building activities as wasted time or fluff. Yet these methods, among other benefits, drive right to the inclusion needs of staff and often the affiliation needs as well. When unmet, these needs can be a definite distraction to the task at hand. We bring our entire beings to work and meetings, regardless of how we define professionalism. A team member who experiences a satisfying sense of inclusion is freer to build relationships that support his work and is going to be more productive and happier than one who is left frustrated in these areas.

### Diversity and inclusion

Inclusion takes on an even deeper meaning when considered in the context of various forms of diversity. With increasing cultural diversity, teams are challenged in communicating effectively, collaboratively using all of the team's knowledge and managing conflict. Differences in approaches, styles of communicating, and values can undermine an otherwise highly competent team. Managers can assist culturally diverse teams by supporting cultural competence—having an awareness of one's own cultural identity and perspectives, while learning about others, and encouraging communication around cultural perceptions.

Diversity can be defined broadly to include differences on many levels including innate differences, such as gender and race; life experiences; and status such as economic, marital, and position in the organization. All of these differences influence our interpretations and behaviors. Inclusion is a key to building effective teams that ultimately use the unique perspectives and experience on the team.

### Phases of group development

In 1965, Tuckman reviewed articles written about work with small groups and proposed a model of group development from his observations of common concepts. This model, which he updated in 1977 after reviewing additional research, defines five stages that a group progresses through as it works together [19]. Tuckman's model has been generalized to work teams, such as task groups, project teams, and committees. Recognizing that there are developmental phases that your team will likely move through can help you target your strategies in productive ways and avoid exacerbating your challenges. A brief summary of his model follows:

- *Phase I—Forming*: Here the team is getting to know one another and beginning to organize. This is a time of testing and dependence, where the members discover what behaviors are acceptable, as well as orient to the required tasks [20].

- *Phase II—Storming*: In this phase, the team experiences some sense of chaos and intragroup conflict. Team members are recognizing differences and there may be some infighting. Similar to responses to change, team members may react emotionally to the demands of the group versus their individual preferences or perspectives, as a form of resistance.
- *Phase III—Norming*: Here we see the group becoming a cohesive entity through acceptance of one another's idiosyncrasies and the group itself. People freely exchange ideas and pursue harmony.
- *Phase IV—Performing*: This phase is marked by energy toward problem solving and efforts toward completing the tasks. Roles are focused more on the group goals and task completion than individually focused.
- *Phase V—Adjourning*: This phase acknowledges the termination of the group, where the group may experience high affection for one another and the group, as well sadly disengage.

## Engagement and the motivational environment

### What is engagement?

In recent years, the conversation about employee motivation has moved from satisfaction to employee engagement. Engagement is defined by Gallup as a psychological commitment to the organization and its goals [21]. An engaged employee is satisfied, but more importantly he is more productive, less likely to leave, and committed to the organization. Why does that matter to you?

Whether you are managing a project team, leading your staff, or participating in a collaborative initiative, the engagement of those you are working with will likely have a significant effect on your success. How frustrating is it to be working on a project team, and the expert in one area suddenly informs you she is leaving for another job? Now you have development delays, heavy or uneven workload issues, and new learning curves, as you transition to a new team member. High performers can be picked off easily if they are not engaged.

So why do some leaders retain their best and brightest people, and others lose their best talent? Kaye and Jordan-Evans [22] find that leaders who change their approaches based on the economy suffer more loss of talent than those who treat their employees consistently well, using engagement strategies. In tough economic times, we want to avoid taking advantage of the circumstances, demanding more and giving less to employees; or communicating that they should put up with more because they at least have a job. In these cases, high performers are likely to be poised to leave as soon as the economy changes. In

*Love'em or Lose'em*, the authors find that the best managers have a "talent-focused mind-set" and employ strategies that influence the engagement of their employees.

## What contributes to engagement?

Gallup has researched and identified core elements of engagement, using questions that are directly correlated with it [21]. Following a pattern that overlays *Maslow's Hierarchy of Needs*, these questions refer to critical areas, including [23]:

1. Basic needs, such as clear expectations and having the materials needed to do the job.
2. Belonging needs, such as the perception of caring coworkers and friendship in the organization.
3. Esteem needs, such as recent recognition of work, valued input, and quality coworkers.
4. Alignment with the organizational mission, encouragement toward development and employee progress.

## Motivational elements

As you weigh the motivational or engaging aspects of your workplace and leadership, there are a few areas you may want to emphasize in your assessment. Wagner and Harter [21] in their book *12: The Elements of Great Managing*, describe 10 elements that Gallup uses in their assessment process, which correlate with increased engagement. These reflect much of the work in the field of organizational motivation. Drawing from their research and others, the following sections explain three critical areas to analyze in your workplace.

### Value and recognition

One study originating in the last century, and replicated several times, asks employees what matters most to them in the workplace, listing multiple options including money, job security, among others. The consistently highest response has been feeling valued or appreciated by the employer for work done. Everyone wants to feel that they have worth and value.

One way that you value people is by listening to what they think and regarding it. Listening is a form of recognition. Additionally, when you encourage input in decision making, feedback, and idea generation, you increase motivation and ownership. You communicate value in the person by acknowledging that what they think matters. The by-product is greater commitment on the part of the employee: if they help make the choice on anything, they are more likely to support it; if they had a say in the process, they are more likely to follow it. If people are given access to communication channels to offer input and glean information, it is a statement of trust and value in them.

Many have written about the need to recognize the accomplishments of employees. Furthermore, there is significant cost to not doing it. According to Wagner and Harter [21], employees who do not feel recognized for their work are more than two times likely to say they will leave the company within a year. Many employees not only do not feel recognized but actually feel invisible.

Recognition needs to be honest and genuine. It can be positive feedback on contributions, or praise for noteworthy accomplishments, but it should be customized to the individual. Preferences for recognition vary from person to person. Some love public recognition or opportunities for superiors to hear about their work, while others prefer a quiet word of acknowledgment. How it is done depends on the leader and the style of the individual. But *that* it is done is essential to engagement.

## Clear expectations

The job description is just the starting point for discussion around expectations with employees. Employees need to clearly see how their work fits in with what the rest of the team is doing. Understanding what they need to do; why they are doing it; where it fits; and what their managers expect them to do can make a significant difference in productivity and profitability of the company. If your company or team is experiencing significant change, expectations can get very unclear. People are not sure where they fit in the new scheme of things. This is a good time to revisit expectations.

## Development

An environment that encourages development means that there are opportunities for natural and informal mentoring, as well as attention to people growing into areas that best fit their skills. Through the years, research supports the importance of mentoring. Wagner and Harter confirm this in their findings that more than two-thirds of those who say that someone encourages their development at work are found to be "engaged employees."

The youngest generation in the workforce today, referred to as "Millenials" are highly focused on advancement [24]. Motivating young workers means supporting their development.

Engagement is linked to being in a position to best use the natural talents and developed skills one brings to the workplace. People are motivated when able to do what it is they enjoy doing. One manager whose team scored highest in employee engagement in their organization pays close attention to the unique abilities and motivations of her team members and assigns work projects accordingly. This gives them opportunity to use their natural motivations for the good of the organization.

Good managers and teams look for strengths among their team members and opportunities to free them to use them. Additionally, letting go of trying to direct *how* work gets done (assuming ethical approaches), and focusing on the results, allows team members to thrive on their own motivation.

Giving team members opportunities to develop in skills, knowledge, access, and to use their abilities will enhance motivation and increase engagement.

### Team engagement

Teams have varying levels of engagement, as do individuals. An effective manager can influence engagement through intentional efforts and strategies. One of the best ways of doing this is to ask the team itself what areas they would like to see improve and then involve the team in the process of planning the improvements.

In general, engagement relates directly to the outcomes you are hoping to produce. You want to hold on to your high performing team members, so attention to engagement is an investment. Engagement's return is productivity, lower attrition, and a more pleasant working environment.

## The power of dialogue

Head down, working hard feels good, but the current professional playing field requires increasing collaboration, particularly due to the complexity of issues and rapid state of change. Conversation has great impact on our ability to influence, solve problems, anticipate the future, and garner agreement. Increasingly, large-scale processes are being designed to draw out meaningful dialogue for improving organizations. Ongoing conversation among your team, stakeholders, leaders, and customers will serve to help you in navigating through the opportunities and issues of your projects, as well as those of the industry. Building off principles of inclusion, collaboration, and involvement, group and interpersonal dialogue is a powerful influence in change and development efforts. Organizations bring in facilitators to support large-scale dialogue. But you can engage your team in productive conversation through effective inquiry and recording of responses.

### Challenging conversations

Your team is in the process of development and your boss, who is heavily invested in the outcome, is overpromising results to his superiors. How do you raise the subject? Or you see a product pushed through that you think has a potential to be destructive, but no one is acknowledging it and you are the newest member of the team. How do you address it?

Encouraging team members to speak up—or even doing it yourself—can be challenging, when you perceive that it may be risky to do so. Yet there are times when silence actually can be more than just costly, but dangerous. Through history, there are tragedies that may have been averted had someone spoken up about what they saw going wrong. Patterson et al. [2] in their book, *Crucial Conversations*, share a model, based on communication concepts, of how to have a conversation when three elements are in place: the stakes are high, there are opposing views, and emotions are present. Through their research we are reminded that there are people in organizations who are able to stay in dialogue even during tough conversations. These people may seem to possess a rare talent, but the skills they employ can be learned. The authors highlight ways of avoiding the conversation from blowing up or shutting down. These include:

- assessing your own motives and focusing on what you really want out of the conversation;
- employing strategies to make it safe for the other person to stay in dialogue;
- checking out your own thoughts about the topic to assure they are based on facts and not stories you have devised in your mind to justify your own behavior;
- remaining open to new data;
- following a pattern of initiating dialogue that includes starting with facts of what you have observed;
- listening effectively;
- establishing win-win solutions and collaborative decisions.

Having the tough conversations that are necessary and staying in dialogue can make or break the success of your team. One team was struggling with tensions among two units of the group. One side thought coworkers in the other unit were overstepping their roles and trying to control them. The other side was frustrated at what they viewed as inaction on the other unit's part. One result of this was increased resistance and slowed progress during a system change. When the team began to share their perspectives, they came to see they had much in common and that their concerns were worthy of the team unifying to work toward overcoming.

Another group was made up of two teams. One required higher degrees than the other to perform their work. The team with less academic experience was assuming arrogance on the part of the other team members and felt their abilities were being underestimated. As a result they felt resentment and were slow to respond and resistant to performing work for the other team. The team with higher degrees felt the other team members didn't appreciate their experience or knowledge and were assuming anyone could do their work; so they viewed the other team as unsupportive. In a healthy dialogue, they came to learn the judgments they had about each other were unfounded and changed their perceptions of the other.

If you have the patience to enter into healthy dialogue with others on your teams, and remain open to changing your conclusions about your observations, you will reduce costly mistakes, conflict, and unproductive relationships. Communication skills and tools are often the key to doing this well. As we will see in the next section, it may take developing more *emotional intelligence*.

## Enhancing success with emotional intelligence

The concept of emotional intelligence was popularized in the mid-1990s, when Goleman [25] wrote his book, *Emotional Intelligence: Why it Can Matter More than IQ*. Goleman had taken psychological and interpersonal research and synthesized it, communicating it in a way that brought emotional intelligence into everyday business vernacular. What the research has continued to point to is that the "soft skills" often dismissed as less important than the technical know-how are distinctly predictive of greater success in work and life—nearly everywhere except the classroom. Research supports that top performers have both.

Think of a person who most people would consider very intelligent by academic standards, but who routinely is passed over for promotions. He might lack the assertiveness to speak up; be overly independent and thus not seen as a team player; or lack the drive to develop additional skills to make him a good candidate. These issues relate to emotional intelligence.

---

### Case Study: "George" Versus "John" and Emotional Intelligence

"George" was the brightest and most knowledgeable in his area of expertise in the company. His manager confided in me that she wished she could put him in a leadership role that was coming open because he knew so much, but his biting comments and negativism, particularly when he was frustrated, disqualified him. In fact his inappropriate directness was affecting the entire team. She was building the nerve to replace him. He was smart but missing the needed emotional intelligence skills to go further.

On the other side of the coin is "John," a blue collar worker with a ninth grade education, who found himself sitting at the executive's boardroom table. He was held in highest esteem by the president of the company and became very successful because of a combination of a powerful positive attitude, continual observation of what others needed and wanted, willingness to help others, and authentic communication that came from a comfort with himself. He had emotional intelligence.

---

### What is emotional intelligence?

Mayer and Salovey [26], early researchers on the topic, define emotional intelligence as *"The ability to monitor one's own and others' feelings and emotions, to discriminate among them, and to use this information to guide one's thinking and action."*

Goleman and Cherniss explain that there are five dimensions to emotional intelligence [27]. Mersino clarifies that they can be categorized as *what you see and what you do* [28]:

1. Awareness of self
2. Self-management

3. Awareness of others
4. Relationship management
5. Motivation.

Another way to describe this is intrapersonal and interpersonal awareness, along with the ability to use those data for increased satisfaction and effectiveness.

### Elements of emotional intelligence

In the 1980s, Reuven Bar-On, in an effort to better understand emotional well-being and why some people—although intelligent—were not as successful, proposed a model of emotional intelligence [29]. Bar-On developed an instrument, similar to the IQ tool, to measure "EQ," or the emotional quotient, based on his model. The instrument, EQi, which has been updated and well validated, includes 16 scales that are skills and attitudes he associated with emotional intelligence. The difference, however, is that this is a subjective tool, with the assumption that the EQ score can change and probably should change as you grow. The 16 elements that are considered skills on Bar-On's EQ 2.0 are batched into six categories that focus on aspects of behavior and well-being [29]. The instrument has been administered to over a million people and used as a source of research for the field of study.

Bar-On's model according to Stein and Book includes the following composites and elements [29]:

*Self-perception*: *Self-regard, self-actualization, emotional awareness*
Rutledge finds this composite reflects how in touch we are with our inner self [30]. How confident are you? How motivated are you to develop yourself? How in touch are you with your present feelings?

*Self-expression*: *Emotional expression, assertiveness, independence*
This area relates to how engaged you are in expressing your inner self to others. How open are you in communicating your emotions? Do you speak up when you need to? Are you willing to go it alone, if necessary?

*Interpersonal*: *Interpersonal relationship, empathy, social responsibility*
This relates to how you tend to trust and establish healthy and mutual relationships. Do you give and receive trust and affection? Are you sensitive to what others are feeling? Do you show concern for and contribute to the communities around you?

*Decision making*: *Problem solving, reality testing, impulse control*
Rutledge describes this area as one which reflects how well someone manages themselves, particularly as it related to everyday difficulties.
Can you easily solve problems, even when upset? How practical are you? How well do you resist temptation?

*Stress management: Flexibility, stress tolerance, optimism*
> This area relates to one's ability to face the challenges of life with adaptability, resilience, and optimism. Do you tend to resist change? How easily does stress make you anxious and overwhelmed? Is the glass half full or empty?

*Well-being: Happiness*
> This area combines four other elements to reflect a general happiness or contentment with life.

A person may not need to be highly engaged in all aspects of EI. For example, if assertiveness is not needed in a person's job or significant relationships, he may not care if he is not strong in that area. Additionally, we can be too engaged in a skill area and that will undermine our effectiveness. Balance is critical to effectively employing emotional intelligence skills.

Many of the skills and activities of the workplace that we rely on for success are tied to emotional intelligence, such as negotiation, team building, sales, running meetings, presentations, decision making, and change management. Determining your own engagement in each of the correlated skills and remaining aware of the impact of emotional intelligence on your team will certainly increase your potential for success.

## Handling conflict

In working with leaders, I can easily say that this is one of most dreaded and avoided areas for managers and team members. Whether it is conflict on the team or their own conflict with a team member or peers, the common impression I get is leaders just want the "noise" to go away. Many managers will avoid facing conflict as long as possible, often leaving it to grow quietly underground, until it explodes or they lose good people. I get the sense that managers often feel a sense of helplessness of what to do with conflict when it arises. I facilitate sessions where two people have had enough conflict to make life so unpleasant that they (or their managers) are frustrated and looking for a lifeline. I help the team members reestablish expectations and move past painful situations.

As a manager, there are two specific types of conflict you will deal with the most: group conflict—or conflict that occurs in meetings and across teams—and interpersonal conflict, usually between two individuals. They are often closely related or intertwined.

Team members have a responsibility to manage conflict, as well. As a team member, disagreement can be healthy and lead to more creative and higher quality outcomes. However, interpersonal conflict can interfere with not only your own productivity but also that of the entire team. I have worked with teams that have become what they called "toxic," yet when distilled down, one negative person or a seething conflict between just two individuals was the source of an entire team's poor morale. Critical to avoiding this dysfunction is to address the conflict, giving the other team member the benefit of the doubt, recognizing the role differences play in conflict, and listening carefully for points of common ground. Challenge your own judgments and assumptions about the other team member and be open to not only changing your view but also apologizing for the role you have played in the conflict. Never vent to members of the same team or negativity will grow like a weed, and your conflict will swell not reduce.

It's helpful, also, to look for other sources in the environment influencing the conflict, such as issues related to recognition of work, competing for a leader's approval or limited opportunities, and work-related pressures that create stress. One team I worked with dealt with negativity in their jobs all day long via their clients and lacked a feeling of being valued by the organization. It was no wonder they were sinking into conflict. Recognizing these influencing elements can assist in reconciliation between you and your coworkers.

## Group conflict

When we consider group conflict that occurs openly in meetings we want to make sure we are viewing it in the proper light. Some conflict is not only acceptable but also preferable to achieve greater creativity and higher degrees of quality. A group that never disagrees or is overly polite may be trying so hard to avoid conflict that honest perspectives and creativity are lost. Groups may agree to solutions that are mediocre or make poor decisions because fear of conflict halts the needed, albeit slightly heated, discussions that may produce better results.

While some conflict is healthy, groups that get entangled in frequent or intense conflict suffer loss of productivity and engagement. How do you address conflict when it reaches this level?

Facilitation practices in meetings are significant tools used to channel or diffuse conflict. If you have not taken a course on facilitation of meetings you should. Running a meeting seems like an everyday skill that most professionals can do; but judging from how people often feel about meetings, it clearly is not. Some techniques used for diffusing unhealthy conflict in meetings include:

- Head off frustrating conflict with collectively determined "ground rules" for which the team can hold each other accountable.
- Assist in summarizing the key points each person is making to clarify the useful information and stop repetitive arguments.
- Determine decision-making processes for specific situations in advance, so that you can move to decision-making approaches to diffuse the conflict.
- Involve the group, asking for some quick response of how others feel or what they think in a structured way. This helps when you have an antagonistic individual who is relentless.

- Protect the team by asking the individuals involved to talk after the meeting and facilitate or even mediate if necessary.
- Take a break in the meeting.
- During a break approach disruptive members and offer them honest but sensitive feedback. Ask them to change their approach. If they are abusive, ask them to not return to the meeting. (Avoid reacting, yourself, as these situations can provoke frustration.)
- Of course, genuine violent responses need to be immediately addressed through either escalation to leaders (or human resources) or met with law enforcement. Never take threats lightly, report them immediately.

Dealing with conflict in groups requires thinking on your feet, which means you need to have knowledge of, and practice in, effective approaches. Training in facilitation skills, as well as handling conflict, is worthy of further investment of time.

For the sake of the team, do not allow conflict that is disruptive to continue unaddressed. Departments, such as organization development, human resources, and training, can assist you when you need support. Don't feel you need to go it alone. But, learn in advance the difference in the approaches of each to determine if the resource is the proper fit for the situation. For instance, contacting HR may escalate the conflict to a level that is in the policy and legal realm. This is appropriate if you see abusive behavior, disruptive patterns, or threats. Organization development may assist in strengthening the trust on the team, conflict resolution, or coaching in a confidential manner. Training can assist if skill development is at the core of the issue.

If your company has no HR department, there are many external resources to access, from independent consultants to organizations that act as outsources to HR departments. Leadership of the company should be made aware of the issues you might otherwise bring to the HR department.

## Interpersonal conflict

When two people on your team become embroiled in an ongoing conflict, it can be exhausting as a manager and deeply distressing for the team. The team may form factions, siding with one or the other in displaced compassion. This is not likely to go away if you ignore it. A manager may think that a stern warning or persuasive appeal to the two involved will suffice. But while conflicts most often arise out of personality differences, they are often provoked or at least inflamed by work-related issues. So, it is essential that you find a way to address the conflict, as well as analyze what else may be impacting this conflict. At times, conflict is influenced by a lack of clarity in goals, ambiguity of roles, lack of leadership, insufficient recognition, ineffective structure, or impacted by a transition in the workplace. The conflict may arise as a symptom of larger issues, somewhat analogous to a child acting out when the family is dysfunctional. Significant conflict should be an impetus for organization assessment.

Consider a department that has just restructured. Teams have been merged and there is new leadership. People in transition (recall Bridge's model from earlier in this chapter) are feeling anxious and somewhat insecure. There has already been some slight racial tension in the department. Suddenly a conflict between two individuals flares up. This conflict has multiple layers. Working through the conflict between the individuals is priority, but you do not want to stop there. It will prove even more beneficial if you pinpoint where the issues may be in the organization and start addressing change management, work processes, team development, and diversity.

## Further development

Managing the development of embedded systems certainly demands the necessary technical expertise. However, applying the knowledge of how to lead and participate effectively in a team may be equally important in activating your success. Achieving the ambitious goals of today's technical world will require a focus on creative solutions and possibilities. These can best be accomplished in ways that stretch past the technical aspects of the work. Technical ability opens the gateway to success, and ability to lead, collaborate, communicate, and resource moves you through it.

## References

[1]   J. Maxwell, The 17 Essential Qualities of a Team Player, Thomas Nelson, Nashville, TN, 2002.
[2]   K. Patterson, J. Grenny, R. McMillian, A. Switzler, Crucial Conversations, McGraw-Hill, New York, NY, 2002.
[3]   <News.executiveboard.com>. CEB Identifies Anatomy of the New High Performer. December 20, 2012.
[4]   American Society for Training and Development, Tapping the Potential of Informal Learning, ASTD Press, Alexandria, VA, 2008.  < http://store.astd.org/Default.aspx?tabid = 167&ProductId = 19946 >.
[5]   J. Kotter, What Leaders Really do, Harvard Business Review on Leadership, Harvard Business School Press, Boston, MA, 1998.
[6]   W. Bridges, Managing Transitions, Making the Most of Change, third ed., DaCapo Press, Philadelphia, PA, 2009.
[7]   E. Kubler-Ross, On Death & Dying, Scribner, New York, NY, 1997.
[8]   D. Kelley, D. Conner, The emotional cycle of change, in: J. Jones, J. Pfeiffer (Eds.), The 1979 Annual Handbook for Group Facilitators, University Associates, San Diego, CA, 1979.
[9]   P. Zagarmi, Leading People Through Change, The Ken Blanchard Companies, Youtube, 2009<http://www.youtube.com/watch?v = DTZEnSvZPqc&feature = related> (accessed 15 July 2013).
[10]   R. Moss Kanter, Managing the human side of change, Manage. Rev. (1985) 52−56.
[11]   K. Bunker, Responses to Change, Helping People Manage Transition, Center for Creative Leadership, Greensboro, NC, 2008.
[12]   B.B. Jones, M. Brazzel (Eds.), The NTL Handbook of Organization Development and Change, Pfeiffer, San Francisco, CA, 2006.

[13]  R. Moss Kanter, 10 Reasons People Resist Change, September 25, 2012. <blogs.hbr.org/2012/09/ten-reasons-people-resist-chang/>.

[14]  J.M. Kouzes, B.Z. Posner, To Lead, Create a Shared Vision, HBR The Magazine, 2009<http://hbr.org/2009/01/to-lead-create-a-shared-vision/ar/1> (accessed 15 July 2013).

[15]  D. Pink, RSA animation, drive—the surprising truth about what motivates us. <http://vimeo.com/15488784>, 2010.

[16]  I.B. Myers, M.H. McCaulley, N.L. Quenk, A.L. Hammer, MBTI Manual: A Guide to the Development and Use of the Myers—Briggs Type Indicator, CCP, Inc., Mountain View, CA, 2003.

[17]  J.R. Gibb, TORI Theory and Practice, in: J.W. Pfeiffer (Ed.), Annual Handbook for Group Facilitators (Jan 15), John Wiley & Sons, San Francisco, CA, 1972.

[18]  W.C. Shutz, FIRO, A Three Dimensional Theory of Interpersonal Behavior, Rinehart, New York, NY, 1958.

[19]  B.W. Tuckman, M.C. Jensen, Stages of small-group development revisited, Group Organ. Manage. 10 (1977) 43–48419–427

[20]  B.W. Tuckman, Developmental sequence in small groups, Psychol. Bull. 63(6) (1965) 384–399.

[21]  R. Wagner, J. Harter, 12: The Elements of Great Managing, Gallup Press, New York, NY, 2006.

[22]  B. Kaye, S. Jordan-Evans, Love'em or Lose'em, Getting Good People to Stay, Berrett-Koehler Publishers, Inc., San Francisco, CA, 1999.

[23]  A.H. Maslow, Motivation and Personality, Harper, New York, NY, 1954.

[24]  R. Alsop, The Trophy Kids Grow Up: How the Millenials Are Shaking up the Workplace, Jossey-Bass, John Wiley & Sons, San Francisco, CA, 2008.

[25]  D. Goleman, Emotional Intelligence: Why it Can Matter More than IQ, Bantam Books, New York, NY, 1995.

[26]  J.D. Mayer, P. Salovey, Emotional Intelligence, Baywood Publishing Company, Amityville, NY, 1990.

[27]  D. Goleman, C. Cherniss, The Emotionally Intelligent Workplace, Consortium on Research in Emotional Intelligence, Jossey-Bass, San Francisco, CA, 2001.

[28]  A. Mersino, Emotional Intelligence for Project Managers: The People Skills You Need to Achieve Outstanding Results, AMACOM, New York, NY, 2007.

[29]  S.J. Stein Ph.D., H. Book, The EQ Edge: Emotional Intelligence and Our Success, Jossey-Bass, A Wiley Imprint, Ontario, CA, 2011.

[30]  H. Rutledge, EQ Workbook, OKA, Fairfax, VA, 2012.

## CHAPTER 3

# Project Introduction

Kim R. Fowler
*IEEE Fellow, Consultant*

## Chapter Outline

Developing and Managing Embedded Systems and Products.
DOI: http://dx.doi.org/10.1016/B978-0-12-405879-8.00003-9

## Overview

"All the really important mistakes are made on the first day" [1]. "Never enough time to do it right, always enough time to do it over" [2]. What you do in the beginning sets precedence for the remainder of the project.

This chapter assumes that the project idea is already in place and that the necessary excitement is already whipped up. Consequently, this chapter is about how people begin the long process of designing and building a successful project. The material is not difficult at all! It just needs to be done correctly.

Proper project introduction and management for that matter are based on integrity and embracing the big picture. It's about facing reality, developing trust through authentic connection, facing the negative, and building for the future. Henry Cloud's book on integrity is an excellent foundation for presenting these principles [3].

Communicate the vision. Communicate often. Establish the "who, what, when, where, why, and how..." Use proven techniques to develop and manage the project. Start on the right foot.

## Establishing the vision, mission, goals, and objectives

This section may sound like "Rah, rah, go team, fight." Regardless, it is very important to get right. Following the admonition of the chapter introduction to start right, this is a first step—to establish or confirm your team's vision, mission, goals, and objectives. Hopefully after the first project, the vision and mission and most of the goals will not need to change. Then this exercise becomes even easier and should give long-term benefits. Many sources on project and program management support my contention of getting this done early and done right. I will not repeat the long list of support evidence here.

These items are summarized in the following four points [4]:

- Vision—where you are heading
- Mission—defines your fundamental purpose(s)
- Goals—specific measurable components aligned with the mission and vision statements
- Objectives—action steps to meet goals

Beyond establishing vision, mission, goals, and objectives, gain consensus and support for them from the entire team. Communicate them often. Measure and track them. See that your organization implements and follows them.

### The lineage and the progression

Every project needs clear vision, mission, goals, and objectives. These should flow from the company's own larger vision and mission. If the company doesn't have these then stop—establish the company's vision and mission first! This is an overarching principle. Vision and mission drive the company and hence your project.

The project's vision and mission should copy or be a subset of the company's vision and mission. The goals and objectives that implement the project's vision and mission should align with the company's goals and objectives.

I will outline each point and assume that you will be applying it to your project.

### Vision—where you are heading

Vision points to the company's future direction. It is a state or set of conditions toward which everyone within the company should strive so that the company is the best at what it does. A *vision* statement encapsulates that direction and state; it *should be the guiding image of ultimate success* [5]. Vision should be owned and believed by everyone in the organization. Sharing vision, though the vision statement, should identify the company to customers and stakeholders (e.g., clients, partners, allied companies, government officials, and neighbors). *The project's vision should align very closely with the company's vision.*

A group's shared vision does the following [6]:

- Communicates a sense of purpose
- Expresses what is important and why
- Focuses on the future
- Reflects the shared values of group members
- Uses pictures, images, and words to bring the vision to life.

---

### Example: Vision for the Society of Naval Architects and Marine Engineers (SNAME) [7]

- SNAME will be the international organization of choice for engineers and other professionals in the marine industry, providing valuable and relevant services to all its members.
- SNAME and its members will be recognized by their peers as the technical leaders in the advancement of the marine industry.
- SNAME and its members will be recognized by the public and by governments as responsible technical authorities and valuable contributors to society.

---

### Example: Vision for the IEEE Instrumentation and Measurement (I&M) society [8]

Be the premier international professional Society in the Instrumentation and Measurement (I&M) fields.

---

---

**Example: Vision for a fictitious embedded controls design team**

Lead the world in designing efficient control systems for sustainable profits and a cleaner environment.

---

So why is vision so important? Vision inspires work ethic and guides the weaving of our efforts together to accomplish our goals. Vision provides clarity in times of confusion or unknown. It is the pinnacle of our effort to control destiny [5].

A good vision statement is "short, simple, and powerful" [5]. Automobile companies work long and hard to produce clear, concise vision statements. In the 1980s, Ford Motor Company touted "Quality is Job 1," which may have been a slogan, but it served as a vision statement, too. Recently that statement has morphed to "One Ford, One Team, One Plan, One Goal" [9]. The web site expands the vision statement into an integrated set of statements for both vision and mission.

## Mission—your fundamental purpose

The mission describes how you and your organization will carry out the vision. It describes what the team or company will do, who it serves, and why it is unique. "A mission statement is simply an organization's reason for existing" [5]. Like the vision statement, a mission statement should be concise in telling the company's stories and ideals in less than 30 seconds. Again, the project's mission should align closely with the company's mission. To prepare a mission statement, ask your team or organization these questions [5].

- What are we going to do?
- How are we going to do it?

---

**Example: Mission for the Society of Naval Architects and Marine Engineers (SNAME) [7]**

The mission of the Society of Naval Architects and Marine Engineers is to advance the art, science, and practice of naval architecture, marine engineering, ocean engineering, and other marine-related professions.

---

**Example: Mission for the IEEE instrumentation and measurement (I&M) society [8]**

- Provide the most *comprehensive* and *high-quality* services to our members and related professionals.
- Serve as the professional *incubator* for the *growth* of all (particularly younger) members.
- Be in the *forefront* of future I&M technological advances.

---

**Example: Mission for a fictitious embedded controls design team**

- Study the latest technology and scientific advances and incorporate them into embedded controls.
- Educate and mentor our employees, customers, clients, and stakeholders in the latest control paradigms.
- Use sustainable processes to design, develop, and manufacture embedded control systems.

---

The primary difference between a mission and a vision is that the mission describes the present state while the vision describes the future state. The mission's present state answers the questions, "Who are we? Why does our organization exist?" The vision's future state answers the questions, "What do we want to achieve? Where do we want the organization to go?" [5].

## Goals—measurable components

Goals describe those components that implement the mission. Goals should be measurable to be effective in tracking and gauging progress. The first example lacks metrics so it is not as effective as maybe the second example.

---

**Example: Some goals for the Society of Naval Architects and Marine Engineers (SNAME) [7]**

- SNAME shall enable the global exchange of knowledge and ideas relative to the marine industry.
- SNAME shall work to further education in engineering as it relates to the marine industry.
- SNAME shall encourage and sponsor research and development in naval architecture, marine engineering, ocean engineering, and other marine fields.
- SNAME shall promote the professional integrity and status of its membership.

---

**Example: Some goals for a fictitious embedded controls design team**

- Attend at least three technical conferences each year to follow advances in the technology of embedded controls.
- Every year prepare and present four whitepapers or trade journal articles on applying the company's embedded controls.
- Every year provide four webinars to potential customers and clients on the company's latest control designs.
- Increase the use of sustainable processes and materials in company products by 25% each year.
- Within 2 years provide the means for recycling of all products.

---

### Objectives—action steps to meet goals

Objectives are discrete steps to meet every goal. Objectives are specific, measurable, possible (realistic and attainable), and defined to a duration of time [5]. You can prepare objectives for either the high-level organizational directives or at a lower-level project basis. The example provides some objectives for a project that updates a process-industry control panel.

---

**Example: Some objectives for a fictitious embedded controls design project**

- Prepare a whitepaper to explain the new user interface on the control panel.
- Present a webinar to potential customers and clients to explain the new user interface on the control panel.
- Use leadless solders, according to Reduction of Hazardous Substances (RoHS) requirements, on the circuit boards.
- Use a water-based conformal coating on circuit boards that cleans up without organic solvents.

---

## Establish the team

Building a team is challenging but critical to the success of your project. You must understand who the stakeholders are and then define the roles within the project. Stakeholders are those people and organizations with an interest in your project succeeding. Stakeholders might include:

- team members
- customers
- upper management
- government regulators
- industrial partners.

Team members might include but are not necessarily limited to:

- Project manager
- Engineers—electrical, software, mechanical, manufacturing
- Manufacturing personnel
- Marketing and sales
- Management, administration, and support staff
- Purchasing and procurement
- Technical support
- Legal department
- Quality assurance (QA).

## Define the roles

A smoothly functioning team is central to successful development. Soft skills have equal footing with technical skills. A project contract is sometimes very effective in establishing boundaries of what is expected, what the team can do, and what the team will not do. Verzuh [10] details all the different roles and efforts to partition the tasks and roles.

All project team members and stakeholders must know and understand their role in the project. Some thoughts follow that emphasize particular issues for specific roles within a team.

### Project sponsor

This is an important role to identify because a project cannot succeed without high-level support and sponsorship. A sponsor often is the stakeholder who will benefit the most from a successful project. Consequently, the sponsor should assume the overall responsibility for the entire project and champion it. The sponsor must grant sufficient authority to the team to ensure success [11].

Giving a team a fixed budget and schedule for a project with features that cannot be developed within those constraints is not granting appropriate authority and responsibility. This is not championing a project.

### Project manager

This quote summarizes some skills of the project manager:

> The project manager is ultimately responsible for developing a cohesive project team motivated toward success making the project managers leadership qualities, interpersonal skills, and credibility are far more important than formal authority. A project manager possessing these attributes can usually find a way to "make it happen" with or without formal authority, but the project owner should give the project manager the level of authority that enables the project manager to successfully accomplish the assigned responsibilities. The project owner should provide the project manager with a formal statement or contract detailing the scope of authority being granted [11].

The project manager communicates the roles to the project stakeholders and manages the project's scope, schedule, and cost. The project manager leads the team through problem resolution by working with the appropriate levels within the organization and escalating to upper management when necessary. The project manager communicates the project status with stakeholders, project owner, and upper management. Finally, the project manager reviews the performance of team members and negotiates resolutions in conflicts between members and for their time.

*Project team members*

Project team members are the "boots on the ground." They are responsible for designing, developing, testing, integrating, manufacturing, and supporting the product. (Remember that team members' disciplines cover more than engineering design.) They are responsible for working together and covering each other, should someone miss a task or insert a design flaw. They are the ones most informed about progress and feasibility; the project sponsor and upper management need to believe them and their assessments.

*Assemble a core project team*

Skills must cover the requirements of the product's architecture. Skills need to be complementary, even if there are overlaps. These sorts of questions need to be asked [11]:

- What skills are needed (e.g., electrical, software, mechanical, manufacturing, user interface, operations)?
- Is additional training needed?
- Do members of the team have relevant experience? If so, how well did they do on previous projects?
- Is each member interested in working on the project? If not, can you find someone who is?
- Will the project team members work well together?
- What is the availability of each person to participate on the team?

"Every person brings a different skill set and personality to the project. We need to accommodate the inconsequential differences and yet be able to confront and question problems. This can only be done through integrity, communication, openness, and trust. Teamwork encompasses these virtues" [12], p. 8.

*Communication is key to the team*

A team is not a team without communications. Team members must function together and communication is central to functioning together. Let's look at the next section of the chapter to expand on what is important about communications.

## Communications

How much time do you think people spend communicating in various ways within a project? Specifically, what percentage of time do people spend in communications? It's a question that I ask my students every semester; I get a wide range of answers from 10% to 80%. The answer is, "Depending on the project it is usually closer to 90% or 95%."

Project communications coordinate and provide status, changes, and concerns to everyone. Verzuh [10], p. 33 writes, "Communication is a primary project success factor no matter what the organizational style. Most organizational structures facilitate vertical (top-down and bottom-up) communication patterns, but your communication requirements may run counter to the prevailing patterns. Crossing organizational boundaries always takes more effort, but you must do whatever is necessary to keep all the stakeholders informed and coordinated."

### Forms of communication

Communications are more than e-mail messages, though those are important. Communications constitute many different forms and situations: verbal discussions, meetings, telephone calls, web meetings, social media, letters, memos, documents— online and paper copy, and yes, e-mail messages. (Chapter 5 goes into detail for various documents.)

### Stakeholders

All stakeholders within the project need appropriate communications. These stakeholders include team members, customers, the sponsor, and upper management.

### Communication plan

Planning "who and what" early in the project is more likely to keep the project on track and stakeholders happy. Table 3.1 lists some potential stakeholders within a project and their information needs along with the frequency and the medium of communication and expected responses.

Having such a plan and following it is just good business. It sets and maintains expectations for stakeholders. Along with careful planning and management, it helps reduce surprises.

### Social contracts

This section refers to the relationships we have or develop within a project. The social contracts are the commitments and promises that we verbalize to one another within the project. Social contracts exist between you and various other people: customers, upper management, your colleagues and staff, and marketing.

You would also be wise to develop a good relationship with your legal counsel, too. Wash your social contracts through your favorite lawyer to gain another perspective (of course, it really helps for the lawyer to be very business minded)—see Chapter 17 on contracts.

Aside from official, legal contracts, customers need frequent assurance that you are working on their problem(s). Good communications are key to assurance. Writing the project goals

**Table 3.1: A potential list of stakeholders and the types of communications they should have**

| Organization | Stakeholder | Information Needs | Frequency | Medium | Response |
|---|---|---|---|---|---|
| **Customer** | Management | bus, prog | a | e, t | Approval, decision making |
| | Project Lead | Tech, bus, prog, QA | w/m | e, t, doc, DES | Approval, decision making |
| | Project Support #1 | Tech, QA, eng | w/m | e, t, doc, DES | Support to decision making |
| | Project Support #2 | Tech, QA, eng | w/m | e, t, doc, DES | Support to decision making |
| | Project Support #3 | Tech, QA, eng | w/m | e, t, doc, DES | Support to decision making |
| **Government or other customer** | Agency #1 | prog | a | doc | Support to decision making |
| | Agency #2 | prog | a | doc | Support to decision making |
| | Agency #3 | prog | a | doc | Support to decision making |
| **Company** | Management— CEO, CFO | bus, prog, mfg | a/w | v, e, t, doc, DES | Approval, decision making |
| | Project Lead | eng, Tech, mfg, bus, prog, QA | d | v, e, t, doc, DES | Approval, decision making |
| | Systems Engineer/ Architect | eng, Tech, mfg, QA | d | v, e, t, doc, DES | Decision making, Execution |
| | Electronic Lead | eng, Tech, QA | d | v, e, t, doc, DES | Execution |
| | Software Lead | eng, Tech, QA | d | v, e, t, doc, DES | Execution |
| | Mechanical Lead | eng, Tech, QA | d | v, e, t, doc, DES | Execution |
| | Sensor Lead | eng, Tech, QA | d | v, e, t, doc, DES | Execution |
| | EGSE Lead | eng, Tech, QA | d | v, e, t, doc, DES | Execution |
| | Manufacturing Lead | eng, mfg, Tech, QA | d | v, e, t, doc, DES | Execution |
| | Test and Integration Lead | eng, mfg, Tech, QA | d | v, e, t, doc, DES | Execution |
| **Vendors or partners** | Company #1 | mfg, QA | a | e, t, doc | Execution |
| | Company #2 | mfg, QA | a | e, t, doc | Execution |
| | Company #3 | mfg, QA | a | e, t, doc | Execution |

eng = engineering issues, mfg = manufacturing issues, Tech = technical compliance, bus = business issues, prog = programmatic issues, QA = quality assurance.
a = as-needed, d = daily, w = weekly, m = monthly.
e = e-mail message, t = telephone, doc = document, DES = design review, v = verbal.

and objectives to align with customers' interests and letting them know is a good first step. Appropriately frequent meetings or even phone calls to update status can do vast amounts of good in maintaining a relationship.

Writing a charter can ease a number of potential problems [10], pp. 63–65. A charter can act as a "contract" with upper management to remind them of assurances of support and authority that they made at the beginning of the project; it can be simple but useful as a social commitment. Sometimes formalizing them to a degree can help set expectations and smooth ruffled feathers when challenges arise later in the project. A "contract" with upper management can also serve to establish your project's legitimacy with other groups within a larger organization, particularly groups who may not be mandated to cooperate with you but you need their services. Examples might be marketing or manufacturing.

Our social contracts with members of the project team are very important. Chapter 2 deals at length with some of these issues.

## Business case

The business case explains the reason for the project. It answers "why?"—it establishes the rationale for the project and justifies its development. The business case covers the economics of the product. It balances cost, schedule, quality, and features. The business case attempts to demonstrate the feasibility of profit (or a positive outcome for a nonprofit effort).

Economics drive most products. If the product isn't economically viable, it won't (or shouldn't) come to market.

Many components play into the economics of product development. Cost figures as a major part of the economics of a product. Many things compose the cost: purchasing components and materials, manufacturing and assembly, distribution, maintenance, and disposal. The cost associated with nonrecurring engineering (NRE) must be amortized over the life of the product. NRE includes all design, test, and qualification performed during the initial development of the product. Then there is the 'cost of goods sold' (COGS), which refers to the ongoing expenses and recurring costs that go into the production of each unit. COGS includes buying components and the labor to assemble and build each unit.

Another concern is the margin between cost and price. Specialized, low-volume products generally have high margins; that is a large ratio (or difference) between the price of the final product and the cost to develop and manufacture it. High-volume products generally have low margins; that is a small difference between cost and the final price" [12], p. 6.

## Understand the market

First, know what problem you are solving. This means that you know where the product is going and what it will do. Here are some questions that you and the business case should answer:

- Who does it help?
- Why does it help?
- What are the current solutions in the market?
- Where do those solutions fall short?
- How can you do better?

---

### Example: Lawn care tractor

A company has developed a new engine controller for a quiet, high-efficiency engine in a lawn care tractor. The company will need to understand which market will be addressed by this engine controller and the quiet, high-efficiency engine: residential, institutions with large grounds to maintain, or commercial lawn care companies. Each market operates differently. Residential use (homeowners) will run the tractor once a week or less to cut the grass during the summer but then they will turn it off for the winter. They might occasionally attach a plow or snow blower for moving snow in the winter; regardless, there will be long periods of sitting dormant in a cold garage or shed and not running. A golf course will use the tractor or a lawn mower for long periods of time every week, probably every day. The company must also ask, "What competitors are in the market?" and "What are the advantages and disadvantages of their solutions?" and "What do we bring to the market that makes for profitable effort?"

---

### Example: Web connected home appliances

A company wants to develop web connected appliances for use at home. A web connected coffee maker could be commanded to turn on for special occasions when a timer would not be capable of anticipating the situation. Web connected lighting could be remotely controlled. Regardless of the appliance or the next bright idea, the company will need to know what is possible, what is feasible, what customers want (or may want once the capability is well known), and who are or might become competitors.

---

### Example: Cubesats for sale

Suppose that you want to build and sell cubesats (little cubical satellites, 10 cm on a side) for various jobs in space. You need to understand the market: where in space (low earth orbit or geosynchronous orbit or interplanetary travel), the environment (radiation fluxes and duration, solar heating, vacuum, launch vibration), the types of "jobs" your cubesat can do, and who might use it (see customers below). You need to know what is possible, what is feasible, what customers want, and who are or might become competitors.

## Understand the customer

Next, understand who will use your product. Markets and customers are inextricably linked. Understanding the market will help you understand the customer but not necessarily completely. Some markets have a variety of customers with slightly different goals and uses. Knowing which of those customer groups your product serves is just as important as knowing the market.

### Example: Lawn care tractor

See the example above for the lawn care tractor. Different uses over different durations drive different concerns for different customers. A golf course wants high efficiency to reduce fuel costs and low acoustic noise to reduce disturbance to golfers. A home residence will not be as concerned with efficiency or noise; cost and ease of use will be more important; furthermore, the engine sits cold for long periods between bursts of use, which makes reliability a challenge (homeowners are not renowned for proper maintenance).

### Example: Web connected home appliances

See the example above for web connected appliances for use at home. Many households use appliances. Not everyone wants a web connected appliance immediately. Who are the "early adopters" of new technology who might buy web connected appliances? Who will buy web connected appliances once they are in vogue? Who are the "late adopters"? How many potential buyers are in each group?

### Example: Cubesats for sale

See the example above for the cubesats. Who will buy them? Will commercial companies use them for maintaining larger satellites? Will commercial companies buy them to handle commercial payloads? Will the military buy them? and why? Will universities buy them both to host their experiments and to avoid the development delay from "reinventing the wheel"?

## *Business plan*

Once you have completed the business case and the project looks feasible, you prepare a business plan. The business plan specifies the framework needed to develop the project. It specifies, but is not limited to, the following concerns:

- Control structures and management
  - Contracts, charters
  - Funding
  - Scheduling
  - Communications, meetings
  - Process controls—company QA, reviews, audits, action items
- Plans
  - Project plan
  - Configuration management plan
  - Risk (and hazard) management plan (see Chapter 4)
  - Concept of operations
  - Technical process plans (e.g., hardware, software, mechanical)
- Documentation (see Chapter 5)
- Archives.

The first management plan you should put in place is the project plan or the project management plan (PMP). The PMP describes who does what and when and where. The project management book of knowledge (PMBOK) details areas of project management life cycle [13]. A large project management life cycle (more than 50 people involved in a project for more than 5 years) with nine PMBOK areas spread across project phases will include the following:

1. Integration management
2. Scope management
3. Communications management
4. Risk management
5. Human resource management
6. Resource management
7. Procurement management
8. Time management
9. Cost management
10. Quality management

Most of your projects, and the focus of this book, are much smaller projects with more limited scopes. Within those smaller projects, you should establish two or three of the most important areas of project management:

- Immediately after preparing the PMP, you should establish the risk management plan. Chapter 4 details risk management. This may include quality management, or a QA plan, which may be a stand-alone document.
- Larger projects (more than 20 people involved in a project for more than 1 or 2 years) may require a resource and procurement management plan.

## Business administration and concerns

Once the business plan is in place, begin monitoring the project according to the various management plans listed above. Redundant communications in various forms and formats are the center of project administration.

Assuming (as most people usually do) that quality and features are already in place, then schedule and budget are always the focus. Numerous articles and studies indicate that many projects end up over schedule and over budget. Let's see what you can do to change that!

### Scheduling and budgeting

The same basic principles hold for both scheduling and budgeting. They are all matters of integrity [3].

1. Be realistic.
2. Be trustworthy; build a reputation for estimating and assessing progress accurately.
3. Communicate the how and why of designing and developing a particular component, subsystem, or product. Maintain communications on the status of the project's progress throughout development and production and delivery.

### Estimating accurately

One method for estimating accurately is to prepare two separate estimates of the exact same development. Prepare a top-down estimate; set dates and deadlines based on your experience and compile the project's timeline. Prepare a bottoms-up estimate; look at each individual component and how long it takes to develop and integrate; this will require knowing when some activities may done in parallel and when one activity will become a bottleneck; add the groups of serial activities to total the time duration. Software tools are available to aid you in bottoms-up estimation.

---

#### Example: A bottoms-up estimate averts an ill-advised project [14]

The US Navy uses hyperbaric chambers with mixtures of gases, such as nitrogen, oxygen, helium, and hydrogen, to train deep-sea divers. Years ago I participated in a proposal effort to automate the controls for a hyperbaric chamber. The proposal called for dual redundant programmable logic controllers (PLCs) for dependable operation and for hot swapping of modules during repair. It required writing software for the human interface. It also required mechanical overrides for failure recovery.

I prepared an estimate of the schedule for the proposal. I used project scheduling software to prepare the estimate, which included design, development, peer reviews, formal design reviews, test, and integration. The bottoms-up calculations for this estimate totaled to a much longer time (years of calendar time and of effort) to implement the design than the original top-down estimate, which called for 4 months of calendar time and much less effort.

This was a safety-critical application with human lives at stake. I was able to defend my estimate from my experience and with logical inference. The sponsor felt that the automation project would take too much time and money and dropped it.

Your estimates should allow time for goof-ups, changes of scope or design, problems, and oversights. Every project has some measure of these problems. Hopefully as you gain experience and wisdom, these problems will shrink. To allow for them requires uncommitted extra time and effort. This is called contingency. Build in contingency, and add even more for the poorly understood processes. Another factor in contingency is the mix of technical and business disciplines in the project. Involving more disciplines usually means that more unknowns exist; consequently these unknowns need more allowance for schedule slips and budget growth. ***If your project cannot have schedule slips or budget growth, then cut back features!***

Another fly in the estimation ointment is the unfounded hope for the miraculous. Do not solely depend on a single tool to solve all the problems. ***There are no silver bullets!***

### Example: A silver bullet it wasn't

Three years ago I interviewed with a company building a portable medical diagnostic instrument. I interviewed for the position of director of engineering, a job where estimating effort, schedule, and budget was important.

Before the interview I discovered that the company was trying to ram through an FDA approval for the device within 3—4 months. During the course of the interview, upper management argued that it had a military application, though the largest market was sports, and that they had a military general on board to urge the FDA to expedite the approval. My experience and understanding of the FDA approval process said that they would be lucky to get FDA approval within a year; in the interest of being honest—trying to build trust—I told them of my reservations.

I did not get the job; probably just as well because the device was still not FDA approved two and a half years later. They were depending on a silver bullet (the military general) to shortcut the process. The silver bullet was not effective.

## A word of warning

If a project began as an ad hoc hobby project or a research proof-of-concept, it's already run its course and needs a complete makeover. Close it out and open a new project with a different name, a new set of ground rules, and a new budget and schedule. Start from the ground up. You can use the concept architecture but you really need to think development now. It's no longer a science fair project. The project team needs a new mindset.

## Effort to introduce a project

Every project introduction varies in its effort, but previous experience and templates can reduce that effort. A small project with a few people working over 6 months may only require 2 or 3 hours to prepare the project introduction. Larger, more involved projects, with 30–50 people working over the course of 4–6 years may require several days to prepare and introduce the project.

Both of these estimates assume that you and your team have had similar experiences and have document templates in place. If not and this is the first time to incorporate these management plans and documents, then expect to spend several weeks preparing these. The second time to introduce a project will then be much, much faster. Chapter 5 has a number of outlines for documents that may serve you to introduce a project or prepare for one.

## Acknowledgement

My thanks to Craig and Diane Silver for their thoughtful and kind comments in reviewing this chapter.

## Recommended reading

H. Cloud, Integrity: The Courage to Meet the Demands of Reality, Collins and imprint of HarperCollinsPublishers, New York, NY, 2006 Dr. Cloud shows how integrity lays the foundation for good preparation, management, and communication in starting and introducing a project.
E. Verzuh, The Fast Forward MBA in Project Management, fourth ed., John Wiley & Sons, Inc., Hoboken, NJ, 2012. Eric Verzuh clearly and concisely describes the concepts of good management and communication within business.

## References

[1]  P. Hawken, A.B. Lovins, L. Hunter Lovins, Natural Capitalism: The Next Industrial Revolution, *second ed.*, Earthscan, London, U.K., 2010. p. 111 (Previously quoted in Little, Brown and Company, Boston, MA, 1999, p. 111).
[2]  Unknown source. Often quoted.
[3]  H. Cloud, Integrity, The Courage to Meet the Demands of Reality, Collins and imprint of HarperCollinsPublishers, 2006.

[4]   Slide show from Colorado State University, Objectives, Goals, Missions, & Visions. <http://www.wcirm. colostate.edu/archive/ObjectGoalMiss.pdf> (accessed 24.04.14).

[5]   Slides by Rizwan Khurram, Lecture 01—Strategic Concepts and Terminology - Missions, Visions, Goals, Objectives, Core Competencies (P1). <http://www.slideshare.net/RIZWANKHURRAM/objectives-mission-and-vision> (accessed 24.04.14).

[6]   Deborah Killam, Extension educator. Bulletin #6107, Vision, Mission, Goals & Objectives...Oh My!. at: <http://umaine.edu/publications/6107e/> 2004 (accessed 24.04.14).

[7]   At: <http://www.sname.org/Membership1/TheSociety/MissionGoalsObjectives> (accessed 24.04.14).

[8]   At: <http://ieee-ims.org/sites/ieee-ims.org/files/documents/IMS%20Strategic%20Plan.pdf> (accessed 24.04.14).

[9]   At: <http://corporate.ford.com/innovation/innovation-detail/one-ford> (accessed 24.04.14).

[10]  E. Verzuh, The Fast Forward MBA in Project Management, fourth ed., John Wiley & Sons, Hoboken, NJ, 2012. p. 33, 63—65.

[11]  Project Management Methodology Guidelines, Project Management Methodology & Step-by-Step Guide to Managing Successful Projects, City of Chandler, AZ. <http://www.chandleraz.gov/Content/PM000PMMethodologyGDE.pdf> (accessed 20.04.14).

[12]  K. Fowler, What Every Engineer Should Know About Developing Real-Time Embedded Products, CRC Press, Boca Raton, FL, 2008, pp. 6, 8.

[13]  A Guide to the Project Management Body of Knowledge: PMBOK(R) Guide, Fifth Edition, Project Management Institute, January 1, 2013.

[14]  K. Fowler, Electronic Product Development, Architecting for the Life Cycle, Oxford University Press, New York, NY, 1996, pp. 52—53.

## CHAPTER 4

# Dealing with Risk

## Kim R. Fowler
*IEEE Fellow, Consultant*

## Chapter Outline

Developing and Managing Embedded Systems and Products.
DOI: http://dx.doi.org/10.1016/B978-0-12-405879-8.00004-0

## Overview

Every project is risky. Some much more so than others. The good news is that risk can be managed.

Risk is the potential inability to stay within defined cost, schedule, performance, or safety constraints. This chapter suggests means for managing various forms of risk.

Risk management encompasses a number of actions. First, identify the various risks and analyze them. Understanding risk generally requires these four activities:

1.  Identify the areas of risk and understand the differences: business, operations, technical disciplines, and security.
2.  Determine the particular problems and consequences of each area.
3.  Determine the associated criticality or severity of each risk within each area.
4.  Understand the various groups of stakeholders—users, operators, owners—who will be affected in each area of risk.

Then control risks by reducing, constraining, or transferring them. Finally assess the state of the risks by analyzing them again. Repeat this cycle throughout the project's development (see Figure 4.1).

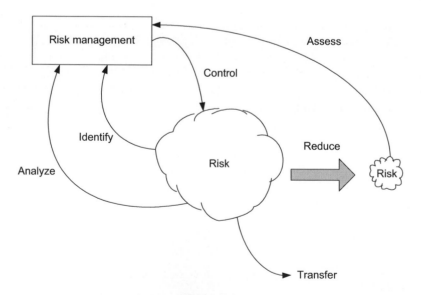

**Figure 4.1:**
General overview of how risk management controls risk. © 2014 by Kim R. Fowler.
*Used with permission. All rights reserved.*

### Example: Business risk—lawn care tractor

In Chapter 3, I introduced an upgraded lawn tractor with a new engine controller and engine. The business risk is whether such an upgrade will generate a profit. Part of that equation for business risk is whether development costs of the controller and engine will stay within schedule and budget.

### Example: Business risk—cubesat

In Chapter 3, I introduced the concept of a standard cubesat for volume sales. The business risk is whether your company will ever sell enough to make a profit. Part of that equation for business risk is whether development costs of the cubesat will stay within schedule and budget.

### Example: Operational risk—military equipment

Military equipment is built for specific purposes. Often during deployment, that equipment is used in unforeseen ways. The operational risk is whether such equipment will adequately serve in those new capacities. If it does not operate as desired in new situations, it could put lives and missions at risk. If its operation is undesirable, it will be jettisoned in favor of other equipment and methods, which affects both the operations and the business case (Chapters 1 and 3).

### Example: Safety risk—medical equipment

Medical equipment is built to sustain human life. The safety risk is if either it does not operate as designed or it fails in unexpected ways, then it could jeopardize patients' lives.

So who performs risk management? When do they do it? How do they do it? And why do they do it?

### *Who performs risk management?*

*Answer: Everyone.* The project leader sanctions the plan for the management of risk. Each member of the team should then contribute to the management of risk. At the very least, every team member should be aware of the project's risks. I worked with a spacecraft program manager who famously said, "If I eat crow, you eat crow." His point was that he was responsible and ultimately accountable for the project's risks, but every team member must share responsibility in handling the risks.

Besides the project team members, other stakeholders in managing risk include the sponsor, customer, user, patient, government, and regulator. These people will have input and suggestions for managing risk; they each have vested interest in reducing risk.

### When is risk management done?

*Answer: Throughout development and production.* Start early in the project—preferably the first day. Develop a plan and follow it. Update the analyses of risks throughout the project. Often the formal design reviews will have a prominent place for managing risk.

### How is risk management done? And why?

*Answer: Tracking repeated analyses to demonstrate trends.* The "how and why" are intertwined. The method is to track risks, make them visible to the team, and follow the trends of the margin. The goal is to decrease over time both the risks and the unknowns that hide risks.

### Hazard analysis within risk management

Hazard analysis and risk management are closely tied within project development. Hazard analysis focuses on safety issues, the damage, and the consequences to people while risk and risk management is a superset that includes safety risk, project risk, and mission risk. *One frequent difference between hazard and risk analyses is whether the focus is safety-critical or mission-critical.* (Manned spaceflight is both, this book does not have the scope to even begin addressing that.)

Proactive analyses help find and prioritize problems before building the system. Reactive analysis finds and fixes problems after building the system. Chapter 7 has more details on analysis. This chapter focuses on the awareness of risk and managing it.

---

**Please remember**

All examples and case studies in this chapter serve only to indicate what you might encounter. Hopefully they will energize your thought processes; your estimates and calculations may vary from what you find here.

## Definitions

One of the biggest problems that I have found recently in performing hazard, risk, and safety analyses is reaching consensus on definitions. Consequently, I have a set of definitions that I use during risk management:

*Failure*—a component, module, or system (implemented in either hardware or software) stops operating; the high-level system may behave in various ways: it might stop or it might malfunction or it might not show any deviation [1]. A failure may also be defined as the inability of a system (component) to perform its intended function for a specified time under specified environmental conditions.

- Example 1: A burned-out bearing in a pump is a failure.
- Example 2: A short circuit in an amplifier is a failure.
- Example 3: Heater coil breaks and no heat is produced, this is a failure.
- Example 4: Data overflows register and causes a wrong command, this is a failure.

*Fault*—higher order abnormal behavior, tends to be seen as a symptom, may be caused by things several steps removed [2]. Superset covers both failures and hazards [1]. NOTE: Some authors and researchers reverse the terms "fault" and "failure." All failures are faults, but not all faults are failures [2, p. 344].

*Degradation*—reduced performance of the system (or component).

*Error*—a design flaw or deviation from a desired or intended state [2]. Error versus failure has these distinctions:

- A failure is an event (a behavior) while an error is a static condition (a state).
- A failure occurs at a particular instant in time; an error remains until removed, usually through some form of human intervention.

*Hazard*—"state or set of conditions of a system (or an object) that, together with a particular set of worst-case environment conditions, will lead to an accident (loss event) where significant risk of injury or damage exists" [2, p. 467]. "Distinguishing hazards from failures is implicit in understanding the difference between safety and reliability" [2, p. 223].

- Example 1: Frayed power wires into a device presents electrical shock hazard.
- Example 2: Unsecure data entry can lead to wrong action by a device.

*Accident*—an undesired and unplanned event that results in a loss (including loss of human life, injury, property damage, etc.) that is important to a stakeholder [2].

- Example 1: Heater "sticks" on and the device's temperature goes above the specified limit.
- Example 2: Incorrect temperature set point entered by a user and temperature is not correct for the application.

*Incident*—(might be considered a near miss), an event that involves no loss (or only minor loss) but with the potential for loss under different environmental circumstances. A particular instance of a hazard [2].

- Example 1: Wrong temperature input by a user, but within limits acceptable for the application.
- Example 2: Battery fails but building power continues to power the device.

*Safety*—freedom from accidents or losses [2].

*Risk*—combination of severity and likelihood:

- *severity*—the worst possible accident that could result from the hazard given the environment in its most unfavorable state.
- *likelihood*—probability of hazard occurrence.

## Risk analysis and management

Risk analysis establishes the likelihood of problems and their severity. Margin management then uses these metrics to follow the risks and manage the system margins to converge to an acceptable solution.

Risk management systematically applies management policies, procedures, and practices to identify, analyze, control, and monitor risk. Risk management includes planning for risk, assessing risk areas, developing risk mitigation plans, tracking risks to determine how they have changed, and documenting the overall risk management program.

Risk management takes the results from reliability calculations, environmental and various analyses, found in Chapter 7, and assigns probability and criticality to every potential event. The probability for a particular scenario is the likelihood that it will occur during a specified operational window. The probability and severity of each problem help the designer and developer to set a priority for fixing the problem.

Hazard analysis is one particular form of analyses that helps reveal the problems. Hazard analysis focuses on the safety and mission-critical aspects of the project. Hazard analysis usually does not provide evidence for the business case or even the operational case in its entirety. Figure 4.2 shows one example of hazard analysis feeding into risk management. Medical devices sometimes combine risk and hazard into one document called the Risk and Hazard Analysis (RHA) (Figure 4.2).

### Margin analysis and management

Margin analysis and management is a part of risk management. It tracks areas of concern and provides trend analysis to show that the concerns are narrowing to acceptable values. Margin analysis and management can help you avoid specification overruns during the system development.

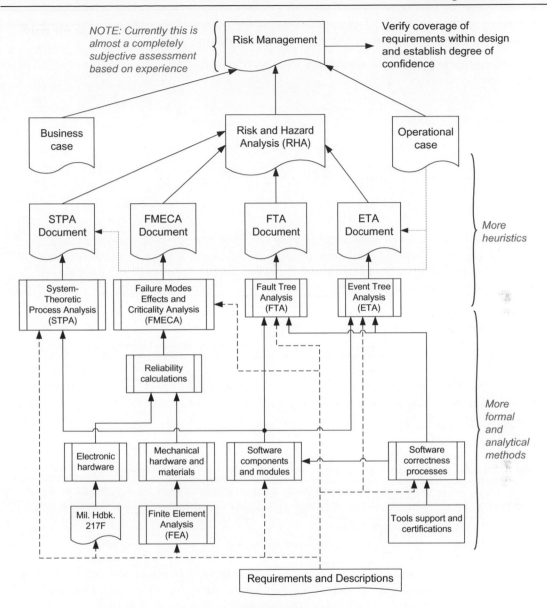

**Figure 4.2:**
How various analyses might feed into the RHA to inform the plan for risk management.
© 2014 by Kim R. Fowler. Used with permission. All rights reserved.

### Example: Margin management in a satellite subsystem

The following parameters are examples of items with margins from an unmanned satellite subsystem:

- Mass
- Power consumption
- Data rate
- Data storage
- Critical CPU computational throughput
- Critical processor memory (code and operational memory) loading
- Critical nonvolatile memory loading
- Data bus throughput

These must be managed to avoid exceeding the specifications. Running over the mass specification clearly makes the satellite subsystem too heavy. If it is too heavy then some other subsystem must be made lighter and the entire satellite needs to be rebalanced to its originally planned moments of inertia.

Overly conservative or excessive margins can make a system more complex and expensive than necessary. Providing a data bus throughput that is four times the necessary bandwidth may mean that you have used unnecessary components or drawn too much power to drive the bus.

Margin analysis has two parts: the *likelihood* of failing to achieve a desired result and the *consequences* of failing to achieve that result. (Both of these can derive from the analyses detailed in Chapter 7.) The likelihood of occurrence ranges from remote to very likely. The consequence of occurrence ranges from noncritical to catastrophic.

Table 4.1 gives the criteria for specific levels that define risk consequence and likelihood. The risk severity is established by combining the consequence and likelihood as given in Table 4.2. These criteria derive from the unmanned aerospace industry. Definitions and values may vary for other industries.

Risks that fall in the light gray (or yellow) boxes (diagonal) and in the dark gray (or red) boxes (upper right corner) of Table 4.2 are tracked in the risk management database. Risks in the clear boxes (lower left corner) are tracked at the discretion of the program manager.

Risk management uses these two tools (Tables 4.1 and 4.2) for each module or subsystem within the system. During each phase of development you then update the risk—likelihood and criticality—of each subsystem. This is a way of monitoring trends to see that the project is converging to a satisfactory solution. The case study lists some potential concerns for designing a hybrid switcher locomotive.

Table 4.1: One example of risk attribute categories

| Attribute | Value | Description |
|---|---|---|
| Consequence (or criticality) | Catastrophic (C) | **Safety-critical:** loss of life or life-threatening injuries. **Mission-critical:** cost impact exceeds project reserves. Schedule slip that affects launch date or start of mission. Loss of mission or failure causes other instruments to fail. Major loss of capability in other instruments or experiments. |
| | Severe (S) | **Safety-critical:** severe injuries. **Mission-critical:** cost impact exceeds planned reserves. Schedule slip affecting critical path but not delivery. Loss of system or instrument that might affect operation of other systems or instruments. |
| | Important (I) (or moderate) | **Safety-critical:** minor bodily injury. **Mission-critical:** cost impact smaller than element cost reserves. Slip reduces slack to one month or <50% of remaining schedule. Minor loss of capability or design/implementation work-around. No impact to cost reserves. Loss of system or instrument but does not affect operation of other systems or instruments. |
| | Noncritical (N2) (or low) | **Safety-critical:** very minor bodily injury. **Mission-critical:** slip but slack greater than 1 month or >50% of remaining schedule. Loss of capability/margin but all mission requirements met. Instrument or system experiences degradation in performance. |
| | Noncritical (N1) and very low impact | **Safety-critical:** inconsequential bodily injury. **Mission-critical:** no slip or change in schedule. Instrument or system might experience a small degradation in performance. |
| | None (N0) | **Safety-critical:** no injury. **Mission-critical:** no slip or change in schedule. DVS continues operating, no performance change, smaller margin of error might be expected. |
| Likelihood | Very likely (v) | >50% chance of occurring |
| | High (hi) | 10–50% chance of occurring |
| | Moderate (mod) | 1–10% chance of occurring |
| | Low (lo) | 0.01–1% chance of occurring |
| | Remote (r) | <0.01% chance of occurring |

## Table 4.2: One example of risk severity

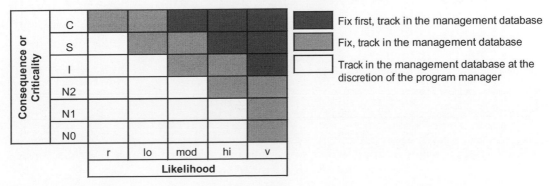

## Case Study: Risk assigned to locomotive subsystems during design

A company will design, develop, build (or buy) subsystems, test, commission, and support a hybrid locomotive. The locomotive will have a genset (motor alternator in a steel frame so that it is removable) and in a separate compartment, batteries. It will be primarily be used for switching in rail yards. The company has decided to purchase the genset, the batteries, and reconditioned motors already in trucks. A contract manufacturer will take old locomotive chassis and trucks and remanufacture them.

The concern is how to manage the risk in designing and developing this locomotive. To start, you could use the tools in Tables 4.1 and 4.2 and assign the likelihood and criticality of failure to each subsystem of the locomotive. Figure 4.3 illustrates the major electrical subsystems that need consideration (this example does not take into account the mechanical structure, a very important part of any locomotive). Table 4.3 lists examples of criticality and likelihood for each subsystem.

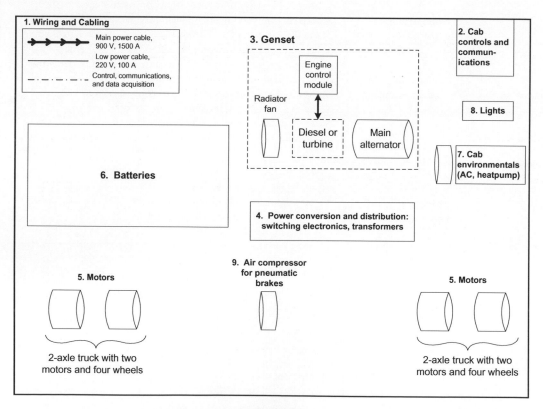

**Figure 4.3:**
An example of the major electrical or data subsystems of a hybrid locomotive. © 2014 by Kim R. Fowler. Used with permission. All rights reserved.

**Table 4.3: Examples of criticality and likelihood for each electrical or data subsystem in a locomotive**

| Subsystem | Criticality | Likelihood | Fix |
|---|---|---|---|
| 1. Wiring and Cabling | | | |
| – control cabling | I or S | r | |
| – low-power cabling | I or S | r | |
| – high-power cabling | S | lo | (shaded) |
| 2. Cab controls and communication | S | r | |
| 3. Genset | I | r | |
| 4. Power conversion and distribution | S | r | |
| 5. Motors (considered one at a time) | I | lo | |
| 6. Batteries | I | mod | (shaded) |
| 7. Cab environmental equipment | N2 | mod | |
| 8. Lights | I | mod | (shaded) |
| 9. Air compressor for pneumatic brakes | S | lo | (shaded) |

The light gray shading in the "Fix" column indicates subsystems that you should focus on making these subsystems more robust. This is only one example and in reality should receive much more detailed attention.

The genset, while important, is not a high criticality item. The batteries could still run the locomotive for a short while. The genset is also a purchased item; it should have a fairly well-refined reliability figure. Losing a motor is not a major concern, because the locomotive can run on the other three motors before fixing the failed motor. Motors also tend to be a fairly reliable component.

You can also perform margin management on each parameter of interest (e.g., mass, power consumption, data throughput, or data storage). You then record and track each value for each parameter of interest in each module or subsystem. The following example from the satellite illustrates tracking margins.

### Example: Margin management in a satellite subsystem-monitoring risk trends

Consider two parameters, mass and power consumption, from the previous satellite example. The final architecture concept has goals for mass and power consumption that work with the rest of the satellite. In this example the subsystem has a mass goal of 2.4 kg and has a goal for average power consumption of 1.5 W. The margin for mass is +15% to begin with and steadily decreases to +4%. The margin for the average power consumption begins at +20% and decreases to +5%.

The two figures (Figures 4.4 and 4.5) show these margins bounding development and the trend of the development for these parameters.

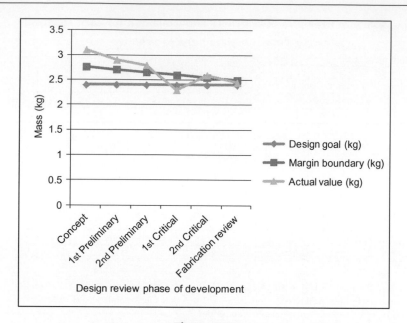

**Figure 4.4:**
Trend of mass margin during development. © *2014 by Kim R. Fowler. Used with permission.*
*All rights reserved.*

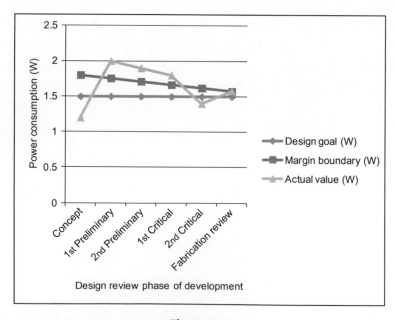

**Figure 4.5:**
Trend of power margin during development. © *2014 by Kim R. Fowler. Used with permission.*
*All rights reserved.*

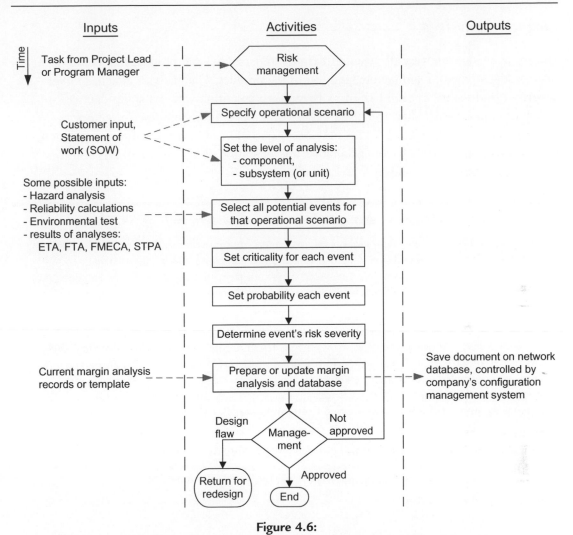

**Figure 4.6:**
The flow of activities within risk management for single iteration or design phase.
These steps should all be repeated as separate tasks in later stages of project development.
© 2014 by Kim R. Fowler. Used with permission. All rights reserved.

Risk and margin management should be performed throughout product development. Figure 4.6 gives the general flow risk and margin management for a single iteration or design phase.

Risk and margin management are accepted forms of product assurance. They are based in considerations accepted by both the military and aerospace communities. Decisions to perform or not perform the analysis depend on either the customer contract or the intent to archive results for future projects.

## *Hazard analysis*

Hazard analysis identifies all potential hazards to personnel and product that could occur during development, manufacture, and test of the product. The analysis attaches either severity or criticality to each hazard. A hazard is a condition where a significant risk of injury or damage exists because of the nature of the operation being performed or of the material being handled. Hazards may occur during installation and during operation; the hazards may differ between installation and operation. The case study that follows derives from the space industry.

### Case Study: Hazards in an unmanned satellite subsystem built by XYZ Company

Typical hazards include electrical shock, explosion, fire, contamination, radiation, temperature extremes, corrosion, and collision. A catastrophic hazard can cause loss of life or damage to adjacent hardware. A critical hazard can cause significant personnel injury or damage to deliverable product. Each potential type of hazard typical to XYZ Company's business is listed below.

#### Electrical shock

Most of XYZ Company's products receive power supplied at 28 VDC and then generate 15, 5, and 3.3 VDC for use within their subsystems. These are below reasonable electrical shock thresholds of 42 V or higher. Furthermore, during most operations the XYZ Company products are sealed within the spacecraft and away from human contact. Consequently, there is typically no hazard from electrical shock. Regardless, the potential for electrical shock is an area of investigation for XYZ Company hazard analysis.

#### Explosion

Typically, the only source of explosive power in XYZ Company's products is 28 VDC, which is fuse protected at the source in the spacecraft. Consequently, there is usually no hazard from explosion. Regardless, the potential for explosion is an area of further investigation for hazard analysis by XYZ Company.

#### Fire

Typically, the only sources of ignition for a fire and material to sustain combustion in XYZ Company's products are the power input and output. Power lines are circuit (crowbar) or fuse protected at the source in the spacecraft. Consequently, there is usually no hazard from fire generated by XYZ Company's products. Regardless, the potential for fire is an area of further investigation for hazard analysis by XYZ Company.

#### Temperature extremes

Typically, the only sources of elevated temperatures in XYZ Company's products are the cabling, circuit boards, and components. These are current limited by design, circuit (crowbar) or fuse protected at the source in the spacecraft. Consequently, there is usually no hazard from an extreme high temperature generated by XYZ Company's products.

Most XYZ Company products have no source of extreme low temperatures, such as cryogenics, thus there is usually no hazard from an extreme low temperature generated by XYZ Company products.

Regardless, the potential for temperature extremes is an area of further investigation for hazard analysis by XYZ Company.

### Radiation

XYZ Company products have no sources of ionizing radiation, consequently there is no hazard.

### Corrosion

XYZ Company products have limited sources of corrosion and corrosion by-products, none of which are hazardous to humans, consequently there is no hazard.

### Collision

Most XYZ Company products have no moving parts; consequently there is usually no hazard of collision. Regardless, the potential for external forces colliding with XYZ Company products is an area of further investigation for hazard analysis by XYZ Company.

### Contamination

Most XYZ Company products have limited sources of contamination, none of which are known to be hazardous in brief or even extended contact with people. Furthermore, all components and circuit boards are protected by conformal coating, which prevents outgassing of these substances. Consequently, there is usually no hazard from contamination during operation. Regardless, the potential for contamination is an area of further investigation for hazard analysis by XYZ Company.

## Criticisms of probabilistic risk assessment

Up to this point in this chapter, RHA primarily describes probabilistic risk assessment (PRA). There are some significant criticisms of PRA that have credence. Some of the most prominent concerns are raised by Nancy Leveson in her textbook, *Engineering a Safer World: Systems Thinking Applied to Safety.* She argues that the chain-of-event concept of accidents typically used for risk assessments cannot account for the indirect, nonlinear, and feedback relationships that characterize many accidents in complex systems. The claim is that current risk assessments do a poor job of modeling human actions and their impact on failure modes [3].

Here are some new assumptions that Leveson believes need to be brought to system design for great safety and more robust operations:

1. "Accidents are complex processes involving the entire socio-technical system. Traditional event-chain models cannot describe this process adequately." [3, p. 31]
2. "Risk and safety may be best understood and communicated in ways other than probabilistic risk analysis." [3, p. 36]
3. "Operator behavior is a product of the environment in which it occurs. To reduce operator 'error' we must change the environment in which the operator works. (This postulated assumption opposes an original assumption that most accidents are

caused by operator error and that we must reward safe behavior and punish unsafe behavior to reduce accidents.)" [3, p. 47]

4.  "Highly reliable software is not necessarily safe. Increasing software reliability or reducing implementation errors will have little impact on safety." [3, p. 50]

5.  "Systems will tend to migrate toward states of higher risk. Such migration is predictable and can be prevented by appropriate system design or detected during operations using leading indicators of increasing risk." [3, p. 52]

6.  "Blame is the enemy of safety. Focus should be on understanding how the system behavior as a whole contributed to the loss and not on who or what to blame for it." [3, p. 56]

From this point, Leveson then goes into the System Theoretic Process Analysis (STPA) in her textbook. Chapter 7 in this book has a section on STPA.

The real question in all this is, "Can we realistically prepare for all events?" The tsunami that flooded the Fukushima nuclear power plant and precipitated a set of failures was unprecedented in magnitude. It topped a 10 m high flood wall with a 14−15 m water level. Most likely PRA or STPA would have predicted the multiple failures (i.e., loss of offsite electrical power to the facility, the loss of fuel for the diesel generators, the flooding of the diesel generators and electrical switchyard, and damage to the cooling water inlets from the ocean) that led to the loss of reactor cooling water if we had known that a 15 m high tsunami was conceivable.

Does digging around for all the very distantly remote and highly severe events really help? Or would that effort mask more likely and realistic problems? In the end, I think a rigorous, ongoing effort to cover the known problems and constraining the "known unknowns" is sufficient for most project developments. The next section of this chapter outlines some problems common to many projects. Dealing with them will be a good start to covering the known problems and constraining the "known unknowns."

---

### Example: NIST encounters Hurricane Isabel

In September 2003, Hurricane Isabel caused power outages in the Washington, DC, area because the wind blew trees down onto power lines. The National Institute of Standards and Technology (NIST) did not have a power outage but did experience 180 VAC overvoltage for 20 minutes that destroyed 1000 seconds of fluorescent lamp ballasts. The protective mechanisms for AC power, which could have prevented the damage, were controlled over telephone lines. Guess what was also knocked down by falling trees? [4]

Clearly, the designers did not realize that power problems and the loss of protective control could have the same source (a common cause problem source). They also may not have realized that power outages in the region would cause extended periods of overvoltage that could damage installed equipment. This is an example of failure not occurring frequently, immediately, or predictably.

The question is, "Could this have been foreseen?" Common cause failure, trees falling on power and telephone lines, is not unusual in this region of the country. More thought should have been applied here . . . but hindsight is always 20—20!

## Types of problems

One way to identify important potential sources of problems to your project is to first identify the consequences, the losses, such as:

- Loss of data
- Loss of personnel
- Loss of project
- Loss of market
- Loss of business.

Next identify the level of detail:

- Component
- Board
- Subsystem
- System to integrate with other systems.

Once these are in place, begin looking at the types of problems that can lead to these consequences. The next three sections of this chapter identify some sources of these problems: failure, disasters and catastrophes, intrusion, sabotage, theft, and destruction.

## Failure

Failure has its primary source in humans and human-related problems. These failures include the following aspects: technical, professional, production, commercial, marketing, and societal change. After identifying these in your project, then build a contingency plan to recover from them.

### Technical failure

One class of failures is *technical failure* in design and management decisions. Choosing an architectural design or subsystem that does not cover all the planned situations is a technical failure. Not performing all the necessary tests or analyses to prove the design is also technical failure (could be a management failure, verging on professional failure, if a manager restricts the designers or developers from doing a full suite of tests and analyses).

### Personal Example: Technical failure

Years ago I helped design a portion of an ultraviolet camera for a satellite instrument. I was responsible for the automatic gain control for an image intensifier that formed the front end of the ultraviolet camera.

Image intensifiers are plates of glass capillaries sandwiched between phosphorus screens. They operate on the photomultiplier principle to intensify the light collected by the optics. They have wide dynamic ranges, with gains up to a million or more, and they are highly nonlinear in their control.

*Problem*: My first design for the control of the image intensifier produced a video picture that repeatedly bloomed (bright light washed it out) and then collapsed to nearly black.

*Background*: My design used discrete logic with up-down counters to record the number of bright pixels within a video frame. Unfortunately, the control for the image intensifier was unstable for bright objects. The short development time (we actually flew wire-wrapped breadboards) rushed me and I did not fully simulate or analyze the control action.

*Should'a*: In retrospect, I "should'a" taken extra time to analyze and simulate the expected video scenes during design of the gain control system. I let the tyranny of the urgent rule me and I did not take time to consider all the aspects of the problem. The good news is that a year later I got the chance to redesign the automatic gain control of the image intensifier for another satellite. This time I did a much better job; the second design controlled the image intensifier appropriately.

Technical failures share some similar characteristics:

1. These failures that stem from problems have confounding complexity. The original designers did not foresee all possible circumstances and that many problems have multiple causes. In the NIST power failure example, no one picked up on the possibility that wind could simultaneously down both power lines and telephone lines, which prevented protective mechanisms from functioning properly.
2. These failures often experience a significant passage of time. Technical failure becomes a major problem when it occurs in fielded units; it is much more difficult and expensive to retrieve and repair fielded units.
3. The third characteristic of these failures is oversight, that is, unintentionally missing the potential problem during design, even if it appears to be a manufacturing problem. Improper routing of wire harnesses or manufacturing problems where wires are "stuffed" into an enclosure only to get crimped by hasty assembly or to chafe against a sharp edge is a design problem.

4. The fourth characteristic of these failures is nonobviousness to the user. The use of the device is improperly labeled or poorly instructed by the user's manual or incomprehensible—knobs are in the wrong place or obscure codes must be programmed or it is awkward to handle. No one foresaw a 15 m high tsunami inundating the Fukushima power plant; it was nonobvious.

5. Finally, failure can be from improper use. Yanking on a power cord repeatedly to unplug it from the wall receptacle, pounding on the keyboard out of frustration, poking the touch screen with a screwdriver instead of a finger, using the wrong lubrication, and your three-year old standing on the toaster as a stepstool are all examples of improper use and abuse.

## Professional failure

A closely related class of failures is ***professional failure***, which is where a character flaw emerges to damage the effort. Not doing all your assigned work is a professional failure. Treating colleagues or clients badly is a professional failure. While not technical, professional failure directly affects and hurts the project development.

---

### Personal Example: Professional failure

Years ago, I was asked to finish programming a project while the original designer moved onto another project. I procrastinated and made a number of false starts on the work. Finally, I removed myself from the work and the original designer had to return and finish the job.

*Problem*: I did not complete my assignment.

*Background*: Management had asked me to take over the project. I quickly realized that this was a no-win situation for me. While I was asked to finish the work and there was a hint that I would be a hero if I did, it seemed to me that I would not see any real recognition of my work. I saw it as setting precedence for helping others out of problems without getting to do significant work myself. I simply lost motivation.

*Should'a*: I "should'a" made another choice in how I handled the job. Either I should not have taken the job in the first place or, if given no choice, I should have plowed through the assignment while finding another job. Muddling through or not completing a project is very bad for progress, your reputation, and the company's reputation.

Muddling along and not completing a project is a professional failure. You can avoid this problem by choosing different projects (or jobs or companies) or by steeling up your motivation and plowing through the project. If you get stuck in this sort of work, you will need to find another line of projects or of work.

---

### Production failure

A class of business failure is *production failure*, which includes inventory, suppliers, machinery, and work force. Anything that disrupts production enough to stop a project is a production failure. The facility of your contract manufacturer burns down; your production is stopped and it costs your company. A long-term worker strike can stop production, too.

### Commercial and marketing failure

Another class of business failures is *commercial and marketing failure*. Building a smoothly running and robust widget that does not meet the market demands is a marketing failure. Getting the wrong features or not enough features or too many features are all marketing failures.

### Failure from societal change

The final class of failure, and the most difficult to circumvent, is *societal change*. A large group of people that move away from your product or service can be a societal change that marketing could not have predicted. Building a large factory in a remote country, which then experiences a coup d'état leaving your work force stranded (or hostages) and stopping payments, is a societal change leading to a business failure.

---

**Personal Example: Business/political failure**

I led the development for a satellite subsystem. NASA, our sponsor, took over the project and out of our control.

*Problem*: Government politics outside the control of my company pulled the project from us.

*Background*: The team spent six months performing trade studies to design the architecture of the subsystem. The work was done thoroughly and well. Meanwhile, several groups at NASA had fallen on lean times and wanted work for their engineers. NASA pulled the project back in-house to keep their engineers employed.

*Should'a*: There is no "should'a." We did a good job and the project was still pulled from us. These things happen.

---

### Steps that you can take

You can do a number of things to reduce failure. Taken together, these activities will not guarantee success but they will help you reduce the frequency and impact of failure:

- Establish a culture of integrity.
- Build trust.

- Continually remain open and communicating.
- Encourage experimentation.

Teamwork is one of the best ways to establish integrity through accountability to one another, building trust, and remaining open. It knits together the organization and its development effort. Honest peer review and code inspections will dramatically improve the quality of your designs. You will also learn from your colleagues. "Iron sharpens iron, so one man sharpens another."[Proverbs 27:17]

Experimentation is planned failure in a controlled environment. Using simulation and then actual systems in test situations will help wring out problems. Doing experiments with field tests often helps locate nonobvious circumstances. Simulation alone is not enough; it is limited to the assumptions of current knowledge.

## Disasters and catastrophes

Various environmental influences can cause major problems that delay your product development. These include:

- power outage
- fire
- flood
- hurricane and tornado
- earthquake
- snow and ice
- environmental changes.

Power outage is the most frequent problem and should be the first thing that you address in contingency planning. Power outage can wipe out memory on your computer and you lose a day's worth of work (or more if you do not have a recent backup). A simple uninterruptible power supply will sustain your computer long enough for you to back up data.

Most people do not realize how frequent fire and flood can be. These disasters are the next most frequent behind power outage. Both fire and flood will destroy your equipment and your paper files.

The remaining types of disasters have similar effects. They can destroy your equipment and your paper files.

The most important thing that you can do is develop an offsite storage site for data and for files. Your data is typically the most important asset in your company. Daily you should back up your computers and their data to an offsite server. There are companies that provide such storage.

Investigate and assess the possibility of flooding—including a burst pipe in your facility. Elevate computers and equipment off the ground. I recently toured a high-technology company with design functions and specialized manufacturing equipment in a building that is situated in the flood plain of a local creek. I wonder if moving facilities to higher ground might not be better than flood insurance?

## Intrusion, sabotage, theft, and destruction

There are both physical and cyber realms for intrusion, sabotage, theft, and destruction. Intellectual property, designs, business plans, and equipment all can be subject to sabotage, theft, and destruction.

The primary defenses are physical boundaries and cyber security. Chapter 12 addresses cyber security.

Restricting physical access to your property is an important defense to reduce physical intrusion by unauthorized people and the subsequent sabotage, theft, and destruction of physical property. Monitor and sign visitors in and out of your facility during business hours. Lock doors and use an alarm system to restrict access after business hours. Secure, off-site storage can also provide some recovery from intrusion, sabotage, theft, and destruction.

## Contingency planning

Contingency plans need to be in place before an emergency, disaster, theft, or destruction actually occurs. Going through the exercise of contingency planning will help have a disproportionate effect of preparing you for a real event. A contingency plan should walk you through steps to deal with an actual event: access the problem, determine its impact, decide on a course of action, and then implement that course of action. Contingency plans should be a part of risk management and referenced in the risk management plan.

### Configuration management

Before any problems occur, put in place a configuration management system. A configuration management system can help you recover more quickly by restoring your company and project data to its state just before the problem occurred.

"Every system of business or technical development needs configuration management because problems, mistakes, and changing circumstances occur frequently, if not continuously. Configuration management helps reduce the consequences of these problems from becoming too disruptive by maintaining an understanding of what is intended and what is the current output.... Configuration management monitors the state of just about everything in a project. Here are some examples of areas within configuration management:

- Software
- Hardware
- Products
- Identification
- Database and network." [5]

You should have off-site storage of all data and business records. You should back up your projects data and records on a daily basis. You should have a version control system to check files in and out. Consult the reference for details [5].

### Assess impact and priorities

Assessing the impact of a loss event requires identifying its impact. Potential areas of impact include the budget, the schedule, the business case, the financial soundness of the company, reconstruction of the design, and safety risks to stakeholders. The amount of impact can range from inconsequential to catastrophic. This is a very similar exercise to margin analysis and management from earlier in this chapter.

### Recovery

Prepare different courses of action to deal with potential problems. Create business continuity plans and procedures; for example, where to move the business should a fire devastate your current facility. Prepare to manage insurance claims, a nontrivial exercise. Maintain effective communications during an emergency.

Institute a configuration management systems and use it rigorously. It will help you recover much more quickly, cutting time lost to recovery. Store your project designs, documentation, and business files to off-site, secure storage. After the loss, you will be able to reconstruct and restore your business and project structures and operations.

Sometimes a proactive approach that prepares redundant prototypes or fault tolerance architectures can allow you and your team to ride through the crisis. Building two or three prototype copies of a product allows you to continue should one fail or be destroyed. Spacecraft instruments often follow this course; the team will build a flight model and an engineering model, which is identical in all respects except that it does not have the final conformal coat on it. Running initial tests with the engineering model reduces the delay and impact of first-time mistakes, for example, reversed power leads, dropping, overheating, and all the sorts of things that you can iron out before testing the flight hardware.

### If recovery is not possible

You have some less desirable choices assuming recovery is not possible. In descending order of desirable outcomes, you could do the following:

- outsource the effort,
- change the course of the project to something that is realizable,
- sell the project or merge the company with another, or
- if nothing else works, your company could file bankruptcy or go through some sort of commercial death.

### Outsourcing

By way of example, let's say your manufacturing facility flooded and the water destroyed expensive equipment. You could outsource the manufacturing to reduce delay in production. Outsourcing may be a greater expense than you planned using in-house facilities, but those are now useless, so at least the product gets out the door.

### Change course

Another example, let's say that your design was stolen and the legal challenge would sink the project. You could change course and develop another project (oh yes, and fix your security so the new project design is not stolen, as well). This is a delay and a hit to your budget and finances, but at least you are still in business.

### Sell or merge

If the loss is significant enough that recovery looks impossible, you could sell the intellectual property and assets to recoup some of the original investment. You will need good legal counsel. Chapters 17 through 23 can give you some insight into some of the things that you will have to do, such as negotiate. Other than bankruptcy this is a last resort.

### Bankruptcy

Sigh, sorry, I don't have any words of encouragement here. See a good lawyer. Chapters 17 through 23 may be a bit of help here, but most likely even they are too little too late.

## Effort to manage risk

Every project varies in its effort but previous experience and templates can reduce the effort. A small project with a few people working over six months may only require one person working for two or three hours to prepare the risk management plan. Larger, more involved projects, with 30–50 people working over the course of four to six years may require that same person to spend several days to weeks to prepare the risk management plan.

Both of these estimates do not include the effort for the analyses, as described in Chapter 7; analyses are considered separate efforts that support the effort to manage risk. Furthermore, both of these estimates assume that you and your team have had similar experiences and have a document template in place. If not and this is the first time to prepare a risk management plan, then expect to spend several weeks preparing the risk management plan. The second time through a risk management plan should be faster.

The effort to manage risk is far more variable and depends on the product and the market. For a spacecraft instrument, risk management takes place alongside the updates of the analyses throughout the project. For this type of project, risk management will occupy a notable portion of someone's time, which does not include the very considerable effort to update the analyses—often the entire team contributes those various analyses. For an appliance model that is similar to previous models, the risk management may be considered almost "boilerplate" because it is done the same way for every new model. Unfortunately, "boilerplate" can breed complacency and problems can creep in to follow-on projects.

So, all this to say, risk management does take time. It is better to spend regular time reviewing and updating your risk management plan rather than cleaning up a mess caused by a risk that ran away from you into a disaster!

## Acknowledgement

My thanks to Craig Silver for reviewing this chapter.

## References

[1]   AS5506/3, SAE Architecture Analysis and Design Language (AADL) Annex Volume 3: Annex E: Error Model Annex, AEROSPACE STANDARD, 2013-09-13, Draft 0.96, SAE Aerospace, An SAE International Group.

[2]   N.G. Leveson, Safeware: System Safety and Computers, Addison-Wesley Professional, Boston, MA, 1995 (Chapters 8 and 9).

[3]   N.G. Leveson, Engineering a Safer World: Systems Thinking Applied to Safety, The MIT Press, Cambridge, MA, 2011, pp. 14−15, 31, 33, 36, 47, 50, 51−53, 56.

[4]   Remarks during presentation by M. Postek, Nanotechnology Research Potentials at the NIST Advanced Measurement Laboratory, Greater Washington Nanotechnology Alliance, Special Topics Symposium, November 25, 2003.

[5]   K. Fowler, Mission-Critical and Safety-Critical Systems Handbook, Design and Development for Embedded Applications, Newnes an imprint of Elsevier, Boston, MA, 2010, pp. 42−45.

## CHAPTER 5

# Documentation

**Kim R. Fowler**
*IEEE Fellow, Consultant*

**Chapter Outline**

## Overview and rationale

Documentation conveys the right information to the right person at the right time. It can be voluminous (Figure 5.1) or not; the most important thing is that documentation contains the right information. Documentation is corporate communication and memory. Every project needs it, otherwise the work cannot be replicated or fully understood.

Developing and Managing Embedded Systems and Products.
DOI: http://dx.doi.org/10.1016/B978-0-12-405879-8.00005-2
© 2015 Elsevier Inc. All rights reserved.

**Figure 5.1:**
Documentation is about communicating information to the right person at the right time. It is not about accumulating "edge inches" of paper as a metric to demonstrate work done. © *2014 by Kim R. Fowler, used with permission, all rights reserved.*

Recall from Chapter 3 that I said communications take 90—95% of our time. Documentation is a form of communication and occupies a sizable portion of that 90—95%. Everyone working on a technical project spends a lot of time preparing documentation.

"Here are three heuristics about documentation that I have found:

- *Documentation is integral to every product.*
- *While good documentation can't help a poor product, poor and inadequate documentation can destroy a good product.*
- *While some good products have poor documentation, I have NEVER found a poor product with good documentation.*" [1]

Actually, I have since heard of one example of a poor product with good documentation. That is one example out of hundreds, if not thousands, of contrary examples that I have encountered over 32 years of technical development.

## *Function*

"Documentation generally serves three purposes:

1. To record the specifics of development (the "who, how, and why"), which include, but are not limited to, engineering notebooks, software source listings, schematics, and test reports. These records help when modifications, upgrades, fixes, and recalls occur.

2. To account for progress (the "what, when, and where") toward satisfying requirements and provides an audit trail of the development. These types of documents include memos, but are not limited to, meeting notes and minutes, review action items, and project plans. An appropriate plan for documentation supports rigorous testing, validation, and verification.

3. To instruct users and owners the extent of functionality of your product. Instructions include, but are not limited to, user manuals, DVDs with instructions, websites, labels and warnings that present concise instructions." [1]

Documentation can have a number of different functions to fulfill its purposes. Figure 5.2 illustrates six different functions of documentation.

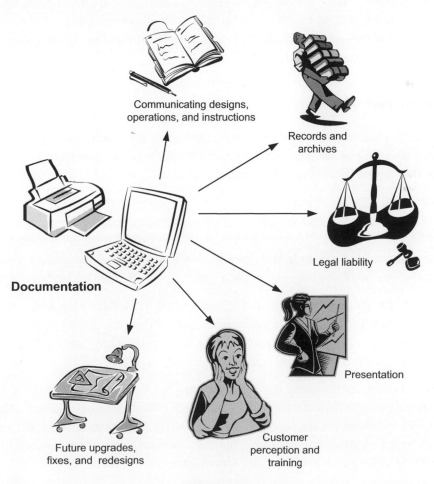

Communicating designs, operations, and instructions

Records and archives

Legal liability

Presentation

Documentation

Future upgrades, fixes, and redesigns

Customer perception and training

**Figure 5.2:**
Documentation serves many purposes. © 2007, 2014 by Kim R. Fowler, used with permission, all rights reserved.

> **Please remember**
>
> All examples in this chapter serve only to indicate what you might encounter. Hopefully they will energize your thought processes; your estimates may vary from what you find here.

## Types and content

The set of project documents should ultimately cover, "who, what, when, where, why, and how." All components of a project's documentation generally fall into one of four primary categories to cover the "who, what, when, where, why, and how": [1]

- Plans (who, what, where, when, why)
- Design documents (how, why)—source listings, schematics, engineering notebooks
- Reviews, reports, and presentations (who, what, where, when, why)
- Instructions (what)—user manuals, webinars, seminar presentations, brochures, training materials, maintenance and repair guides.

## When, who, and what

In addition to the four categories of documents, documentation serves two primary arenas: the company and the project. Typically everyone within a project contributes specific documents that represent their area of work.

That said, the project leader or program manager has responsibility for the documents found in Table 5.1. These documents tend to be plans and management functions for running the project. Other members of the team may contribute material to any one or all of these documents in the Management Sector. Most certainly team members will supply important content to the technical plans; in some cases they may write a specific plan and then the manager approves it.

Members of the design, development, and manufacturing team have responsibility for the documents found in Table 5.2. These documents include designs, descriptions, analysis reports, test plans, test results, and manufacturing documents. They tend to be the technical descriptions necessary to replicate the product.

Any number of company personnel have responsibility for the documents found in Table 5.3. These documents are the face of the company and deal with the customer, stakeholders, and legal aspects of the project. Usually upper management, marketing, business staff, and legal counsel handle the various documents in Table 5.3. The technical staff members and project team typically provide much assistance in preparing these documents.

Other documents that your team might develop if your project is safety or mission critical include, but are not limited to: [1, pp. 50–82]

# Table 5.1: Primary documents prepared to manage a project

| Project Sector | Description of Documents | Concept Design | Preliminary Design | Critical Design | Test and Integration | Release to Manufacturing | Commissioning |
|---|---|---|---|---|---|---|---|
| **Management** | Project charter | ✓ | | | | | |
| | Project plan | ✓ | | | | | |
| | Configuration management | ✓ | | | | | |
| | Risk management | ✓ | | | | | |
| | Control board plan | ✓ | | | | | |
| | Quality assurance plan | ✓ | | | | | |
| | Communications plan | ✓ | | | | | |
| | Documentation plan | ✓ | | | | | |
| | Review plan | ✓ | | | | | |
| | Technical plans | | | | | | |
| | — Electronic development plan | | ✓ | | | | |
| | — Software development plan | | ✓ | | | | |
| | — Mechanical development plan | | ✓ | | | | |
| | — Manufacturing plan | | ✓ | ✓ | | | |
| | — Support equipment plan | | ✓ | | | | |
| | Records—action items, meeting minutes, problem reports and corrective actions (PRCA), engineering change notice (ECN) | ✓ | | | | | |

Each check mark indicates the phase when you initiate the particular document. Most documents are updated in succeeding phases up to Commissioning.

**Table 5.2: Primary technical documents prepared during development of a project**

| Project Sector | Description of Documents | Concept Design | Preliminary Design | Critical Design | Test and Integration | Release to Manufacturing | Commissioning |
|---|---|---|---|---|---|---|---|
| **Project-specific documents** | Architecture design | √ | | | | | |
| | Requirements | √ | | | | | |
| | Analyses and trade-offs | √ | | | | | |
| | Risk and hazard analysis (RHA) | √ | | | | | |
| | Design descriptions | | | | | | |
| | – Electronic description | | √ | | | | |
| | – Software description | | √ | | | | |
| | – Mechanical description | | √ | | | | |
| | Test and integration plan | | √ | | | | |
| | Test results | | √ | | | | |
| | Manufacturing transfer | | | | √ | | |
| | Manufacturing tests | | | | √ | | |
| | Commissioning plan | | √ | | | | |
| | Schematics | √ | | | | | |
| | Source listings | √ | | | | | |
| | Bill of materials (BOM) | | | √ | | | |
| | Manufacturing travelers | | | | | √ | |

Each check mark indicates the phase when you initiate the particular document. Most documents are updated in succeeding phases up to Commissioning.

Table 5.3: Primary documents prepared to communicate a project and its business concerns

| Project Sector | Description of Documents | Concept Design | Preliminary Design | Critical Design | Test and Integration | Release to Manufacturing | Commissioning |
|---|---|---|---|---|---|---|---|
| **Customers, stakeholders** | User manual | | | ✓ | | | |
| | Installation instructions | | | ✓ | | | |
| | Training, webinars, seminars, DVDs | | | | | | ✓ |
| | Presentations | ✓ | | | | | |
| **Business and legal** | Concept of operations (CONOPS) | ✓ | | | | | |
| | Business case | ✓ | | | | | |
| | Marketing plan | ✓ | | | | | |
| | Contracts, MOUs | ✓ | | | | | |
| | Statement of work (SOW) | ✓ | | | | | |
| | Work breakdown structure (WBS) | ✓ | | | | | |
| | Purchase order | | ✓ | | | | |
| | Quotes | ✓ | | | | | |
| | Licenses | ✓ | | | | | |
| | Patents, trademarks, copyrights | ✓ | | | | | |

Each check mark indicates the phase when you initiate the particular document. Most documents are updated in succeeding phases up to Commissioning.

- Interface control documents (ICDs)—these documents describe the content, conditions, and specifications of the interfaces between subsystems. The specifications include things like power levels, signal levels, connector pin designations, current limits, signal frequencies and waveforms, data bandwidth, signal timing and synchronization, and mechanical alignments.
  - Electrical ICD
  - Mechanical ICD
  - Data ICD
  - Optics ICD
  - Sensor ICDs
  - Support equipment ICDs
- Approved materials, parts, and processes
- Materials, parts, and processes plan
- Prohibited materials verification plan
- Contamination plan
- Metrology and calibration plan
- Vendor qualification plan
- Software style guide
- Highly accelerated stress testing (HAST).

## Document formats

Tailor each document to the audience. Provide an appropriate level of detail for reading comprehension with an intuitive format or instructional flow in the layout of the text and the graphics. A technical document writer, or even a graphic artist, can help you with layout. Every document should have the following attributes: [1]

1. Correct
2. Complete
3. Consistent
4. Clear
5. Concise.

Most documents should have a stereotypical format. Often company policy defines the format. A general outline might have the following pages:

- Title and identification for the project and company—this can include the company logo, the project name, the document title and identification number, the company address, and the author's name.

- Copyright, ownership, and disclaimers—here is sample information that you might include:
  - **Proprietary Information**: This document contains information considered proprietary by COMPANY NAME.
  - **Copyrights and Intellectual Property**: **Copyright © 2014 COMPANY NAME Corporation. All rights reserved.** This document is protected by US and international copyright laws, as well as by other intellectual property laws and treaties. No part of this document may be reproduced in any manner whatsoever without the express written permission of COMPANY NAME. Permission is granted to make copies of this document as necessary to facilitate execution of the project for which this document was prepared.
  - **Disclaimers**: This document could contain typographical errors or technical inaccuracies. The reader is requested to bring these to the attention of the editor: Your name, COMPANY NAME Corporation, [e-mail address], COMPANY_website_here · com.
- Signatures—here are example signatures that you might include, each with a date and signature line:
  - author
  - editor
  - reviewer, product assurance
  - reviewer, customer representative
  - authorization to release
- Change log—with dates, signatures, and a very brief description of the changes
- Table of contents
- References
  - COMPANY processes and documents
  - Industrial and commercial standards
  - Books and white papers
- Overview and background
- Purpose and scope
- Ground rules and acronyms
- Body of material—contains the important value in all documents, the following section gives examples.
- Appendices
- General notes—headers, footers, page numbers, document identifier. These are important to identify the document should a page become separated from the main body or as a quick reference.

## Document contents

This section of the chapter has a several document outlines that can serve as structures for document templates. These outlines include a Project Plan and a Design Plan, Requirements, Design Descriptions, a Test Plan, and a User's Manual. Chapter 3 indicates what a Communications Plan might contain. Chapter 4 goes into considerable depth for Risk and Hazard Analysis and Risk Management. Chapter 13 has examples of action items, minutes, and meeting agendas.

### Project plan outline

1. *Management and Staffing*—list the staff, their titles, and their primary area of responsibilities.
2. *Schedule and Budgeting*
    a. Project schedule
    b. Project resources
    c. Project budget.
3. *Work Products*
    a. Design inputs—are the **sources of information** that inform the requirements, such as client surveys, government regulations, commercial standards, company policies, technical papers, and expert opinion. Design inputs are NOT the actual design details.
    b. Design outputs—are the **instruments that contain the design information**, such as schematics, source listings, design descriptions, bill of materials, manufacturing transfer document, and white papers. Design outputs are NOT the actual design details.
    c. Documents required—these are the documents that you need for both design inputs and design outputs.
4. *Standards and Practices*—elaborate the commercial or government standards that may regulate your product. Many appliances require an Underwriter's Laboratory (UL) label, which means that your product must undergo compliance testing. A medical device must have Food and Drug Administration (FDA) approval before it can sell in commercial markets.
5. *Reviews*
    a. Peer reviews requirements—defined by company policy, but regular review is very important to successful development.
    b. Formal reviews—many different names for these, but generally a formal review completes the end of a development phase. Figure 5.3 illustrates just one set of phases and design reviews at the end of each.
    c. Action item management—defined by company policy, it does not need to be complex.

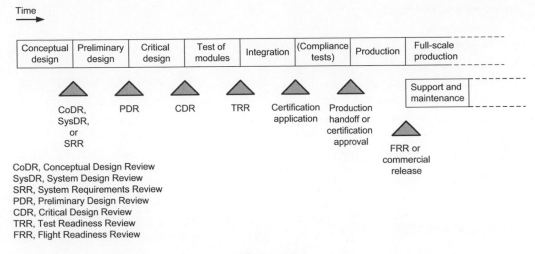

**Figure 5.3:**
An example timeline of when some design reviews might occur. © 2013–2014 by Kim R. Fowler. *Used with permission. All rights reserved.*

6. *Analyses and Tests*—you may want to give some rationale here but in most situations you only need to cite the RHA (or Analysis Plan) and the Test Plan.
7. *Tools, Techniques, and Methodologies*—this is primarily reserved for the unusual or extraordinary technical approaches that the project may need.
8. *Risk Management*—give some rationale or cite the Risk Management Plan
9. *Configuration Management*—(cite the Configuration Management plan)
10. *Documentation Plan*—(cite the Documentation Plan)
11. *Operations Plan*—give the concept of operations or cite the Operations Plan
    a. Installation
    b. Training
    c. Logistics
    d. Maintenance and repair
    e. Disposal
12. *System Architecture Development Process*—(cite the Architecture Plan)
13. *Software Development Process*—(cite the Software System Plan)
14. *Electronics Development Process*—(cite the Electronic System Plan)
15. *Mechanical Packaging Development Process*—(cite the Mechanical System Plan)
16. *Glossary*
17. *Technical Appendices*

### Design plan outline

This is an example design plan. It is very similar to the project plan; the primary difference is that a Design Plan goes into specific details for its technical area.

1. Management and Staffing
2. Work Products
    a. Design Inputs
    b. Design Outputs
    c. Documents Required
3. Standards and Practices
4. Reviews
5. Analyses and Tests
6. Tools, Techniques, and Methodologies
7. Risk Management

## Requirements' examples

The notes that follow are only some examples of requirements and some parameters that you might consider. Requirements group under four primary headings: operations, functions, performance, and manufacturing.

1. *Operational Requirements*—have to do with the scenarios within which the product will operate. Operational requirements usually cover the following concerns (*here are suggested areas, but they do not include of all possibilities*):
    • Mission profiles
    • How an operator uses the system—the human interface
    • Environment
        − Mechanical shock
        − Vibration
        − Corrosion (possibly salt spray or toxic environments)
        − Humidity
        − Temperature range
        − Vacuum
        − Dust
        − Radiation (only for space environments or nuclear power)
    • Infrastructure needed
    • Logistics and maintenance
    • Responsible party for generating the requirements
2. *Functional Requirements*—have to do with the physical plant and situations within which the product will reside. Functional requirements usually cover the following concerns (*here are suggested areas*):
    • Interfaces (perhaps from ICDs)—human, mechanical, electrical, software, special (e.g., optics)
    • Mechanical—size, shape, weight, volume, density

- Electrical—power sources, distribution
- Responsible party for generating the requirements

3. *Performance Requirements*—have to do with the metrics and parameters that describe the product's capability. Performance requirements can cover, but *are not limited to*, the following concerns:

- Responsible party for generating the requirements
- Sensor parameters (*here are just some possible considerations, these may or may not apply to your project*):
  - Measurand
  - Speed of transduction—samples per second
  - Span
  - Full scale output
  - Linearity—%, SNR
  - Threshold
  - Resolution—ENOB
  - Accuracy—SNR
  - Precision
  - Sensitivity—%
  - Hysteresis
  - Specificity
  - Noise—SNR, % budget
  - Stability
- Data throughput
  - Bytes or samples per second
  - Data transmission protocol
  - Data storage
  - Control
- Operation
  - Electrical—power consumption, efficiency, signal integrity
  - Mechanical—strength, motion required
  - Structural—capability to withstand environments in mission profiles
  - Optical (*probably not needed in many projects*)
- Calibration
- Dependability
  - Reliability
  - Maintainability
  - Testability
  - Fault tolerance
  - Longevity
- Power consumption

- Dissipation and cooling
- Electromagnetic compatibility (EMC)
    - Conducted susceptibility
    - Radiated susceptibility
    - Conducted interference
    - Radiated interference

4. *Manufacturing Requirements*—have to do with the metrics and parameters that describe the manufacturing of the product. Here are some examples—these are ONLY examples and are very incomplete at that:

- Logistics of materials
    - raw stock shall be aluminum 6160 alloy
    - received components shall be stored in antistatic bags
- Fabrication
    - enclosures must be machined from aluminum 6160 alloy
    - circuit boards must be FR4 for fabrication within six business days or less
- Assembly
    - subassemblies must each assemble within 45 seconds by one person
    - subassembly enclosures must be fastened together with screws of type X, size Y, and thread pitch Z
    - subassemblies must mate to the final system frame within 150 seconds by a team of three people
    - the wiring harness must have connectors with connector savers that remain on until final integration test
- Test
    - assembly line functional test shall use the QXB-10 test assembly
    - functional test shall complete within seven seconds
- Inventory
    - assembled circuit boards shall be stored in antistatic bags
    - The system shall be wrapped with an antistatic bubble wrap
- Distribution
    - The system shall ship in a wooden crate.
    - Posts with mechanical shock mounting shall mechanically isolate the system from the shipping container.

### Outlines of design descriptions with examples

A design description is the document that explains what a subsystem does. It also describes how and why it works. Examples of three different types of design descriptions follow. Each has some suggested components for describing the particular technical effort.

*Operations, data flow, and software design description*

1. Functional and architectural overview
   - What it does from an operational viewpoint
   - What the software does—basic description
   - Block diagram—partitioning of modules
   - Why—rationale
2. Standards compliance
3. Data flow ICD
   - Input signals—types, timing, synchronization
   - Output signals—types, timing, synchronization
4. User interface and operations
   - Display operations
   - Buttons and switches operations
   - Use cases—cite use cases and the User Manual
5. Software processing
   - General flowchart or state diagram of operations
   - Specialized processing (if needed, e.g., graphics processor or FPGA)
   - Mechatronic operations
   - Memory operations
   - Handling I/O and peripherals (e.g., ADC, DAC, timing module, interrupts, PWM)
   - Power cycling
6. Manufacturing concerns (should be short for software)
   - Installation
   - Test points manufacturing quality and for diagnostics—might cite the test plan

*Electrical and electronic design description*

1. Functional and architectural overview
   - What it does from an electrical and signals viewpoint
   - How the product does it—basic description and cite the Operations, Data Flow, and Software Design Description document
   - Block diagram—partitioning of modules
   - Why—rationale
2. Standards compliance
3. Electrical ICD
   - Input signals—types, timing, synchronization
   - Output signals—types, timing, synchronization
   - Signal integrity in communicating with outside subsystems
4. User interface
   - Display
   - Buttons and switches

5. Processing
   - Controller (e.g., microcontroller, microprocessor, multiple processors)
   - Specialized processing (if needed, e.g., graphics processor or FPGA)
   - Mechatronics support
   - Memory
   - I/O
   - Peripherals (e.g., ADC, DAC, timing module, interrupts, PWM)
6. Power
   - Type—for example, batteries or AC line input or solar panel
   - Conditioning
   - Conversion
   - Distribution
   - Range of loads and available margin
   - Safety concerns—leakage, energy density—potential problems (e.g., explosions, fire)
7. I/O
   - Input signals
   - Output signals
   - Signal Integrity internal to system
8. Manufacturing concerns
   - User interface—fabrication and test
   - Printed circuit boards—fabrication, assembly, attachment
   - Wiring and cabling
   - Test points manufacturing quality and for diagnostics—might cite the test plan

*Mechanical and materials design description*

1. Functional and architectural overview
   - What it does from a mechanical viewpoint
   - How it does it—basic description and cite the Operations, Data Flow, and Software Design Description document
   - Block diagram—partitioning of modules
   - Why it does it—rationale
2. Standards compliance
3. Mechanical or structural ICD
   - Supporting structures and mechanical connections
   - Materials compatibility—for example, electrochemical corrosion between dissimilar metals
   - Cables
   - Connectors

4. Mechanisms
   - Propulsion
   - Manipulators
   - Miscellaneous actuators
   - Mechatronics—electric motor drives, gearbox, linkages, chains, belts
   - Kinematics
5. Weight, volume, shape
   - Electronics
   - Power supplies or batteries
   - User interface
   - Mechanisms
6. User interface
   - Display
   - Buttons and switches
   - Materials—wear and resistance to wear to UI and to enclosure
7. Manufacturing concerns
   - Fabrication and test
   - Wiring and cabling
   - Test points manufacturing quality and for diagnostics—might cite the test plan

### Test plan example outline

An example outline of a test plan follows. It has some suggested types of tests for different areas within the project. Appendix A has examples from a test plan for a spacecraft instrument. Remember this is a plan; it specifies *who* does *what* and *when* they do it; test procedures give the specific *what* and *how*.

1. Types of tests
2. Code inspections
3. Software unit tests
4. Electronic unit tests
5. Mechanical unit tests
6. Field or prototype tests
7. Integration tests
8. Acceptance test

<u>**NOTE**</u>: Manufacturing tests are not the same as these tests in this test plan. Manufacturing tests are for quality control and assurance. The tests in this test plan verify and validate the product design and development.

### Examples of test procedures

An example of a set of test procedures follows in Appendix B, which has examples of test procedures for a spacecraft instrument. It has some suggested types of tests for different areas within a project. Remember these are test procedures that give the specific *what* and *how*. (A test plan specifies *who* does *what* and *when* they do it.)

### User's manual example outline

Most projects require a User's Manual. Probably every appliance that you have ever purchased had a User's Manual, even if it was very short and just a folded sheet of thin paper. Regardless, a good User's Manual is an important part of project development and delivery.

Tailor the User's Manual to the audience. Provide enough detail, at the appropriate level for comprehension for the reader. Try to get the services of a technical document writer and graphic artist to help with preparing the document. Here are basic sections in a User's Manual:

1. Title page
2. Company identifications, labeling, and disclaimers
3. Table of contents
4. Introduction
5. How to use (the product)
6. Installation—only for a simple installation; for complex equipment or processes, reference both the Installation Plan and the procedures to explain installation
7. Maintenance—for example, refills, lubrication, and battery changes
8. Repair (or replacement)
9. (if needed. . . ) Training
10. Index

## Summary and parting thoughts

Documentation is difficult, tedious, and boring. It's also incredibly important to the project. A project is not finished until the documentation is finished. It communicates the design for replication and archives it for later modification. It instructs other team members about integration. It informs users and clients about the product and trains them to use it.

From my experience, documentation takes between 50% and 80% of your time on a project. That time commitment is true for every member of the project team.

## Recommended reading

K. Fowler (Ed.), Mission-Critical and Safety-Critical Systems Handbook, Design and Development for
Embedded Applications, Newnes an imprint of Elsevier, Boston, MA, 2010. Chapter 1, pp. 45–82. This
section of the textbook contains a full example of documents written during the course of development of a
spacecraft subsystem.

## References

[1] K. Fowler (Ed.), Mission-Critical and Safety-Critical Systems Handbook, Design and Development for
Embedded Applications, Newnes an imprint of Elsevier, Boston, MA, 2010, pp. 45–48, pp. 50–82.

# Appendix A: Examples from a test plan

## Development tests to verify design and development

### Electrical and electronic test procedures

Set up the tests for electrical components (e.g., cables, connectors, switches, resistors,
LEDs) and electronics (e.g., circuit boards, modules). Use the general Test Plan T01 to
prepare the tests.

Run the verification tests on the electrical and electronic components, modules, and
subsystems. Use the general Test Plan T02 and 03 to run the tests.

These electrical and electronic tests may encompass many different types of tests. These
tests may include, but are not limited to, the following:

- Continuity checks of wires and circuit board traces
- HiPot (high potential) tests of wires, cables, and connectors
- Power on and measuring voltage and current levels
- Monitoring power consumption and dissipation
- Monitoring signal presence and levels
- Measuring signal integrity—ringing and damping on pulse edges, drive capability
- Measuring frequency response, filtering, and spectral harmonic components
- Testing for EMC/EMI
- Testing for radiation tolerance

These tests need to consider the following assignments, conditions, and issues.

1. Personnel: either electronic technician or electrical engineer
2. When: during the Test Phase, and often for the Preliminary and Critical Design Phases
   in the laboratory

3. Where: laboratory benches and environmental chamber, simulations at desk computer; specialized tests (e.g., EMC or radiation tolerance will be performed at contractor facilities)
4. Conditions: simulate operational environment, depends on test—for example, cycling temperatures in environmental chamber, vacuum and atomic bombardment in cyclotron for radiation tolerance
5. Types of stimuli: power input (e.g., ranges of voltages), signal levels and frequencies, signal sequences, environmental changes, radiation bombardment
6. Types of results: ranges of output voltage and current levels, signal drive capabilities, proper sequences of signals, proper responses to overvoltage or overcurrent

### Software test procedures

Set up the tests for software modules. Use the general Test Plan T01 to prepare the tests.

Run the verification tests on the software modules. Use the general Test Plan T02 and 03 to run the tests.

These software tests may encompass many different types of tests. These tests may include, but are not limited to, the following:

- Measuring response to data value underflow and overflow
- Determine that code recognizes input data correctly
- Monitoring memory loading
- Measuring response times to specified data inputs
- Monitoring processing for correctness and timeliness
- Monitoring interrupts for correctness and timeliness
- Monitoring output values for correctness and timeliness
- Measuring data queuing times and depth of queues to ensure against overflows
- Monitoring data sets for correctness and avoidance of corruption
- Testing software for resistance to conditions and stimuli that are out of operational bounds
- Check that interlocks function correctly
- Measuring memory margins
- Measuring timing margins

These tests need to consider the following assignments, conditions, and issues.

1. Personnel: either technician or engineer
2. When: during the Test Phase, and often for the Preliminary and Critical Design Phases in the laboratory

3. Where: laboratory benches, simulations at desk computer
4. Conditions: simulate operational environment—for example, variety of memory loadings, variety of operational loads
5. Types of stimuli: data inputs—for example, variety of data inputs and timings, input loading with multiple inputs
6. Types of results: ranges of output values and timings

### Mechanical—structural test procedures

Set up the tests for mechanical structural components (e.g., enclosures, platforms, modules). Use the general Test Plan T01 to prepare the tests.

Run the verification tests on the mechanical components, platforms, and modules. Use the general Test Plan T02 and 03 to run the tests.

These mechanical tests may encompass many different types of tests. These tests may include, but are not limited to, the following:

- Measuring mass
- Check for clearances and mechanical tolerances for size and volume verification
- Measuring ground fault impedance
- Measuring resistance to scratching, collision, and puncturing
- Measuring resistance to corrosion
- Measuring resistance to humidity
- Measuring stress and strain of materials
- Measuring responses to mechanical shock in various axes
- Measuring responses to vibration in various axes
- Testing for EMC/EMI shielding effectiveness
- Testing for radiation shielding effectiveness

These tests need to consider the following assignments, conditions, and issues.

1. Personnel: either mechanical technician or mechanical engineer
2. When: during the Test Phase, and often for the Preliminary and Critical Design Phases in the laboratory
3. Where: laboratory benches and environmental chamber, simulations at desk computer; specialized tests (e.g., EMC or radiation shielding will be performed at contractor facilities)
4. Conditions: simulate operational environment, depends on test—for example, cycling temperatures in environmental chamber, vacuum and atomic bombardment in cyclotron for radiation tolerance

5. Types of stimuli: mechanical power input—for example, ranges of mechanical shock, vibration, and metal impingement for scratch and puncture resistance, cycles of environmental changes (temperature, mechanical shock, vibration), radiation bombardment
6. Types of results: ranges of movement and creep, levels of corrosion spread, levels of conducted EMI, levels of radiation penetration

### Mechanical—mechatronic test procedures

Set up the tests for mechatronic components (e.g., motors, solenoids, squibs, mechanisms). Use the general Test Plan T01 to prepare the tests.

Run the verification tests on the mechatronic components. Use the general Test Plan T02 and 03 to run the tests.

These mechanical tests may encompass many different types of tests. These tests may include, but are not limited to, the following:

- Measuring mass
- Check for clearances and mechanical tolerances to avoid collisions
- Measuring response times and variations
- Measuring resistance to corrosion
- Measuring resistance to humidity
- Measuring stress and strain of materials
- Measuring responses to mechanical shock in various axes
- Measuring responses to vibration in various axes
- Measuring metal fatigue from movement
- Measuring metal flexibility and cycles of useful operational life

These tests need to consider the following assignments, conditions, and issues.

1. Personnel: either mechanical technician or mechanical engineer
2. When: during the Test Phase, and often for the Preliminary and Critical Design Phases in the laboratory
3. Where: laboratory benches and environmental chamber, simulations at desk computer
4. Conditions: simulate operational environment, depends on test—for example, cycling temperatures in environmental chamber, vacuum
5. Types of stimuli: mechanical power input—for example, ranges of mechanical shock, vibration, and metal impingement for resistance to movement, cycles of environmental changes (temperature, mechanical shock, vibration)
6. Types of results: ranges of movement and creep, levels of corrosion spread

## Optical test procedures

Set up the tests for optical components, modules, and subsystems. Use the general Test Plan T01 to prepare the tests.

Run the verification tests on the optical components, modules, and subsystems. Use the general Test Plan T02 and 03 to run the tests.

These optical tests may encompass different types of tests. These tests may include, but are not limited to, the following:

- Measuring focal length
- Measuring aperture
- Measuring environmental effects on imaging or light sensor
- Measuring AGC response to changes in light levels
- Monitoring autofocus capabilities
- Measuring image quality
- Counting "dead" pixels
- Testing for range of light levels

These tests need to consider the following assignments, conditions, and issues.

1. Personnel: either technician or optics engineer
2. When: during the Test Phase, and often for the Preliminary and Critical Design Phases in the laboratory
3. Where: laboratory benches and environmental chamber, simulations at desk computer; specialized tests (e.g., optical bench performed at contractor facilities)
4. Conditions: simulate operational environment, depends on test—for example, cycling temperatures in environmental chamber, vacuum
5. Types of stimuli: light levels, different test images
6. Types of results: ranges of output voltage and current levels, signal drive capabilities, proper image size, frame rate, shape, and focus, range of responses to different light levels, numbers of "dead" pixels

## Support equipment test procedures

Set up the tests for electrical components (e.g., cables, connectors, switches, resistors, LEDs) and electronics (e.g., circuit boards, modules, desktop computer). Use the general Test Plan T01 to prepare the tests. Set up the tests for software modules. Use the general Test Plan T01 to prepare the tests.

Run the verification tests on the electrical and electronic components, modules, and subsystems. Use the general Test Plan T02 and 03 to run the tests. Run the verification tests on the software modules. Use the general Test Plan T02 and 03 to run the tests.

These electrical, electronic, and software tests may encompass many different types of tests. These tests many include, but are not limited to, the following:

- Continuity checks of wires and circuit board traces
- HiPot (high potential) tests of wires, cables, and connectors
- Power on and measuring voltage and current levels
- Monitoring power consumption and dissipation
- Monitoring signal presence and levels
- Measuring signal integrity—ringing and damping on pulse edges, drive capability
- Measuring filtering and spectral harmonic components
- Testing for EMC/EMI
- Testing for radiation tolerance
- Measuring response to data value underflow and overflow
- Determine that code recognizes input data correctly
- Monitoring memory loading
- Measuring response times to specified data inputs
- Monitoring processing for correctness and timeliness
- Monitoring interrupts for correctness and timeliness
- Monitoring output values for correctness and timeliness
- Measuring data queuing times and depth of queues to ensure against overflows
- Monitoring data sets for correctness and avoidance of corruption
- Testing software for resistance to conditions and stimuli that are out of operational bounds
- Check that interlocks function correctly
- Measuring memory margins
- Measuring timing margins
- Measuring end-to-end timing responses

These tests need to consider the following assignments, conditions, and issues.

1. Personnel: either technician or engineer
2. When: during the Test Phase, and often for the Preliminary and Critical Design Phases in the laboratory
3. Where: laboratory benches and desktop, simulations at desk computer
4. Conditions: simulate operational environment, depends on test—for example, cycling temperatures in environmental chamber
5. Types of stimuli: power input (e.g., ranges of voltages), signal levels and frequencies, signal sequences, environmental changes, data values and loading from telemetry link

6.  Types of results: ranges of output images and displays, timeliness of display, quality of display

## Integration test procedures

Integration is the combination of subsystems into a functioning system. The goal of integration tests is to confirm proper function of the whole system. The most important part of integration is confirming functional behavior. Integration can also test system responses to mechanical shock, vibration, thermal, vacuum, condensation, corrosion, and EMI/EMC.

Integration often is a combination of verification and validation. Field or engineering model or prototype tests can help validate the translation of customer intent into a functional system. Some or all of the previously stated verification tests might help fulfill parts of the integration plan.

### Functional integration procedures

Functional integration tests primarily confirm that the subsystems function as a system as planned. Functional integration tests focus on the conditions in electrical, mechanical, and software domains; they tend not to include the extremes of environment or anomalous conditions. Integrating the subsystems for the first time is completed on a testbed, which comprises prototype subsystems and breadboards; sometimes engineering models (near flight quality units) may be used for integration instead of prototypes and breadboards.

Set up the integration tests for electrical and mechanical components (e.g., enclosures, cables, connectors, switches, displays) and for the electronics, mechatronics, and software (e.g., modules and subsystems). Use the general Test Plan T04 to select the necessary integration tests. Use the general Test Plan T05 to prepare the necessary integration tests.

Add each module or subsystem in turn and run the integration tests on the system. Use the general Test Plan T06 to run the integration tests.

These integration tests may encompass many different types of tests. These tests may include, but are not limited to, the following:

*   Continuity checks of wires and circuit board traces
*   Power on and measuring voltage and current levels
*   Monitoring power consumption and dissipation
*   Monitoring signal presence and levels
*   Measuring response to data value underflow and overflow
*   Determine that code recognizes input data correctly

- Monitoring memory loading
- Measuring response times to specified data inputs
- Monitoring processing for correctness and timeliness
- Monitoring interrupts for correctness and timeliness
- Monitoring output values for correctness and timeliness
- Measuring data queuing times and depth of queues to ensure against overflows
- Monitoring data sets for correctness and avoidance of corruption
- Testing software for resistance to conditions and stimuli that are out of operational bounds
- Check that interlocks function correctly—electrical, mechanical, and software
- Measuring memory margins
- Measuring timing margins
- Measuring end-to-end timing responses
- Measuring challenge tests for data throughput overloads
- Measuring mass
- Fit check for mechanical clearances and tolerances to avoid collisions
- Measuring mechatronic response times and variations
- Measuring stress and strain of materials
- Measuring optical sensor AGC response to changes in light levels
- Monitoring optical autofocus capabilities
- Measuring optical image quality
- Counting "dead" pixels in image
- Testing for optical range of light levels

These tests need to consider the following assignments, conditions, and issues.

1. Personnel: either technician or engineer
2. When: during Integration Phase
3. Where: laboratory benches
4. Conditions: simulate operational environment
5. Types of stimuli: power input (e.g., ranges of voltages), signal levels and frequencies, signal sequences, data values and loading from telemetry link, various images into sensors
6. Types of results: ranges of output images and displays, timeliness of display, quality of display, timeliness of mechatronic operation, data throughput, memory loading

### Engineering model testbed integration procedures

Sometimes a program requires an engineering model or testbed. The engineering model is essentially a flight quality system without conformal coating or some of the extreme environmental testing. It can be used in place of prototype units for the functional integration tests.

### System integration procedures

System integration is the full suite of test procedures to fully qualify the system before delivery. It includes functional integration, plus additional tests. Some of these additional integrations may include, but are not limited to, the following:

- HiPot (high potential) tests of wires, cables, and connectors
- Measuring signal integrity—ringing and damping on pulse edges, drive capability
- Measuring filtering and spectral harmonic components
- Testing for EMC/EMI
- Testing for radiation tolerance
- Challenge testing with system overload to determine robustness and degradation responses
- Environmental extremes (see the next section)

### Environmental test procedures

System integration often includes tests of the entire system to environmental extremes to qualify the system before delivery. Some of these environmental tests may include, but are not limited to, the following:

- Measuring resistance to mechanical shock
- Measuring resistance to vibration
- Measuring resistance to thermal cycling between temperature extremes
- Measuring resistance to vacuum and outgassing
- Measuring resistance to thermal-vacuum cycling
- Measuring resistance to condensation and humidity
- Measuring resistance to corrosion

## Some test plans have a manufacturing section—here is an example

Manufacturing tests may be performed as portions of a unit are fabricated, assembled, or produced. This means that manufacturing tests are not necessarily a onetime event. They may occur multiple times as components, modules, and subsystems are added to a manufactured unit. They may even be repeated, as in regression testing, to confirm that formerly tested portions are still functioning correctly.

Manufacturing tests require test jigs and fixtures and test equipment. These tests are unique to the project. Manufacturing tests may include, but are not limited to, the following:

- Highly accelerated stress screening
- Inspections for fit, spacing, and component orientation

- HiPot (high potential) tests of wires, cables, and connectors
- Measuring resistance to mechanical shock
- Measuring resistance to vibration
- Measuring resistance to thermal cycling between temperature extremes
- Measuring resistance to thermal-vacuum cycling
- Measuring resistance to condensation and humidity
- Measuring resistance to corrosion
- Continuity checks of wires and circuit board traces
- Power on and measuring voltage and current levels
- Monitoring power consumption and dissipation
- Monitoring signal presence and levels
- Boundary scan testing of functional operation

## Acceptance test procedures

Acceptance test procedures are the suite of tests that a customer requires before accepting delivery of a completed system. These acceptance test procedures are performed in Integration or Acceptance Phase. Many acceptance procedures comprise the system integration procedures and environmental test procedures from sections above.

The company and the customer define the acceptance test procedures as part of the project contract.

## Installation test procedures

Installation tests verify the functionality of a system as it is installed. Installation tests are not integration tests or acceptance tests, because they do not have the range, depth, or completeness of integration or acceptance tests. These installation tests are performed during Installation Phase; they may, however, be prepared before Installation Phase.

Use the general Test Plan T80 to select the necessary installation tests. Use the general Test Plan T81 to prepare the necessary installation tests. Use the general Test Plan T82 to run the installation tests.

> **Please note**
>
> Installation tests may be performed as portions of a unit are installed. This means that installation tests are not necessarily a onetime event. They may occur multiple times as components, modules, and subsystems are added to an installation. They may even be repeated, as in regression testing, to confirm that formerly tested portions are still functioning correctly.

Installation tests require test equipment. These tests are unique to the project. Installation tests may include, but are not limited to, the following:

- Inspections for fit, spacing, and component orientation
- Continuity checks of wires and circuit board traces
- Power on and measuring voltage and current levels
- Monitoring signal presence and levels
- Boundary scan testing of functional operation
- Functional tests for system operational behavior

# Appendix B: Examples of test procedures

## Introduction

This example document has some acronyms:

> DBX—data handling subsystem
> GSE—ground support equipment
> TBD—to be determined

## Mechanical, packaging, and cabling test scripts

The test scripts for the mechanical requirements break into four main areas:

- Size, volume, and weight.
- Connector policies—guidelines for keying connectors.
- Cabling policies—guidelines for labeling wires and cables.
- Shielding—describes the electromagnetic environment and the guidelines for shielding.

### Size, volume, and weight

2.1.1.  The DBX vehicle package, which includes circuit board enclosures and cabling, shall not exceed a mass of (TBD) kg.
   *Procedure: weigh each component and sum every recorded value to the final total mass.*

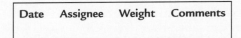

| Date | Assignee | Weight | Comments |
|------|----------|--------|----------|

2.1.2.  The DBX vehicle package, which includes circuit board enclosures and cabling, shall not exceed a volume of (TBD) m$^3$.

*Procedure: measure each component and sum every recorded value to the final total volume.*

| Date | Assignee | Volume | Comments |
|------|----------|--------|----------|
|      |          |        |          |

2.1.3. The enclosure for circuit boards in the DBX vehicle package shall have the following linear dimensions (TBD) m.

*Procedure: measure and record the linear dimension of each component.*

| Date | Assignee | Linear Dimensions | Comments |
|------|----------|-------------------|----------|
|      |          |                   |          |

2.1.4. The enclosure for circuit boards in the DBX vehicle package shall be capable of attaching to (TBD) in the vehicle.

*Procedure: measure, record, and compare the linear dimensions of each component to the point of attachment within the vehicle.*

| Date | Assignee | Attachment | Comments |
|------|----------|------------|----------|
|      |          | Yes ____ No ____ |    |

2.1.5. The ground system in the DBX, the GSE, shall be of a size, volume, and mass that is typical of desktop computers.

*Procedure: measure, record, and compare the linear dimensions of each component of the DBX and its GSE to a typical desktop computer.*

| Date | Assignee | Size | Comments |
|------|----------|------|----------|
|      |          | Yes ____ No ____ | |
| Date | Assignee | Volume | Comments |
|      |          | Yes ____ No ____ | |
| Date | Assignee | Mass | Comments |
|      |          | Yes ____ No ____ | |

## Connector policies

3.1.1. The connectors within the DBX vehicle package shall be configured or keyed in a manner to avoid accidental connection in the wrong orientation.

*Procedure: compare the configuration of each connector of the DBX to its drawing and to its mate, note variances or differences.*

| Date | Assignee | Connector | Correctly Keyed | Comments |
|------|----------|-----------|-----------------|----------|
| | | | Yes _____ No _____ | |
| | | | Yes _____ No _____ | |
| | | | Yes _____ No _____ | |
| | | | Yes _____ No _____ | |
| | | | Yes _____ No _____ | |
| | | | Yes _____ No _____ | |
| | | | Yes _____ No _____ | |
| | | | Yes _____ No _____ | |
| | | | Yes _____ No _____ | |
| | | | Yes _____ No _____ | |

3.1.2. The connectors within the DBX vehicle package shall be configured or keyed in a manner to avoid accidental connection in the wrong location.

*Procedure: compare the configuration and location of each connector of the DBX to its drawing and to its mate, note variances or differences.*

| Date | Assignee | Connector | Correct Location | Comments |
|------|----------|-----------|------------------|----------|
| | | | Yes _____ No _____ | |
| | | | Yes _____ No _____ | |
| | | | Yes _____ No _____ | |
| | | | Yes _____ No _____ | |
| | | | Yes _____ No _____ | |
| | | | Yes _____ No _____ | |
| | | | Yes _____ No _____ | |
| | | | Yes _____ No _____ | |
| | | | Yes _____ No _____ | |
| | | | Yes _____ No _____ | |

3.1.3. The connectors within the DBX vehicle package shall be strained relieved in a manner to survive the specified mechanical shock, vibration, and temperature environment.

*Procedure: compare the configuration and location of the strain relief on each connector of the DBX to its drawing, note variances or differences.*

| Date | Assignee | Connector | Strain Relieved | Comments |
|------|----------|-----------|-----------------|----------|
| | | | Yes _____ No _____ | |
| | | | Yes _____ No _____ | |
| | | | Yes _____ No _____ | |
| | | | Yes _____ No _____ | |
| | | | Yes _____ No _____ | |
| | | | Yes _____ No _____ | |
| | | | Yes _____ No _____ | |
| | | | Yes _____ No _____ | |
| | | | Yes _____ No _____ | |
| | | | Yes _____ No _____ | |

3.1.4.   The ground system in the DBX shall have connector policies typical of desktop computers.

     *Procedure: compare the configuration and location of each connector in the DBX, the GSE, to the typical configuration for a desktop computer, note variances or differences.*

| Date | Assignee | Connector Policies Typical | Comments |
|------|----------|----------------------------|----------|
|      |          | Yes _____ No _____         |          |

## Cabling policies

3.1.5.   The cabling in the DBX vehicle package shall be clearly labeled with (TBD) cable tags that attach to cables at (TBD location).

     *Procedure: record and compare the labeling of each cable of the DBX to its drawing, note variances or differences.*

| Date | Assignee | Cable | Correctly Labeled | Comments |
|------|----------|-------|-------------------|----------|
|      |          |       | Yes _____ No _____ |          |
|      |          |       | Yes _____ No _____ |          |
|      |          |       | Yes _____ No _____ |          |
|      |          |       | Yes _____ No _____ |          |
|      |          |       | Yes _____ No _____ |          |
|      |          |       | Yes _____ No _____ |          |
|      |          |       | Yes _____ No _____ |          |
|      |          |       | Yes _____ No _____ |          |
|      |          |       | Yes _____ No _____ |          |
|      |          |       | Yes _____ No _____ |          |

3.1.6.   The cabling in the DBX vehicle package shall be attached to the vehicle at (TBD location).

     *Procedure: record and compare the attachment of each cable of the DBX to its drawing, note variances or differences.*

| Date | Assignee | Cable | Correctly Attached | Comments |
|------|----------|-------|--------------------|----------|
|      |          |       | Yes _____ No _____ |          |
|      |          |       | Yes _____ No _____ |          |
|      |          |       | Yes _____ No _____ |          |
|      |          |       | Yes _____ No _____ |          |
|      |          |       | Yes _____ No _____ |          |
|      |          |       | Yes _____ No _____ |          |
|      |          |       | Yes _____ No _____ |          |
|      |          |       | Yes _____ No _____ |          |
|      |          |       | Yes _____ No _____ |          |
|      |          |       | Yes _____ No _____ |          |

3.1.7.  The cabling within the DBX vehicle package shall be strained relieved in a manner to survive the specified mechanical shock, vibration, and temperature environment.

    *Procedure: compare the configuration and location of the strain relief on each cable of the DBX to its drawing, note variances or differences.*

| Date | Assignee | Cable | Strain Relieved | Comments |
|------|----------|-------|-----------------|----------|
|      |          |       | Yes ____ No ____ |          |
|      |          |       | Yes ____ No ____ |          |
|      |          |       | Yes ____ No ____ |          |
|      |          |       | Yes ____ No ____ |          |
|      |          |       | Yes ____ No ____ |          |
|      |          |       | Yes ____ No ____ |          |
|      |          |       | Yes ____ No ____ |          |
|      |          |       | Yes ____ No ____ |          |
|      |          |       | Yes ____ No ____ |          |
|      |          |       | Yes ____ No ____ |          |

3.1.8.  The ground system in the DBX, the GSE, shall have cabling policies typical of desktop computers.

    *Procedure: compare the configuration and location of each cable in the DBX, the GSE, to the typical configuration for a desktop computer, note variances or differences.*

| Date | Assignee | Cabling Policies Typical | Comments |
|------|----------|--------------------------|----------|
|      |          | Yes ____ No ____         |          |

## Shielding

3.1.9. The DBX vehicle package shall conform to good EMC design practice.

    *Procedure: examine the circuit boards, cables, and enclosures for good EMC processes: the circuit boards (PCBs) use multilayer construction and minimize the trace lengths, signal traces and wires are properly impedance matched and terminated or are short enough not to allow pulse edge reflections, and enclosures are isolated by 1 M$\Omega$.*

| Date | Assignee | Power Supply PCB | Comments |
|------|----------|------------------|----------|
| | | Yes _____ No _____ | |
| Date | Assignee | Analog Telemetry PCB | Comments |
| | | Yes _____ No _____ | |
| Date | Assignee | Video Compression PCB | Comments |
| | | Yes _____ No _____ | |
| Date | Assignee | Digital Serial PCB | Comments |
| | | Yes _____ No _____ | |
| Date | Assignee | Radiometer Cable | Comments |
| | | Yes _____ No _____ | |
| Date | Assignee | Spectrograph Cable | Comments |
| | | Yes _____ No _____ | |
| Date | Assignee | IR Camera Cable | Comments |
| | | Yes _____ No _____ | |
| Date | Assignee | Visible Camera Cable | Comments |
| | | Yes _____ No _____ | |
| Date | Assignee | FTS Cable | Comments |
| | | Yes _____ No _____ | |
| Date | Assignee | Housekeeping Wires | Comments |
| | | Yes _____ No _____ | |
| Date | Assignee | Enclosure 1 MΩ Isolation | Comments |
| | | Yes _____ No _____ | |
| Date | Assignee | Enclosure is electrical conductive | Comments |
| | | Yes _____ No _____ | |

3.1.10. Circuit boards (PCBs or PWBs) within the DBX vehicle package shall contained ground (signal return) planes that are continuous and uninterrupted with slots.

*Procedure: examine circuit board drawings for good layout processes—incorporation of a ground/return plane without slots or large open areas.*

| Date | Assignee | Power Supply PCB | Comments |
|------|----------|------------------|----------|
| | | Yes _____ No _____ | |
| Date | Assignee | Analog Telemetry PCB | Comments |
| | | Yes _____ No _____ | |
| Date | Assignee | Video Compression PCB | Comments |
| | | Yes _____ No _____ | |
| Date | Assignee | Digital Serial PCB | Comments |
| | | Yes _____ No _____ | |

3.1.11. Circuit boards (PCBs or PWBs) within the DBX vehicle package shall use good design practices to minimize concerns for EMC.

*Procedure: examine connection drawings for good layout processes—every signal should pair with a return line or run over an uninterrupted ground/return plane, differential signals have traces that route together, signal traces are impedance matched and terminated or are short enough not to allow pulse edge reflections.*

| Date | Assignee | Power Supply PCB | Comments |
|------|----------|------------------|----------|
|      |          | Yes _____ No _____ |          |
| Date | Assignee | Analog Telemetry PCB | Comments |
|      |          | Yes _____ No _____ |          |
| Date | Assignee | Video Compression PCB | Comments |
|      |          | Yes _____ No _____ |          |
| Date | Assignee | Digital Serial PCB | Comments |
|      |          | Yes _____ No _____ |          |

3.1.12. Electrical connections between circuit boards (PCBs or PWBs) within the DBX vehicle package shall use either multilayer backplanes with continuous signal return planes or short, wire conductors that pair each signal with a return line connected to the ground (signal return) plane.

*Procedure: examine connection drawings for good layout processes— incorporation of a ground/return plane without slots or large open areas, differential signals have traces that route together, signal traces are impedance matched and terminated or are short enough not to allow pulse edge reflections.*

| Date | Assignee | Power Supply PCB | Comments |
|------|----------|------------------|----------|
|      |          | Yes _____ No _____ |          |
| Date | Assignee | Analog Telemetry PCB | Comments |
|      |          | Yes _____ No _____ |          |
| Date | Assignee | Video Compression PCB | Comments |
|      |          | Yes _____ No _____ |          |
| Date | Assignee | Digital Serial PCB | Comments |
|      |          | Yes _____ No _____ |          |
| Date | Assignee | Backplane PCB | Comments |
|      |          | Yes _____ No _____ |          |

3.1.13. Electrical connections between circuit boards (PCBs or PWBs) and connectors within the DBX vehicle package shall be either direct, solder fits on the PCB or PWB to the connectors or short, wire conductors that pair each signal with a return line connected to the ground (signal return) plane.

*Procedure: examine wires, connectors, and cables for short runs—all wire pairs are short enough not to allow pulse edge reflections.*

| Date | Assignee | Power Supply PCB | Comments |
|------|----------|------------------|----------|
|      |          | Yes _____ No _____ |        |
| Date | Assignee | Analog Telemetry PCB | Comments |
|      |          | Yes _____ No _____ |        |
| Date | Assignee | Video Compression PCB | Comments |
|      |          | Yes _____ No _____ |        |
| Date | Assignee | Digital Serial PCB | Comments |
|      |          | Yes _____ No _____ |        |
| Date | Assignee | Backplane PCB | Comments |
|      |          | Yes _____ No _____ |        |

3.1.14. The cabling in the DBX vehicle package shall pair each signal line with a return line to form a continuous circuit.

*Procedure: examine wires and cables for good shielding processes— incorporation of a ground/return wire with every signal and differential signals are paired together ( + and − ).*

| Date | Assignee | Radiometer Cable | Comments |
|------|----------|------------------|----------|
|      |          | Yes _____ No _____ |        |
| Date | Assignee | Spectrograph Cable | Comments |
|      |          | Yes _____ No _____ |        |
| Date | Assignee | IR Camera Cable | Comments |
|      |          | Yes _____ No _____ |        |
| Date | Assignee | Visible Camera Cable | Comments |
|      |          | Yes _____ No _____ |        |
| Date | Assignee | FTS Cable | Comments |
|      |          | Yes _____ No _____ |        |
| Date | Assignee | Housekeeping Wires | Comments |
|      |          | Yes _____ No _____ |        |

3.1.15. Paired signal and return lines within cabling in the DBX vehicle package shall be twisted at 6 turns per meter.

*Procedure: examine wires and cables for good shielding processes—all wire pairs twisted together at 2 twists per foot, differential signals are paired together ( + and − ).*

| Date | Assignee | Radiometer Cable | Comments |
|------|----------|------------------|----------|
| | | Yes _____ No _____ | |
| Date | Assignee | Spectrograph Cable | Comments |
| | | Yes _____ No _____ | |
| Date | Assignee | IR Camera Cable | Comments |
| | | Yes _____ No _____ | |
| Date | Assignee | Visible Camera Cable | Comments |
| | | Yes _____ No _____ | |
| Date | Assignee | FTS Cable | Comments |
| | | Yes _____ No _____ | |
| Date | Assignee | Housekeeping Wires | Comments |
| | | Yes _____ No _____ | |

3.1.16. The cabling in the DBX vehicle package shall have a continuous, electrically conductive capacitive shield over all wire conductors.

*Procedure: examine wires and cables for good shielding processes—confirm through visual inspection and electrical continuity tests that a conductive capacitive shield covers all wire conductors.*

| Date | Assignee | Radiometer Cable | Comments |
|------|----------|------------------|----------|
| | | Yes _____ No _____ | |
| Date | Assignee | Spectrograph Cable | Comments |
| | | Yes _____ No _____ | |
| Date | Assignee | IR Camera Cable | Comments |
| | | Yes _____ No _____ | |
| Date | Assignee | Visible Camera Cable | Comments |
| | | Yes _____ No _____ | |
| Date | Assignee | FTS Cable | Comments |
| | | Yes _____ No _____ | |
| Date | Assignee | Housekeeping Wires | Comments |
| | | Yes _____ No _____ | |

3.1.17. Capacitive shields surrounding the cabling in the DBX vehicle package shall follow Orbital Sciences Corporation's (OSC's) directions for electrical connection to the vehicle; if no directions from OSC, then capacitive shields shall be electrically connected to metallic enclosure that houses the circuit boards.

*Procedure: examine wires and cables for good shielding processes—confirm through visual inspection and electrical continuity tests that a conductive capacitive shield covers all wire conductors.*

| Date | Assignee | FTS Cable | Comments |
|------|----------|-----------|----------|
| | | Yes _____ No _____ | |
| Date | Assignee | Housekeeping Wires | Comments |
| | | Yes _____ No _____ | |

3.1.18. The ground system in the DBX, the GSE, shall have EMC and shielding policies typical of desktop computers.

> *Procedure: compare the configuration of the DBX, the GSE, to the typical configuration for a desktop computer, note variances or differences.*

| Date | Assignee | EMC Shielding Policies Typical | Comments |
|------|----------|-------------------------------|----------|
| | | Yes _____ No _____ | |

## Software processes test scripts

The Software Requirements will break into three main areas:

1. Development processes
2. Development metrics and rates
3. Error rates and defect records

### Development processes

4.1.1. The software development team(s) shall follow good industry practices to develop, generate, debug, test, and integrate software for the vehicular portion of the DBX.

> *Procedure: examine company records for good software development processes: code inspections and action items to correct deficiencies are on file, formal reviews and action items to correct deficiencies and concerns are on file, version control is in use.*

| Date | Assignee | Code Inspections Records | Comments |
|------|----------|-------------------------|----------|
| | | Yes _____ No _____ | |
| Date | Assignee | Formal Review Records | Comments |
| | | Yes _____ No _____ | |
| Date | Assignee | Version Control in Use | Comments |
| | | Yes _____ No _____ | |

4.1.2. The software development team(s) shall follow established software style guidelines when developing code.

*Procedure: examine company records for software development following style guidelines: code inspections and action items to correct deficiencies are on file and formal reviews and action items to correct deficiencies and concerns are on file.*

| Date | Assignee | Code Inspections Records | Comments |
|------|----------|--------------------------|----------|
|      |          | Yes ____ No ____ |          |
| Date | Assignee | Formal Review Records | Comments |
|      |          | Yes ____ No ____ |          |

### Development metrics and rates

4.2.1.  The software development team(s) shall maintain a record of lines of code (LOC) generated, debugged, tested, and integrated into the vehicular portion of the DBX.
  *Procedure: examine company records for development progress: formal reviews and action items to correct deficiencies and concerns are on file.*

| Date | Assignee | LOC–Formal Review Records | Comments |
|------|----------|---------------------------|----------|
|      |          | Yes ____ No ____ |          |

4.2.2.  The software development team(s) shall maintain a record of the time/effort expended to generate, debug, test, and integrate each software module in the vehicular portion of the DBX.
  *Procedure: examine company records for maintenance progress: formal reviews and action items to correct deficiencies and concerns are on file, and version control is in use.*

| Date | Assignee | Formal Review Records | Comments |
|------|----------|-----------------------|----------|
|      |          | Yes ____ No ____ |          |
| Date | Assignee | Version Control in Use | Comments |
|      |          | Yes ____ No ____ |          |

4.2.3.  The software development team(s) shall calculate the rate of development by dividing the LOC for each module by the time/effort to develop the module.
  *Procedure: examine company records for development progress: formal reviews and action items to correct deficiencies and concerns are on file.*

| Date | Assignee | Rate–Formal Review Records | Comments |
|------|----------|----------------------------|----------|
|      |          | Yes ____ No ____ |          |

4.2.4.  The GSE software developed for the ground portion of the DBX shall have the following metrics—(TBD).

### Error rates and defect records

4.3.1.  The software development team(s) shall maintain a record of errors/defects in the code found during development (debugging, testing, and integration) of the vehicular portion in the DBX.

> *Procedure: examine company records for a defects database: confirm that it is current.*

| Date | Assignee | Defects Database current | Comments |
|------|----------|--------------------------|----------|
|      |          | Yes _____ No _____       |          |

4.3.2.  The software development team(s) shall calculate the rate of errors by dividing the errors/defects discovered in each module by the time/effort to develop the module.

> *Procedure: examine company records for defect error rates: code inspections and action items to correct deficiencies are on file, formal reviews and action items to correct deficiencies and concerns are on file.*

| Date | Assignee | Code Inspections Records | Comments |
|------|----------|--------------------------|----------|
|      |          | Yes _____ No _____       |          |
| Date | Assignee | Formal Review Records    | Comments |
|      |          | Yes _____ No _____       |          |

4.3.3.  The GSE software developed for the ground portion of the DBX shall have the following metrics—(TBD).

## Hardware test scripts

The Hardware Requirements will break into three main areas:

1. Performance—describes types of logic families and component blocks.
2. Memory size.
3. Download and test ports.
4. Power—describes the input power, power dissipation, and supply voltages.

### Performance

5.1.1.  The electronic circuitry within the DBX vehicle package shall have sufficient capability to run the necessary software to compress and multiplex the data at the performance specified above for the system.

*Procedure: perform laboratory bench tests end-to-end, from sensors, through the DHU, to the GSE, for negligible delay. Examine company records for operation of the DBX and GSE: formal reviews and action items to correct deficiencies and concerns are on file.*

| Date | Assignee | Bench tests—delay | Comments |
|------|----------|-------------------|----------|
| Date | Assignee | Formal Review Records Yes ____ No ____ | Comments |

5.1.2.   The electronic circuitry within the DBX vehicle package may use a variety of components, such as digital signal processors and field programmable gate arrays, to accomplish performance with power consumption at or below the specification below.

*Procedure: perform laboratory bench tests of the DHU and sensors for power consumption; sum the individual values of consumption and compare to end-to-end bench tests of the DHU and all the sensors. Examine company records for operation of the DBX: formal reviews and action items to correct deficiencies and concerns are on file.*

| Date | Assignee | Power Consumption (W) | Comments |
|------|----------|-----------------------|----------|
| Date | Assignee | Formal Review Records Yes ____ No ____ | Comments |

5.1.3.   The ground portion of the DBX shall have performance typical of a desktop computer.

*Procedure: compare the performance of the DBX, the GSE, to the typical performance for a desktop computer, note variances or differences.*

| Date | Assignee | Performance Typical Yes ____ No ____ | Comments |
|------|----------|--------------------------------------|----------|

### Memory size

5.2.1.   The memory within the DBX vehicle package shall have sufficient memory capacity to run all the code without reducing the performance to below those already specified.

*Procedure: perform laboratory bench tests end-to-end, from sensors, through the DHU, to the GSE, for negligible delay due to memory constraints. Examine company records for operation of the DBX: formal reviews and action items to correct deficiencies and concerns are on file.*

| Date | Assignee | Bench Tests | Comments |
|---|---|---|---|
| Date | Assignee | Formal Review Records<br>Yes _____ No _____ | Comments |

5.2.2.   The memory within the DBX vehicle package shall have 30% or greater additional margin in memory capacity at the time of final integration at JHU/APL.

*Procedure: inspect the compiled code for size in memory, compare it to the size of the physical memory.*

| Date | Assignee | Code Size: Memory Size | Comments |
|---|---|---|---|
| | | | |

5.2.3.   The ground system in the DBX shall have memory capacity typical of desktop computers and sufficient to run the GSE.

*Procedure: compare the memory size of the DBX, the GSE, to the typical memory size and configuration for a desktop computer, note variances or differences.*

| Date | Assignee | Memory Typical<br>Yes _____ No _____ | Comments |
|---|---|---|---|
| | | | |

## Download and test ports

5.3.1.   The electronic circuitry within the vehicle package in the DBX shall have connections or a port available to download software code into the circuitry.

*Procedure: inspect the DHU for a connection port for downloading software. Examine company records for design of the download port: formal reviews and action items to correct deficiencies and concerns are on file.*

| Date | Assignee | Inspection<br>Yes _____ No _____ | Comments |
|---|---|---|---|
| Date | Assignee | Formal Review Records<br>Yes _____ No _____ | Comments |

5.3.2.   The electronic circuitry within the vehicle package in the DBX shall have connections or a port available to test the circuitry.

*Procedure: inspect the DHU for a connection port for testing the circuitry. Examine company records for design of test port: formal reviews and action items to correct deficiencies and concerns are on file.*

| Date | Assignee | Inspection | Comments |
|---|---|---|---|
| | | Yes _____ No _____ | |
| Date | Assignee | Formal Review Records | Comments |
| | | Yes _____ No _____ | |

### Power

5.4.1.   The DBX vehicle package shall draw input power from the vehicle.

*Procedure: inspect the DHU for a power connection port to the vehicle. Examine company records for design of the power connection port: formal reviews and action items to correct deficiencies and concerns are on file.*

| Date | Assignee | Inspection | Comments |
|---|---|---|---|
| | | Yes _____ No _____ | |
| Date | Assignee | Formal Review Records | Comments |
| | | Yes _____ No _____ | |

5.4.2.   Input power on the supply bus will be supplied to the DBX vehicle package at a nominal voltage of 28 VDC. The variation in voltage could be $\pm 20\%$.

*Procedure: perform laboratory bench tests end-to-end, from sensors through the DHU, for operation at 28 VDC; vary the input voltage and monitor operation.*

| Date | Assignee | Operation at Various Input Voltages | Comments |
|---|---|---|---|
| | | 28 VDC: | |
| | | 22.4 VDC: | |
| | | 33.6 VDC: | |
| | | $(-)$ varying input voltage: | |
| | | $(+)$ varying input voltage: | |

5.4.3.   The maximum input current on the supply bus supplied to the DBX vehicle package will be 3A.

*Procedure: perform laboratory bench tests end-to-end, from sensors through the DHU, for operation at 28 VDC; vary the input voltage and measure current.*

| Date | Assignee | Current (A) at Various Input Voltages | Comments |
|------|----------|---------------------------------------|----------|
| | | 28 VDC: | |
| | | 22.4 VDC: | |
| | | 33.6 VDC: | |
| | | ($-$) varying input voltage: | |
| | | ($+$) varying input voltage: | |

5.4.4.　The DBX vehicle package shall consume a maximum of 90 W.

　　　*Procedure: perform laboratory bench tests end-to-end, from sensors through the DHU, for operation at 28 VDC; vary the input voltage, measure current, and calculate power.*

| Date | Assignee | Power Consumption (W) at Various Input Voltages | Comments |
|------|----------|------------------------------------------------|----------|
| | | 28 VDC: | |
| | | 22.4 VDC: | |
| | | 33.6 VDC: | |
| | | ($-$) varying input voltage: | |
| | | ($+$) varying input voltage: | |

5.4.5.　The ground portion of the DBX shall draw standard, residential power typical of a desktop computer at 120 VAC.

　　　*Procedure: inspect the operation of the DBX, the GSE, to determine if it runs from 120 VAC, note variances or differences.*

| Date | Assignee | 120 VAC Operation | Comments |
|------|----------|-------------------|----------|
| | | Yes _____ No _____ | |

5.4.6.　The ground portion of the DBX shall consume power typical of a desktop computer.

　　　*Procedure: monitor the operation of the DBX, the GSE, and measure voltage and current to calculate power consumption; compare consumption to a typical desktop computer, note variances or differences.*

| Date | Assignee | Power Consumption Typical | Comments |
|------|----------|---------------------------|----------|
| | | Yes _____ No _____ | |

## CHAPTER 6

# System Requirements

### Robert Oshana

*Director, Software Research and Development, Digital Networking, Freescale Semiconductor*

**Chapter Outline**

## Definitions

System requirements engineering aims to deliver the right product to the right customer at the right time for the right price. System requirements engineering determines the

Developing and Managing Embedded Systems and Products.
DOI: http://dx.doi.org/10.1016/B978-0-12-405879-8.00006-4

true customer needs and specifies them completely and correctly. This approach is intended to:

- achieve high customer satisfaction,
- reduce rework and improve productivity within development,
- control scope creep and requirements changes in the project, and
- reduce maintenance and support costs of the delivered system.

The IEEE Definition of a Requirement is [1]:

- a condition or capability needed by a user to solve a problem or achieve an objective,
- a condition or capability that must be met or possessed by a system or system component to satisfy a contract, standard, specification, or other formally imposed document,
- a documented representation of a condition or capability as in definition (a) or (b).

Fred Brooks, in his article "No Silver Bullet: Essence and Accidents of Software Engineering" states, "The hardest single part of building a software system is deciding precisely what to build. No other part of the conceptual work is as difficult as establishing the detailed technical requirements, including all the interfaces to people, to machines, and to other software systems. No other part of the work so cripples the resulting system if done wrong. No other part is more difficult to rectify later" [2].

---

**Editor's note**

While this quote is for software, it holds true for systems!

---

We can further refine the definition of requirements engineering as a set of activities to identify and communicate the purpose of a proposed system, and the contexts in which it will be used. Requirements engineering acts as the bridge between the real world needs of stakeholders, including users, customers, sponsors, administrators, and other constituencies affected by a system, and the capabilities and opportunities afforded by these technologies.

## *Developing and managing requirements*

Requirements engineering consists of a requirements development phase and a requirements management phase, as shown in Figure 6.1. Requirements development can further be classified into:

- elicitation—techniques for determining the requirements,
- analysis—techniques for assessing the requirements,
- specification, and
- validation.

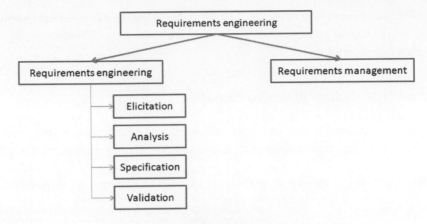

**Figure 6.1:**
Requirements Engineering approach [3].

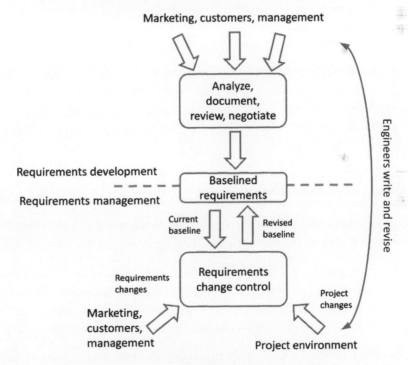

**Figure 6.2:**
Requirements development and management.

Figure 6.2 shows another view of requirements development versus requirements management. Requirements development has key stakeholders, such as marketing, customers, and users, provide input as to the system function and then engineers create the requirements. The team then makes these requirements the baseline. After this, they analyze, document, review, and negotiate further changes to the requirements through a disciplined change control process with input from the same key stakeholders, as well as other key project personnel.

## Customer interpretation of requirements

It is important to understand how a customer will interpret documented requirements. Table 6.1 summarizes the expectations from a customer on having documented requirements.

**Table 6.1: Spoken versus unspoken inputs from the customer during requirements development**

| Requirement | Spoken | Unspoken (Functional) | Unspoken (Delights) |
|---|---|---|---|
| If present | Customer is pleased | Customer expects it to be there, takes it for granted | Customer is pleasantly surprised |
| If absent | Customer is dissatisfied | Customer is dissatisfied | Customer is unaffected |

## Requirement categories

There are three primary categories of system requirements: functional, nonfunctional, and architectural. Functional requirements are "what" the system should do. Nonfunctional requirements are "how well" the system should perform in one or more areas. Architectural requirements are more descriptive of "connections" between the subsystems to form the final system.

### Functional versus nonfunctional requirements

Functional requirements specify what the system has to do. They are traceable to a specific source, often to Use Cases or Business Rules. They are often called "product features."

Nonfunctional requirements are mostly quality-related requirements which include the areas of performance, availability, reliability, usability, flexibility, configurability, integration, maintainability, portability, and testability. This category may also include implicit requirements for modification and upgrades, reusability, and interoperability.

Table 6.2 shows the importance of nonfunctional requirements to different stakeholders.

*Availability* is an indicator of the planned up time where the system is available for use and fully operational. Availability is the mean-time-to-failure (MTTF) for the system divided by the sum of the MTTF and the mean-time-to-repair the system after a failure occurs. Availability encompasses reliability, maintainability, and integrity [4].
*Efficiency* measures how well the system uses processor capacity, disk space, power, memory, or input/output (I/O) communication bandwidth [5]. It also measures the system's capability to convert mechanical and electrical energy.

**Table 6.2: Nonfunctional requirements and their importance to users and developers**

| Important Primarily to Users | Important Primarily to Developers |
|---|---|
| Availability | Maintainability |
| Efficiency | Portability |
| Flexibility | Reusability |
| Integrity | Testability |
| Interoperability | |
| Reliability | |
| Robustness | |
| Usability | |

*Flexibility* measures how easy it is to add new capabilities to the product. You might refer to it as augmentation, extensibility, extendability, or expandability.

*Integrity* relates to security. This quality requirement category describes what is required to block unauthorized access to certain system functions, how to prevent information loss, how to ensure that the system is protected from virus infection, and how to protect the privacy and safety of data entered into the system.

*Interoperability* defines the ease in which the system can exchange data or services with other systems.

*Reliability* defines the probability of the system operating without failure for a specific period of time. Ways to measure system reliability include the percentage of operations that are completed correctly and the average length of time the system runs before failing [6].

*Robustness*, or " fault tolerance," is the degree to which a system continues to function properly when presented with invalid inputs, or defects in connected software or hardware or mechanical components, or unexpected operating conditions. System robustness can be viewed as an aspect of reliability.

*Usability*, or " ease of use," refers to the many factors that constitute what users often describe as user-friendliness.

*Maintainability* measures how easy is it to correct a defect or modify the system. It depends on how easily the system can be understood, changed, and tested.

*Portability* addresses the effort required to migrate a system component from one operating environment to another. Portability may include the ability to internationalize and localize a product and has aspects similar to reusability.

*Reusability* defines the effort involved to convert a system component for use in other applications. Lack of reusability leads to systems that are more costly to develop. Reusable systems are modular, well documented, independent of a specific application and operating environment, and somewhat generic in capability.

*Testability* or verifiability is an important class of requirement. This class of requirement addresses the ease with which system components or the integrated product can be tested for defects. This is very important for complex algorithms and logic, as well as subtle functionality interrelationships.

Performance requirements define how well or how efficiently the system must perform specific functions. Examples of performance might include:

- Speed—e.g.,
  - database response time, or
  - how fast your motorcycle with a computerized engine management system goes down the road.
- Throughput—e.g.,
  - transactions per second, or
  - how much water a sump pump with an embedded motor driver pulls out of a basement.
- I/O—e.g., frames per second through a video port.
- Capacity—e.g.,
  - concurrent usage loads or channels of execution, or
  - how much load your power jack with an embedded controller can hold without dropping a car on your head.
- Timing—e.g., hard real-time demands.
- Power—e.g., battery life or milliwatts/transaction.

Figure 6.3 shows the relationship between functional and nonfunctional requirements.

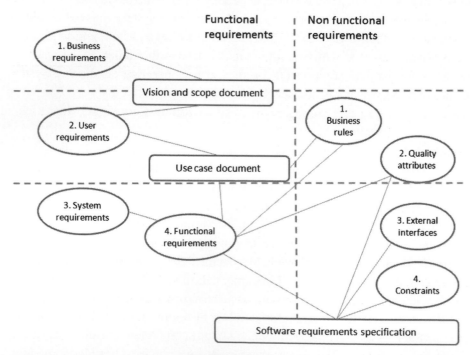

**Figure 6.3:**
Relationship between functional and nonfunctional requirements.

## System architecture requirements

Another important class of system requirements relates to system architecture. Examples of this class of requirements include:

- *Component-naming*—be consistent in naming components (mechanical, hardware, or software). This makes systems easier to understand.
- *Compatibility*—many systems need to be compatible with other systems. A hotel reservation system, for example, needs to be compatible with several financial systems.
- *Ease of interfacing*—this relates to coupling. Systems should be loosely coupled, which means that system components can operate properly without having intimate knowledge of the workings of other system components.
- *Upgradability*—systems that have a high probability of later revisions should be designed to support easy upgrades.
- *System building*—the activity of creating new systems by using tools to build new mechanical, hardware, or software definitions. For systems that will change often, this process needs to be made easy.

Systems live within the context of the real world, and the real world has constraints. Otherwise, with no constraints, there are an infinite number of ways to solve the problem. Reducing the number of available options often eases the job of designing systems by limiting the degrees of freedom when creating a new system. Table 6.3 shows several classes of system constraints and some examples of each.

**Table 6.3: Several classes of system constraints**

| Source of Constraint | Examples |
| --- | --- |
| Schedule and Resources | • System must ship by April 1 to be ready in time for Christmas<br>• Must use existing engineers with no expansion<br>• Can only use outside labor from approved vendors |
| Regulatory, legal | • Must support environmental and regulatory constraints for all states<br>• Must support appropriate security requirements for the country<br>• Must support legal requirements for the states/countries used |
| System | • New system should be based on architecture of existing system<br>• New system must be compatible with existing systems<br>• New system must support both Windows and Linux |
| Technical | • Must use technologies approved by customer<br>• Freedom to use commercial operating systems and graphics software<br>• Cannot use technology from unapproved outside vendors |
| Environmental | • Must operate over a span of temperatures, from $-40°C$ to $+75°C$<br>• Must endure a vibration profile<br>• Must operate over a span of humidity, from 10% to 98% relative humidity |
| Political | • Work must be performed from the design center of excellence within the company<br>• Must adhere to local city regulations |
| Economic | • Must pay required licensing fees for use of software packages<br>• Must meet certain margins for profit for the program |

## Requirements' attributes

System requirements also have key attributes that can measure how well each requirement is written. Table 6.4 lists these key attributes. Please note: there are many nonfunctional requirements and not all of them can be a primary focus due to system constraints.

**Table 6.4: Key attributes of requirements**

| Requirement Attribute | Definition |
|---|---|
| Correct | According to the customer or stakeholder representative. |
| Necessary | Requirement must be feasible. Tools, techniques, resources, budget must be able to satisfy the requirement. |
| Prioritized | Should be prioritized by developers and stakeholders. Performed during the interview or elicitation phase. Should be able to trace back to a use case. |
| Unambiguous | There is only one interpretation. The requirement is easy to read and understand. |
| Concise | The requirement only contains the information needed to proceed to the next development step. |
| Verifiable | The requirement must be testable and measurable. A person or machine is able to confirm that the system meets the requirement. |

As shown in Figure 6.4, nonfunctional requirements have impacts on each other. Increasing the focus on one nonfunctional requirement, such as reliability, has a positive impact on other nonfunctional requirements such as availability. On the other hand, increasing the focus on flexibility, for example, may adversely affect the nonfunctional requirement of efficiency. This needs to be considered when formulating and prioritizing nonfunctional requirements.

|  | Availability | Efficiency | Flexibility | Integrity | Interoperability | Maintainability | Portability | Reliability |
|---|---|---|---|---|---|---|---|---|
| Availability | ▓ |  |  |  |  |  |  | + |
| Efficiency |  | ▓ | − |  | − | − | − | − |
| Flexibility |  | − | ▓ | − |  | + | + | + |
| Integrity |  | − |  | ▓ | − |  |  |  |
| Interoperability |  | − | + | − | ▓ |  | + |  |
| Maintainability | + | − | + |  |  | ▓ |  | + |
| Portability |  | − | + |  | + | − | ▓ |  |
| Reliability | + | − | + |  |  | + |  | ▓ |

**Figure 6.4:**
Impact of nonfunctional requirement types on each other. " + " indicates that increasing the attribute in that row has a positive effect on the attribute in the column. " − " indicates that increasing the attribute in that row adversely affects the attribute in the column.

## Common risks in setting requirements

When specifying system requirements, there are some common risks that must be considered:

- *Insufficient User Involvement*: this leads to unacceptable products.
- *Creeping User Requirements*: this contributes to budget and schedule overruns and leads to a decline in overall product quality.
- *Ambiguous Requirements*: this results in ill-spent time and rework.
- *Inappropriate Features*: usually this is done by developers or users and leads to unnecessary features to address extraneous or low-priority desires or concerns.
- *Minimal Specification*: this leads to missing key requirements.
- *Overlooked User Classes*: which leads to dissatisfied customers.
- *Inaccurate Planning*: this is often caused by incompletely defined requirements.

## Process and QA

Chapter 1 introduces processes and standards for Quality Assurance (QA). In this chapter, I will only focus on Capability Maturity Model Integration (CMMI) to serve as an example process and standard under which requirements might develop.

The CMMI is a process improvement framework and appraisal program. The CMMI is required by many projects for the Department of Defense, as well as commercial companies, for system development (but it can be used for general system development, too). CMMI can be used to guide process improvement across a project, division, or an entire organization. CMMI processes are rated according to their maturity levels, which are initial, repeatable, defined, quantitatively managed, optimizing as described in Chapter 1. The five maturity levels of CMMI are based on a strong foundation of requirements at Levels 2 and 3 as fundamental processes to support the standard.

At process area Level 2 (Managed), Requirements Management is a key process area. Requirements Management is the capability that helps an organization achieve more predictable and less chaotic projects. One key activity within Level 2 is collecting and documenting requirements. This doesn't mean that an organization should wait until achieving Level 2 before collecting and documenting requirements. Collecting and documenting requirements is profitable at any level and will help the organization climb through the levels of CMMI.

At process area Level 3 (Defined), Requirements Development is required. Requirements Development is performed after an organization has adopted the discipline of managing requirements changes and tracking status. The focus is on developing high-quality requirements.

The key themes of the CMMI relating to requirements include:

- The development team understands requirements and resolves issues with customers.
- Manages changes.
- Maintains traceability and bidirectional traceability of requirements.
- Records the source of lower level or derived requirements.
- Traces each requirement downward into derived requirements and its allocation to functions, object, processes, and people.
- Establishes horizontal links from one requirement to another of the same type.

There are three sets of practices to perform during requirements analysis:

1. Identify a complete set of customer requirements to develop product requirements. Elicit stakeholder needs and transform those needs and constraints into customer requirements.
2. Define a complete set of product requirements. The three key steps are establish product components, allocate requirements to product components, and identify interface requirements.
3. Derive child requirements, and understand and validate those requirements. The five key steps are establish the operational concepts and scenarios, define the required system functionality, analyze the derived requirements, evaluate product cost, schedule, and risk, and validate requirements.

## Domains and properties

You can analyze a system from the perspective of an application domain, as well as from the perspective of a machine domain. We write the requirements in the application domain. We then implement the system in the machine domain. Figure 6.5 illustrates these domains in abstract form.

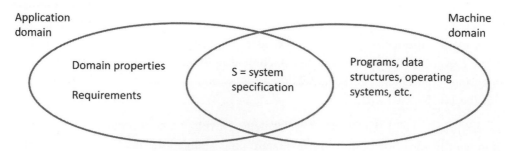

**Figure 6.5:**
Application domain and machine domain.

In Figure 6.5, the domain properties define the things in the application domain that are true whether or not we ever build the system.

The requirements are those things in the application domain that we wish to be made true by delivering the proposed system. These may also involve phenomena to which the machine has no access. For example, a trucking navigation application that must control where trucks drop off and pick up cargo does not have direct access to the environment the truck is in but must control the truck navigating in that environment.

The specification is the description of the behaviors that the system must have to meet the requirements. These can only be written in terms of shared phenomena of the system. The shared phenomena for the trucking example mentioned above would be the user interface to the computer system. This interface would include capabilities to be able to navigate the truck in the environment.

## Setting boundaries

When eliciting requirements, the first thing is to set the boundaries. For example, how will the system interact with the world? Asking that question will help set a boundary. The requirements engineer must decide what phenomena are shared in the application domain. One approach is to use the four-variable model, developed by Parnas and Madey, as shown in Figure 6.6 [7]. Here we decide the boundaries by designing the input/output devices. Then we use I/O data items as proxies for the monitored and controlled variables for the system we are specifying.

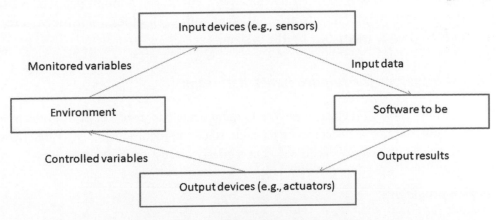

**Figure 6.6:**
The four-variable model of Parnas and Madey [7].

Van Lamsweerde describes this as an extension to the four-variable model when it comes to describing requirements as shown in Figure 6.7 [8].

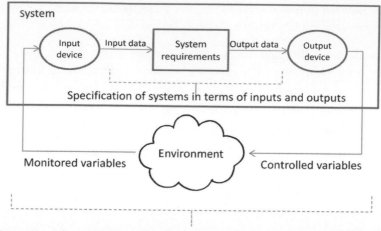

Requirements; what control actions the system must have under what circumstances
Domain; what domain properties exist that constrain how the environment can behave

**Figure 6.7:**
Extension of the four-variable model for system requirements definition [8].

Van Lamsweerde's satisfaction argument for system requirements is then presented as:

$$\{SOFREQ,\ ASM,\ DOM\}| = SysReq$$

and defined as:

***if*** **the software requirements in set SOFREQ are satisfied by the software, the assumptions in set ASM are satisfied by the environment, the domain properties in set DOM hold and all those statements are consistent with each other,** ***then*** **the system requirements SysReq are satisfied by the system.**

## Framing the system for requirements definition

Ben Kovitz, in his book, "Practical Software Requirements" describes a set of problem frames used to help frame a system to better understand the requirements [9]. He defines five problem frames as given in Table 6.5. Let's briefly discuss each of these.

### Information problems

This type of problem primarily answers queries about a certain part of the real world and the requirements describe types of information requests to be satisfied. The requirements need to address how the system can get access to that part of the real world. This means describing those relevant parts of the world, the queries necessary to get that information, as well as the people or things that initiate those queries.

**Table 6.5: Five problem frames used for system requirements elicitation and definition [9]**

| System Requirement Type | Description | Problem Frame |
|---|---|---|
| Queries | Requests for information from the application domain | Information |
| Behavioral rules | Rules according to which the problem domain is to behave | Control |
| Mappings | Mappings between data input and output | Transformation |
| Operations on realized domains | Operations what users can perform on objects that exist inside the software | Workpiece |
| Correspondence between domains | Keeping domains that have no shared phenomena in corresponding states | Connection |

In these types of systems, the requirements should be able to satisfy the queries initiated by information requestors. These types of information systems report on the state of the world but not change its state. In other words there is no causation. Examples of information systems include most inventory control systems, search engines, and Global Positioning System (GPS) mapping systems.

Information problems can be categorized as dynamic systems which query and report information that changes, for example, stock quotes, and static systems where the information does not change often, such as a city map.

Information systems can be passive systems where the queries are initiated by the user, such as an audio system providing explanations for museum displays, or active systems where the information is supplied without being asked, such as a burglar alarm system. If a query is triggered by an event, such as inventory running low or a warning light, then we need to document further the requirements of what triggers these notifications, what kind of performance (e.g., response time) is required, and how to determine that the information was received properly.

In an information system, the requirements specification must describe the model of the real world for which you are making inquiries. Figure 6.8 shows a simple frame diagram of an information problem. This diagram shows:

- The computer system, M
- The requirement category, R (queries in this case)
- The target domain (in this case the real world we are attempting to model)
- The entity querying for information (the information requestors).

Most information systems require some method to relay information from the real world to the software, for example, people performing data entry. This leads to inefficiencies that you must consider as well. For example you may need additional requirements that check for

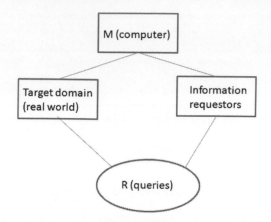

**Figure 6.8:**
Information system model as an example of a model used in developing requirements.

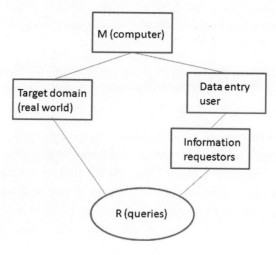

**Figure 6.9:**
Information problem with a connection domain in the model.

errors in data entry from users. This is shown in Figure 6.9. The data entry user "connects" the computer system to the information requestors. Whenever you have a connection domain like this (very common in computer systems), this introduces two inefficiencies:

1.  Distortion, the data may be entered incorrectly, and
2.  Delay, measurable time required to enter the data.

You may need to formulate additional requirements to address this distortion and delay introduced by this connection domain.

## Control problems

Unlike information systems, control systems focus on causation. This means that control systems interact with the world around them rather than just extracting information from the environment.

Control systems require that you document what is responsible for ensuring that some part of the real world behaves in accordance with a set of rules. Your requirements must describe those things that inhabit that part of the world that the rules obey. Requirements for control systems describe how the system will monitor that part of the world and initiate causal chains that cause the specified rules to behave correctly.

Because control systems interact with the world, the focus of requirements for control systems should describe the following:

- The causal properties of those relevant parts of the world applicable to what you are trying to control.
- The rules that the system components follow to achieve this control.
- Any important and relevant phenomena shared between the computer system and the problem domain (the real world).

Examples of control problems are heating ventilating and air conditioning (HVAC) systems and traffic lights. Figure 6.10 shows a simple frame diagram of a control system.

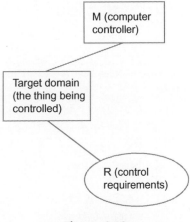

**Figure 6.10:**
A control system.

### Examples: HVAC and traffic lights

In an HVAC system, we set the rules for temperature control on the user panel. The system "controls" the real world in the sense that it heats or cools the real world according to the rules.

Traffic lights are also examples of control problems handled by a system. The "rules," how the lights transition from red to yellow to green, define the control actions. The system controls or affects the real world by stopping and starting cars at a traffic intersection.

Control systems can also have connection domains. For example, control systems can direct people to perform certain types of activities like taking a receipt to a manager or calling in a prescription. This leads to the same forms of distortion and delay that a control system may solve with additional requirements and constraints.

Requirements for control systems need to specify the behavioral rules for the shared phenomena. In addition, timing rules must be well documented.

### Transformation systems

Transformation systems convert some form of input data to output data based on a set of rules or algorithms. The requirements for transformation systems must describe the entire set of all possible inputs and the mapping rules that indicate, for each possible input, the correct output. Figure 6.11 is a frame diagram of a transformation problem.

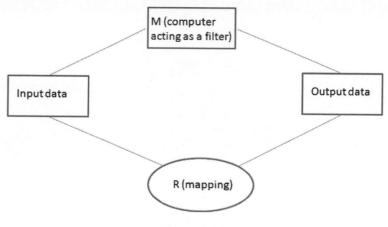

**Figure 6.11:**
Transformation system.

The requirements for transformation systems describe all possible inputs, the possible outputs, and the rules relating each possible input to its corresponding output. The rules are referred to as the "mapping." Examples of transformation systems include bar scanners at supermarkets that read bar codes and translate into product and price (e.g., bar codes to numbers), and image processing software that takes a video frame and enhances the image or performs edge detection on the image. The algorithms that perform these functions are the primary requirements that must be documented.

## Workpiece

Workpiece systems are systems where the application serves as a tool for creating objects such as files of databases that exist within the software in the system itself, like memory or a hard disk.

Requirements for workpiece systems describe the objects that must exist within the computer, as well as the operations that users can perform on these objects.

The system helps users create the objects of interest. These can be documents, designs, databases, and other forms of electronic media that exist inside the computer. There may also be requirements that allow these objects to be printed, so in a sense, the objects can "exist" outside the computer, as well.

These workpieces are usually created using a user interface, so requirements for workpiece problems must also describe the user interface operations required to produce the workpieces.

## Connection

Connection domains "connect" application domains that, for some reason, cannot be connected directly. The goal is to make them appear to be connected. An ideal connection is rarely possible so a connection domain would set an upper limit as to how well the requirements can be fulfilled. Each connection domain adds a level of distortion and delay between the two domains that you are trying to connect. In Figure 6.9, for example, the data entry clerk is a form of connection domain. The distortion in the data entry clerk is entering the wrong data into the system. The delay component represents the finite amount of time to enter the data into the system.

Other examples of connection domains include a video conferencing system that connects two domains of interest in different cities. There is obvious distortion (video signal not ideal) and delay (in the transmission of video and audio between the sites). The requirements engineer must be able to identify one or more connection domains in the system and, for each one, determine the distortion and delay, their upper limits, and then determine if additional requirements must be derived to determine if the connection domain is not operating properly (e.g., the data entry clerk is entering the wrong data), and be able to correct (e.g., provide range checking).

## Use cases

Use cases provide an important technique for eliciting requirements. A use case describes, at a high level, "what" the system will do, not "how" the system will achieve a particular behavior. More to the point, a use case is not a design description of the system. A use case has a user focus for the purpose of scoping the project and giving the application some structure. Use cases can be detailed in "scenarios" that identify a particular thread of system usage in the form of a written narrative. Doing so shifts the perspective to what users need to accomplish (in contrast to what users want the system to do).

A use case is a discrete, stand-alone activity that an actor can perform to achieve some outcome of value. Use cases capture functional requirements of a system from the perspectives of the different users of the system. Use cases are textual narratives describing the different kinds of scenarios in which the system will be used. A use case diagram helps to visualize which actors are involved in which scenarios. Figure 6.12 is an example of a use case diagram for an Automated Teller Machine (ATM).

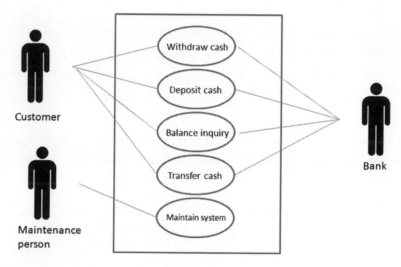

**Figure 6.12:**
Example of a use case diagram for an ATM.

Collaborations among objects in the design realize the functional requirements described in use cases. Use cases do not depict nonfunctional (or qualitative) requirements. A use case is a scenario that describes how the system works in a given situation and is defined from an "actor's" perspective.

An actor is a role played by a person or device as it interacts with the system. An actor is anything that communicates with the system and that is external to the system itself. An actor is outside the system model but interacts with the system. An actor can be a person, machine, or an information system.

By creating use cases, you keep the focus of development on the user and the user's perception of the system. A nontrivial system will often have many actors, each involved in multiple scenarios.

The first step in identifying use cases is to identify the different people or devices that will interact with the product. Consider what each actor hopes to accomplish by using the product. This can lead to the discovery of new requirements. There can be overlap, in that different actors may expect to use the same system facilities. Be consistent and look for opportunities to represent two similar scenarios with a more general one.

The key questions answered by a use case are:

- What main tasks or functions are performed by the actor?
- What system information will the actor acquire, produce, or change?
- Will the actor have to inform the system about changes in the external environment?
- What information does the actor desire from the system?
- Does the actor wish to be informed about unexpected changes?

A use case scenario is essentially a narrative text describing the sequence of events of an actor, using a system. It can be considered a short story. Think of this as a static model of functionality that describes what happens when an actor interacts with a system. A scenario represents a function that makes sense to an actor. An actor initiates interactions with the system; and *vice versa* but there is no temporal sequencing with a scenario.

The steps to develop use cases are:

1. Identify who will use the system directly.
2. Pick one of those actors.
3. Define what the actor wants to do with the system, this becomes a use case.
4. For each use case, define what happens.
5. Describe the "basic course."
6. Consider alternatives.
7. Review for commonality.
8. Repeat steps 2—7 for each actor.

A scenario is a use case instance. It contains the same steps as the use case. A scenario typically describes just one execution path (one outcome) through the use case. It describes a sequence of actions and interactions of objects. The objects in the scenario are from the classes in the use case (e.g., a particular set of objects acting to cancel a catalog order).

A template for a use case scenario is shown below for an ATM. This template follows the outline given in the DOT/FAA *Requirements Engineering Management Handbook* [10].

---

**Example use case for an ATM interaction**

**Use Case**:

 < Enter a short name for the Use Case using an active verb phrase.

e.g. Withdraw Cash, Deposit Cash, Balance Inquiry, etc. >

**Id**:

< Enter a unique numeric identifier for the Use Case. e.g. UC-001 >

**Description**

< Briefly describe this use case.

e.g. Customer inserts teller card, types in PIN, and is presented with various financial selections by the ATM system. The customer selects "withdraw cash", enters the amount to withdraw, and the ATM system provides the money to the customer and updates customer account information accordingly. >

**Level**:

< Enter the goal level of this Use Case. Specify whether the Use Case level is - High Level Summary, Summary, User Goal, etc >

**Primary Actor**

< List the Actor whose goal is being satisfied by this Use Case and has the primary interest in the outcome of this Use Case.

e.g. ATM customer >

**Supporting Actors**

< List the Actors who have a supporting role in helping the Primary Actor achieve his or her goal.

e.g. Bank >

**Stakeholders and Interests**

< List the various entities who may not directly interact with the system but who may have an interest in the outcome of the use case. Identifying stakeholders and interests often helps in discovering hidden requirements which are not readily apparent or mentioned directly by the users during discussions. >

**Pre-Conditions**

< List the system state/conditions which must be true before this Use Case can be executed.

e.g. ATM machine has adequate amount of money in the system, customer has adequate funds in her checking account. >

**Post Conditions**

Success end condition

< Enter the successful end condition of the Use Case where the Primary Actor's goal is satisfied.

e.g. customer has her money, customer account is updated, ATM machine is now back at the welcome screen. >

Failure end condition:

< Enter the failure end condition of the Use Case if the Primary Actor's goal has not been achieved.

e.g. Customer is unable to withdraw cash, account information is not changed. >

Minimal Guarantee

< The guarantee or assurance that this Use Case provides to all Actors and Stakeholders to protect their interest regardless of whether the Use Case ends with success or failure.

e.g. For Withdraw Cash (ATM Use Case), minimal guarantee could be, Customer is logged out of the ATM system. This minimum guarantee ensures that the system will ensure that no unauthorized withdrawals can be made from the ATM thus protecting the interest of the Bank Customer as well as the Bank's stakeholders. >

**Trigger**

< The event that starts this Use Case.

Example; For *Withdraw Cash* Use Case - Customer inserts the bank card into the ATM machine. >

**Main success scenario**
1. Customer inserts ATM card into the ATM machine
2. ATM system validates ATM card and prompts customer for PIN
3. Customer enters PIN
4. ATM system validates PIN and customer credentials and prompts for financial options
5. Customer selects "withdraw cash"
6. ATM prompts for checking or saving account
7. Customer selects "checking"
8. ATM prompts for amount
9. Customer enters amount
10. ATM validates account information and outputs bills
11. Customer takes bills
12. ATM prompts for additional transactions
13. Customer selects "none"
14. ATM prompts for receipt
15. Customer selects "yes"
16. ATM outputs receipt

< Enter the Main flow of events. i.e. The steps narrating/illustrating the interaction between Actors and the System. Describe Actor's actions/stimuli and how the system responds to those stimuli. Describe the 'happy path/day' scenario, meaning the straight and simple path where everything goes 'right' and enables the primary actor to accomplish his or her goal. Main flow/path should always end with a success end condition. >

**Extensions**
< Enter any extensions here. Extensions are branches from the main flow to handle special conditions. They also known as Alternate flows or Exception flows. For each extension

reference the branching step number of the Main flow and the condition which must be true in order for this extension to be executed.

Example of an Extension in ATM Use Case:

3a. In step 3, if the customer enters the wrong PIN

1. System will prompt for reentry of PIN and errors message

2. Customer enters PIN again

4. Use Case resumes on step 4.

> >

**Variations**
< Enter any data entry or technology variations such as — different methods of data input, screen/module invocation, etc.

e.g. 3'. In step 3, instead of entering PIN by typing it in, the customer may enter it directly using voice recognition

**Frequency**:
< How often will this Use Case be executed. This information is primarily useful for designers.

e.g. enter values such as 20 per hour, 500 per day, once a week, once a year, etc. >

**Assumptions**
< Enter any assumptions, if any, that have been made while writing this Use Case.

e.g. For *Withdraw Cash* Use Case(ATM system) an assumption could be: The Bank Customer understands either English or Spanish language. >

**Special requirements**
< Enter any special requirements such as Performance requirements, Security requirements, User interface requirements, etc. Examples:

Performance

1. The ATM shall dispense cash within 15 seconds of user request.

User Interface

1. The ATM shall display all options and messages in English and Spanish languages.
2. The height of letters displayed on the display console shall not be smaller than 0.5 inches.

Security

1. The system shall display the letters of PIN numbers in a masked format when they are entered by the customer. i.e. Mask the PIN with characters such as ****. Rationale — This is to ensure that a bystander will not be able to read the PIN being entered by the customer.

2. The ATM system will allow user to Cancel the transaction at any point and eject the ATM card within 3 seconds. Rationale — In case the customer is under duress or in fear of own security he/she needs to quickly get away.

3. The ATM system shall not print the customer's account number on the receipt of the transaction.

>

**Issues**
< List any issues related to the definition of the use case.

Example

1. What is the maximum size of the PIN that a customer can have? >

**To do**
< List any work or follow-ups that remain to be done on this use case.

Example

1. Need to ensure that we have covered all parties under the 'Stakeholders and Interests' heading. >

Requirements traceability focuses on documenting the life of a requirement and providing bidirectional traceability between various attributes of a requirement. It enables stakeholders to find the origin of each requirement and track changes to those requirements and other important attributes. Table 6.6 is an example of requirements traceability. It shows the user requirement, described in a use case, and traces it to functional requirements, design elements, code modules, and test cases.

**Table 6.6: Example of a traceability matrix for requirements**

| User Requirement | Functional Requirement | Design Element | Code Module | Test Case |
|---|---|---|---|---|
| UC-1 (use case) | Catalog<br>Sort | Class catalog | Catalog.<br>Sort() | Search.7<br>Search.8 |
| UC-2 (use case) | Catalog<br>Query<br>import | Class catalog | Catalog<br>Import()<br>Catalog.<br>Validate() | Search.12<br>Search.13<br>Search.14 |

## Prioritizing requirements

Requirements prioritization is done to make sure the product contains the most essential functions and to provide the greatest product at the lowest cost. You must balance project scope against the constraints of schedule, budget, staff resources, and quality goals and drop or defer low-priority requirements to a later release. Establishing priorities early allows for more options. Doing this balances the business benefit of each function against cost. The customers can determine value while considering cost, technical risk, and other trade-offs. Table 6.7 suggests some prioritization levels to be used for prioritization of system requirements.

**Table 6.7: Prioritization levels to be used for prioritization of system requirements**

| Names | Meaning |
|---|---|
| **High** | A mission-critical requirement; required for next release |
| **Medium** | Supports necessary system operations; required eventually but could wait until a later release if necessary |
| **Low** | A functional or quality enhancement; would be nice to have someday if resources permit |
| **Essential** | The product is not acceptable unless these requirements are satisfied |
| **Conditional** | Would enhance the product, but the product is not unacceptable if absent |
| **Optional** | Functions that may or may not be worthwhile |

**Table 6.8: The relationship between urgency and importance in setting priorities for system requirements**

| | Important | Not Important |
|---|---|---|
| **Urgent** | High priority | Don't use these |
| **Not urgent** | Medium priority | Low priority |

Table 6.8 shows the relationship between urgency and importance in the prioritization of system requirements.

*High-priority requirements* are both important (the user needs the capability) and urgent (the user needs it in the next release). Contractual or legal obligations might dictate that the requirement must be included, or there might be compelling business reasons to implement it promptly.

*Medium-priority requirements* are important (the user needs the capability) but not urgent (they can wait for a later release).

*Low-priority requirements* are not important (the user can live without the capability if necessary) and not urgent (the user can wait, perhaps forever).

Requirements in the fourth quadrant appear to be urgent but they really aren't important. Don't waste your time working on these. They don't add sufficient value to the product.

Prioritizing system requirements should be based on value, cost, and risk. The project manager should perform the following steps to set priorities:

- Leads the process.
- Arbitrates conflicts.

- Adjusts input from the other participants, customer representatives, product champions, marketing staff, or development representatives, such as team technical leads.
- Supplies benefit and penalty ratings.
- Provides the cost and risk ratings.

When prioritizing requirements use these steps as a guide:

1. List all features, use cases, or requirements.
2. Customer representatives estimate the relative benefit on a scale of 1 (not useful) to 9 (extremely valuable).
3. Estimate the relative penalty if the feature were not included, on a scale of 1 (no one will be upset if it's excluded) to 9 (a serious downside).
4. Calculate the total value = benefit + penalty (weights can change).
5. Developers estimate relative cost of implementation on a scale of 1 (quick and easy) to 9 (time consuming and expensive).
6. Developers estimate relative degree of technical or other risks on a scale of 1 (easy to program) to 9 (serious concerns).
7. Calculate priority = value%/((cost%*weight) + (risk%*weight)).
8. Sort features in descending order by calculated priority.

## Recommendations to reduce requirements' risks

Here is a summary of the key risks for requirements with some recommendations on how to reduce or eliminate these risks.

1. Risk: Product vision and project scope
   - Recommendations: Early on, write a vision and scope document that contains business requirements; use it to guide decisions about new or modified requirements.
2. Risk: Time spent on requirements development
   - Recommendations: Record how much effort is actually spent on requirements development for each project; use it to judge whether it was sufficient and improve planning for future projects.
3. Risk: Completeness and correctness of requirements specification
   - Recommendations: Elicit requirements by focusing on user tasks. Devise specific usage scenarios, write test cases from the requirements, have customers develop acceptance criteria. Create prototypes; elicit feedback. Have customer reps inspect specs and analysis models.
4. Risk: Requirements for highly innovative products
   - Recommendations: Emphasize market research, build prototypes, use customer focus groups for feedback early and often.

5. Risk: Defining nonfunctional requirements
   - Recommendations: Query customers about performance, usability, integrity, reliability. Document in the System Requirements Specification with acceptance criteria.
6. Risk: Customer agreement on product requirements
   - Recommendations: Determine the primary customers; use a product champion.
7. Risk: Unstated requirements
   - Recommendations: Use open-ended questions.
8. Risk: Existing product used as the requirements baseline
   - Recommendations: Document requirements discovered through reverse engineering; have customers review them for relevance and correctness.
9. Risk: Requirements prioritization
   - Recommendations: Prioritize every functional requirement, feature, and use case; allocate each to a specific system release. Evaluate the priority of new requirements against the body of work remaining.
10. Risk: Technically difficult features
    - Recommendations: Evaluate the feasibility of each requirement. Track and watch for those falling behind schedule. Take corrective action early.
11. Risk: Unfamiliar technologies, methods, languages, tools, hardware
    - Recommendations: Don't underestimate the learning curve. Identify high-risk requirements early; allow time for false starts, experimentation, and prototyping.
12. Risk: Requirements understanding
    - Recommendations: Conduct inspections and include developers, testers, and customers. Create models and prototypes; represent requirements from multiple perspectives.
13. Risk: Time pressure to proceed despite To Be Determined (TBDs) still remaining in the document
    - Recommendations: Record the person responsible for closing and the target date for resolution.
14. Risk: Ambiguous terminology
    - Recommendations: Create both a glossary and a data dictionary.
15. Risk: Design included in requirements
    - Recommendations: Make sure requirements emphasize <u>what</u> needs to be done rather than <u>how</u> to do it.

---

**Editor's note**

Rob Oshana has laid out detailed and rigorous groundwork for preparing and managing requirements. Mike Gard has a case study and a complementary perspective to Oshana's development that follows.

## Mike Gard: thoughts on developing requirements

Participation by the design team begins at the very earliest stages of the project effort. Design team effort is needed before and during proposal generation, where even nascent product and project ideas are being explored. The idea is not to do a preliminary design at the proposal stage—although that is a wonderful thing to do if circumstances permit. Rather, the very earliest work should identify possible pitfalls, identify design unknowns, carefully assess and rank design priorities, and, if necessary, identify those unknowns that fall into the category of *inventions needed* ("we must find a way to . . ."). This is necessary to avoid surprises, of course, but also to make sure budgets for time, money, materials, and manpower are well thought out.

Design trade-off information is especially important in the concept definition stage of embedded systems because project definition must do the following:

- make and clarify expectations about the functions and features to be embedded in circuit hardware,
- clarify expectations about the complementary or mating functions and interfaces external to the embedded design,
- clarify expectations about implementation and partitioning between the mechanical, hardware, software, and firmware components.

Failure to establish these matters early in concept development will often result, at best, in over specification or functional redundancies in the design. At worst, failure to establish these matters will plant the seeds of a catastrophic omission or fundamental, and sometimes fatal, flaw.

In September 1999, the Mars Climate Orbiter (formerly the Mars Surveyor '98 Orbiter) disintegrated after making an improper low entry into the upper Martian atmosphere because ground-based computer software produced thruster force output in Imperial pound-seconds (lbf*s) rather than the specified newton-seconds (N*s). One section of the final investigation is worth quoting in entirety:

> *The MCO MIB has determined that the root cause for the loss of the MCO spacecraft was the failure to use metric units in the coding of a ground software file, "Small Forces," used in trajectory models. Specifically, thruster performance data in Imperial units instead of metric units was used in the software application code titled SM_FORCES (small forces). The output from the SM_FORCES application code as required by a MSOP Project Software Interface Specification (SIS) was to be in metric units of Newton-seconds (N-s). Instead, the data was reported in Imperial units of pound-seconds (lbf-s). The Angular Momentum Desaturation (AMD) file contained the output data from the SM_FORCES software. The SIS, which was not followed, defines both the*

*format and units of the AMD file generated by ground-based computers. Subsequent processing of the data from AMD file by the navigation software algorithm therefore, underestimated the effect on the spacecraft trajectory by a factor of 4.45, which is the required conversion factor from force in pounds to Newtons. An erroneous trajectory was computed using this incorrect data.*

*The discrepancy between calculated and measured position, resulting in the discrepancy between desired and actual orbit insertion altitude, had been noticed earlier by at least two navigators, whose concerns were dismissed. A meeting of trajectory software engineers, trajectory software operators (navigators), propulsion engineers, and managers, was convened to consider the possibility of executing TCM-5, which was in the schedule. Attendees of the meeting recall an agreement to conduct TCM-5, but it was ultimately not done [11].*

**If there is a single best guarantor of success, it is the opportunity for designers and system architects to spend time in the customer's shoes**. It is one thing to design to a set of specifications—that is expected. The difficult thing is to make sure the design is being done to the correct specifications; this is to say, that the specifications accurately reflecting what the customer wants and needs. This is not always what the customer's agent, the contracting authority, or even the customer, originally says is desired. The need to root out the true requirements is most acute when the intended end user is not a technical individual or organization; in those cases the customer often describes needs based on what already exists rather than what is possible. In such cases, the design team needs the insights that come only from actually experiencing, or at least observing, the task requirement at hand. This is not always possible, but it should be a design team priority whenever circumstances allow it. Hands-on experience should be obtained as early and as often as possible during the project.

On occasion, design team recommendations may be articulated but overruled by the customer, the customer's agent, or other folks within the design team's organization. This is unfortunate, but it happens. In the commercial marketplace, the marketing strategy may emphasize cost over performance to obtain greater market penetration in the low end (euphemistically called the "value" tier) of the market. In other cases, the ultimate design option may be something simple, intuitive, reliable, extremely durable, and nearly foolproof (military hardware for combat infantry comes to mind), whereas the engineering team (and, more often, the marketing team) sees opportunities to produce something elegant and capable of satisfying very difficult, but relatively rare, problems. In these cases, document everything and forget nothing—there will be other projects and other opportunities to apply what has been learned.

Design requirements often evolve as design effort progresses and functional prototypes get their first customer use. **Editor's Note:** *See Tim Wescott's section in Chapter 8 titled, "The Importance of Early Prototypes."* The later the customer begins to experiment with the product, the greater the risk to the schedule and budget. Even the problem known as "creeping elegance" often can be traced to the tendency for large projects to be handled by specialized contracting groups which, themselves, have little or no direct experience with the needs of the true end user. Specification creep, cost overruns, schedule delays, creeping elegance, and poor final performance are the almost inevitable (although predictable and preventable) results.

Military projects are often cited for these issues; they are very expensive and highly visible examples because of their size and complexity. Military projects are hardly unique in their occasional ability to get it wrong. My own work experience encompasses missile, aircraft, ophthalmic ultrasound, geophysical (seismic) systems, computer aided tomography scanners, and construction equipment. Despite differences in technologies and applications, the problems have much in common. More often than not, the ultimate causes of most problems are failures to completely understand the end user's true needs coupled with vague, high-level, nonspecific definition that cover only a few critical functions. As it inevitably becomes obvious that the specifications are inadequate, there are specification changes to correct the deficiencies. Just as often, these changes in specification are strongly driven by the influence of schedule or budget (because the project is now well advanced) rather than technology or good engineering judgment. All too often, the results of this situation are an unrealistically compressed schedule, design compromises to preserve schedule and money, and an end product that is not all it could or should have been. ***Invest time and effort up front to develop good specifications—it is much better for all concerned. Good specifications are the essential first step of the design process***.

## Oshana's Maxim—estimating requirements' efforts

In my experience, requirements engineering is a complete lifecycle process that takes between 15% and 25% of the total effort of a complex engineering project to do properly. Nonfunctional requirements formulation (performance goals, usability goals, scalability, etc.) are much harder than functional requirements because analysis, simulation, modeling, research, and benchmarking are required to create well defined, quantifiably correct, measureable, and testable requirements. Multiple iterations of discussion, clarification, and review are needed to drive clarity into the requirements process. Globally distributed development teams and customers exacerbate the problem.

## Acknowledgments

I thank Geoff Patch and Tim Wescott for their reviews of this chapter.

## References

[1]  IEEE 610.12-1990. IEEE Standard Glossary of Software Engineering Terminology. Software Engineering Standards Collection. Institute of Electrical and Electronics Engineers, New York, NY.

[2]  F.P. Brooks Jr, No silver bullet: essence and accidents of software engineering, Computer 20 (4) (1987) 10−19, Washington, DC.

[3]  K.F. Wiegers, Software Requirements: Practical Techniques for Gathering and Managing Requirements Throughout the Product Development Cycle, Microsoft Press, Portland, WA, 2003.

[4]  T. Gilb, Principles of Software Engineering Management, Addison-Wesley, Boston, MA, 1988.

[5]  A.M. Davis, Software Requirements: Objects, Functions, and States, Prentice Hall, Upper Saddle River, New Jersey, 1993.

[6]  J. Musa, A. Iannion, K. Okumoto, Software Reliability: Measurement, Prediction, Application, McGraw-Hill, New York, NY, 1987.

[7]  D. Parnas, J. Madey, Functional Documentation for Computer Systems Engineering (Version 2), Technical Report CRL 237, McMaster University, Hamilton, ON, 1991.

[8]  A. van Lamsweerde, Requirements Engineering: From System Goals to UML Models to Software Specifications, Wiley, Hoboken, NJ, 2009.

[9]  B.L. Kovitz, Practical Software Requirements: A Manual of Content and Style, Manning Publications, Co, Cherry Hill, NJ, 1998.

[10]  D.L. Lempia, S.P. Miller, Requirements Engineering Management Handbook, Report No. DOT/FAA/AR-08/32, June 2009.

[11]  A.G. Stephenson, L.S. La Piana, D.R. Mulville, P.J. Rutledge, F.H. Bauer, D. Folta, et al., Mars Climate Orbiter Mishap Investigation Board Phase I Report, NASA, 1999. Available from: http://sunnyday.mit.edu/accidents/MCO_report.pdf.

## Recommended reading

D.L. Lempia, S.P. Miller, 2009. Requirements Engineering Management Handbook, Report No. DOT/FAA/AR-08/32, June 2009. A well written, didactic but easily followed set of guidelines on how to prepare and maintain requirements. In spite of being a government document, it is very readable and very practical.

## CHAPTER 7

# Analyses and Tradeoffs

## Kim R. Fowler
*IEEE Fellow, Consultant*

## Chapter Outline

Developing and Managing Embedded Systems and Products.
DOI: http://dx.doi.org/10.1016/B978-0-12-405879-8.00007-6

## Introduction

This chapter reveals a real chicken-and-egg problem—which comes first, design creation or analysis? Can you even separate the two subjects? What if you modify the design? Most of you realize that some form of analysis is a part of design creation and modification. You cannot complete a circuit schematic, for example, without calculating the currents and voltages or defining signal paths and reviewing their operations. The same is true for software; you cannot complete the design of a code module without reviewing its function to see if it performs as you expect. Consequently, analysis tightly couples with design creation and modification—or should.

Chapter 1 gives definitions for design, analysis, tradeoffs, and synthesis. This chapter describes various analyses, portions of which are necessary to the successful development of product and embedded systems. ***Analysis may not be sufficient to guarantee success, but it is necessary!***

### Why analysis?

***Analysis forms the feedback between design and requirements.*** The right types of analyses, which depend on the system and the market, will gauge how well a design meets the requirements. If a design does not match the requirements, then either the design or the requirements need modification (Figure 7.1).

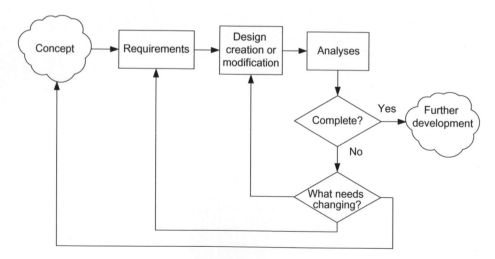

**Figure 7.1:**
Analysis forms the necessary feedback between design and development. © 2013 by Kim R. Fowler.
*All rights reserved. Used with permission.*

## When to analyze?

Analysis is used from the first stages of the concept to the final evaluation of the design. It can also be used after a design is built to explain behaviors or problems. (Chapter 1 provides some examples of design processes and where analyses fit into each phase of those processes.) Figure 7.2 illustrates the timing of various types of analyses during development.

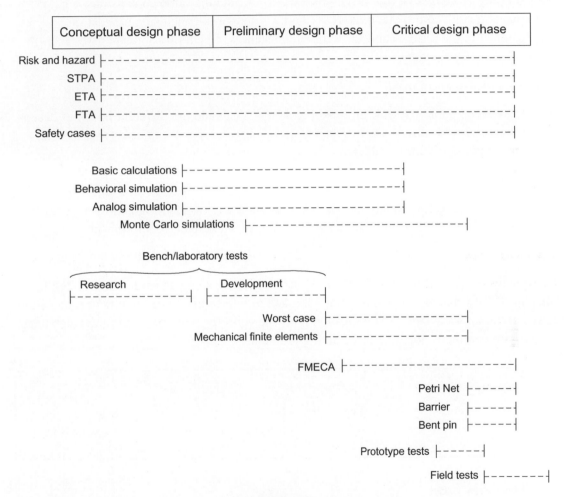

**Figure 7.2:**
Timeline for performing various analyses during a project. © *2013 by Kim R. Fowler. All rights reserved. Used with permission.*

## How is analysis used?

There are two primary types of analyses: proactive and reactive. Proactive analyses help with the design of a system by informing designers of the expected operations and behaviors of the system. Reactive analyses can guide you in debugging and finding problems in a functioning system; then you can fix them after a design has solidified into the built system.

Some analyses use a bottom-up approach, while others use a top-down approach, while even others might be considered "sideways-in" approaches. All the analyses that I describe are a mixture of ***deductive*** (the facts will always generate the same conclusions, e.g., a mathematical calculation) and ***inductive*** (the facts support a conclusion or a set of conclusions but the results can vary with circumstances, e.g., a probabilistic assessment) forms of investigation.

### Where are analyses performed?

Analyses can take many forms. Engineering analyses can address different levels of design from components to subsystems to systems. In this chapter, I will indicate the preferred levels for applying various analyses.

### Who analyzes?

Many people can perform analyses. This book is about the engineering and management concerns of embedded systems, so I focus on engineers and managers in this chapter.

### Risk management

These analyses can and should feed forms of risk analysis and management described in Chapter 4. Risk management is another approach to informing the development process to converge on an acceptable design. It highlights the concerns that arise during development so that you can focus attention on reducing problems.

---

**Please remember**

All examples and case studies in this chapter serve only to indicate what you might encounter. Hopefully these examples and case studies will energize your thought processes; your estimates and calculations may vary from what you find here.

---

## The business case

The business case for product development drives most products. (We just cannot get into all the altruistic reasons for other products; we will focus on business here.) If a good idea looks like it could make money, then someone in your company should perform the appropriate analyses to bolster the perception.

A product needs to make profit, which means selling it generates more money than you spend developing, producing, and distributing it. You need to understand that you can feasibly design and build the product and then manufacture and sell it profitably. The business case always includes schedule, budget, quality, features, and performance. Hopefully these aspects of the business case will lead to a successful and profitable system. Unfortunately, many products fail because the business case was done poorly.

Analysis supports the business case by connecting the evolving design with the requirements. Requirements, in turn, should express the system function, which should derive from the business case.

### Time and money

It is true—time is money! Your company pays very real money in salaries for staff to spend time developing a product. Time figures directly into some less obvious expenses, such as the cost of interest (the cost of getting investment funds to support development) and the cost of inventory. Maybe even less obvious is the opportunity cost in pushing a product into commercial sales, called time to market, economists and marketers understand that the sooner a finished product gets to market, the better chance it has for larger sales than a later introduction of the product. Finally, innovation that reduces the time to do something, whether for you, your company, or the customer, is often valuable in saving dollars for someone in some way.

Many companies have the concept of the three-legged stool; the legs represent features, schedule, and budget. Changing one affects the other two. *Most tradeoffs get down to features, schedule, and budget, which ultimately translate to time and money*.

### NRE and COGS

NRE and COGS are two basic terms that you need to know in designing, developing, and manufacturing electronic products. NRE is the cost associated with "nonrecurring engineering" effort. COGS is the "cost of goods sold." Figure 7.3 is a pictorial representation of the definitions of NRE and COGS.

*NRE is the cost to design, test, and qualify a product during development*. NRE includes the cost of the professional and administrative time consumed during development, the cost of tools and assets to prepare the design, and the overhead items used to design products, such as rent, heating, lighting, furniture, and office equipment. You amortize (an accounting term for "spread out") the cost of NRE over the sales life of the product to recoup the expense of designing a product.

*COGS includes the ongoing expenses and recurring costs that go into manufacturing each unit*. The procurement of large quantities of components are recurring costs and figure into COGS. The labor to assemble and build each unit is a part of COGS. COGS also includes the energy, materials, and contract services to produce. The manufacturing

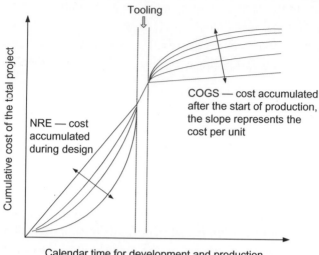

**Figure 7.3:**
Pictorial indications of NRE and COGS. © 2003–2013 by Kim R. Fowler. All rights reserved. *Used with permission.*

overhead, such as facilities, rent, and machinery, are a part of COGS, but at this point you need a good accountant to explain the details.

---

### Example: Motor controller

A company that builds and sells industrial motor controllers is willing to devote several full-time engineers for two years to design the controller and program it. They will sell tens of thousands of these units each year during production. Management's paramount demand is to reduce component costs. If functions can reside in software, then they do. They simply will not add more components to upgrade a circuit board during the market life of the motor controller. They would rather modify the software. This is a case where the NRE of the engineers' time is quickly overcome by the COGS in producing the circuit boards.

---

### Please note

This explanation of NRE and COGS is simpler than most real situations. NRE can often continue into the production phase of an item, which means the amortized cost of development will continue to rise. COGS, too, can sometimes go down during production because your company buys bigger quantities of components or the price of components goes down with maturity. Furthermore, tooling costs are not included here. Chapter 8 gives more detail about NRE, tooling, and COGS.

**Example: Core competency drives NRE and COGS (contributed by George Slack)**

A company's core competency often drives NRE and COGS. In the 1980s, Xerox designed their own processor chips for rapid image processing. With the advent of specialized components from companies like AMD, Intel, and Sun, Xerox scrapped this portion of their core competency and purchased processor chips to save money and accelerate time to market.

In this example, Xerox's core competency could not drive NRE and COGS low enough to compete with other companies' components.

## Tradeoffs

Tradeoffs are critical to design. The example of the industrial motor controller, above, weighs the NRE of the design against the COGS of the produced item.

You use tradeoffs to compare different approaches to design or different manufacturing concerns or various logistics. First specify the parameters and then develop the different approaches according to the chosen parameters, e.g., a particular circuit for signal processing and another circuit or software module to do the same thing. Then compare the approaches directly parameter-by-parameter to determine which approach is the best for you. The example, below, illustrates a tradeoff that you might encounter.

**Example: Buying a software tool**

Suppose that your company wants to reduce the time to design software. Several tools may be available to help with the design process that you can purchase. Table 7.1 gives one possible set of tradeoffs that might afford you a clear tradeoff to select a tool.

**Table 7.1: Example set of comparisons to make a tradeoff analysis for selecting a product.**

| Parameter | Vendor A | Vendor B | Vendor C | Open Source |
|---|---|---|---|---|
| Cost ($) | $10,000 one time purchase, no license renewal | $5000 purchase with $2000 yearly license renewal | $1100 one time purchase | $0 |
| Features | 5 regularly used 8 occasionally used 12 not needed | 4 regularly used 6 occasionally used 8 not needed | 5 regularly used 4 occasionally used 6 not needed | 4 regularly used 10 not used |
| **(0—bad, 10—excellent)** | | | | |
| Ease of use and training | 6 | 6 | 3 | 0 |
| Vendor technical support | 4 | 8 | 2 | 0 |
| Comments from users | 5 | 7 | 1 | 3 |
| Reviews | 7 | 6 | 1 | 2 |

From this table (often called a decision matrix or Pugh diagram), Vendor B might be the best fit for your company. Bottom-line cost should not be your only determining factor. How a vendor supports their product will be an important indicator of how well they will support you and actually meet your needs. Comments and reviews can be a source of insight into choosing a tool, as well.

Think critically about comparing between the prices and lost engineering time to get a tool working. Note that it takes only about two weeks of engineering time, or less, to waste the dollar savings of a cheaper tool versus a more functional, better operating tool (assuming that an engineer costs about $70–100 per hour, 80 hours equals somewhere between $5000 and $8000 over two weeks).

## Design tradeoffs

Design tradeoffs can be important considerations to form the concept and design of the system, which in turn feed the requirements. There are many different types of tradeoffs for many different concerns, such as (and these are only a sampling):

- Power type, conversion, and distribution
- Materials
- Weight, size, and volume
- Software versus electronic hardware versus mechanical mechanisms
- Number of features
- Reliability versus fault tolerance versus availability
- Buy versus build
- Design for X (often called DFx), where X is
    - Manufacturing
    - Assembly
    - Disassembly
    - Test diagnostics
    - Repair
    - Availability.

### Case study: Design tradeoffs for the LWS-SET

The US National Aeronautics and Space Administration (NASA) has a program called Living With a Star (LWS) that began in 2000. The stated mission of LWS is, "The LWS Program provides missions to improve our understanding of how and why the Sun varies, how the Earth and Solar System respond, and how the variability and response affects humanity in Space and on Earth" [1]. Within the LWS program is the Space Environment Testbeds (SETs) Project, which performs flight and data investigations with the goal to improve the performance of electronic hardware in the space radiation environment [2].

In 2002, I was the program manager for the Johns Hopkins University Applied Physics Laboratory (JHU/APL) to guide the basic development of the SET system [3]. My team developed a concept for a support platform to host or carry low-cost university experiments; the SET support platform would fit various different satellites and would supply power, data, and control services to the university experiments. These university experiments were to investigate space radiation effects on electronics. Figure 7.4 illustrates the basic diagram of the SET platform [4].

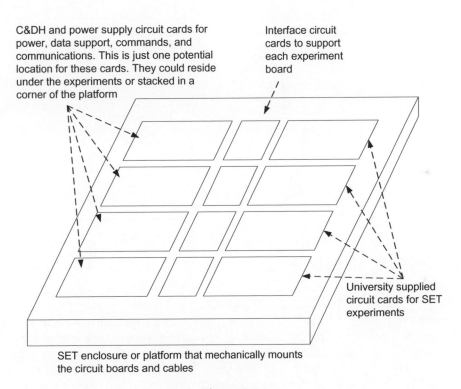

**Figure 7.4:**

Basic diagram of the SET platform. © *2003–2013 by Kim R. Fowler. All rights reserved. Used with permission.*

My team performed a number of tradeoffs to design an appropriate system for the SET platform. One set of tradeoffs that we made was for the cabling harness configuration, which was to conduct power, data, and commands. We could use either a centralized "star" configuration or a distributed network for the cabling. Figure 7.5A,B illustrates the basic diagrams of the two possible cable configurations [3]. Note: POL in Figure 7.5B refers to "point of load" power converters. POL converters take a higher voltage input, e.g., +28 VDC, and output multiple lower voltages, e.g., +5 and ±15 VDC.

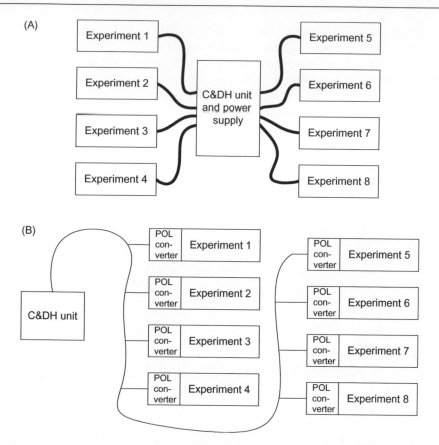

**Figure 7.5:**
(A) Centralized star configuration for command, data, and power distribution. (Note: C&DH is an acronym for command and data handling. This configuration generates power centrally and then distributes the regulated DC voltages, e.g., +5 and ± 15 VDC, via large gauge, heavy cables; each cable is point-to-point and dedicated to a single experiment.) (B) Distributed configuration for command, data, and power distribution. (Note: This configuration distributes a higher voltage, less-well regulated DC voltage, e.g., +28 VDC, via smaller gauge, lighter cables and then generates power locally via POL converters; a single cable connects all the experiments.)

Table 7.2 gives the parameters and considerations made to prepare the tradeoff analysis in selecting the configuration. The centralized configuration is traditional in spacecraft and well understood; however, considering all the parameters, the distributed configuration was better suited for LWS-SET and we chose that design.

We performed a number of tradeoffs for this project. Unfortunately, we did not complete this project for business and political reasons beyond our control. Ultimately, NASA and JHU/APL agreed to dissolve JHU/APL's involvement in the project for administrative reasons

**Table 7.2: Example set of comparisons to make a tradeoff analysis for selecting a product.**

| Parameter | Centralized | Distributed |
|---|---|---|
| Fault tolerance—power distribution | − − − | + + + |
| Cable weight reduction[a] | − − − | + + + |
| Cable complexity reduction in numbers of conductors | − − − | + + + |
| Ease of cable design | − | + |
| Ease of power supply design | − − | + + |
| Ease of command and communications design | + | − |
| Ease of assembly, freedom from mistakes | − | + |
| Ease of diagnostics for repair during development | − | + |
| Reliability | 0 | 0 |
| System mass (weight)[a] | 0 | 0 |
| Cost to implement | 0 | 0 |

*Note*: + is better, + + is good, + + + is outstanding; − is worse, − − is bad, − − − is really bad; 0 is neutral.

[a]Note that heavier cables to distribute centrally generated power means that you do not have POL converters needed by a distributed power system; the tradeoff in weight is inconsequential to the total system weight.

other than technical or management performance. NASA has continued the development and plans a first launch in 2015.

### Control: Software versus electronic versus mechanical

Control, particularly of physical movement, involves tradeoffs between various configurations of software, electronics, and mechanisms. Tradeoffs should ask, "How much of the control resides in algorithms in the software? How much in the electrical configuration? And how much in the mechanism and its range of motions?"

These tradeoffs often simplify to comparing the NRE with the COGS for the final configuration of the embedded system. A simple example is the flashing LEDs in children's shoes—these could be designed with a simple transistor circuit but use a microcontroller circuit instead! A more apropos example for most of you would be developing the control of a filter wheel for a scientific imager on a satellite; it can make sense to use a motor control IC or you might perform all the control in software algorithms—as discussed in the example below.

### Example: LEDs in children's shoes

While the flashing LEDs in children's shoes could be implemented with a simple transistor circuit, they use a microcontroller circuit instead. Microcontrollers are very cheap, they can reduce the parts count in the manufacturing bill of materials, and, most importantly, they can be reprogrammed between production runs for new flash patterns to satisfy fashion trends. For LEDs in shoes, amortizing the NRE of programming the flash patterns over millions of shoes is much less expensive than the COGS of the circuitry in the shoe soles.

## Example: Filter wheel control in a satellite imager

Scientific imagers in satellites sometimes incorporate a filter wheel mechanism to select different ranges of the optical spectrum. These wheels contain different optical filters and some sort of electric motor drive. Each filter has a specified position on the wheel. The wheel cuts through the optical path in front of the imager to insert a filter. Sometimes, the filter wheel has slot detents to dwell at each position and a drive mechanism that moves the filter wheel from one detent slot to the next. Or you might use a photodiode detector and alignment holes in the wheel rim to center each filter in the optical path. Spacecraft instrument designers typically use stepper motors to move filter wheels and software to count the steps to indicate position; designers then rely on a highly reliable combination of a mechanical position indicator and an electronic sensor to confirm the wheel position.

## Example: Radar front-end circuitry

Most radar systems now use a digital front-end with very high-speed analog-to-digital converters to digitize the received signal for processing. These digital front-ends replace original analog electronics front-ends to down convert the received signal into intermediate- and low-frequency signals for processing. The digital front-ends tend to be more reliable and they do not drift with temperature or environmental changes. This is an example of a tradeoff between analog circuitry and software operating on digital electronics.

### Number of features

The number of features can lead to an exponential expansion in effort. If each feature takes the same amount of time to design and test as all the others and if every feature interacts with every other feature, then the time to develop the system is:

$$\text{Time} = Y \cdot N \cdot (N - 1) = Y \cdot (N^2 - N)$$

where $Y =$ time per feature and $N =$ number of features. Figure 7.6 illustrates this quadratic relationship. This relationship gives the upper maximum of time required to develop features. One way around this expansion of effort is to decouple the features and limit their interactions. Regardless, you will still devote significant effort to verify correct operation. The point is that adding features is not free—it takes time to develop and test them carefully.

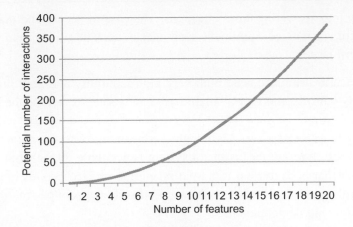

**Figure 7.6:**
Potential number of interactions between features or modules as they increase in number.

### Buy versus build

One tradeoff that we encounter in most projects is whether to buy particular components (or modules or subsystems) or to build them. The conundrum is whether they will solve your design need neatly and immediately or they will need to be "shoe-horned" into the system. A designer's natural inclination is usually to build something, rather than buy it, because you see so many ways to do it better or fit your project better. Chapter 8 has more details on the decision to buy or build.

### Dependability—reliability versus fault tolerance versus availability

Other tradeoffs that most of us encounter include determining the reliability versus fault tolerance versus availability. Questions that you need to ask to make these tradeoffs include the following:

- What is the specified reliability? Reliability means that a system continuously operates without fail for a specified duration. All functionality remains as designed with no degradation in capability.
- What is the specified fault tolerance? Fault tolerance means that the system can operate with diminished capacity, but acceptably, until it recovers from the fault, if it ever does recover.
- What is the specified availability? Availability is the measure of "uptime" of the system. A system may not be very reliable, i.e., it operates continuously for shorter time durations, but it might be easily and quickly repaired and brought back to full functionality. An example would be swapping circuit boards in and out of a rack of equipment.

Many issues play into answering these questions. The ease of access and diagnostics is one issue. The need for specialized equipment is another issue. The skill of maintenance personnel is a third issue.

Some systems, such as satellites, need both very high reliability and very high fault tolerance over any consideration for availability. Once a satellite is launched, repair, and hence, availability, cannot be done. Other systems, such as a server farm with massive numbers of subsystems, need a very high degree of availability to remain operating. The current vernacular of "five 9s" means that a system is available 99.999% of the time; over the space of a year that is equivalent to a total downtime of 315 seconds or a little over 5 minutes! Regardless, the higher any of these specifications becomes, the more effort you will spend in design and development.

These terms, reliability, fault tolerance, and availability, are subsets of dependability. Dependability encompasses much more; here are all the components of dependability:

- Reliability—"... is the probability that the system operates correctly throughout a complete interval of time... Reliability is most often used to characterize systems in which even momentary periods of incorrect performance are unacceptable, or it is impossible to repair the system" [5]. Complexity, vibration, mechanical shock, wide temperature variations, corrosion, and material aging all reduce reliability [6].
- Availability—"... how frequently it [the system] becomes inoperable but also, how quickly it can be repaired" [5]. It is the probability that a system is operating correctly at a specific instant in time. A system can be highly available even though it experiences frequent periods of nonoperation, which are extremely short in duration [6].
- Fault tolerance—"... a system continues operating in the face of a component failure or operational fault, but the performance may decrease. The difference between reliability and fault tolerance (performability) is that reliability defines the likelihood that *all* of the functions perform correctly, while fault tolerance (performability) defines the likelihood that a *subset* of the functions performs correctly" [6].
- Testability—"... defines the ease of test for certain attributes within a system. Testable architectures generally do not provide continuous monitoring and they usually aren't automatic either—either they must periodically trigger testing or an operator must initiate the testing" [6].
- Maintainability—"... defines the ease of system maintenance, as well as the repair for a failed system. Quantitatively, it is the probability that a failed system or one down for maintenance will restore to operation within a set period of time... Testability, BIT, diagnostics, and repairability all are components within maintainability. Furthermore, completing maintenance and repairs more quickly means that the system is more available than a similar system that is not maintainable. Ultimately this means that the system is more dependable" [6].

- Safety—". . . is the probability that a system will either perform its functions correctly or will discontinue the functions in a manner that causes no harm" [5].

While considering these components of dependability, the design of embedded systems also involves other characteristics, such as fidelity (defined by accuracy, precision, and resolution), longevity, and highly technical support.

### Explicit versus implicit

Explicit versus implicit could also mean the obvious versus the hidden. Some system operations are immediately obvious as necessary or functional; other operations may not. Being implicit or nonobviousness does not mean any less difficult; in fact, it usually indicates greater difficulty for development. One example of the implicit or nonobvious is the user interface; it may seem simple and straightforward but seldom is. I worked for a small medical device company where we developed a basic instrument for programming implanted devices with a notepad-type computer; we spent 95% of the software effort on developing the GUI, not on the calculations, communications, or database.

Another area of explicit versus implicit that can confound development, and even acceptance of the produced system, is the tradeoff between manual and automatic operation. Manual operation sometimes can reveal internal states that automatic operation would conceal. In situations, such as engine controllers, where operation necessarily must be automatic, you will spend considerable effort to reveal the appropriate conditions and states.

An ancillary concern for the tradeoff between manual and automatic operation is the documentation necessary to explain the operations versus intuitive operation. You will find a large portion of your career will be involved in documentation—schematics, source code listings, reviews, papers, presentations, and communications.

### Manufacturing tradeoffs

Chapter 15 goes into the many details of manufacturing and producing an embedded system. Some of the tradeoffs discussed in Chapter 15 and mentioned here for emphasis include:

- In-house versus contract manufacturing versus offshore manufacturing
- Automated versus manual assembly
- Ease of manufacturing versus design effort.

Manufacturing in-house requires buying, owning, and operating machinery and equipment. You also have to train and then retain skilled operators and assemblers. To use your assets and human resources to the optimum advantage means that you must keep them busy with a continuous stream of products on the assembly line.

Contract manufacturers, on the other hand, can do smaller production runs efficiently because they fill in with jobs and work from other customers. Contract manufacturers often have expertise that you might find difficult to locate, such as medical or military or aerospace markets. Furthermore, they can inventory components to relieve you of concern and storage costs.

Automated assembly can be very fast and accurate, thereby increasing quality in manufacturing. For assembling circuit boards, pick and place machines are quick and easy to reprogram and load. Unless you are manufacturing high volumes (hundreds or thousands of systems per month) and can afford automated manufacturing equipment, you will need manual labor to build, place, and attach the larger piece, such as subsystems, cables, and connectors. Test for quality can be automated if the production volume warrants; otherwise, this is another manual check. Automated versus manual assembly all comes down to the ease of manufacturing versus the design effort; it is a straight COGS versus NRE tradeoff. Your question will always be, "Where is the breakeven point?" because the breakeven point strongly indicates whether you will have a return on investment and be profitable or not.

### Logistics and support tradeoffs

Chapter 16 goes into the many details of distributing, maintaining, and supporting an embedded system. One of the first tradeoffs is, "Who will keep the inventory and distribute the product?" Inventory and distribution require facilities and personnel; does your company have enough product shipping to keep staff busy? Like contract manufacturing, there are logistics companies that can provide various aspects of the warehousing, shipping, and even technical support—if you wish.

Another tradeoff is whether the embedded system will be disposable versus repairable. Computer printers and smart phones are disposable, whereas vehicles tend to be repaired. Clearly, more costly and more complex embedded systems will tend toward being repairable. Computers are a funny kind of "in-between" with repairs for memory tune-ups and drive replacements versus disposing computers that need a new processor chip. Maintenance skill plays into the consideration for this tradeoff—how skilled does the maintenance personnel need to be to perform the repairs and are they worth the expense?

Another tradeoff is customer training; do you train customers or rely on the user manual to inform customers? If you decide to train customers, what type of training is better, an in-house, week-long seminar or a webinar, or on-site at customers' facilities? Can you get by with a DVD or a web site and telephone support?

## Use cases

Use cases supply scenarios of the embedded system's operation when people interact with it. Each use case lays out the detailed steps of the interactions. A use case starts with a template that asks the journalistic standard of "who, what, when, where, why, and how." You fill out the template for each operation to prepare a use case. Collecting together all the use cases should represent all the possible operations of the embedded system that interact with people. Figure 7.7 provides an example template for a use case. There are a variety of ways to generate use cases from commercial software to simple post-it notes on a flip chart; the important thing is just to do it.

---

Use Case Title: _____

Use Case #: _____     Revision #: _____

Creation date: _____     Modification date: _____

**Who** is involved in this use case?

_____

**How** are they involved in this use case?

_____

**When?** _____

**Where?** _____

**What?** Describe the steps of the operation, include any pre- or post-conditions:

**Why?** Describe the rationale for the operation:

---

**Figure 7.7:**
Example template for a use case. © 2013 by Kim R. Fowler. All rights reserved. Used with permission.

Examples of "who" doing "what" include:

- Users and operators—interact with the embedded system when it is installed
- Maintainers—perform diagnostics, replenishment, repair, and upgrades

- Customers—people who do the purchasing (may not be the users or maintainers)
- Influencers—provide persuasive input and insight for the purchase and use of the system.

Use cases should drive requirements. After the fact, use cases help frame the extent of operations for analysis. While thorough preparation of use cases costs time and effort, it more than makes up with better requirements and less time in targeting analyses later.

## Design analyses

Design analyses refer to mathematical calculations, modeling, and deductive forms of analysis used to support the synthesis of a component, module, or system. Their purpose is to define the boundaries of operation and estimate the expected operations. Here are some major types of design analyses and their areas of application:

- Mathematical calculations and modeling
  - Circuit design and behavior—signal generation, reception, control, filtering, data bandwidth, control operations
  - Power consumption of the system
  - Heat dissipation and cooling of the system
  - Reliability—electrical, mechanical, materials
  - Probability assessment—types of operations, faults and failures, errors
- Worst case—linear and root-sum-square combinations for power dissipation, cooling, bandwidth
- EMC—circuit components, PCBs, cables, connectors, subsystems, enclosures and gaskets
- Derating margins—circuit components, mechanical stress and strain, kinematics tolerances and clearances.

## Physical forms of analysis

### Simulations

Simulations very closely relate to design analyses. Simulations contain more assumptions than physical models and, consequently, create looser constraint boundaries. The main purpose of simulations is to show what will not work; usually simulations do so in a manner more quickly and for lower costs than physical models. Simulations are less convincing at demonstrating physical operation; they can support but do not prove that a system or device will operate in a particular way. Simulations necessarily require simplifying assumptions, which can limit their utility in supporting and proving operation before actually building a physical component.

Types of simulation include:

- Mechanical—finite element analysis for fit, vibration, mechanical shock, materials failures, corrosion
- Electrical—SPICE, commercial packages
- EMC—fields, generation, reception, interference
- Cooling
- System dynamics.

---

### Case study: Simulating plant controls and flows

A wide variety of manufacturing, processing, and refining plants use embedded systems for process measurement and control. These systems control motors, valves, pumps, fluid flows, and material movement. These systems also use sensors, which themselves are often embedded systems, to monitor and determine the current state of the process.

These embedded systems work together to adjust for environmental compliance, efficiency, and safety. Companies strive to optimize these plants and their processes to achieve these disparate goals and still make money. These networks of embedded systems implement a succession of controls from supervisory to loop level to maintain set points and minimize disturbances. The highly nonlinear behaviors of the controlled materials compound the complexity of the optimization problem.

"Trial-and-error approaches to improve performance can adversely affect plant operations and safety. Setting less aggressive controller gains as a quick fix to control system instability problems often leads to suboptimal performance. As an alternative to these approaches, engineers can perform dynamic simulations of the control system (controllers and process) to gain insight into the system dynamics, understand what is causing instabilities, tune controllers, design and validate a better control architecture, and achieve better plant performance" [7].

"During plant design, simulation enables engineers to optimize processes, formulate the plant control system architecture, and study steady-state capacity. Once in operation, plant simulations let engineers identify the root cause of inefficiencies and fine-tune the process. Often, problems can be resolved by tuning isolated control loops with a single controller. Loop tuning uses an empirical model based on process data during a predefined set point change, helping calculate new gain coefficients such as in PID controllers" [7].

"When plant conditions change and no longer match those used in the original design simulations, the problems can become more complex. Multiple control loops may interact with each other, and systems may have unforeseen coupled dynamics that cause oscillatory behavior or uncontrollable instabilities. Multivariate control techniques are often used to address this class of problem" [7].

"Once a new control system strategy is verified via simulation, the distributed control system (DCS) can be reconfigured or supplemented with a supervisory control system, as in the case of a model predictive controller. Before plant start-up, a company may optionally verify the DCS configuration by connecting the control system to the process simulation..." [7].

### Bench (or laboratory) tests

Bench (or laboratory) tests are a good start to physical realization and confirm the results of design analyses and simulation. Types of bench tests include:

- Circuits—confirming signals and controls, trying out concepts
- Mechanisms—confirming movements, trying out concepts
- Comparing to simulation results for
  - power consumption
  - heat dissipation
  - signal generation
  - control movements
  - component variations.

### Prototypes

Prototypes perform similar functions to bench tests for proving out operations. Prototypes, however, tend to be more realistic because they operate in the actual operating environment with more features than units in bench tests.

Prototypes are good for field tests where you can subject an embedded system to real-world operations, realistic operator interactions, and environmental challenges. You can put prototypes through extensive laboratory tests to shake out problems and confirm design analyses and simulations; these tests can include tests for EMC, thermal cycling, mechanical shock, vibration, condensation, dust, salt spray, pressure, and vacuum.

## Formal analysis techniques

### Types and when used

Hazard analysis and risk management are closely tied in project development. Hazard analysis focuses on safety issues, while risk is a superset that includes safety risk, project risk, and mission risk. Chapter 4 goes into more detail on the high-level considerations for hazard analysis and risk. This chapter will look at the details of specific forms of analysis.

*Proactive analyses find and prioritize problems before completing the system*. Proactive analyses use multiple different analyses to cover potential areas of fault and failure and increase the probability of finding them.

*Reactive analyses find and fix problems after building the system*. Root cause analysis (RCA) is the primary form of reactive analysis. A thorough set of proactive analyses will provide a strong foundation for RCA and will reduce the time to complete an RCA should the need arises.

## Proactive analyses

There are a number of different proactive analyses, each with a different set of strengths and weaknesses. These analyses include:

- Failure modes effects criticality analysis (FMECA)
- Fault tree analysis (FTA)
- Event tree analysis (ETA)
- System theoretic process analysis (STPA)
- Dependability—reliability, fault tolerance, availability, maintainability
- Safety cases.

FMECA, FTA, ETA, and safety cases all examine aspects of design and design faults or flaws. Dependability focuses on material wear-out and failures and does not highlight design flaws.

None of these analysis techniques are sufficient on their own. Each covers problem areas the others do not. You should consider preparing each analysis to end up with a suite of results that provide a better chance of finding potential problems in both design and materials. Performing these analyses can consume a lot of time. In some situations, they can double or triple the initial design effort—but that investment in time and effort can more than repay the project later in robust designs that function nearly flawlessly.

Remember, "it's pay me now or pay me later." Paying me later (i.e., delaying analyses) will most likely cost you a LOT more. The NASA study in Chapter 1 showed how costs skyrocket as problems are found later in development. Do the analyses early and you can avoid many of those exorbitant later costs.

### Failure modes effects criticality analysis

FMECA is a technique that arose in the military and aerospace industries in the 1950s and 1960s. It examines single-point failures to determine their extent and effects. FMECA helps developers see if a design must change to improve its reliability, safety, or operation. FMECA can also estimate the overall reliability of a system from the base component reliabilities [8].

To develop an FMECA, you determine the failure effects from various levels of detail: functions, components, assemblies, modules, or subsystems. The idea is to look at how the failures appear at interfaces to other subsystems and then see how they propagate and affect those subsystems. Your effort in preparing the FMECA is both qualitative and quantitative in nature and you use a tabular, bottom-up approach. You identify failures occurring within each component (or assemblies or modules or subsystems) and then determine its effect on the component and then on the connected components. Finally, you assign a criticality to each effect.

FMECA is a manual form of analysis that requires knowledge and expertise of the developer. FMECA becomes significantly more complex and lengthy to prepare as you descend into levels of the system, moving from subsystems to components. A good compromise is to look at the failure effects of modules or subsystems with other modules or subsystems; usually this means that you need to look at each connection and conductor between the subsystems and consider their failure effects, which is considerably less effort than examining the failure effects of each component within the module.

Once you have prepared a complete FMECA, you can answer these questions:

- How can each component (or assemblies or modules or subsystems) fail?
- What are the effects of each failure?
- What are the consequences of each failure?
- (If reliability data are available:)
  - How frequently can it fail?
  - How does it affect system reliability?

Knowing the answers to these questions can give you insight into how you might redesign or reconfigure the system to mitigate problems. FMECA helps a developer address the priority risks during design. FMECA also helps enforce a measure of discipline and rigor during development.

FMECA has some strong points. Engineers can easily learn it. FMECA gives rigor and focuses analyses. It can be relatively inexpensive (in terms of effort) to prepare, but that does depend on the level of detail that you examine. FMECA can predict reliability. Commercial software is available to aid developers in preparing FMECA [8].

FMECA has several weak points. It only considers single-point failures; it does not consider combinations of failures. It does not identify hazards unrelated to failure or interactions that lead to failure. It provides little or no examination of human error, external influences or interfaces, or software or operations. Finally, and most importantly, FMECA requires knowledge of the system and extensive engineering expertise with the product [8].

---

### Example: Simple FMECA for spacecraft imaging system

Consider a spacecraft imaging system, as diagrammed in Figure 7.8, that has several imagers, a processing section, a power supply, and telemetry support. The system also has a ground support element, a basic desktop computer, in mission control. Figure 7.9 illustrates a simple

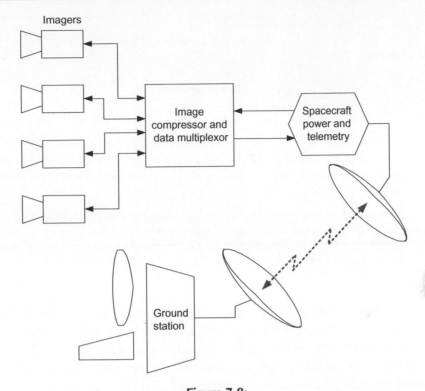

**Figure 7.8:**
A spacecraft imaging system with both spacecraft subsystems and a ground-based subsystem.
© 2013 by Kim R. Fowler. All rights reserved. Used with permission.

| Component | Failure | Immediate effect | Systemic effect | Criticality | Failure rate (failures/year) | Prob. detect. | RPN |
|---|---|---|---|---|---|---|---|
| | Fails open | No video from that camera | No data received from that camera, no influence on other cameras or systems | 3 | 0.0876 | 0.99999 | 0.263 |
| | Fails shorted | No video from that camera | No data received from that camera, no influence on other cameras or systems | 3 | 0.0876 | 0.999 | 0.263 |
| Camera video line to compressor | Occasional dropouts, loose signal pin | Video stops from that camera occasionally | Occasional data drops from that camera, no influence on other cameras or systems | 2 | 0.46105 | 0.85 | 1.085 |
| | Frequent dropouts, camera heating | Video stops from that camera frequently | Frequent data drops from that camera, no influence on other cameras or systems | 2 | 0.64899 | 0.75 | 1.730 |
| | Noise on signal | Video signal from that camera is noisy | Noisy data from that camera, no influence on other cameras or systems | 1 | 0.03982 | 0.998 | 0.040 |

**Criticality:**

| | |
|---|---|
| None | 0 |
| Inconsequential or very low | 1 |
| Low to moderate | 2 |
| Serious | 3 |
| Severe | 4 |
| Catastrophic | 5 |

**Figure 7.9:**
A simple example of one page from a FMECA for the spacecraft imaging system. © 2013 by
Kim R. Fowler. All rights reserved. Used with permission.

example of one page of a much more extensive FMECA. This FMECA contains probabilistic failure rates for each type of failure and the risk priority number (RPN). The RPN is calculated as follows:

$$RPN = \frac{(\text{Criticality})(\text{Failure rate})}{\text{Probability of detection}}$$

Once an engineer understands the imaging system and the types of failures, which could take several days at a minimum, this sort of FMECA could take 5 to 10 minutes per line to analyze and develop the numbers for calculation. This simple little exercise alone took me about 30 minutes.

Extrapolating this effort, at this level of detail for this imaging system, you could take up to a week of full-time effort to analyze this imaging system.

### Fault tree analysis

FTA is a technique that arose in the aerospace industry in the 1960s. It examines high-level faults and traces the sequence of events to lower-level causes. FTA helps developers see if a design must change to improve its safety or operation. FTA can also educate designers to potential problems and it provides proactive, foundational documentation should a fault occur and someone must perform an RCA [8, pp. 183–221].

An FTA helps a designer to determine the sources, or root causes, of potential faults. FTA is a graphical, top-down approach that uses Boolean algebra, logic, and probability. FTA is both qualitative and quantitative in nature; it can handle multiple failures and can support probabilistic risk assessment.

You identify all potential and possible causal paths from the high-level fault to lower-level components (or assemblies or modules or subsystems). If the lower-level fault or failure adequately explains at least one root cause, then terminate that pathway. Otherwise, iterate to lower-levels of components (or assemblies or modules or subsystems) in that pathway until you find all the root causes. Repeat the same iterative effort for all possible paths. Each level and pathway has its own probability of occurrence; this provides the mechanism for assessing the probabilistic risk.

Like FMECA, FTA is a manual form of analysis that requires knowledge and expertise of the developer. FTA becomes significantly more complex and lengthy to prepare as you descend into more detailed levels of the system, moving from subsystems to components. A good compromise is to look at the failure effects of modules or subsystems with other modules or subsystems, which is considerably less effort than examining the failure effects of each component within the module.

Once you have prepared a complete FTA, you can answer these questions:

• What are the root causes of failures?
• What are the combinations and probabilities of causal factors in undesired events?
• What are the mechanisms and fault paths of undesired events?

Knowing the answers to these questions can give you insight into how you might redesign or reconfigure the system to mitigate problems. FTA helps a developer address potential root causes for failure during design. Like FMECA, FTA also helps enforce a measure of discipline and rigor during development.

FTA has some strong points. It is structured and rigorous. Engineers can easily learn and understand its visual format. It combines hardware, software, environment influences, and human operations. It can be relatively inexpensive (in terms of effort) to prepare, but that does depend on the level of detail that you examine. FTA can assess risk probability. Commercial software is available to aid developers in preparing FTA [8, pp. 183−221].

FTA has several weak points. Depending on the level of detail, preparing an FTA can be very time-consuming. It does not identify hazards unrelated to failure. It provides little or no examination software, timing, scheduling, intermittent faults, or injected noise. Finally, and most importantly, FTA requires knowledge of the system and extensive engineering expertise with the product [8, pp. 183−221].

### Example: Simple FTA for spacecraft imaging system

Consider a spacecraft imaging system, as diagrammed in Figure 7.8, that has several imagers, a processing section, a power supply, and telemetry support. The system also has a ground support element, a basic desktop computer, in mission control.

Figure 7.10A−C briefly explains the symbols used in FTA. Figure 7.10D illustrates a simple example from one page of a much more extensive FTA. GSE stands for Ground Support Equipment and is the computer that records and displays images transmitted from the spacecraft.

Once an engineer understands the imaging system and the types of failures, which could take several days at a minimum, this sort of FTA could take 30 to 60 minutes per column to analyze and develop the numbers and calculations. This simple little exercise alone took me about an hour.

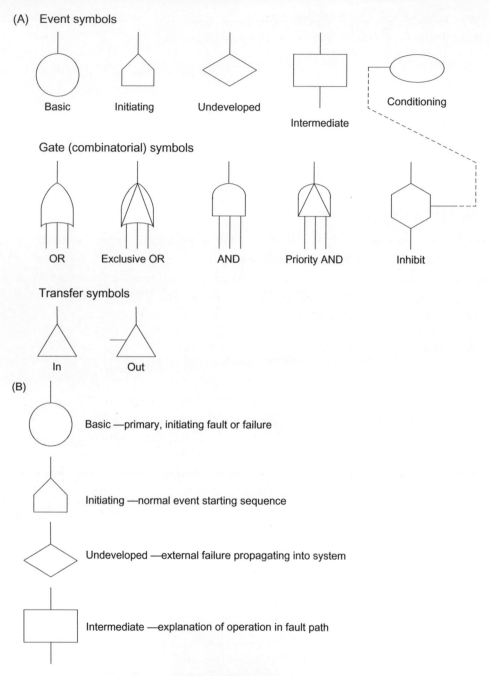

**Figure 7.10:**
(A) Symbols used in FTA. (B) Explanation for the termination symbols used in FTA. (C)
Explanation of the symbols and probabilities used in FTA. (D) A simple example of one page
from an FTA for the spacecraft imaging system. © 2013 by Kim R. Fowler. All rights reserved.
*Used with permission.*

(C)

Out

OR gate — output occurs if one or more inputs occur

$P_{out} = P_A + P_B + P_C - (P_A P_B + P_A P_C + P_B P_C) + (P_{ABC})$
*[Note: that the even cross-terms in probability subtract while the odd cross-terms in probability sum]*

A B C

Out

AND gate — output occurs only if all inputs occur together

$P_{out} = P_A \cdot P_B \cdot P_C$

A B C

Out

Exclusive OR gate —— output occurs *ONLY* if one input occurs but *NOT* any others. May need to attach a condition statement to the gate to explain the rationale.

$P_{out} = P_A + P_B + P_C - 2(P_A P_B + P_A P_C + P_B P_C) + 4(P_{ABC})$
*[Note: that the even cross-terms in probability subtract while the odd cross-terms in probability sum and that cross-terms have even coefficients]*

A B C

Out

Priority AND gate — output occurs *ONLY* if all inputs occur together but *ONLY* in sequence, first A, then B, then C. May need to attach a condition statement to the gate to explain the rationale.

$P_{out} = P_A \cdot P_B \cdot P_C / N!$ *[PLEASE NOTE: this expression is only true for equal rates of faults, i.e., $\lambda_A = \lambda_B = \lambda_C$ and N = number of inputs to the gate, 3 in this example; otherwise the probability derivation becomes more complex]*

A B C

(D)

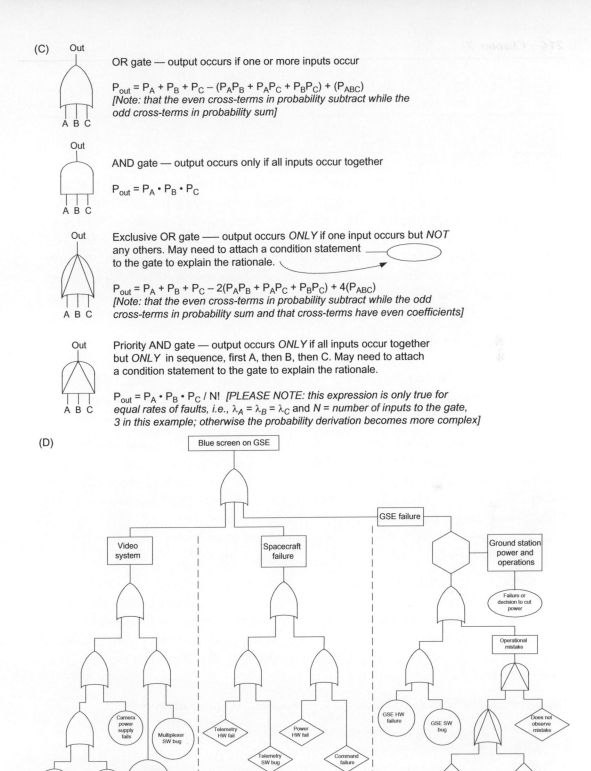

**Figure 7.10:**
(Continued)

Extrapolating this effort, at this level of detail for this imaging system, you might take more than a week of full-time effort to analyze the entire imaging system.

*Event tree analysis*

ETA is a technique that arose in the nuclear power industry in the mid-1970s. It examines any event, successful, fault, or accidental, and branches to all possible outcomes. ETA can provide varieties of consequences for many different operations and conditions. ETA helps developers see if a design must change to improve its safety or operation. Like FTA, ETA can also educate designers to potential problems and it provides proactive, foundational documentation should a fault occur and someone must perform an RCA. ETA is better, however, than FTA at elucidating intermittent and unusual modes and conditions [8, pp. 223–233].

An ETA helps a designer to determine the sources, or root causes, of potential faults and problems. ETA is a graphical, "sideways-in" approach that can assign probability; it can evaluate all possible outcomes from any initiating event and not just faults, failures, or undesirable events. ETA is both qualitative and quantitative in nature; it can handle multiple failures, degraded operation, and can support probabilistic risk assessment.

You identify all potential and possible events that include both normal and abnormal situations; then you identify any subsequent conditioning (or pivotal) events. After establishing the events, you map all possible outcomes through each sequence of possible events. This gives you a branching tree structure that lies on its side. You can conduct this "sideways-in" analysis at different levels of abstraction for various levels of subsystems. Each branch has its own probability of occurrence; this provides the mechanism for assessing the probabilistic risk.

Like FMECA and FTA, ETA is a manual form of analysis that requires knowledge and expertise of the developer. ETA becomes significantly more complex and lengthy to prepare as you descend into detailed levels of the system, moving from subsystems to components and into other possible events. A good compromise is to look at the events at the level of modules or subsystems, which is considerably less effort than examining the failure effects of each component within the module.

Once you have prepared a complete ETA, you can answer these questions:

- What are all the possible outcomes of each initiating event?
- What are the systemic consequences of any event?
- What are the combinations of causal factors and probabilities in any event?
- What are the mechanisms of any event?

ETA has some strong points. It is structured and rigorous. Engineers can easily learn and understand its visual format. It combines hardware, software, environment influences, and human operations; ETA can analyze any event—it is not restricted to failures or faults as are FMECA and FTA. ETA can analyze multiple events at the same time. ETA can also assess risk probability, assuming those probabilities are available. Commercial software is available to aid developers in preparing ETA. Assuming engineering expertise in the product, ETA can distinguish partial degradation or intermittent faults (contrary to Ericson's claim on p. 233), but it does require domain expertise to do so [8, pp. 223–233].

ETA has several weak points. Preparing an ETA can be very time-consuming. It provides little or no examination software, timing, or scheduling. Finally, and most importantly, ETA requires knowledge of the system and extensive engineering expertise with the product [8, pp. 223–233].

### Example: Simple ETA for spacecraft imaging system

Consider a spacecraft imaging system, as diagrammed in Figure 7.8, that has several imagers, a processing section, a power supply, and telemetry support. The system also has a ground support element, a basic desktop computer, in mission control.

Figure 7.11A shows a template for ETA that includes the format for calculating probabilities assessment. Figure 7.11B illustrates a simple example of one page of a much more extensive ETA.

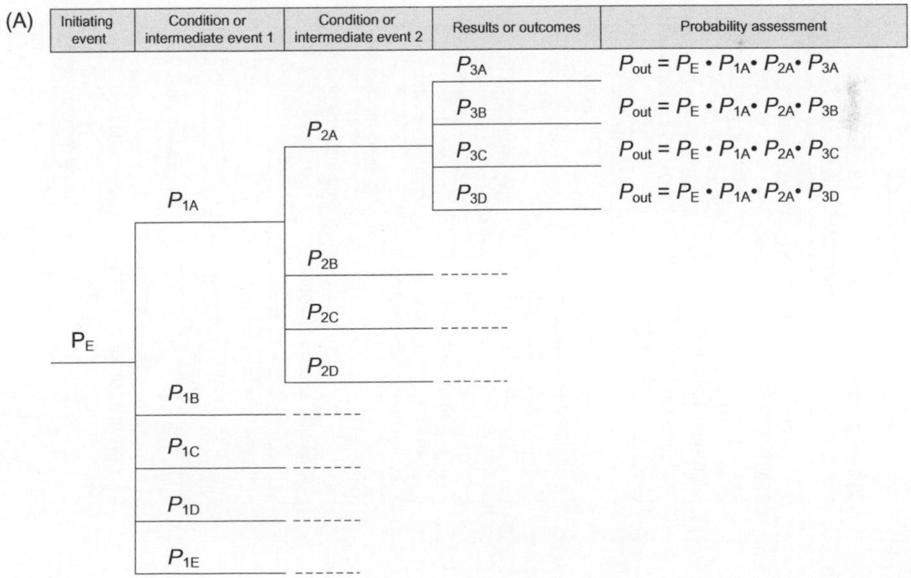

**Figure 7.11:**
(A) A template for ETA. (B) A simple example of one page from an ETA for the spacecraft imaging system. © *2013 by Kim R. Fowler. All rights reserved. Used with permission.*

(B)

| Initiating event | Pivotal event or intermediate condition | Outcomes or results | Hazard, consequence, or comment |
|---|---|---|---|
| Select camera from GSE | Cameras, multiplexer, GSE, and operator function correctly | System functions as designed, desired camera selected | |
| | Camera fails, other systems function correctly | No video from selected camera, GSE indicates problem | (1) No data from camera, switch to another |
| | | No video from selected camera, GSE does not indicate problem | (2) No data from camera, operator might recognize problem and switch cameras |
| | Camera fails intermittently, other systems function correctly | No video from selected camera, compressor cannot synchronize with intermittent video from camera, GSE indicates a problem | (1) |
| | | Intermittent video from selected camera, compressor synchronizes with camera, GSE indicates a problem | (1) |
| | | Intermittent video from selected camera, compressor synchronizes with camera, GSE does not indicate a problem | (2) |
| | Multiplexer fails, other systems function correctly | No video from any camera, GSE indicates problem | (1) |
| | | No video from any camera, GSE does not indicate problem | (2) |
| | Multiplexer fails intermittently, other systems function correctly | No video from the selected camera, GSE indicates a problem | (1) |
| | | Intermittent video from camera, GSE indicates a problem | (1) |
| | | No video from selected camera, GSE does not indicate a problem | (2) |
| | | Intermittent video from camera, GSE does not indicate a problem | (2) |
| | GSE fails, other systems function correctly | No video from any camera, no communication with spacecraft | No data from system. Diagnostics in GSE should indicate problem. Switch out GSE |
| | GSE fails intermittently, other systems function correctly | Intermittent video from selected camera, GSE indicates a problem | (1) |
| | | Intermittent video from camera, GSE does not indicate a problem | (2) |
| | Operator selects wrong camera, other systems function correctly | Video data streams from wrong camera, operator corrects and selects desired camera | (1) This is an operator concern. The entire system is operating as designed. Finding and fixing this problem relies on an expert operator, which is typical in space missions |
| | | Video data streams from wrong camera, operator does not notice | (2) |

**Figure 7.11:**
(Continued)

Once an engineer understands the imaging system and the types of failures, which could take several days at a minimum, this sort of FTA could take 5 to 10 minutes per branch path to analyze and develop the numbers and calculations. This simple little exercise alone took me about an hour.

Extrapolating this effort, at this level of detail for this imaging system, you might take more than a week of full-time effort to analyze the entire imaging system.

*Dependability*

Dependability quantifies several dimensions to predict how well and how long a system will operate. Dependability focuses on material strengths and failure modes; dependability reveals little about design flaws, unless a specific calculation reveals that the design violates a physical principle.

Dependability considers both wear-out and random latent failures in materials and components. Historically, the study of reliability has explained and modeled mechanisms of wear-out and random latent failures reasonably well. Models include mechanical systems and electronic semiconductors.

The calculations for reliability, time-to-repair, and availability have limitations. These calculations are standard formulas that serve very well in comparing different design approaches. These calculations, however, are seldom accurate for absolute predictions of reliability because they apply to steady-state conditions and there are many unknowns, such as stresses and susceptibility factors. Dependability, and specifically reliability, cannot predict failure from situations where environmental extremes, operating stresses exceed design limits, and similar types of abuse inflict the system.

See Appendix A for a basic description of the mathematics of dependability.

*System theoretic process analysis*

Nancy Leveson and colleagues developed STPA over the past 10 years after reviewing a variety of industrial, avionic, and medical accidents that have occurred since 1980. STPA is fundamentally different than more traditional forms of analysis, such as FMECA and FTA. It identifies potential hazards, accidents, losses, model variables, environmental variables, and process control variables. STPA also alerts and guides you in identifying the process model for the controlled system or subsystem. Then from these identified quantities, variables, and the system configuration, STPA allows you to infer behaviors and identify potential causes of hazardous control actions. STPA can point out where developers might improve safety or operation by revealing potential problems and providing proactive, foundational documentation before a fault occurs [9].

STPA provides a template method to identify faults based on control loops and explicit process models. (A process model is the explanation of the controlled process that the system controller relies on to generate appropriate actions. A process model resides within the controller either as an explicit mathematical description or as an implicit behavior within the software code.) STPA's primary strength appears to be coverage of many potential problems within many systems. It is a generalized control diagram with guide words and a database of common actions and problems; with these attributes and resources, STPA can describe a great many different systems—potentially the vast majority of interesting systems.

STPA helps a designer to determine the sources, or root causes, of potential faults with a graphical, control theory approach. Leveson claims that STPA can identify all potential causes of high-level fault from lower-level modules or components. STPA is qualitative in nature but it can analyze multiple hazards and failures [9].

Like FMECA, FTA, and ETA, STPA is a manual analysis. Currently the form, functions, capability, and automation of STPA are still developing through intense research by a number of different researchers [10]. STPA requires domain knowledge and expertise from the developer but possibly less so than FMECA or FTA when the analyses are performed at higher levels of the system (e.g., an electronic circuit analyzed in an FMECA requires an electrical engineer to perform the detailed analysis; STPA generally does not get to that level of detail unless an electrical engineer dives into component-level design).

Once you have prepared a complete STPA, you can answer these questions:

- What are the potential causes of hazards?
- What are the potential causes of accidents or faults or failures?
- How do hazards trace to accidents?
- What are the combinations of causal factors in undesired events?
- What are the mechanisms and fault paths of undesired events?

Knowing the answers to these questions can give you insight into how you might redesign or reconfigure the system to mitigate problems. Like FMECA, FTA, and ETA, STPA also helps enforce a measure of discipline and rigor during development.

STPA has a number of strong points. Engineers can quickly learn how to prepare STPA. It combines all aspects of hardware, electronics, software, environment influences, and human operations. STPA is a more comprehensive analysis for various hazards than FMECA or FTA or ETA because it can capture and evaluate software, timing, scheduling, intermittent operations, injected noise, timing problems, environmental inputs that contribute to hazardous operations, and multiple interactions of apparently correctly designed subsystems that have hazardous results. It is not tied to single-point component failures like FMECA. In particular, STPA can identify hazards unrelated to failure because it embraces the discovery of hazards based on control loops, which may cause faults that are not caused by failures. Generally, STPA does not greatly expand in complexity or length to prepare, as much as FMECA or FTA or ETA does, when you descend into lower or more detailed levels of the system, moving toward subsystems, until you reach component or module levels; then STPA can require significantly more effort.

STPA has several weak points. As of the Fall 2013, it is still under active research development and automated tools are not available. STPA is not yet quantitative and does not yet conform to formal mathematical methods or proofs of operation. Manual preparation of STPA can be a bit of a learning curve (but I expect that to become much easier in the near future as more research, tools, and familiarity develop within the research community). It does become a bit cumbersome, requiring significant domain expertise, as the analysis focuses down to the detailed component level.

Another important, current concern is that it is unclear if there are hazards unassociated with control loops. It is possible that the process of defining accidents and hazards covers some of these hazards and makes them observable; it may also put these accidents and hazards into newly defined control loops.

Finally, STPA does not strongly connect stakeholders to the development and use of a system. Actually none of the techniques explicitly connect stakeholders to the development and use of a system.

---

### Case study: Simple STPA for infant isolette

Figure 7.12 shows a simple STPA diagram for an infant isolette. Several feedback loops can reveal the operation of the isolette—the current flow in the tell-tail of the heater element and the temperature of the air. The tell-tail signal from the heater element indicates (but does not prove) that the heater is functioning. The temperature sensor gives a more direct indication that the heater is warming the air. Many problems in a particular sensor could be covered by the operations of the other sensors. Inappropriate actuation through the electronic drive module will drive the heater incorrectly, which the feedback paths will help elucidate. An incorrect process model for the heater will cause the controller to generate inappropriate commands or misinterpret the feedback.

The general step-by-step method for STPA follows the procedure in an EPRI paper by Torok and Geddes: [11]

1. Identify the system boundary
2. Identify accidents or losses
3. Identify system-level hazards
4. Draw the control structure
5. Create the process model—list the process model variables
6. Identify hazardous control actions

a. Identify control actions
b. Postulate control action behaviors—control action is provided, not provided, provided too soon, provided too late, or stopped too soon
c. Determine if control action behaviors are hazardous in various contexts expressed by the process model variables
7. Identify the potential causes of the hazardous control actions
8. Remove or mitigate hazards.

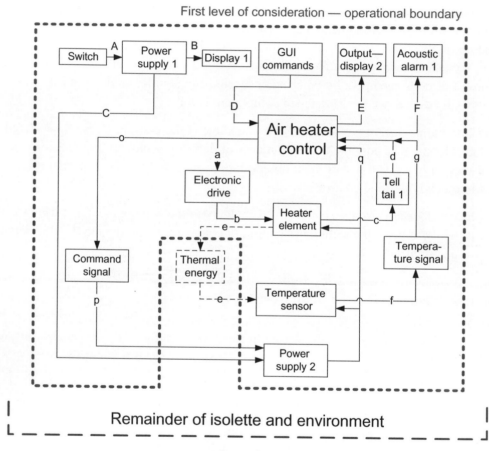

**Figure 7.12:**
A simple example diagram used in an STPA for control of an infant isolette. © *2013 by Kim R. Fowler. All rights reserved. Used with permission.*

This particular STPA does not look at all issues for an isolette. It provides some initial analysis of the operations around the isolette. You can, of course, shift the focus of the STPA by moving it to a higher system level to look at other operations or by moving it to a lower level to "drill down" into specific details of a subsystem. Figure 7.13 illustrates another step in expanding the STPA to include more details within the analysis.

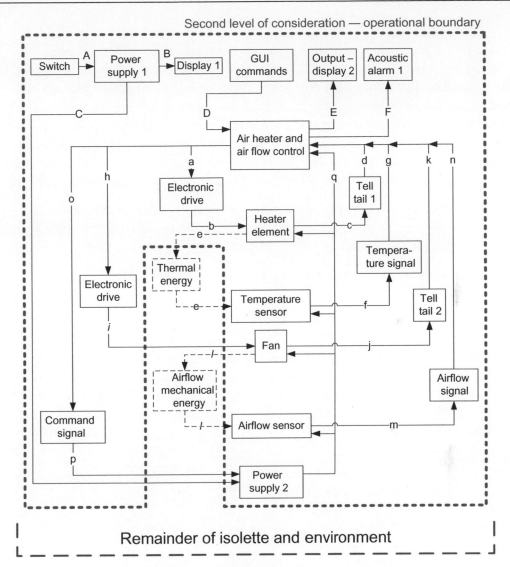

**Figure 7.13:**
An expanded diagram following Figure 7.12 used in an STPA for control of an infant isolette.
© 2013 by Kim R. Fowler. All rights reserved. Used with permission.

Note that the air of the isolette is not included in the analysis. For these particular analyses, it is not directly controlled; the heater is directly controlled, the air temperature is a secondary result. If the isolette cover lid is open, then the heater would try to heat the entire room air. Once a follow-on analysis includes a lid latch and tell-tail for the latch, then the air within the isolette can be included in the analysis.

Also note that Figure 7.13 includes a fan. The heater system might need a fan to mix the air within the isolette to distribute the heat for more even temperature distribution.

*Safety cases*

The safety case is an inductive argument that claims a facility is as safe as can be reasonably expected. The safety case contains all the necessary information to support the safety arguments; that information contains volumes of evidence to show that risk and hazard have been reduced to an acceptable level. Safety cases are usually restricted to large, complex industrial systems such as nuclear power plants, petrochemical plants, offshore oil or gas platforms, pipelines, mining, and railways. One basic definition of a safety case is, "A documented body of evidence that provides a demonstrable and valid argument that a system is adequately safe for a given application and environment over its lifetime" [12].

The operator and regulator determine the nature of the safety case for their particular situation; it is usually not governed entirely by prescriptive regulations. The basic principles for a safety case are the following:

- Those creating the risks are responsible for controlling those risks.
- Safe operation comes from setting and achieving goals NOT by following prescriptive rules.
- Reduce risks below a threshold of acceptability.

Safety cases are prepared for new installation or major changes and for ongoing services in these industries. Concerns for safety cases are that they require time-consuming amount of paperwork and the safety case itself does not ensure or improve safety. Its implementation requires ongoing commitment and participation from management, key personnel, and all employees [12].

A safety case has the following features:

- Duty-holder responsibility
- Participation and commitment
- Information availability
- Nonprescriptive and performance-based
- Risk management system
- Management systems
- Living document
- Auditor/assessor responsibility.

A safety case is organized along these lines:

I—Executive Summary
II—Introduction
III—Policies, Objectives, Regulations, and Standards
IV—Facility Description
V—Safety Management System

VI—Formal Safety Assessment

VII—Audit and Review

The European Organization for the Safety of Air Navigation outlines a slightly different organization for the safety case [13]. It also outlines the types of evidence used in supporting the safety argument:

- Analysis
- Design
- Simulation
- Test
- Previous usage
- Standard compliance
- Rigor of evidence appropriate to the associated risk
- Relevance so that the argument refers to the correct configuration of system.

*Comparison between analyses*

People have used these five different types of techniques, FMECA, FTA, ETA, STPA, and safety cases, to examine the effects of failures, faults, and hazards in safety-critical and mission-critical systems. Table 7.3 compares these five techniques on a line-item by line-item basis.

STPA is the most comprehensive form of analysis and covers nearly all of the line-items; the only item it does not cover is probabilistic assessment. FMECA is the least capable (but the oldest and best known) technique; it is limited to single-point failures of components and their subsequent effects. The safety case also focuses on large, industrial systems and plants and is of limited utility for embedded systems. Each technique needs a fair amount of expertise and domain knowledge to be useful.

STPA is similar to Chaos theory in that for each level of abstraction it has an identical form of detail. The highest level can be a system of system view. You can then descend into the next level of detail and examine a system using the standard control loop diagram. Then you can descend into the next level of detail and examine a subsystem using the standard control loop diagram. You can continue this drilling down effort until you arrive at the basic component level.

To cover for STPA's one lack of probabilistic assessment, you can use one of the other three techniques, FMECA, FTA, and ETA, to generate the appropriate numbers.

The basis for the comparison among techniques has several components:

1. A recent thesis by Vincent Balgos compares FMECA and STPA for analyzing a particular accident. Note that CAST, which is the acronym for causal analysis based on

**Table 7.3: Comparison of analysis techniques—FMECA, FTA, ETA, STPA, and safety cases.**

|  | = (blank) does not address |
|---|---|
|  | − − = very poor to completely inadequate |
|  | − = poor |
|  | 0 = neutral or OK |
|  | + = good |
|  | + + = very good to excellent |

| Issue | FMECA | FTA | ETA | Safety Case | STPA |
|---|---|---|---|---|---|
| **Addresses potential problems leading to hazards** | | | | | |
| Architecture |  | + | + | + + | + + |
| Electronics | + | + | + | 0 | + + |
| Mechanics | + | + | + | 0 | + + |
| Materials | + | + | + | 0 | + + |
| Software |  | − | − | 0 | + + |
| Production |  |  | + | + + | + + |
| Maintenance |  | − − | + | + + | + + |
| Probabilistic assessment | + | + + | + |  |  |
| Common cause problems |  | 0 | 0 | + | + + |
| Timing |  | − − | − − |  | + + |
| Design flaws |  |  | 0 |  | + + |
| **Identifies failures and faults** | | | | | |
| Component | + + | 0 | 0 |  | + + |
| Stuck at 1 or 0 | + + | + + | + + |  | + + |
| Intermittent faults | + | + | + |  | + + |
| Operational drift (e.g., op amp circuit sensitivities) | 0 | + | + |  | + + |
| Degraded faults | − | − | − |  | + + |
| Overheating | 0 | + | + |  | + + |
| Mechanical jams and binding | − | + | + |  | + + |
| **Identifies interactions: (requires significant technical and domain expertise)** | | | | | |
| Operator |  | 0 | 0 | 0 | + + |
| Between correctly operating subsystems |  | − | 0 |  | + + |
| Between incorrectly operating subsystems |  | − | 0 |  | + + |
| **Identifies assaults: (requires significant technical and domain expertise)** | | | | | |
| Heat and thermal cycling | 0 | 0 | 0 |  | + |
| Corrosion | 0 | 0 | 0 |  | + |
| Dust | 0 | 0 | 0 |  | + |
| Condensation | 0 | 0 | 0 |  | + |
| Salt spray | 0 | 0 | 0 |  | + |
| Light, UV degradation | 0 | 0 | 0 |  | + |
| Pressure | 0 | 0 | 0 |  | + |
| Fatigue and stress fracturing | 0 | 0 | 0 |  | + |
| Collision | 0 | 0 | 0 |  | + |
| Fungus | 0 | 0 | 0 |  | + |
| Animal invasion | 0 | 0 | 0 |  | + |
| EMI/EMC | 0 | 0 | 0 |  | + |

*(Continued)*

Table 7.3: (Continued)

| Issue | FMECA | FTA | ETA | Safety Case | STPA |
|---|---|---|---|---|---|
| **Power problems** *(requires significant technical and domain expertise)* | | | | | |
| High-voltage surges | 0 | 0 | 0 | 0 | + |
| Low-voltage sags | 0 | 0 | 0 | 0 | + |
|   Noise and harmonic ripple | 0 | 0 | 0 | 0 | + |
| Transients and spikes | 0 | 0 | 0 | 0 | + |
| Dropouts | 0 | 0 | 0 | 0 | + |

STAMP, and STAMP are very similar analyses to STPA. In comparing FMECA and STPA (CAST) Balgos found, "... the 175 hazards found [by CAST] were only for the six identified control loop(s) of almost 20 possible control loops. The CAST analysis produced significantly more system level hazards than the 70 found via the industry standard FMECA methodology in less than half of the identified loops in the control structure. The sheer voluminous findings indicate that the systems thinking model in the CAST methodology was more effective in discovering hazards. Furthermore, in addition to the number of hazards, the type hazards identified were significant. This CAST analysis was able to identify single component factors (inability to detect foreign material on sensor) that were a critical factor in the case accident. Moreover, this non-reductionist approach allowed the analyst to consider upstream failures, and how it affected the downstream dynamics with the potential to migrate the system to an unsafe state... able to identify additional hazards that did not include a component failure, a limitation of the FMECA practice. The control loop template in Figure 19 provided a guideline for easily recognizing system level hazards and conflicting constraints amongst the various control loops. Viewing safety as a control problem help elucidated many system level hazards. Therefore, the CAST application to risk analysis can provide a more rigid evaluation for a variety hazards that occur with and without failures" [14].

Balgos concludes, "... CAST analysis... [uses a]... systems approach was superior to the industry standard FMECA practice in identifying hazards. It was able to detect significant contributors to the case accident in form of failures (foreign material on the sensor), and non-failures (a conflict in controlling actions)... The CAST approach was able to distinguish more than twice the number of system level hazards with considerable less time and resources. Multiple failure hazards and hazards that occurred without component failures due to conflicting system control actions were confirmed. The CAST methodology was able to increase not only the quantity of hazards found, but identify complex and nonlinear hazards. The current FMECA is incapable of producing these results based on its inherent design and structure" [14, p. 85].

2. John Thomas, in his recent dissertation on automating STPA for requirements generation, points out the following, "Systems Theoretic Process Analysis (STPA) is a

powerful new hazard analysis method designed to go beyond traditional safety techniques—such as Fault Tree Analysis (FTA)—that overlook important causes of accidents like flawed requirements, dysfunctional component interactions, and software errors. Although traditional techniques have been effective at analyzing and reducing accidents caused by component failures, modern complex systems have introduced new problems that can be much more difficult to anticipate, analyze, and prevent. In addition, a new class of accidents, component interaction accidents, has become increasingly prevalent in today's complex systems and can occur even when systems operate exactly as designed and without any component failures" [10, p. 5].

Thomas goes on to say, ". . . traditional hazard analysis techniques assume accidents are caused by component failures or faults and oversimplify the role of humans. Attempts have been made to extend these traditional hazard analysis techniques to include software and cognitively complex human errors, but the underlying assumptions do not match the fundamental nature of systems we are building today. For example, most software-related accidents can be traced to incomplete or flawed software requirements, however current hazard analysis methods like Fault Tree Analysis (FTA) analyze component failures and easily overlook unsafe requirements. In addition, new technology is changing the role of humans in systems from followers of procedures to supervisors of automation and high-level decision makers. New models of accident causation and hazard analysis techniques are needed to address these issues" [10, p. 17]. This supports the need to use and to automate STPA.

3. STPA is a template that represents a large proportion of many interesting systems. It contains the basic subsystem blocks, control variable flows, and sensed variable flows of these systems. It also recognizes and includes the process model for the controlled subsystem/process within the controller.

4. STPA has guide words that drive the examination of the system and its interactions. While generic, they are quite powerful for covering many concerns within control loops and controlled processes.

Considering these components, the comparisons in Table 7.3 are not entirely subjective. There is mounting evidence for STPA's coverage of many potential hazards in many different types of systems.

## Further analyses for specific applications

The following analyses are particular to specific industries and markets, such as aerospace or process control. Where safety is paramount, you may need to perform these analyses.

### Sneak circuit analysis

"A sneak circuit is a latent path or condition in an electrical system that inhibits a desired condition or initiates an unintended or unwanted action. This condition is not caused by component failures but has been inadvertently designed into the electrical system to occur as normal operation. Sneak circuits often exist because subsystem designers lack the overall system visibility required to electrically interface all subsystems properly" [8, p. 291].

Sneak circuit analysis identifies pathways in electrical, hydraulic, pneumatic circuits or in command and control functions that cause inappropriate or undesired operation or inhibit operations. It has been used extensively in military and aerospace applications. Sneak circuit analysis is difficult to learn, requires extensive experience, and only applies to a system once significant design effort has been completed. Ericson details the operation of sneak circuit analysis [8, pp. 291–306].

Sneak circuit analysis requires a list of clues that staff develop from experience. Companies have proprietary lists of sneak circuit clues, which they consider intellectual property. Peer review of logic and data flows often catch the same problems as sneak circuit analysis [8, pp. 291–306].

I was involved in a military project years ago where a consulting firm performed a sneak circuit analysis of my company's system. The system had four different types of circuit boards comprising a total of 11 circuit boards and several hundred electronic components. This consultation was considered a subset of the system verification. The analysis took them about three weeks to complete. Commercial software operating on CAD schematics may be much faster than the manual analysis I experienced.

### Petri net analysis

"Petri Net analysis (PNA) is an analysis technique for identifying hazards dealing with timing, state transitions, sequencing, and repair. PNA consists of drawing graphical Petri Net (PN) diagrams and analyzing these diagrams to locate and understand design problems. Models of system performance, dependability, and reliability can be developed using PN models. PNA is very useful for analyzing properties such as reachability, recoverability, deadlock, and fault tolerance. The biggest advantage of Petri nets, however, is that they can link hardware, software, and human elements in the system" [8, p. 307].

You can perform PNA early in design. PNA is limited, however, because it only manages and analyzes timing and transitions in simple circuits or networks with few elements. PNA does not provide root causes. It is also difficult to learn [8, pp. 307–316].

### Barrier analysis

Barrier analysis identifies hazards from energy sources, such as batteries. It evaluates the possibility of energy flow from energy sources to people or other systems by examining barriers designed to prevent that flow. Barrier analysis supports portions of hazard analyses but is not sufficient to account for all circumstances [8, pp. 335–351].

While barrier analysis can be learned fairly quickly, it does not identify all energy sources automatically. Identifying energy sources requires the expertise of the designer or analyst.

### Bent pin analysis

Bent pin analysis identifies hazards from bent pins within cables and connectors. This is a very limited type of analysis that looks at open, short, or bridged circuits between adjacent pins. Bent pin analysis has found use within the military and aerospace markets. Bent pin analysis supports portions of hazard analyses but is not sufficient to account for all circumstances [8, pp. 353–364].

While bent pin analysis can be learned fairly quickly, it does not automatically identify bent pins or their effects. Identifying bent pins or their effects requires the expertise of the designer or analyst.

### Markov analysis

Markov analysis performs some of the same functions that PNA does. Markov analysis models state transitions and calculates the probability of reaching system states, dependability, and safety [8, pp. 317–333]. A Markov process is a stochastic process that can predict its future based on knowing its present state and needs no memory of its history; specifically, a Markov process is conditional on the present state of the system, while its future and past states are independent [15]. A Markov chain is a system that transitions from one state to another state that is in a set of finite or countable number of possible states; a Markov chain is stochastic and memoryless—the next state of the system depends only on the current state and not on the sequence of events that preceded it [16].

While you can apply Markov analysis early in design to identify issues, it does not identify hazards or root causes. Performing a Markov analysis requires the mathematical expertise of the designer or analyst. Other techniques presented in this chapter tend to be more useful for system analysis than Markov analysis.

## Root cause analysis (RCA)

"Root cause analysis is a structured investigation that aims to identify the cause of a problem and the actions necessary to eliminate it" [17]. RCA is a systematic method for reactive analysis to debug and to solve problems and then generate fixes or design changes to prevent the problem from occurring again. RCA is a toolbox of techniques used by a small team of investigators, which have been organized for a short duration to solve a particular problem. Figure 7.14 illustrates the general process of RCA.

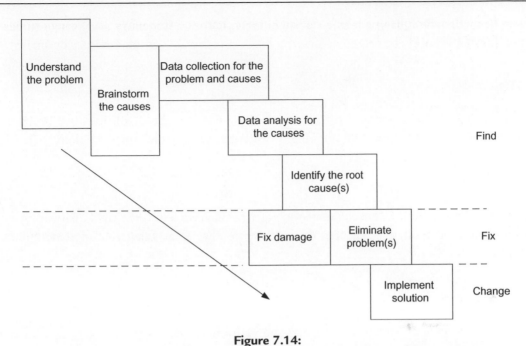

**Figure 7.14:**
Schematic of the flow of activities within RCA. © 2010–2013 by Kim R. Fowler. All rights reserved. Used with permission.

"RCA is often considered to be an iterative process. . . . RCA is typically used as a reactive method of identifying event(s) causes, revealing problems and solving them. Analysis is done *after* an event has occurred. Insights in RCA may make it useful as a preemptive method. In that event, RCA can be used to *forecast* or predict probable events even *before* they occur. . . Root cause analysis is not a single, sharply defined methodology; there are many different tools, processes, and philosophies for performing RCA. However, several very-broadly defined approaches or 'schools' can be identified by their basic approach or field of origin: safety-based, production-based, process-based, failure-based, and systems-based" [18].

## RCA find

The first step is to understand the problem through techniques such as flowchart, critical incident, spider chart, and performance matrix [17]. The next step is to brainstorm potential causes through techniques such as "is—is not" matrix, nominal group, and paired comparisons. Then collect data resulting from the problem and any information that you might find on the causes [17].

With data in hand, perform the analysis through techniques such as histogram, Pareto chart, scatter chart, and affinity diagram. Finish this section by identifying the problem cause or causes through techniques such as cause-and-effect chart, the five whys, and results from both FTA and ETA [17].

### RCA fix

Fix the problem. Use techniques such as six thinking hats, the theory of inventive problem solving, and systematic inventive thinking. Ensure that you fix the entire problem [17].

### RCA change

Implement the solution. Use techniques such as tree diagram and force-field analysis. Ensure that unintended consequences do not arise in other areas when you implement the solution [17].

## Final case study

I consulted for a small aerospace company and performed some of the analyses described in this chapter. The product was an imaging subsystem very similar to the one shown in Figure 7.8. I performed a worst case study of the circuit boards in the processing subsystem, and then I performed FMECA, FTA, and ETA of the system's connection to the spacecraft power and telemetry channels. The worst case analysis of the passive electrical components took me about 50 hours over two weeks.

I generated three different FMECAs: one for the imagers to the processing unit, one for the processing unit to the spacecraft power and telemetry channels, and one for the test equipment connection to the processing unit:

1.  The FMECA for the test equipment connection had 100 separate lines for failures across 50 connector pins (many functions shared redundant pins) with as many as 9 different failures per set of pins.
2.  The FMECA for the spacecraft channels had 61 separate lines for failures across 25 connector pins (many functions shared redundant pins) with as many as 8 different failures per set of pins.
3.  The FMECA for the processing unit had 55 separate lines for failures across 31 connector pins (many functions shared redundant pins) with as many as 9 different failures per set of pins.

The document describing the FMECAs had 56 pages with 29 pages of written description and diagrams. Each of the remaining 27 pages had a single FMECA chart of between 22 and 47 lines. The effort took about 70 hours of my time.

I generated three different sets of FTAs: one for the data from the imagers through the processing unit and spacecraft to the GSE, one for the memory storage units in the processing unit, and one for the RF telemetry. The document describing the FTAs had 25 pages of written description and diagrams. The effort took about 50 hours of my time.

The document describing the ETAs had 90 pages with 22 pages of written description and diagrams. Each of the remaining 67 pages had a single ETA chart. The effort took about 120 hours of my time.

## Acknowledgment

My thanks to George Slack for his careful and thoughtful review, critique, examples, and suggestions in this chapter.

## References

[1]  Accessed on April 17, 2013: <http://lws.gsfc.nasa.gov/>.
[2]  Accessed on April 17, 2013: <http://lws-set.gsfc.nasa.gov/>.
[3]  K.R. Fowler, L.J. Frank, R.L. Williams, Space environment testbed (SET): adaptable system for piggybacked satellite experiments, IEEE Trans. Instrum. Meas. 53(4) (2004) 1065–1070.
[4]  B. Sherman, M. Cuviello, G. Giffin, NASA's LWS/SET technology experiment carrier, Proceedings of the 2003 IEEE Aerospace Conference, Big Sky, MT March 8–15, 2003, vol. 1, pp. 427–435.
[5]  D.K. Pradhan, Fault-Tolerant Computer System Design, Prentice Hall PTR, Upper Saddle River, NJ, 1996 (pp. 4–6, 104, 143, 159–166).
[6]  K. Fowler, What Every Engineer Should Know About Developing Embedded, Real-Time Products, CRC Press, Boca Raton, FL, 2007 (pp. 1, 15, 76–78, 80).
[7]  T. Lennon, Using modeling, simulation to optimize plant control systems, ISA White Paper, July/August 2010.  <http://www.isa.org/InTechTemplate.cfm?utm_medium = kimf@ieee.org&utm_source = Eloqua& utm_campaign = COSYEmail_Existing-15JAN13-AW&Section = Control_ Fundamentals1&template = % 20/ContentManagement/ContentDisplay.cfm&ContentID = 82972&elq = de8f4194193145d0ac1787ec1ee 148b6&elqCampaignId = 400> (accessed 15.01.2013).
[8]  C.A. Ericson II, Hazard Analysis Techniques for System Safety, Wiley-Interscience, Hoboken, NJ, 2005 (pp. 235–259).
[9]  N.G. Leveson, Engineering a Safer World, The MIT Press, Cambridge, MA, 2011.
[10]  J. Thomas, Extending and automating a systems—theoretic hazard analysis for requirements generation and analysis, Dissertation for Doctor of Philosophy, Massachusetts Institute of Technology, April 2013, p. 5.
[11]  R. Torok, B. Geddes, Systems theoretic analysis (STPA) applied to a nuclear power plant control system, EPRI Paper, MIT STAMP Workshop, Cambridge, MA March 26–28, 2013.
[12]  Sutton Technical Books, —Safety Cases, Sutton Technical Books 2007–2012, Ashland, VA. <http://www.stb07.com/management/safety-cases.html> (accessed 29.09.2013).
[13]  Safety case development manual, European Organisation for the Safety of Air Navigation, DAP/SSH/091, November 13, 2006.
[14]  V. Balgos, A systems theoretic application to design for the safety of medical diagnostic devices, Thesis for Master of Science in Engineering and Management, Massachusetts Institute of Technology, April 2013, p. 83.
[15]  Accessed on July 30, 2013: <http://en.wikipedia.org/wiki/Markov_process#cite_note-1>.

[16]  Accessed on July 30, 2013: <https://en.wikipedia.org/wiki/Markov_chain>.

[17]  Andersen, Fagerhaug, Root Cause Analysis, second ed., ASQ Quality Press, Milwaukee, WI, 2006. p. 12.

[18]  Accessed on July 23, 2013: <http://en.wikipedia.org/wiki/Root_cause_analysis>.

## *Recommended reading*

C.A. Ericson II, Hazard Analysis Techniques for System Safety, Wiley-Interscience, Hoboken, NJ, 2005 (The specific chapters on analysis techniques are the best. The first five chapters of introduction tend to be applicable only to large-scale military systems during or before the 1990s. The chapters devoted to specific analysis techniques are well organized.)

N.G. Leveson, Engineering a Safer World, Systems Thinking Applied to Safety, The MIT Press, Cambridge, MA, 2011. This book will introduce you to STPA.

Andersen, Fagerhaug, Root Cause Analysis, second ed., ASQ Quality Press, Milwaukee, WI, 2006.

## CHAPTER 8

# The Discipline of System Design

## Tim Wescott

*IEEE Senior Member, Owner, Wescott Design Services*

**Chapter Outline**

Developing and Managing Embedded Systems and Products.
DOI: http://dx.doi.org/10.1016/B978-0-12-405879-8.00008-8

Areas of particular focus in this chapter include:

- how to communicate effectively with nonengineering personnel,
- how to partition a large system into manageable subsystems,
- how to partition system requirements into requirements for individual disciplines (i.e., software, electronics, and mechanical engineering),
- how to control costs for projects of various sizes and production volumes.

## What to expect in this chapter

A successful product is far more than a collection of correctly working parts. The parts of a successful product must not just work, they must work together smoothly to deliver the value that the user or buyer of the product expects from it.

Moreover, a successful product must, itself, become a working part of some greater whole.

System design is the action of designing the product as a whole so that all of its parts work together, it achieves the desired ends, and it works for users to help them achieve their desired ends. System design ensures that you don't build a vehicle with the front end of a bicycle and the rear end of a dump truck. System design ensures that you build a product that will fit into economical packaging for shipping and sales. System design ensures that a piece of electronics designed to plug into a wall power socket does not require the building owner to rewire his building before things will operate correctly.

This chapter on system design is placed before the chapters on electronics, mechanics, and software design because system design always comes first in the design process. Your system design effort may be so automatic and brief that you do not notice it or it may take years of effort. But until you know how your electronics, mechanics, and software designs fit together, you cannot start these design processes. The organization of these chapters echoes this organization of your project design.

The thrust of this chapter—Learn to design before you build. ...

> **Please remember**
>
> All examples and case studies in this chapter serve only to indicate what you might encounter. Hopefully these examples and case studies will energize your thought processes; your estimates and calculations may vary from what you find here.

## Basic definitions

### What is a system?

There are probably an infinite number of different ways that the term "system" can be defined. You find the term "system" defined in physics, medicine, chemistry, politics, and for all that I know, religion as well.

For the purposes of this chapter, I will define "system" in a narrower engineering sense: *a system is a collection of parts that must work in concert to achieve some desired behavior.* While this definition may be narrow, it still has plenty of breadth. Practically anything can be defined as a system. Indeed, you can even invent definitions of "systems" that bring the status of a "thing" to collections of objects, ideas, people, and actions that had never been viewed as connected.

The breadth of application of this definition of "system" carries with it opportunities for systematic study and design. It also brings many pitfalls and sometimes even terrors (if only of a narrower, engineering sort). See Chapter 1 and the discussion and definition of system, as well.

The most immediate realization from this definition to product design is that if your team is designing a product, then your team is designing a system. This means that methods of system design can be applied to any product design effort.

Two useful applications of this definition to product design are one, that it will become part of some larger system, and two, that it will, itself, be composed of lesser systems—subsystems—that enable it to work.

### What is system design?

If the definition of a system can cover almost anything, then how can we derive a definition of system design? There are two ways. One definition of the term "system design" (and, by extension, "system engineer") is to start with the assertion that "all engineering is system design," then remove all known disciplines of engineering from that. Another more useful definition of the term "system design" is the act of defining a system's place in the larger system in which it works, while simultaneously defining exactly what the system's formal description is in terms of its own subsystems.

The first definition of "system design"—to start with everything and eliminate all the knowns—leads to a definition of "systems engineer" that roughly reads, "They are really smart and really useful, but we have absolutely no clue what they actually do." While this sort of definition does lead to the ability to hire useful people into engineering teams, it can lead to trouble down the road—a "systems engineer" who is really an optical designer, for instance, cannot easily step into the shoes of a "systems engineer" who is a specialist at shepherding completed products through regulatory processes or customer compliance testing.

The second definition of "systems design" is the more useful. It can lead us directly to an idea of what the job actually entails.

Systems design involves making sure that for all points in the design cycle the design team has their arms wrapped around the whole problem and that they are driving rapidly and efficiently toward a good, feasible solution to the design problem. This means that the system itself is well described as a part of the larger system to which it belongs, while at the same time the system is partitioned into smaller bits that make sense. These smaller system definitions should be:

- specified in a rational manner,
- apportioned out to individual designers or groups of designers that have the necessary skills to carry out their design,
- scheduled so that the earliest results can be verified in isolation and, in turn, support the verification of subsequent results,
- designed so that the system as a whole achieves its purpose in a robust manner with the least cost.

Possibly the biggest pitfall to any design effort, and one that you can easily fall into more than once in any given project, is the problem of getting the project painted into a corner. It is nearly impossible to get a project underway without putting details aside to resolve later. It is easy to then overlook these details as the system design evolves. As a consequence, you can end up with a final design that works altogether correctly except for some vital missing piece.

### How does system design fit into a project?

In a sense, system design can be viewed as "design of everything." So how does it fit into a project? The answer is "early and often." At the very beginning of a project, the design task is solely one of system design. You have a goal, often inadequately stated, and you have constraints, again, often inadequately stated or even unstated. Starting from this goal and this set of constraints, you must produce an outline of the product to be designed, as well as an estimate of the time, budget, and manpower required to complete it.

So the project manager must work closely with whomever is doing the system design. As more is found out about how the system is actually going together, the system design—and often the project schedule—must be continually revised and updated to reflect the real results of the product development.

## Who should do the system design?

There are three common choices for who drives the system design. These choices are the project manager, a committee of the lead engineers on the project, or a designated systems engineer (or systems engineering team).

Each of these three personnel choices has an impact on the project. None of these choices are necessarily the best for any given team and product development effort nor are all necessarily best or worst for your company culture. You should neither reject any approach out of hand nor automatically embraced any approach.

### Project manager

The benefits to having the project manager do the systems design are twofold. First, the project scheduling and staffing decisions are closely tied to systems design decisions. Second, the systems engineer often needs to have considerable authority to gain acceptance for a system design that is optimal overall, but various engineering disciplines sometimes may not understand or accept that authority. When the systems designer and the project manager are the same person, all of these needs are satisfied at once.

As a guideline, a project manager can be a reasonable choice for very small projects, typically with a team of less than 6 or 8 people. The caveat is that the project manager _must_ be technically capable _and_ well rounded.

There are downsides, however, to having the project manager do the system design. System design is a discipline unto itself; for an individual to do it effectively requires an ability to span engineering disciplines. Asking one person to be facile at this great breadth of technical ability, while at the same time possessing the business acumen and organizational ability to be a good project manager is asking quite a lot. In addition, system design takes time; it involves a considerable amount of cross-checking to make sure that the whole mission of the product has been accounted for in the various subsystems; it requires spending a lot of time communicating with various people from various disciplines and not just within the engineering department; it requires synthesizing this information and communicating it in a way that can be understood by all of the people involved in the project. These demands on both ability and time may be too much to ask from one person, particularly if the project is large.

### Committee

A solution to the problems of having the project manager perform the system design is to hand the system design over to a committee of the lead engineers on the project, preferably with the project manager moderating the proceedings of the committee. In this model, each of the engineering disciplines (e.g., electronic, software, or mechanical) involved are all headed by their own designated technical lead engineer. These technical leads convene to make the systems design decisions.

---

### Sidebar: Definition of lead engineer

There are many possible definitions for lead engineer. In this chapter, a lead engineer is the governing engineer within a particular discipline. An example could be the Mechanical Lead Engineer—the person responsible for technical direction of the team designing and developing the mechanisms and packaging within a project. The lead engineer is not necessarily a business manager nor a personnel manager, though could be either or both depending on the company and project.

---

The advantage of the committee system is twofold. It relieves the project manager of the need to have the skill and the time to do all of the systems engineering tasks. It also means that each of the people who are doing the system design has a very good idea at least of what their discipline can do, which can significantly reduce the possibility of asking one or another discipline to do the impossible.

The potential set of disadvantages to the committee system is manifold, in no small part because the design is, indeed, by committee. If the members of the committee do not stay focused on the project goals, then the system design can get lost in politics. Skill gaps can exist on the team, either because an engineering discipline was left out or because the chosen leads cannot communicate. Even without skill gaps, culture clashes between engineering disciplines or lead engineers who are too territorial about the role of their chosen discipline can cause difficulties. In both the committee approach and the project manager as systems engineer approach, the systems engineering can be shortchanged. Finally, there is a twisted form of design by committee that I call "serial design" that can cause great difficulties.

Skill gaps can take two forms in the committee system. First, entire disciplines may be left out. Second, the technical leads on the committee may not be able to communicate issues about their discipline or understand the points made by people outside of their discipline.

The danger of leaving out an entire discipline happens when the job of one discipline seems trivial or when there is an unrecognized need to have the design visited by an expert in some field. An example of this is a product that has strong electrical and software components but has mechanical content that is "only packaging." If a mechanical engineer is not brought on board early such a product may end up with severe difficulties when this "trivial" mechanical design is not given the attention it deserves.

The case of technical leads not being able to bridge the gaps in their disciplines is more insidious. This form of skill gap happens when two or more technical leads cannot reach outside of their own discipline to communicate their needs or to appreciate what other members of the committee are saying about their needs or what solutions they can deliver.

The solution to this problem is to choose technical leads who have breadth of talent. The ideal candidate to be a technical lead is not necessarily someone who is the best designer in their field. Rather, the ideal candidate will be someone who is good at what they do, but who is able to poke their nose into someone else's specialty and come away enlightened rather than baffled (Figure 8.1).

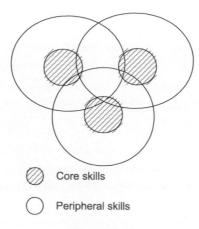

Core skills

Peripheral skills

**Figure 8.1:**
A pictorial representation of the overlapping skill sets needed in the members of a design committee. No one has to be able to do someone else's job, but they must be able to reach outside of their own discipline to understand how their own design work can help or hinder that of other disciplines.

It is entirely possible for an otherwise competent project manager or team of lead engineers to perform their system design tasks without ever realizing that they are performing system engineering or that a need for formal system engineering may exist. In such cases, the system design task can get shortchanged, to the ultimate detriment of the project.

As a final note, it should be realized that design-by-committee has a very pathological variant that should be avoided if at all possible. This is the "serial committee," where a design is passed from hand to hand, with information only flowing from the prior holder of the design torch to the next, without the team of engineering leads ever meeting at one time to hash out the design. In extreme cases, the principals in such "serial design" may not even be employed on the project at the same time or even meet each other.

**Case Study: Serial committee design**

I worked on a project where design-by-serial-committee occurred. The project needed to design mechanical packaging, mechanisms, and electronics, and it needed software to make all the parts play together. Indeed, this was exactly how the project was carried out. First, the mechanical engineers designed their mechanisms and attached the various new auxiliary equipment to the system as required. As part of their effort, they selected motors and sensors and made what space they could for the electrical engineering team to put circuit boards. At the end of this, they handed the design effort off to the electrical team. The electrical team designed the circuit boards to fit into the available space—and if they found that their estimates of the space needed was off, or if the increased electrical load necessitated more functionality than they had expected, they were out of luck for getting more space for boards. Finally, the electrical team proudly presented their efforts to the software team, along with a long list of requirements for getting the whole assemblage to work.

As a consequence of using this "serial committee" approach, there was a phenomenon whereby the mechanical designers would do their best job to make life possible (and sometimes even easy) for the electronics designers—but they did this in ignorance of what the electronics designers could easily do. The electronics designers did the same thing to the software engineers. Finally, the software engineers made do with what they had. As a consequence, each discipline, to some extent, painted the next discipline into a corner. A rational, comprehensive system design step at the beginning of the project would have avoided this problem.

*Systems engineer*

In some ways, the most rational way of doing system design is to have a designated specialist to do the systems design or to be the team lead on the systems design. Equip this person with enough authority to make decisions, place him close to the project manager so that everyone stays well informed, give him a word processor and some scientific computing software; then stand back and watch technology happen!

Indeed, having one individual do the overall system design does have advantages. The project manager is relieved from the detailed engineering involved in systems design. Each technical lead can shine at leading engineers to do their individual jobs well. And you avoid the difficulties inherent in design-by-committee.

Having a designated systems engineer to take up the whole-system design, however, can have its own set of problems. If the chosen engineer does not understand the issues and solutions to the problems from a wide variety of engineering disciplines and specialties, he or she will fail. If the chosen engineer is not strong enough to truly drive the design to completion, then you end up with the design driven by a strong personality outside the systems engineering office or by a hidden committee. If the chosen engineer is too full of himself, then he will micromanage and take on too many of the design decisions upon himself which will either generate resentment in the group or force unwise decisions on the individual engineers who must carry out his wishes. Successfully finding a systems engineer who is wise enough to know when to defer to team members, yet strong enough to stand firm when challenged, pays off well.

On large enough design efforts, it may even be necessary to have more than one systems engineer or even an entire team of systems engineers. Such projects are generally very large efforts that span multiple companies, such as the design of a bomber or fighter aircraft, or a tank or other weapons system, or a large commercial aircraft.

## Human elements in system design

### The systems designer's skill set

So what should you look for in a systems designer? Do you need a person who can do the job of any member of the team? Do you need someone who could design the whole system alone, if need be? Or do you just need someone to quietly sit in a corner and maintain the requirements documents as a checklist and quietly bring any discrepancies to the project manager's attention?

While it can be nice to have a "superman" systems designer, able to leap over any engineering problem in a single bound, such people can be hard to find. (And the attempt can be dangerous: if he thinks he is "super," then he is probably deluded!) The minimum skill set that a systems designer needs to have to deserve the name is the ability to find the right people, ask them the right questions, boil the answers down to the parts that are essential to the problem at hand, integrate all of the received answers into an understandable whole, and to communicate this to the team. Now doing that is "super!"

What a systems designer does not need is the ability to do everything. In fact, if you have a systems designer that can do a wide range of tasks, you need to make sure that he has the self-control not to try to solve every problem himself, or you need to be ready to take problems away from him and keep him on the task of arranging the forest, not planting the individual trees.

A good systems engineer and designer also needs to have considerable diplomacy. In some ways, good system engineering isn't a matter of not stepping on other people's toes—it is a matter of mashing people's toes on a regular basis and making them happy that you have done so. It means being able to:

- present decisions to one team or another that may not be popular,
- hold to important technical decisions in the face of stiff opposition,
- realize when interdisciplinary politics trumps a technical decision and accept a slightly suboptimal technical decision if it will keep the peace and keep the project moving forward,
- mediate between the technical disciplines and project management.

These are all abilities that a systems engineer must have if the project is going to proceed smoothly, without resentment and other "people problems" crippling an otherwise technically sound approach.

## Language differences

Humans have this remarkable propensity to invent new languages and dialects as we go through daily life. This colors not only communication with people in different parts of the world, but across generations (when bad—or rather "baaaad"—is good, for instance), and across technical disciplines.

The consequence for this linguistic inventiveness for a systems engineer is that terms or words that are perfectly sensible in one technical discipline may be complete nonsense in another. Worse yet are terms or words that have a very specific and concrete meaning in one discipline may have an altogether different meaning in another discipline—while still being quite concrete and specific.

Consequently, whoever is doing systems design on a project must be on the lookout for such collisions in terminology. Cases where one engineer's term is another engineer's nonsense are benign but potential time-wasters. Cases where one engineer's perfectly sensible term is another engineer's perfectly sensible term with an entirely different meaning can have consequences as benign as wasted time or as dire as being the center of major personality conflicts, or as latent problems that do not crop up until late in a project when they are difficult, time consuming, and expensive to correct.

### Example: Quirk in language "state"

The first time this quirk of the language came to my attention was the use of the word "state." As I have spent a good portion of my engineering career, switching gears between control systems engineering and software design I should have known, yet when I first started teaching software engineers the fundamentals of control system design my students and I were ambushed by this.

In short, a control systems designer uses the word "state" to mean "an aspect of the system's dynamic behavior that can be represented by an integrator." Sometimes a control guy may use state to mean "the values of all the individual states at a given point in time."

A software engineer, on the other hand, usually uses the word "state" to mean "one of a finite number of discrete states taken on by a state machine." Moreover, while a control designer usually thinks of a system "state" as being a continuous value that is best modeled as a real number, a software designer often thinks of a "state" as being best modeled as a collection of enumerated values, each of which takes on a linguistic meaning for which it is not sensible to attach to a numeric value (i.e., "start," "go," "stop").

Similar terminology differences abound: the civil engineer's "permeability" means something vastly different from an analog circuit designer's "permeability." A mechanical engineer's "resistance" may well refer to a physical force where an electrical engineer's "resistance" probably refers to an electrical effect. "Head" to a software engineer refers to the part of a computer system that interacts with a person, while "head" to a hydraulic engineer refers to the amount of pressure available in a tube. The possibilities are, unfortunately, endless.

The bottom line is—be aware of these language differences, or at least of their possibility, and be ready to recognize a potential misunderstanding and nip it off in the bud.

## Case Study: Language differences

This happened to me during a discussion about one of my hobbies, but it could easily have happened in a professional context.

The discussion came about as I was working on this chapter and was about using Kalman filters to aid model airplane performance. I was having a rather frustrating exchange with a systems engineer for an aerospace company about how one estimates vehicle position and velocity.

He kept using the term "observer" in a way that made no sense at all to me. Normally, I would have been very tempted to discount what he was saying as being due to a lack of expertise, but from previous conversations with him I know the fellow is very competent in a number of different fields.

In control systems practice and literature, the term "observer" means a filter that is designed specially to estimate the states of a system for the commands that are being given to the system and the measured outputs of the system. Observer design is not trivial, because the system states are not directly accessible, but must be deduced by the difference between the system's observed behavior and the observers predictions of that behavior.

After several exchanges where I, at least, was getting more puzzled and frustrated, it finally dawned on me that I was dancing on the edge of a language pitfall. So I asked him to please define the term "observer" in his context.

It turns out that in his world, an "observer" is the measuring device used to report on vehicle position, such as a GPS device, or an actual human navigator sighting landmarks or determining locations by astronomical sightings. This use of the word is totally different from the usage that I am used to.

Had I not realized that there was a discrepancy, then I might have either abandoned the conversation, or much worse, attempted to discredit what he was saying. Had this happened in the context of a project it could have been damaging to the business interests of all involved. Once I realized that the misunderstanding was revolving around the usage of one word, however, I was able to ask for clarification and get an answer that not only let me learn an interesting linguistic tidbit but also clarified several hundred words worth of conversation.

This exchange is an example of what might happen to you when you talk to an expert outside of your field and stumble across one of these terminology collisions. I was fortunate in that I was able to recognize what was happening and to take my own advice in dealing with the situation.

### System design flow steps

Whether you are designing a keyless remote transmitter or an entire aircraft carrier, the system design process is the same. While smaller designs do not formally need most of the steps, if you dig deep you'll still find all of them implicitly embedded in the process.

I present this design flow as if it were a waterfall model. For small systems, proceeding in known territory, the waterfall model works. In most system design cases, however, you will find that your design flow is more of a spiral model; you need to make assumptions about later steps to complete prior steps, but on reaching those later steps you find that your assumptions were wrong, which requires you to go back and refine your original design. (Chapter 1 discusses these processes in more detail.)

### Business concerns

Product-design engineering does not happen in a vacuum. If you are designing a product, then you are almost certainly doing so as part of a business that intends to make a profit, and your product is intended to make a profit. Even if you designing a product for a nonprofit organization, or designing a product that is intended to be sold at a loss, or are working for a government entity to develop a product, your organization is still going to have some aim in mind to get the maximum positive effect for the money and resources spent.

The tools for tracking the true costs of an engineering effort are business tools, so it makes sense to pay some attention to them. The tools to estimate the profit to be made from a product are also business tools; while you may nominally not be involved in that side of the computation, it is good to know the language, out of self-defense if nothing else.

We trained, or are training, as engineers to think in terms of the physical world, and how to analyze physical phenomenon and to produce desired physical phenomena from stuff that we can get. Engineering, in this sense, is applied science, and as such there is always a rational touchstone based on mathematics and physics.

Business does not have this same touchstone. In the end, a business does not succeed or fail based on its leaders ability to understand and affect the physical world. Instead, a business succeeds or fails based in large part on its leader's ability to understand and affect the people around them and to keep control of the financial issues. While the quality of the product that a business sells has a great impact on the success of the business, financial management and customer relations have a greater impact. You can tell that this is so because the world is full of successful businesses that sell poor products and failed businesses that sold great products.

Product managers, and whoever is acting in the role of systems engineer, constantly interact with people whose concerns are those of the larger business. As such, it is essential that they understand at least some of the concepts and language.

This one section of one chapter of one book does not replace the sum total of knowledge available. If you did not benefit from a microeconomics or an engineering economics course, then take one, or at least get a good book, and study up.

### The cost of money

Money isn't free. When a company funds an engineering effort, they have to find the money to do so. This money either comes out of the company's reserves, or the company has to borrow the money from a bank, or the company must attract investors willing to spend money on the effort. All of these sources have two things in common: first, they will expect to get paid back and second, they will expect to get more than they put in.

The sensible businessman, banker, or investor looks at an opportunity to put money into an endeavor with an eye to getting more money in return than they invested. At the top level, this return is measured in two ways: return on investment (ROI) and risk. The goal of anyone putting money into a project is to, on average, get more money out than they put in. That means that the ROI, with risk factored in, is a positive number.

An alternative way of looking at ROI is the amount of time it takes to get your value out of the investment: the ROI on a house loan can be up to 30 years, while the ROI on a restaurant meal can be measured in minutes or hours.

This desire to earn a return on their investment leads any lender or investor into, essentially, charging a fee for its use. In typical situations, when you borrow money this fee is the interest on the loan for a car or a house, or when we let a balance build up on our credit cards. When you or your company attracts investment, this "fee" is expected to be returned either in the form of increased value of the investor's stake in the company (as in rising stock prices) or in the form of direct cash payments back to the investor (as in dividends).

The cost of money is still there even if a company is working from cash reserves. The company will always have a finite amount of cash, and when the company spends money on one project, that money becomes unavailable to spend on another. Thus, a wise company will not just fling its money around randomly. Instead, it will seek to only spend the money on projects that have an ROI that is not only positive, but that is better than it could get from other projects, or by putting the money into the bank or by buying stock in another company.

This is all important to the project manager, because (in rational companies, at least) projects and their modifications get approved or rejected based on their perceived ROI. While a project manager may not have much control over the return part of the ROI, he should have a rock-solid knowledge of costs and schedule times, and he should at least have an idea of how changes to the product may affect that return.

Even when a project is being done for nonmonetary reasons, by a nonprofit or by a government, there will still be some perceived benefit such as feeding the hungry, defending the country, and improving the environment, and the organization's or government's leaders will still want to get the maximum benefit for the minimum cost.

### The cost of time

If the discussion about ROI has not made it apparent, the timing of money spent also has an impact on the cost of spending the money. Because ROI is based on an interest rate, spending an amount of money early costs more than spending the same amount of money later.

Financial analysts have ways of translating these costs to make an apples-to-apples comparison. They can either translate all future costs to the present value of spending or they can translate all costs to some specified future date. Sometimes they can do other useful manipulations. This sort of financial accounting for costs in the presence of interest rates is covered in a good course in engineering economics.

### The cost of opportunity

You will hear sales people and product line managers talking about "opportunity cost." This is the cost of waiting to sell something. When you choose to stretch a project out by three months or six months to sell something better at the end, you are losing the opportunity to sell something lesser, sooner.

Opportunity cost comes in two parts. The nice concrete part is simply the cost of getting the same amount of money at a later date and can be calculated with financial tools. The squishy part is the cost of coming onto the market after a competitor has established a product.

Fortunately, it is not usually the systems engineer's or project manager's job to decide whether to ship a four-button widget on Tuesday or a six-button widget on Thursday. But it is important to know that these concerns can weigh heavily on the sales and product line managers and to be ready to support whatever decision they make (and perhaps even to be ready to lend one's own opinions to the mix, at least if you can defend them with data).

## NRE, tooling, and COGS costs

Nonrecurring engineering costs (NRE costs, or just NRE; sometimes defined as nonrecurring expenses), tooling costs, and cost of goods sold (COGS) are terms that one quickly encounters in the workplace at any manufacturing company. They are terms that you need to understand in designing and producing product. NRE is the cost associated with the engineering effort to design and prototype a system. COGS is the cost associated with building each item during manufacturing that follows development. Tooling costs, which sometimes get lumped into NRE, are the cost of manufacturing setup that occurs between engineering development and regular production where COGS dominates.

The NRE cost is the amount of money necessary to get a design ready for production and any support needed by the manufacturing group to get spun up to speed. You can think of NRE as "one-time" fee that the engineering group must charge the company to deliver a working design. NRE includes all design, test, and qualification performed during development of the product. NRE cost is important and cannot be ignored. The NRE costs must be amortized over the life of the product.

### Sidebar: Amortization

"Amortize" is an accounting term that design engineers must understand when they go to calculate costs.

If you spend $500,000 to design a system and tool up to make it, then sell 10,000 units that cost $1.00 to make, then in the absence of one really enthusiastic early adopter, you cannot plan on charging $1,000,000 for the first unit sold and $2.00 for every unit thereafter. Instead, you take that $500,000 of nonrecurring costs, spread it out over the 10,000 units, and charge $102.00 for each one (well, perhaps $99.95).

This action of dividing the $500,000 up 10,000 ways is called "amortizing," and it is how accountants and businesspeople sensibly spread the NRE, the cost of spinning up a product, over the lifetime of the product.

No products can be built by throwing the components into a bag, giving it a shake, and having the finished parts fall out. Managers can wish otherwise, but if it is to be built, it must be built with tools. Moreover, many products require that specialized tools be designed and built, specifically to manufacture that product. These tools, and particularly these specialized tools, are known as "tooling." The money necessary to build molds, assembly fixtures, software used in manufacturing, and any other efforts associated with actually putting a design into production are tooling costs.

Tooling costs, like NRE, must be amortized over the life of the product. Unlike NRE, the volume of production has a direct impact on the tooling costs. If a particular tool can make

10 components a day and you need to make 50 components per day to meet production schedules, then you need five tools or you need a different approach to manufacturing. Engineering costs (at least to a first-order approximation) do not follow this rule; once designed, the engineering money is spent and does not need to be spent again.

The COGS includes the ongoing expenses and recurring costs that go into the manufacture of each unit. A calculation of the COGS for a product should include all of the money that is spent to build each item, including components, labor, and overhead costs. In addition, COGS should reflect the cost of any parts that are built and rejected: if you spend $2.00 to build each gizmo, but you can only ship one of every ten built, then the true COGS for that gizmo is $20.00, not $2.00.

Figure 8.2 is a pictorial representation of the definitions of NRE, tooling, and COGS.

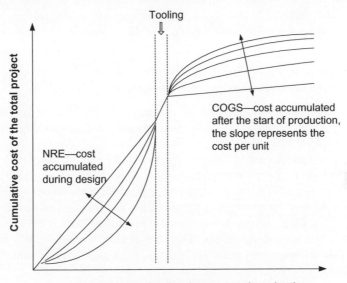

**Figure 8.2:**

Pictorial indications of NRE, tooling, and COGS. © 2003–2013 by Kim R. Fowler. All rights reserved. Used with permission.

### Case Study: Two motor controllers

I know of a company that is contemplating an upgrade to an existing motor control board, to work with a wider range of motors. Two upgrade paths are being contemplated. The "simple" path mimics the existing board by using a simple microprocessor that translates commands from the outside world into analog command voltages, which are then applied to all-analog motor control and drive circuits. The "complicated" path will have much simpler hardware but will require more software content; it will have a more capable processor whose output directly drives the output amplifiers of the board, with all the same feedback from the motor that used to go to the analog circuitry. Instead of the control loops being closed in analog hardware, they will be closed in software.

It is estimated that the "simple" path to upgrade the motor control board will take a year of calendar time and approximately one and a half man-years of effort from a mechanical designer, a circuit designer, a software engineer, and an engineering technician. The cost of the engineering effort is estimated to be $270,000. (For the sake of argument, I'm assuming that the engineers in this case study earn around US$90,000 per year and that their loaded cost equal about twice their annual pay. I think the final figure is right for some parts of the United States, but certainly not for high-dollar places like the San Francisco bay area.)

It is estimated that the "hard" path to upgrade the motor control board will take the same year of calendar time, but due to the higher software content, it will take approximately two man-years of effort from a similar team. The cost of the engineering effort is estimated to be $360,000.

The board is going into an existing product, which enjoys a sales volume of 10,000 units a year. This sales volume is expected to continue unchanged.

The "simple" board has an estimated component cost of $15. This picture is complicated, however, by the fact that many of these components are precision parts on which the performance of the board depends. The board suffers from a 10% rejection rate, largely due to unanticipated component variations. It has a documented support cost of $25,000 a year as the manufacturing engineering team constantly chases component obsolescence and quality issues to keep the rejection rate down. Because the motor controller's characteristics are defined by analog components, the board must be run through a functional test that costs $7.50 per board. Finally, the board is used in two versions of the product, and while the base design is the same, the tuning of the control system is different. Thus, two different part numbers must be built and stocked, which adds 10% to the overall costs of the board.

The "hard" board has an estimated component cost of $15, because the cost of the analog components is offset by the needed analog-to-digital converters (ADCs) and more powerful processor. Because the circuit of the new board is simpler, and because its performance is not dependent on the performance of each individual component, it is estimated that it will be produced with a rejection rate of 5%. Manufacturing engineering has looked at the proposed design and has estimated that by virtue of its lack of critical analog components it will only require $10,000 a year in support. Moreover, the required functional test is estimated to cost only $2.50 per board. Finally, as the control system characteristics are defined in software, only one part number must be built (Table 8.1).

While the component costs of the two boards are identical, the less tangible contributions to COGS push the cost of the "simple" solution to be $10.43 more than the "hard" solution. Over 1 year of production this adds up to a $104,300 (remember, 10,000 units, each saving $10.43), which is greater than the $90,000 extra in engineering time for the new software.

**Table 8.1: Comparison of total COGS cost between a simple and more difficult board design**

| Description of Costs | "Simple" | "Hard" |
|---|---|---|
| Components | $15.00 | $15.00 |
| Support | $2.50 | $1.00 |
| Testing | $7.50 | $2.50 |
| Rejects | $2.50 | $0.93 |
| Extra stocking | $2.75 | $0.00 |
| **Total** | $30.25 | $19.43 |

This shows that neither the NRE nor the COGS of a product tells the whole story. In this case study, the balance is tipped toward the more modern solution in the absence of other determining factors. The "simple" older-style solution, however, might make sense for a product with a lower production volume.

## Overhead

Any time you calculate costs, you must factor in overhead. Overhead is the amount of money it takes to get something that is not directly related to the cost of the thing itself.

For instance, when you show up at work, you know how much you are being paid to show up—that is your direct cost. But in addition to paying you directly, your employer pays for the space in the building where you sit, they pay some of your taxes and your various benefits, they pay for the water cooler and coffee machine, the restrooms and break rooms and meeting rooms, they pay for the IT people that maintain your computer and the network, they pay for the HR people that hired you and take care of your concerns,... and the lists go on. ...

All these hard-to-define, but still very real, costs are lumped into "overhead." This overhead must be taken into account when you compute project costs. Thus, if you want to take a project plan that is expressed in terms of man-hours and material costs, you must find out the overhead factor for each man-hour, and the overhead for the material costs, and you must add these into the costs of the project in order to get a realistic dollar amount.

The common business term for direct costs plus overhead costs is "loaded cost." For most employees in the United States, it's a good rule of thumb to assume that the loaded cost is twice their direct pay. This can vary significantly, however, so if you can do so, you want to find out your company's estimate of loaded costs for the personnel you contemplate using.

## Human factors

Because business decisions can be made to sound rational but are ultimately based on squishy human interactions, the manner in which information is presented can easily be subject to the individual desires or ambitions of the people involved. People will always be inclined to fudge their numbers. Anyone championing a project will naturally tend to make it look shorter and less expensive and have higher customer acceptance. Anyone opposing a project will naturally tend to do the opposite.

I cannot advocate any particular approach to dealing with this phenomenon when making estimates. Personally, I try to always be as honest with myself as I can and to take as objective a view as possible—but I still regularly find my own enthusiasms or lack thereof coloring my estimates.

Often project managers need to be salesmen for their projects. This spills over onto whoever is doing the systems engineering, because it is often the systems engineering team upon which the management team of a company will rely for direction to make their business decisions. If you are doing systems engineering, expect to provide information about what the system will do and to give informative answers to people without technical backgrounds.

You can't force the world to be a rational place—but you can at least be aware of the forces involved and try to deal with them.

## The art of system design

### Understand system purpose and requirements

The first step in system design, which you cannot avoid without inviting disaster, is to understand the purpose and requirements of the system that you are designing. In essence, what you are doing is deciding exactly what it is that you are building; get this wrong, and you will build the wrong thing. (Chapter 6 discusses developing requirements; this section ties into that material with specific examples and issues.)

You must understand, at a top level, what your system needs to do, the operating environment, and, if it is a part of a larger system, how it fits into that system. You must understand how much it will cost to build and deploy the final system. If you get any of these wrong, then you may design the best system in the entire world, and still have a product that is an abject failure.

The critical aspect of understanding a system's purpose and requirements is that you must be able to describe the system requirements clearly, in an engineering sense. Management directives to "just build it," or "just build a <insert name here>," or "make it work" are all well and good and are a necessary component of interacting with people whose yardsticks are marked out in dollars, but leaving such descriptions as they are is not systems design! A large part of formal systems design is flowing requirements down from the top level into every nook and cranny of the design. "Works good" is not even correct English, let alone a reasonable requirement for a system; the systems engineer must translate such phrases into specific, testable requirements, written in formal language, and quantified in engineering units. Doing so allows the team to track requirements and show how they are met in the finished product. Proper requirements don't stop there, they allow anyone to review how the system is broken down into smaller pieces, and see how the behavior of each element of the system combines to meet the requirements for the system as a whole.

Note that there is an apparent exception to this rule, which applies far more commonly to commercial development than military. At times it makes business sense that some of the requirements be left flexible, with some minimum performance standard being stated yet some enhanced performance standard being regarded as highly valuable. In this case, you will see a requirements breakdown that lists a performance level as "required" and higher performance levels (or additional features) as "desired." Being able to identify which of such "desire-ments" can be met without objectionable increases to the project cost is one of the areas where systems design is more art than science.

### Business as a foreign language

You may find a tremendous amount of reward and frustration in pinning down requirements. Often the impetus for a product comes from management, and management usually comprises nontechnical people. So your project is being driven forward by people who have—or think they have—a clear picture of what the product must be but are unable to communicate their vision in engineering terms. Since we were not created with the ability to see into each other's minds, the process of developing that product vision into a formal form written in "Engineer-speak" is not necessarily straightforward.

I cannot stress enough, that one of the opportunities to produce a disaster out of a system design is to allow this part of the process to slide by without giving it enough attention. Moreover, you cannot assume that the responsibility for the translation of the system description from English[1] to Engineer-speak rests with anyone but you and your engineering team. You and your group speak English, you and your group speak Engineer. Management speaks English and Business, but for the most part doesn't speak Engineer at all. It is your responsibility to handle the translation from your common language into the language of engineering design.

A great deal of the frustration involved in this process comes about if your management contacts do not have much engineering acumen. They may well believe that they have already described the system fully; if you go back to them for more detail, they will be frustrated with you for needing it, and you will be frustrated with them for not furnishing it. Be prepared to explain why you need the extra information and how things may go wrong

---

[1] (or in your native tongue)

if you don't get it. Furthermore, be diplomatic, in the event that your management doesn't understand why you are asking the questions.

---

### Example: Engineer-speak versus management-speak

There is a problem with communications between engineers and managers that crops up again and again. The problem can seem to be linguistic in nature, but the underlying problem is that while engineers and managers inhabit the same planet, we often inhabit different worlds.

Manager, "I want this working." (Translation: now or close-of-business today, tomorrow is OK, end of the week is acceptable.)

Engineer, "I'll get it working." (Translation: probably in 6 weeks but maybe 12 or more weeks.)

Manager, "I want this working." (Translation: complete, finished, out the door, in production.)

Engineer, "I'll get it working." (Translation: the prototype will have most of the features but it will need a complete set of verification tests, and then more features will need to be added and a full suite of regression tests run.)

---

*Prepare use cases*

Be sure that when you are talking to management that you cover every possible use case. If you cannot, make it clear that things are missing and that you will get back to them. Do not collect only the "normal" use case. Capture every action from the time that a user unpacks it until the time that they finish with it and put it back on the shelf (hopefully with a smile). If the product can be customized or adjusted by the user, capture those use cases. If the product must have batteries charged or replaced, capture those use cases, too. Think about whether the product needs nonvolatile memory; can it lose calibration or user-sourced customizations when the batteries run down or must such information be preserved even when the electronics have no power source at all?

Depending on the product, not all of the use cases will come from the same source. Probably, the most common examples of this would be for a product that either needs factory calibration or is designed for repair. In this case, your management contact will probably view the system operation from a user perspective, but their take on production and repair will be "well, of course it needs to be calibrated and fixed." In this case, you need to coordinate with manufacturing and service personnel to capture (or invent) those use cases.

---

### Example: Manufacturing use case

We should all know how to write use cases for the end user. Such a use case may have a high-level description as, "User opens the box, throws the directions on the floor, and starts pushing buttons until the power comes on." This use case would then be elaborated to investigate that particular user's out-of-box experience with the product.

Manufacturing and service use cases can be just as important, but they must start with a different set of assumptions. For most consumer products, you have to assume a variety of levels of training and knowledge in the end user and a wide variety of approaches to problem solving. In the case of manufacturing personnel, however, you should have a much better idea of how they will approach the job, and (hopefully) you will be able to count on a much firmer commitment to doing the job as specified.

Consider a calibration use case for a hypothetical gyroscope-based product that measures angles. In this case, the end-user use case would be to turn the device on, wait for it to stabilize, then rotate it while watching the angle readout. The calibration use case, on the other hand, might be a detailed set of instructions:

- Attach the device(s) to the calibration fixture (inventory asset # xxx).
- Turn the devices(s) on and put them into calibration mode.
- Rotate the devices through a predetermined angle (in the event of a well set-up factory, this will be defined by the calibration fixture).
- Instruct the devices that the calibration cycle is done (presumably the devices would then calculate some parameters to be stored in nonvolatile memory as their calibrations).

As you can see, the use case above is almost a procedural definition; it requires a considerable level of training and skill from a technician. This means that you must not only determine what must be done, but you must decide how much you can ask the manufacturing technicians to do and know.

As you flesh out your collection of use cases, you will not only clarify the level of skill required at each manufacturing or servicing step, but you will be able to generate a list of tools that must be purchased or built. This will make for a smooth transition to manufacturing, instead of a "throw it over the wall" event.

There are times that you simply cannot get the attention of manufacturing or service to help you with these use cases. You may not be allowed to talk to them, they may not be in the same physical location as you are, or they may not be interested in talking to you.

All of these obstacles are, to some extent, understandable. Manufacturing and service operations are generally focused on much shorter time scales than engineering groups. Furthermore, they may not be rewarded for the time they spend working with engineering. Your "immediately" may be a month or two out; for a production line or a service department, that same month or two may equal "forever." Except for the extreme case where you cannot, by company policy, talk to manufacturing or service, your best bet is to try to convince someone in each group that talking to you will make their future jobs easier, and that they have an opportunity to affect the future ease of their jobs now, by helping you out.

*Managers and company leaders please take note: Preventing people from talking is far more problematic in the long run than freeing folks to communicate. Strangling communications adds bottlenecks to information flow that stretches calendars and budgets. Information and time are indeed money! Strangling communications will ultimately cost you more!*

If you cannot talk to manufacturing or service, it is important that you try to put yourself in their shoes. One of the many ways that a design team can make a useless "lab queen" design is to leave the design in a state that each and every article that goes out the door must be touched by a design engineer at some point in manufacture. One of the ways that a design team can make a product that suffers from a bad reputation in the field is to make it so that it cannot be taken apart and put back together again by ordinary human beings—meaning that the customer's service department can't fix things that go wrong. So make sure that when you design a product, you have a clear idea of how it is to be manufactured and serviced. Address every single step of both manufacture and service. Realistically assess the skill of the people who will manufacture and service it. And clearly specify the precision of the equipment needed for manufacturing and servicing.

## The importance of early prototypes

When you get to the point where you have translated your requirements into formal engineering language, you are not done. You must translate these requirements, and the important characteristics that they imply, back into English[2] and perhaps Business-speak and run them by management.

In particular, you must make a point to capture any places where you were forced to change the nature of the system from that agreed on in your initial conversations with management. Those changes are where you had to revise (or chop, or mangle, or violently contort) their descriptions to come up with a realizable design.

Don't just depend on verbal or written descriptions. Language is a great tool for communication, but it is not the only one that we have been granted. Any time that you have an opportunity to make your examples concrete, do so. Sketches, physical prototypes, and computerized simulations are all very important ways to communicate how your system is going to end up looking and feeling to the end user. Indeed, you may find that such concrete examples are not only useful to you for communicating to your management team about the nature of the project, but they will be useful for the management team to communicate to investors or customers as well. Figure 8.3 is an example of a product designer preparing concepts for a new product.

Even during your initial meeting, when you are given a physical description try to generate a sketch. It doesn't have to be pretty—you aren't trying to win awards for your art, you are

---

[2] (or in your native tongue)

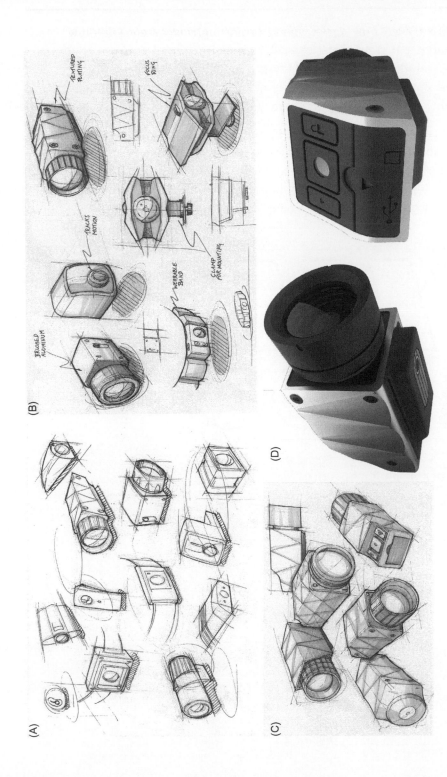

**Figure 8.3:**

A series of sketches and models prepared by a product designer in considering concepts for a new product, in this case a rugged consumer athletic camera. (A) Rough sketches of initial concepts. (B) Some refinement of the initial concepts. (C) Selection of the final concepts. (D) Final rendering of the selected concept. © 2013 by Seth Fowler. All rights reserved. Used with permission.

trying to communicate an idea. If the device you are sketching is small enough, draw it full size. If the device is large, put some object in (such as a stick figure of a person or a rough sketch of a car) to give an idea of scale. If your system includes a graphical user's interface (GUI), sketch the most important user screens. If your system includes blinking lights or small character-only displays, then sketch the content of those as the system is operated.

Figures 8.4 gives an example sketch for a hypothetical product with a simple seven-segment LCD display and three buttons for user input, showing how one might be able to use such a simple display to implement an interface with a multilayer menu. Note how crude these sketches are—yet they have been included in a book with international distribution, because they can effectively get a point across (Figure 8.5).

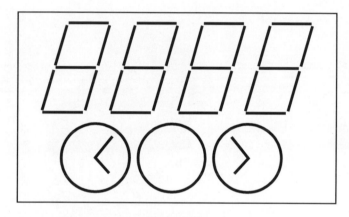

**Figure 8.4:**
A sketch of a simple front panel. It's ugly, but it gets the point across.

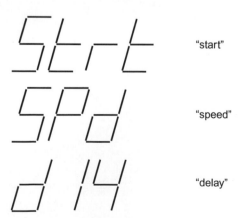

"start"

"speed"

"delay"

**Figure 8.5:**
Some menu choices, proving that it can be done with seven-segment displays.

While snazzy 3D solid-printed models are always nice, you don't have to staff up with mechanical designers and find a 3D printing service. A sufficient model of a system might be nothing more than a cardboard box with a front panel drawn on with a pen and rocks taped inside to give it weight. A model doesn't have to be pretty (although you do want to avoid giving anyone wood splinters from a really rough model). Try to match the projected size and weight of the final product—remember that a weight requirement on paper doesn't mean nearly as much as someone picking up a physical object and hefting it. You are trying to elicit, "Wow, this is pleasantly light!" or "No, this is just too heavy." responses early in the process, when things are cheap to change, rather than late in the process when significant amounts of money must be spent on revision.

### The importance of frequent prototypes

Don't think that you can rest after you get the initial product described. As you refine your product, be sure to make prototypes of the user interface. Even when the product is a small hand-held device with a "buttons and lights" user interface, you can still use computers to simulate a "buttons and lights" user interface. Have someone on your team code up software with a GUI that is representative of the end product's GUI (or have them code up a picture of your hand-held product, with buttons and lights simulated—Figure 8.6 is an example). Make sure that you can use the software to walk your management team through any critical parts of the functioning of the device.

Don't confuse prototype GUIs with functioning software. It is often unnecessary, or even distracting, to try to connect anything up to the GUI. If you have broken functionality

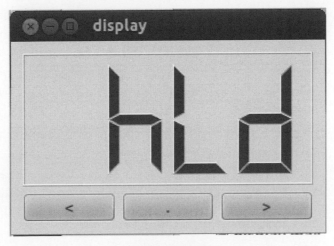

**Figure 8.6:**
Here is an example of a simulated display.

connected to a GUI, then the conversation will be about the functionality—not about the look and feel the user will experience.

Do not be surprised if you find that your early prototypes generate requests for changes. For that matter, do not be disappointed or upset if you get such requests. The reason that you are doing such early prototyping is to find the need for change in a timely manner. You want to make sure that any, "I wish it didn't do this." or "I wish it did do that." issues crop up when it is still cheap to change. (See Figure 8.6—changing in the concept stage is cheaper than changing in the preliminary stage, which is cheaper than changing the design in the final stage. See the notes from the NASA study mentioned in Chapter 1 that quantified the cost of finding and correcting errors in the various stages of product development.)

### Analyze requirements for feasibility and cost

Once you understand what your system is going to be, you must figure out whether it can be built at all and how much it will cost. If it can't be built at all, or if it will just cost too much, then it would be foolish to launch your company on an expensive design effort that will, in the end, go nowhere.

Feasibility is about knowing whether what you are doing is possible and likely to work. Setting aside considerations of cost, you need to ask yourself if the product that you are to build is physically possible, and whether the product as envisioned has a chance of meeting the stated goals. If you are a moral and law-abiding engineer working for a moral and law-abiding company, you must also weigh in on whether the product you are to build stays within the bounds of the law and does not violate your principles.

Feasibility does not stop at knowing whether the product can work. You need to design a product that can be reliably built by people that your company already employs or can reasonably hire. It is easy to design a product that can only be assembled, tested, and tuned by members of its design team. It can be much more difficult to build a product that can be built economically by people with the skills and education that you find on a normal assembly line.

Generally, the system cost has at least two, but possibly more, major dimensions. The two ever-present costs are the amount it will cost to develop the system (this is NRE) and the amount it will cost per deployed system (or per unit sold) to manufacture (this is COGS). For many systems, the cost to equip the factory to build the product (tooling costs) may also be significant. Less often, but still significant, some systems will also need to have an estimate of any warranty repairs or replacements.

In general, figuring out this feasibility and cost is a matter of extrapolating from known systems that you have experience with and taking any new advances in technology into account. Starting from what you know, you first extrapolate to what is possible, then you can decide on mundane things like estimated bill of materials cost, size, power consumption, and so on.

---

### Sidebar: BOM costs

As with the terms COGS and NRE, you will see the term "BOM" flung around conference rooms with regularity during the design phase of any physical product. "BOM" stands for "bill of materials"—it is the list of all of the items that must be assembled into the physical part of a product. As the bill of materials is the most visible part of the COGS, it gets considerable attention.

---

If you are building a system design that is a complete departure from what you have done before, just having the system requirements in hand may not be enough to estimate costs. In that case, you must accept the fact and at least take a first cut at the balance of the system design, so that you can come back and make reasonable estimates.

In truth, this step of analyzing for feasibility and cost should be monitored throughout the project. As you build the system, you will learn more about what you are doing and the feasibility of building the individual modules. With this additional knowledge in hand, you can go back and refine your feasibility and cost estimates and raise an alarm if the latest estimates differ from expectations. (Remember the NASA study in Chapter 1—finding and correcting problems later in a project is disproportionately more expensive.)

### Partition system design into modules

At its most obvious, breaking a system down into modules means deciding what the physical parts of the system will be and what they will do. There are many examples of this partitioning in systems that we are used to in everyday life:

- The cell phone system (which is a module of the telecommunications system) partitions roughly into cell phones and base stations.
- Highway transportation can roughly partition into automobiles and trucks, roads, bridges, tunnels, gas stations, and repair shops.
- Air transportation can be roughly partitioned into airplanes and helicopters, the air, and airports (which serve the dual functions of fueling and repair, with facilities for getting airborne and back on the ground, to boot).

As an example of how a system is partitioned, consider various ways that you may leave your house and go visit your favorite aunt in a distant city. You might drive or take a bus, you might take the train, or you might fly. (In reality you might do some combination of these. For the sake of argument, let's assume that you and your aunt each live next to an airport that has a train station).

If you take a train to your aunt's house, the two most important components of the system are the train and the railroad. The train can move itself along the tracks, and it can stop, but it can only go where the tracks go. In this case, the train is responsible for providing the motive power, but the steering comes from the rails, as does the support of the train against gravity.

If you take a car to your aunt's house, the two most important components of the system are the car and the roadway. Your car is held up against gravity by the roadway, but it (with your help) determines where it goes, and it provides the motive power. Thus, in this case, the car is responsible for both the motive power and the steering, but the support against gravity again comes from the road.

If you take an airplane to your aunt's house, the two most important components of the system are the airplane and the airports from which it departs and at which it arrives. Unlike a car or a train, an airplane must take an active part in supporting itself against gravity, and if it fails to do so then bad things happen. Like a car, an airplane can steer itself. Like both cars and trains, the airplane must provide its own motive power.

In each of these three cases, the demands placed on the various modules in the systems shape their economics and utility. Trains are more fuel efficient and less labor intensive per ton of cargo than either trucks or airplanes, but train tracks are expensive and more rare than roads. Automobiles are more fuel efficient than airplanes, but they travel much more slowly. Airplanes can (in theory) fly over any spot on earth, but because they must support themselves they must be maintained to a much higher standard than cars or trains, and the more economical an airplane gets to operate over long distances, the more expensive and extensive an airport it requires to be able to safely land and take off again.

How a system partitions into modules can have a significant effect on the amount of time and effort it takes to develop it, on the amount of money it takes to deploy the system, and on the systems eventual utility. This dependence on partitioning is very strong in systems that have a "public" component, such as telecommunications, transportation, and broadcasting, or systems that have some modules that must be duplicated (and paid for) numerous times, but it also impacts an "in-house" system with one module of each type.

In general, you want to partition your system into modules with the following goals in mind:

- to minimize the number of physical and logical connections between modules,
- to minimize the overall cost of the system,
- to make sure that modules fit in the available space, and where possible,
- to put any sensitive components into modules that will experience the mildest environmental conditions.

Depending on your system design, you may have other constraints. You may need to control access to trade or military secrets or you may want to load all of the patentable intellectual property into one module. You may want some modules to be very low power dissipation. Finally, you may want to divide a bunch of tasks among several processors to reduce the programming burden rather than take more time to fit all of the tasks into a single, monolithic processor; reducing time directly translates into quicker time to market.

---

### Sidebar: Intellectual property

Intellectual property (IP), such as patents, copyrights, and trade secrets, is the "secret sauce" that everyone tries to protect because it sets their products apart as a competitive advantage—remember the lawsuits that have been fought over screen operations on smart phones, those were all issues about IP protection. See Chapter 20 for more information on IP.

---

## Requirements budgeting

When you design a system, you will generally be presented with a set of fixed constraints to the system and a set of performance requirements. Your job is to meet these performance requirements while not exceeding the fixed constraints. A tool that works very well to help you to manage these dual and contradictory tasks is the concept of budgets. Find a way to express each constraint and performance requirement as a single number and keep track of each module's contribution to each of these numbers. In the end, you should be able to tot up each module's contribution to each budget and verify that the overall system budget is not exceeded. Examples of numbers that you can budget this way are overall system power consumption, system cost (including NRE, bill of materials, and COGS), overall weight, and total end-to-end delay.

You should make a requirements budget as early as possible. This budget should identify all of the requirements that are needed, then it should identify all of the modules that impact the budget for each of these requirements. Finally, it should assign a number to each of the modules as a minimum or maximum acceptable amount for that parameter.

When you budget this way, you realize two significant advantages. First, it helps you ensure feasibility. If you can show that your system will work by demonstrating that each module meets its requirements, and if you can show that it is feasible for each module to meet its requirements, then you can make a much stronger case that the system as a whole is feasible. Second, it gives you a way to ensure feasibility as you progress; as each module nears completion, you can test it to make sure that it meets its requirements.

An advantage that is less obvious, but still significant, is the help that a requirements budget gives you when modules fail to meet their requirements, either during the analysis stage or when they are actually built. Having a requirements budget, lets you plug in the actual numbers that you are achieving and lets you see how closely your system meets its overall requirements. In the case where you have some modules that are coming in below budget and others that are coming in above, a requirements budget can tell you if you are still on track to have a viable system or if your system is failing. Having a requirements budget also makes it easier to juggle requirements midstream or to assess just how hard you have to push on a subpar module to make the entire system perform as needed.

## Example: A simplified requirements budget

Suppose that you wish to build an audio recording system to transfer recordings from vinyl records to digital. The product line manager wants to maintain a 90 dB signal-to-noise ratio from the magnetic pickup to the digital audio. This magnetic pickup has an output voltage of 10 millivolts peak-to-peak at maximum. The frequency response of the system ranges from 30 to 20,000 kHz (ignore equalization for the purposes of this example).

After considering the possible circuits involved, you determine that the main drivers to the final product signal-to-noise ratio are the ADC and the preamplifier for the magnetic pickup. There are opportunities for noise to creep in at other points, but you decide that these can be dealt with by exercising due diligence in your design, paying attention to grounding and shielding.

You select an ADC that has a 93 dB signal-to-noise ratio (SNR). When this noise is reflected back to the equivalent noise at the preamplifier, you find that it is equal to 224 nanovolts, RMS. To meet your 90 dB signal-to-noise specification, you must have a total noise voltage of no more than 316 nanovolts RMS. This total noise voltage is your noise budget.

Noise power adds as the sums of the squares of the voltages. Total system noise is the square root of the sum.

$$V_{noise} = \sqrt{V_1^2 + V_2^2 + V_3^2 + \ldots + V_n^2}$$

$$SNR = 10 \log_{10} \frac{V_{signal}^2}{V_{noise}^2}$$

where $V_{signal}$ is the full-scale amplitude of the signal. The combined noise from the ADC and preamplifier, then, is

$$V_{noise}^2 = V_{ADC}^2 + V_{preamp}^2$$

$$V_{preamp} = \sqrt{V_{noise}^2 - V_{ADC}^2}$$

Solving this equation for the preamplifier noise, we find that the noise from the preamplifier must be 224 nanovolts, RMS or less, this maximum being the same as the equivalent noise from the ADC. Thus, we tell the preamplifier designer this noise level, knowing that it will result in the correct system SNR.

*Elaborating and tracking requirements*

Some requirements simply cannot be broken down numerically in a budget. Yet these requirements cannot be left to chance. You still need a means of ensuring that your overall systems requirements are met, even if the behavior of any one module cannot be quantified.

In this case, you need to elaborate your requirements. You must break them down into requirements for each module and verify that the sum of individual requirements for the modules satisfies the parent requirement. In essence, you have ask two questions, to ask, "Is this top-level requirement going to be satisfied?" and "Why must I meet this derived requirement on my module?" (Or is that, "Why, why, WHY must I meet this requirement???....").

An example of a split requirement would be a user interface. While a user interface can be a well-integrated entity in the software world, in the physical world it is often split into two distinct portions: there will be a display or other set of indicators (most likely visual), there will be at least one input device that may not be attached to the display, and there will be software. So to meet the requirements for the user interface, it may be necessary to specify two or three separate modules.

Take, for instance, a system that has a game-like user interface: there is a display, which gives the user information from the system, there is a hand-held input device with any combination of buttons, knobs, or joysticks, and there is a processor somewhere that is processing user and system input and deciding how to present the information that shows up on the screen. Fulfilling the overall user interface requirements means that the hand-held device must register its inputs and communicate them to the processor. Then the processor must register the communications from the input device and respond appropriately. The processor must also send the correct graphical information to the display, and that the display must be large enough, clear enough, and bright enough so that the user can actually see what the processor is trying to present.

You must be able to trace backward from any requirement and find it in the top-level requirement. This is true for any input device, display device, processor, or software so that no piece of the requirement is lost that might provide a clear and usable user interface.

The simple way to trace requirements is with a software traceability matrix. While a software traceability matrix can get unwieldy if you try to tie too many requirements together (which can easily happen if your project has a large software content), it is the acknowledged starting point for requirements traceability.

You and your company should seriously consider purchasing a software package to manage requirements. A number of companies provide good tools for generating, tracking, testing, modifying, and archiving requirements. These tools help you to avoid those embarrassing lapses or, worse, gaps in preparing and handling requirements.

## Design for this and that

"Design for manufacturing," "design for usability," and "design for quality"—the two words "design for," followed by a noun, seem to be ubiquitous. While you might like to dismiss these phrases as meaningless and overused, you should not. The phrases may arguably be overused, but they aren't meaningless. The reason for their ubiquity is that the "design fors" are often ignored, which leads to trouble. What you are designing for should have a significant impact on what you are actually designing because if you design for the wrong things, then your product will fail. Keep in mind the various "design fors," including volume, anticipated life cycle, maintenance, reliability, ease of manufacturing, longevity of the product, and all other aspects of the product that make a difference to how you prioritize the various parts of your design.

These "design for" issues can be slippery ones to address, primarily because there are so many disparate details to take into account, from so many different disciplines. I have seen "design for" issues successfully handled by small teams without a great deal of conscious effort on the part of the team, but in these cases the teams have been made up of experienced hands who had been through the pain of getting it wrong in the past. However, I have seen teams of smart people fail at the "design fors" when they are new to either a product or a market segment, or when they are trying to do an improved job in one of the areas they are designing for. If you have reason to believe that you or your organization are working in such unfamiliar territory then it helps to consciously think through the issues to explain or modify your behavior.

### Design for life cycle

Design for the anticipated lifetime of your product and for the anticipated reaction of a user to the product wearing out. A product that is intended to last for a month and be thrown away should not be designed with the same mindset as a product that is expected to run for 20 years. Design for too long a life and your product costs too much to buy. Design for too short a life, and either your customers reject it before they buy or you get a bad reputation in the field.

Products that you might expect to have long life cycles are such things as automobiles, large home appliances, capital equipment on a factory floor, aviation hardware, military hardware, some medical devices, and (possibly) personal computers.

Some products are designed to be disposable and inexpensive. These products are such things as inexpensive radios, wrist watches, and other small consumer items that would cost more to repair than to replace.

Other products are designed to be disposable but are by no means inexpensive or "cheap." They are impractical to reuse or are destroyed in the normal course of use, such as implantable ID "chips" for pets, single-use medical devices, and "high-tech" military munitions (e.g., rockets and cruise missiles).

### Design for production volume

Design for the anticipated production volume of your product. Missing the target on this score can have a significant impact on the production economies.

I touched on this issue earlier. A device whose production volume can be counted on the fingers of one hand will have a price that strongly reflects the NRE cost. On the other hand, a product device whose production volume runs into the millions should have its price dominated by COGS; a much lower portion of its price should depend on NRE cost.

Misplaced effort can be hugely costly. In general if you are building a few of anything, then you should buy a component when it costs less than you can spend to design it. More subtly, intentionally over-specifying components (e.g., buying $1000 worth of extra processing power to save yourself a man-week of engineering) makes far more sense than sweating the last penny out of the bill of materials. Finally, sometimes it makes sense to overengineer from the start. Spending a huge amount of time trying to find exactly what is good enough is a waste if you can identify a candidate design that is obviously more than good enough, and will let you proceed.

On the flip side of this coin, when production volumes are high, it can make sense to spend a lot of time sweating the details. If you're building a million of something, saving a dime on each one saves $100,000 over the lifetime of the product; that is a savings worth some engineering effort to pursue.

### Design for maintenance

The decision to design for maintenance is, to some extent, a part of designing for the life cycle of the product. Deciding that the product should be fixed when it breaks, rather than thrown away and replaced, is a life cycle decision. Once the decision is made to make a product maintainable, however, the goal of making the product maintainable must extend to every facet of the system design.

You must scrutinize each module in the system for its place in the maintenance schedule, and how easily it is maintained. For each module in the system, you must ask what level of repair operation is necessary to fix faults with that module, how easy it is to remove and replace that module, and what, if any, diagnostic aids should be built into the module. With those questions answered, the module can then be designed to facilitate repair.

Aerospace and the military applications commonly determine the facility that will repair a system module. Modules may be specified to be reparable on the flight line, in the avionics shop, at a specialized repair depot, or at the factory. Preference is given to systems that maximize the ease of repair, by minimizing the work of replacing any particular module, by maximizing the ability to use automatic diagnostics (BIT or Built-In-Test), and by minimizing the number of system problems that must be addressed at the depot or the factory level.

Other industries are not so formal, but repair can still be an issue. Large, complex consumer goods such as automobiles, major appliances, and built-in systems in homes are expected to be repairable. Brands that gain a reputation for being expensive or difficult to repair are seen as less desirable by the buying public, and goods that are sold under warranty are less expensive for the manufacturer when they can be easily repaired.

Capital equipment in factories is generally expected to be repairable. Typically, the amount of time it takes to repair a piece of factory equipment is just as important as the fact that it can be repaired. When a factory has a piece of equipment go down then failure is not just an inconvenience; every minute the factory cannot be productive is money lost.

---

### Case Study: Design for repair

A friend of mine was trained as an analog circuit designer. His first job out of college was as a service engineer for a semiconductor equipment manufacturer. These service engineers were sent out to job sites to fix equipment, usually with a satchel of tools and spare circuit boards.

He arrived at a semiconductor plant to repair a sick machine. He immediately started to follow his college training, by carefully going over the symptoms, measuring and poking and tweaking to try and find just which board needed repair.

After he had been at this for half an hour, the factory manager came up to him.

"Do you know how much longer you'll be?"

"No, I don't. I need to find the problem before I can fix it."

"Well, if you're going to take more than a day it'll be cheaper for me to have that wall over there knocked down, and bring in a new machine from your competitor."

My friend looked into his satchel full of new boards, replaced every one that he could, verified that the machine was working correctly, and got out of there.

He was able to do this because the company that built that machine not only designed it so that it could be repaired quickly by swapping boards, but because they had made sure that their service engineers were equipped to do so (if, perhaps, not trained strenuously enough to understand just how quickly they should act in the field).

---

### Design for upgrade

If you buy a digital thermometer or a clock radio, you do not expect to ever replace a component within it with something else that will make it better. If, on the other hand, you buy a component stereo system, then you are almost by definition starting from scratch and "making it better." Some products are sold with the intention that at some point in its life the user will add parts, or replace parts, to improve its performance; other products are sold with the assumption that they will not be substantially modified between the time that they come out of the box until the time that they have reached their end of life.

Designing a system to allow upgrades is similar in many ways to designing it to be maintainable. The requirements for replacement, upgrade, or modification are close enough to those of maintainability that it is difficult to do one without doing the other. As an example of this, consider that most automobiles are ostensibly designed to be used as is, yet there is an extensive industry devoted to various enhancements—for example, performance packages—that you can purchase and bolt on to your car.

What is different between upgrading a system and maintaining it is that maintenance makes it work as it did when it came out of the factory. Upgrades make it work better (at least for their purposes). If you are contemplating building a system that is upgradeable then you place a new set of constraints on the system design, which is particularly true if you want to throw the doors open to allow upgrades by third parties. Even if you never intend for the upgrades to be anything but "in house," you must take upgradability into account.

In general, you must consider the module definitions and their ramifications for upgrade issues. When designing a system to be easy to upgrade, you must be much more strict about avoiding each module's functions "bleeding" into another module.

---

### Example: Automobile dashboard

You design a car dashboard. You have a choice: you could save a few pennies and use the processor in the radio to manage the dashboard gauges, or you could make the radio a completely separate unit, but require that the rest of the dashboard be run from its own processor. If you choose the first path, you simplify vehicle production, but you make it extremely difficult for the eventual owner to upgrade the sound system. Furthermore, you make it impossible for your company to produce a "stripped-down" version with no radio.

---

### Example: Furnace

As a second example, consider what might happen if you were designing a home furnace. You see how you can save some money by making the furnace with its own unique thermostat, rather than an industry standard one. On the surface this makes sense—but doing so makes it difficult for a homeowner to use an existing thermostat, and it makes it more difficult to replace a furnace in an existing building, because he must also replace the thermostat. Thus, this "money-saving" idea ends up making the whole system of furnace, existing housing base, and independent contractors more expensive than the "more expensive" alternative.

The world of personal computers offers a wealth of examples of upgradeable systems. At one end of the spectrum is the common desktop PC, where not only is the system entirely modular, but each component piece of the system (e.g., motherboard, power supply, and case) has industry standard electrical, mechanical, and software interfaces. Because of these industry standard interfaces, each piece can be purchased separately, allowing technicians and even consumers to mix and match parts to meet their unique needs. With only a bit of suspense and drama, you can assemble parts from disparate manufacturers into a working whole, on a kitchen table and with simple hand tools.

At the other end of the spectrum are the newest innovation in personal computers, tablet PCs, which offer practically no internal upgrade paths at all. When you buy one, you are buying a closed appliance with no opportunity to change anything. Even these machines, however, are designed with external ports that allow you to expand on their capabilities by plugging in peripherals.

### Design for part obsolescence

Do you have a 6SH7 in your parts box? Do you have any equipment that uses one? Do you know what it is? The 6SH7 was a "must-have" vacuum tube for high-performance radio receivers in the 1940s. Yet today, unless you have an interest in old technology (or you are, perhaps, an old technologist), then you wouldn't even know what one is nor would you know where to get one.

You don't have to reach back into the 1940s to find parts that are no longer available. Anyone who has worked around electronics for more than a decade knows that just because you can get a part today, you may not be able to get that same part next year. As designs age, parts can go obsolete. When parts go obsolete, production teams have a far more difficult time building systems that depend on them.

For some of us, parts obsolescence may never be a serious issue. Cell phones, PC motherboards, and other high-volume consumer products can have life cycles that last for less than a year from conception to final ship dates, even though they may stay on store shelves for longer. When you are designing for such markets, dealing with part obsolescence is simply a matter of buying enough parts, early enough to build an anticipated production run.

Some systems, however, must be designed to last far beyond any reasonable initial buy. Automotive modules may change with the turning of a model year, but if the car that they go into is to be maintained over decades then those modules must be available for that same period of time. Military, industrial, and medical capital equipment can have lifetimes that exceed a generation, and yet must be kept going for their entire useful lifespan.

Designing systems to be "future proof" against parts obsolescence requires a many-pronged line of attack. To do this, you must:

- select manufacturers that are committed to supporting their components for the long term,
- make sure that circuits and mechanical systems are designed—wherever possible—to not rely too heavily on specialized parts,
- make sure that your software is written to minimize the effort in porting it to different processors,
- scrutinize your overall system design to make it easy to redesign entire circuit boards or modules if such a step becomes necessary when critical parts going obsolete.

Finally, where possible, it is advisable to write contracts with vendors of critical subsystems to supply these subsystems for some defined amount of time, or to give warning in plenty of time (generally a year or two) if they are going to obsolete a given subsystem or module. Of course, it is not always possible to get these commitments from vendors. When you are faced with needing to buy a part in such a circumstance, it is wise to make sure that there is a feasible backup plan in place for when—not if—the part becomes unavailable.

---

### Case Study: Fickle vendor (contributed by Kim R. Fowler)

A small medical products company developed an interface device for programming implanted devices. They selected a commercial-off-the-shelf (COTS) pen-tablet computer (precursor to the smart phones or tablet computers today) as the basis for their programming device.

The company spent months of engineering time to select a pen-tablet computer, all the while receiving assurances from the vendor that the computer would not become obsolete within several years. Well, surprise, surprise, the COTS vendor stopped making the particular model of pen-tablet computer within 4 months after its selection and made it obsolete (yes, in spite of their previous assurances). The small medical products company had to switch to another pen-tablet computer in midstream development.

The problem was that pen-tablet computers had a design and market cycle of 6 months to a year (smart phones or tablet computers today have even shorter cycles of around 2–4 months). Another confounding problem was that pen-tablet computers served much smaller markets than consumer products such as laptop computers; therefore, manufacturers don't maintain inventories of obsolete product—it was not profitable for them.

The only two markets where manufacturers maintain long-term inventories of older products are industrial automation and military equipment. But these devices are not necessarily COTS because they need to be rugged to endure harsh environments. This type of equipment is more appropriately called rugged-off-the-shelf (ROTS). ROTS tend to be three to five times more expensive than comparable COTS products.

The small medical products company could have purchased ROTS computers if it felt that it could afford the much higher costs. Doing so, they would have been able to write a contract that required the vendor to inventory sufficient numbers of ROTS computers to ensure future supplies. The other avenue the small medical products company could have taken was to design and build their own tablet computer; if they had done so, they could have contracted for long-term supplies of subsystem components, much the same as they would have done with a ROTS vendor. The tradeoff was between the NRE of designing their own device and the COGS of purchasing a COTS computer component.

---

## Case Study: A different fickle vendor story—contributed by Tim Wescott

A company I used to work for made a very high-end industrial camera. At the time that they were designing this camera, it was clear that building the computing engine around an ISA-bus[3] compatible "PC motherboard" circuit was quite feasible. No matter how the project team shined a light on the idea, no problems showed up.

Because of the nature of the design, the software and even much of the hardware could be prototyped simply by making first-article boards and plugging them into standard PCs.

The advantages proved too hard to resist. Indeed, as the camera wended its way through the design and development process, all seemed well. The software ran perfectly well on a PC and was nearly done by the time the hardware was ready, and the critical hardware elements had been already tested in ISA-bus PCs. Even the actual digital design was simplified, because the design team used a chipset that came with complete applications documentation that included schematic files that were compatible with our board layout software.

The fly landed squarely in the ointment, however, when the camera transitioned from engineering to production. The circuit engineers had verified that the chipset was available in quantity before they designed it in. By the time that the camera made it into production, however, the chipset had gone obsolete and was no longer available. By a combination of obsolete-part buys, a crash program to redesign the digital board to a newer chipset, and an up-front lifetime buy of the chips to build the board, the product was saved—but the mistake was costly and traumatic.

---

Designing the system so that it is easy to replace modules with newly designed ones is a similar discipline to designing the system for maintenance or for upgrade. In fact, designing for maintenance and upgradeability often builds in a certain degree of insurance against obsolescence even when it is not intended.

As with designing for maintenance or upgrade, the modules need to be well thought out, without too much overlap between the function of the module and the functions of the other modules around it. The only real additional work when doing such design is to take care

---

[3] The "Industry Standard Architecture", or ISA bus, was the first-generation IBM-compatible PC bus specification.

that any communications interfaces that you use are themselves "future proof." Basing a system on a communications interface that seems hot, but which fade from the market in a year, can be unfortunate.

---

### Example: Digital video interface

In cahoots with a colleague of mine, I designed a system to use IEEE-1394 as a transport medium for digital video in the late 1990s, only to see the standard fade from the market at about the same speed that we got our system ready for production. We had knowingly taken the risk that IEEE-1394 would remain a strong force in the industry, based on the fact that it seemed popular, that the roadmap for USB did not indicate nearly as much speed as we needed, and the ease of implementing the IEEE-1394 interface.

For all of our care in deciding what transport medium to use, we fell under the Microsoft steamroller: IEEE-1394 died in the market and was replaced by USB.

Fortunately for us and for our company, we had chosen a version of IEEE-1394 that persisted on the market long enough for our product to be a success. Because IEEE-1394 never realized its full roadmap, however, subsequent versions of the product turned to different technologies to transmit video.

---

Selecting manufacturers for support can be a difficult task if you aren't familiar with the market. Many manufacturers will be quite frank about not supporting a product over long periods of time, others will freely claim such support and then not have the will or the ability to carry through in the fullness of time (see the example above). Finally, some companies will put a design into escrow or contract to inventory some minimum number of units—but it will cost you more. At best you can select electronics components that are as generic as possible, that is, so-called "jelly bean" parts. When you cannot do this, I have found that selecting large manufacturers that stress automotive markets and other long-term applications give me the best chance at selecting long-lived part lines.

---

### Sidebar: Design escrow

The term "escrow" means putting money or documents into the hands of a third party, to be released to the new owner or returned to the original owner depending on whether contracted conditions have been met.

In industry, putting a design in escrow means that the manufacturer of a subsystem will put a copy of all of their design documents into safe keeping. Then, in the event that the subsystem manufacturer goes out of business or discontinues production of their design without providing a suitable replacement, their customer can use those design documents to put the design into production.

Properly done, having a design in escrow can limit the damage to the customer when that product is no longer available from the original manufacturer.

In designing electronic circuits, you can "future proof" the circuit to some extent by using "jelly bean" parts. A "jelly bean" part is one that is readily available and has been around for decades and is either directly replaceable by a part with the same part number from a different manufacturer or is very similar to a great number of other parts from a variety of manufacturers.

Examples of such "jelly bean" parts are the various comparators available as the "LM393" or the wide variety of operational amplifiers that are designed to drive high-impedance loads and have gain-bandwidth products of no more than 10 MHz or so. In the digital world, choosing simple 74xx gates in the variety of available CMOS technologies, or the "one-gate" equivalent chip, is often a very safe bet.

Using a "jelly bean" part, where they will do the job, will go far to future proof your circuit. This is often valuable, even if it is at the expense of needing more components. Choosing an operational amplifier that can be replaced by any one of a thousand other operational amplifiers gives you a very good chance at being able to fix an obsolescence problem by simply selecting another part. Choosing an operational amplifier that is at the edge of the performance envelope for operational amplifiers in general ties you to that part number from that manufacturer, unless it happens that other manufacturers jump on that bandwagon and try to compete with it.

Of course, you cannot always use this expedient. There are times when the value of using some unique part is so compelling that it is hard to say no. In that case then you should scrutinize the reputation of the manufacturer involved. If you are lucky enough to have high enough volumes, then getting contractual guarantees of continued supply should not be ruled out. (See escrowing the design and the case study above.)

Finally, remember that for all of your effort in designing for obsolescence, your system will still be susceptible. Companies that maintain product lines for years or decades have numbers of engineers dedicated to sustaining those product lines. Some companies may have entire departments titled "sustaining engineering," while others will draw their sustaining engineering expertise from their new product developers, as a continuing "tax" on the new product design process. Wise companies may even use sustaining engineering as a training ground for new product developers, to season them to the realities of that company's particular market segment and customer base.

As with all other engineering decisions, you cannot make a system entirely future proof. You can only try to find a balance between the present cost of decreasing the chances that part obsolescence will affect your product and the future cost of dealing with the obsolescence when it does finally happen.

### Design for manufacturing

Consider two systems. Both systems are modules of a larger systems. Both systems have similar all-around requirements for weight, performance, size, connectivity, and power consumption. Both systems have a stack of half a dozen circuit boards. In the interests of saving weight and engineering time, and increasing reliability, system A has tie-points on all of the circuit boards, and before the system can operate at all, the boards in the stack must be connected from point to point by soldering on wires. System B, by contrast, has that same stack of half a dozen circuit boards but each one has a card-edge connector, which plugs into a backplane board that carries signals from one board to another.

Which system is going to be easier to manufacture? Which system is going to be easier to maintain? Which system is going to be easier to upgrade?

(You may think this is an unrealistic, contrived example—it isn't. System A was inflicted on some engineering friends of mine by a strong-willed project manager. System B was actually a subsequent revision of system A, once the project manager was forcibly made to see the light of building an unrepairable system.)

This example is an extreme case of how bad design decisions can affect manufacturability (and, for that matter, serviceability). But it helps to highlight, by negative example, some of the issues you must consider when designing a manufacturable system.

As with other aspects of system design, how you design for manufacturability depends on the production volumes you anticipate, as well as the level of maturity of your manufacturing operation. It is not uncommon, where overall production volumes are less than a dozen systems a year, for the design engineers to be intimately involved in the manufacture of the systems (indeed, in such cases, it is not uncommon for each system to be unique). At the other end of the spectrum, when production volumes are high enough it is not uncommon for the design engineers to be on an entirely different continent from the production line. There is no fixed line to divide these models, but as production volumes increase, you generally see a shrinking role of the design engineers in the day-by-day production of the systems, and you see the manufacturing operation taking an increasing role in those aspects of the design that most impact them.

A system with very small production volumes does not need nearly as much accommodation to the demands of manufacturing as one that is being produced in large quantities. Even if you are building a one-off, making a system manufacturable will save money on the manufacturing floor—but if you spend more than you save during the engineering phase, you have not saved overall. So, if your volumes are small then when all else is equal you should come down on the side of manufacturability, but you should be prepared to trade off manufacturability for design cost wisely.

Do not shirk your responsibility to have a manufacturable system when you staff up for a project. As volumes increase, the importance of getting manufacturing engineers on board increases. In some companies this is a given, with project managers having manufacturing engineers on staff. In other companies, the manufacturing engineers may report through an entirely different management chain, and it may be difficult to get them assigned. If it is at all possible to do so, however, it is worth the effort.

Here are some of the properties that a manufacturable system must have:

1.  It must be transparent. Each module, subsystem, and board must be well specified as to how it must work. Ideally, each subassembly will be designed in such a way that it can be tested individually, apart from the rest of the system.
2.  It should be modular. A technician should be able to remove a module or circuit board, replace it with an equivalent one, and have the system still operate correctly. Even if the system is a snap-together throwaway, this modularity can assist in diagnosing production line problems.
3.  This modularity should extend up and down the chain. In a complex system, there should be test fixtures for each individual board and mechanical assembly, as well as test fixtures (or at least test procedures) to verify entire assembled systems.
4.  No part should have to go into its assembly untested. Each part should have a way of being checked before it is committed to its assembly.
5.  If all the pieces work, the whole system should work. This is, to some extent, an extension of modularity and transparency. It guarantees that an assembly of tested and functioning modules will work together.
6.  Calibration should be minimized and simplified. Calibration takes money and can require skills that are hard to train for and replicate. If a system must be calibrated, try to make it so that it calibrates itself or (assuming sufficient production volume) can be calibrated automatically.

## Design for anything

The biggest impact that "design for something" has on your design efforts is that, in general, the "design fors" cannot be deferred until the end of the product design phase. Doing so means that the success of the "something" that you should have designed for, whatever it was, is not a matter of design but of luck. You may find that you hit the mark, but if you did it was by random chance, not by your intent or skill.

Perhaps one of the biggest mistakes made by young engineers who have not yet grown into wisdom (and by older engineers who are immune to it) is to assume that as soon as you have a working example of your system sitting in a lab, functioning at room temperature on artificial tests, that you have completed a product design. Even after a prototype system has proven itself in the field, the actual production versions of the system must be deployed and must work correctly.

In general, leaving out a "design for" until the end of the process means that you have built the wrong thing! Pounding every penny out of from the bill of materials cost when you are building one item costs you time and wastes NRE. Specifying any component that will work in a design, regardless of price, costs you if you go to volume manufacture. Designing in a component that has the market lifetime of a mayfly leaves a ticking time bomb in the product if you must make the product available for years. The list of potential errors goes on and on, but the consequences are similar in all cases. When you design to inappropriate assumptions, or do not consider the assumptions that you are designing to, your efforts will lead to costs and schedule problems that can prove to be insurmountable.

So you must identify your "design fors" early. Preferably, you will identify them when the project still only exists on paper. These additional requirements must be laid on to all aspects of the system design, from the top level down through each module. As the system is being designed and built, you (and your project manager, if you are not one and the same person) must constantly check to make sure that the "design for" constraints are being observed and followed.

## System design choices

When you start with a blank sheet of paper and launch a system design, you will find yourself coining metaphors to juggling, or fog, or snake-infested jungles on dark nights. For any given system goal, there is nearly always a huge number of different ways of attaining that goal; trying to find the very best solution can be (to draw another metaphor) akin to wandering in a maze in the dark, hoping that there are no bears.

Part of the problem with system design is that engineers are taught that there are always hard and fast answers to every question. By implication, our training tells us that if we submit a question to enough analysis, we will get an answer.

Indeed, analyzing the workings of a system, whether real or hypothetical, can usually be done in this manner. But systems design is an entirely different matter. There are always innumerable ways to proceed on a system design task, with each method providing advantages and disadvantages that are dissimilar enough that it is not possible to enumerate every possible path, much less to analyze each one to try to find some "optimum." Moreover, the advantages and disadvantages are not always easy to pin down in some tidy numerical way, with engineering units attached to each element of your cost—benefit tradeoff.

Hence, juggling. And, since some of the tradeoffs are unknown prior to trying them, juggling in the dark. Moreover, since some of the choices, if made incorrectly, can be exceedingly expensive, juggling in the dark with knives. Suddenly, the idea of walking barefoot through a snake-filled jungle doesn't sound so bad.

In some cases, the nature of the system to be designed takes many of the design choices out of your hands. If the new system is to be an incremental upgrade of an existing one or if it is to be customized in some manner from an existing, designed, system, then much of the system design work has already been done, and the amount of work is inherently limited.

There are times, however, when you must make a significant departure from existing systems. You may be faced with a product that has reached the logical end of its design lifetime and must be replaced. Or there is an existing design that ought to fulfill the role but whose requirements have made such a leap that it cannot be used. Or you are simply working on an entirely new product.

Even in the case where you are designing an entirely new product, you should not get caught up too strongly in the natural desire to innovate. Engineers are often accused of "re-inventing the wheel," and as often as not this accusation is fair. Engineers do tend to innovate unnecessary new solutions, and you should fight this desire where it is inappropriate. Even when you are working on an entirely new product, there should be plenty of room for using old and boring technology.

It is good to remember that one of the definitions of "boring" is "predictable." Making sure that those parts of the system that can be done in some old and proven way is "boring" because they work exactly as expected. It is exactly this sort of "boring" that will save you energy and your company money. You can then use this energy and money to focus on those parts of the system that must be new and will need the extra work to function correctly.

### Build versus buy—contributed by Kim R. Fowler

In designing a product, you have to make a fundamental decision—to build the components or to buy them. Custom design allows you to optimize for a narrow set of requirements. Purchasing components, on the other hand, usually reduces the time to develop the product. These can be good reasons but many more reasons should go into the tradeoff between custom design and COTS components.

#### Definitions

Understand the relationships among components, subsystems, custom design, and COTS to begin the tradeoff between build and buy [1]. A component is any module, hardware, or software that you choose to connect with other modules to form a system. Any component is a system of interconnected, smaller, and simpler components. You need to understand at what level your design resides and then define the "components" accordingly.

If your product is a power supply, then the components are capacitors, inductors, resistors, transistors, and integrated circuits; the power supply is the final system. If your product is a data acquisition and processing system within a chassis, then the power supply is a component in your product and your data acquisition and processing system is the final system. If you are designing, building, and integrating a large system with lots of equipment (perhaps a weapons system within a navy cruiser or a product manufacturing line), then that data acquisition and processing system is merely a component to you—it is a large and expensive component, to be sure, but a component all the same.

As systems become more complex, so do components. The function that was yesterday's system, is today's subsystem, and will be tomorrow's component. The COTS revolution reflects historical trends of systems becoming components. Discrete transistor designs gave way to logic gates in the late 1960s and early 1970s. Then in the late 1970s and early 1980s, designs with logic gates and small-scale integration moved to large-scale integration and microprocessors. In the 1990s, boards and modules replaced custom designs based around microprocessors. Now boards and modules are integrating into systems-on-chip. In each transition, the design cycles became shorter and packed more functionality into smaller spaces.

### Custom design

Custom design may include mechanical hardware, electrical hardware, software, or any combination of the three. Designing a custom system allows you to optimize the design for a narrow and specific set of requirements. In general, custom-designed systems make sense when the advantages of meeting those specific requirements override the disadvantages and expense of the custom design.

If you are manufacturing thousands or millions of units, then a custom design can make sense because you can amortize the cost of an extensive custom development effort over the entire production run. If you are building something that has unique requirements that cannot be satisfied by any existing system, then a custom design makes sense because it is the only answer. Sometimes these unique requirements are partially satisfied by a collection of off-the-shelf parts but can be done so in a much more satisfactory manner with a custom design, such as a hand-held instrument that could be replaced by a PC, 20 pounds of equipment, and a rats nest of cables, but that just looks sharper as a single unit and is easier to cart around and use.

Building your own custom design can be a satisfying experience—or it can be a harrowing, mind-numbing catastrophe (remember juggling knives in the dark with snakes in a jungle?). Problems usually arise from unintended consequences in unforeseen circumstances.

### COTS

If production quantities are low or the market demands a short delivery time then you should lean toward buying COTS components. COTS components can be hardware or software; examples include cases, motherboards, standard interface modules, or commercial operating systems. COTS components are also recommended when you have a large, complex system or you need good technical support.

One example of COTS is buying software, particularly a real-time operating system (RTOS). Buying an RTOS is usually much cheaper than "rolling your own." But you need to run the numbers to be sure. It takes time to learn the features and operations of an RTOS. An RTOS also requires some overhead in processing time to schedule tasks; a general-purpose commercial RTOS may have so many features that this task-switching overhead is excessive. Finally, you have to ask, "How important is the technical support for this RTOS? Is it really maintainable? Does this vendor have a good reputation for supporting customers?"

---

### Case Study: Arc fault detector—both custom and COTS designs

Arcing faults are, in essence, high-impedance short circuits in AC power switching systems. They conduct sufficient current to sustain an arc but can remain below the trip threshold of circuit breakers. Arcs typically start as inline short circuits around dirty or loose connections and then jump between electrical phases on the bus bars to generate white-hot heat that melts and consumes the metal in switchgear in 1 or 2 seconds.

The Arc Fault Detection (AFD) system monitors and extinguishes potentially deadly arcing faults within the power switchboards on ships. It uses two different types of sensors to detect arc faults: a photodiode detector and a fast-acting pressure sensor. The first version of the system was custom built to military standards in the 1980s. The Navy built and installed such 175 systems, which totaled to more than 10,000 photosensors, 4000 pressure sensors, and 175 control units [2].

In the late 1990s, some colleagues developed a second generation of the AFD system. The redesigned system incorporates COTS components and uses a high-level software language. The computer board is PC/104 format and purchased from a commercial vendor. The enclosure is a commercial enclosure for industrial applications. The sensors have been modified and simplified with plastic housings and COTS components so that cost per sensor dropped from over US $2,500 for the custom design to about US$600 per unit for a design using COTS. The total cost per protected switchboard has dropped from US$26,600 for the first generation, custom system to about US$4,500 for the second generation, COTS system [3].

---

*Tradeoffs: parameters of build versus buy*

First define the parameters of your decision. Here are seven important ones:

1. Cost
2. Quantity
3. Time
4. Longevity

5.  Specifications
6.  Resources
7.  Technical support.

### Cost

Cost affects nearly all projects. You need to weigh the cost of buying system components versus building them. Building can be cheaper but you have to consider how long it takes to ramp up manufacturing; this decision affects time to market for the product.

Development time can be costly! NRE quantifies the development costs, which must be amortized over the production life of the product. Tooling for manufacturing can drain your budget as well. Sometimes buying system components can be far less expensive than building a custom design. Many times it can also save a lot of time. At other times, however, the COTS solution is more expensive because it offers too many features, because it requires modification to meet very precise specifications, or because it does not fit well in the system and must be designed around.

Your next consideration is the expense of components. This can be nonintuitive. If components within a product are cheap, then chances are that you can build the system cheaply (Figure 8.7). As components get very expensive, you should seriously consider buying them and leverage the buying power of the component vendor who can build them more cheaply than you can. The component vendor can spread costs around many customers; you can, well, spread your costs around yourself, unless you have many customers for your product.

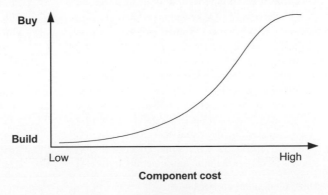

**Figure 8.7:**
An indication of how component costs can influence the decision to build or buy. © 2003–2013 by Kim R. Fowler. All rights reserved. Used with permission.

## Case Study: DSP board analysis—build or buy? (contributed by Kim Fowler with comments by Tim Wescott)

Assume that your company wants to build a system that requires a board with multiple digital signal processing (DSP) processors. The question is whether you and your team will design and build the board or will you buy it.

For this particular study, here are some of the basic assumptions and considerations that you might make to design and build the board:

1. The initial cost of the components and circuit board is US$870. (Large quantities of components will be discounted.)
2. It takes 2.2 hours of skilled, technical labor to assemble one board. (Assume a pick-and-place machine for some components and a solder reflow oven for soldering the board.)
3. You will consume about 300 hours of engineering effort to select and manage the component vendors.
4. You will consume about 2000 hours to design the hardware.
5. You will consume about 4000 hours to design and program the basic software.
6. Finally, it will take about 2000 hours to test the board and integrate it into the system.

**Table 8.2: List of pertinent expenses in building or buying a multiprocessor DSP circuit board**

|  |  | Hours | Costs ($) |
|---|---|---|---|
| COGS | Initial component costs for custom build = |  | $870 |
|  | Initial COTS costs to buy a board = |  | $6,000 |
| NRE | Assemby costs for custom build (effort in hours) = | 2.2 | $176 |
|  | COTS assembly costs (effort in hours) = | 0.2 | $16 |
|  | Selection of component vendors (effort in hours) = | 80 | $10,400 |
|  | Selection of COTS vendor (effort in hours) = | 200 | $26,000 |
|  | Management of component vendors (effort in hours) = | 200 | $26,000 |
|  | Management of COTS vendor (effort in hours) = | 200 | $26,000 |
|  | Custom hardware engineering effort (individuals × hours) = | 2016 | $2,62,080 |
|  | Custom software engineering effort (individuals × hours) = | 4032 | $5,24,160 |
|  | COTS hardware engineering effort (individuals × hours) = | 168 | $21,840 |
|  | COTS software engineering effort (individuals × hours) = | 2016 | $2,62,080 |
|  | Custom build test (effort in hours) = | 672 | $87,360 |
|  | Custom build integration (effort in hours) = | 1344 | $1,74,720 |
|  | COTS build test time (effort in hours) = | 168 | $21,840 |
|  | COTS build integration (effort in hours) = | 672 | $87,360 |

COGS, cost of goods sold; NRE, nonrecurring engineering.
No factor for inflation or for changing labor rates with time.
Support and assembly staff cost ($/hour) = $80
Engineering staff cost ($/hour) = $130

Here are some of the basic assumptions and considerations that you might make to buy a COTS board:

1. The initial cost of a board is US$6,000. (Quantity purchases of COTS boards will be discounted.)
2. It takes 0.2 hours of skilled, technical labor to insert one board into a system and test it.
3. You will consume about 400 hours of engineering effort to select and manage the COTS vendor.
4. You will consume less than 200 hours to design the system to hold the hardware.
5. You will consume about 2000 hours to design and program the basic software. (TW comment—This would only apply if, for some reason, the manufactured board came with software that happened to hit a sweet spot in your design. In my experience, the software

**Table 8.3: Per unit costs for building or buying a multiprocessor DSP circuit board**

| | Units Produced | % Discount on Components | Total Cost ($) | Time to Market (Months of Effort)* | Per Unit Cost ($) |
|---|---|---|---|---|---|
| Build | 1 | 0 | $1,085,766 | 50 | $1,085,766 |
| COTS | | 0 | $451,136 | 20 | $451,136 |
| Build | 2 | 0 | $1,086,812 | 50 | $543,406 |
| COTS | | 0 | $457,152 | 20 | $228,576 |
| Build | 5 | 0 | $1,089,950 | 50 | $217,990 |
| COTS | | 0 | $475,200 | 20 | $95,040 |
| Build | 10 | 0 | $1,095,180 | 50 | $109,518 |
| COTS | | 5 | $502,280 | 20 | $50,228 |
| Build | 20 | 0 | $1,105,640 | 50 | $55,282 |
| COTS | | 10 | $553,440 | 20 | $27,672 |
| Build | 50 | 0 | $1,137,020 | 50 | $22,740 |
| COTS | | 10 | $715,920 | 20 | $14,318 |
| Build | 100 | 10 | $1,180,620 | 50 | $11,806 |
| COTS | | 20 | $926,720 | 20 | $9,267 |
| Build | 200 | 10 | $1,276,520 | 50 | $6,383 |
| COTS | | 30 | $1,288,320 | 20 | $6,442 |
| Build | 500 | 20 | $1,520,720 | 50 | $3,041 |
| COTS | | 40 | $2,253,120 | 20 | $4,506 |
| Build | 1000 | 30 | $1,869,720 | 50 | $1,870 |
| COTS | | 50 | $3,461,120 | 20 | $3,461 |
| Build | 2000 | 30 | $2,654,720 | 50 | $1,327 |
| COTS | | 50 | $6,477,120 | 20 | $3,239 |
| Build | 5000 | 50 | $4,139,720 | 50 | $828 |
| COTS | | 50 | $15,525,120 | 20 | $3,105 |
| Build | 10,000 | 50 | $7,194,720 | 50 | $719 |
| COTS | | 50 | $30,605,120 | 20 | $3,061 |

COTS, commercial-off-the-shelf.
* Effort in time to market does not translate directly to calendar time.

(A)

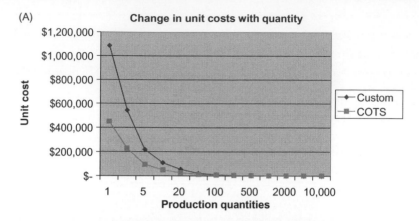

(B)

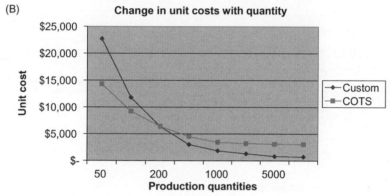

**Figure 8.8:**
Graph of the crossover point for Table 8.3. (A) Consideration of all unit costs between 1 and 10,000 units. (B) Highlighting the crossover of costs just below 200 units. © *2003–2013 by Kim R. Fowler. All rights reserved. Used with permission.*

savings in such off-the-shelf hardware efforts is about a wash. You spend less time getting the thing running initially, then you spend the rest of the project maneuvering around the fact that the hardware isn't quite exactly what you wanted.)

6. Finally, it will take a little over 800 hours to perform both the initial tests of the COTS board and the integration into the system.

For either situation, assume that support staff and assembly costs US$80 per hour and that engineering time costs US$130 per hour. (TW comment—These labor cost rates include overhead in some parts of the United States, they are too low in other parts of the United States, and they are too high in many other parts of the world.) Also note that the hours of effort do not translate directly into calendar time; some efforts can be done in parallel, thus reducing the calendar time; other efforts will be both sequential and performed on a part-time basis, thus seriously extending the calendar time.

See Table 8.2 for the listing of pertinent expenses. Eventually, the cost per unit produced crosses at about 200 units (Table 8.3 and Figure 8.8). This may be the cost crossover but there is no crossover point for the time to market. It is nearly always quicker to buy rather than to build a custom design.

**Case Study: COTS—a cautionary tale from the AFD system**

Remember that case study about the Arc Fault Detector system? Well, the redesign with COTS components was not quite as clean a divide from custom design as you might think. One of the designers later commented that though they bought COTS components, those components required considerable effort to design a custom interface to connect the different subsystems. The processor board, a PC104 format circuit card, did not connect directly to the sensors. The designers didn't use most of the functions of the PC104 interfaces, they had to perform custom design to develop the interfaces.

That's not all, folks. The vendor changed the card format of the PC104 card part way through their development cycle and moved the connectors 90° around the board! This necessitated another design cycle to move the interfaces to connect to the new board.

*Quantity*

Closely coupled to cost is the number of units manufactured. Quantity can be either a per year amount or the cumulative total for a production run. High production volume can amortize a custom design's NRE over the production run. What is not obvious is what determines the crossover from buying to building—a major consideration is the number of units produced.

Larger, more complex systems tend to have higher profit margins (meaning net income after expenses). Quantities of these larger systems, when less than 10s (for really large, complex systems in multiple, 7 foot tall, 19 inch racks, like specialized test systems for industry) or less than 100s per year (for a control console used in process industries), tend to indicate COTS; volumes above these might warrant custom design.

Even in such "full-custom" designs, however, closer examination often reveals a considerable level of COTS content. Our hypothetical, complex system in relay racks is likely going to be built in purchased relay racks—not ones that we custom built. Moreover, when someone pulls one of the custom-built modules out of that relay rack, you probably will find that it is in a COTS enclosure and contains many COTS modules, such as power supplies, keyboards, monitors, and possibly even entire COTS industrial PCs.

Conversely, smaller, less complicated, and cheaper devices tend to have smaller profit margins. Quantities of these smaller systems, when less than 1000s or 10,000s per year, tend toward purchasing COTS components. Depending on complexity, a custom design might be appropriate; more complex systems that cost more would indicate a crossover into custom design at lower quantities than for simpler systems that cost less. Figure 8.9 illustrates the general trends you might encounter in making a build versus buy decision.

In the absence of other compelling reasons, production volume can be a good barometer for choosing between custom and COTS design (but not always—see the counterexample below). High production volume, in the thousands (1000s) or millions of units, usually means that you can amortize the development cost of custom design over the production

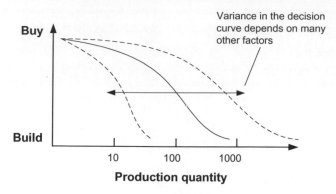

**Figure 8.9:**
The general curve for deciding to build or buy based on the final, manufactured quantity.
Quantity is an obvious component in the decision to build all the subsystems or buy the major
ones; what is not obvious is the point of decision between buying and building, which depends on
complexity of design, as well. © 2003–2013 by Kim R. Fowler. All rights reserved. Used with permission.

run. When production quantities dip into hundreds (100s) of units or less, then COTS
components may be preferable.

---

## Counterexample from Tim Wescott

All this is nice but I've got two active customers as I write with custom designs and markets
in the single or low two digits.

Being a responsible design contractor, I discussed off-the-shelf alternatives with them, using
many of the arguments that are presented in this section, but in both cases they decided to
go with full-custom boards.

They have a variety of reasons for doing this, mostly having to do with size, customer
impression, and concern over making life easy for potential imitators.

In both cases, they wanted to keep their products small. COTS modules tend to be large,
rectangular, and require external equipment to function. All of these characteristics stem
from perfectly reasonable causes, but in the case of these customers it meant that the final
product would have been too bulky if we had used COTS modules.

They wanted to keep their products unique. It is difficult to put together a product from
COTS parts without it being obvious that you have done so. Selling a product that is
obviously a collection of COTS modules can reduce your image in your customers eyes.

They wanted to prevent reverse engineering. Potential competitors can find it far easier to reverse
engineer a product that is made up of COTS modules than one that is custom designed.

That having been said, both of these "100% custom" products contain circuit boards that
feature BlueTooth radio modules—because they needed BlueTooth, and at the production
levels involved it was far easier and less expensive to put a module on the board than it
would have been to go through the extensive certification process required by BlueTooth.

---

Generally larger, more complicated systems have a larger profit margin; their breakeven point between build or buy is in the 100s of manufactured units. For smaller, cheaper products, the breakeven point moves out to 1000s of units manufactured each year. Figure 8.9 illustrates the general shape of the decision curve. This is the primary curve for deciding between building or buying. The other factors influence the sharpness of its transition and its translation along the *x*-axis, production quantity.

---

### Example: Power supplies within equipment box [4]

One manufacturer of electronic equipment has large circuit boards in products that use point-of-load (POL) power converters. Their business model requires the following criteria to design a custom POL converter:

- If the quantity of converters needed exceeds 1,000,000 and the output capacity of the converter is less than 20 W, then they will design a custom POL converter and build it in-house.
- If either the quantity is less than 1,000,000 or if the power capacity must be greater than 20 W, then they will buy the POL converter for their product.

---

*Time*

Time to market affects most projects. Time is money. Whatever shortens the design cycle should reduce the cost (assuming no bad or stupid decisions) and generally leads to increased market share. Usually buying the system components takes much less time than building them (Figure 8.10). The caveat in buying COTS components occurs when optimization complicates the design. Sometimes forcing a square COTS peg in a round application hole takes more time than starting from ground zero with a custom design. This is where your engineering judgment plays an important role.

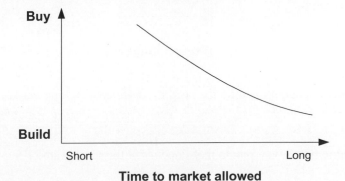

**Figure 8.10:**
How time to market can affect the decision to build all the subsystems or buy the major ones.

In the DSP spreadsheet case study above, the breakeven point for cost was about 200 units. Below 200 units, it is less costly to purchase COTS. Above 200 units, COTS may be more expensive. The personnel effort, however, was considerably different between build, which could take 50 months, and buy, which required 20 months. More time spent in custom development often means less market share for the product—this is the opportunity cost mentioned previously. To shorten development time, you might argue that adding people to work in parallel will shorten the schedule, but time expended does not decrease proportionately with more people thrown at a problem. It follows the old adage that "you can't make a baby in one month with nine women." [5] All other things being equal, the additional people will add expense to the project. Please give careful consideration when you trade more development time for the increased cost of components.

*Product longevity*

The longevity of a product can also drive the buy versus build decision. For many markets, vendors usually don't like to keep inventory of old designs. Product obsolescence is a fact of life. Figure 8.11 illustrates the general trend for making the build versus buy decision according to product longevity. The breakeven point can be fairly subjective but the longer a product stays in the field (or in the kitchen), the more likely you are to build a custom design.

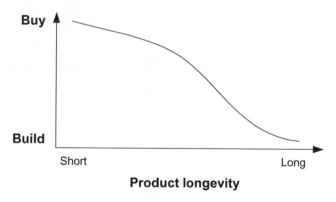

**Figure 8.11:**
Tend to buy the components, if the product has a short life cycle; a long market life often dictates a build decision. © 2003—2013 by Kim R. Fowler. All rights reserved. Used with permission.

Here are some vague breakeven points that depend on the market and life cycles:

- Military equipment: 3—15 years, some systems go for 30 or even 50 years. The airframe of B-52 bombers may even reach 100 years in 2052—this is not likely, but it's possible.
- Medical products: 7—12 years.

- Industrial products and process controls: 4—10 years, some systems go for 20 or 30 years.
- Common household appliances: 1—3 years for items such as coffee makers.
- Large appliances (washers, dryers, ovens): 5—10 years, though some appliances, such as ovens, might last as long as 40 years.

### Specifications and product complexity

Specifications describe what requirements the design should fulfill. Specifications don't tell you how to design the instrument, they just give the metrics of the design (e.g., weight, volume, power consumption, and performance). Once you know the specifications then you can begin the creative process of developing designs that perform accordingly. When you choose between building and buying COTS, you are deciding between refining an application in a custom design and buying the basic building blocks.

For very narrow and highly optimized specifications—higher speed, lower power, more severe environments, or other unique considerations—the decision tilts toward custom design. Examples of systems with very tight specifications include those used in space, undersea, or down-hole oil exploration; these sort of products exceed the market's expertise and capability to deliver COTS solutions.

As the specifications widen, COTS designs tend to be more flexible in a wider variety of applications. They are less optimized but generally have more features and are more modular than custom designs. Finally, as the specifications become very loose, you may swing back to considering building a custom design with fewer features but it will be good enough (Figure 8.12).

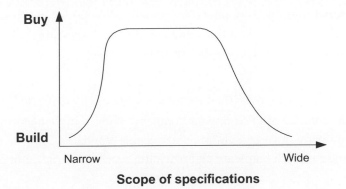

**Figure 8.12:**
How specifications can affect the build versus buy decision. © 2003—2013 by Kim R. Fowler. *All rights reserved. Used with permission.*

Product complexity affects your decision between build and buy similarly to specifications. The two are related. Some simple instruments may be easily built. As complexity increases, buying the components will save you a lot of development time and effort. Finally, as the complexity of the product becomes very high, COTS components may not have the necessary features, which means that you must build the component modules and subsystems (Figure 8.13).

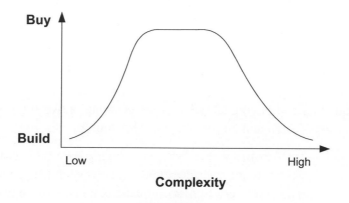

**Figure 8.13:**
How product complexity can affect the build versus buy decision. © *2003—2013 by Kim R. Fowler. All rights reserved. Used with permission.*

You need to know who will use your product; information about users should be implicit in the specifications. If only a small group of people will use the instrument, then custom design might closely fit their needs and capabilities. Conversely gurus, who are using a complex instrument and clearly understand its operation, may want the large variety of features in a COTS-based design. These same features, however, may stymie someone less familiar with the instrument. Otherwise, COTS might offer a well-known standard, such as a GUI, that may serve untrained operators but actually slow down and frustrate a highly experienced user. So good luck with that convoluted line of reasoning! Obviously, you will need to spend serious time considering the tradeoff.

### Resources

Custom design can demand a lot from your company, both in manufacturing assets and in expertise. Your company must be current on applicable standards. Your team must have the necessary knowledge in both hardware and software; recent experience in a similar project helps (Figure 8.14).

COTS components can circumvent some of these concerns. Vendors of COTS components typically have surmounted the difficulties in applying expertise to standards, functionality, environment, and manufacturing. They are able to amortize their NRE over many units or

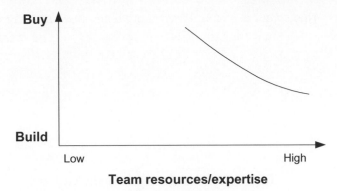

**Figure 8.14:**
Buy the components, if your team does not have the capability or capacity to build.

"design wins," whereas, you and your company can amortize your NRE over a single application with many fewer units than the vendor.

If the application requires expert knowledge, radiation hardened robots for example, then custom design may be your only avenue. If you and your team have specialized expertise in the desired application, then building makes sense. If the product is both very expensive and very specialized, like a space-based instrument on a satellite, then custom design may be your only choice.

### Technical support and training

Technical support has two very important and very different dimensions:

1. support needed by you from the component or subsystem vendor,
2. support that you must provide to customer for the final product.

If you have simple components, such as capacitors, resistors, or transistors, then you probably won't need much vendor support. (Regardless, a good relationship with a component vendor will help you circumvent many problems—problems still crop up even with "simple" components [6].) Conversely, vendors of complex components or subsystems, such as single-board computers, can provide valuable expertise that will save you time and effort from having to cultivate it within your company (Figure 8.15).

Providing technical support is like NRE at the end of the development cycle. If your product is simple or the customer does not need any technical support, this will influence your decision toward building the components and saving some money. If your product is very complex and customers require a lot of hand holding to operate and use it, then you probably will want to develop the expertise to design, develop, and then support it. For specific expertise of a complex product, you may not want to pay the overhead for COTS

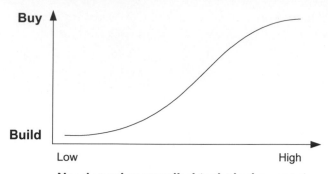

**Figure 8.15:**
The general curve for deciding to build or buy based on the vendor-supplied technical support.

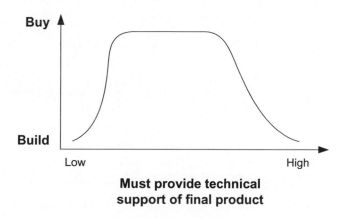

**Figure 8.16:**
An indication of how final technical support can influence the decision to build or buy.

components and extra technical support for features that don't matter to you or your customers (Figure 8.16).

*Other issues*

A number of issues revolve around these decision parameters. Some "-ilities," like maintainability, modularity, expandability, and interoperability, will favor COTS solutions. COTS vendors often have incorporated the appropriate standards in their products and have considered the necessary government regulations. Design maintenance, such as documentation and technical support, can be easier with COTS components. Technical support, in particular, is usually better with COTS components because more trained personnel are available to service them. (Clearly, COTS vendors cannot provide the system support, that is your job. They can and should help with the components and subsystems that you purchase from them. At least, they can help with part of the support problem.) That said, COTS vendors are not all equal; some are top-notch technical providers; others are scum of the earth. Read on.

### How to pick a COTS vendor

A COTS vendor should serve and support you. You really don't want to be left out in the cold with a complex component and no technical support—that would be the "snakes in the jungle" part.

You can evaluate a vendor several ways:

1. Read their user's manual. Get a user's manual from the vendor—a good manual means that it is most likely a good product. *I have never found a bad product with a good* manual (OK, with one exception out of hundreds of examples). Occasionally, you will find a good product with a poor manual, but a reputable vendor recognizes that the manual is an integral part of the product and will strive to produce accurate and useful documents. (OK, I have heard of one instance where the manual was outstanding and the product was a dud—it was for a complex application and the vendor must have gotten a good technical writer and then made up stories.)

   As a corollary to this, if the prospective vendor refuses to provide a user's manual, worry. If the prospective vendor has a user's manual and does not want to share it, then you've just learned that they will be tight with information through the lifetime of the relationship. If the prospective vendor does not have a user's manual, then they either don't have a product, or their market is very small, or they are not a mature company. All of these issues are causes for serious concern, and should be deal killers unless there is both something compelling about the product, and significant mitigating factors about the lack of this first-tier support.

2. Monitor their responsiveness to your requests for literature and information. See how accurately they fulfill your requests. Do they send you the information that you requested? I have had both bad and good experiences here. If they don't return your telephone calls or forget to send literature, that is a very bad sign. It indicates that they are either stretched too thin to serve customers adequately or they are neither careful nor businesslike.

3. Gauge their expertise in answering your questions. Does the vendor understand your application? Do they understand their product? Do they answer your questions honestly and reliably? Are their responses timely? A yes to each of these questions indicates a reputable COTS vendor who will stand by you (caution: does not prove, only indicates!).

4.  Ask colleagues for their experiences with COTS vendors and find out about the vendors' reputations. A good reputation is gained for a reason.

For processors, benchmarking is one method to help select a COTS component. Benchmarking runs the component or subsystem through a standard battery of tests or operations; then it records a single performance number. Hopefully, this gives you a basis to compare components from different vendors. "Benchmarks attempt to abstract and simplify complex systems so you can better perform apples-to-apples comparisons. . . . It is incumbent on you to analyze benchmark disclosures to determine a given score's relevance to your situation." [7] Benchmarking should be only one out of many activities that helps you select a suitable component or subsystem. Unfortunately, vendor-supplied benchmarks are usually marketing pieces skewed to their product; often, they are worth less than the paper (or PDF file) on which they are printed. Therefore, run your own benchmark tests tailored to your application.

Buying an evaluation system from the vendor will answer all these questions. Yes, buy it. That is the fastest way to uncover all the interesting "features" in a product, and a way to get to know parts of the sales channel that your manufacturing operation will have to deal with later. Getting a loaner system for evaluation may not give you a complete or accurate picture of the product or vendor. For one thing, getting a loaner is usually a hassle. For another, you may not have it long enough or be allowed to use it the way that you would like.

So what about cost? It is a *very* low priority in my experience. You may be disappointed with a product when you base a significant evaluation on cost. It's true—you get what you pay for.

### Marketing hype

Hype—virtual reality, vaporware, bugs—is out there. Some COTS vendors may claim that their products do more than they really can. A favorite trick is to play "specmanship" and advertise performance in unusual ways. This misleading tactic makes difficult the product comparison between vendors. Another trick is to announce capabilities that are not yet been developed. Another smoke screen is to advertise third-party alliances that provide support to the products and extend their usefulness. This is particularly true for hardware vendors who use third-party software. Again, that software may or may not yet work with the specific hardware.

You can debunk a lot of hype by doing what I suggested above. Check the veracity of the vendor. Get their manuals and read them. Buy evaluation systems and study them.

Finally, some vendors will hype certain aspects of performance but then fail to mention that you just can't use the product that way. If you do your homework, you will probably avoid this problem.

**Case Study: Marketing hype**

I once had an interesting encounter with marketing hype. I was involved with design using some large multiprocessor boards that had between 8 and 24 processors on each board. One company offered two different boards, one with 8 processors and another with 16 processors. Three other companies offered boards with 12, 16, 20, or 24 processors. These processors were identical between all the vendors in architecture, capabilities, and instruction sets. The hype was computational performance—imagine what kind of number crunching you can do with 24 processors on one board! Here is the catch—except for the first company, they didn't give any values for the power consumption per board.

The first vendor provided detailed power dissipation for both types of boards. The eight-processor board ran at 27 W and the 16-processor board ran at 31 W; both were worst case figures. The vendor carefully explained that the board with 8 processors ran at 5 V while the board with 16 processors ran at 3.3 V, thus reducing the average power consumption per processor. They also explained that either board would consume less power and dissipate less heat than the maximum specified for the chassis slot occupied by the board.

The other three vendors conveniently omitted the specifications for power consumption and heat dissipation from their literature. Furthermore, all their boards only used 5 V processors, so their worst case figures could easily range from 45 to 80 W per board. This far exceeded the power and cooling specifications for any slot within a VME chassis. While boards with 16, 20, or 24 processors could provide amazing amounts of computational horsepower, they would only do so until they melted!

In this case study, the hype of computational performance overlooked power consumption. *Remember the big picture. Do your homework.*

## Approaching a design

Figure 8.17 shows a general truth that I have noticed in systems design of all sorts during my career. The approach that you take to solve a given problem must be a good fit to the complexity of the problem you are trying to solve. Approaches that are appropriate and simple when you applied them to simple problems may become unbearably complex when you apply them to complex problems. On the other hand, approaches that work well to solve very complex problems may incur such severe start-up costs that while they can also solve easy problems, you should not use them.

As an example, you could choose programming languages for some embedded application. In the figure, line "A" might correspond to an assembly-language program (assuming that you are familiar with assembly language). For a simple-enough problem, it often takes less effort on a tiny processor to write and debug several hundred lines of assembly code than setting up a C run-time environment and verifying that it works. Similarly, as the problem becomes more complex, line "B" might correspond to using C to do some task. It is harder to get a working C environment going on a given processor than to use assembly, but once

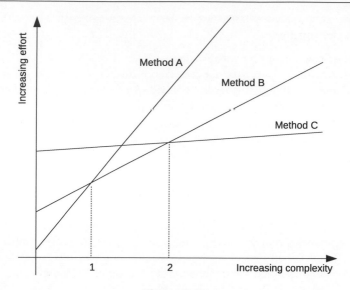

**Figure 8.17:**
Choose appropriate tools for the job at hand: tools that are easy to start using may be too unwieldy for large jobs; tools that handle high complexity well may be too costly for small jobs.

you do, a program that takes several thousand lines of C code would supplant tens of thousands of lines of assembly. For the right problem, using C is less effort. Now let your problem grow even more complex—say that you find yourself in charge of a large project team trying to manage tens of thousands of lines of code. In this case, you may find that the modularity and reusability of C++ or Ada starts to edge out the ease of coding in C, and then you progress to line "C."

This way of dealing with complexity not only applies to the choice of programming language but also applies to doing some purely mechanical tasks, where:

- line "A" would correspond to using only mechanical engineering,
- line "B" would correspond to using mechanical and electronic engineering,
- line "C" would correspond to using mechanical, electronic, and software engineering.
  (Editor's note: Chapter 9 on mechanical engineering supports this progression.)

In each case, you can always find a compromise between a simple approach that may blow up in your face and a complicated approach that covers all possible bases but has a steep start-up cost. Some examples include: choosing COTS when custom design is better; choosing an approach that is clearly best for dealing with the worst-case complexity that may have too great an entry cost, for example, using radiation-hard components in a coffee maker for terrestrial applications. Of course, there is always some approach that is just wrong no matter how you slice it, for example, using a steam engine to drive your autonomous drone.

So when you approach a system design, you should consider the available choices, think about where the problem may lie on the complexity scale, and then make a correspondingly appropriate choice of the proper approach to take.

Take note! Sometimes you might get by with choosing a less than optimum approach. If, however, you choose an approach that is entirely wrong, then you will cripple your product. Choosing approach "C" when your project complexity is to the left of point "1" in Figure 8.17 means that you have saddled your engineering staff with unnecessary work, and possibly your manufactured design with an unnecessary bill of materials cost. You may find that size, power dissipation, or reliability simply become unacceptable. Sometimes, the way to solve a problem is to pound on it with a rock. But by the same token, if you have a problem that lies to the right of point "2" on the complexity scale and you try to solve it with a "pound with a rock" solution, you will expend excess engineering effort, possibly burden your manufacturing group with unnecessary cost, and generally fail just as badly as if you tried to solve a simple problem with a too-complex solution.

All of your design choices should keep this these approaches and problems in mind. Moreover, you should be on guard against applying approaches simply because they worked well in the past. You should consider why these approaches worked, ask yourself if the situation matches what you have done before, and ask if that approach is the best one for the problem at hand. Make the hard decision! Look closely. Look carefully. Look objectively.

### Processor

Most of this section is covered under Processors in Chapter 10, Electronic Design. Both here and in Chapter 10, I assume that you are selecting board-level components or that you choosing a microprocessor or microcontroller chip that will be incorporated into a circuit designed by your team. The same considerations, however, still apply if you are limiting your design to the chassis level and you are choosing a processor board, or even if you are going in the other direction and incorporating a processor core into an ASIC or an FPGA.

The task when choosing a microcontroller or a processor board is complex. Not only are you selecting a processor core, which must be adequate to run your software, but you are selecting the amount (and speed) of memory and a number of peripherals. It is easy to fall into a trap of concentrating just on the core or on the peripherals, and failing to ensure that the neglected half of the choice is truly good enough. See Chapter 10 for more details on choosing a processor.

### Algorithms

The choices of the algorithms used in a system often don't bubble up to the top-level systems design. This can be a serious mistake. More and more with embedded system, the algorithms are the system or, at least, defining part of it. In these cases, the mechanical and electrical portions of the system may be nothing more than the carrier that lets the algorithm happen. In these cases, the systems engineer should be prepared to give the system's algorithms a great deal of attention.

When the algorithms do have a large impact on the system performance, then you will also often find that you cannot make the choice of algorithm in a vacuum. Your algorithm choice often will drive your choice of processor and its peripheral components. Conversely, choosing one processor over another may impact what algorithm you can use to good effect. Sometimes you simply cannot separate the processor choice from the algorithm choice; you have to determine the best tradeoffs among several possible combined systems to make an intelligent decision.

Here are some factors to consider when you are generating or choosing algorithms. Some of these considerations conflict with one another. You cannot have it all, so you must sort out what is necessary from what is merely desirable. Then where there are still conflicts, choose what is most desirable out of all that is necessary. Ask yourself these questions:

- Is it sufficient? Does it get the job done that needs to get done?
- Is it robust? If the situation changes or if you mistook what was required from the algorithm will it still work with little or no engineering? Or will you need to completely revamp it?
- Is it too much? Is your algorithm unnecessarily "gold plated?" Having any system component, algorithm or not, that is more than sufficient isn't a bad thing unless you have to pay for it—if there is a price to having an excessive solution, then is it a price that you and your team can afford?
- Do you understand it? Does anyone else understand it? At some point, any algorithm that you generate, buy, or have generated for you by a colleague will pass into someone else's hands. When it does, the algorithm will be judged not only by how well it works, but by how well other people can maintain it. At times it is better to use a less-optimal algorithm that anyone can understand than it is to use the World's Best Solution that is clear only to you—or worse, to some consultant.
- How many processor resources does it demand? Are they sufficient? Having a perfect algorithm that is clear and powerful and always works is a great thing, unless it cannot be executed on the hardware that you are planning on using. This part of the algorithm decision goes hand in hand with the choice of processing hardware; you can't choose one and then the other; you must be reasonably confident that you will have an overall solution that works before you can proceed.

### Signal processing chains

One of the possible areas of contention in a system design revolves around the signal processing chain. In the dawning days of the industrial revolution, humans or mechanical devices performed both signal processing and control. Today the available choices for signal processing technologies, whether strictly for signal conditioning or for the purposes of implementing a control rule, suffer from an embarrassment of riches.

Given exactly the same overall signal processing specification, you, the system designer, and your team are faced with an almost overwhelming number of ways in which the signal processing can be carried out. In some cases (but is becoming ever more rare as time goes on), the most sensible place for this signal processing is in the mechanical world. In some other cases, this signal processing can be best done in analog circuitry. In still other cases, this signal processing can be done digitally—but often only after the decision has been made as to whether it will be done in an FPGA, a DSP chip, or a general-purpose processor.

As with any other system design choice, the decision should rest on which approach minimizes the overall system cost while at the same time retaining a good fit with the available talent.

In general, the characteristics of the the locations for signal processing are:

- Doing signal processing entirely in the mechanical world tends to be very bulky and inflexible, but in a system that has no other reason for electronics it can make the system less complex, "low-tech," and easier to work on in the field. These advantages are particularly apparent where maintenance, manufacturing, and repair facilities are primitive. Examples of such mechanical signal processing include the flush toilet, automatic speed governors on small engines, carburetors, mechanical thermostats, and electromechanical ignition systems.
- Doing the signal processing entirely in the electrical world is more flexible and smaller. It often consumes less volume than a mechanical component. But remember that actuators and sensors add complexity, cost, and size. Electronics have the widest bandwidths of any approach. Feedback loops, for example, that are closed with electronics can have bandwidths into the gigahertz range; such bandwidths cannot be accomplished in any other way. On the other side of the coin, signal processing generally requires memory. The analog electronic components that provide memory are capacitors and inductors, and both of these devices do poorly when the "memory" has to extend to minutes or longer.
- Performing signal processing digitally in a processor or an FPGA has the best potential for flexibility, and often for size. You can't download resistor and capacitor values into analog electronics—but you can download entirely new control or signal processing algorithms at the touch of a button. When you put a processor into a system that did not have one before, you force an increase in complexity, but if the system requirements are complex, then using a processor gives you the best chance at dealing with those complex requirements. If those system requirements are subject to change, or if you are treading new ground in your system design, then having a processor that is correctly incorporated into your signal processing chain allows you to change the system behavior easily, either to match requirements or to cope with surprises as your design matures.

## Apportioning among disciplines

It is necessary to slice your system design into virtual or physical modules. But it is also necessary to identify problems that must be solved and to apportion the solution of those problems among the engineering disciplines involved in the project.

Slicing up problems by disciplines is no more clear-cut a practice as slicing a system into modules. Considering and anticipating the skills and training of your team members further complicates this slicing. Slicing the system up perfectly in a global sense may still lead to disaster for your team if you require the wrong skills from the wrong hands.

To compound your troubles, you would like to partition any particular module's design into as few different disciplines as possible, yet somehow the big three, mechanical, electronic, and software, manage to creep in despite your best efforts.

No matter what you do, if you are dealing with physical modules then some amount of mechanical engineering is going to be unavoidable. Any physical module must be packaged and packaging is a mechanical engineer's job. Even if the "module" in question is just a circuit board, someone must design that board to fit into some sort of space. If someone doesn't design the board for the space, then someone will have to design a space around it, which will increase expense, frustration, and project time.

To some extent you can establish a causal relationship among the disciplines. Electronics live in mechanical assemblies and software executes on electronic packages. Thus, if your project involves electronics design then it involves mechanical design. Even if you, or some electrical designer, does the mechanical design, it is still mechanical design. Similarly, if your project involves software content, then it must, perforce, include electronics design, and thus, mechanical design.

Possibly the only exception to this is if you happen to be in charge of some software-only project, where your team is given space on an existing processor in an existing mechanical assembly. Even here you are not totally isolated from electrical and mechanical design decisions—rather, you are a beneficiary of them (or, perhaps, a hapless victim).

In general, when I am thinking about how to partition a system function into mechanical, electronic, and software components, I find that there are two questions that always rise to the top:

1. What components or subsystems does the system need for interacting with the physical world?
2. Where should the system keep its brains?

The question of what components or subsystems does the system needs to interact with the physical world is the more concrete one. The system specifications will tell you how large the system must be, where it must fit, how it is to be handled, what it will be doing to the physical world outside of itself, what the physical world will be doing to it, and how it should respond to changes in the physical world around it.

Where the system performs its decision making or signal processing, that is, where it keeps its brains, is the more difficult question. This is because, in theory, decisions and signal processing can be done in nearly any engineering realm that you name. It is easy to see that it can be done in any of the Big Three: it can be done with mechanical parts, in analog circuits, in dedicated digital logic, in a processor running software, or in any combination thereof. But the possibilities are not limited to the Big Three. Theoretically at least, you could do your signal processing chemically, or acoustically, or optically. Indeed, historically each of these realms have hosted signal processing. To complicate your life further, each of these realms can be divided down into more options. For instance, mechanical signal processing can and has been carried out using linkages, hydraulically, pneumatically, and by other means. Indeed, the first use of control theory recognized in the literature concerned the use of mechanical devices to perform the signal processing necessary to close speed-control loops around steam engines [8].

It is exactly because the world is so versatile at doing signal processing that deciding just where the signal processing should be done can be so difficult. Often, there simply is no answer to what is clearly best. Worse, for nearly any signal processing task, the world abounds with examples of that task having been done badly, or well, in a wide variety of strange ways. And to cap off your misery, technology is always advancing and not always in the direction of just putting everything into a processor.

Consider the lowly ON-OFF switch. You can make your ON-OFF switch entirely mechanical, with a rotary, toggle, or push-button action. Or you can make an ON switch that uses a bit of circuitry to latch the power on when it is pressed, and an OFF switch that removes power when it is pressed. Or you can make an ON switch that turns power on but requires a processor to hold the power on. Finally, you can design your system so that any time it has power applied (or batteries installed) a processor that starts running in the background, which then monitors the ON switch and decides if and when it should power up the rest of the system.

Each one of these choices carries consequences with it. Mechanical switches that can implement an ON-OFF function can be astonishingly bulky and expensive. The more complex your mechanical assembly, the more unreliable it is. Switching schemes that depend on circuitry require the circuitry to be reliable. Finally, switching schemes that depend on the processor are vulnerable to software bugs. Users do not like products that are otherwise perfectly good but that cannot be reliably turned on (or off). For all that, a product can often be made to look much slicker and polished, and its cost can be reduced, if the ON and OFF switches are simple and match all the other switches in the system.

## Case Study: Optical system design considerations

Consider a high-performance optical system that must work over a wide range of temperatures. Optical systems must hold their lenses to tight tolerances in every direction. Yet optics benches expand and contract with temperature; furthermore, some lens materials change their optical properties with temperature.

A very desirable feature of such an optical system is for the focus (and perhaps the zoom and other lens characteristics) to hold steady even as the temperature changes. This allows the user (or the factory) to set the lens up once and then use the system without constant adjustment. This quality of being able to hold a steady focus in the face of changing temperatures is called "athermalization."

Optical systems are generally athermalized in one or a combination of three ways: fully mechanically, electromechanically with analog electronics, or electromechanically through digital processing hardware or software. In a fully mechanical design, the optics bench itself is designed to hold the lenses in the correct positions as the temperature changes. If the optical adjustments are motor driven, then the circuitry that controls the optics drive motors can monitor the ambient temperature and adjust the positions of the optics accordingly. Finally, if the optical adjustments are ultimately controlled by an embedded processor, then the processor can, with the aid of the circuitry, monitor the temperature and adjust the optics accordingly.

Each of these choices has both good and bad consequences for the life cycle cost of the system. Because of this, there is no one "best" choice.

An all-mechanical athermalization system means that your optics do not require any electronics or software content to work. But doing the athermalization mechanically involves extra complexity in the optics bench. This extra complexity increases design cost, manufacturing cost, fragility, and weight. Moreover, athermalization is always a fussy, error-prone process. You almost never get your athermalization correct the first time around. This means that an all-mechanical athermalization scheme will require a number of design passes to get the athermalization tuned and working correctly. Then, once your mechanical design is done, if it is changed for some reason other than athermalization, there is a good chance that the entire athermalization exercise will need to be repeated. Thus, if all-mechanical athermalization is chosen, there will be an impact on the engineering cost, bill of materials cost, and ongoing sustaining engineering to the lens assembly.

Taking a purely mechanical optical assembly and introducing motors, sensors, and electronics simply for the purpose of athermalization is probably not a good choice. Any savings in mechanical engineering grief that you enjoy from not having to athermalize mechanically will be matched, and topped, by the mechanical engineering grief that will result from trying to maintain your tight tolerances while moving lenses around. Then, when you are done complicating the mechanical engineers' lives, you will still need electronics.

Assuming, however, that you have a need to drive the optics electrically anyway, performing athermalization electrically or in software has a positive impact on the mechanical assembly costs. The NRE cost, the sustaining engineering, and the BOM materials cost will all see a savings by leaving out mechanical athermalization. Because the optics bench can leave out all of the mechanical parts related to athermalization it will be simpler and easier to make it strong and mechanically robust. However, you have now taken a task away from the mechanical designer and dropped it into the lap of the electronics designer, and possibly the software designer, as well.

Putting the "brains" of the athermalization process into the electronics will require both extra circuitry and extra mechanical work. It will require more sensors—if you wish to athermalize your optics, you must at minimum add a temperature sensor and, depending on your actuation, you may need to add a lens position sensor, as well. If you do not add a lens position sensor then you will be constrained to using an actuator that has positive motion steps, such as a stepper motor. Mechanically, homes will have to be found for these sensors or extra room found for the stepper motor (stepper motors are often bulkier than gear motors with equivalent torque output). Finally, the electronics will have to have the smarts to monitor the temperature and change the lens position as the temperature changes by just the right amount so that the focal distance stays constant over changing temperature.

Putting the "brains" of the athermalization process into software has all the drawbacks of doing athermalization electrically. Compared to having the signal processing done in analog hardware, however, it does have some obvious attractions. As long as your analog-to-digital conversion is accurate enough, you can easily use difficult sensors and linearize their measurements in software. If you are already controlling the motors in software, it is easy to augment your lens position loops with thermal information. In general, if the software already has full control over your lenses, then also making it responsible for athermalization is a good common-sense approach.

### Case Study: Apportioning among disciplines gone wrong

A product that I once worked on had a very odd feature. Signal 1 came out of a digital-to-analog converter (DAC), controlled by the microprocessor, and fed into an analog multiplier. The second input to the analog multiplier was signal 2, which was available from an ADC. The result was used in a motion control loop.

On inspection of the code in the microprocessor, it became apparent that the signal out of the microprocessor was derived by performing some trigonometry on yet another signal, signal 3, which came to the microprocessor as a digital signal.

There was no earthly reason why that multiplication could not be performed inside the microprocessor. Indeed, the multiplier in question was an ongoing source of trouble for the manufacturing engineering staff, because even though it should have been a ubiquitous "jelly bean" part, the control loop was sensitive to variations. Whenever the manufacturer changed their process, manufacturing would see their reject rate go up, and the design engineering staff would have to tweak component values to restore correct performance.

I asked one of my colleagues who had worked on the software, if she knew why this multiplication was performed "out there." It turns out that the circuit was designed by a couple of very good, but very pure analog circuit designers who did not know what microprocessors could do for them. They had determined the math needed to be performed on the signal and had designed a circuit to do everything but the trigonometry.

The trigonometry stumped them, however. So they asked the software engineer if she could provide signal 3, since she already had signal 1.

"Sure." she replied, and wrote the code—and a totally unnecessary circuit block was born.

If there had more system oversight, or even if the three engineers had gotten together and discussed why the circuit designers needed the trigonometry, then everyone would have realized that the multiplication could have been carried out in software. Doing so would have saved board space, cost, and ongoing headaches for manufacturing.

## Finding parts

### Build versus buy tradeoffs

As you break your system down into smaller and smaller modules, the inevitable question arises, "Who should design and build the parts?" This aspect of system design can be the most critical. It is often one that is fraught with inter-company politics, and there is often no clear path to take. (The earlier section on build versus buy tradeoffs focused on larger modules and subsystems. Here I am looking at the finer grain problem of components and small modules.)

Just to illustrate, let us look at the two extremes to the build versus buy tradeoff.

At one extreme, you can "design" your system by choosing your favorite product from the market, and having your company arrange with the manufacturer to build that exact product with your nameplate replacing theirs. At the other extreme, your company could mine ore, drill for oil, collect sand, and build all of the product elements by themselves, from the screws and semiconductors up to the paint on the case.

The "slap on a label" case above is not really a design effort, at least not in the engineering sense. It is sensible from a business perspective, and I have seen many companies do it successfully, but there is little or no engineering carried out in such a case. The "total vertical integration" case above is theoretically possible, but in this day and age it's almost impossible to imagine that a company could manage to be a successful resource-extraction company, and a components company, and a top-level product manufacturing company all at the same time without splitting into its component parts.

These two cases do, however, define the limits between which you and your company must operate. In a sense, each build versus buy decision is a decision between one of the above extremes, in a small way.

Defining the level of integration at which you are operating can be problematic. If you can buy contacts, shafts, and switch bodies separately is a switch a component part or an assembly? Is a purchased real-time operating system a component part when you could, in theory, roll your own? Is a BlueTooth radio that you solder onto a circuit board count as a component, because you treat it as one, or is it a module, because it contains several parts itself?

In general, the guidelines that I find most useful in making the build versus buy tradeoff is to answer the following questions:

1. Can this system function be realized at all by a purchased module, or am I on my own?
2. Do I really need to take this approach to realize this system function?
3. Is there some arrangement of purchased parts that will do the job in a satisfactory manner?
4. Does this system function involve skills that my company has or wants to develop?
5. Do the revenues and savings garnered from building this module offset the expense of designing, manufacturing, and maintaining it?

Clearly, if the answer to questions 1 through 3 are, "I am on my own and there is no alternative," then you need to implement that system functionality without recourse to a purchased module. Indeed, most new product development comes about because someone asked those questions, and came up with those answers. In many cases, you are doing what you do so that someone else can answer question 1 with "yes, there is a part."

The answer to question 4 should color your decision. In general it is very risky to try to design and implement something that is beyond your company's competency. However, there are times that you must simply bite the bullet and implement something that is new to your company. On the other hand, there may be times when your company is perfectly capable of doing the task, yet it will still make economic sense to farm out a module for another company to build.

Note, too, that question 4 can be a political hot potato. It is not uncommon for various divisions within a company to disagree about what the company's core competency or preferred direction should be. At times there will even be disagreement about whether a particular system function lies within the company's competency at all.

The question that counts the most is probably question 5. Restated, this question boils down to, "Is this the best way for my company to make the most money?" In a sense, it is the question that you are trying to answer throughout the entire system design process.

---

### Example: Question 5 and spending the right money

Question 5 is also one that is highly sensitive to the production volume of your product. To understand this point, ask yourself this question, "Is it a good trade to spend $30,000 of engineering time to save $400 in the bill-of-materials cost per system?" If you think about it, this is an incomplete question. Why? Because if you are building one system, then you are comparing a $30,000 expense with a $400 savings, and you come out ahead by a factor of

75:1 if you used the purchased part. But if you are building 10,000 systems, then you are comparing a $4,000,000 expense with a $30,000 savings, and the advantage of build (and design) versus buy is well over 1:1000, the other way!

## Buying off the shelf

While making the build versus buy tradeoff, you will want to consider all of the factors involved. In my example above, I present a clear-cut choice between a large amount of engineering time versus buying a part and using it. But that example assumed that there was:

- an exact one-to-one (1:1) correspondence between the purchased part and the module that you might design,
- there was no engineering work involved in selecting and using the purchased part,
- there was no risk to using the purchased part.

In practice, just as in the COTS discussion, none of these assumptions are entirely true. It is easy to underestimate the difficulty of using a purchased part. You would like to think that a purchased part can just be peeled off of its backing and pasted into your design. This is not so. You must sweat a bit over any purchased part before you selected select them for use. You must select with care even seemingly mundane parts, like bolts or resistors, or you risk having a $10,000 failure over a $.01 part. By the time you get up to the module level, you must be very careful that all of your system's design needs are truly satisfied by what the module delivers.

This process of vetting a module for inclusion into a system does not come for free. Vetting modules takes time, and it is not a task that can be glossed over. In general, any time you design a part into your system (even our hypothetical, lowly bolt, or resistor), you need to carefully compare the capabilities of the part with the demands that you are going to place on it.

Misleading, incomplete, or just wrong data sheets often complicate the vetting process. Furthermore, as modules get more specialized not only do they got more expensive, and riskier, they generally are designed for lower production volumes; lower production volumes in turn mean that their data sheets will tend to be even less informative and trustworthy than those of more mundane, commonplace parts.

You should approach the data sheets of any candidate subsystem with skepticism and care. You should scrutinize them just as if you were designing that subsystem yourself, from a blank sheet. You should make sure that all of your systems requirements for that part are truly satisfied. As you do this, you cannot leave anything to chance. Data sheets are published to sell products as well as to describe them. Any missing specifications could mean an oversight by the manufacturer, but could also mean that the manufacturer knows perfectly well that its product is a lousy performer in a particular area, and is trying to gloss over that fact. If you must have some feature or aspect of performance that is not mentioned on the data sheet, then ask the manufacturer to guarantee that aspect of their product in writing. If it isn't written down and signed, then it doesn't mean anything—it is equivalent to the feature not existing.

You should make a conscious decision about whether you want to extend your vetting of the module beyond just scrutinizing the data sheet. If it makes sense to do so, get a test article for evaluation. In the case of electronic or mechanical components, this may be in the form of an evaluation board or assembly; in the case of software components, there may be code available under an evaluation license.

This decision on whether to proceed on the data sheet alone should depend on how critical the module is to your design and how easy it is to replace it with an equivalent module. If the module is difficult to fully test outside of the system, but if there is an inexpensive and low-risk alternative that can be swapped in, then it may make sense to just build it in and try it later. Usually you will find that it is a good idea to at least poke at the module, to make sure it behaves as advertised.

Once you've thoroughly vetted the candidate module for inclusion in your design, you may still find that it doesn't do exactly what you wanted it to do. In some cases, this will lead to a decision to remove the part from your system and use an alternative. In other cases, this will force you to adjust the design of the surrounding parts of the system to accommodate that the module's particular characteristics. This is a risk that you should account for in your project planning, and it is a risk that you should plan on mitigating as much as possible by identifying not only how to test the module early, but how to test the module again and again as the prototype system goes together and starts to function as a whole.

Finally, when you buy a module there is always an element of risk. It is inevitable that you should be less familiar with a purchased module and that you may make some error when you include the module in your design that you wouldn't have if you had designed your own module. The manufacturer may have made a mistake in their predicted performance for the module and it may not live up to its published specifications. At worst, the manufacturer may have outright lied on their data sheet. Finally, purchased parts can always suffer from quality problems, and manufacturers can obsolete parts or just go out of business.

For all your care in analyzing the module's benefits, you may have missed some essential need of your system that the module cannot fulfill. Worse than that, you may find that the physical parts that you receive may not live up to their data sheets. In general, if you choose a module and it does not work to specification, the most that you can get out of the manufacturer is your purchase price. When such disaster strikes, all of the money that your company spent on design and development to fit that module into your system simply gets flushed down the drain.

## *Repurposing of parts*

Sometimes the functionality that you want isn't quite available in the form that you want it—perhaps the product you're looking at would work just fine if it were stripped out of its case and mounted in yours, or if its circuit boards were rearranged, or if it were sawn into two pieces and patched together with 20 feet of cable.

Buying subsystems from the consumer market, or buying subsystems that are almost what you need but must be modified is an attractive notion, particularly if you are building a system in small production volumes. You can get exactly what you want from something that is cheap and readily available, with what seems to be a minimum of effort.

I have direct experience with this practice, at two different companies, with hand-held page scanners and video cameras. The notion paid off each time, but there were troubles, trials, and tribulations that went beyond even the problems that you run into with buying COTS parts.

In general, such modifications can range from something as simple as taking the case off of the product and embedding the workings into your own case, to doing modifications as extensive as replacing PC boards or sawing boards in half. It can be a very tempting thing to do. It can, in fact, be made to be profitable. It has, however, many drawbacks.

First, it is almost impossible to get authorization from the original manufacturer to repurpose their parts. Product support is already a hugely expensive proposition when the manufacturer knows that the customer's interaction with the device is limited to the outside. When you, the customer, start delving inside the box, then the product support costs grow dramatically. Furthermore, the original manufacturer is going to be highly uncomfortable with the idea of helping you, a third party, to get familiar with their product. Even if your company makes promises not to go into competition with the manufacturer, you will be creating a route that you, or some of your company's employees, might use to become, or carry information to, one of the manufacturer's competitors.

Second, even in those cases where the manufacturer is willing to work with you in the use of their parts, your use of their product will leave you at their mercy when they make design changes. Any changes that they make other than, maybe, the exterior paint color will impact your product. Since product changes that do not alter the external character of the piece are often not announced even with a version change, this means that such product changes will come as a surprise to you—and they will often not be apparent until you have systems failing on the production line.

Finally, even if all you are doing is embedding the part into your system, without any overt changes at all, unless the purchased part is specifically designed for this use you may find that changes to the product—from something as simple as a change in the exterior design that introduces mechanical fit issues, to the product going out of production—will bring down your manufacturing line.

In general, unless the manufacturer has specifically agreed to selling you the product for some fixed period of time, and supporting you as it changes, then you will get little or no support from the manufacturer as these changes happen. This means that when you use such a product, you are signing up to making constant engineering changes to your product to keep it functioning.

### Case Study: Sawing a camera in half

I worked at a company that made airborne imaging equipment. This equipment was used for movie making, TV news, and airborne surveillance. The general architecture of the system was to have a camera on a gimbaled mount on the outside of the aircraft, communicating with one or more electronics boxes inside the aircraft.

The company had a product that was built around a very high-performance, broadcast-quality visible-light video camera. The gimbal necessary to fit the entire camera with the desired telescopic lens would have been enormous. Such a system would have been large, heavy, and expensive. The company decided that they would get far more market share if they could squeeze the required video capabilities into a much smaller gimbal.

The company could have built their own cameras. But building a professional-quality video camera is not a trivial task. Attempting to do so would have increased the amount of engineering necessary by several factors and would have required people with expertise far outside of what was available in-house.

There was no suitable solution available off the shelf. While it is obviously possible to separate a modern TV camera into parts separated by cables, there was not a suitable camera available on the market at that time.

The solution to this dilemma was to saw the camera in half. The light-sensing and low-noise analog electronics (the camera head) went into a box that was about a quarter of the size of the whole camera. The rest of the camera (the camera body) went into the central electronics box. The outputs of the camera head plugged into a custom board that buffered and amplified all of the signals, and those signals were transmitted by coaxial cable to the camera body in the central electronics box. That central electronics box needed even more custom interface boards between the cables and the electronics in the camera body.

The system worked very well, for many years. But every time a problem cropped up in the video chain, in-house engineers had to solve it, because the original camera manufacturer would not touch the product. Moreover, when the camera manufacturer made changes to their product they would inevitably come as a surprise to the production people, who would have to go back to the original system designer so that the various customized parts could be modified to accommodate the changes. These changes, in turn, had to be tracked as part of the main product. Ultimately, the camera vendor discontinued the camera. Because the technology had changed so much in the intervening time from the original design, the engineering effort to split a new camera was as involved, expensive, and lengthy as the effort to do it the first time around.

The bottom line is that in this case, modifying a readily available product worked very well. But the effort was fraught with difficulties that did not end until a camera became available from the manufacturer that offered an already-split unit.

### Buying custom-made subsystems

Buying custom-made subsystems can seem to be a project manager's dream. You have a problem that you desperately need to solve, and you have a white knight at hand (granted, one with his hand outstretched for cash) who will fight your dragons for you. As with any other buy versus design decision, however, there are downsides that are not immediately apparent. These downsides can range from mild to severe; you should approach such purchases of custom-made subsystems with care.

All of the problems that exist with buying ready-made modules exist when you set out to buy custom-made modules. In general, the best guidance that you, a seasoned project manager or technical lead, can get in relation to buying custom-made modules is to consider all of the ways in which your own company has fallen short when dealing with customers, and assume that any company you deal with may turn around and do that to you.

At best, when you buy a custom-made module you can clearly define the system needs that a module can satisfy, and you have candidate companies that have already designed and manufactured similar products. Ideally, you would be asking a company to make some simple modification to a standard product or variation on a custom module they may for someone else. Or you would choose a company that has a long history of designing similar custom-made products for customers in a similar industry to yours.

The process of using a custom-made product ends up being the same as the process for using ready-made parts or for purchased and modified parts. You have a hole to fill in your product, and you have a module with a data sheet that fills the hole. The biggest difference is that, where a ready-made module demands that you vet the data sheet and possibly modify your system to match, a custom-made module demands that you write the data sheet and negotiate its details with the manufacturer.

Negotiating a data sheet for a custom-made module is as much a design task as any other. It is not only common, but almost entirely expected, that what you initially ask for cannot or should not be done. There will almost inevitably be some feature that you ask for that drives the module price too high or that makes the candidate vendor flatly refuse to pursue the contract. When these cases crop up, you must adjust what you are asking for at the same time that you adjust your system design so that the hole you provide in your system fits something the vendor can reasonably supply.

The work involved in negotiating data sheets for custom modules expands exponentially if you are trying to source from more than one vendor. Either you will try to get exactly the same thing from both vendors, in which case you will be constrained by the limitations of both, or you will be trying to make an "A or B" tradeoff, which means that you will essentially be juggling two different system designs, one to accommodate each module, until you make your decision.

The most easily overlooked downside to custom-made modules is the amount of time it takes to negotiate the data sheet, the amount of engineering effort that needs to be made to accommodate the module design that is ultimately agreed upon, and the inherent risks involved with buying custom-made modules.

The inherent risks with custom-made modules are, for the most part, the same as for ready-made, writ larger. The module vendor may not provide exactly what you want, you are at the mercy of the module vendor, and the module vendor may decide to terminate production or may go out of business. Unfortunately, all of these risks are magnified in the case of custom-made modules.

When you buy a ready-made module, you are—at least nominally—buying an established product. You are buying something that has been made and will be made for other customers. This means that you are part of a larger customer base, which makes it harder for the manufacturer to ignore problems with the product. Even if you are the first customer for a new product, if any problems crop up with that product the manufacturer will be anticipating that they will be able to amortize their engineering cost over more than one customer when they fix the problem.

This is not the case when you buy a custom-made product. In essence when you buy custom-made, you are buying an orphan product that has no other buyer but you. If problems crop up with the product, the manufacturer knows that the only money that they are going to make from fixing those problems are going to come from you. In general, manufacturers of custom-made products will budget a certain amount of engineering effort to your module; you will find that as long as that budget has not been exceeded, they will be quite willing to help you with any problems. Should you have persistent problems, however, their profit/loss tradeoff for that module will slide into the red, at that which point, vendors can get very reluctant to continue doing business with you.

This problem is mitigated when you buy semicustom. Many companies are in the business of building products with customized parts around some standard core. For example, you may be able to buy an industrial pressure sensor that has the same actual sensor, but which comes with whatever sort of mounting provisions or connectors that you specify. In this case, the vendor does know that they can amortize any engineering that goes into the core over a great many customers. The only fully custom parts are the ones you specify. If you are wise, you will try to make those parts as easy for the vendor as you can.

Another mitigation to this problem occurs when you buy from a vendor that is established in the business of providing custom modules. Most such companies understand that their customers will have all the concerns that I just laid out, and they understand that their customers talk to one another, and that an engineer working at your company may be working for another potential customer tomorrow. These established companies will generally bend over backward to help you out with problems, as long as the problems do not get too severe.

---

### Case Study: DC motor drive

As a case in point, I was involved in a situation where I consulted with a customer to replace a drive module. The existing module had a DC brushed motor and a gearbox. My customer wanted to replace the brushed motor with a DC brushless motor. The decision was made to replace the entire module with a new motor, gearbox, and added drive electronics to accommodate the brushless motor.

The vendor chosen to build the replacement had a long-established history of doing such work successfully. We chose them with confidence that both the relationship and the product would be a smashing success.

Originally, my job was just to work with my customer's engineers to make some minor changes in the electronics and software of the product to accommodate the new interface to the drive mechanism and to do any necessary debugging and tuning of the related control loops.

What my customer did not know, however, was that the vendor had recently had a change of personnel. Their long-time control engineer had retired, leaving them with a hole in their staffing. The module ended up being designed solely by one of their sales engineers, with little review from their main-line engineering staff. As a consequence, the module that they built had a number of problems. It simply did not work in a satisfactory manner. After many trials and tribulations with trying to get the vendor's drive electronics whipped into shape, my customer was forced to throw out the vendor's drive electronics entirely and to spend a considerable amount of engineering time to develop their own drive electronics and software.

---

At best, designing in a custom-made module can be a slam dunk. You have a need and your vendor has the ability to fill it for far less money, time, and effort than you would have spent to make things work. At worst, designing in a custom-made module can be a system killer. In between are those cases where things work but not as well as you would like, or with far more money, time, and pain than anyone anticipated.

## Counterfeit parts

A growing problem in the industry is supply of counterfeit parts. A recent article indicates the breadth of the problem.

"The number of fake tech products floating around in the market quadrupled from 2009 to 2011, according to IHS—and they're sneaking into some high-profile places.

In September 2010, the Missile Defense Agency found that the memory in a high-altitude missile's mission computer was counterfeit. Fixing the problem cost $2.7 million. Had the bomb launched, it most likely would have failed, the agency said.

Two years earlier, the FBI seized $76 million of counterfeit Cisco (CSCO, Fortune 500) routers that the Bureau said could have provided Chinese hackers a backdoor into U.S. government networks. A number of government agencies bought the routers from an authorized Cisco vendor, but that legitimate vendor purchased the routers from a high-risk Chinese supplier.

... Counterfeit parts aren't just a government problem. Consumer electronics topped the Homeland Security Department's pirated goods list last year. ... They're popping up in every segment of the market, including wireless devices, PCs and even automobiles. Common problems include short circuits, failures in unusually hot or cold temperatures, and systems that don't boot up." [9]

There are ways to avoid counterfeit chips. The US government and NASA are active in anticounterfeiting efforts by vetting suppliers to pick the lowest risk vendors. They require vendors to show proof that they have government and vendor certifications. They regularly survey vendors and suppliers. Finally, they are cataloging vendors in a database and requesting developers and manufacturers to report counterfeit components to other buyers and to criminal investigative authorities (GIDEP, ERAI) [1,10].

# System analysis and test

System design is an art. There are too many different approaches to be considered and too many variables to juggle to be able to codify system design. System analysis and testing are much more science than system design, although you are still stuck with some uncertainty and in need for human judgment. With care you can, however, make definitive statements about a system design without building the entire system.

## System modeling

### Modeling from theory

Before you can analyze a system's behavior or even test it in a meaningful way, you must have a model for how the system behaves. Without a model you cannot even begin to calculate or even guess at the system's behavior; without a model the only system tests you can perform with any degree of certainty is full, final, field tests. System models are a necessary part of the system design.

A system model is a description of the system behavior, pertinent to the problem at hand. A system model is a simplification of the true system behavior by necessity. These two qualities, simplicity and pertinence, define the difficulty of coming up with a system model—not only is the full system behavior not known, but even if you could capture a complete behavioral description of the system's behavior, the resulting model would be too unwieldy to calculate or to guide tests. This leaves the system analyst treading a sometimes fine line between two extremes. If the system model is too simple, then it will not take important effects into account. On the other hand, if the analyst works too hard to take every possible effect into account, the resulting model will be too unwieldy to be useful.

In general, when given a the choice between building multiple small models that each illuminate some portion of the system behavior, or building one Grand Unified Model of the system behavior, you should go with multiple small models. The "large model" approach only makes sense if you have a need to study the ways that the various parts of the system may interact. Even then, the "large model" approach should only be attempted if you have high confidence that you can build a model that will be sufficiently accurate. You will have to decide how to construct your models based on the problem at hand, but remember that any system modeling effort has pitfalls to trap the unwary. As the model gets larger the number of pitfalls increases exponentially.

When you model a system's behavior, you need to start with a clear idea of several points. You need to know:

1. exactly what behavior that you are trying to model, for example, thermal, mechanical motion, electrical, control rules, or some combination of these,
2. if you can model the system in steady state, or if you must model the system's transient, dynamic behavior,
3. the factors that come into play in the behavior that you are trying to model,
4. how the various facets of the system's behavior and environment will affect the validity of the model,
5. just how firm your grasp of the true behavior of the system is, so that you can assess how close your model will be to reality.

---

### Case Study: Stabilizing a mechanical load

As an example of the choices involved in modeling, consider a pair of closed-loop motion control systems. You would like to develop a model to use for designing a stable control loop. Like all such systems, they must include an actuator, a mechanical load, a sensor, and a controller. For the sake of our illustration, give each of the systems an identical load that weighs 10 pounds, with a known low-loss (i.e., high Q) resonance at 100 Hz, but give each entirely different actuators and load supports. In the first of these systems, the load is floating on an air bearing with extremely low friction, with a speaker-coil actuator for low friction and fast response, and assume that the anticipated loop closure frequency is not limited at all by the actuator or the position sensor. On the second, the load is supported by rollers on ball bearings and is actuated with a rack and pinion attached to a motor (Figures 8.18 and 8.19).

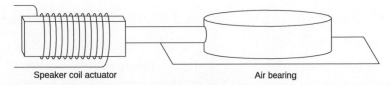

Speaker coil actuator                                    Air bearing

**Figure 8.18:**
Schematic diagram of a high-precision locating mechanism using an air bearing and a speaker-coil actuator.

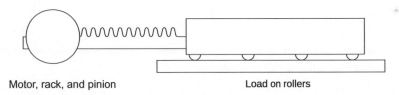

Motor, rack, and pinion                          Load on rollers

**Figure 8.19:**
Schematic diagram of a locating mechanism that uses a motor, rack, and pinion gear and a set of roller bearings.

Now ask the question, "Should the resonance of the load be in the model?" The answer is almost certainly different for the two different systems because of the circumstances of the load's suspension and drive.

For the speaker-coil and floating load, the answer is almost certainly yes. Air bearings and speaker-coil drives are designed to be low friction and will have almost no affect on the Q of the resonance. Moreover, air bearings and speaker-coil drives imply that the precision of the control is important, which in turn means that the control bandwidth should be as wide as possible. In this case, it is very likely that the control bandwidth will be limited by the system designer's ability to dodge the effects of the resonance. In this case, the resonance will have a significant effect on the system behavior, and it must be modeled.

For the rack and pinion with a rolling load, however, the answer is almost certainly no. Rolling suspension and geared drives will add considerable friction, which will, in turn, tend to damp the resonance significantly. Moreover, geared drives and rolling suspension will limit the precision of the mechanism and will also by themselves limit the attainable bandwidth. Both the anticipated damping from the mechanism and the implied bandwidth limitations mean that you can safely ignore the resonance.

Does this mean that we can automatically ignore the resonance in the geared-and-rolling case? Possibly not. Depending on the strength of the resonance and what is mounted on the load, it is possible that while the resonance wouldn't interfere with the behavior of the control loop, it may severely affect the correct operation of whatever is mounted to the load. So it is possible that the model that you develop for the purpose of assessing the control loop will turn out to be insufficient for the purpose of assessing the suitability of the mechanism to support its device.

Does this mean, then, that you should always model the resonances of a system? To answer the question, consider two things:

1. Any rigid mechanical assembly will have innumerable resonances, and the characteristics (including frequencies) of the resonant modes will be difficult to calculate (and hence model) accurately.
2. A mechanical assembly that includes any bolted joints, bearings, or other attachment points can get exceedingly difficult to model thoroughly. In theory you could use up the entire project budget for mechanical engineering on trying to chase the illusion of a fully accurate listing of the resonances of a part.

### Sidebar: Air bearings and speaker coils

Air bearings and speaker-coil actuators are mechanical elements used in control systems where a load must travel in a straight line and have very precise control.

An air bearing consists of a flat plate, with many small holes drilled in it. This flat plate supports the load, which is smooth and flat on the bottom. Air is pumped into the space beneath the plate, so that the air is forced between the plate and the load. The plate floats on a thin layer of air, making for a nearly friction-free bearing.

A speaker-coil actuator is a magnet and coil assembly. It works similarly to a motor, however, it has a small stroke, without any contact between the magnet and the coil. The force exerted between the magnet and the coil is proportional to the coil current, with no dependence on the relative position or velocity of the coil and magnet.

### Testing and refining models

The task of developing a system model is complex. Deciding what to include and what to ignore introduces an element of art. Furthermore, you could easily be wrong about your model or spend too much money and time generating a fully accurate model.

Practical systems engineers want to expend the least amount of effort and money on a system model, yet still end up with a model that is sufficiently detailed and accurate. Part of how you do this is through experience. Over time, you will learn just what is necessary and what is not. But part of how you do this, while also taking normal human error into account, is to test your model.

Model testing simply means that as you get physical parts you find a way to test these parts in isolation, look at how they respond to real-world inputs, and compare these real-world responses to what your models predict. If the modeled results deviate from the results of the physical parts, then something is wrong. When this happens you either have a conceptual error in your model which you must find and correct, or your model is structurally correct but needs its parameters tweaked. In extreme cases, you may find it best to discard your model and find a way to use the measured part's behavior in its place.

Whichever action you take—correcting, refining the parameter values, or discarding the model in favor of measured data—depends on:

- the importance of the accuracy of the final model,
- the ease with which measured data can be substituted,
- the amount of certainty that you have that the articles that you test are representative of the population of articles to be manufactured.

An example of wholesale substitution of collected data for modeled behavior is the frequency response method of control system design. In this method, you express control system goals in terms of its frequency response, and you use a modeled frequency response of the controller. During preliminary designs when you have not measured the plant characteristic, you use a mathematical model of the plant. Once you have a chance to measure the actual plant's response, you substitute that measured response for the modeled response.

By way of example, if you have a control system that has nonlinearities that are significant enough to rule out using frequency domain design methods, then you could not use a wholesale substitution of collected data for the modeled behavior to do your design. In this case, you would start with whatever model you had and use it to decide what time-domain measurements would make sense for the plant. Then you would take these measurements and compare them with the results from exciting the model of the plant in the same way. In the event that there were significant discrepancies between measured and modeled behavior, then you would tweak the model as necessary to match the actual behavior.

The process of modeling a system, testing its behavior, and refining the model can be easy and straightforward or it can be long and drawn out. There are times when you test the system and find out that your model is either too simplistic to match the real system, that you neglected to take some behavior into account, or made some other mistake that means that your model is just plain wrong. In these cases, you may have to go all the way back to the modeling phase with a clean sheet of paper and take another try at it, with the effort illuminated by the actual system behavior.

## Analysis

When I use the term "analysis" in this book, what I mean in general is "predicting the system's characteristics from the system's design." Thus, if I have a complete set of mechanical prints for a system, I should, in theory, be able to determine the system weight, its ability to bear weight, the amount of temperature rise that its interior will experience in response to power being dissipated in its electronics, etc. Similarly, if I have a set of code for a system and I know what processor is being used, I should be able to determine how much memory is being consumed, how heavily the processor is loaded, and other important factors from a software engineering perspective.

I'm not going to try to cover how to analyze every possible facet of a system's behavior and construction in just a few pages of one book. I'm going to assume that you already know how to analyze systems in your area of expertise, or at least that you know how to analyze pieces of such systems, and can figure out how to analyze the whole. Moreover, I'm going to assume that while you almost certainly don't know how to analyze every facet of a system's behavior, that you're modest enough to get help where needed, flexible enough to listen and understand the analysis done by others, and wise enough to know when you've asked too much of someone, and need to find someone else to do that bit of analysis.

The biggest challenges to determining a system's behavior by analysis are to make sense of your results and to ensure that your analytical results are real. Both of these problems can crop up and both can delay or derail your project. You should be sure that you are correctly interpreting what your analytical results mean and that your analytical results are correct.

### Static and dynamic analysis

When you analyze a system you often need to aim for two different targets. On the one hand, it is often sufficient to know the unchanging characteristics of the system, such as weight, memory requirements, strength, and power-handling capability. On the other hand, many systems must respond to a changing environment, changing demands, or both; in this case, you may need to understand the dynamic behavior of the system in response to commands or to environmental influences.

In general, analyzing the unchanging, static aspects of a system is easier than analyzing in detail how the system changes. It is easier, for instance, to analyze a set of code to determine the maximum stack usage for a task than it is to predict, from that same set of code, what that task's stack usage will be from moment to moment. Similarly, given a mechanical design, it is far easier to determine how much temperature drop there will be from some heat-dissipating component to the outside world after that component has been in use for a good long time than it would be to compute the exact temperature–time profile of that component.

Considering the challenges in analyzing your system, you should limit the scope of the analysis. One important place to restrain yourself is choosing between dynamic analysis and static analysis. Dynamic analysis of a system gives you answers that are more complete than just static analysis, but dynamic analysis can be far more costly. Thus, you should pick and choose when to do dynamic analysis (or other detailed analysis) when some simpler form of analysis will do.

## Case Study: Dynamic versus static analysis in software

You are designing an embedded system. This system has a control loop that samples at 10 kHz. Your system must meet two goals:

1. Sample the ADCs every 100 microseconds then update the DACs with a new command no more than 25 microseconds later.
2. Your embedded system is a module in a larger system. Your system receives a status request on its communication port every one millisecond. To meet requirements, the system must respond to this status request within 250 microseconds.

You have picked out a candidate processor. Through static analysis of the code, backed up by a bit of benchmarking on evaluation boards, you determine that the processing required to respond to the ADC with a command on a DAC requires about 10 microseconds to complete. The processing required to respond to a status request takes about 100 microseconds to complete.

You need to use analysis to determine two questions:

1. Is there enough raw processing speed to operate this embedded application with your candidate processor?
2. Can you use a cooperative (blocking) multitasking operating system without putting constraints on the software that responds to the communications port?

Static analysis is enough to answer the first question. Each of these two tasks only takes ten percent of the total available processing power. Thus, the candidate processor will answer our needs.

Static analysis is not enough to answer the second question. Some simple dynamic analysis involving drawing out the timelines of the two tasks, however, quickly tells you that if your "answer the status request" task is allowed to block the processor for a full 100 microseconds, then your system is absolutely guaranteed to miss its control systems deadlines. Thus, you must either use a preemptive multitasking operating system that can wrest the processor from the communications task to service the control loop or you must design the communications task to yield control to the operating system at intervals of no more than 15 microseconds. Either way, you give the task for servicing the control loop enough time to meet its deadlines.

## Case Study: Static versus dynamic analysis on the border between mechanical and electrical

I recently consulted with a company that wished to build high-performance hand-held photo flash devices. They wanted to be able to discharge capacitors at currents up to 600 amps, for up to 5 milliseconds at a time, using IGBT (Insulated Gate Bipolar Transistor) devices.

An IGBT drops between one and two volts when it is conducting. At 600 amps, this results in a power dissipation of between 600 and 1200 W.

The initial static thermal analysis of the situation indicated that we needed IGBT devices that were capable of dissipating 1200 W continuously. Such devices did exist—but they were larger than the intended size of the entire photo-flash unit.

All was not lost, however. Further consultation with the customer indicated that the anticipated flash rate of the photo flash unit would be no more than three full-strength flashes per second. At this rate, the average power dissipated by the IGBT devices drops to 9 W. Dissipating 9 W in the desired package size does present some thermal challenges, but it is entirely feasible.

One IGBT device, in a readily available package, can easily dissipate 9 W. It would seem, from the enhanced static analysis of the situation, that I was done and had got off easy. Did I?

No. Devices such as IGBTs cannot conduct heat infinitely fast from their junctions to their cases, and even if they could this heat could not be conducted from the IGBT case to the environment. Details of the IGBT's thermal behavior must be known. Fortunately, most IGBTs that are designed for high-current transients are specified in a way that makes the dynamic analysis easy. Such devices have a parameter known as the "Effective Transient Thermal Impedance." Looking at the graphs of the effective transient thermal impedance allowed my customer and me to determine that connecting several IGBTs in parallel, in the manufacturer's recommended fashion, would result in a circuit that met the electrical requirements without burning up our transistors.

## Case Study: Unnecessary dynamic analysis

You are designing a conveyor belt system using a brushless AC gear motor and a brushless drive that has built-in speed control. Based on the design of the conveyor, you know that the output shaft of the gear motor cannot exceed 400 rpm without the conveyor going too fast and jamming. The conveyor is in a machine that is designed in such a way that the conveyor never "helps" the motor. The motor is always pushing on the conveyor, never the other way around.

Your boss comes to you and asks you to do a detailed design of the control system for the conveyor, so that you can analyze its behavior and ensure that the conveyor will never go over speed.

You answer "Wait! I may not have to do that much work!" How can you say this?

Brushless AC motors have a fixed proportional relationship between their maximum unloaded speed and the voltage applied to them, called the "speed constant." Most brushless drives have a fixed relationship between their supply voltage and the maximum voltage that they can supply to the motor. If you know the motor's speed constant and if you know the maximum voltage that your drive will ever be able to supply, then you can calculate the maximum speed that the motor will ever go. If this speed is less than the 400 rpm speed limit on your motor, then the dynamic analysis (much less the premature control system design) is unnecessary.

### Finding symbolic solutions

Getting a symbolic solution to our analysis problem is how we often think of "system analysis." It is certainly the method that is commonly taught in engineering school. In this sort of analysis, you take your system model, you do a bunch of mathematics (or logical thinking, depending on the system), and you try to come up with an "exact" answer, often in symbolic form, to the question "how does this system behave?" (or weigh, etc.)

Finding a symbolic solution has the greatest potential to give you general answers, which are nice and succinct. To give a very simple example of such analysis, if you have a nice rectangular, sharp-edged brick, you can compute the volume of that brick by multiplying its height, depth, and width together. You can, in fact, easily write a formula to compute the volume of any such brick. Then you can use that formula over and over again (to, for instance, do a case study on the utility of different sizes of brick).

---

**Sidebar: Examples of symbolic solutions**

In signal processing, the transfer function of a filter or other physical system is a symbolic answer to the question "how does this system respond to input?"

In classical mechanics, if a solid shape is regular, you can use calculus (or tables of solutions) to answer the question "what is the volume, centroid, and moment of inertia of this shape?"

In software engineering, the equations in Rate Monotonic Analysis are symbolic answers to the question "can my RTOS schedule these tasks?"

---

On the down side, trying to derive a symbolic solution has the greatest potential to not give you an answer at all. Paradoxically, while symbolic solutions are often nice succinct, clear answers, they can also be horribly complex and opaque. Worse, it is often just plain impossible to arrive at a general solution to an analytical problem.

Other drawbacks to finding symbolic answers is that it is often difficult, which either means that it is possible but no one on your team can do it or that one member of your team can do it, but no one can double-check the analysis for validity.

Often when deriving symbolic solutions you can ease the task of analysis by modifying the system model. If you can find a system model that is close to the real one, but which makes the computation easier during analysis, then you can make such general solutions into a viable option where it was not before.

Modifying a problem to derive symbolic solutions is a common practice. It is very prevalent when solving systems of differential equations, because linear differential equations are very easy to solve, while nonlinear ones are often impossible. However, it is easy for practitioners in some fields (especially control theory and analog circuit design) to perform this "problem modification" step so routinely that they do not realize exactly what they are doing or why. This means that it is very important, when you are simplifying a system model to make the analysis easier, that you do so consciously and ensure that your simplifications are valid.

When you are faced with a task that requires analysis, but it is impossible or impractical to derive a general solution, then you must choose another method. The two prevalent methods to do this are either to let a computer crunch a lot of numbers for you or to construct some analog of the system and test that analog.

### Numerical analysis

If you have a model of your system that accurately represents your system, but you cannot analyze by computation, then a second-best solution that often works is to analyze numerically. Numerical analysis is used primarily when the system analysis requires the solution of some differential equation that does not allow for a convenient symbolic solution. This can happen in static analysis when a system must be described with a partial differential equation. It can also happen in dynamic analysis when a system must be described with a nonlinear differential equation that cannot be adequately approximated by a linear differential equation.

Numerical analysis is often done by application-specific tools. Analog circuit designers will use circuit simulators, mechanical engineers and magnetics designers might use finite-element analysis programs, and control and communications systems engineers may use general-purpose numerical analysis packages such as MATLAB or Scilab.

It is difficult to make many sweeping statements about numerical analysis of system problems, other than this: be careful. It is very easy to make a very high-fidelity simulation of a model that does not describe what one then actually builds or to perform a numerical analysis that is based on completely unrealistic assumptions. When this happens, you can easily lead yourself to believe that your system performs far better than it really does.

When you perform numerical analysis, you should have a checklist, and you should make sure that every item in this checklist is covered before you accept the results of a simulation or finite-element analysis program. It is not a bad idea as a project manager or technical lead to keep this list in mind, too, and to ask your team members if they have covered all of the bases in their numerical analysis.

The minimal checklist should consist of these items:

- Have you verified that your analysis is correct for the entire range of environmental variation (temperature, pressure, humidity, etc.) over which the system must function? Physical systems will change their characteristics over as the environment changes. Making sure that everything works under laboratory conditions is just the beginning, not the end, of your efforts.
- Have you tested your analysis with component variation? No two "identical" mechanical parts are ever truly identical, no electronic component ever exactly matches its data sheet values. You (or your designer) should have an idea of how variations in parts will impact the system performance. If it is possible, you should simulate the system with these variations.
- Have you done "sanity checking?" While you wouldn't be doing numerical analysis if symbolic analysis was easy and tractable, you can often use symbolic analysis, or alternative numerical approaches, to find limits on the numbers that you see from your "big" simulation. If the numbers are grossly different, then you should figure out why, and correct the problem as necessary.

### Testing physical models

You may be wondering why I put this section under "Analysis." It is because if you are testing some physical model of a subsystem, then you are really performing a simulation—you're just doing it the old fashioned way. (Editor's note: Both Chapters 7 and 9 do the same thing by including notes on physical testing under Analysis.)

Testing physical models or prototypes for the purpose of finding out something about your eventual system should not be confused with testing your actual system, or subsystems, for the purposes of verification. In the latter case, you are making sure that your analysis, and possibly your design, is correct. In this case, you are using testing of a physical article as another means of predicting the behavior of the system that you are eventually going to build. Examples of testing physical models include:

- Getting evaluation boards for a chip or a processor and testing it.
- Building scale models of aircraft to test in a wind tunnel or in flight.
- Building scale models of some mechanical assembly to test its strength or weight or other physical characteristic.
- Building a full-scale mockup of some part of your system to check fit or to test your user's reactions to it.

The important thing to remember about this testing, and the reason that it is in a section about analysis, is that you are doing the testing for all the reasons that you do analysis. Thus, you should put the results of this testing in the same place as you put your analytical results. You should treat the results of your physical tests with the same skepticism as you treat any other analysis that may be using an imperfect model of your eventual real system.

### Worst-case, nominal, and statistical analysis

So far I have talked about static analysis versus dynamic (or other detailed) analysis. Another distinction that one can make in one's analysis is worst-case analysis versus nominal analysis versus statistical analysis. These three types of analysis can be summed up as follows:

1.  Nominal analysis assumes that every component behaves as specified, and that the system's behavior does not change with changing environmental conditions.
2.  Statistical analysis assumes that certain component or environmental characteristics will vary in a known way, and attempts to find a family of system responses to these changing conditions by using statistical methods.
3.  Worst-case analysis assumes the same component or environmental effects as statistical analysis, but instead of trying to find a family of behaviors, it tries to pick out the very worst behavior.

Nominal analysis is useful to understand how a system behaves, but it fails to predict how the system may respond to undesirable effects. Generally, a nominal system analysis is not sufficient, unless the analysis indicates that the system is exceeding specifications by a good margin, and if the engineer doing the analysis has a good idea that it will continue to do so as factors change.

Statistical analysis attempts to predict how the system will respond to variations. To do statistical analysis, you identify all of the factors that you believe will contribute to the system meeting (or not meeting) its specifications, and you perform multiple analysis (or simulation) runs as you vary these factors. When you vary the contributing factors randomly and do many many runs, this sort of analysis is called "Monte Carlo" analysis.

Worst-case analysis tries to answer the same questions that statistical analysis does, but rather than varying parameters randomly, you identify what the worst possible values of these parameters may be, and then perform your analysis to see if the system meets its specifications under such circumstances.

In general, statistical analysis is most useful when you are building a product that can be "tested into quality" after production. In that case you may intentionally design the product such that if the preponderance of the variables fall in a particular direction it will fail to work. Then, on the production line, you test each instance off the line to make sure that the product as a whole works correctly, and you reject the ones that do not pass testing.

Statistical testing is also useful to find bad combinations of factors that are not easy to anticipate. In this role it is a good backup to worst-case analysis.

Worst-case analysis (or testing) is most useful when you have a few parameters that are known to be troublesome, and when you can expect that the system will be at its worst when these parameters are at one end or another of their anticipated range. For instance, CMOS circuits generally respond their slowest when they are at the high end of their temperature range, while mechanical assemblies are generally most likely to bind up and refuse to move at the cold end of their range. Both mechanical and electrical systems are most likely to suffer from mismatches between their various parts at both ends of the temperature range. Thus, a worst-case analysis of a piece of electronics may check the behavior at room temperature plus both extremes of temperature. A worst-case analysis of a mechanical system would do the same thing.

### Types of analysis to perform

For the most part in this section, I have spoken as if the only important sort of analysis to do is to analyze the system's intended behavior against its actual behavior. This is just a part of the overall analysis.

In addition to analyzing and testing a system for intended behavior, you should pay attention to the following:

- Thermal considerations. Electronics generate heat, and in most systems that heat must be dissipated in a way that ensures that every part of the system works correctly. The system analysis should ensure that every part of the system remains in its working temperature range. If this analysis indicates a failure, then parts must either be redesigned or selected to work over a wider temperature range, or the system must be designed with better heat-handling capabilities.
- Electromagnetic compatibility (EMI/EMC). Your system must play well with others in the electromagnetic realm. It must not emit too much electromagnetic radiation, and it must not be too susceptible to electromagnetic radiation that is emitted by other nearby devices.
- Reliability. Particularly if you are working on systems for military, or medical, but often if you are working on a system for any application where long life or reliable operation is a selling point, you should do formal reliability analysis. Knowing what mean time between failure and mean time to failure are is essential, although calculating them is often not only an engineering specialty but often uses methods that are determined by regulation or contract terms.

- Safety. Electronic and electromechanical systems can create dangerous situations for the human beings that use them. They must not electrocute people who touch them, they must not burst into flame and catch houses or businesses on fire. Particularly, in the case of industrial equipment and power tools they must not present an undue danger of physical injury to users or bystanders. In cases where there are applicable safety standards or laws, it is incumbent on whoever is taking on the systems engineer's role to understand the issues and not only make sure that a safe and legal product is designed, but that it can be shown to be safe and legal.

## References

[1] A Special Report, Counterfeit Parts: Increasing Awareness and Developing Countermeasures, Aerospace Industries Association of America, Inc. <http://www.aia-aerospace.org/assets/counterfeit-web11.pdf>, March 2011 (accessed on 26.08.13).

[2] H.B. Land III, Sensing switchboard arc faults, IEEE Power Eng. Rev. April (2002) 18−2027.

[3] H.B. Land, C.L. Eddins, L.R. Gauthier, J.M. Klimek, Design of a sensor to predict arcing faults in nuclear switchgear, IEEE Trans. Nucl. Sci. 50(4) (2003) 1161−1165.

[4] D. Strassberg, Distributed power: taming the dragons, EDN Mag. July(24) (2003) 40−42.

[5] The Mythical Man-Month, Fred Brooks. <https://en.wikipedia.org/wiki/Brooks%27s_law>, 1975.

[6] K.R. Hancock, Build vs. buy—the great debate, PC/104 Embed. Solut. (Summer 2002) 35−39.

[7] R. Cravotta, Uncovering the truth in benchmarks, EDN Mag. October(2) (2003) 58.

[8] J.C. Maxwell, On Governors, 16, Proceedings of the Royal Society of London, 1868, pp. 270−283

[9] D. Goldman, Fake tech gear has infiltrated the U.S. government, by David Goldman@CNNMoneyTech. <http://money.cnn.com/2012/11/08/technology/security/counterfeit-tech/index.html>, November 8, 2012 (accessed on 9.11.12).

[10] B. Hughitt, Counterfeit Electronic Parts, NASA/ESA/JAXA, Trilateral Safety and Mission Assurance Conference. <http://www.hq.nasa.gov/office/codeq/trismac/apr08/day2/hughitt_NASA_HQ.pdf>, April 2008 (accessed 26.08.13).

## CHAPTER 9

# Mechanical Design

Steve Zeise

*Aerospace Electronics Industry*

## Chapter Outline

Developing and Managing Embedded Systems and Products.
DOI: http://dx.doi.org/10.1016/B978-0-12-405879-8.00009-X

## What to expect from this chapter

I will start with a review of the basics of mechanical design, establishing a common language and foundation. Then I'll address packaging and thermal design, which are closely related. I have concentrated on the fundamentals leaving the details to several of the excellent references. The more exciting topic of mechanisms is next where I focus on robust mechanism design and methods for calculating loads and forces. Finally, I'll look at analysis and test, discussing how to use Finite Element Analysis (FEA) effectively and then end with a simplified approach to vibration problems.

The basics of mechanical design: materials, fasteners, fabrication, and finishes. These topics can be daunting if you're not a practicing mechanical engineer because there are so many choices. I will describe a simple approach but I'll also give you the rationale behind it so that you can look back and check your mooring as you move past the simple approach.

## Materials

To create a top-notch product, you need to go deeper than thinking "I want a gray plastic case with rounded corners" or "I want a shiny chrome cover with black lettering." The material choices you make must be based on the material properties along with the aesthetics. So when I hear the word plastic, for instance, immediately a "property card" pops up in my mind and I begin to consider how these material properties may affect the operation of the part. Although there are thousands of "plastics," they all share common properties that you should know to make a good decision. There are four broad categories of materials: metals, plastics, ceramics, and composites. Figure 9.1 shows a top-level comparison of their material properties.

The material properties that drive our decisions are typically strength, stiffness, weight, electrical conductivity, and thermal conductivity. Two additional properties, coefficient of thermal expansion (CTE) and resistance to corrosion, are very important but can be managed with proper design.

Properties of composite materials are difficult to pin down because, by their very nature, they result from combination of several other materials. The composites used for the construction of embedded systems include reinforced plastics such as fiberglass-reinforced epoxy and carbon graphite-reinforced epoxy. The properties of the reinforcement fibers are dramatically different from the plastic that holds them together. By orienting the fibers in a specific direction, you can get dramatically different properties in that direction. The popular FR-4 material used in the construction of printed wiring boards (PWBs) is a fiberglass-reinforced epoxy composite created by weaving strands of fiberglass perpendicular to one another. For this type of construction, the lengthwise direction is noted

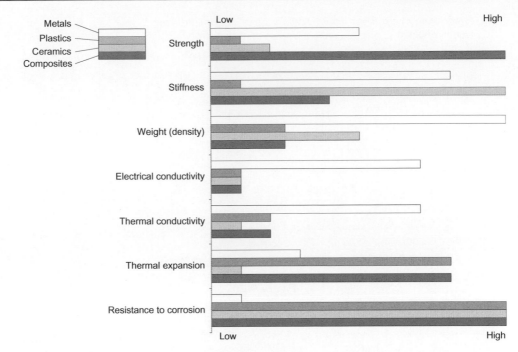

**Figure 9.1:**
Comparison of material properties.

**Table 9.1: Properties of FR-4 PWB composite material**

| Mechanical Properties | LW | CW | Z | |
|---|---|---|---|---|
| CTE [ −40°C to +125°C] | 12 | 15 | 50 | ppm/°C |
| Modulus of Elasticity | $3.5 \times 10^6$ ($24.1 \times 10^6$) | $2.8 \times 10^6$ ($19.3 \times 10^6$) | NA | lb/in.$^2$ (N/m$^2$) |
| Poisson's Ratios | 0.13 | 0.11 | NA | |
| Thermal Conductivity | | 0.525 | | W/m K |
| Specific Heat | | 1.22 | | J/g K |
| Density | | 1.97 | | g/cm$^3$ |

Typical values for Nelco® N4000-7 FR-4 50% resin content from Park Electrochemical Corp. Rev 7–12 [1].

as "LW" and the crosswise direction is noted as "CW" and the material properties are different depending on the direction. As can be expected, the properties perpendicular to the thickness of the sheet (direction "Z") are dramatically different, particularly the CTE as shown in Table 9.1.

If you are like me, a list of properties will mean nothing to you unless it is applied to a real life example. Let's consider how these properties would be applied in the selection of an enclosure material. Figures 9.2 and 9.3 illustrate how the enclosure material choice may affect the performance of an embedded system.

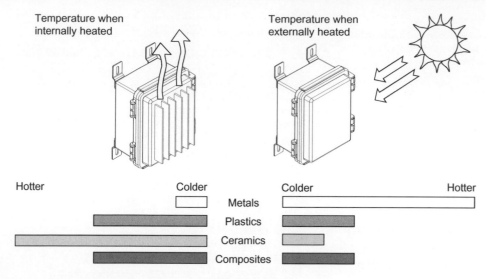

**Figure 9.2:**
Thermal effects of material choice on enclosure design.

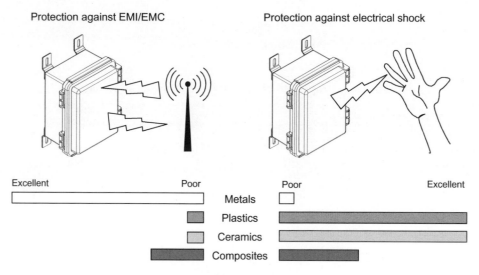

**Figure 9.3:**
Electrical effects of material choice on enclosure design.

Which material is the best for an enclosure? The choice is not obvious until you know more about your environment and the embedded system within the enclosure.

The first three properties: strength, stiffness, and weight are tightly related so we'll consider them together. The terms of strength and stiffness are very distinct properties though sometimes incorrectly interchanged. It's easiest to illustrate this with a diagram. Figure 9.4 shows the difference between the two properties as applied to a simple structure.

The stiffness shown in Figure 9.4 is the amount of deflection for a given amount of force. Strength is the amount of force before the material breaks or ends up permanently

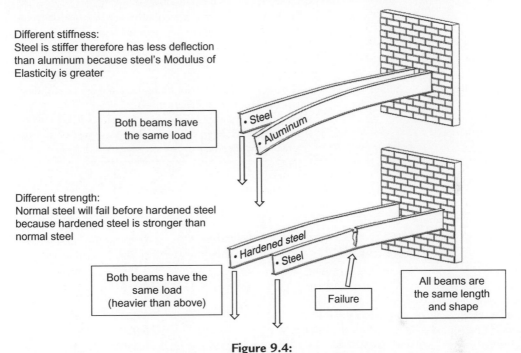

**Figure 9.4:**
The difference between stiffness and strength.

deformed. An important distinction is that stiffness is related to the shape of the material and the Modulus of Elasticity, an inherent material property that is virtually unaffected by processing. Contrast that to the strength of the material which can be altered by forming and heat treatment. For example, all steels have essentially the same Modulus of Elasticity but with processing, such as heat treatment, high-strength steel can have a higher strength than ordinary steel.

The choice of materials must consider both strength and stiffness but usually one or the other will dominate the decision. When I start a new design, I try to determine if it is "stiffness driven" or "strength driven." Some examples are listed in Table 9.2.

The property listed as weight is actually "mass density," in other words, the material mass per unit volume. I grouped this together with strength and stiffness because most designs are forced to trade-off one against the other. If I want to reduce weight, I will most certainly reduce both the strength and stiffness of my parts. Because of this challenge,

#### Table 9.2: Examples of stiffness versus strength designs

| Design | Type | Comments |
|---|---|---|
| Enclosure lid | Strength driven | Lid must be strong enough to resist impacts and loads |
| Fasteners | | Fasteners rely on being flexible yet strong |
| Ball bearings | | Ball bearings require high-strength steel to resist damage |
| Adhesives | | Adhesives are evaluated and selected based on shear and tensile strength |
| Mounting brackets | | The mounting brackets of an embedded system are typically evaluated for strength |
| PCB and stiffeners | Stiffness driven | PCBs must be stiff to resist vibration and shock and may require bolt-on stiffeners |
| Mechanism links | | Mechanisms rely on stiff linkages to maintain alignment |
| Optical structures | | Structures holding optics or mechanisms must be stiff to maintain alignment |

#### Table 9.3: Specific stiffness and specific strength [2]

| Specific Stiffness | | | | |
|---|---|---|---|---|
| Material | Specification | Modulus of Elasticity (Mpsi) | Density (lb/in.$^3$) | Specific Stiffness (in.) |
| Ceramic | Aluminum oxide 98% | 20.3 | 0.141 | 144 |
| Titanium | 6Al-4V, heat treated | 16.5 | 0.160 | 103 |
| Aluminum | 6061-T6 | 10 | 0.098 | 103 |
| Stainless steel | 304, 60% Cold rolled | 29 | 0.285 | 102 |
| Carbon fiber/Epoxy | Cytec Thornel® P-650/42 | 5.51 | 0.056 | 99 |
| Nylon 6 | 50% Glass filled | 3.84 | 0.056 | 69 |
| Copper | Annealed | 16 | 0.323 | 50 |
| ABS plastic | 40% Glass fiber filled | 1.6 | 0.059 | 27 |
| **Specific Strength** | | | | |
| Material | Specification | Yield Strength (kpsi) | Density (lb/in.$^3$) | Specific Strength (in.) |
| Carbon fiber/Epoxy | Cytec Thornel® P-650/42 | 430 | 0.066 | 6515 |
| Titanium | 6Al-4V, heat treated | 140 | 0.160 | 875 |
| Steel | 4140, heat treated | 175 | 0.285 | 614 |
| Aluminum | 6061-T6 | 34.8 | 0.067 | 519 |
| Nylon 6 | 50% Glass filled | 40 | 0.098 | 410 |
| Ceramic | Aluminum oxide 98% | 22 | 0.059 | 376 |
| ABS plastic | 40% Glass fiber filled | 34.8 | 0.141 | 247 |
| Copper | Cold rolled 1/8 hard | 20 | 0.323 | 62 |

strength and stiffness are often categorized as a ratio of mass density as shown in Table 9.3. These ratios are defined as "Specific Stiffness" also known as "Specific Modulus" and "Specific Strength." An interesting fact appears when you view the table of Specific Stiffness, the common metals are the same! What does this mean? In practice, it means that

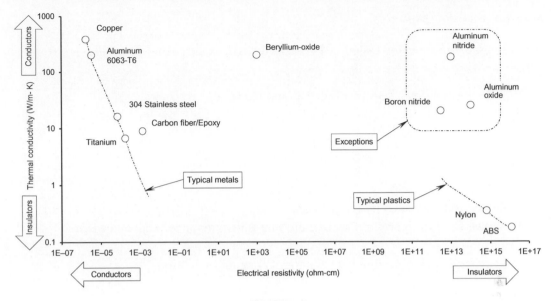

**Figure 9.5:**
Electrical and thermal conductors [3].

if your part is metallic and you are trying to improve the stiffness without increasing the weight, look at optimizing the structural design before changing the materials. Strength to weight is a whole different ball game where there are clear winners. Steel is ten times the specific strength of copper while carbon fiber is another ten times the specific strength of steel!

Electrical and thermal conductivity are the next two properties; they are also related. In the vast majority of materials, good electrical conductors are good thermal conductors; likewise, poor electrical conductors are poor thermal conductors. As you select materials for your embedded system, these broad generalizations can guide your decisions. There are exceptions that form a group of materials you can use to solve difficult problems such as dissipating heat from electronics while maintaining electrical isolation. Figure 9.5 shows how these materials differentiate.

## Fasteners

Threaded fasteners often hold the assembly of parts together in embedded systems, especially for industrial markets. Products that are consumer oriented with heavy emphasis on cosmetics and style will use subtler techniques, such as snap fits and hidden clips. For the context here, I will focus on making a good choice of a threaded fastener. Don't be overwhelmed by the sheer volume of fastener designs on the market. It is a daunting task to make a comprehensive survey for an optimum fastener. Instead I will focus on specific choices and if you want to search further I list some references at the end of the chapter.

## Goals for choosing fasteners

Before I talk about threaded fasteners I need to discuss exactly what you will want from them:

- Hold the device down during all operational loads.
- The fastener will sustain extreme loads and not fail even if the load is several times the weight of the device.

And you don't want:

- a loose assembly,
- the device to be crushed by the fastener,
- the device held so tightly to the structure that it distorts under heat and thermal expansion issues,
- the fastener to loosen during temperature cycling and vibration.

Finally, you will want the ability to remove and reinstall the fastener without damaging or compromising any of the above. Once you list all that you expect from the fastener, you can begin to appreciate how complex a threaded fastener can be.

## Fastener types

First let's indulge in some name calling. Figure 9.6 shows the correct names of the most common types of fasteners and head styles. Make every attempt to use the correct names

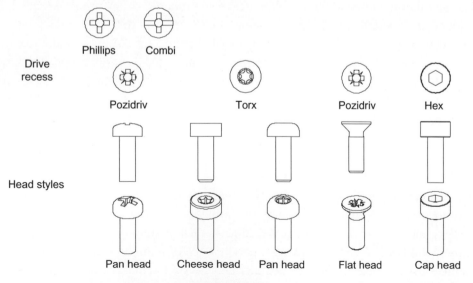

**Figure 9.6:**
Fastener head styles.

since there are subtle differences between the various fasteners. The main choices you have to make are the head style, the drive recess, and the thread type.

The popular choice for head style is the pan head with a Phillips drive recess. Many industrial enclosures use a Phillips/slot combo for the convenience of using either a Phillips or straight screwdriver. If you anticipate that your unit will be assembled with automated equipment or power screwdrivers, use the pan head with a Pozidriv recess. The Pozidriv recess is a new version of the Phillips drive recess. For higher rate production, the pan head or cheese head with the Torx drive recess is an even better choice. The Torx drive recess offers higher torque and doesn't get damaged as easily and the cheese head provides more material for the drive recess.

Use a flat head screw with a Phillips or Pozidriv recess when a flush appearance is desired. The hex drive recess is not recommended because it is easily damaged at higher torques.

---

### A little history

When you look at the different drive recesses, you are actually studying the gradual refinement of threaded fasteners. Starting with the slotted recess, designers looked for a way to avoid the tool sliding or jumping out of the slot. The Phillips drive recess was developed because it allowed power drivers to be used. One design feature of the Phillips was that it releases the tool, or cams out, when the proper torque is reached. The downside is that the Phillips recess requires a significant force pushing in as you are turning. With the advent of torque-limiting power drivers, the Pozidriv head was developed to eliminate the problem with cam out. Pozidriv screws can be driven with a standard Phillips driver but only realize the full improvement when driven with a Pozidriv tool.

The next design is the hex recess. The hex recess screw can be tightened without having to push the tool into the screw. But the hex recess is limited by the amount of torque before the tool or the head fails. The Torx head evolved to improve the amount of torque before failure and is widely considered to be the best choice.

---

The flat head screw offers a nice flush appearance but it requires attention to several details for it to fit properly. Since the head of the screw fits snugly in the countersunk hole, the fit between parts must be excellent or you will end up with the result shown in Figure 9.7. Both the cover and the base need to be precision machined to avoid this type of mismatch. A second alternative shown is to use a floating nut on the base which allows each screw to align to the countersink.

The next decision you have to make when choosing a fastener is the thread style and there are as many as there are head styles! This is one decision that you leave until last because the choice of a thread must wait until the material of the parts has been selected. For example, if you select an enclosure fabricated from aluminum with tapped holes you will

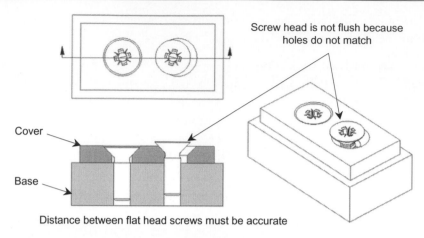

**Figure 9.7:**
Possible mismatch of multiple flat head fasteners.

select a standard machine screw but if the enclosure is a soft thermoplastic with just a plain hole molded in, you will choose a thread forming screw with threads designed specifically for plastic. One very important point to remember is that the screw and the hole work together. Each type and diameter of screw has a specifically designed hole to receive it with very specific tolerances.

For screws threading into a metallic material, you have to decide whether to pre-thread the holes or use a self-tapping fastener. For machined parts, it is an easy decision to thread the holes but for die-cast or injection-molded parts the decision is tougher because it will require a secondary machining step just for the threads. A self-tapping fastener can generate debris when the fastener is inserted or removed; so, if you cannot live with debris, you will have to pay the extra expense of having the holes pre-threaded.

For screws threading into plastics, the choice of thread style is based on the properties of the plastic. The basic guidelines are shown in Table 9.4 and are based on the Modulus of Elasticity in bending which is called the Flexural Modulus. You use these as a starting point but as soon as you have the exact material properties you consult the fastener supplier. Selecting the wrong screw type will result in damage to the screw or to the plastic.

For example, if you choose a screw for a polycarbonate enclosure, which is a thermoplastic, an excellent choice for the thread type is the Delta PT® from EJOT or the REMFORM® from Research Engineering & Manufacturing Inc. Both manufacturers offer further information and guidance for other plastic types.

The length of screw engagement is very important; too little and the thread will strip out of the material. Table 9.5 shows the minimum recommended screw engagement for various materials.

**Table 9.4: Recommended screw threads for plastic materials [4,5]**

| Type | Material Name | Flexural Modulus (kpsi) | Delta PT® | PT® | REMFORM® | DST* | Plastite® | | | |
|---|---|---|---|---|---|---|---|---|---|---|
| | | | | | | | 48° Twin Helix | 48° Single Helix | 45° Single Helix | Duro PT® |
| Thermoplastic | Polyethylene (PE) | 150 | • | • | • | • | | | | |
| | Polypropylene (PP) | 200 | • | • | • | • | | | | |
| | Polycarbonate (PC) or Lexan | 340 | • | • | • | • | | | | |
| | ABS, 0–20% glass | 350 | • | • | • | • | | | | |
| | Polyamide 66 (PA) or Nylon | 350 | • | • | • | • | | | | |
| | Acetal (AC) or Delrin | 400 | • | • | • | • | | | | |
| | Polystyrene (PS) | 430 | • | • | • | • | | | | |
| | Polypropylene, 40% talc (PP40) | 500 | • | • | • | • | | | | |
| | Polyphenylene sulfide | 550 | • | • | • | • | | | | |
| | ABS, 20% glass | 650 | • | • | • | • | | | | |
| | Polyamide 66, 12% glass | 800 | • | • | • | • | | | | |
| | Polycarbonate, 20% glass | 850 | • | • | • | • | | | | |
| | Polycarbonate, 30% glass (PC30) | 1100 | • | • | • | | • | • | • | |
| | Polybutylene terephthalates 30% glass (PBT30) | 1100 | • | • | • | | • | • | • | |
| | Polyamide 66, 30% glass (PA30) | 1200 | • | • | • | | • | • | • | |
| | Liquid crystal polymer (LCP) | 1400 | • | | | | • | • | • | |
| Thermoset | Polyphenylene sulfide, 40% (PPS40) | 1700 | | | | | | | • | • |
| | Phenolic, 20% glass | 1750 | | | | | | | | • |
| | Polyester, 50% glass | 2100 | | | | | | | | • |

Delta PT®, Duro PT®, and PT® are registered trademarks of EJOT Verbindungstechnik GmbH & Co. KG.
REMFORM® and Plastite® are trademarks licensed by Research Engineering & Manufacturing Inc.
*DST, dual-spaced thread.

**Table 9.5: Minimum recommended screw engagement for various materials**

| Material Name | Recommended Full Thread Engagement ( × Screw Diameter) |
| --- | --- |
| Steels | 1 |
| Aluminum | 1.5 |
| Plastics | 2–3 |

**Table 9.6: Common panel nuts**

| Panel Nut | Manufacturer | Website |
| --- | --- | --- |
| Perma-Nut® | Stanley Engineered Fastening | www.stanleyengineeredfastening.com |
| PEM® | PennEngineering | www.pemnet.com |
| WELL-NUT® | Stanley Engineered Fastening | www.stanleyengineeredfastening.com |

For thinner materials, if you can't add a boss to thicken it locally, you typically employ panel nuts to create a threaded hole. Specially designed "press nuts" are one good design choice. An alternative for sheet metal enclosures is a rubber WELL-NUT®. These are only intended for lower strength applications but are great as a rattle-free method for attaching sheet metal covers in an industrial, high-vibration environment. Table 9.6 shows common panel nuts with their manufacturers.

A washer distributes the load of the screw over a larger area and protects the part from the spinning and scarring action of the screw head. But a loose washer is a liability that should be avoided. Anytime the underlying material is soft or easily damaged, such as a PWB, use a washer but only if it is captive to the screw. This can be done in two ways. The washer can be built-in to the head. This works well if the purpose is to distribute the load. But if you are worried about scarring or scratching the part, then use a flat washer that is attached to the screw. Another type of washer that is used with a flat head screw is the crown washer. This combination gives a nice finished look to a design. Figure 9.8 shows these combinations.

### Fastener sizes

Now that you have considered types of fasteners, what size fastener and how many should you use? Table 9.7 lists the common ISO Metric and American Standard thread sizes used with embedded systems.

Consider the industry and surrounding equipment when choosing metric versus standard— maintaining a common standard helps when installing or servicing the equipment. Choosing the size of the fasteners is all about proportions—larger enclosures, larger PWBs, and larger components will require larger fasteners or a greater quantity of smaller fasteners. I've included the approximate screw proof strength in Table 9.7 so you can attempt to choose fasteners based on the weight of the component and the expected loads on that component,

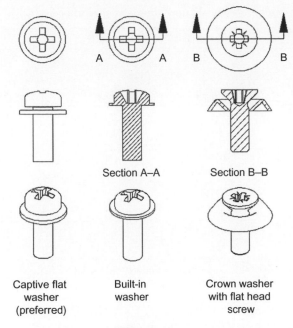

Section A–A        Section B–B

Captive flat
washer
(preferred)

Built-in
washer

Crown washer
with flat head
screw

**Figure 9.8:**
Different types of washers.

but that can get fairly complex. Just for reference, the proof strength is the point above which the screw begins to stretch permanently and is usually considered to have been ruined as a fastener. I base the fastener choices on a process similar to the flow diagram shown in Figure 9.9.

After you work through the process and the design starts to take shape, you back up your choice with a calculation of loads versus proof strength of the fasteners. If necessary, higher strength versions are often available for each of the fastener sizes.

### Preload

I will end the discussion of fasteners with the topic of fastener preload. The main purpose of a threaded fastener is to hold a component or cover in place and preload is the mechanism by which that happens. Simply put, preload is the stretching of the fastener by applying a torque to the head or the nut. The fastener can develop a preload force of up to and slightly beyond its proof strength but typically I shoot for a value of about 70% of the proof strength. I say "about" because the process of calculating torque versus preload is dependent on many variables. Fastener suppliers publish typical torque values that can be

**Table 9.7: Metric and standard thread sizes [6,7]**

| Common Metric Threads | | | | |
|---|---|---|---|---|
| | Pitch (mm) | | Approximate Screw Proof Strength[a] | |
| Nominal Diameter (mm) | Coarse | Fine | N | lbs |
| 1.6 | 0.35 | | 260 | 58 |
| 2 | 0.4 | | 420 | 94 |
| 2.5 | 0.45 | | 690 | 160 |
| 3 | 0.5 | | 1000 | 220 |
| 3.5 | 0.6 | | 1400 | 310 |
| 4 | 0.7 | | 1800 | 400 |
| 5 | 0.8 | | 2900 | 650 |
| 6 | 1 | | 4100 | 920 |
| 8 | 1.25 | 1 | 7500 | 1700 |
| 10 | 1.5 | 1.25 or 1 | 11,900 | 2700 |

| Common Standard Threads | | | | | |
|---|---|---|---|---|---|
| | | Pitch (threads/in.) | | Approximate Screw Proof Strength[a] | |
| Number | Nominal Diameter (in.) | Coarse | Fine | N | lbs |
| #0 | 0.06 | 80 | | 240 | 54 |
| #2 | 0.086 | 56 | | 490 | 110 |
| #4 | 0.112 | 40 | | 800 | 180 |
| #6 | 0.138 | 32 | | 1200 | 270 |
| #8 | 0.164 | 24 | 32 | 1600 | 360 |
| #10 | 0.19 | 24 | 32 | 2400 | 530 |

[a]The approximate screw proof strength is the approximate load which a common, low strength, screw can support without damage. The strength is calculated using a tensile yield strength of 30 kpsi and a tensile stress area based on a diameter midway between the minor and pitch diameters. These values are intended to be used as a guide but should not be used for critical load calculations. Consult screw manufacturer for exact values.

used for most situations. A complete description of the calculations of preload based on torque is given by Shigley [8].

## Fabrication

Understanding how custom mechanical parts are fabricated saves time and money during the development of the project and allows you to effectively navigate the process from concept through prototyping, from low rate production into final production. Changes are inevitable but how you manage them, and at what point in the process changes are made, can make or break a product.

Before I dive into the various methods of fabrication, there are two basic concepts to touch on. The first is economies of scale; simply put, quantity drives the price down. With any

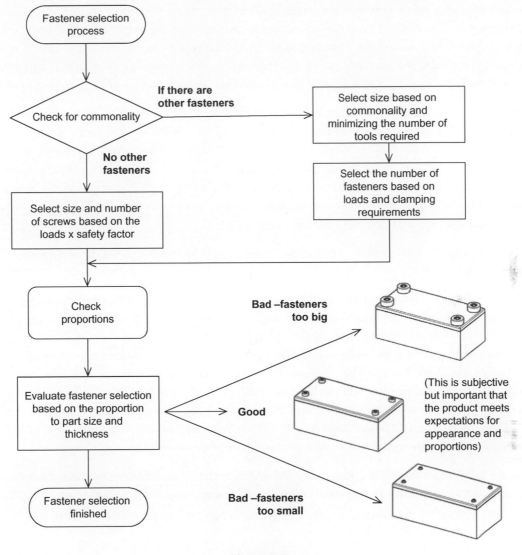

**Figure 9.9:**
Fastener selection process.

manufacturing process, there is always a nonrecurring expense (NRE) and a recurring expense (RE). (*Editor's note*: See Chapter 8 for more details on NRE and RE. RE is typically identical to COGS.) NRE includes machine setup fees, molding dies, and tooling. RE includes the cost of material and the actual fabrication time for each part. Mass produced parts are always cheaper than one-of-a-kind custom parts because the NRE is distributed over a large number of parts. Higher quantities justify the more expensive fabrication processes, so the higher the projected quantities, often the higher the NRE. This is identical to the electronic world: when you pick out a new low-cost processor, you don't see the very high fabrication cost associated with the development of it because the extremely high volume has absorbed the NRE. As for me, I attempt to minimize the cost by using an off-the-shelf solution if it exists. If I am forced to design my own, I try to design a common part that can be used multiple times. This technique can increase the quantity and yield a nice cost savings.

The second concept is often called front-end loading and is basically the attitude of pushing for design verification as early as possible in the product life cycle. The fundamental principle in the concept of front-end loading is that design decisions made in the early stages of a product are inexpensive but have the greatest effect on the performance and per piece cost of the product. In contrast, changes that are made in the later part of the design process and especially in the production phase are very expensive and have marginal effect on the performance and per piece cost of the product.

When you begin a new project, keep those two concepts in mind and set the following goals for the project:

- *Strive to pay NREs once* by committing to production tooling only after all design verification has taken place. Design changes that happen after production commences will force you to pay the NRE again and again.
- *Build prototypes* for the purpose of gathering information on the front end to make effective design decisions. For example, a plastic model of an aluminum enclosure can be painted to look like the real part but will not have the same thermal performance. A simple sheet metal prototype may have the correct thermal characteristics but may not look anything like the end product. But if it helps speed design verification, build a sheet metal prototype to verify the thermal performance and build a plastic model to show the sales guys.

The six fabrication methods common to embedded systems are additive manufacturing (rapid prototyping), machining, sheet metal bending, sheet metal stamping, plastic injection molding, and metal die casting. Figure 9.10 shows examples of each method with their attendant strengths and weaknesses.

## Finishes

The final stage of part design is selecting the finish. Most parts require or are enhanced by a surface finish. Plastics can have color and texture added as part of the molding process but all other materials will require a surface finish. Even the natural look of polished stainless steel or brushed aluminum is actually a purposeful surface finish and requires an additional manufacturing process.

How do you select a surface finish? Start with the environment. A more aggressive environment, such as seaside in the tropics, will require a more intensive surface finish than a benign office environment. The second factor is appearance. Is a particular color

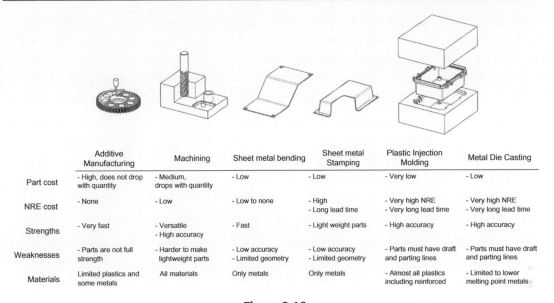

| | Additive Manufacturing | Machining | Sheet metal bending | Sheet metal Stamping | Plastic Injection Molding | Metal Die Casting |
|---|---|---|---|---|---|---|
| Part cost | - High, does not drop with quantity | - Medium, drops with quantity | - Low | - Low | - Very low | - Low |
| NRE cost | - None | - Low | - Low to none | - High<br>- Long lead time | - Very high NRE<br>- Very long lead time | - Very high NRE<br>- Very long lead time |
| Strengths | - Very fast | - Versatile<br>- High accuracy | - Fast | - Light weight parts | - High accuracy | - High accuracy |
| Weaknesses | - Parts are not full strength | - Harder to make lightweight parts | - Low accuracy<br>- Limited geometry | - Low accuracy<br>- Limited geometry | - Parts must have draft and parting lines | - Parts must have draft and parting lines |
| Materials | Limited plastics and some metals | All materials | Only metals | Only metals | - Almost all plastics including reinforced | - Limited to lower melting point metals |

**Figure 9.10:**
Fabrication methods.

required? Should the surface be polished or dull? Does it need a texture for grip or to hide scratches that may inevitably appear? Lastly, do you need to maintain or refresh the appearance of the part during its lifetime?

The most versatile surface finish is painting. Aside from cleaning, this is the only viable finish that can be maintained. Paints fall roughly into two categories: paints that are wet applied and dry powder coatings.

For wet applied paints, use polyurethane paints for the vast majority of industrial applications. The thickness of polyurethane paint is typically 0.002 inches and can be applied in any color with finishes from a glossy smooth to a flat, matte appearance. Special additives can be used to create a pleasing, textured appearance that hides imperfections and dirt.

Use epoxy paints for extreme chemical resistance. If you need resistance to acetone and other cleaning solvents, and want to stay with a wet applied coating, epoxy paints are the only way to go. Epoxy paints are the same thickness at 0.002 inches and can be formulated with the same properties as urethane. The disadvantages of epoxies are that they are more expensive and most will yellow and fade with sunlight exposure.

Powder coating is method of applying paint as a dry powder then baking it to create a smooth finish. The same paint chemistries, such as polyurethanes and epoxies, are available, as well as thermoplastics that literally melt and flow as the finish is baked on. Powder coating can easily create thicker films from 0.020 inches to as much as 0.120 inches if desired. The thicker coatings have a richer appearance and hide superficial scratches better.

But the thicker coatings tend to chip more easily and the coating thickness varies more around corners than wet paints. The appearance can be glossy finish or flat and some texture processes are available. Because of thickness variations, a glossy powder coated finish will never look quite as good as glossy wet paint finish. An advantage of powder coating is lower environmental waste than wet paint with the possibility to recycle leftover powder.

All paint systems include a primer that is specific to the part material. *I rarely deviate from the suggestions of the paint manufacturer and only if I have extensive testing to support a different solution*. A good relationship with a skillful painter is worth developing. When the appearance of the part is important, I have found that working directly with the painter is the best approach. The same paint applied with different techniques can produce completely different appearances.

Chemical and electrochemical surface finishes on metallic parts open up a new set of options. Aluminum, magnesium, and titanium can be treated to produce anodized and conversion coatings. These coatings are thin, corrosion resistant, and part of the metal. Alternatively, most metals can be plated to add a thin coating of another metal, such as nickel, that will resist corrosion and wear or just change the appearance cosmetically. Stainless steel is a whole different animal and while it doesn't require a coating to resist corrosion, it does require specific surface preparation for best performance. In general, these surface finishes leave the surface texture unchanged. Polished surfaces before anodizing will look polished afterwards and brushed surfaces will stay brushed.

As mentioned, the surface of aluminum parts can be anodized or conversion coated. Aluminum will corrode quickly and should not be used without a coating. The oxides of aluminum are not as obvious as the red rust oxides of iron and steel but can be just as present. If you clean a piece of aluminum that has been left untreated, even in an office environment, you will see a black oxide residue on the cleaning cloth. This oxide layer adversely affects electrical conductivity and adhesion of paints and structural adhesive bonds.

By far the most popular coating for aluminum is anodizing. The most common process is sulfuric acid anodizing that is performed by a plating company; anodizing creates a layer of aluminum oxide ranging up to 0.001 inch thick. The oxide layer is the same composition as aluminum oxide sandpaper so it is very hard and resistant to wear. The coating is also an electrical insulator. If you need an electrical connection, you must mask the anodize from specific surfaces of the part. Masking is a manual process and will increase the cost but it is sometimes necessary. The thickness of the anodize coating does affect tightly fitting parts. I specify on my drawings that dimensions apply after anodizing so the machinist and the plating company have to work together to create a part that meets the dimensions. I also regularly mask threaded holes and small holes because the anodize coating is often irregular in these holes anyway. My machinist will usually choose to control the masking process himself and not rely on the plater for their interpretation of the drawing.

Several other anodizing processes are available. One option for close fitting parts is a thinner anodize process using chromic acid. These coatings are only about 0.0005 inches thick. For a thicker coating, "hard coat" anodize is a refrigerated sulfuric acid process and creates a hard thick layer up to 0.006 inches. I have used this extensively for wear surfaces and when extreme corrosion resistance is required. The hard coat anodize can also be impregnated with Teflon®* for a good low friction coating. As the hard coat anodize gets thicker, it can get a fuzzy appearance. When I need a smooth good quality hard finish, I specify hard coat anodize to a specific thickness but then return the part to the machine shop for a final grinding operation to bring the surface to a smooth condition.

Anodized coatings are microscopically porous and can be dyed with a variety of colors. Beware—dyes do fade with prolonged sunlight and are difficult to color match from one plating company to another.

The second coating for aluminum is a conversion coating. The most popular is the chromate conversion process which creates a very thin coating that is electrically conductive and resists corrosion. The chromate conversion process can be used for those surfaces that were masked in the anodize process for a complete corrosion solution. The chromate conversion coating can also be used as a surface preparation for painting and bonding when corrosion protection is desired. Whenever using conversion coatings, specify an RoHS (Restriction of Hazardous Substances) compliant process to minimize environmental impact.

Stainless steel parts do not require a coating but should be chemically treated to improve corrosion resistance. The process is called passivation and creates a stable oxide layer on the surface of the part removing any free iron particles that might have been introduced during the machining process. Passivation should not change the dimensions of the part but I have had some tight tolerance parts ruined by an overly aggressive passivation process. I always specify that tolerances apply after passivation on the drawing.

Other steels will require a protective coating such as zinc plating followed with a clear chromate. This combination is about 0.0005 inches thick and provides excellent corrosion protective in all but extreme environments. Galvanizing is a different, hot-dip, zinc plating process that adds a thicker layer of zinc between 0.0014 and 0.0039 inches thick. Galvanizing adds much more corrosion protection but obviously the thickness must be accounted for in the design of the parts. Because of the thicker coating, galvanizing adds a sometimes undesirable texture to parts and a distinctive speckled appearance.

## Packaging

Now that I have prepared a foundation in materials, fasteners, and finishes, I will discuss packaging, the first of my responsibilities as a mechanical engineer. The term packaging,

---

* Teflon is a registered trademark of DuPont.

for this chapter, will refer to a broad range of topics from the mounting and attachment of the processor and PWB to the design of the overall chassis and the exterior. The thermal design, although covered in the next section, must proceed simultaneously with the packaging design since each heavily influences the other.

### Enclosures

The enclosure of the embedded system provides protection and support, as well as fulfilling any cosmetic requirements. Two popular standards define the level of protection for enclosures and electrical connections. In the United States, the National Electrical Manufacturers Association created NEMA Standards Publication 250, a comprehensive product standard addressing ingress and corrosion protection, environmental requirements, and testing. The International Electrotechnical Commission (IEC) created IEC 60529, which is popular worldwide although less comprehensive; it specifies testing only for ingress protection. The benefit of these standards is that they make the selection of an enclosure a simple matter of identifying the level of protection then choosing an enclosure that meets that level of the specification. If you are designing your own enclosure, still go through the process of identifying the level of protection you need and then use an off-the-shelf enclosure as a starting point for your custom design. This strategy leverages all the experience and design time that was invested into one of these off-the-shelf enclosures for a fraction of what it would cost you to develop and design one of your own.

A second benefit of these specifications is that they serve as a guide to help ask, and answer, the questions that you should be asking. For example, "Do I expect my enclosure to encounter water being sprayed at it?" If so, a quick check of the construction of a NEMA 250 Type 4 or Type 6 compliant enclosure will demonstrate how to protect your embedded system from hose sprayed water. "What about corrosion and water spray?" NEMA 250 Type 4X or Type 6P adds corrosion protection. Table 9.8 lists each type of NEMA 250 enclosure and the degree of protection that is intended.

Table 9.9 lists each type of IEC 60529 enclosure and the degree of protection that is intended. Keep in mind that neither the NEMA specification nor the IEC specification requires testing but rely on the manufacturer to conduct self-testing.

Before embarking on a custom enclosure design, search through the multitude of off-the-shelf designs. Table 9.10 lists several major enclosure manufacturers but don't limit the search to these. Also included are manufacturers that are in the business of creating custom enclosures.

### Connectors and cabling

*I cannot overemphasize the importance of well-designed connectors and cabling*. Connectors are the one part of the equipment that gets touched by many hands, during the

**Table 9.8: NEMA 250 enclosure types [9]**

| | NEMA 250 Enclosures—Nonhazardous Installations | | | | |
| Type | Intended Location | Protection Against | Watertight | Ventilated | Comments |
| --- | --- | --- | --- | --- | --- |
| 1 | Indoor | Limited amounts of falling dirt | Not required | Possible | General purpose, not dust tight |
| 2 | Indoor | Limited amounts of falling dirt, dripping/light splashing of liquids | Not required | Possible | |
| 3 | Outdoor | Rain, sleet, windblown dust, and damage from external ice formation | Not required | Possible | |
| 3X | Same as 3 with corrosion protection | | | | |
| 3R | Outdoor | Rain, sleet, and damage from external ice formation | Not required | Possible | No dust protection |
| 3RX | Same as 3R with corrosion protection | | | | |
| 3S | Outdoor | Rain, sleet, windblown dust, and to provide for operation of external mechanisms when ice laden | Not required | Possible | |
| 3SX | Same as 3S with corrosion protection | | | | |
| 4 | Indoor or outdoor | Windblown dust and rain, splashing water, hose-directed water, and damage from external ice formation | Yes | No | |
| 4X | Same as 4 with corrosion protection | | | | |
| 5 | Indoor | Falling dirt, dripping/light splashing of liquids | Not required | Possible | General purpose, dust tight |
| 6 | Indoor or outdoor | Hose-directed water, the entry of water during occasional temporary submersion at a limited depth, and damage from external ice formation | Yes | No | |
| 6P | Indoor or outdoor | Hose-directed water, the entry of water during prolonged submersion at a limited depth, and damage from external ice formation | Yes | No | Prolonged submersion |
| 11 | Indoor or outdoor | Corrosive effects of liquids and gases | Not required | Possible | General purpose |
| 12 | Indoor | Circulating dust, falling dirt, and dripping noncorrosive liquids | Yes | No | General, purpose, not oil-tight |

(Continued)

**Table 9.8: (Continued)**

### NEMA 250 Enclosures—Nonhazardous Installations

| Type | Intended Location | Protection Against | Watertight | Ventilated | Comments |
|------|-------------------|--------------------|-----------|-----------|----------|
| 12K | Same as 12 with knockouts | | | | |
| 13 | Indoor | Dust, light spraying of water and noncorrosive coolants | No | Possible | General purpose |

### NEMA 250 Enclosures—Hazardous Installations

| Type | Intended Location | Type of Hazard (As Defined in the National Electrical Code) | Comments |
|------|-------------------|-----------------------------------------------------------|----------|
| 7 | Indoor | Class I, Groups A, B, C, or D | |
| 8 | Indoor or outdoor | Class I, Groups A, B, C, or D | |
| 9 | Indoor | Class II, Groups E, F, or G | |
| 10 | Mining applications | Meets the applicable requirements of the Mine Safety and Health Administration | |

**Table 9.9: IEC 60529 enclosure types [10]**

| International Protection Rating (IP Code) | | | |
|---|---|---|---|
| **Examples** | | | |
| IP | X | 6 | IPX6 protects against powerful water jets |
| IP | 4 | 7 | IP47 protects against anything larger than a wire and good for shallow immersion |

| Code | Solids | Code | Liquids |
|---|---|---|---|
| X | Not specified | X | Not specified |
| 0 | Not protected | 0 | Not protected |
| 1 | >50 mm (hand) | 1 | Dripping water |
| 2 | >12.5 mm (finger) | 2 | Dripping water when tilted up to 15° |
| 3 | >2.5 mm (tools) | 3 | Spraying water |
| 4 | >1 mm (most wires) | 4 | Splashing water |
| 5 | Dust protected | 5 | Water jets |
| 6 | Dust tight | 6 | Powerful water jets |
| | | 7 | Immersion up to 1 m |
| | | 8 | Immersion beyond 1 m |

**Table 9.10: List of enclosure manufacturers**

| Enclosure Manufacturers | | | | |
|---|---|---|---|---|
| **Company Name** | **Website** | **PWB Templates** | **Type of Construction** | **Comments** |
| Polycase | www.polycase.com | Yes | P | |
| Bud Industries | www.budind.com | No | P, M, S | |
| Protocase | www.protocase.com | No | S, E | Custom enclosures in two to three days—no minimum order |
| METCASE | www.metcase.com | No | M, S | |

Type of construction: P, molded plastic; M, metal die cast; S, sheet metal; E, extruded aluminum.

assembly, testing, end use by the customer, and service. I pay attention to every person that touches the connector especially during the prototype and test phase. Was it obvious how to mate and de-mate the connector? Did they plug it in wrong or misaligned? Never ignore a connector failure with the excuse that it was due to a clumsy technician or engineer. If it happened once, it should be addressed and if necessary the design changed to use a more durable connector. I always regret it when I don't address it the first time!

***Failures due to connector or cabling damage are common but can be minimized with proper design***. There are many excellent sources for design practices; I have found the industry standard IPC-A-620 a good start to both understand the problems and also to help find solutions and specify components [11]. Also enlightening is the "Aircraft Electrical Wiring Interconnect System (EWIS) Best Practices" from the FAA [12]. This guide from the FAA is rich with examples of what to do and what not to do to ensure the ultimate reliability in connectors and cabling.

Selecting a connector can be a daunting task. I defer the electrical characteristics of voltage and current to the electrical engineer but I participate by following the thought process below:

External connectors:

- Identify the environment and use either the IEC 60529 standard or another to narrow the choices.
- Separate power and signal into different connectors. Power connectors should always be arranged to protect the live electrical contacts. Usually this means the live power uses the recessed socket side of the connector to minimize electrical shock hazards.
- Consider the user and make the connections foolproof by selecting a unique connector for each cable.
- Avoid using industry standard or readily recognized connectors for nonstandard signals or purposes. For example, don't use a nine pin D-shaped connector for anything but a serial link because users will try to plug in a serial connection.
- Get an accurate signal count, and the voltage and current requirements before starting the connection selection. You would be wise to have extra connector pins for growth in new designs should changes occur in the design.
- Trade-off packaging, ease of use, and durability. These are usually conflicting requirements and this is where the hard decisions have to be made.
- Remember to consider other factors, such as "If the user is wearing gloves in cold weather, can the connector be removed with gloves on?"

Internal connectors:

- As stated above, get accurate requirements and leave room for growth to add pins.
- Trade-off pin density, packaging, ease of use, and durability.
- Trade-off the ease of mating and demating the connection with the connector retention force; the connector must remain mated during vibration, mechanical shock, and temperature cycling.

Blind-mate and docking connectors:

- Include sufficient guides and tactile feedback to make sure the connector can be mated.
- Carefully dimension and tolerance the placement of both connectors.

**Table 9.11: Cable and wire routing—what to do and what not to do [11,12]**

| Do | Do Not |
|---|---|
| Segregate signal wires along one side of the enclosure and power wires down the other side | Run sensitive signal wires and power in the same bundle |
| Add a service loop at each connector and going into each grommet | Run wires in tension or pull wires to one side of a connector or grommet |
| Tie wires down, more often for higher vibration environments | Allow wires to be unsupported for long lengths |
| Arrange connectors and grommets on the sides or bottom of an enclosure and add a drip loop before the connector or grommet to shed water and moisture | Run wires vertically down into a connector or grommet mounted on the top of an enclosure because it allows water or moisture to run down into the connector |
| Remove sharp edges from corners where wires cross, adding protective plastic edging if necessary. When wires cross an edge or another bundle, add a wire tie to prevent movement and chafing. | Run wires over sharp edges or wires over top of one another |
| Use a bend radius of 10 times the outside diameter of the largest wire. If the wire is tied down firmly this bend radius can be minimized | Bend wires sharply |
| Choose a conduit or passage 25% larger than the wire bundle to allow for movement and thermal expansion | Completely fill a conduit or wire passage |
| Add drain holes to the lowest point of a conduit or wire passage or slope the conduit to create a natural drain | Allow moisture to collect in conduits or wire passages |

Cable and wire routing is usually the last thing on my mind when I start a new packaging design but it always turns out better when I consider cable routing from the beginning. Table 9.11 lists some reminders of what to do and what not to do.

### Vibration and mechanical shock

The next function of the packaging design is to protect the embedded system from vibration and mechanical shock. First, how do vibration and shock failures occur? *I have observed the following failures firsthand*:

- The forces caused by vibration or shocks can cause one component to hit another—I had a hall-effect component fail after vibration testing and an examination with a microscope revealed that it was hitting the sensing magnet ever so slightly. These impacts caused an internal electrical failure in the hall effect. Stiffening the PWB eliminated the deflection, the contact, and the failure.
- The forces overstress the material of a component and the component separates or retains a permanent distortion rendering it inoperative—I was examining a large mechanical assembly after a shock test when I noticed that the mechanism was

jamming up. Close examination revealed that the end-of-travel stop had bent under the shock testing and was allowing the mechanism to travel too far. The stop was redesigned and the entire assembly passed without further problems.

- Repetitive motion caused by vibration creates rapid wear that renders the parts inoperative—Numerous times I have failed to adequately secure cables and the insulation wears through as the cable rubs against a harder surface.
- Repetitive motion can generate significant heat in heavily loaded joints and can cause thermal softening of plastics and rubber—I have observed corrosion in stainless steel joints in a matter of minutes which has been traced to loose parts rubbing at high frequency and generating extreme local heating. I tightened up the fit of the parts or tightened up the preload of the bolted joint and solved the problem.
- Repetitive stresses can cause fatigue failures in structures, solder joints, and component wires or leads—I have experienced reliability problems with specific PWBs that were a result of PWB traces cracking due to flexing of the PWB. I improved the mounting of the PWB and the reliability problems disappeared.

In retrospect, all of these failures were the result of poor designs. That is why vibration testing is often used as a design check even for office products that may never see vibration in operation. Later in the chapter, I will explain how to calculate the effect of vibration.

### Thermal cycling

Another load that I always evaluate is thermal cycling. Recalling the property sheets, every material has a characteristic CTE which is the expansion of the material per degree Celsius or degree Fahrenheit for each unit of length. Because of CTE differences, thermal cycling creates structural stress that causes parts to buckle or move. I had one optical assembly that would be misaligned after every extended temperature cycling. Careful measurements taken after each cycle revealed that one of the parts would creep like an inchworm in one direction because of CTE differences. In another unfortunate design, a $6000 optical prism would crack at very low temperatures not because of the CTE difference in the aluminum–glass joint but because the CTE difference between an adjacent steel screw and the aluminum housing created a distortion in the housing.

Larger parts and extreme temperatures aggravate the problem. Depending on the material, the CTE can change value with temperature expanding usually less proportionally at cold temperature and more at hot. Plastics are notorious for nonlinear behavior of the CTE while metals will have virtually the same CTE over normal working temperatures. My goal when I design an assembly is to avoid joining parts that have drastically different CTEs. Figure 9.11 is a comparison of CTEs for common materials. Selecting materials close to one another will minimize the stresses from thermal changes.

Coefficient of Thermal Expansion

| Material | Specification | CTE (in./in./°F) | CTE (mm/mm/°C) | Length @ 30°C dt (originally 10 in.) | "D" (in.) |
|---|---|---|---|---|---|
| Nylon 6 | Unfilled | 9.44E-05 | 1.70E-04 | 10.051 | 0.440 |
| ABS plastic | Unfilled | 5.04E-05 | 9.07E-05 | 10.027 | 0.202 |
| Nylon 6 | 50% Glass filled, average, transverse | 4.00E-05 | 7.20E-05 | 10.022 | 0.145 |
| Aluminum | 6061-T6 | 1.31E-05 | 2.36E-05 | 10.007 | -- |
| Stainless steel | 304, 60% Cold rolled | 9.61E-06 | 1.73E-05 | 10.005 | −0.019 |
| Copper | Annealed | 9.11E-06 | 1.64E-05 | 10.005 | −0.022 |
| ABS plastic | 40% Glass filled, average | 8.78E-06 | 1.58E-05 | 10.005 | −0.023 |
| FR-4, Crosswise | Nelco® N4000-7 FR-4 50% resin content | 8.33E-06 | 1.50E-05 | 10.005 | −0.026 |
| FR-4, Lengthwise | Nelco® N4000-7 FR-4 50% resin content | 6.67E-06 | 1.20E-05 | 10.004 | −0.035 |
| Titanium | 6Al-4V, Heat treated | 4.78E-06 | 8.60E-06 | 10.003 | −0.045 |
| Ceramic | Aluminum oxide 98% | 4.50E-06 | 8.10E-06 | 10.002 | −0.046 |

Resulting bow when various materials are bolted together with aluminum

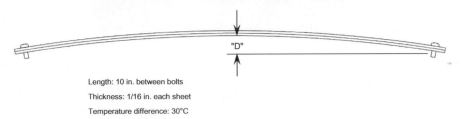

"D"

Length: 10 in. between bolts

Thickness: 1/16 in. each sheet

Temperature difference: 30°C

**Figure 9.11:**
Comparison of CTEs for common materials [13].

## Thermal design

Address the thermal design of an embedded system early in the concept phase and revisit it constantly as the design progresses. Once prototypes are available, conduct room temperature tests to validate thermal predictions. If the prototype is exceeding predictions at room temperature, it will fare no better at high temperature. *I have found that early and decisive action to solve a thermal problem before production will save time and money*.

The maximum thermal power that the embedded system can safely dissipate is dictated by the shape and size of the device and by the surrounding environment. For many embedded systems, most of the electrical power going in will be dissipated as heat with the exception of a small amount of work that may be performed through mechanical actuation or signal transmission. Since these parameters are established early on in the design process, it is important that the anticipated electrical power be balanced with ability of the device to reject the waste heat.

The embedded system, by definition, is part of a larger host product. This host product creates and defines the thermal environment within which the embedded system must function. I attempt to calculate the thermal impact of the embedded system on the host

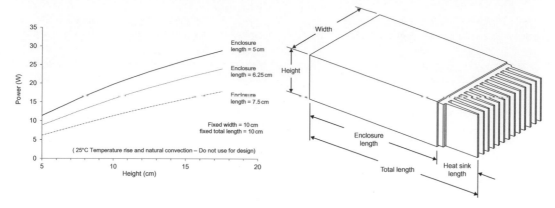

**Figure 9.12:**
Example of a thermal parametric study.

product at an early stage of the design. For example, a desktop PC graphics card is an example of an embedded system inside a host PC. But a high-power graphics card may require the addition of an extra cabinet fan in the host PC. If this is impractical, the power dissipated may need to be reduced, resulting in a rewriting of the graphics card specification and often a reduction in capability.

My role as a mechanical engineer is to define the maximum thermal power dissipation and provide the electrical engineers with a budget. This budget should help guide the processor selection and system architecture. When I'm faced with loose definitions for physical size and shape, I create a parametric study of various sizes and power projections as shown in Figure 9.12. I use this as a high level "choose this, not that" type of decision guide.

One decision that dramatically improves heat dissipation is the use of fans or forced convection. Forced convection "can provide heat transfer rates in electronic systems that are 10 times greater than those available with natural convection" [18, p.157]. The improved cooling comes with the cost of complexity, increased noise, reduced reliability, and additional waste heat from the fan itself. The optimal fan will maximize the cooling while adding a minimum amount of waste heat. Selecting too large a fan might actually raise the temperature of the device instead of reducing it and waste electrical energy.

The detailed thermal design can begin once the initial architectural and design decisions have been made. The goal of the detailed thermal design is to quantify the temperature rise of each major heat source and to optimize each thermal path out to the ambient environment. This path always starts with thermal conduction through the case of the electrical component and possibly into an attached heat sink. Depending on the amount of power, thermal convection through air may be used to transmit heat to the exterior of the device or to the outside air itself. For cases where the power is very high, you can use a heat pipe, which effectively reduces losses from thermal resistance.

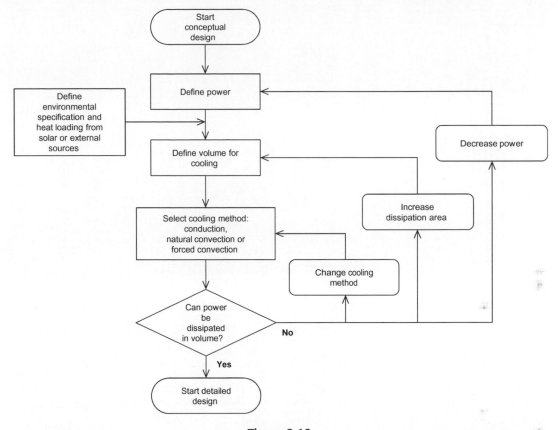

**Figure 9.13:**
The iterative nature of thermal design.

When the temperature of a component must be closely controlled or be held below ambient, I have used thermoelectric coolers. The efficiency is low, however, and may rarely be justified in an embedded system. Exceptions include the temperature control of laser diodes since the wavelength of the light is directly related to the component temperature.

### Thermal design during the concept phase

***Thermal design in the conceptual phase is iterative*** as shown in Figure 9.13. The details that need to be established are listed below:

- Define environmental specification
- Identify external heat sources (solar loading or engine compartment)
- Define external shape along with dedicated heat rejection areas
- Identify power sources and power components

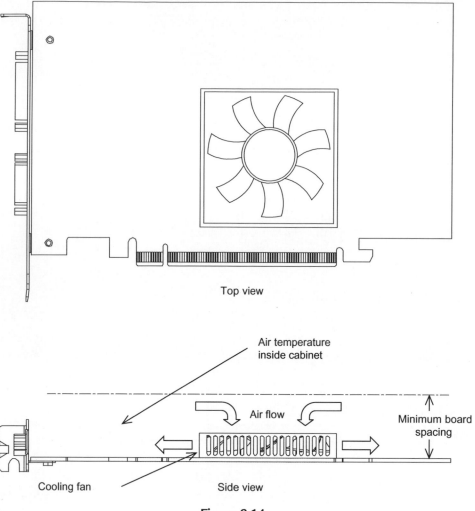

Top view

Air temperature inside cabinet

Air flow

Minimum board spacing

Cooling fan

Side view

**Figure 9.14:**
Cooling diagram for PC graphics process or card.

- Estimate overall power dissipation
- Select overall method for cooling: conduction, natural convection, or forced air cooling
- Develop a plan for each thermal heat path.

Most likely the host product will define the overall method for cooling the embedded system. Handheld products are typically intended to cool themselves through natural convection to the surrounding air. In that case, the embedded system would cool conductively through the case to the outside air. Larger products are typically cooled using natural or forced convection. The graphics card in the desktop PC is an example of a forced convection cooling. Figure 9.14 shows a typical diagram for the cooling of a graphics card.

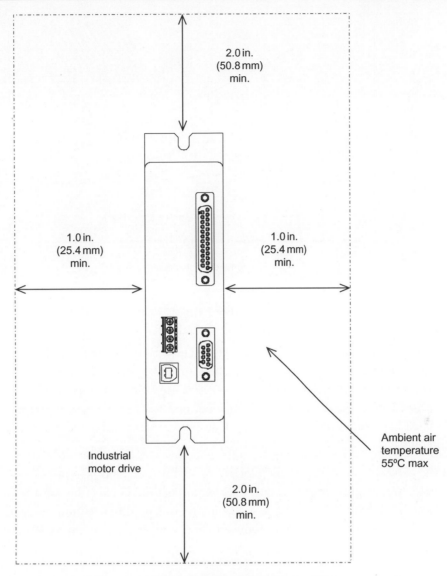

**Figure 9.15:**
Motor drive specification.

Carefully define each interface with the expected temperature and power transmitted through that interface. Figure 9.15 shows the definition of the cooling parameters for an industrial motor drive, which could eventually be used as an installation guide.

### *Define the external shape and estimate maximum power dissipation*

During the concept phase, try to understand the relationship between the external shape of the device and its ability to reject heat. Figure 9.16 presents guidelines for estimating the

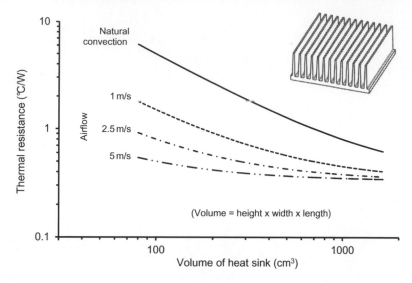

**Figure 9.16:**
Temperature rise based on power, external shape and size.

temperature rise for a given shape and power [14]. These rough guidelines can be used early in the design process to guide the partitioning of space between the electronics and the cooling of the electronics. ***An even more effective method is to construct a mockup of the device, reproducing the air flow and heat load, and test the temperature rise directly***. (*Editor's note*: I have mocked up the thermal load for a system and agree that this is a very useful technique!)

As part of the external shape definition, identify areas for dedicated heat rejections, such as heat sinks or air inlets and outlets. Products relying on natural convection must allot a significant amount of space due to the low efficiency of natural convection. Natural convection is also dependent on gravity requiring heat sink fins and possibly PWBs to be oriented vertically for maximum cooling.

---

### Case study: Thermal runaway

Few issues can ruin a product as fast as poor thermal design. Years ago during the infancy of 3D computer graphics I blew the entire capital budget on a high-performance graphics processor designed by then industry leader, Evans and Sutherland. I purchased the unit after a stunning demonstration but unfortunately I didn't pay enough attention to the computer crash midway through the demo. I replayed the event in my mind for several years after the purchase. The two salesmen shot each other a look of despair but recovered with a contrived story of thermal problems that would be fixed by the time the product shipped. Unfortunately, the crash was a harbinger of trouble to come. Once our new system was installed, the graphics processor crashed every two hours and required a half hour downtime

**Table 9.12: Example of a power source table**

| Mfgr PN | Object Description | Design | Package | Length (in.) | Width (in.) | Maximum Temperature (°C) | Power Dissipation (mW) |
|---------|-------------------|--------|---------|--------------|-------------|--------------------------|------------------------|
| SC8544CPXANG | IC—MPC8544E, POWER QUICC III PROCESSOR | UI | PBGA | 0.33 | 0.30 | 85 | 5660 |
| ICS841003AKI-02 | IC—PCIE CLOCK SYNTHESIZER | U2 | VFQFN-32 | 0.20 | 0.20 | 85 | 580 |
| VIVO105THJ | CONVERTER—DC/DC, 48 TO 5, VICOR | U5 | VI Chip | 0.87 | 0.65 | 100 | 4000 |
| VIB0101THJ | CONVERTER—DC/DC, 48 TO 12, 120W, VICOR | U3, U4 | VI Chip | 0.87 | 0.65 | 100 | 2000 |

to cool off. Despite several returns and replacements, the problems subsisted. Without money for an alternative, we struggled on valiantly for two years. Needless to say, as soon as the money materialized, we purchased a lower performance, higher reliability graphics processor, and left Evans and Sutherland in the rear view mirror. The lesson I learned as a young engineer was that thermal issues, once designed into a product, cannot be removed unless you completely redesign the product.

### Estimate overall power and identify power sources

I have found it is a challenge to estimate the power that will be dissipated especially with brand new designs. Often prototypes or breadboards of the circuits exist and the electrical power can be measured directly. The law of conservation of energy assures us that after a careful accounting of the power in and out, the remainder will be dissipated as heat. If the embedded system drives an actuator or performs mechanical work, the amount of work per unit time is equal to power and is included in the energy equation and not dissipated as heat.

When it comes to CPUs and MPUs, the amount of power dissipated is a function of operating voltage, clock frequency, software code, use of I/O, and ambient temperature. But the main two factors are operating voltage and frequency. When in doubt, bracket the problem by obtaining worst case estimates along with best case estimates. As the power sources are identified, compile a table similar to Table 9.12.

### Develop plan for thermal heat paths

Design a thermal heat path for each of the components listed in Table 9.12. Schematically this path can be shown as a series of resistances, is analogous to an electrical circuit, and can be solved as such if you establish the equivalent quantities shown in Table 9.13.

**Table 9.13: Electrical analogies for thermal analysis**

| Electrical Quantity | Electrical Symbol | Thermal Quantity | Thermal Symbol |
|---|---|---|---|
| Current | I (A) | Rate of heat transfer | $q$ (W) |
| Voltage | V (V) | Temperature | $T$ (°K) |
| Resistance | R (V/A) | Thermal resistance | $R_T$ (°K/W) |
| Steady State Equations | | | |
| $\Delta V = IR$ | | $\Delta T = qR_T$ | |

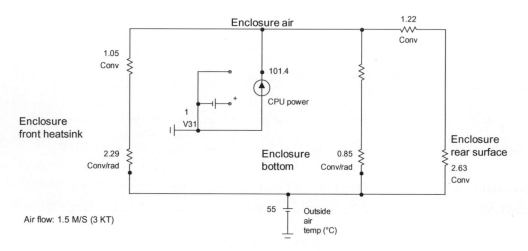

**Figure 9.17:**
Typical thermal circuit.

Figure 9.17 shows a typical thermal circuit. The circuit can be solved using Excel or with circuit simulation software such as Micro-Cap by Spectrum Software. The electrical circuit analogy can be extended to the transient response. Consult Steinberg for a detailed explanation [18, p.355].

## RFI/EMI shielding

One question I ask while planning out the thermal path "What are the radio frequency interference (RFI) and electromagnetic interference and compatibility (EMI/EMC) issues with this circuit?" Embedded systems can produce RFI/EMI or be susceptible to it. Often a section of the PWB will be shielded but for extreme cases the entire embedded system will require a complete metallic or metallic-coated case that will function as a Faraday shield against RFI and EMI. This requires a completely different approach to thermal design because it adds several more resistances to the thermal path.

**Table 9.14: Thermally conductive dielectric interface materials [16,17]**

| Trade Name | Manufacturer | Website | Thickness Range (mm) | Thermal Impedance Range $°C − in.^2/W$ ($°C − cm^2/W$) |
|---|---|---|---|---|
| THERMATTACH® | Parker Chomerics | www.chomerics.com | 0.005−0.010 (0.127 − 0.25) | 0.6−1.2 (3.7−7.7) |
| CHO-THERM® | Parker Chomerics | www.chomerics.com | 0.003−0.018 (0.08−0.46) | 0.33−0.64 (2.1−4.1) |
| Sil-Pad® | The Bergquist Company | www.bergquistcompany.com | 0.005−0.012 (0.127 − 0.30) | 0.23−1.45 (1.48−9.3) |
| Gap Pad® | The Bergquist Company | www.bergquistcompany.com | 0.010−0.250 (0.25−6.35) | 0.30−2.48 (1.93−15.9) |
| Bond-Ply® | The Bergquist Company | www.bergquistcompany.com | 0.005−0.011 (0.127−0.279) | 0.50−1.03 (3.21−6.61) |

If the embedded system is susceptible or is radiating at frequencies below 100 kHz, high permeability magnetic shielding materials, such as MμMetal from MμShield may be required. MμShield publishes an excellent design guide for RFI/EMI shielding design [15].

Shielding complicates the thermal design because the internal electronics must maintain electrical isolation from the shield while still dissipating the thermal energy through the shield. This paradox is solved by selecting a thermally conductive dielectric material such as listed in Table 9.14. Parker Chomerics is an excellent source for these interface materials.

### Cooling of CPU/MPUs

Effective cooling of a high-power CPU or MPU requires attention to detail. Start with a detailed thermal circuit of the CPU as installed on the PWB. Heat will be dissipated through the top of the case, the bottom of the case, and through the leads. In the case of a ball grid array part, heat is dissipated through the solder balls' grid. The heat dissipated through the bottom of the case and the leads or balls will be conducted through the board. This conduction can be improved with added copper layers or dedicated vias in the PWB.

The top of the case often provides the best opportunities for heat dissipation. The important details are surface finish, thermal interface materials, and pressure. Manufacturing tolerances will also strongly affect the thermal performance and manufacturability.

### Air cooled heat sink design

Some embedded systems will reject heat to the ambient air. A properly designed heat sink minimizes the temperature rise. The first choice should be a stock design from a company such as Aavid. Aavid publishes an excellent guide entitled "How to Select a Heat Sink" [14].

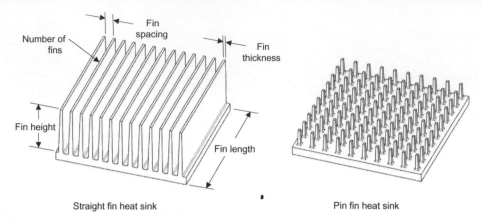

**Figure 9.18:**
Heat sink design parameters.

Heat sink companies also provide design services and can design a custom heat sink to meet my specifications.

The heat sink design minimizes thermal resistance by increasing the surface area available for heat transfer while at the same time ensuring that the air will flow through the heat sink. Figure 9.18 shows the typical constructions of fins or pins. The design parameters include fin spacing, fin thickness, fin dimensions, and the number of fins. The optimum fin spacing is strongly dependent on the air velocity; as the velocity increases, the fin spacing can decrease. The air velocity for natural convection is low but not zero. Air moves during natural convection because of buoyancy forces generated as the air heats.

### Natural convection heat sinks

Because air velocity is dependent on buoyancy forces when air heats, the fins on a natural convection heat sink must be oriented vertical. The low air velocity requires much larger fin spacing than with forced convection. Steinberg recommends that the fin spacing should be a minimum of 0.60 inches (1.52 cm) in perfectly still air [18, p.113]. In a typical office environment, the ventilation system creates more air flow allowing the spacing to be reduced to 0.50 inches (1.27 cm). I have heard it said that if you can't fit your little finger into a natural convection heat sink, the spacing is too tight. Since the fin spacing is predetermined, the remaining variables of fins depth, height, and number of fins must be optimized.

### Selecting a cooling fan

Figure 9.19 shows the process to select a cooling fan as recommended by most fan manufacturers.

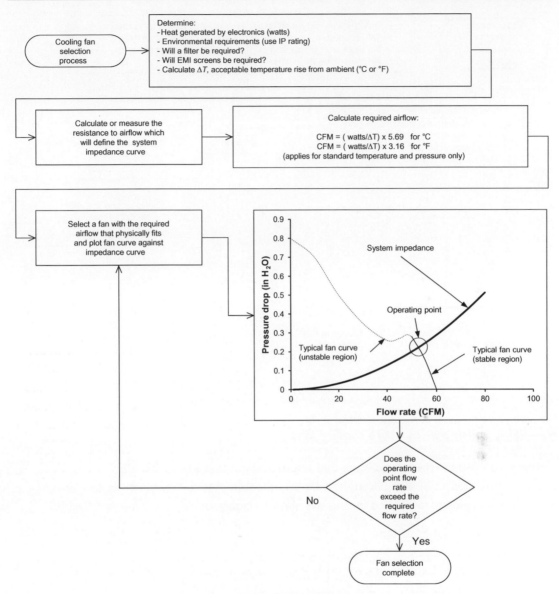

**Figure 9.19:**
Cooling fan selection process.

Finding the optimum solution can be a time-consuming process so an abbreviated, experimental process is shown in Figure 9.20. Table 9.15 lists the common risks in the fan selection process and ways to mitigate them.

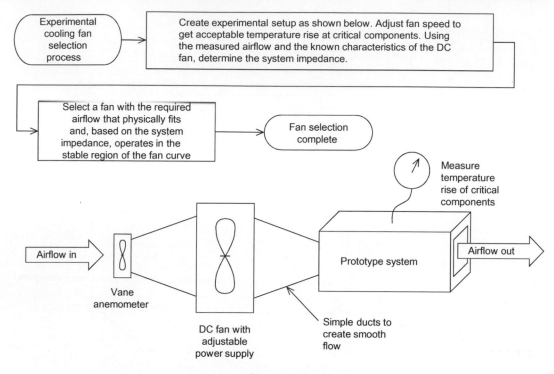

**Figure 9.20:**
Experimental fan selection process.

## Reliability considerations with cooling fans

Cooling fans reduce the temperature and improve the reliability of the electronics. But cooling fans often represent the highest failure rate components in an embedded system. Some simple principles applied during the fan selection process can improve reliability:

- Reduce the number of fans.
- If given a choice select the fan with a lower speed.
- Consider alternative designs for bearings, such as magnetic levitation or ceramic sleeves.
- Select a high quality fan.

## Minimizing noise from fans

Some embedded systems will have noise requirements. If noise from the fan is the problem, the following tips will help to optimize the thermal performance while reducing noise.

- Design for the lowest air velocity.
- Select a larger, low-speed fan if possible.

### Table 9.15: Common fan selection risks and mitigation

| Decision | Risks | Mitigation |
|---|---|---|
| General fan selection | Selecting a fan that is too small | Select an appropriately sized fan since operating at too low a flow, in the unstable region of the flow curve, results in flow pulsations, increased noise, low efficiency, and shorter life |
| | Selecting a fan that is too large | Select an appropriately sized fan since too large a fan will have greater electrical current draw and higher noise |
| | Extreme conditions result in choosing oversized fan for normal conditions | Include a fan speed controller that reduces speed during normal conditions and increases speed for extreme conditions |
| Use one large fan or several smaller fans | Poor performance with high-pressure drop enclosures | For densely packed electronics with high-pressure drop, one large axial flow fan is more efficient than using several smaller axial flow fans |
| | Poor performance with low-pressure drop enclosures | For lightly packed electronics with low-pressure drop, using several smaller cooling fans instead of one large fan will lower the noise and electrical current value |
| Placement of air inlet and air outlet | Air recirculates from outlet back to inlet | Move outlet further away from inlet or add baffles to redirect air |
| | Poor efficiency due to outlet lower than inlet | Outlet should be physically higher than inlet to take advantage of hot air rising as it moves through the enclosure |
| Size of air inlet and air outlet | Poor efficiency due to outlet smaller than inlet | Outlet should be equal to or greater in size than the inlet |
| Placement of fan—at the inlet or the outlet | Fan at inlet pushing air through enclosure adds fan heat to electronics | Pushing air through enclosure is more efficient but to reduce heat use as small a fan as possible and reduce air flow restrictions |
| | Fan at outlet pulling air through enclosure is very sensitive to leaks | Use gaskets for fan mounting and gaskets on covers. Components that require cooling must be in the direct air path—parts outside the flow path will not receive as much cooling as when pushing air |
| Fan obstructions due to space constraints | Increased noise and lower performance | Where possible, relocate the fan to obstruct the outlet not the inlet of the fan |

- Remove unnecessary airflow interruptions and move them as far away from the fan as is practical.
- Operate the fan at its peak efficiency point.
- Allow a less conservative, higher component temperature rise which reduces the airflow requirement.
- Finally, add a soft isolation mount to the fan.

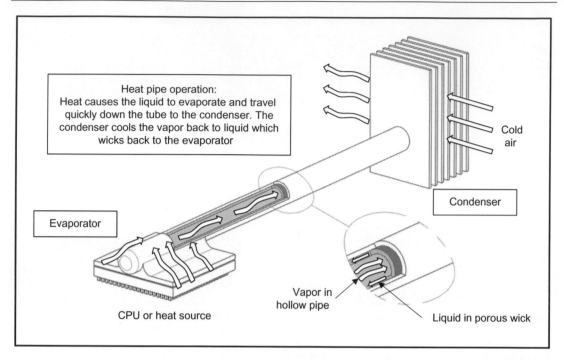

**Figure 9.21:**
Heat pipe schematic.

According to Comair Rotron, a preeminent fan supplier, the primary cause of noise is disturbances to the air flow [19]. In the case of a forced convection, it is necessary for the heat sink to interrupt the flow but careful design can move items such as finger guards and sheet metal edges, screws away from the fan inlet and outlet.

## Heat pipes

The use of heat pipes has become commonplace as an effective method to transport heat from component to heat sink. A heat pipe seals a small amount of liquid under vacuum in a sealed tube as shown in Figure 9.21. The liquid boils on the hot end and moves rapidly to the cold end as a vapor where it condenses, effectively transferred the heat. The condensed vapor, now liquid, travels back to the hot end usually by capillary action to complete the cycle.

The advantage of a heat pipe is that it can improve thermal conductivity by a significant amount, from 10 to 10,000 times, over the thermal conductivity of copper (depending on the length). A correctly sized heat pipe will have a thermal resistance that is much lower than the other resistances in the path. Heat pipes benefit from the effects of gravity by having the heat source lower than the heat sink. The efficiency will decrease as the heat source is elevated above the heat sink [20].

Heat pipes are typically a custom designed assembly. I have incorporated several heat pipe with excellent results. The best way to design a heat pipe is to work with a supplier such as Thermacore. They provide on-line design tools as well as expert designers [21].

## Mechanisms

Now let's focus on one of the more exciting parts of mechanical design—mechanisms. A mechanism is a set of parts working together to change an input motion, usually from an electric motor, into a different, more useful output motion. I will discuss robust mechanism design and methods for calculating loads and forces. I like to capitalize on the power of embedded systems by exchanging complex mechanisms for simpler mechanics with more complex controls. For strict mechanism design, I've listed several excellent sources at the end of the chapter.

Like most design processes, mechanism design starts with an understanding of the requirements then proceeds through synthesis, analysis, and finally testing. But mechanism design can get very complex unless, right from the beginning, we create a manageable design space consisting of simple mechanisms that we know how to analyze which can be combined to solve even complex motion problems. The theoretical, simplified mechanism is used to calculate the loads and forces, then the loads and forces are used to design a mechanism that complies with our simplification. This classical method of mechanism design is shown in Figure 9.22.

The first step of simplification is to separate the mechanism into the functional parts of joints and links as shown in Figure 9.23. The joints are specifically designed parts of the mechanism that are intended to move in a precise manner. Figure 9.24 shows the more common types of joints and includes the degrees of freedom of each one. The concept of "degrees of freedom" is important in mechanism design. Any object in space that is free to move has 6 degrees of freedom and the object can be described by six parameters. The first 3 degrees of freedom describe the position and traditionally are given variable names $x$, $y$, and $z$. The last 3 degrees of freedom are the angular orientation of the object and traditionally given variable names $\alpha$, $\beta$, $\gamma$. The purpose of a joint is to specifically restrain the degrees of freedom of the connecting links so that the mechanism functions in a particular manner.

Links connect the joints together and maintain the desired geometry. Links connected to a slider or screw joint will have a variable length. The complete assembly of joints and links is called a linkage. The specific design of the linkage, the number of links the type of joints and the length of each link, will determine the motion of the mechanism. I start most mechanism designs by selecting from a short list of classical linkages to see if one of these will solve the problem. The characteristics of these linkages are well known and documented.

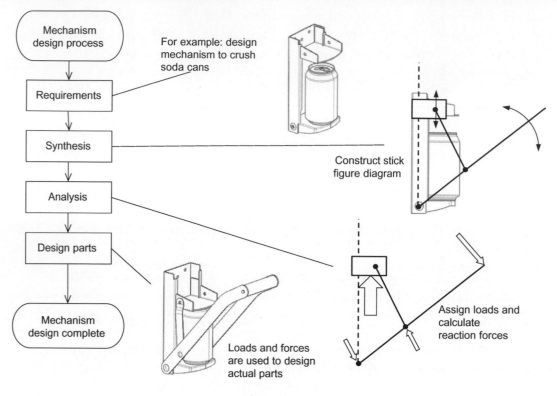

**Figure 9.22:**
Method for mechanism design.

The simplest motion is rotational motion driven directly from a motor or through a gear, belt, or worm gear as shown in Figure 9.25. Any point on the rotating link travels in a circular arc. Also shown in Figure 9.25, the second simple motion is a straight-line motion driven by a screw, a belt, or a rack and pinion gear. Any point on the sliding link travels in a straight line.

If you connect the simple rotating motion to the simple straight-line motion with a third coupler link, you have created the popular slider crank three-bar mechanism shown in Figure 9.26. This mechanism converts rotary motion to straight-line motion and vice versa. The slider crank has some interesting and very useful attributes. As the crank rotates, there are two positions where the coupler link becomes parallel to the crank and the motion of the slider stops and changes direction yielding a rough sinusoidal motion. This change of direction generates a "toggle"-type mechanism that can effectively clamp or hold a heavy part in a precise position as shown in Figure 9.27.

This toggle action occurs in many mechanisms and the mechanical advantage of the mechanism becomes infinite at these reversals. A small motor driving the input link can

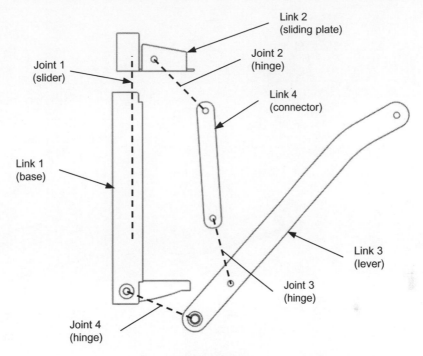

**Figure 9.23:**
Links and joints of a mechanism.

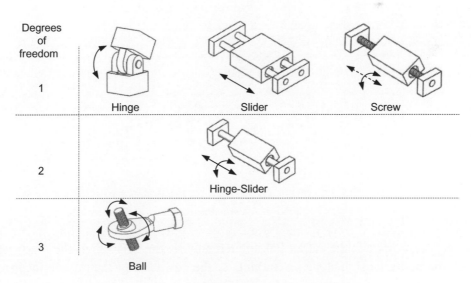

**Figure 9.24:**
Types of joints with degrees of freedom.

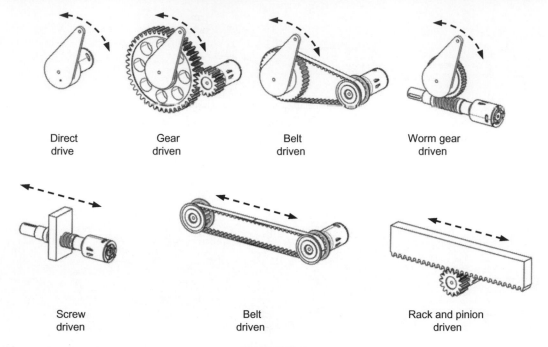

**Figure 9.25:**
Mechanisms for simple rotary and linear motion.

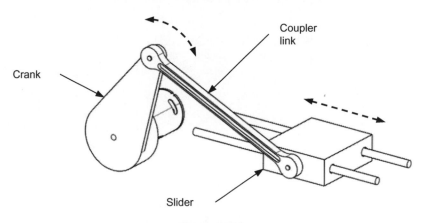

**Figure 9.26:**
Slider crank mechanism.

create a very large force at the output link limited only by the stiffness of the mechanism. Of course you don't get something for nothing, so the downside of the infinite mechanical advantage is that the range of motion of the output link is small at the toggle position.

Adding one more link generates the classical four-bar linkage as shown in Figure 9.28. In the simplest form, this linkage is composed of four hinge joints connecting four links.

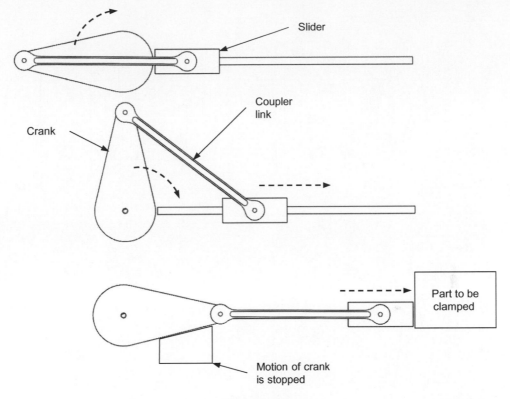

**Figure 9.27:**
Toggle clamp mechanism.

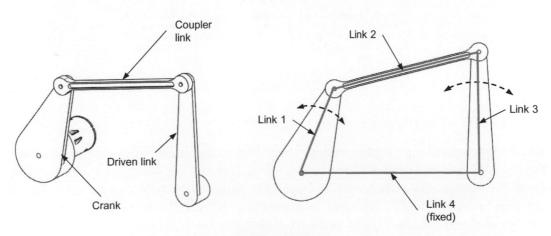

**Figure 9.28:**
Four-bar mechanism.

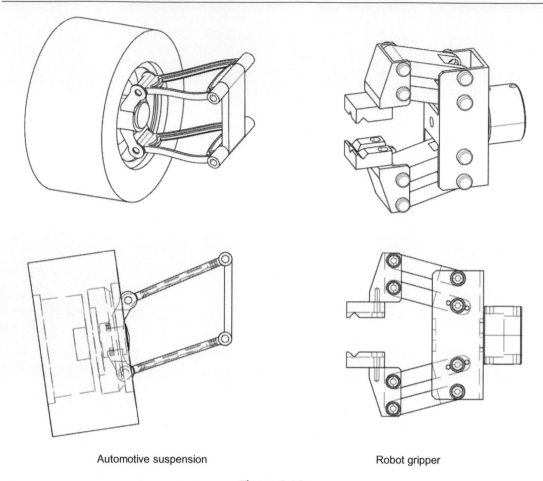

Automotive suspension                    Robot gripper

**Figure 9.29:**
Examples of four-bar mechanisms

When opposite links are of equal length, the linkage forms a parallelogram and the linkage enforces parallel motion. This linkage has been used extensively everywhere from car suspensions to robot grippers as shown in Figure 9.29.

The main benefit of a parallelogram four-bar linkage is that it simplifies the motion and therefore the design. Figure 9.30 shows how three- and four-bar linkages are combined to create the linkage of a delta printer. Although the equations of motion are complex, the equations are simplified because each four-bar enforces parallel motion and therefore the moving platform remains parallel to the fixed platform. Note that the four-bar linkages in the delta printer require ball joints, not hinge joints. Hinge joints would over constrain the moving platform.

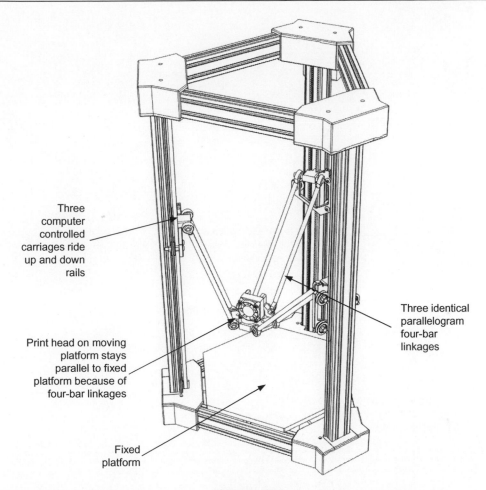

Three computer controlled carriages ride up and down rails

Three identical parallelogram four-bar linkages

Print head on moving platform stays parallel to fixed platform because of four-bar linkages

Fixed platform

**Figure 9.30:**
DeltaMaker personal 3D printer using four-bar linkages (www.deltamaker.com) [22].

Modifying the link lengths of the four-bar linkage produces a variety of motions that can be useful. Figure 9.31 shows how a four-bar linkage with unequal link lengths can create a "toggle" mechanism similar to the slider crank toggle mechanism.

Before leaving the three-bar and four-bar mechanisms, please note a unique attribute of the coupler link. It is obvious there are an infinite number of combinations of link lengths, each creating a unique mechanism, but what is not obvious is that the coupler link of each unique mechanism can unlock an infinite number of motion paths. Figure 9.32 shows the motion of the input and output links; any point on those links is simple circular motion. But if you extend the size of the coupler link, the motion of any one point on the coupler link is distinct and very useful. Complex motions such as walking-type motions can be achieved by selecting the proper point on the coupler link along with specific link lengths.

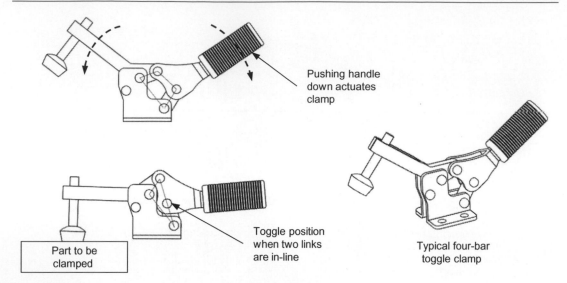

Pushing handle down actuates clamp

Toggle position when two links are in-line

Part to be clamped

Typical four-bar toggle clamp

**Figure 9.31:**
Toggle four-bar mechanism.

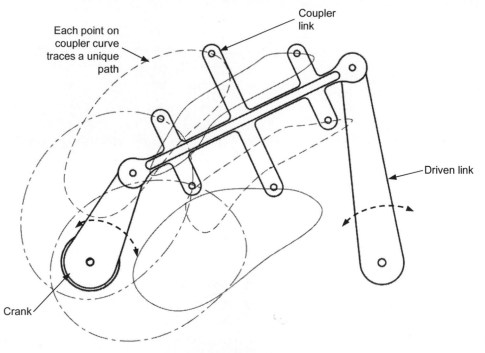

Coupler link

Each point on coupler curve traces a unique path

Driven link

Crank

**Figure 9.32:**
Unique motions of four-bar coupler link.

The process of determining the link lengths is called mechanism synthesis. I like the graphical method for mechanism synthesis. This method plays right into the strength of parametric sketchers that are included in most CAD software. In case you're not familiar with it, a parametric sketcher is a CAD sketch software that allows you to assign parameters to the sketch such as the length of lines or the angle between lines. Changing these parameters causes the software to redraw the sketch geometry with the new parameters. The case study for the car trunk lock mechanism demonstrates the graphical method by synthesizing a car trunk lock mechanism.

## Case study: Car trunk lock mechanism

Simple mechanisms can amplify the power of a small electric motor and create useful motions. One application is a mechanism to latch the trunk of a car. Common in many luxury cars, this mechanism starts its motion once the trunk lid has been lowered to the closed position. The mechanism reaches up and grabs the hook on the trunk lid and pulls it down compressing the trunk lid gasket. At the end of the motion, the mechanism must lock in position so that even after power is removed the mechanism cannot move backward and allow the trunk lid to reopen.

For the purpose of this case study, I am choosing to synthesize a four-bar mechanism using the graphical method as shown below in Figure 9.33. Step 1 is to draw the start and finish positions of the mechanism.

I want the hook arm of the mechanism to reach out and grab the trunk lid hook and pull it down so I draw the desired path of the hook arm. This type of motion is only going to be

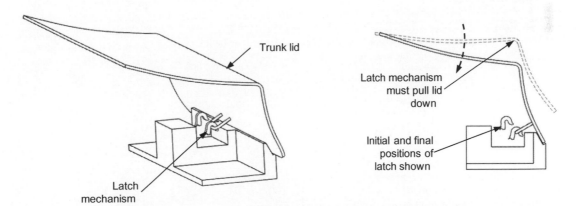

**Figure 9.33:**
Car trunk lock mechanism overview.

found somewhere in the middle of the coupler link since the two ends of the coupler link only travel through simple arcs. Therefore, I am going to sketch the hook arm extending from the coupler link since I know from experience that this will have that type of motion. To create the locking motion, I am going to draw the crank of the four-bar linkage as starting and finishing the motion in a toggle position. This toggle action will create enough mechanical advantage to compress the gasket on the trunk lid. The electric motor will drive this crank from the start to the finish position. What remains is to find a rotation point for the third bar of the four-bar link, which we will call the rocker arm (Figure 9.34).

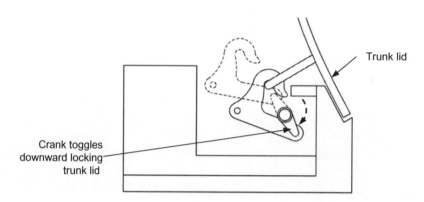

Trunk lid

Crank toggles
downward locking
trunk lid

**Figure 9.34:**
Car trunk lock mechanism with Crank defined.

Step 2 of the process is to use simple rules of sketching to synthesize the rocker arm. Starting with the two desired positions of the coupler link, I draw the one arc that will fit through the end points. This describes the motion of the rocker arm and defines its center of rotation (Figure 9.35).

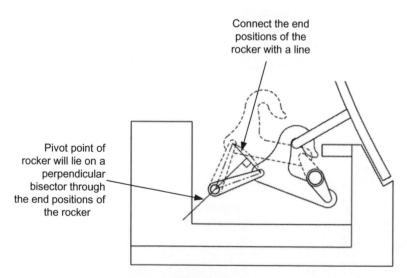

Connect the end
positions of the
rocker with a line

Pivot point of
rocker will lie on a
perpendicular
bisector through
the end positions of
the rocker

**Figure 9.35:**
Car trunk lock mechanism with Crank and Rocker defined.

The final Step 3 of the process is to use the power of the parametric sketcher and change the dimensions of the link lengths to investigate and optimize the motion and size of the mechanism (Figure 9.36).

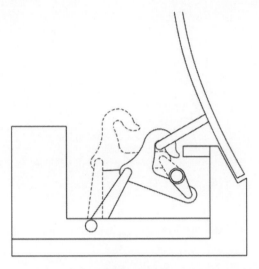

**Figure 9.36:**
Completed car trunk lock mechanism.

Once you have a linkage model you can calculate the forces on the mechanism. What I am going to describe is called a static analysis and it assumes that the mechanism is moving slow enough that you can ignore the dynamic forces. Again a graphical method is the easiest to understand, so I will illustrate using the car trunk lock mechanism from the case study. As shown in Figure 9.37, I've added force vectors to the linkage diagram. A vector represents magnitude and direction which is expressed by the length and angle of the arrow, respectively. In addition to the linear forces, the graphical analysis will also include twisting forces called "moments," which are represented by a curved arrow. It is not practical to represent a moment graphically so the length of the curved arrow has no meaning. The typical convention is that positive moments act in a counterclockwise direction and the magnitude must be indicated with a number on the graphical diagram.

Moments are also called "couples," because any moment can be replaced in the graphical diagram with a couple, or two, parallel forces.

The key points of a static analysis, shown in Figure 9.37 also, are as follows:

• Any rigid body is static, or nonmoving, if the sum of forces acting on the body is zero and if the sum of moments is zero.

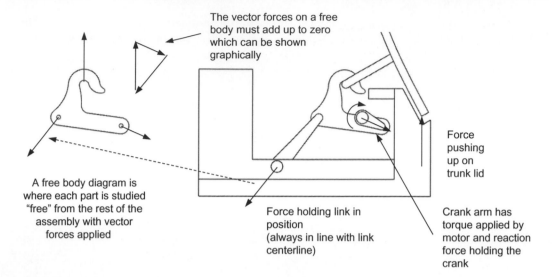

The vector forces on a free body must add up to zero which can be shown graphically

A free body diagram is where each part is studied "free" from the rest of the assembly with vector forces applied

Force pushing up on trunk lid

Force holding link in position (always in line with link centerline)

Crank arm has torque applied by motor and reaction force holding the crank

**Figure 9.37:**
Car trunk lock mechanism with force vectors.

- The forces in the joints reflect the degrees of freedom of the joint type. For example, a hinge joint is free to pivot and will only transmit forces through the center of the joint; no moments can be transferred through the joint because the joint pivots freely.
- Each joint can be analyzed in isolation as long as all forces and moments from the neighboring links are included. A "free-body diagram" of each link can be drawn as long as the forces and moments sum to zero.

I've drawn what I believe will be the two important positions of this mechanism along with the trunk lid force vector at each position in Figure 9.38. In the real design, this force vector would be measured from the amount of force it takes to pull the trunk lid downward against the gasket. What I want as the output of this static force analysis is the torque required by the electric motor at each position of the mechanism. A free-body diagram of each link shows the forces and moments that exist at each position. The solution can be calculated graphically by choosing an appropriate scale for the length of the vector or using trigonometry. The example I described was a flat, two-dimensional, type of mechanism but the principles can be extended to three dimensions [23].

The simplified analysis described above is only valid if you enforce the simplifying assumptions with actual design. (I have assumed I can neglect friction in the joints.) Now that you know the applied force, you must go through joint by joint and design for the specified motion and low friction. The methods to accomplish this are going to vary by industry so I am going to aim my comments at mechanisms that are lightly loaded and low speed. These types of mechanisms can accomplish low friction and long life with good design and material

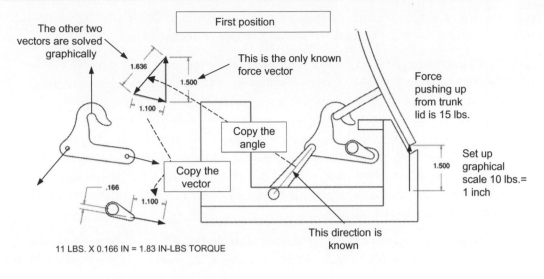

First position

The other two vectors are solved graphically

This is the only known force vector

1.636

1.500

1.100

Copy the angle

Force pushing up from trunk lid is 15 lbs.

.166

Copy the vector

1.500  Set up graphical scale 10 lbs.= 1 inch

1.100

This direction is known

11 LBS. X 0.166 IN = 1.83 IN-LBS TORQUE

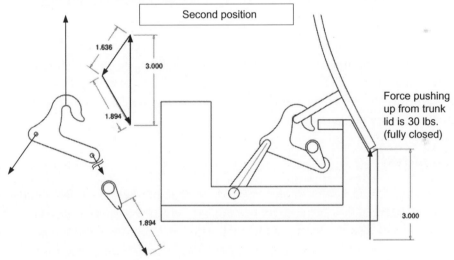

Second position

1.636

3.000

1.894

Force pushing up from trunk lid is 30 lbs. (fully closed)

3.000

1.894

The mechanism is in the toggle position so 18.9 lbs. x 0 in = 0 in-lbs (there is no torque)

**Figure 9.38:**
Car trunk lock mechanism with free-body diagrams.

selection but minimal lubrication. If the market is heavy or high-speed machinery, this requires constant and dedicated lubrication which is beyond the scope of this discussion.

The design of joints is one area where I stick to two basic principles: (i) always attempt to use a lower order of joint whenever possible and (ii) try to purchase joints from off-the-shelf. The principle of choosing a lower level joint is based on my experience that a simple hinge joint can be made to last longer and operate more accurately than a higher order ball joint. Likewise, a simple slider is preferred over a hinged slider. Obviously, the higher order joints are required for some designs but ***starting simple is always better***.

The second basic principle is to ***purchase joints off-the-shelf whenever possible*** because you are also purchasing a wealth of engineering and design. For hinge joints, use ball bearings or sleeve bearings. The manufacturers of these items list the load capacity, the amount of free play in the joint, and the wear characteristics. If you keep your loads within the guidelines, you will be confident that the mechanism will function as designed with good reliability. The same goes for slider, screw, slider hinge, and ball joints. All can be purchased with known attributes. Every time I have designed my own version of these joints, it has taken several iterations, with a good deal of testing, before the mechanism would function correctly and reliably.

These two principles help improve the reliability of the joints, which is key because most mechanism failures can be traced back to failure of one of the joints. I expect a joint to move smoothly with a minimum amount of friction. But when the joint friction increases, or worst case the joint freezes, all the calculations change and the mechanism will often fail.

Once the motor torque has been calculated it is possible to select a motor. I always resist the desire to create something from scratch and first search for existing solutions. Especially if I have space, integrated motor drives can reduce the design time dramatically. Integrated motor drives can come complete with feedback, speed control, torque limiting, and can communicate via Ethernet or RS422.

## Analysis and test

I have chosen to group the topics of analysis and test together because I believe there should be no actual boundary between them. (*Editor's note*: I heartily agree.) Though the analysis or the testing may be conducted by two different individuals or by two different departments, ***both analysis and test are part of the larger verification task that should be performed before a product is completed***. Verification usually starts with mathematical analysis and progresses to experimental testing as the product matures and as the hardware becomes available. But there is value in analysis only when it accurately predicts the performance of the eventual product. This goal is best accomplished by merging mathematical analysis and experimental testing in a continuum as shown in Figure 9.39.

As you proceed you should acknowledge the weaknesses of both analysis and testing and guard yourself against them. The weakness of analysis is that you will get an answer, instead of getting to the truth. All the pretty pictures are worthless if they aren't rooted in a continual drive to get to the truth. The weakness of test is that you simplify the real world, so you can actually do the test, but in the process of simplification you make the problem too easy to pass. The other weakness of tightly controlled testing is that you will not get a reasonable scattering

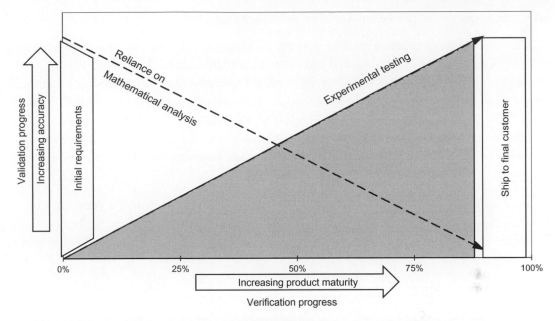

**Figure 9.39:**
The product verification continuum.

of results because the inputs are too tightly controlled. That is why it is important that the verification continuum shown in Figure 9.39 include the initial rollout to the customer.

My purpose in analysis and test is two fold: predictive and preventative. Predictive in that you will want to understand how your designs are going to perform and preventative in that you will want to know if your design will not fail given a set of circumstances. There are also several fundamental questions that you should ask of any design:

- How hot will the device get in operation?
- Are there any structural weaknesses?
- During operation, is it being stressed in any extraordinary manner?
- Will it be exposed to shipping drops, vibration, or mechanical shock?
- Are the requirements to continue operation after significant shock events like an automotive or aircraft crash? Since we are relying more and more on embedded systems, we must define operation through and after failure of other systems. This is the basis for fail-safe design.

Mechanisms have a different set of questions:

- How long does it have to last?
- What behavior is acceptable if the mechanism is jammed? Outright failure or should the motor stall and the rest remain intact?
- How will manufacturing tolerances affect the mechanism performance?

All of these questions can be answered by analysis and by test.

Start analysis first by creating a set of general guiding equations by hand or with the help of an FEA tool, Excel or Matlab, and run theoretical experiments. The results are completely mathematical and may require some interpretation but ultimately try to answer the question "Will it operate in the real world?" If you are honest, the answer is "I still don't know but it operates in my mathematical world." So it is imperative, but certainly not comfortable, that as soon as hardware is available, go back and confirm the analysis with test. In the process, you are building a knowledge base that will improve future products.

Even with FEA, Excel, Matlab, and other automated solvers, there is still a need for closed-form expressions that can be solved by hand. Having equations I can solve in my head is an advantage over computerized methods because I can see trends immediately where it may take many FEA runs to establish trends.

Figure 9.40 shows the general guiding equations useful for embedded systems.

---

### Case study: The benefit of general guiding equations

I started my career at Westinghouse Electric Corporation in Baltimore, MD, working on surveillance cameras for jet aircraft. I was responsible for the design of complex mechanisms and there were many times as a young engineer when my designs would fail in testing and I had no solution to the problem. For these unsolvable problems, Westinghouse would call in retired engineer Frank Rushing. Frank at the time was 80 years old and when he would hobble in, all conversations would drop to a whisper. I remember the first time when I sat down with him. I had my stack of computer calculations complete with FEA color stress plots and a carefully prepared presentation. After all, Mr. Rushing's time was not to be wasted! Instead of poring over my presentation and calculations, after listening to my verbal introduction, he opened his notebook and in a shaky hand he wrote: $F = ma$ (the most basic of all mechanical engineering equations!).

Step by step he asked "How much does this part weigh?" then "What acceleration are you calculating for the part?" Within 15 minutes, he systematically calculated the forces and accelerations for all the major parts in my mechanism. One part was questionable with high forces disproportionate to its size. He tapped his pencil and with a smile he said "Check this part again; I think this is where your problem lies." He stood up and shuffled out of the room.

Frank was right and this was the first of many problems he helped me solve over the years. Each time regardless of the complexity of the problem, maybe just as a reminder to himself, at the top of the page in a shaky hand he would write: $F = ma$.

---

I talked about thermal analysis previously so I will focus only on structural and vibration problems. First some terms: I am going to use the word "loads" to refer to any externally

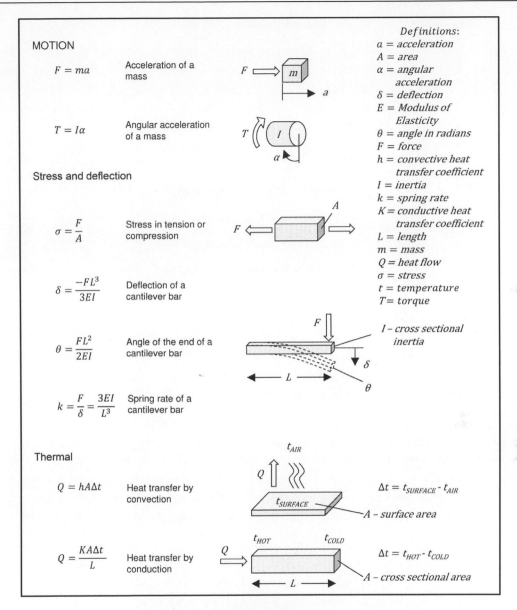

**Figure 9.40:**
General guiding equations useful for embedded systems.

applied force to my parts and the word "stress" as referring to the internal forces inside of my parts. Therefore, the external loads create internal stress. "Deflections" are the movements of the parts in response to the internal stress. At a microscopic level, I will refer to the deflections as "strains." These microscopic strains are actually stored energy that can be capitalized in the design of springs, fasteners, and even ordinary structures. In some

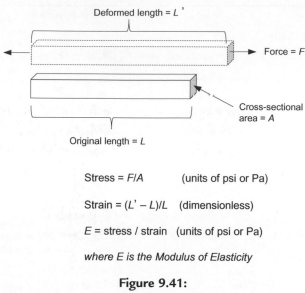

Stress = $F/A$          (units of psi or Pa)

Strain = $(L' - L)/L$    (dimensionless)

$E$ = stress / strain   (units of psi or Pa)

*where E is the Modulus of Elasticity*

**Figure 9.41:**
Stress versus strain.

cases, I may not know the load but I observe or can calculate the deflection. In that case I can calculate the stress and in turn the load that is required to create that stress. Figure 9.41 shows this relationship between loads, stress, deflection, and strain.

Every material whether it is a metal, plastic, ceramic, or rubber has a unique response to stress and strain. The stress versus strain curve is a material characteristic and is often published by the material manufacturer. A typical stress–strain curve for a metal is shown in Figure 9.42. The curve is created by applying a load to a material sample and measuring the deflection. The stress is calculated as the load per unit area and is plotted on the vertical axis. The strain is calculated as deflection per unit length and is plotted on the horizontal axis.

A small load applied and released will cause most materials to stretch and then spring back to the original length. This is "elastic behavior"; it shows up as a straight line on the stress–strain curve. The slope of this part of the curve is called the Modulus of Elasticity and denoted by the capital letter "E." At some point, the load becomes great enough that the material permanently stretches and once the load is removed, the material will not spring back completely. We mark this point on the stress–strain curve as the yield point. Increasing the load beyond the yield point will eventually lead to breaking the part and we mark this point as the ultimate stress of the material. From the yield stress point to the ultimate stress point is called the plastic region.

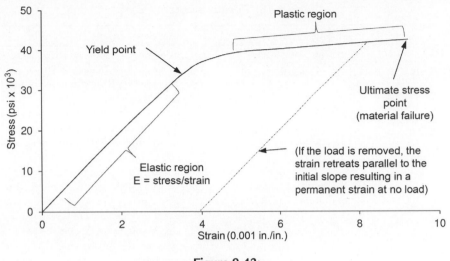

**Figure 9.42:**
Typical stress–strain curve.

Each material has a unique yield stress and ultimate stress. The terms yield stress and yield strength are used interchangeably as well as ultimate stress and ultimate strength. The goal of stress analysis or structural analysis is to make sure that the operating loads do not cause stresses that exceed the yield strength. For fail-safe designs, you must also analyze extreme loads such as crash loads. In these cases, it is often acceptable for the stresses to exceed the yield as long as they do not exceed the ultimate strength.

The yield and ultimate strength for most materials is published by manufacturers though I must advise caution when designing close to the yield or ultimate stresses. The actual yield and ultimate strength for all materials will be subject to a statistical distribution based on the manufacturing process. If you do design close to the yield or ultimate, research carefully the quality of the published specifications for the material. For parts that are safety critical, add material testing requirements to the drawing to ensure that the part will perform as expected.

### Finite element analysis

The technique of FEA has become the tool of choice for structural analysis. FEA software programs will provide excellent results provided you follow specific rules as shown in Table 9.16.

## Table 9.16: Tips for accurate FEA

| FEA Task | Tips | Comment |
|---|---|---|
| Constraining attachment points | Constrain realistically | This is the source of most inaccuracies. Study carefully how the actual structure will be attached and work to constrain it accordingly |
|  | For bolted connections, limit constraints to the area around each bolt hole | Based on the bolt size, constrain an area equal to the size of a small washer. Do not constrain the entire bolt flange or surface, just the area around each bolt |
|  | Bracket the problem when the exact attachment condition is unknown | Bracketing is a powerful technique for spotting high-risk areas; model the attachments in several different ways ranging from very conservative to very optimistic. Study the results to decide how sensitive the results are to the attachment technique. For example, a conservative assumption is that bolted joints are frictionless and joints will slip in the plane of the attachment |
| Modeling bolted connections that will stay preloaded | Do not model actual bolt but attach the area around each bolt hole from one part to the other | Based on the bolt size, attach an area equal to the size of a small washer. Do not attach the entire bolt flange or surface, just the area around each bolt |
| Modeling bolted connections that will unload | Model actual bolts with appropriate elements and setup contact regions between the parts | Use flexible elements that reflect the stiffness of the actual bolt. Pretension element with the anticipated preload. Setup contact regions between the two parts so the surfaces will separate as loads are applied but will not penetrate when loads are lower |
| Modeling the mass of nonstructural parts | Use mass elements carefully to avoid stiffening structure | Mass elements can be used to simulate nonstructural parts but care must be used when attaching the mass element. If rigid elements are used, the rigid elements may inaccurately stiffen the structure. If the nonstructural part is weak by itself but attaches across a large area, consider breaking the mass up into several mass elements that add up appropriately or attach the mass with flexible elements. Bracket the problem, as necessary, by trying several techniques |
| Meshing the parts | Eliminate small incidental features such as small chamfers and radii but increase density in high stress area or thin sections | Use at least two elements across thin sections or use shell elements. Study stress plots and increase density of mesh in high stress regions |

*(Continued)*

**Table 9.16: (Continued)**

| FEA Task | Tips | Comment |
|---|---|---|
| Checks to perform on every model | Run a 1 G analysis in each direction X, Y, and Z | For each analysis, apply a gravity load (1 G acceleration) and check the deflections and stresses. This is a good check because most analysts have an good sense of what effect gravity should have on their parts. If the deflections are very high for 1 G or the material is overstressed then solve the problems before analyzing with actual loads |
| | Check the reaction forces at the attachment points | After solving each analysis, whether the 1 G or the actual loads, check the reaction forces at the attachment points. The forces at the attachments must equal the applied loads and must agree with any conservative assumptions you have made. For instance, if bolted joints were modeled as frictionless, the reaction forces should be zero in the plane of the attachment |
| | Check the forces between parts and at specified contact regions | After solving each analysis, whether the 1 G or the actual loads, check the forces between parts. The forces must equal the applied loads. Contact forces should always be appropriate and never be in tension, the parts should separate instead |
| Interpreting stress results | Do not ignore high stress areas or discount locally high stresses as "singularities" | Study carefully high stress areas and try to understand if the simulation is accurate. Sharp internal corners can create stress singularities and refining the mesh just increases the stress. Don't ignore these areas but replace the sharp corners with fillets or radii if the part will actually have them. If the actual part has a sharp corner, it is quite possible the high stress will exist in the actual part; I know this from unfortunate firsthand experience |
| Interpreting frequency results | Be cautious and bracket the problem with conservative modeling | Frequency results are often optimistic and rarely obtainable with actual structures. Temper the results by bracketing the problem with conservative attachments and part connections. It is better to report the results as "no lower than XX Hz based on conservative modeling but we expect more like XX Hz" |

## Case study: Finite Element Analysis

My first FEA was on an airborne camera structure. I had installed a brand new revision of popular FEA software and excitedly began to generate beautiful color stress plots. The older experienced analyst looked at the results, turned up his nose, and stated "Nope, I don't believe it!" As I pressed him, he confided that he didn't believe any new revision of software until it was tested against real results. With his help I ran a couple of analyses, with known results, and wouldn't you know one of them was wrong! We continued digging until we found the problem: when we entered the inertia matrix of a lumped mass, the program incorrectly inverted the matrix. Sheepishly the software company issued an update that fixed the problem.

I learned a valuable lesson that any analysis tool, from hand calculations to the most sophisticated computer analysis, is only as accurate as you demand it to be. Never be satisfied with a convenient answer or a pretty graphic. Demand accurate prediction of reality and that requires testing with real problems for which you know the right answers. I tell the new mechanical engineers "FEA is only useful when you already know the answer!" In other words, if you are a novice, use FEA only when you have a set of similar problems that you have verified with actual tests. Once you have built up a body of experience and know the limitations inherent in FEA, you can begin to trust the results although I still attempt to verify whenever possible.

## Vibration analysis

When you get a set of vibration requirements take time to look it over and calculate the magnitude of the vibration. Usually, the requirement is in the form of a random vibration power spectral density (PSD) chart as shown in Figure 9.43. While the purpose and meaning of this chart and its units may seem odd, it is actually a powerful and concise method to communicate random vibration requirements. If the requirements are given in a different form, such as sine sweeps or dwells, please consult the recommended reading for more information. Since the PSD is the most common vibration specification, I'll cover it here.

The first step with a new specification is to calculate the overall G's root mean square (RMS) value. The vibration is random; random in frequency and random in level but it will follow the frequency distribution of the specification. The overall G's RMS is simply the area under the log—log curve calculated using a piecewise integration as shown in Figure 9.44. Since the units are in frequency (Hz) along the *x*-axis and $G^2$/Hz along the *y*-axis, the area units are $G^2$. The square root of the area is the overall vibration energy level in G's RMS.

The vibration test equipment is designed to create random vibration with the frequency content specified and with a normal distribution. Therefore, the peak G level at any given time is subject to the normal distribution, shown in Figure 9.45, designated as multiples of the overall G's RMS. Looking at the figure, $6\sigma$ peak G levels seem to occur infrequently. But in real life applications, which can have a long duration, I have seen $6\sigma$ events.

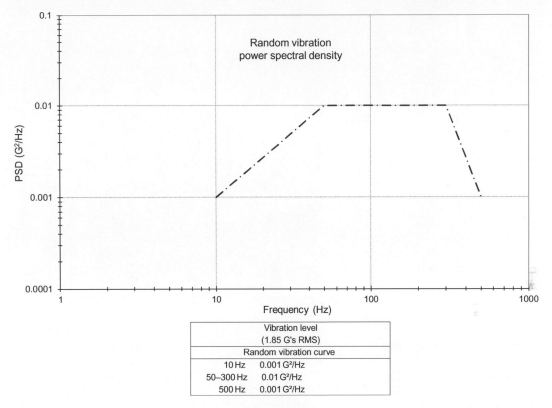

| Vibration level |
|:---:|
| (1.85 G's RMS) |

| Random vibration curve | |
|:---:|:---:|
| 10 Hz | 0.001 G²/Hz |
| 50–300 Hz | 0.01 G²/Hz |
| 500 Hz | 0.001 G²/Hz |

**Figure 9.43:**
Vibration PSD specification.

However, the test lab environment is a completely different story. Many specifications, including military specifications, allow for the clipping of events at the $3\sigma$ level. Therefore, if you are designing just for a laboratory test, use 3 or $4\sigma$ multiplied times the overall G's RMS as your expected peak value.

The second step is to understand how your particular equipment will respond to the random vibration. In the simplest situation, the equipment will have a single resonance at which it will freely vibrate. This is the resonant frequency or natural frequency of the part or equipment and the variable "$f_n$" represents it. Often, there are several significant resonances each with a unique preferred direction and shape of vibration. At the resonant frequency, the equipment can be lightly damped and ring like a bell or heavily damped and barely resonate at all. One of your challenges is to quantify this damping for it represents the amplification that the structure creates at the natural frequency. The variable "$Q$" is typically used to represent the amplification and therefore a structure with a $Q = 10$ will amplify energy at resonance by a factor of 10. The response of a simple system with one resonance can be plotted as shown in Figure 9.46 and is called the structural transfer function.

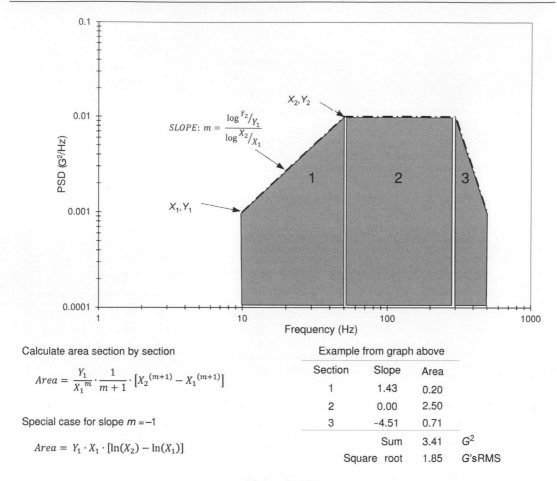

**Figure 9.44:**
Integration of the PSD.

Calculate area section by section

$$Area = \frac{Y_1}{X_1^m} \cdot \frac{1}{m+1} \cdot \left[ X_2^{(m+1)} - X_1^{(m+1)} \right]$$

Special case for slope $m = -1$

$$Area = Y_1 \cdot X_1 \cdot \left[ \ln(X_2) - \ln(X_1) \right]$$

Example from graph above

| Section | Slope | Area | |
|---------|-------|------|---|
| 1 | 1.43 | 0.20 | |
| 2 | 0.00 | 2.50 | |
| 3 | -4.51 | 0.71 | |
| | Sum | 3.41 | $G^2$ |
| | Square root | 1.85 | G'sRMS |

The interpretation of the transfer function is that for frequencies well below resonance, the structure will follow the input vibration at 1:1. As the frequencies approach resonance, the structure begins to amplify the input vibration, finally reaching a peak of $Q$ at the resonance. Above the resonance, the vibration falls off sharply crossing the 1:1 point at $1.414 \times f_n$. Above this point, any vibration is attenuated and the structure is isolated from the input vibration. The electrical equivalent of this is a low pass filter.

Once you have characterized the vibration and the equipment, the final step is to predict the response of your equipment to the specified vibration. The response of the equipment is equal to the specification PSD multiplied at each frequency by the square of the transfer function at that frequency as shown in Figure 9.47. Once you have the equipment response, calculate the overall G's RMS *at the equipment*. This equipment response follows the same

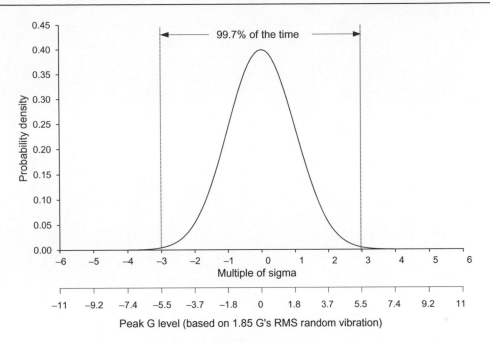

**Figure 9.45:**
Normal distribution of random vibration.

statistical distribution as the input but will be dominated by the resonant frequency of the equipment. Using the resonant frequency, you can estimate the distribution of the vibration and also estimate the maximum deflection also shown in Figure 9.47. This is a first-order estimate and works only for simple structures dominated by a single resonance but I have found it to be very useful.

Understanding the structure or equipment and finding the resonate frequencies is the most important step in the process above. Here are some practical approaches to find the resonate frequencies.

- For new designs:
  - Use FEA to determine the natural frequency. (The FEA only calculates the natural frequency not the amplification.)
  - Calculate the natural frequency using the examples in Steinberg's book "*Vibration Analysis for Electronic Equipment, 3rd ed.*" which addresses a variety of structures and assemblies including PWBs [24, p.35].
  - Estimate the amplification $Q$. From my experience, $Q$ ranges from 100 for monolithic metallic structures, 40 for rigid bolted structures, and 10 for loosely bolted assemblies. Steinberg advises that for PWBs, $Q$ can be estimated using the natural frequency $Q = \sqrt{f_n}$ [24, p.35].

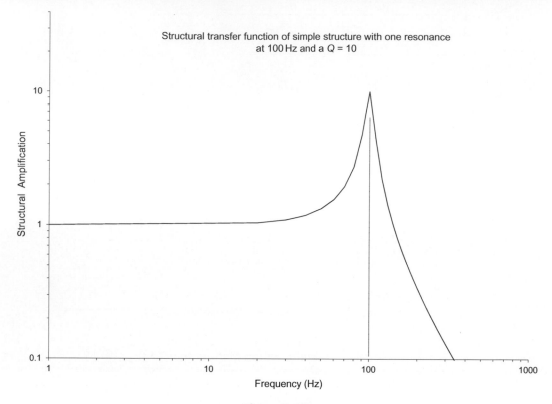

**Figure 9.46:**
Structural transfer function.

- For existing equipment, run a low-level sine sweep at a vibration test lab and ask the test lab to measure the transfer function.
- Without access to a vibration test lab, sophisticated users will conduct a hammer modal test which uses an instrumented hammer and an accelerometer to measure the response of the structure to an impact. With just an accelerometer, I have "bump" tested a structure and measured the natural frequency and even estimated the damping by examining the decaying time waveform. Please consult Tom Irvine's excellent website and extensive library of vibration topics for tips on how to conduct this type of testing [24].
- Always mount the equipment the same way it will be mounted in operation. The natural frequency can change dramatically if the equipment is bolted to a heavy rigid structure versus bolted to a light structure. Likewise, any FEA calculations should replicate as much as possible the same attachment method.

Once you have characterized the predicted response of the equipment, use this information to improve the design. The peak G's predicted above will not occur everywhere in the structure. Near the mounting points, the input vibration with its peak G's will be dominant.

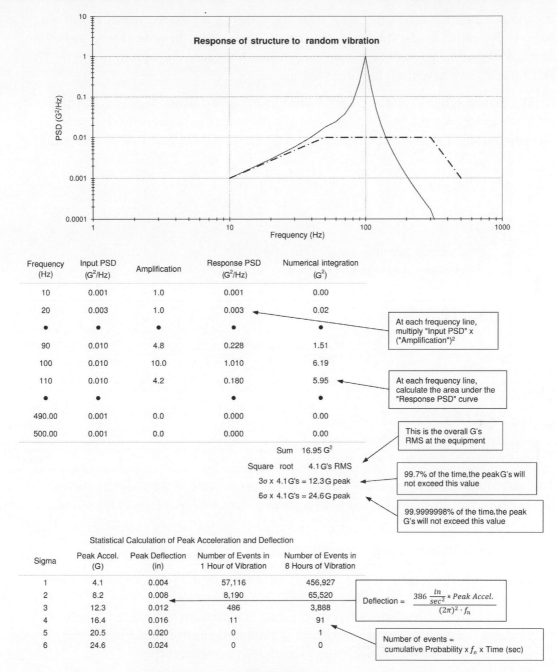

**Figure 9.47:**
Simple structural response Excel worksheet.

As you move farther and farther away from the mounting points, the vibration will typically become dominated by the structure's resonant frequency with the peak G's as predicted above. Usually, the point of maximum acceleration is the furthest point from the attachments but I have been fooled several times.

Since vibration failures can be caused by one part impacting another, a useful equation is the relationship between the peak displacement and the peak G's. The equation below is a first-order estimate of the peak displacement based on the natural frequency of the structure [24].

$$\delta_{\text{peak}} = \frac{G_{\text{peak}} \times 386}{(2\pi f_n)^2}$$

where

$\delta_{\text{peak}}$ = Peak displacement in inches (one-sided)
$G_{\text{peak}}$ = Peak acceleration in G's
$f_n$ = Resonant frequency in hertz.

For displacement in millimeters:

$$\delta_{\text{peak}} = \frac{G_{\text{peak}} \times 9807}{(2\pi f_n)^2}$$

where

$\delta_{peak}$ = Peak displacement in millimeter (one-sided)
$G_{peak}$ = Peak acceleration in G's
$f_n$ = Resonant frequency in hertz.

This estimate of peak displacement is with respect to the base of the structure. Internal components may deflect more or less with respect to each based on the shape of the structure as it vibrates.

Let me finish with some tips for designing vibration proof assemblies:

- Avoid loose items such as wire bundles and connectors. Fasten everything and use the peak G estimation to decide how much preload is necessary.
- Avoid cantilever structures wherever possible. If a cantilever structure is necessary, evaluate the peak G's and allow space for the peak deflection of the end of the cantilever. Use a conservative value for the $Q$ of the cantilever.
- Instead of cantilever structures, attempt to create attachments that are outboard of the center of gravity. In other words, the center of gravity should be near the geometric center of the fastener pattern.
- Preload all bolted joints by specifying a screw torque and use a thread locking adhesive or other locking means.

In those situations where the clearance between parts is less than the predicted peak displacement, add a plastic or elastomeric bumper to stop or snub the motion before damage occurs.

## References

[1] Nelco® N4000-7. Park Electrochemical Corp., Rev 7–12, <www.parkelectro.com> (accessed 03.10.13).

[2] MatWeb, Your Source for Materials Information. <www.matweb.com> (accessed 03.10.13).

[3] MatWeb, Your Source for Materials Information. <www.matweb.com> (accessed 05.09.13).

[4] Threaded fasteners for plastics. STANLEY Engineered Fastening—Formerly Emhart Teknologies & Infastech, <www.stanleyengineeredfastening.com> (accessed 28.10.13).

[5] The REMFORM® Screw, Research Engineering & Manufacturing Inc., <www.taptite.net> (accessed 28.10.13).

[6] ISO General Purpose Metric Screw Threads, ISO 262.

[7] Gages & Gaging For Unified Inch Screw Threads, ASME B1.2.

[8] J.E. Shigley, Mechanical Engineering Design, third ed., McGraw-Hill, New York, NY, 1977. p. 245.

[9] Enclosures for Electrical Equipment (1000 Volts Maximum), NEMA 250, 2003.

[10] Degrees of Protection Provided by Enclosures (IP Code), IEC 60529, Edition 2.2, August 2013, International Electrotechnical Commission.

[11] Requirements and Acceptance for Cable and Wire Harness Assemblies, IPC/WHMA-A-620, 2002.

[12] Aircraft Electrical Wiring Interconnect System (EWIS) Best Practices, FAA, Revision 2.0. <www.faa. gov> (accessed 12.06.13).

[13] MatWeb, Your Source for Materials Information. <www.matweb.com> (accessed 01.11.13).

[14] S. Lee, How To Select A Heat Sink. Aavid Thermalloy, LLC., <www.aavid.com> (accessed 18.02.13).

[15] Design Guide: Useful Information for Designing Your Own Magnetic Shielding. <http://www.mushield. com/design-guide.shtml> (accessed 03.01.14).

[16] Thermal Interface Materials for Electronics Cooling. Chomerics a Division of Parker Hannifin Corporation, <www.chomerics.com> (accessed 19.11.13).

[17] Sil-Pad® SELECTION GUIDE. The Bergquist Company, <www.berquistcompany.com> (accessed 19.11.13).

[18] D.S. Steinberg, Cooling Techniques for Electronic Equipment, second ed., John Wiley & Sons, Inc., 1991.

[19] Acoustic Noise: Causes, Rating Systems, and Design Guidelines. Comair Rotron, <www.comairrotron. com/acoustic-noise-causes-rating-systems-and-design-guidelines>, 2013 (accessed 01.09.13).

[20] Heat Pipe Technology: Passive Heat Transfer for Greater Efficiency. Thermacore, Inc., <www. thermacore.com/thermal-basics/heat-pipe-technology.aspx>, 2013 (accessed 29.09.13).

[21] R. DeHoff, K. Grubb. Heat Pipe Application Guidelines. Thermocore <www.thermacore.com> (accessed 30.11.13).

[22] Graphic of DeltaMaker Personal 3-D Printer used by permission. DeltaMaker LLC, <www.deltamaker.com>.

[23] J.E. Shigley, J.J. Uicker Jr., Theory of Machines and Mechanisms, McGraw-Hill, 1980, pp. 384–399.

[24] T. Irvine, Response of a single degree-of-freedom system subjected to a unit step displacement. <www.vibrationdata.com/unit_step.pdf> (accessed 30.11.13).

## Suggested reading

D.S. Steinberg, Vibration Analysis for Electronic Equipment, third ed., John Wiley & Sons, Inc., New York, NY, 2000. This is an excellent text for detailed vibration analysis of electronic equipment. Written assuming no knowledge of vibration theory, Dave Steinberg derives all the math and equations needed to predict vibration response for typical vibration and shock environments. The book also contains many "rule of thumb"-type solutions and is an excellent companion for vibration testing and troubleshooting.

D.S. Steinberg, Cooling Techniques for Electronic Equipment, second ed., John Wiley & Sons, Inc., New York, NY, 1991. A comprehensive guide for thermal design and analysis specifically geared toward electronics. Dave Steinberg derives all the math and equations needed for thermal analysis. The book contains many "rule of thumb"-type solutions. The focus is more geared toward theoretical with no content on the specifics of temperature testing and instrumentation.

T. Kordyban, Hot Air Rises and Heat Sinks, ASME Press, New York, NY, 1998. This light-hearted approach toward thermal design by a practicing thermal engineer is a quick and informative read. Although humorous in the approach, it is packed with practical lessons and once you sift through the allegories, Tony Kordyban points you to the thermal equation that applies. This is an excellent book for technicians as well as degreed engineers. The importance of testing and practical experience is highlighted, therefore, this book should be required reading before embarking on the test phase of a project, especially for novices.

J.E. Shigley, C.R. Mischke, R.G. Budynas, Mechanical Engineering Design, seventh ed., McGraw-Hill, New York, NY, 2004. This is the definitive mechanical engineering desk reference now in the ninth edition. The seventh edition is recommended by engineers because that is the last edition Joseph Shigley was directly involved with. I personally use my 3rd edition on a daily basis as a practicing mechanical engineer. The basics of mechanical design are covered from stress and strain to bearings, fasteners, and gear trains.

J.P. Holman, Heat Transfer, tenth ed., McGraw-Hill, New York, NY, 2009. An excellent introductory text on heat transfer, Jack Holman presents the theoretical basis for conduction, convection, and radiation. The text addresses the physics and equations of heat transfer and applies to all industries.

M. Silverman, How Reliable is Your Product, Super Star Press, Cupertino, CA, 2010. A very readable text written by a practicing reliability engineer, Mike Silverman, covers all aspects of reliability and describes how it should be applied whether you're in the concept phase or in the manufacturing phase of product development. Every concept is illustrated with multiple case studies from his experience of running a reliability test lab.

V. Adams, A. Askenazi, Building Better Products with Finite Element Analysis, OnWord Press, Santa Fe, NM, 1998. This is an entry-level text addressing FEA from the perspective of practicing engineers. Adams and Askenazi focus on how FEA is applied to everyday design problems. The examples are useful and the practical advice regarding setting up boundary conditions and FEA in general is excellent.

T. Irvine, <www.vibrationdata.com>. Tom Irvine has compiled a vast knowledge base of vibration calculations, software, and tutorials on his website. For almost any vibration problem I have encountered, Tom has written a tutorial explaining the mathematics and the solution.

## CHAPTER 10

# Electronic Design

## Michael F. (Mike) Gard
*Senior Product Design Engineer The Charles Machine Works Perry, OK, USA*

## Chapter Outline

Developing and Managing Embedded Systems and Products.
DOI: http://dx.doi.org/10.1016/B978-0-12-405879-8.00010-6

## *Overview of electronic design*

Electronic circuit design involves the selection and interconnection of physical devices in a variety of topologies to meet performance specifications, environmental requirements, power and cost budgets, operating-life requirements, and a host of other design constraints in agreement with an overall schedule. Embedded systems are an advanced extension of this general paradigm for circuit design because some of the physical devices are processors, FPGAs, and semi-custom or full-custom integrated circuits with software and firmware functions included as part of the hardware package.

The first step in circuit design is not designing circuits. The first step is to establish, identify, and harmonize the specifications and standards the design is expected to satisfy. This is a nontrivial exercise because it is not unusual to find some requirements in stark conflict with others. Resolution of such conflicts is itself a high practice of the engineering arts. It is imperative that the circuit designer (who may be a single individual on a small project) or the design team (which may consist of multiple individuals with specialist skills in the various design subdisciplines) have a fundamental understanding of the user, the user's environment, the user's operational expectations, and other aspects of the end user's world which define the many dimensions of the design assignment. ***Whenever possible, the design team should be involved in specification definition, cost estimation, and schedule formation.***

### Requirements

Chapter 6 explored the challenges of setting and iterating requirements. The distinction between "gotta haves," "should haves," and "wanta haves" within requirements is certainly a major issue at the system level, but their effects are felt acutely at the level of circuit design. Embedded systems are particularly vulnerable to the effects of poor design trade-offs and poor product definition. Large systems often have sufficient resource margins to incorporate newly defined (or newly emergent) software elements into existing computing resources; in more severe cases, it may be possible to shift part of a new software burden from one computing platform to another. Embedded systems usually have no appreciable intrinsic resource buffer. It is a serious matter—if not catastrophic—when the project team learns the system's processor must have additional ADC (analog-to-digital) inputs, additional I/O lines, or higher clock frequency.

### General processes and procedures

The general design flow for embedded systems requires specification accuracy and system integration because your team will almost always develop hardware (electronics and mechanics), firmware, and software in parallel. You must have modules from each of these domains in some functional, if preliminary, form before meaningful product testing can begin. Embedded systems have the added complication that pieces from these major domains are not readily separable; firmware and software ordinarily must be available in rudimentary form before testing hardware; firmware and software, in turn, require well-defined interface specifications and control structures in the hardware. For example, early development of the graphical user interface (GUI) is indispensable for hardware development.

### Specific requirements

Parameters in the requirements will determine the difficulty of the component selection process for most commercial projects. These parameters include operating-temperature range, circuit-board area, and volume. Other considerations may come into play in more demanding applications or extreme environments, including but not limited to:

- operating and storage temperature requirements for certain components (e.g., capacitors and batteries),

- power consumption and heat generation for active devices, because they drive the size of the power supply, cooling, and packaging designs,

- weight specifications can be a very severe design constraint, it is often a major consideration for packaging materials, component selection, and even battery selection. In certain unusual cases—for instance, equipment and subsystems on the rotating gantry of a computed tomography machine—both weight and center of gravity may be specified,

- vibration, mechanical shock, and drop—often associated with anticipated conditions of use or simply conditions of storage and transportation—can impact component selection, circuit-board layout, packaging, and thermal management (especially cooling) in many ways,

- humidity, immersion, salt spray, fungus, and corrosion, when specified, usually affect the selection of fasteners, enclosure materials and finishes, and wiring details (dissimilar metals and humidity are a bad combination that can lead to corrosion and form a chemical half-cell reaction that contributes offset error to precision DC circuits),

- dust, humidity, and altitude can complicate selections of components operating at high voltages or which depend on sealed reference compartments for their operation, for example, some types of absolute pressure transducers,

- electrical noise specifications are usually implicit in the performance specifications; if given explicitly, they can be dominant design drivers,

- acoustic noise specifications, if given, can influence the selection of fans, motors, and power transformers,

- radiation, primarily specified for space, nuclear power, or radiological applications, can severely limit the choice of semiconductors and can also affect packaging decisions,

- outgassing specifications, normally given for space or high-vacuum applications, will limit the available selection of semiconductors and may restrict the availability of certain passive components.

An operating-life requirement, typically specified as mean-time-between-failure (MTBF), can be difficult to meet. It is often difficult to establish what the design's MTBF calculation really is: newer semiconductor devices rarely have established reliability numbers and the design team must rely on comparisons with devices of a similar nature having established reliability records. Rigid compliance with an MTBF specification will restrict the number of qualified active devices available for design use. (A frequent lamentation is that established MTBF numbers are available only after the device has become obsolete.)

Many consumer and laboratory device environments are relatively benign. Ordinarily, temperature requirements are not extreme. Mechanical shock and vibration are concerns during transportation but rarely during use. Line power is readily available. MTBF expectations are ordinarily not specified.

Automotive, truck, rail, and other vehicular applications are more demanding than consumer environments—consider the engine firewall temperature or the interior of a sealed automobile passenger compartment on a hot sunny day. Construction and agricultural equipment and other devices used in an uncontrolled and unprotected outdoor environment are even more demanding—because of the environment in which they are used and stored; the forces, mechanical shocks, and vibrations encountered in normal operation; and because the increasing use of engine and electrohydraulic electronics can result in potentially fatal RFI/EMI/EMC problems (machinery used in underground construction has very strict EMI/EMC requirements which are continually upgraded as cell phone operating frequencies increase—every operator will simply place his/her active cell phone on the control console while at work). Aerospace, submarine, satellite, and defense applications are more demanding yet.

Embedded systems contain processing capability and require a significant programming effort. We tend to overlook the analog circuit functions, which often hide behind displays, operator interfaces, soft keys, and other aspects of the design. If a consumer project intended for a stationary room-temperature environment requires rudimentary filtering, limited power (especially if battery operated), a few switch interfaces, and performance is not critical, it may be appropriate to be casual when selecting passive components and semiconductors—price is often the major criterion in such applications. If another application in the same environment requires preamplification and signal conditioning for a critical process control variable—for example, the temperature controller of a hospital blood bank or a security system intrusion alarm—component selection should be done with great care.

## Circuit design

Discussion of the general aspects and processes for electric design begins with the assumption that design specifications are known and accurately represent performance requirements. I will start with basic components and work "bottom-up" through circuit boards to the final electronic subsystem.

## Components

It is not reasonable to attempt an encyclopedic discussion of device types, performance characteristics, configurations, and specialized uses. Most embedded systems use fairly common components with well-known characteristics and, for the most part, these devices are readily available from multiple sources. A brief review of the most salient component types and characteristics is appropriate, if only to serve as a reminder to you. Excellent general discussions of components are found in Refs. [1,2], which also provide very helpful filter design information.

## *Resistors*

Resistors are the unsung heroes of the electronics business. They are indispensable passive circuit elements which, in most cases, are plentiful, reliable, accurate, generally inexpensive, and stable over a wide range of frequencies in most environments. The degree of refinement is such that some resistors closely approximate theoretically-ideal components over many decades of frequency (Figure 10.1).

I will focus on general-purpose resistors, those individual resistive-circuit elements generally available in small axial-leaded cylindrical packages for through-hole applications or in chip form for surface-mount applications. Precision resistors, available in most of the general-purpose packaging options, are sometimes available in rectangular packages with radial leads for through-hole applications. Resistance values are available from tens of milliohms to tens of megohms in well-defined standard resistance values. A component tolerance of ±1% is very common for general-purpose resistors, although tolerances of ±0.5% and ±0.1% are readily available and are represented in standard resistance tables (Table 10.1). Power-dissipation rating determines the physical size of general-purpose resistors.

Surface-mount and axial-leaded cylindrical devices are usually rated at 1 W or less. Although axial-leaded cylindrical resistors are an older technology and require significantly greater circuit-board area than their surface-mounted equivalents, they are readily available and are still used in large quantities, especially for low-volume production runs and prototyping purposes. The production economics of automated pick-and-place machinery increasingly favors surface-mount devices. Furthermore, manual assembly becomes very difficult for the very smallest devices (0402 and smaller). Typical temperature coefficients of resistance are one or two hundred parts per million per degree Celsius (ppm/°C), although the temperature coefficient varies by construction and from manufacturer to manufacturer; individual parts could have a temperature coefficient twice this typical value.

Most general-purpose resistors are either thick-film or thin-film construction. In thick-film construction, the manufacturing process involves screening a controlled conductivity paste or ink on a substrate (typically alumina or other ceramic) which is fired to form a resistive element. The geometry of the paste pattern and the size of the substrate determine the resistance and power-dissipation rating [3]. The manufacturing process for thin-film elements deposits a thin metal layer on a substrate in carefully controlled patterns which can be trimmed by laser etching if needed [4]. As with thick-film construction, the geometry and substrate size in thin film determine the resistance and power-dissipation rating. Axial-leaded cylindrical resistors can be internally constructed of either thin-film or thick-film technology enclosed by a molded outer case and axial lead wires. Typical temperature coefficients of resistance are one or two hundred parts per million per degree Celsius (ppm/°C).

(A)

(B)

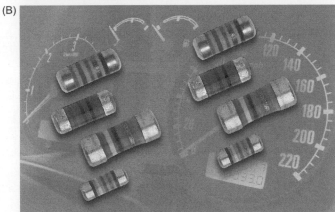

(C)

**Figure 10.1: (A)–(C): Resistors.**
The most common resistors used in general electronics work are (A) through-hole (axial-leaded)
resistors, (B) MELF resistors, and (C) chip resistors. These devices are available in a wide variety
of compositions, tolerances, power-dissipation ratings, and temperature coefficients.
*Image courtesy of Vishay Intertechnology, Inc.*

# Table 10.1: Standard resistance values in a standard resistance decade

**Resistance Tolerance (%)**

| E192 | E96 | E24 | E12 | E6 | E192 | E96 | E24 | E12 | E6 | E192 | E96 | E24 | E12 | E6 | E192 | E96 | E24 | E12 | E6 | E192 | E96 | E24 | E12 | E6 |
|---|---|---|---|---|---|---|---|---|---|---|---|---|---|---|---|---|---|---|---|---|---|---|---|---|
| 0.10% 0.25% 0.50% | 1% | 2% 5% | 10% | 20% | 0.10% 0.25% 0.50% | 1% | 2% 5% | 10% | 20% | 0.10% 0.25% 0.50% | 1% | 2% 5% | 10% | 20% | 0.10% 0.25% 0.50% | 1% | 2% 5% | 10% | 20% | 0.10% 0.25% 0.50% | 1% | 2% 5% | 10% | 20% |
| 10.0 | 10.0 | 10 | 10 | 10 | 15.8 | 15.8 | - | - | - | 24.9 | 24.9 | - | - | - | 39.2 | 39.2 | 39 | 39 | - | 62.6 | - | - | - | - |
| 10.1 | - | - | - | - | 16.0 | - | 16 | - | - | 25.2 | - | - | - | - | 39.7 | - | - | - | - | 63.4 | 63.4 | - | - | - |
| 10.2 | 10.2 | - | - | - | 16.2 | 16.2 | - | - | - | 25.5 | 25.5 | - | - | - | 40.2 | 40.2 | - | - | - | 64.2 | - | - | - | - |
| 10.4 | - | - | - | - | 16.4 | - | - | - | - | 25.8 | - | - | - | - | 40.7 | - | - | - | - | 64.9 | 64.9 | - | - | - |
| 10.5 | 10.5 | - | - | - | 16.5 | 16.5 | - | - | - | 26.1 | 26.1 | - | - | - | 41.2 | 41.2 | - | - | - | 65.7 | - | - | - | - |
| 10.6 | - | - | - | - | 16.7 | - | - | - | - | 26.4 | - | - | - | - | 41.7 | - | - | - | - | 66.5 | 66.5 | - | - | - |
| 10.7 | 10.7 | - | - | - | 16.9 | 16.9 | - | - | - | 26.7 | 26.7 | 27 | 27 | - | 42.2 | 42.2 | - | - | - | 67.3 | - | - | - | - |
| 10.9 | - | - | - | - | 17.2 | - | - | - | - | 27.1 | - | - | - | - | 42.7 | - | - | - | - | 68.1 | 68.1 | 68 | 68 | 68 |
| 11.0 | 11.0 | 11 | - | - | 17.4 | 17.4 | - | - | - | 27.4 | 27.4 | - | - | - | 43.2 | 43.2 | 43 | - | - | 69.0 | - | - | - | - |
| 11.1 | - | - | - | - | 17.6 | - | - | - | - | 27.7 | - | - | - | - | 43.7 | - | - | - | - | 69.8 | 69.8 | - | - | - |
| 11.3 | 11.3 | - | - | - | 17.8 | 17.8 | 18 | 18 | - | 28.0 | 28.0 | - | - | - | 44.2 | 44.2 | - | - | - | 70.6 | - | - | - | - |
| 11.4 | - | - | - | - | 18.0 | - | - | - | - | 28.4 | - | - | - | - | 44.8 | - | - | - | - | 71.5 | 71.5 | - | - | - |
| 11.5 | 11.5 | - | - | - | 18.2 | 18.2 | - | - | - | 28.7 | 28.7 | - | - | - | 45.3 | 45.3 | - | - | - | 72.3 | - | - | - | - |
| 11.7 | - | - | - | - | 18.4 | - | - | - | - | 29.1 | - | - | - | - | 45.9 | - | - | - | - | 73.2 | 73.2 | - | - | - |
| 11.8 | 11.8 | - | - | - | 18.7 | 18.7 | - | - | - | 29.4 | 29.4 | - | - | - | 46.4 | 46.4 | - | - | - | 74.1 | - | - | - | - |
| 12.0 | - | 12 | 12 | - | 18.9 | - | - | - | - | 29.8 | - | - | - | - | 47.0 | - | 47 | 47 | 47 | 75.0 | 75.0 | 75 | - | - |
| 12.1 | 12.1 | - | - | - | 19.1 | 19.1 | - | - | - | 30.1 | 30.1 | 30 | - | - | 47.5 | 47.5 | - | - | - | 75.9 | - | - | - | - |
| 12.3 | - | - | - | - | 19.3 | - | - | - | - | 30.5 | - | - | - | - | 48.1 | - | - | - | - | 76.8 | 76.8 | - | - | - |
| 12.4 | 12.4 | - | - | - | 19.6 | 19.6 | - | - | - | 30.9 | 30.9 | - | - | - | 48.7 | 48.7 | - | - | - | 77.7 | - | - | - | - |
| 12.6 | - | - | - | - | 19.8 | - | - | - | - | 31.2 | - | - | - | - | 49.3 | - | - | - | - | 78.7 | 78.7 | - | - | - |
| 12.7 | 12.7 | - | - | - | 20.0 | 20.0 | 20 | - | - | 31.6 | 31.6 | - | - | - | 49.9 | 49.9 | - | - | - | 79.6 | - | - | - | - |
| 12.9 | - | - | - | - | 20.3 | - | - | - | - | 32.0 | - | - | - | - | 50.5 | - | - | - | - | 80.6 | 80.6 | - | - | - |
| 13.0 | 13.0 | 13 | - | - | 20.5 | 20.5 | - | - | - | 32.4 | 32.4 | - | - | - | 51.1 | 51.1 | 51 | - | - | 81.6 | - | - | - | - |
| 13.2 | - | - | - | - | 20.8 | - | - | - | - | 32.8 | - | - | - | - | 51.7 | - | - | - | - | 82.5 | 82.5 | 82 | 82 | - |
| 13.3 | 13.3 | - | - | - | 21.0 | 21.0 | - | - | - | 33.2 | 33.2 | 33 | 33 | 33 | 52.3 | 52.3 | - | - | - | 83.5 | - | - | - | - |
| 13.5 | - | - | - | - | 21.3 | - | - | - | - | 33.6 | - | - | - | - | 53.0 | - | - | - | - | 84.5 | 84.5 | - | - | - |
| 13.7 | 13.7 | - | - | - | 21.5 | 21.5 | - | - | - | 34.0 | 34.0 | - | - | - | 53.6 | 53.6 | - | - | - | 85.6 | - | - | - | - |
| 13.8 | - | - | - | - | 21.8 | - | - | - | - | 34.4 | - | - | - | - | 54.2 | - | - | - | - | 86.6 | 86.6 | - | - | - |
| 14.0 | 14.0 | - | - | - | 22.1 | 22.1 | 22 | 22 | 22 | 34.8 | 34.8 | - | - | - | 54.9 | 54.9 | - | - | - | 87.6 | - | - | - | - |
| 14.2 | - | - | - | - | 22.3 | - | - | - | - | 35.2 | - | - | - | - | 55.6 | - | - | - | - | 88.7 | 88.7 | - | - | - |
| 14.3 | 14.3 | - | - | - | 22.6 | 22.6 | - | - | - | 35.7 | 35.7 | 36 | - | - | 56.2 | 56.2 | 56 | 56 | - | 89.8 | - | - | - | - |
| 14.5 | - | - | - | - | 22.9 | - | - | - | - | 36.1 | - | - | - | - | 56.9 | - | - | - | - | 90.9 | 90.9 | 91 | - | - |
| 14.7 | 14.7 | - | - | - | 23.2 | 23.2 | - | - | - | 36.5 | 36.5 | - | - | - | 57.6 | 57.6 | - | - | - | 92.0 | - | - | - | - |
| 14.9 | - | - | - | - | 23.4 | - | - | - | - | 37.0 | - | - | - | - | 58.3 | - | - | - | - | 93.1 | 93.1 | - | - | - |
| 15.0 | 15.0 | 15 | 15 | 15 | 23.7 | 23.7 | 24 | - | - | 37.4 | 37.4 | - | - | - | 59.0 | 59.0 | - | - | - | 94.2 | - | - | - | - |
| 15.2 | - | - | - | - | 24.0 | - | - | - | - | 37.9 | - | - | - | - | 59.7 | - | - | - | - | 95.3 | 95.3 | - | - | - |
| 15.4 | 15.4 | - | - | - | 24.3 | 24.3 | - | - | - | 38.3 | 38.3 | - | - | - | 60.4 | 60.4 | - | - | - | 96.5 | - | - | - | - |
| 15.6 | - | - | - | - | 24.6 | - | - | - | - | 38.8 | - | - | - | - | 61.2 | - | - | - | - | 97.6 | 97.6 | - | - | - |
|  |  |  |  |  |  |  |  |  |  |  |  |  |  |  | 61.9 | 61.9 | 62 | - | - | 98.8 | - | - | - | - |

Resistors are commonly available in standard values within a standard resistance decade, known as EIA standard values because they were originally developed by the Electronic Industries Alliance. The number of available values in a resistance decade depends on the tolerance. In this table, the E24 series represents the 24 standard values available in a resistance decade having resistors with ±5% tolerance, for example, standard values in the 100–1000 Ω decade are 100, 110, 120, 130, 150 Ω, and so on. Likewise, standard values in the 10–100 kΩ standard decade are 10, 11, 12, 13, 15 kΩ, and so on. E96 represents ±1% standard values, and E192 represents standard values with tolerances of ±0.5%, ±0.25%, and ±0.1%. Standard resistance decades spanning milliohms to megaohms are readily available as stock items from many manufacturers.

*Image courtesy of Stackpole Electronics, Inc.*

Metal foil is the other readily available resistor category. Metal-foil construction attaches a resistive metal foil to a ceramic substrate in controlled geometries, with particular attention being given to the thermal-expansion coefficients of the metal foil and substrate material. The linear expansion coefficients of the materials are different by design, with the metal foil having a higher coefficient of expansion than the substrate. As the resistor's temperature changes the metal foil, which also has a tightly controlled temperature coefficient of resistance, is stretched or compressed by mechanical expansion and contraction of the substrate. These dimensional changes may be in response to either changing environmental conditions (the ambient environmental temperature) or internally generated heating produced by current flow in the resistive element (self-heating). Careful geometrical design and control of the substrate's linear expansion coefficient produces minute length and thickness changes in the metal foil which compensate the metal foil's resistance change produced by thermal effects. Metal-foil resistors are precision parts with tightly controlled variances around the nominal value and a temperature coefficient of only a few parts per million per degree Celsius (ppm/°C). Current manufacturing processes are sufficiently well controlled to render these parts nearly ideal resistive elements [5].

These construction details are important to the systems designer, particularly the analog designer, because amplifier gains are almost always determined by resistor ratios in a feedback arrangement. At the reference temperature (usually 23°C, 25°C, or room temperature), the ratio of two resistors will determine the gain of a typical voltage-type operational amplifier stage in simple inverting or noninverting configurations. Departure of the actual resistances from their nominal values will determine the room-temperature gain error (assuming an inverting gain configuration, the actual gain variation from nominal using two independent ±1% thick-film resistors could be as much as ±2%). You must add the effect of different temperature coefficients over the anticipated range of operating temperatures to the basic gain-tolerance error, because no two parts from different production runs and different manufacturers can be expected to exhibit exactly the same temperature coefficient of resistance. This difference in temperature coefficients cannot be ignored, because thermal effects are significant in amplifiers expected to operate over wide temperature ranges, for example, the −55°C to +125°C temperature range typical of many military systems. The same effects are evident in many filtering applications; actual cutoff frequencies of conventional active RC filters are influenced in exactly the same way by the production tolerances and differential temperature coefficients of the resistors and capacitors from which they are constructed (Figures 10.2 and 10.3).

One solution to this difficulty is to specify resistors with tight resistance tolerances (e.g., those with ±0.1% tolerance). This certainly will improve the room-temperature gain error for a slight increase in component cost, but it does nothing to improve the relative gain error associated with differential temperature coefficients. A better, but more

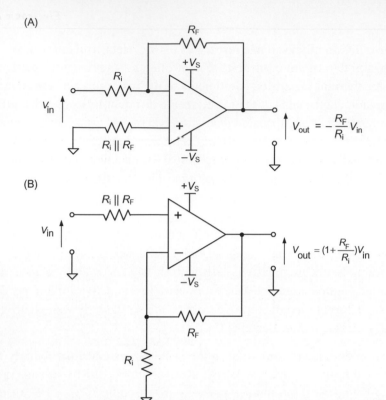

(A)

$$V_{out} = -\frac{R_F}{R_i}V_{in}$$

(B)

$$V_{out} = (1+\frac{R_F}{R_i})V_{in}$$

**Figure 10.2: (A) and (B): Basic inverting and noninverting amplifiers.**
Basic operational amplifier connections for (A) inverting and (B) noninverting amplifiers.
The resistors marked $R_i \| R_F$ are often omitted if amplifier bias current errors may be neglected. Note
the assumption of bipolar power supply operation. See text for discussion of unipolar operation.

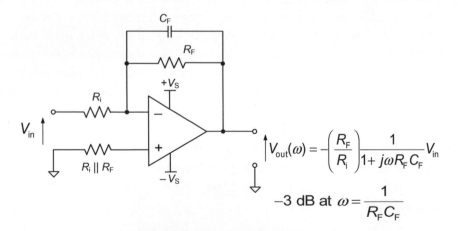

$$\left|V_{out}(\omega)\right| = -\left(\frac{R_F}{R_i}\right)\frac{1}{1+j\omega R_F C_F}V_{in}$$

$$-3 \text{ dB at } \omega = \frac{1}{R_F C_F}$$

**Figure 10.3: Simple single-stage active lowpass filter.**
The simple single-stage lowpass filter illustrates the consequences of component
tolerances and temperature coefficients on both the low-frequency stage gain
and the corner ($-3$ dB) frequency.

expensive, solution is to purchase ratio-matched resistor sets, which (in the case of a thick-film implementation) are parts screened, formed, and fired on the same substrate. In the ratio-matched solution, construction geometry establishes the resistor ratio. The resistors' temperature coefficients of resistance are now virtually identical because individual resistances are part of a single structure formed by the same operation during manufacturing. Better-quality, hand-held digital multimeters (DMMs) typically have ratio-matched, thick-film resistors in their scale-selector networks. An even better solution involves the use of ratio-matched metal-foil resistors (Figure 10.4).

Many designers and systems architects overlook the importance of noise performance in individual components, possibly because noise performance is not always explicitly specified. In many high-performance sensing and measurement applications, the noise floor of the channel establishes the system's ultimate sensitivity limit. The channel noise floor, in

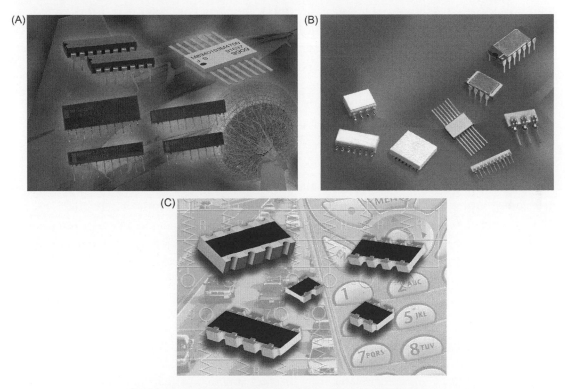

**Figure 10.4: (A)–(C): Resistor networks and RC arrays.**
Resistor networks and resistor–capacitor (RC) arrays are used for a variety of reasons. Resistor arrays in single in-line packages (SIPs) and dual in-line packages (DIPs) are commonly used to provide ratio-matched sets, tapped voltage divider arrays in instruments such as DMMs, attenuator arrays, and anywhere it is necessary to reduce circuit-board area. Networks and arrays are available in many package styles and in many standard values and divider ratios. Custom arrays make economic sense for even modest production volumes. *Images courtesy of Vishay Intertechnology, Inc.*

turn, is very often determined almost entirely by the noise performance of the preamplifier or "front end." The ultimate theoretical limit is the thermal noise or Johnson noise, given by the well-known equation

$$e_{\text{noise}} = \sqrt{4kT(\text{NBW})R}$$

where $k$ is Boltzmann's constant ($1.38 \times 10^{-23}$ J/K), NBW is the channel's noise bandwidth (in Hz), $T$ is the absolute temperature of the channel resistance (in degrees Kelvin), and $R$ is the channel resistance (in ohms) [6]. One commonly cited point of reference is that a 50 $\Omega$ resistor with 1 Hz noise bandwidth at room temperature (298 K) produces 1 $\text{nV}_{\text{rms}}$ of noise. Unfortunately, other noise mechanisms rise above the theoretical minimum and often contribute to the channel's thermal noise. Shot noise is one of these mechanisms and is associated with certain types of resistor construction.

Shot noise was especially troublesome when pressed carbon composition resistors were the standard resistor construction. These devices, rarely seen nowadays, consisted of compressed carbon (usually graphite) granules tightly compressed to form a resistive slug forming the core of an axially leaded cylindrical package. The noise-generating mechanism was associated with the minute grain boundaries of the carbon granules. Thick-film resistors also exhibit shot noise effects (although they are far less noisy than the old carbon composition resistors) because the paste from which the resistance element is formed consists of a colloidal suspension of conductive particles in a binder material. For this reason, metal-film or metal-foil resistors are preferred for low-noise applications.

The ultimate preamplifier arrangement would contain a low-noise amplifier with gain determined by a standard or custom-made, ratio-matched, metal-foil resistor network. The ratio-match feature (frequently used ratio sets are available as standard items; custom units are available if need be) provides excellent gain tolerance. Metal-foil construction provides very low resistance temperature coefficients; the ratio-matched approach provides highly accurate tracking of the already-low temperature coefficients. Finally, the metal-foil resistance element, having no grain boundaries, is an intrinsically quiet (hence, low-noise) construction.

### Wirewound resistors

Wirewound resistors derive their name from their construction, which typically consists of many turns (windings) of resistance wire or ribbon (usually Nichrome or a similar alloy) along or over a ceramic frame. In some constructions, an aluminum extrusion encases the composite ceramic resistance element; the aluminum case has integral mounting feet for attachment to a metal chassis or heat sink. In other cases, a hard durable vitreous enamel or similar protective material coats the entire assembly to exclude moisture and to afford a measure of mechanical protection to the resistance element. This type of resistor is used primarily for high-power

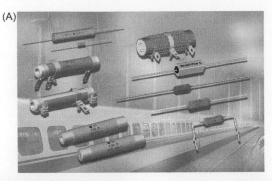

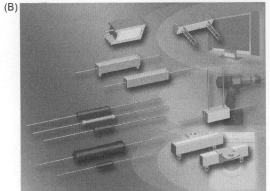

**Figure 10.5: (A)–(C): Power resistors.**
Power resistors are available in a variety of packages reflecting their high power dissipation and the need for adequate air circulation. They are available in a number of special mounting arrangements to facilitate mounting to heat sinks. Shown are a variety of wirewound devices (A) and (B) and power thick-film devices (C). *Images courtesy of Vishay Intertechnology, Inc.*

applications. The resistors themselves are often large and bulky. Wirewound resistors are something of a specialty item in most embedded systems although they have been in existence for a very long time and can remain in service for a very long time if well supported and mechanically protected. Wirewound resistors are most frequently used in applications related to motor control, power supplies, and high-voltage applications (Figure 10.5).

In addition to their size and weight, most wirewound resistors are inductive (they are essentially large pitch, single-layer solenoid windings) which limits their use in higher frequency applications. They are used in power-handling situations and generate considerable heat. This heat is a problem for surrounding components; it can also force a resistance change in the power resistor itself because the resistance wire has a temperature coefficient of resistance. This effect is known as self-heating. If the resistor's primary function is simply to dissipate power, as in a dynamic brake application for a motor-control circuit, self-heating may not be objectionable. In certain other applications involving higher frequency signals or networks with sensitive tuning requirements, the resistance change can be a serious problem [5]. Wirewound resistors with higher power-dissipation ratings are often constructed in the form of cylindrical tubes meant to be mounted on brackets which hold them above the chassis or circuit board for improved air flow and improved convective cooling. However, the resistor's ceramic is brittle; the resistors are vulnerable to breakage when exposed to mechanical shock and vibration; elevating them above the circuit board or chassis often increases the likelihood of breakage.

Table 10.2 summarizes a number of trade-offs you can make when choosing resistors for design.

**Table 10.2: Summary comparisons between resistor types**

| Resistor Type | Relative Accuracy | Relative Cost | Relative Reliability | Application |
|---|---|---|---|---|
| Thick film | Good<1% | Low | High | General purpose |
| Thin film | Good<0.5% | Low | High | General purpose |
| Metal foil | Highest<0.1% | Highest | High | Precision applications |
| Wirewound | Lowest>2% | High | Risk of breakage | Power applications |

### The lowly pullup and pulldown resistor

Of all the applications requiring resistors, pullup and pulldown applications are undemanding yet vital. The classic pullup/pulldown application uses a resistor to terminate an unused logic input so that it is either fixed at logic low (pulldown) or logic high (pullup). The resistance of the pullup/pulldown resistor is relatively unimportant as long as it is consistent with termination requirements of the particular logic family and the power supply voltage used to operate the part. The actual resistance value is less important than the need for the resistor simply to be present—if not present, the gate will float randomly and can result in erratic operation and enormous mischief.

Other pullup resistors are typically used on the outputs of logic gates which have open-collector or open-drain outputs. Once again the actual resistance of the pullup resistance is less of an issue than the fact that it needs to be present.

Less obvious are the biasing resistors needed in certain analog applications. One application in particular—involving instrumentation amplifiers—requires biasing resistors to provide a path to circuit common for the amplifier's input bias currents. Failure to do so results in erratic problems similar to those associated with unterminated logic gates in digital circuits.

### Potentiometers and digital potentiometers

Potentiometers and related devices are very nearly as old as the science of electricity. Structurally, they are an exposed resistive element across which a third electrical contact (the wiper) is moved by rotation of a control shaft. They are very familiar and are commonly used as speed and volume controls, instrument controls, and tuning mechanisms. Although still used in large numbers, potentiometers have drawbacks: they are large and expensive; they are vulnerable to mechanical damage due to wear, mechanical shock, dust, and dirt; resistance tolerances and adherence to their nominal resistance tapers are not especially good; their MTBF (reliability) numbers are poor; and their temperature coefficients of resistance are large and significantly different from those of standard discrete resistors. Even the very best adjustment mechanisms exhibit some hysteresis, and potentiometers are notoriously vulnerable to movement, resistance change, and damage when mechanically shocked or vibrated. You can immobilize the potentiometer shaft with adhesives or a locking mechanism to minimize disturbance of a critical setting (Figure 10.6).

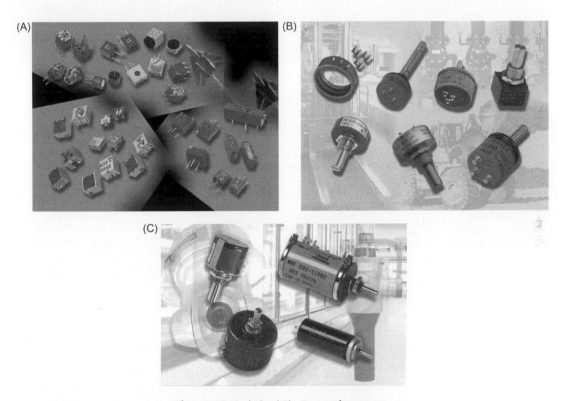

**Figure 10.6: (A)–(C): Potentiometers.**
Potentiometers are continuously variable resistance elements used for a wide range of circuit applications. They are most often used to provide performance adjustments, as rotary position sensors in control applications, and as control elements for such common adjustments as speed, volume, position, and voltage and current settings on laboratory power supplies. Small trimmer resistors mounted on a circuit board (A) are widely used for infrequent performance adjustments and calibration purposes; they are available in a remarkable number of sizes and configurations to insure access to the adjustment mechanism. Other devices (B) and (C) are single-turn or multi-turn panel-mounted devices typically used for control applications. *Images courtesy of Vishay Intertechnology, Inc.*

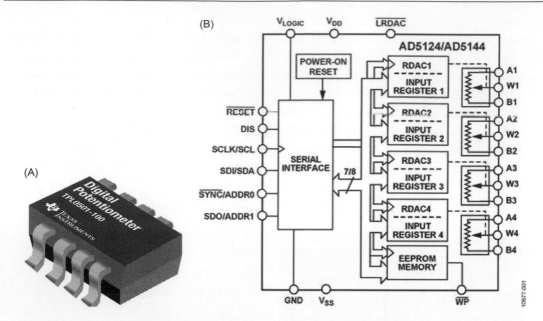

**Figure 10.7: (A) and (B): Digital potentiometers.**
Although not true continuously variable analog control elements, digital potentiometers provide the functionality of conventional potentiometers in a digital structure largely immune to the reliability, thermal sensitivity, and mechanical damage issues associated with many analog potentiometers. Contemporary digital devices with 128 or more taps offer a high-quality approximation to a continuously variable traditional analog device. Because they are true digital devices with no moving parts, they occupy little circuit-board real estate and are ideally suited for embedded applications requiring adjustments made under software control. Many contemporary devices offer nonvolatile memories in multichannel devices, making them especially useful when power must be duty cycled for energy conservation purposes. *Photos courtesy of (A) Texas Instruments Inc. and (B) Analog Devices Inc.*

Digital potentiometers, typically incorporating a resistor network, analog switches for tap selection, and a digital interface, are steadily supplanting true electromechanical potentiometers in applications. Embedded systems, which often must function autonomously or under the control of a remote processor, cannot use conventional potentiometers for many applications other than for system setup and occasional calibrations. Digital potentiometers are ideally suited for processor-driven adjustments. As is seen in Figure 10.7, multichannel digital potentiometers are readily available and a great many digital potentiometers are configured with a serial interface making their use in autonomous digital systems especially convenient.

## Capacitors

Capacitors, like resistors, are widely used passive components in embedded systems. Capacitors have benefited greatly from continuous improvements in dielectric materials and manufacturing techniques for over a century and are reliable components. Certain capacitor constructions closely approximate theoretically ideal components over many decades of frequency, although possibly to a lesser extent than resistors. Capacitors have a wider variety of constructions, ratings, capabilities, and limitations when compared to the constructions and ratings for the different resistor technologies. Capacitors are optimized for different applications. Limitations in operating and storage temperatures frequently restrict the capacitor types available for a given application. I discuss briefly, in the following paragraphs, the most common types of capacitors typically used in embedded systems, but this limited coverage is misleading. This discussion is a simplification of the rich and varied array of technologies all devoted to the storage of energy by means of an electric field.

### Ceramic capacitors

If there were such a thing as a truly general-purpose capacitor, ceramic capacitors would be heavily favored contenders for the title. Ceramic capacitors derive their name from the ceramic dielectric that forms the barrier between two conducting plates or surfaces. This capacitor construction has been available for many years in the familiar radial-leaded disk and rectangular molded epoxy forms. Ceramic dielectrics readily lend themselves to surface-mount construction, where the robust nature and small physical size of a ceramic dielectric is especially useful. Ceramic capacitors are available in a wide range of package sizes, voltage ratings, and capacitance values. Ceramic capacitors have the highly desirable property of being nonpolar, thereby allowing them to be used in many analog applications in which a bipolar power source is desirable. Another important characteristic of ceramic capacitors is their ability to tolerate wide environmental extremes without damage. Storage and transportation temperature specifications can be more severe than the associated operating-temperature requirements, and ceramic devices are particularly useful in such situations. Some other capacitor types, particularly those made with polyester, polystyrene, and similar film dielectrics using wrap-and-fill construction, can be damaged simply by exposure to elevated storage temperatures. Certain electrolytic capacitors can be damaged by cold exposure which causes the electrolyte to freeze. Ceramic capacitors are generally immune to these difficulties (Figure 10.8).

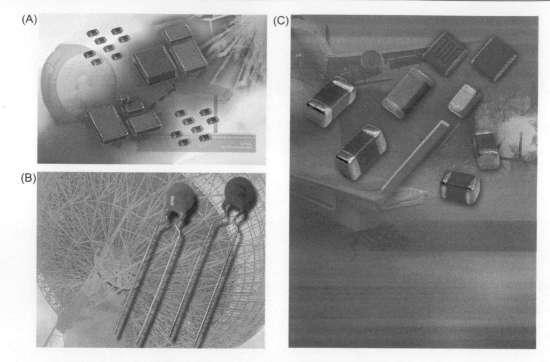

**Figure 10.8: (A)−(C): Ceramic capacitors.**
Ceramic capacitors are available in a wide range of dielectrics and packages. They are suitable for most applications and are reliable, inexpensive, and volumetrically efficient. Surface-mount devices (A) are available in almost any standard device size, and familiar disk through-hole devices (B) have been industry standards for decades. In recent years, multilayer ceramic chip construction (C) has added additional height to standard surface-mount footprints to provide additional capacitance for small circuit-board footprints. *Image courtesy of Vishay Intertechnology, Inc.*

Many types of ceramic dielectric are in general use. The detailed technical differences are many, but the major differences between dielectric types are reflected in specific critical performance characteristics of the resulting capacitor. A standardized descriptive scheme helps make sense of an otherwise bewildering array of ceramics used as capacitor dielectrics. The four major ceramic types, available from virtually any component vendor, derive their names from this coding scheme: these are the types NPO (C0G), X7R (and the closely related X5R), Y5U, and Z5U.

Type NPO, still commonly called C0G from an earlier coding scheme, is a temperature-stable ceramic formulation useful across the entire military temperature range ($-55°C$ to $+125°C$). The temperature coefficient (tempco) is not zero, but it is quite small and is usually the lowest tempco available from a common commercial capacitor. The dielectric material is stable and changes little over temperature, but its dielectric constant is relatively low. Consequently, NPO (C0G) capacitors are physically large compared to comparably rated devices made with other ceramic dielectrics. NPO (C0G) capacitors are most frequently used where temperature stability of the part is of paramount importance (e.g., in oscillators and RF circuits). While relatively large by ceramic capacitor standards, NPO (C0G) capacitors are typically smaller than constructions using film or mica dielectrics. NPO capacitor high-frequency performance is good across a wide temperature range.

Type X7R, called a semi-stable ceramic formulation, is a versatile ceramic still useful across the entire military temperature range ($-55°C$ to $+125°C$), although the X7R temperature coefficient is inferior to that of NPO types at the extremes of temperature. X7R dielectrics have a higher dielectric constant than NPO, making X7R capacitors volumetrically efficient while remaining useful for many applications over a wide temperature range. I consider X7R the dielectric of choice for most applications requiring a considerable temperature range, particularly when circuit-board area is limited. Depending on the required voltage rating, capacitance value, and the need to avoid self-resonance effects, X7R is usable from DC to many MHz and is an excellent choice for most analog filtering applications and other analog functions where modest temperature sensitivity is acceptable. A closely related dielectric formulation, X5R, is readily available and can extend the larger capacitance values available in some voltage ranges. There are technical differences, but as a practical matter X7R and X5R capacitors are virtually interchangeable.

Type Y5V is a dielectric formulation with a more pronounced temperature characteristic than X7R (X5R) and, as a consequence, is ordinarily used over a more restricted temperature range, often ($0-70°C$). Volumetric efficiency is good, and Y5V devices will have higher maximum capacitance values than X7R for the same physical size and voltage ratings. This capacitance range is adequate for most embedded applications used in controlled indoor environments. Performance over frequency is commensurate with other common ceramic types.

Z5U is the last, and most temperature sensitive, of the common ceramic capacitor dielectrics. The capacitance change over temperature can be quite large, on the order of 30% or more, but it is a very good choice when the highest possible capacitance in a given ceramic package size and voltage rating is needed. Many applications require bulk capacitance and can tolerate relatively wide production tolerances and relaxed temperature performance, making Z5U a suitable candidate. For example, Z5U capacitors would be reasonable candidates for power supply decoupling applications in digital, analog, and embedded processing circuits used in consumer electronics.

### Electrolytic capacitors

Aluminum electrolytic capacitors (electrolytics) are widely used in power supply applications requiring high capacitance in energy-dense, small-volume packages having very low equivalent series resistance (ESR). Electrolytics are polarized types with limited

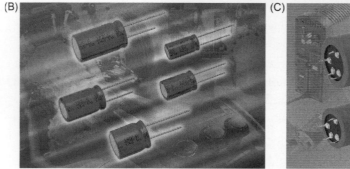

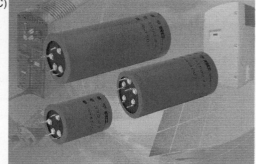

**Figure 10.9: (A)–(C): Electrolytic capacitors.**
Electrolytic capacitors are commonly employed when large capacitance in a small physical volume
is needed. Small surface-mount devices (A) are widely used, particularly when relatively low
operating voltages are involved. Many consumer and industrial applications use familiar
radial-leaded can electrolytics (B). Electrolytics are very common in power supplies and other
line-operated equipment, where larger electrolytics with higher operating voltages (C) are used.
Note the special terminal arrangements available to help secure the devices to a circuit board.

frequency response, somewhat restricted temperature range (approximately $-20°$ to
$+85°C$), and a relatively limited operating life (many devices are rated for 3000 hours at
full rated voltage, although longer life devices are available). Whereas ceramic capacitors
usually survive storage temperatures outside their specified operating-temperature range,
electrolytics can be damaged by unpowered storage at low or high temperatures
(Figure 10.9).

Most electrolytic capacitors are available in a cylindrical aluminum "can" with radial wire
leads, solder tabs, snap-in prongs, or screw terminals on the bottom plate. The electrical
connections are reliable, but external support is usually needed if these "can" structures are
used in a high mechanical-shock and vibration environment. Larger devices have special
circular mounting hardware, typically clamping devices with feet that bolt to a circuit board
or chassis and hold the capacitor securely in place. Smaller devices are often mounted

vertically and held in place with epoxy or even hot glue; this is especially useful when multiple devices can be clustered together and glued in place for mutual mechanical support. Many devices have no preferred mounting orientation; if a particular radial-leaded capacitor is considerably taller than its diameter, you can often mount the device on its side and secure it to the circuit board by epoxy, glue, lacing cord, or a cable tie. The horizontal mounting option is especially useful when circuit assembly headroom is restricted, such as when a device must be mounted inside a low-profile flat enclosure (e.g., a 1U relay rack chassis), a pipe, or a tube.

***Respect the voltage ratings of electrolytic capacitors!*** Larger devices are manufactured with scored tear lines and, occasionally, pressure vents; if exposed to a sufficiently long overvoltage, the electrolyte can produce high-pressure gas with sufficient pressure to explode the device's case. Circuit designers dealing with high-voltage or high-power devices should give thought to the possibility of catastrophic capacitor failure when doing the packaging design and incorporate appropriate safeguards. For example, a perforated-metal safety cage would be appropriate if an open-frame power supply containing high-energy, high-voltage electrolytic capacitors operating near their maximum ratings is used in a rack assembly with unprotected computer or communications boards—the safety cage might very well prevent damage to the more expensive circuit elements in the event of catastrophic capacitor failure.

### Tantalum capacitors

Tantalum capacitors are another polarized capacitor type widely used when high capacitance is needed in a small volume. Operating voltages are usually low, generally 50 volts or less, but operating lifetimes are far longer than those of electrolytic capacitors. Most tantalum capacitors are available in epoxy-dipped radial lead or molded epoxy surface-mount packages, although certain types of tantalum capacitor are available in cylindrical axial-leaded packages. Tantalum capacitors are poor performers in AC applications; they should be used only for DC and certain unipolar filtering applications (e.g., to attenuate noise and ripple in the output of switching voltage regulators, often being useful up to frequencies of 100 kHz) and should *never* be operated in violation of their polarity marking. They are very useful in the power supply and voltage regulator sections of embedded systems because of their small size, low ESR, and compatibility with the operating voltages typical of many portable battery-powered devices. Tantalum capacitors will burn (producing actual flame) if installed with reversed polarity, so you would be wise to include reversed-polarity protection in the power input section of a battery-operated system.

### Film capacitors

Film capacitors include devices with film dielectrics made of materials such as mylar, polystyrene, polyester, polycarbonate, and polypropylene. Film capacitors are typically

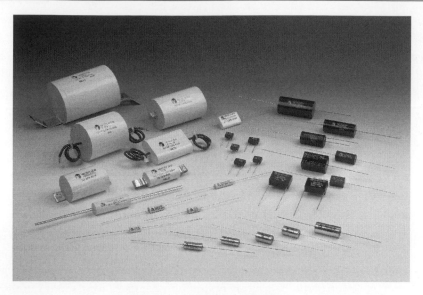

**Figure 10.10: Film capacitors.**

Film capacitors are available with many different dielectrics, case geometries, and sizes. As seen in the photo, special lead arrangements are possible to facilitate mounting to interior structures. The "poly" dielectrics—polycarbonate, polyphenylene, polypropylene, and polystyrene—are especially useful in AC, pulse, low ESR, and filtering applications where good thermal performance, tight tolerances, and low losses are important design considerations. Other dielectrics in this type of construction include mylar and kraft. High-temperature and high-voltage characteristics are very good. Their primary disadvantage is relatively large physical size. *Used with permission of Electrocube, Inc.—Photo by Juan Tallo.*

bipolar and are capable of high-voltage ratings (hundreds or a few thousand volts in some cases). They have good high-frequency performance (e.g., 100 kHz or better) and come in a variety of packaging options. Many of them are rated for the full military temperature range ($-55°C$ to $+125°C$), although some film capacitors can be damaged if exposed to elevated storage temperatures. Their temperature stability characteristics are good and they exhibit low leakage and low dielectric loss, making them useful for such applications as peak detectors, sample-and-hold circuits, low-power resonant circuits, and general analog filtering applications. Polypropylene capacitors are good choices for many DC and AC applications, and special polypropylene safety capacitors are preferred or required for across-the-line and line-to-ground AC power applications. They are large compared to other types, which restrict their usefulness in volume-limited applications (Figure 10.10).

## Silver mica capacitors

Silver mica capacitors are often considered a specialty capacitor ordinarily used for RF and other high-frequency work. The dielectric is mica, a naturally occurring mineral which is an excellent insulator. These devices are very stable over temperature and are frequently rated at hundreds to over a thousand volts. Their low-loss characteristic is greatly appreciated for RF work. They are available in surface-mount or radial-leaded packages. Available capacitance values are relatively low, typically no more than 10 nF with most devices being 1000 pF or less. They are relatively heavy and large for their capacitance ratings.

Table 10.3 summarizes a number of trade-offs you can make when choosing capacitors for design.

**Table 10.3: Summary comparisons between capacitors**

| Capacitor Type | Typical Accuracy | Relative Cost | Relative Reliability | Application |
|---|---|---|---|---|
| NPO (C0G) | ±5% or better | Medium | High | General purpose |
| X7R | ±10% | Low | High | General purpose |
| Y5V | ±10% | Low | High | General purpose |
| Z5U | ±20% | Low | High | General purpose |
| Electrolytic | ±20% +80% or −20% | Low | Limited life | Power supplies |
| Tantalum | ±10% | Medium | Medium—high | Power supplies |
| Film | ±5% or better | Medium | High | Filtering, timing, sample/hold, AC line |
| Silver mica | ±5% or better | High | High | RF |

## Inductors

Inductors of interest in embedded systems are generally those used for power converters, RF sections, signal coupling, galvanic isolation, noise control (common mode chokes), impedance matching, a limited number of filtering applications, and small ferrite beads or surface-mounted devices used to control high-frequency transients in power supplies (Figure 10.11). Power converter inductors can be the power line transformers in line-operated equipment, but most are used in the switch-mode power converters typical of embedded applications, for example, in boost and buck regulators. Other applications are relatively specialized, such as:

- RF subassemblies used for data, voice, and video applications.
- Common mode chokes and other specialty devices used for noise control.
- Telephone line coupling transformers.
- Baluns and impedance matching transformers.
- Certain types of sensors (those involving moving cores such as LVDT devices).
- Inductive devices used for such tasks as such as measuring motor current, object detection by eddy current, and similar specialized sensing applications.

General-purpose inductors, once a mainstay of classical passive LC filter design, are still available for signal processing applications but have been largely replaced by active RC

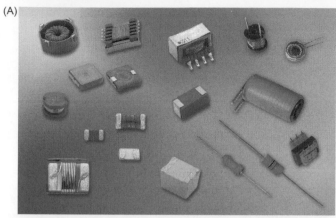

**Figure 10.11: (A) and (B): Inductors.**
Inductors are frequently overlooked. They are commonplace in embedded systems, which often require highly efficient power conversion from AC line or battery sources, making inductors critical power conversion components. Other applications include noise control (often with ferrite beads), signal coupling transformers, common mode chokes, and baluns for impedance matching. Active RC filters have largely replaced classical passive LC filters for analog signal processing tasks, although discrete inductors remain popular and necessary for high frequency and power-handling RF tasks. *Image courtesy of Vishay Intertechnology, Inc.*

filter designs unless the application involves power transfer. ***Other system devices have inductive elements: relay coils, solenoids, miniature AC and DC motors, and stepper motors.***

## Semiconductors

Semiconductors—transistors in particular—are the enabling elements for almost all nontrivial embedded systems. There are a variety of types from bipolar to field effect.

They can be quite small and low power or very large and handle hundreds to thousands of amps of current.

Someone having a casual acquaintance with electronic hardware developments, as from trade press releases, can easily form the impression discrete semiconductors—transistors and diodes in particular—are things from the past, largely relegated to repair and replacement of legacy systems but rarely used in new designs. This impression is incorrect. It is true technology increasingly squeezes dense functional capability into ever-smaller pieces of silicon. Discrete semiconductors are very much alive and well in new hardware design. Densely packed semiconductors must obey certain design rules which often constrain operating voltage and power dissipation. When the latest miraculous integrated device can't handle external sensors, input signal excursions, input transients, and load requirements the hardware designer turns to discrete semiconductors. If you question the accuracy of this assertion, consult a comprehensive catalog from a major part supplier to learn the variety of discrete semiconductors available, new part numbers appear every year in response to new application challenges.

Discrete semiconductors are commonly employed in interfaces between sensors and processors, between integrated processors and controllers and their loads, in power-handling circuits (especially load switches or motor drivers), and in transient protection applications. The applications are remarkably diverse, and it would be foolish to attempt a useful presentation in only a few paragraphs. The system engineer/ planner needs to be aware that fully integrated solutions may not be possible. Applications, well suited for the use of discrete semiconductors, are those involving voltages greater than the processor's power supply voltage (which, in some cases, might be 3.3 volts or less), steady-state current drive applications greater than a few mA, unusually high or low temperatures, applications having inductive or heavily capacitive loads, or applications with demanding requirements subject to transient power supply and load disturbances. Anything deriving operating power from the commercial AC power grid or a battery system is a likely candidate for using at least a few discrete devices.

Proper device selection is a classic exercise in hardware design that requires knowledge of operating voltages, steady-state and transient load current requirements, thermal environments, volume limitations, EMI/EMC requirements, radiation exposure, regulatory requirements, likely failure modes, and anticipated operating-life requirements. Certain emerging technologies (e.g., silicon carbide) hold the promise of extending electronic performance in environments characterized by extreme heat.

Table 10.4 summarizes a number of trade-offs you can make when choosing transistors for design (Figure 10.12).

**Table 10.4: Summary of most common transistor types**

| Transistor Type | Typical Uses | Control |
|---|---|---|
| Bipolar | General purpose, power | Current controlled base |
| MOSFET | General purpose, power | Voltage controlled gate |
| MESFET (GaAsFET) | High-frequency FET | Voltage controlled gate |
| Silicon carbide | High-temperature operation | Bipolar or FET structure |
| Insulated Gate Bipolar Transistor (IGBT) | Large motor drives | Voltage controlled gate |

**Figure 10.12: Typical discrete transistors used in embedded systems.**
Although traditional leaded transistor packages are available in abundance, embedded system assemblies are commonly limited in physical size. In addition, circuit-board fabrication is widely optimized for surface-mount parts and machine assembly rather than through-hole technology. Accordingly, discrete transistors selected for most embedded applications will be the surface-mount package types illustrated here. Discrete power transistors for low- and medium-power applications are readily available. *Photo used with permission of Diodes Incorporated.*

## Visual displays

### Lamps and LEDs

Incandescent lamps and certain specialized vacuum tubes (e.g., the "magic eye" tube) are among the very earliest visual displays. Vacuum tube technology should not be dismissed as a quaint anachronism, because the cathode ray tube is the basis for the oscilloscopes, television screens, and monitoring equipment of various sorts that dominated display electronics through most of the twentieth century. Among the early devices manufactured specifically for display and status indications were a remarkable variety of small incandescent instrument bulbs still available from commercial sources (consider the turn indicators and tail lights of many automobiles still on the road). Such display devices were ubiquitous in status displays on industrial machinery, truck and automotive applications, aircraft and marine applications, computers, radios, and virtually all electronics applications. Color-status indicators were most often realized by using colored transparent lenses or shutters. Neon indicators were also used extensively, because they could take their power from an AC power line and they had no filament to fail. The very first alphanumeric displays, primarily used for numeric displays, were Nixie tubes, devices with multiple numeral-shaped cathodes inside a gas-filled tube and a mesh anode. These devices were later modified to create the earliest 7-segment displays.

Incandescent bulbs had their drawbacks—they produced heat (sometimes painful amounts of it) and the filaments failed through use or mechanical shock. Development of the light emitting diode (LED) was immediately adopted by the instrumentation community, which took advantage of the LED's lower power, longer life, and exceptional durability. Other early applications were 7-segment LED alphanumeric displays. The widespread application of digital semiconductors soon resulted in a number of integrated digital devices directly related to display applications (BDC coded counters and BCD-7-segment display drivers) which, with continuing integration, resulted in a digital takeover of the instrument display business. LED displays remain true workhorses of the instrument industry. Continuous development effort over many decades has resulted in LED devices emitting at a great many frequencies (colors) ranging from the infrared to the ultraviolet (UV), and white or pseudo-white LED assemblies are now challenging the familiar Edison-based light bulb for interior illumination applications.

## Display devices

***Display devices are the heart of the system's human interface***, with display devices now being capable of presenting images of great sophistication and permitting direct human interaction. The system designer's options are many, but there are a few pitfalls to be avoided.

- What is the environment? Operation in high or low temperatures, low light or no light conditions, severe environmental conditions, and in equipment subject to high mechanical shock and vibration can be especially challenging. For example, display devices for construction and other outdoor equipment must be capable of working in a wide and demanding range of environments. The designer must select an option appropriate to the system's intended use.

- What is the actual display requirement? It has become seductively easy to provide exceptionally detailed graphics on a small device. Consider whether high graphics density is a necessary feature or a distraction. It is often sufficient and desirable to keep the display as simple as possible, because the system operator does not need (nor usually wants) to know every system parameter in exquisite detail. Intuitive displays require little operator training and interpretation, and intuitive displays are simple displays. The system designer or design team has the responsibility of knowing what content, in what detail, and in what circumstances the user actually requires information. There is a tendency, usually driven by schedule pressure, to defeature a relatively complex engineering GUI used during product development and call it a user interface. The design team easily falls into this trap because they are used to the engineering GUI and are very comfortable with it. This often succeeds only in burying important information in a mass of unnecessary detail. Automotive gas gauge presentations have remained largely unchanged over the decades, despite massive changes in technology since the first vehicles were produced. Why?—because they don't need to be complicated. If you were told there is 87.3% of maximum fuel capacity remaining, is that really helpful? (7%, possibly, because the operator is nearly out of fuel. 87.3%, no.) Make it easy for the operator to get required information, but don't overload the operator with too many details. The design team must know the system's functions and the operator's needs and expectations uncommonly well to design a really useful display.
- Consider mimicking (or, as old-fashioned as it sounds, using) an analog gauge. A visual image obtained in a glance is often sufficient for most operating parameters. The system's logging and feature extraction software may benefit from frequently updated information accurate to $\pm 0.1\%$, but the human operator usually only wants to know if things are operating where they normally operate. Status displays in red, yellow, or green are readily understood, simple to implement, and comfortable for the operator.

## Integrated circuits

Semiconductors are at the very heart of embedded system design. The notion of an embedded system encompasses very simple devices and very complex structures alike, making it difficult to generalize in any meaningful way. I cannot possibly go into all types of integrated circuits with any depth, because the architectures and tools vary widely. I will discuss a few of the more important concepts.

### Processors and controllers

Almost any embedded application will contain a processor or controller executing some sort of program code. While details are strongly dependent on the application, a few general thoughts are in order.

- Software should include a bootloader allowing code changes and software updates to be loaded by a simple serial line during design and debug, and by telephone connection once the device is in production and in the field.
- Select a processor/controller with a generous amount of excess program and data memory beyond initial project estimates. Some projects recommend or require 100% excess capacity even after software is in production form.
- Do not confuse the desire for memory margin with the need to execute at the fastest possible clock rate. Once a range of devices having acceptable temperature, architectural features, and performance ratings have been identified, make a serious effort to minimize power consumption. Select a clock frequency that is adequate for the task at hand, with a reasonable margin, but not excessively fast— excess operating speed is almost always accompanied by excess power consumption.
- Select devices with built-in data communication (e.g., RS-232, RS-485, Ethernet, USB, or Bluetooth) capabilities when possible.
- Do not overspecify A/D converters. Do not confuse resolution (the smallest detectable change) with accuracy. Few systems actually require, or achieve, true 16-bit A/D conversion accuracy, although it may take a 16-bit converter to obtain 12 or 14 truly accurate bits over the entire dynamic range. Carefully analyze performance specifications to determine actual accuracy requirements.
- Seriously consider isolating all processor inputs and outputs, particularly when dealing with processors for 3.3 or 3.0 volt systems or when designing embedded systems which include or must work with high-power switching devices (e.g., IGBT-based motor controllers for multi-horsepower, or even multi-kW, electric motors).
- Is the system to be an all-digital effort, or will the system contain analog sensors which must detect and process low-level analog signals? Low-level signals are often bipolar signals; not all microprocessors have bipolar A/D converters. Level-shifting a small bipolar signal to fit a converter's unipolar input range often introduces offset and noise problems.
- Determine and respect the project's production facility capabilities and limitations when making component selections. Some devices have very small pin spacings which require specific production capabilities to handle and adequately solder the devices. For example, a device which is available only in a ball-grid array may be desirable for performance reasons, but not all small- and medium-sized production operations have the production equipment needed to fabricate with ball-grid arrays. Know what your production facility can do, or will be able to do, when the system goes into production and select devices accordingly. Failure to do so will either (i) increase cost and schedule risks by forced outsourcing of some critical device assembly operations or (ii) increase cost by forcing acquisition of specialized production equipment which will remain idle for significant periods of time.

### Case Study: Choosing a processor (by Tim Wescott)

Factors involved in processor choice are legion, but here are some you should consider:

1. **Product longevity**: What is your product's expected life cycle? How long do you expect to be able to get processors from the manufacturer? If you are making an automotive product that must be available to provide spares for a decade or more, then choosing a part designed for the 6-month lifetime typical of a cell phone will lead to an expensive product redesign early in its life.

### Example: Problems with semiconductors in large appliances (by Kim Fowler)

Large appliances, such as ovens, clothes washers, and refrigerators, tend to have very long lifetimes. Some might last up to 40 years. At least one manufacturer of these large appliances has encountered a unique problem with selecting microcontrollers for the appliances. Most, if not all, microcontrollers with EPROM memory are specified by semiconductor vendors for operation out to 20 years. There is no indication how long these microcontrollers might operate beyond 20 years. This leaves a maintenance and support issue decades into the future for the manufacturer.

Another, more immediate problem is the obsolescence of semiconductors. Even if the manufacturer contracts with a semiconductor vendor to keep producing a component for years into the future, that component still might fail EMC specifications years down the road. What happens is that the vendor maintains the instruction set, pinout, and clock frequency for a microcontroller component but somewhere along the way the vendor changes its fab line and invariably the design rules for the transistors get smaller. This means that now the transistors switch faster and have steeper digital edges. Steeper edges mean higher harmonics for the same clock frequency; the controller board then exceeds its EMC specifications. This means that the manufacturer must redesign its controller board to meet the new EMC environment.

2. **Cost**: What is your budget for cost of goods sold (COGS) or bill of materials (BOM)? How big of a bite does the processor take out of it? While you are pondering that, what is your production volume? If you can shoehorn the entire application into a one-dollar microprocessor at the expense of man-months of engineering time you'll be a hero—as long as your production volume ranges upward of a million. Sometimes the least expensive choice is the one that costs the most per board (because of NRE; see Chapter 8 for more discussion about NRE).
3. **Level of integration**: All else being equal, if you can choose a processor that does everything you need it to do by itself, without requiring external peripherals, you may well save money, engineering time, and board space by choosing that processor. I prefer to ask myself what combination of processor and peripherals is the least expensive, which generally leads to a highly integrated processor.
4. **Processing power**: This factor isn't just about clock speed, because a 32-bit processor will make more use out of each clock cycle than an 8-bit processor. Make absolutely sure that you have more processing power than you need. If you are working on a low-volume

product, make sure that you have more processing power than you can ever imagine needing. If you are not sure, then get some evaluation hardware and test it. The end product almost always uses more clock ticks than you think.

5. **Memory**: See the comments about processing power. You can start a project with what you think of as oceans of memory and end up feeling like you're trapped in a small box. More is nearly always better, unless you are working on a very high-volume product. If possible, choose a processor that has an upward migration path, so that should you run out of memory you can fix the problem with a part-number change.

6. **Size**: Does the processor physically fit in the space you have available? If you have a processor choice you dearly love but that pushes all the other components off of the board, then it's a no-go. Unfortunately, size choices are difficult, because it seems that you can always—for more engineering and materials cost (NRE and COGS)—pack more parts onto any given board. Be prepared to do some soul searching and negotiating if you are size constrained.

7. **Power**: What impact does the processor have on your board's (and, hence, your system's) power budget? Does it work stand-alone, or does it require power-hungry (and perhaps space-hogging) support electronics to function?

8. **Voltage levels**: A processor is just one component of a larger system, and it must talk to other components on its board. The more voltage supplies you have on a board, the more complex, expensive, and time consuming the design will be. The best processor may be one that is otherwise lackluster, but happens to have I/O voltages that are the best fit to whatever else you have on the board.

9. **Commonality**: In some systems, you may have a number of different modules, each needing a processor. In these cases it rarely makes sense to have a number of dissimilar processors. Think hard about whether it might be less expensive to go with just one processor, or perhaps with a related family of processors, rather than a diverse group of processors optimized to particular applications. Keep in mind the advantages aren't just in the commonality of the BOM. Should the product have a number of similar processors, you can leverage one electrical design and one base software design across all the boards that use a single processor type, giving you more slots over which to amortize the engineering expense—the NRE.

10. **Prior experience**: In those rare cases when all else is equal, a significant amount of engineering time can be saved by using a processor that you have used before. If you have a processor that meets all your requirements and you and your team (both software and hardware folks) like it, then stick with it. While you should be ready to abandon an existing processor in a moment if the situation warrants it, you should never change processors on a whim or for some reason that isn't traceable directly back to a project requirement.

There are no hard and fast rules to choosing a processor. Be sure to consider all of the requirements above. Expect to spend time juggling the choices (or knives).

## Power semiconductors

Power semiconductors are undergoing a renaissance at this time, with exciting new design options (such as silicon carbide) for very hot and difficult environments and higher power ratings and gallium arsenide (GeAs) for very high frequencies. Be aware of the consequences of these devices to RFI/EMI/EMC and enclosure cooling requirements. Some examples of power semiconductors in the embedded world follow:

•   Solid-state relay (SSR) technology slowly continues to displace the electromechanical relay in general-purpose applications. Control signals commanding state changes can be made compatible with the supply voltage of most logic family supply voltages, SSR control signal inputs typically consume less power than relay coils, and switching speeds are as fast or faster than most electromechanical devices. Most SSR devices are fabricated with MOS transistors, and the power-handling capability, current ratings, and voltage ratings of SSR devices parallel developments in power FET technologies. Most SSR devices require assertion of the control signal to establish the conduction channel, equivalent to a normally open relay configuration, although a recent new product introduction provides one of the very first normally closed channel configurations. SSRs have size (both circuit-board footprint and total volume) and weight advantages, which make them highly desirable design options for embedded systems. SSR reliability is excellent (although conventional electromechanical relays frequently operate flawlessly for decades), and their costs are increasingly competitive with electromechanical devices. The SSR's lack of moving parts and totally enclosed construction is a highly desirable feature in high-mechanical-shock, high-vibration, high-dust, high-humidity, and similarly demanding environmental conditions which may otherwise require precision, high-reliability, or hermetically-sealed relays (Figure 10.13).

**Figure 10.13: Solid-state relay.**
Low- and medium-power solid-state relays (SSRs) are available in dual in-line and single in-line packages (DIPs and SIPs, respectively). The device illustrated here is a relatively new type which provides optically isolated single-pole, normally closed operation. Most SSRs provide normally open operation. The control circuit is optically isolated from the load side, which gives the designer great flexibility in application options. The output pins can be configured for AC/DC or DC-only operation, allowing a single type to be used in multiple applications. *Photo used with permission of Ixys Integrated Circuits Division.*

- Embedded systems may include motor-control functions for DC or AC motors. A recent proliferation of motor-control semiconductors and high-power motor driver devices has made the design of embedded-motor applications much, much easier. The availability of integrated control and drive solutions is highly desirable for the designer, because they greatly simplify many motor-control design tasks. These controllers are also important because they typically result in very efficient design solutions. Greater efficiency results in lower heat loss, which has desirable effects of reducing semiconductor costs and increasing semiconductor reliability. Quite aside from these considerations, the US Department of Energy asserts that "Over half of all electrical energy consumed in the United States is used by electric motors."[7] This is a trend that can be expected to spread worldwide as developing economies take increasing advantage of available technology. Thus, even a few percent improvement in motor drive electronics can have a significant effect overall. The most recent devices implement motor tuning on a per-unit basis, which effectively normalizes a motor-control design and makes it largely reusable on motors of different sizes. This is a very significant advance for certain sampling and control applications requiring greater sophistication than ON/OFF commands to a motor (Figures 10.14 and 10.15).

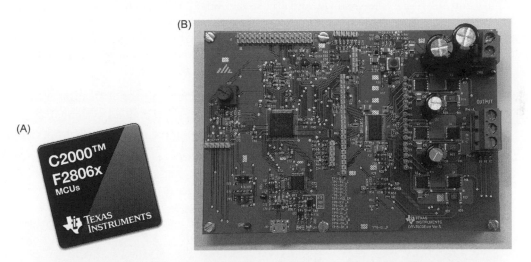

**Figure 10.14: (A) and (B): Optimized motor microcontrollers and evaluation boards.**
The TMS320F2806xF (A) is representative of microcontroller integrated circuits optimized to provide integrated motor-control solutions. This controller family, and similar motor controllers available from a number of other manufacturers, provide flexible speed and torque control for medium-voltage DC motors via three-phase power stage outputs. The DRV8303EVM, shown at (B), is representative of companion evaluation boards and other development tools available to the embedded system developer. The small size of the development board in (B) is testimony to the exceptional design options available to the system developer in embedded applications.
*Images used with permission of Texas Instruments, Inc.*

**Figure 10.15: (A)—(C): Integrated motor driver devices and evaluation board.**
Embedded system developers having existing (and familiar) controller solutions at their disposal do not need to abandon their hardware development investment to take advantage of advanced motor-control options. This figure shows two representative examples of an emerging class of sophisticated motor driver integrated circuits now available from a variety of manufacturers. The DRV8301 and DRV8832 (shown at (A) and (B), respectively) are driver chips readily adaptable to existing hardware of varying capabilities. Evaluation boards such as the DRV8301EVM (C) allow the designer to develop motor-control software and firmware on a stable platform which also provides useful design guidance during circuit-board layout. *Images used with permission of Texas Instruments, Inc.*

- Specialized LED-driver chips are proliferating in numbers and sophistication. LED lighting is often the primary motivation for these devices; embedded systems can readily take advantage of these devices for remotely operated or unattended operation in precision lighting situations (Figure 10.16).

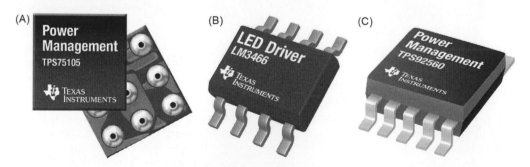

**Figure 10.16: (A)—(C): LED drivers.**
LED drivers are specialized constant current source devices which compensate the operating characteristics of individual LED devices to obtain greater uniformity in display and lighting applications. They may be used for banks of individual devices (A), strings of multiple devices (B), or high-power drivers for lighting applications (C). *Photo used with permission of Texas Instruments, Inc.*

### Analog semiconductors

While many embedded systems are almost completely digital, every one of them has an analog element or elements to interface with the analog world in which we live. For those systems needing more than a resistive divider, here are some typical examples of analog circuits and their concerns that you may encounter:

- Carefully consider the nature of all analog signal inputs during early design effort. If a sensor provides a bipolar output, I strongly recommend a bipolar power supply for all front-end analog electronics. This avoids DC errors introduced by level shifting and noise introduced by the use of a midrail signal common.
- Amplifiers should have the slew rate necessary to reproduce the highest frequency component in the channel passband at the desired channel gain, but excessive speed is to be avoided. Excessive speed (equivalently, excessive bandwidth) requires extra power from the power source and merely passes along undesired high-frequency noise components which must be filtered out later to limit channel noise.
- Analog filters, especially lowpass filters used in preamplifiers and anti-alias filters, should be based on a Bessel prototype whenever possible. Of the three canonical, all-pole, lowpass prototypes (Butterworth, Chebychev, and Bessel), the Bessel filter is the closest approximation to a linear-phase filter and has the best-behaved impulse and step responses (exhibits the least overshoot and ringing), making the Bessel filter an excellent choice for sampled-data systems.
- Similarly, a Bessel filter prototype is an excellent choice for bandpass or band-reject (notch) filters because of its relatively limited response to impulsive noise. A Sallen–Key implementation is a very good choice for second-order filters or filter sections because it provides relatively uniform response despite variations in the production tolerances of its components.
- Band-reject (notch) filters, particularly filters with high Q, can be difficult to tune; they sometimes exhibit instability problems. Band-reject filters are typically based on twin-tee structures; I suggest an alternative form (Figure 10.17) which is very helpful if a gain-phase analyzer is available to do the tuning. This approach is very stable and can achieve exceptional attenuation at the center frequency. A unity-gain bandpass section is designed to provide the desired Q and center frequency (a Bessel prototype is recommended) and is tuned to provide 180° of phase shift at the center frequency of the notch. The phase-shifted bandpass signal drives an inverting summing amplifier with the other summing amplifier input being the original input to the bandpass filter. Trim the gain of the summing amplifier to be exactly the same as the bandpass filter gain at the center frequency. This arrangement is capable of 100 dBV or better attenuation at the center frequency and is excellent for removing the effects of noise at the power line frequency and low-order harmonics.

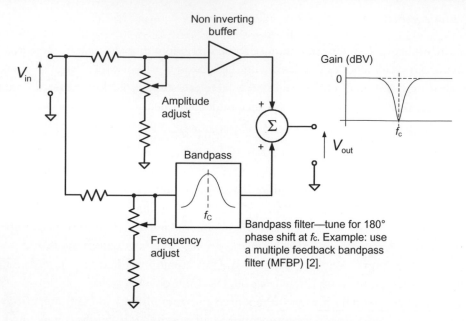

**Figure 10.17: Band-reject (notch) filter implementation.**
The circuit shown is a stable band-reject (notch) filter implementation which provides simplified tuning, making it easier to use than conventional twin-tee implementations in many applications. Notch characteristics are determined by the bandpass filter design. Depth and center frequency of the notch are affected by tuning accuracy and thermal effects. It is desirable to have independent tuning for amplitude and frequency. Although the tuning elements are shown as conventional potentiometers, this circuit approach is a classic opportunity to improve performance of an embedded system using a dual digital potentiometer to provide the tuning elements.

- A Schmitt trigger is an analog comparator using positive feedback to provide hysteresis around a switching threshold. Schmitt triggers are especially useful as comparators or threshold detectors because they avoid the problem of output switching chatter if the input signal hovers or dithers very near the switching threshold. Two similar structures are possible, depending on whether the designer wishes the signal to go high or low when the input signal exceeds the reference level. The transition thresholds determine the width of the hysteresis window. The reference voltage is the center of the hysteresis window only if the amplifier's positive and negative output saturation voltages are equal in amplitude, which typically requires a bipolar power supply and an amplifier with rail-to-rail output response. If this is not the case, you simply define the desired hysteresis window by identifying the desired transition voltages and then select an appropriate reference voltage knowing the feedback and input resistor values and the amplifier's output saturation voltages (positive and negative). Defining a hysteresis window is especially important if the amplifier is operating from a unipolar power supply. The amplifier output will attempt to swing rail-to-rail (some amplifiers allow true rail-to-rail output; others saturate before reaching the power supply positive and negative rail voltages). Actual output voltages and the input and feedback resistor values determine the Schmitt trigger's switching points. Since Schmitt trigger outputs are essentially digital signals and usually provide a binary input signal to the embedded system's control logic, designs often attenuate and diode clamp the amplifier's output to closely approximate any desired digital logic family (Figure 10.18).

- Noninverting summers are useful circuit elements for combining signals in a system having only a unipolar power supply. This situation often arises in certain control loop applications. As seen in Figure 10.19, two resistors implement the summing action at the noninverting input while the inverting input and feedback resistances determine the gain.

### Digital and mixed-signal semiconductors

When it comes to digital and mixed-signal semiconductors, do the following:

- Pick devices with adequate speed and a reasonable margin, but without significantly excess speed. Power consumption increases as operating speed increases. Emissions also increase as operating speeds increase.
- Select devices minimizing power consumption whenever possible.

## Circuit boards

Circuit boards are often (and incorrectly) considered passive substrates that support and interconnect the system elements. Circuit boards certainly have these functions, but they are **engineered components in every sense of the word.** This is especially true of circuit boards for RF and microwave assemblies, and it is painfully evident when the system must meet RFI/EMI/EMC requirements. Circuit-board layout is equally demanding even at lower and intermediate frequencies where part placement, parasitic capacitances, stray inductance, and noise coupling can often degrade weak signal performance.

The vast majority of circuit boards are fiberglass-epoxy formulations, FR4 or G10, available in various thicknesses with or without the copper cladding that is etched to form the circuit traces. Conventional circuit-board construction is satisfactory for most embedded systems, although special board materials and special solders are sometimes used for extreme environments exemplified by petroleum well logging, spacecraft instruments, and engine firewall applications. Fabrication techniques and equipment are highly advanced.

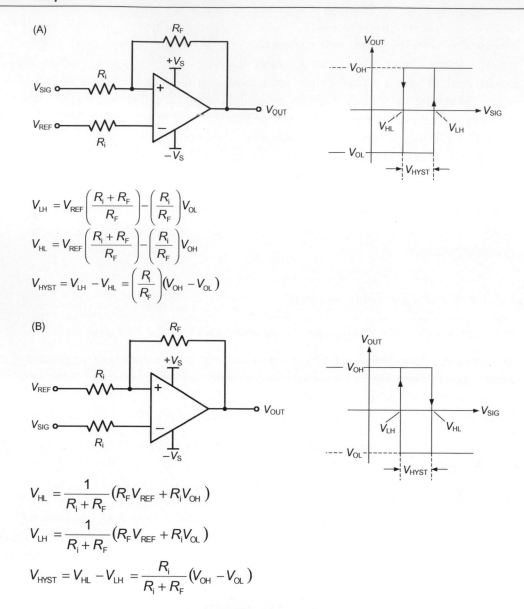

(A)

$$V_{LH} = V_{REF}\left(\frac{R_i + R_F}{R_F}\right) - \left(\frac{R_i}{R_F}\right)V_{OL}$$

$$V_{HL} = V_{REF}\left(\frac{R_i + R_F}{R_F}\right) - \left(\frac{R_i}{R_F}\right)V_{OH}$$

$$V_{HYST} = V_{LH} - V_{HL} = \left(\frac{R_i}{R_F}\right)(V_{OH} - V_{OL})$$

(B)

$$V_{HL} = \frac{1}{R_i + R_F}(R_F V_{REF} + R_i V_{OH})$$

$$V_{LH} = \frac{1}{R_i + R_F}(R_F V_{REF} + R_i V_{OL})$$

$$V_{HYST} = V_{HL} - V_{LH} = \frac{R_i}{R_i + R_F}(V_{OH} - V_{OL})$$

**Figure 10.18: (A) and (B): Schmitt-trigger implementations.**
In (A), the reference voltage is applied to the inverting input, whereas in (B) the reference
voltage is applied to the noninverting input. Notice the effect of output voltage on the
switching voltages. This effect is especially pronounced when a unipolar power supply is used
and $-V_s = 0$.

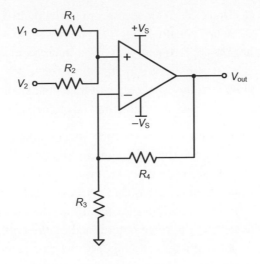

$$V_{out} = \left(\frac{1}{R_1 + R_2}\right)\left(\frac{R_3 + R_4}{R_4}\right)(R_2 V_1 + R_1 V_2)$$

If $R_1 = R_2$, then $\quad V_{out} = \left(\frac{1}{2}\right)\left(\frac{R_3 + R_4}{R_4}\right)(V_1 + V_2)$

If $R_1 = R_2$ and $R_3 = R_4$, then $\quad V_{out} = V_1 + V_2$

**Figure 10.19: Noninverting summing amplifier.**
Circuit diagram of a noninverting summing amplifier useful for unipolar and bipolar supply operation. Circuit analysis is readily accomplished by using superposition.

There is abundant literature treating circuit-board layout, fabrication techniques, and assembly techniques [8–11].

## Connectors, cables, and conductors

*It is astonishing how frequently beautifully-designed and fabricated pieces of equipment fail because connectors, cables, and conductors are poorly selected and specified—more often than not in a misguided effort to save a tiny fraction of the equipment's cost.* Even more disheartening are the number of times a suitable connector, cable, and conductor were selected and incorrectly installed—oftentimes in a misguided effort to save a slight production expense by scrimping on installation tooling and training of assembly personnel. Be prudent when you select and install a connector—do it right.

### Connectors

Connectors, as their name indicates, exist to establish electrical connections between two different assemblies or between an assembly and an associated cable set, two of the very oldest tasks in the electrical and electronic arts. Connectors are available in a staggering number of configurations, sizes, materials, and constructions. Despite this long experience and an abundance of options, connectors are frequently involved in problems at the system level. It is possible to misapply a good connector, to be sure, but most connector problems arise from selection of an unsuitable type. For convenience, I will treat connector requirements as interior board-to-board applications and exterior housing-(firewall-)to-cable applications.

By board-to-board applications, I mean those connectors contained within a protective product enclosure and thus having relatively benign environmental requirements. These connectors are installed to simplify manufacturing interconnections, troubleshooting and repair. They mate and demate only infrequently. Under ideal conditions, they may very well remain undisturbed in a mated condition for the life of the product. Many connector types are available, and the selection is often determined by the connector's physical size, number and type of conductors, diameter (or gauge) of the conductors, type of connection means (crimp or solder), constraints (if any) on the housing material, security of engagement, and exposure to mechanical shock and vibration. Cost is always a consideration, especially in consumer applications, but cost should never be the dominant consideration in connector selection. A poor connector choice can cause problems that far outweigh the anticipated cost savings.

- Number and type of conductors—insure that the connector pins and sockets are appropriate for the conductors; sloppy connections will fail. If circumstances permit, conductors should be stranded wire rather than solid wire to resist stress fractures of handling, mechanical shock, and vibration. When possible, allow space for three or four spare contacts, especially for a project in the early stages of development. In many connector families, contacts are made for only certain selected gauges: for example, #4 AWG, #8 AWG, #12 AWG, #16 AWG, #20 AWG, and #24 AWG are common choices. Use only those wire diameters appropriate for the available contact pins. When in doubt, go with the next larger diameter—for example, if a #18 AWG conductor would be the preferred choice in a benchtop environment, use a #16 AWG conductor and contact if a #18 AWG set is not available for the connector.
- Type of connection means—will the conductors be connected to the connector pins and sockets by crimping or soldering? I do not trust insulation displacement connections and cannot recommend them, despite manufacturing economics and salesmen's claims. Crimped connections are reliable *provided* the insulation is correctly stripped from a correctly sized conductor, the conductor is crimped in a correctly sized barrel by the proper crimping tool, and the conductor—connector pin assembly is installed in the connector cavity with the proper insertion tool. Solder connections are reliable *provided* the insulation is correctly stripped from a correctly sized conductor, the conductor is soldered in a correctly sized solder cup using the correct solder brought to the correct heat by a satisfactory soldering iron, and, if appropriate, the conductor—connector pin

assembly is then installed in the connector cavity with the proper insertion tool. For extra reliability on some connectors—for example, solder-type D-sub connectors with exposed solder cups—it is good practice to slip a short length of heat-shrink tubing over the solder joint to reduce the likelihood of an accidental short circuit during troubleshooting or repair. Be sure to heat the shrink tubing so that it fits securely. Some individuals crimp a connection and then solder it in the belief this improves reliability. I do not subscribe to this practice, for sufficiently long exposure to mechanical shock and vibration will eventually produce a stress fracture immediately above the end of the solder column, particularly on smaller diameter wire.

- Diameter (gauge) of conductors—if mixed-diameter (gauge) conductors share the same cable bundle, as when power conductors and signal conductors share the same connector, the connector will be more expensive and the designer will have fewer connector options. It is often preferable to use two different connectors, each with a single size contact, if space allows.

- Housing material—the most inexpensive connectors generally have nylon or thermoplastic housings. In special cases, particularly those involving very low signal currents (as when photodiodes are the signal source), you may have to purchase connector bodies containing minute amounts of graphite or carbon fiber to provide very slight conductivity to avoid issues with static charge and discharge.

- Security of engagement—all connectors, without exception, should have a positive locking mechanism to insure mated connectors remain mated during shipment and use. In the case of board-to-board connectors, a ratchet and pawl mechanism usually is sufficient. Mounting ears secured by small screws are acceptable on D-sub connectors. Embedded systems are often deployed in remote field situations; it is extremely annoying to go on a trouble call to a remote location and, after a complete teardown, discover the problem was caused by a loose connector.

- Exposure to mechanical shock and vibration—failure to provide a positive locking mechanism can allow a connector to loosen and produce intermittent contact. The likelihood of such problems increases with exposure to mechanical shock and vibration, even if these conditions arise only during transportation from place to place before installation. Mechanical shock failures can occur when small-diameter solid wire is used in a connector pin or socket—prolonged exposure to low-level vibration, in particular, can result in a fatigue fracture immediately above the crimp or solder column on a connector pin. Such problems can be extremely difficult to identify. The likelihood of these problems is greatly reduced by providing a suitable mechanical strain relief for the conductor bundle—even though the connector and associated conductor bundle are assumed to be inside an overall enclosure, mechanical shock and vibration are readily transmitted to the connector assembly through mechanical coupling.

- Connector pin and socket materials establishing so-called dry contacts—those carrying very little current—should be gold plated to prevent contact oxidation and eventual contact failure. Such problems can be very difficult to identify and often manifest themselves for no apparent reason after the connection has been working for months or years. Connections carrying a few amperes of current are not a problem, for the contact's current density will burn the oxide layer away. When in doubt, opt for gold plated contacts [12].

Exterior housing-(firewall)-to-cable applications are more demanding, because in addition to mechanical shock and vibration during transportation and use, these connectors are generally exposed to environmental conditions such as humidity, rain, snow, dust, UV light, and extremes of temperature. They also experience other less readily anticipated perils such as those presented by rodents, insects (especially ants), curious animals and humans, and vandalism [13,14].

Whereas electrical connections inside the system's enclosure can remain undisturbed for long periods of time, exterior connections may be disturbed as the device is moved from place to place or during routine calibration, field maintenance, repair operations, and routine use. Many of the general principles presented for board-to-board applications still apply, but *the severity of the installation environment makes the selection of a quality connector even more important and often more difficult*.

A lesser number of suitable connector types are available for external applications. Many of the better options are military connectors or their commercial equivalents, although industrial connectors designed expressly for heavy industries (e.g., mining, construction, and petroleum) are every bit as rugged as military connectors. Selection criteria will involve not only the number and type of conductors, diameter (or gauge) of the conductors, type of connection means, and security of engagement but also exposure to a whole host of environmental considerations. External connectors, cables, and conductors must be chosen with particular care.

- Use stranded conductors to resist stress fractures from handling, continuous bending and flexing (e.g., the cables in an industrial robot), mechanical shock, and vibration. Conductor insulation should be tough, capable of remaining flexible for a prolonged period of time, resistant to damage by abrasion and UV light, and without a tendency to creep over time. Conductor insulation and the cable's exterior jacket have a tendency to crack or break where they enter the connector shell, thus proper strain relief is very important.
- An environmentally exposed cable is a poor place to use tiny individual conductors. Provide room for three or four spare contacts. Use only wire diameters which match the available contact sets. When in doubt, use a larger diameter. Insulation selection will be a critical consideration for cables and conductors used in certain extreme locations—for example, those with very high temperatures and pressures, exposure to intense UV, high radiation exposure (the containment building of a nuclear power

plant), installations requiring direct cable burial, and installations intended for use in or near water (especially brackish or marine environments).

- The connector backshell's cable entry is an especially troublesome location. There is considerable stress and flexing in this area, which makes it difficult to provide a truly tight seal. In my experience, this is the most likely location for cable failure or for moisture entry. Consider how many molded power line cords you've seen with failed cable jackets immediately behind the plug body or the overmolded strain relief.

- Conductor diameter (wire gauge)—RFI/EMI/EMC considerations and environmental requirements usually dictate a system enclosure should have the fewest possible penetrations. This directly affects external connector selection and often forces connectors to contain a mix of different conductor diameters. Military and heavy industrial connector inserts (the interfacial bodies with holes for the pins and sockets) often permit a mix of conductor sizes, but the combinations are limited. Some, not all, connector families for military and heavy industry allow a limited number of coaxial or fiber-optic connections with special pins and sockets.

- Housing material—suitable connectors are usually aluminum (often with chrome or cadmium plating), although some thermoplastic housings are intended for rugged service. The choice of housing material will depend on the system's intended use. Nylon and other lightweight thermoplastic alternatives are satisfactory for relatively protected environments. Heavy industrial, marine, or military systems are subject to far greater abuse and should be metal or material approved for the application. Circular connectors are usually available with a choice of bayonet lock or threaded shells. I prefer bayonet lock connectors; bayonet locks have a positive detent to confirm engagement, require only fractional rotation to engage and disengage, and are relatively easy to use even when wearing gloves. Threaded shells can be clumsy to start, require many turns for full engagement, and can be extraordinarily difficult to connect if the threads are dirty or the shell has been deformed (which, in my experience, is a common problem—you would be surprised by the number of times a fielded connector is stepped on or run over). Difficulties with threaded connectors often frustrate system users, which result in the connectors being only partially engaged. Partial engagement, in turn, invites water, humidity, and dust intrusion, which then results in intermittent contacts. Dust intrusion and humidity also combine to provide likely paths for high-voltage arcing failures. Some specialized connector families have sealing glands and interfacial seals to exclude water, oil, fuels, solvents, and hydraulic fluids.

## Cabling

External connections are ordinarily established using cable assemblies rather than a loose bundle of individual wires. For most purposes, *cables can be loosely described as being*

*coarial or controlled impedance types, power cables, and instrumentation cables.*
There are many exceptions, and the cable industry has a wealth of special constructions for applications such as marine geophone cables, twinax, direct burial cables, well logging cables with special strength members, high-temperature insulations, and highly flexible shielded cables with unusually small bend radii for use on industrial robots. Applications having sufficiently high volume can justify using custom cables fabricated to satisfy the exact job requirements, and these constructions can be virtually any combination of mixed-diameter single conductors, mixed insulations, custom cable markings, and a mix of single-conductor, coax, and optical fibers. Typical cable construction will include the conductor bundle itself (with or without optical fibers, if a custom cable), electrically insulating filler materials to provide mechanical protection and a substantially circular exterior cross-section, a woven braid shield or a spirally wound foil shield with a bare drain wire for noise mitigation, and a tough outer jacket to protect the interior cable structures and to prevent problems arising from exposure to moisture, chemicals, and mechanical abuse. Some manufacturers also offer flexible spirally wound steel armor inside the cable's outer jacket to provide crush resistance and special strength members to allow the cable to support substantial mechanical loads.

Generalizations are difficult because embedded systems have almost unlimited configurations and purposes. Some common concerns will be:

- Estimate the maximum length of cable required for any reasonable application of the system. Copper has low but still finite resistance per foot. In extremely long-line applications (e.g., petroleum well logging), the cable must be treated as a lossy transmission line. Conductor resistance is responsible for load-dependent line losses and a downhole voltage that is effectively modulated by load current. Cable capacitance and inductance severely limit the realizable bandwidth of very long cables. Line loss issues can be minimized by using high-voltage power connections, although increased source voltage affects the type and thickness of insulation permitted in the conductor bundle. Coping with variations in the long-line load voltage variations can significantly impact the entire system architecture.
- Itinerant measurement systems—for example, seismic instruments and geophone strings deployed during seismic exploration—are deployed in the wild for many days at a time, but move frequently from place to place. Cable handling by the field crew is a routine and frequent task. Perspiration from the field crew's hands accumulates on the cable jacket, and small animals will eat the cable jacket to get the salt. Marine streamer cables constantly in salt water when in use. The cable jacket demands careful consideration and selection for these and many other examples, because the cable jacket is the primary protection for the conductor bundle it encloses.
- There are a great many noise sources in the world. Cables external to the system enclosure should be shielded. The availability of a drain wire is very useful to help

connect the cable shield to the system signal common, although RFI/EMI/EMC considerations could demand more elaborate terminations.

• Quality strain reliefs are needed to mechanically join cables to the system enclosure and to protect the cable bundle from damage through use and weathering.

## Conductors

Embedded systems occasionally require the design team to specify individual conductors. This happens most often when the embedded system is located a significant distance from the power source (which occurs in a variety of industrial plant installations) or when the embedded system involves relatively high-power loads, loads at a distance, or loads with individual wire terminations (e.g., motors and relays).

Selection of individual conductors requires knowledge of the current the conductor must carry and the operating voltage the conductor's insulation must withstand. Inductive applications, which often involve switching transients, require additional knowledge about the peak current surges and voltage transients that may arise. Applications with unusual strength requirements—for example, long-wire antennas and well logging applications—may require a conductor having copper cladding over a steel core (one such arrangement is a commercial product known as Copperweld®).

Certain instrumentation applications require long power and signal conductors (which may or may not be carried in a cable) from the source to a remote instrument package. This arrangement requires a reasonable amount of analysis, because conductor resistance is no longer negligible. Voltage loss in a long power line causes voltage at the load end of the line to be a function of load current. This makes it important for you to insure the remote end of the line has an acceptable line regulation response. The other issue, which can force a significant change in the remote package's architecture, is to note that the low-voltage end of the load is not zero volts when voltage is measured with respect to the power source common. This gives rise to common mode concerns, electrical safety issues, and possible data corruption problems if a data signal is referenced to the low side of the remote package's power supply (which is often the case). The easiest and sometimes least expensive solution is to oversize the supply cable conductors, but this adds cost and weight to the installation and may not be a completely effective solution to the data corruption problem. Optical isolation of digital lines and fiber-optic data transmission may be necessary in exceptional cases.

## Connections

I have addressed board-to-board connections between various internal subsystems and housing-to-cable electrical connections between the system and the outside world.

Unfortunately, other electrical connections are easily overlooked until they present problems. Grounding and bonding practices can cause many difficult noise problems. Noise issues are especially likely to arise in systems applications involving low-level analog signals and high-power switching. While each system application is unique, several common issues are usually present to some degree. They are discussed for completeness.

- If the system is powered by an AC line, should it have a ground-fault circuit interrupter (GFCI)? If so, should the GFCI be included in the equipment itself, or is the system part of a fixed installation allowing the GFCI to be included in the power line's branch circuit breaker or power outlet?
- If the system contains high-voltage DC, should it have a DC-GFCI? If so, where should it be located? DC-GFCI equipment is relatively difficult to find; the design team may have to develop their own especially for the project.
- To the extent possible, partition the system to approximate an equipotential grounding arrangement. In simple systems, you can do this using a bus bar as the single common voltage reference. Be advised: an equipotential ground arrangement may not be the optimal solution for RFI/EMI/EMC problems at some frequencies and signal levels.
- Ground (or common) connections to chassis metal should be established by a strong metallic fastener and a toothed lockwasher. If a connecting wire is needed between chassis metal and a circuit assembly, the connecting wire must be short, of sufficiently heavy gauge to present low impedance and carry all possible fault currents, and must be terminated in a suitable crimp connector, solder lug, or other hardware intended for the purpose. You must securely engage the connection with bare metal.
- Metallic enclosures involving multiple panels should be designed to insure good electrical contact at all seams. In some situations, sheet metal fasteners used to attach lids or sides to other structures may be sufficient. In other situations, especially those subject to stringent RFI/EMI/EMC requirements, conductive gaskets or metal fingers will be needed between all panel joints.

## *Operating life (MTBF)*

MTBF calculations can assess the anticipated operating life of equipment. These calculations use established failure rates for the components comprising any piece of electronics. Calculations also take into consideration such information as anticipated ambient temperature ranges. A great many devices, particularly passive devices, connectors, common standard semiconductors, and the like have established MTBF life figures which are simply applied to the BOM. Unfortunately, the semiconductors that most strongly impact MTBF calculations—voltage regulators, power components, processors and controllers, micromachined accelerometers, operational amplifiers, and any relatively new devices—are often too new to have established reliability values. While MTBF for new semiconductor devices sometimes can be estimated from accelerated life tests, the reliability of semiconductors is such that even accelerated life tests require a surprising amount of time before statistically meaningful life estimates are available. More often, you will calculate MTBF by analogy with older parts having similar operating characteristics, roughly similar physical characteristics, similar manufacturing processes (if possible), and established MTBF values. While inconsistencies are evident in this approach (the significance of manufacturing process differences, for example), *a good-faith effort will result in a MTBF estimate that is debatable but more enlightening than no estimate at all*. The governing document for military reliability calculations in the United States is MIL-HDBK-217 [15]. Detailed information about the rules and procedures for MTBF calculations and associated data also are available in Ref. [16].

## Power and power consumption

*Power supplies are associated with a great many problems*. Many people, especially experienced designers, know this. It, therefore, is very puzzling to me that the part of the design receiving the least attention, and the part of the design where people often attempt to cut corners to save production costs, is the power supply. An expensive power supply does not guarantee good performance, but a cheap one often performs poorly. A power supply is more than a box of volts; it is a system component that touches every other part of the system. A good power source is indispensable; a poor power supply will corrupt the performance of an otherwise acceptable design. If the power supply is a purchased part, be prepared to specify and select it carefully and pay a reasonable price for it. If the power supply and associated voltage converters are designed in-house as part of the project, be prepared to invest enough effort and money to do it correctly.

### An aside about power consumption

Reduced power consumption should be a design priority through the entire design cycle. While it is important to insure all components meet the technical requirements of the project, it is also important to insure the entire system uses the minimum amount of power needed to satisfy the technical requirements with a reasonable margin for production tolerances. Selection of amplifiers, processors/controllers, logic devices, voltage references, display devices, communications devices, and all other system elements should be made with the objective of saving every possible milliwatt. *Think of power as money and spend it sparingly*.

Speed typically comes with a power penalty. Use as much as you need, but no more than you need. Amplifiers with excessive slew rates and processors with unnecessarily high clock frequencies consume more power than necessary with no demonstrable benefit to system performance. Pay attention to quiescent current of power converters, amplifiers, and data handling devices. ***Discipline yourself to pay attention to power consumption rather than efficiency***—a 75% efficient power supply for a 2 W load is a better system option than an 85% efficient power supply for a 5 W load, because the former option produces less heat. Consider using a low-power logic family. (The old 4000-series CMOS family is slow, but it can operate over a wide range of supply voltages, has good noise immunity, and requires little power. It is available from several major manufacturers in tiny, single-gate configurations for those occasional small glue logic requirements.)

Why does this matter? The most obvious answer has to do with operating life. Many embedded systems are meant to be field-deployable and they are often battery operated. The less power it takes to operate the system, the longer the battery's operating life will be. Other reasons are less obvious.

- One of the heaviest items in many line-operated systems is the power supply transformer. Reduced power consumption permits the use of a smaller transformer. Similarly, associated power semiconductors, capacitors, inductors, and circuit breakers can be smaller, leading to reduced weight. Size reductions are beneficial because less volume requires less size and weight in the enclosure. Power savings result in weight savings.
- If you generate too much internal heat, you must get rid of it. Low-power systems require little or no heat sinking, fewer or smaller fans for convective cooling, and reduced thermal stress on components (thereby increasing MTBF and reliability).
- If special cooling is necessary, reduced heat generation gives the system designer a wider range of options.
- Reduced heat generation inside the system enclosure reduces the thermal excursions experienced by the system elements, which somewhat reduces problems arising from temperature coefficients of critical components.
- Reduced power consumption in a battery-operated system means increased operating life for a given battery. In some cases, the difference can allow an acceptable operating life with less battery, thereby saving size, weight, and cost. Occasionally, reduced power consumption will increase the number of battery chemistry options available to the system designer.
- Reduced power consumption often has RFI/EMI/EMC benefits. If the system radiates less power, it more easily meets the emissions standards. Likewise, reduced clock frequencies not only reduce power consumption in processors, controllers, and logic devices, they also reduce the frequency of associated emissions.

### Line-operated power supplies

Embedded systems in a line-operated environment have the luxury of unlimited operating time under normal circumstances. If battery backup is necessary for critical line-operated systems, the preceding discussion about power consumption applies in its entirety. Contemporary commercial single-phase line-operated power supply designs often include expanded input voltage and frequency ranges (85–264 Vac and 47–63 Hz, the single-phase "universal input" ranges, are commonly available). Higher power units often incorporate power factor compensation. Interfaces for software control are often standard or are available as inexpensive options. Multiple outputs, excellent protection networks, and similar features are commonplace. One significant decision facing the systems designer will be to choose between a smaller, more efficient switching power supply and a larger, heavier, less efficient, but electrically quieter, linear power supply. The decision will often depend on the nature of the system. Systems intended for use with very low-level analog signals, or for signal channel operation at or very near the noise floor, will often require a linear power supply because they have far less noise and ripple. It is possible to obtain electrically quiet performance from a switching power supply, but high-quality signal channels often require additional filtering or linear post-regulation of the switching supply's output, which negates some of the advantages otherwise realized by selecting the switcher in the first place.

### Battery-operated and alternative energy systems

Embedded systems are often used for portable field applications with little or no access to commercial AC power or AC generators. In such cases, power is usually provided by batteries or such technologies as solar panels, fuel cells, and occasionally wind-driven generators [17]. The dominant technologies, electrochemical and solar energy conversion in particular, are fundamentally producers of DC electricity, and this discussion will treat all such sources as if they were batteries.

Some battery technologies are very mature, having been in use for more than a century. Newer technologies (e.g., lithium-iron phosphate and zinc-air batteries) are entering the commercial product stream in increasing numbers. More exotic specialty chemistries are expensive and are used only for the most demanding applications (such as space vehicles). Battery chemistry will determine the nominal terminal voltage of an individual cell; the volume of reactants will determine the total energy capacity and physical size of an individual cell. Batteries may be composed of either primary cells (single-use batteries which are discarded when the cell reaches the end of useful life, represented by common alkaline batteries) or secondary cells (multiple-use batteries which may be recharged and reused, represented by lead-acid gel cells, nickel-cadmium, nickel-metal hydride, and lithium-ion polymer cells). As a rule of thumb, a secondary cell will have roughly half the energy storage capacity of a primary cell in the same chemistry and size. The nominal energy capacity of a cell is a rating obtained at a reference temperature (usually around room temperature) and at a specified discharge rate. Useful energy storage capacity is strongly temperature dependent, with many cell chemistries unable to deliver a substantial fraction of nominal useful energy when the cell temperature is below 0°C. The operating life of a secondary cell is determined by the cell's history, with the number of charge/discharge cycles being limited and influenced by the charging profile.

It is possible to use individual cells for small projects, but multiple series-connected cells (stacks) are normally used to provide a higher source voltage for the system electronics. Internal system power converters normally raise (boost conversion) or lower (buck conversion) the battery stack voltage (which decreases as the battery discharges) into the regulated voltages needed by system electronics. Switch-mode power converters are overwhelmingly favored for these applications because of their high efficiencies and small circuit-board footprints. Switch-mode power conversion and regulation is generally satisfactory, especially for digital systems, which are relatively tolerant of the output ripple commonly found on switch-mode converter outputs. Precision analog applications, especially those requiring high resolution and good low-noise performance, require very quiet power rails for the amplifier chain. In such cases, switch-mode power supplies may not be the best choice for the application without considerable care being taken to filter their inevitable output ripple. A quiet battery-operated power supply section, including bipolar power supplies, can be made using switched-capacitor techniques which often outperform conventional switch-mode power converters.

### Notes about unipolar and bipolar (single-ended and double-ended) power supplies

It is very tempting, especially in battery-powered systems, to use a unipolar (single-ended) power supply—which is to say, a power supply with only a single output polarity. This may be entirely adequate if the system is completely digital. However, a single-ended power supply introduces design complications if the system requires analog gain and signal processing. Power supply selection is often made early in the project; an unfortunate choice will cause problems through the remainder of the design cycle and has the very real possibility of compromising performance in production. I strongly suggest using a bipolar power supply for all but the very simplest analog structures.

Many first-time designers of equipment having a unipolar supply, especially those designers who do analog work only occasionally, are puzzled when a simple inverting analog amplifier with a first-order pole performs badly. They eventually recognize the source of the error—the inverting structure's output cannot go negative if the only power supply connections to the amplifier are circuit common (ground) and the single positive power supply connection. Two choices present themselves: change the power supply to a bipolar structure, or define a midrail reference (usually half the power supply voltage) and use the midrail as an artificial signal common. While I urge the first course, let's explore the consequences of the second.

First, the dynamic range of the signal is effectively reduced. A bipolar power supply of $\pm5$ VDC with a rail-to-rail amplifier having a gain of $\times10$ (20 dBV) can readily handle an input signal approaching $\pm500$ mV. The same amplifier operating from a unipolar 5 VDC power supply with 2.5 VDC midrail cannot apply the same $\times10$ (20 dBV) gain to the same $\pm500$ mV signal, because the amplifier's output will saturate beyond $\pm250$ mV.

Second, a great many analog structures apply signal to the inverting input, with the circuit reference being tied to the noninverting input. In a bipolar supply, the circuit common is a low-impedance connection intended to be the common reference for the supply outputs. It is the common point against which all voltage measurements are made. A unipolar supply's midrail connection, on the other hand, is usually formed by a simple voltage divider between the single supply output and the supply common. Any noise (e.g., output ripple) on the supply output will appear on the midrail connection, attenuated by the same voltage divider. This noise-corrupted midrail signal is now applied to the noninverting input of an amplifier, typically having gain and, consequently, amplified noise appears on the output.

Third, a typical analog system will contain more than one amplifier. The same midrail voltage can be applied to multiple amplifiers, often resulting in layout complexities and the equivalent of a fan-out problem arising from using a single midrail, or multiple resistive dividers can be used to simplify layout. This latter alternative produces DC errors across the amplifier chain; no two sets of voltage divider resistors will be exactly the same. Consequently, any signal channel having a DC response will now comprise multiple amplifiers, each with slightly different midrail reference voltages. The accumulated DC errors contributed by multiple midrail connections will be indistinguishable from signal. In effect, the channel's output offset error is increased, sometimes significantly, by the use of multiple midrails.

Fourth, if the designer persists in using a midrail connection, the next step usually is to filter and buffer the midrail signal to provide a lower impedance midrail signal with reduced noise content. (Commercial devices are available to do this sort of thing; one such integrated device is manufactured by Texas Instruments, the TLE2426 "railsplitter.") Use of a buffer amplifier merely succeeds in increasing the power consumption of the analog chain—although the increase may be slight—and the filtering action of the buffer stage reduces (but does not eliminate) the noise component introduced by the midrail divider in the first place. It is much better to avoid introducing the noise. The number of active devices increases, and, even if only one midrail is used, the circuit-board complications remain.

Note the above arguments considered only power supply output ripple as a noise source. In addition to supply output ripple, the midrail injects power supply line and load transients into the analog chain. These effects can be severe, especially in embedded systems involving motor control, relay drivers, and similar inductive loads or heavily capacitive loads. The resulting complications can be very difficult to identify and resolve, and you begin to understand a unipolar analog supply with midrail reference was not the promising economic alternative it was thought to be. Everything depends on particulars of the design and application, of course, but in most cases a slightly more expensive power supply with bipolar outputs will provide a quieter analog channel with fewer design complications and reduced performance compromises. If only a few amplifiers are needed in an otherwise heavily digital design having a single 5 VDC supply, the example network discussed in Figure 10.20 will produce a high-quality (quiet and tightly regulated) bipolar supply capable of driving 20 mA supply current per side at approximately $\pm 5$ VDC. This is more than enough power for a limited number of analog devices in most applications.

## Cooling

Cooling issues are best resolved by avoiding unnecessary heat generation; typical heat control and cooling solutions add cost, complicate packaging design, increase weight, and often increase power consumption. Reducing power consumption has to be done as a matter of routine in all phases of circuit design. Certain circuit topologies (e.g., use of Class D audio amplifiers instead of analog audio amplifiers) can be helpful in reducing thermal management problems. Other schemes, as already discussed, reduce clock speeds, reduce amplifier slew rates whenever possible, and specify reasonable (yet essentially minimal) reserve capacity in processor and memory performance. *If you don't generate heat, you don't have to get rid of it*.

If an embedded application presents a cooling issue, a handful of design options are available. Some design options have the effect of increasing operating life in battery-operated systems, making them desirable in their own right at the expense of somewhat more complex circuit operation and system operating rules. Other useful options may mitigate a heat problem but compromise or complicate the mechanical aspects of packaging design. Design options include the following:

•   Duty-cycle the heat-generating parts. This is traditionally used when dealing with the transmitting elements of an RF data link or the link's entire RF section, but it can be successfully applied to virtually any part or even all of an embedded system. RF-controlled seismic systems, for example, routinely implement a "sniff" mode whereby the entire instrument is turned off with the exception of a small watchdog timer circuit in the receiver. The watchdog turns on the receiver for a few hundred milliseconds every 3 or 5 seconds—if no control carrier is present, the receiver returns to sleep mode. If control carrier is present, the receiver remains active, receiving and decoding data, until commanded to turn off. If the receiver's instrument is involved in the seismic

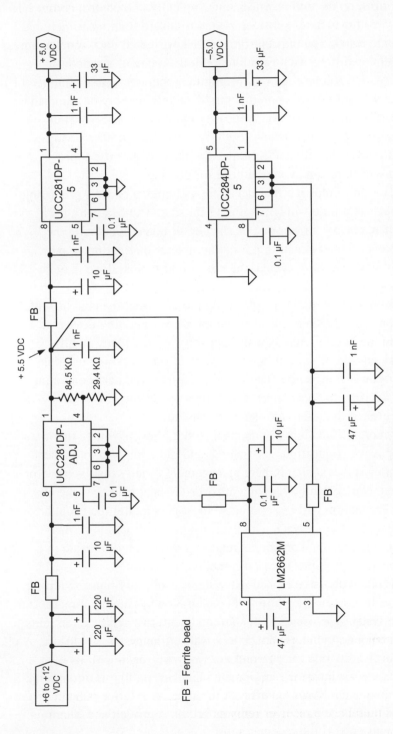

**Figure 10.20: Unipolar/bipolar converter for low-power analog loads.**

This figure illustrates a compact bipolar ±5 VDC power section for modest analog loads in battery-operated embedded systems. This arrangement has proved to be a reliable and efficient way to provide a bipolar power supply for analog front ends in circuit assemblies drawing their power from two lithium cells or three alkaline cells. *Used with permission of The Charles Machine Works, Inc.*

experiment, the receiver turns on the entire instrument. Typically, the control carrier has a minimum duration of three times the basic sleep period to insure each instrument has at least two opportunities to respond to the control carrier. Distributed data systems are readily structured to acquire and store data for a long period—a day or a month, for example—for later download to a central unit according to a operator-issued command or a time-based operating schedule. Most contemporary data systems rely on sampled rather than continuous data acquisition, so you can easily implement proper power cycling of the system's various elements, which results in a dramatic power savings. Because system units are on only briefly and infrequently, they spend little time producing heat. Thus, duty-cycling can avoid the need for cooling.

- Devices which produce heat for prolonged intervals—for example, a resistor producing 1 W of heat for five minutes at a time—may require a heat sink. The heat sink itself might be nothing more than a heavy copper pad on the circuit board containing the device, or it might be a larger finned structure presenting greater thermal mass and surface area. If the heated device is duty cycled, the use of a heat sink alone may be sufficient to control the heat problem. If the system is intended for use in a residential, office, or laboratory environment, it can be very effective to construct the chassis with finned heat sink projections outside the enclosure with the heat-generating device in good thermal contact with the interior chassis wall opposite the fins. Many small laboratory power supplies capable of several hundred watts of output power use this arrangement to cool their power transistors. Heat is removed from the heat-generating parts by thermal conduction (into the heat sink), which then leaves the heat sink through convection and infrared radiation into the environment.

- Higher power devices or more difficult environments may not allow adequate thermal management using only passive conduction, convection, and radiation to remove heat from the part and enclosure. The next step in heat management usually involves one or more fans providing forced convection by blowing air over the heat sink fins, thereby improving heat transfer. This option will be familiar to most people, but the peculiar nature of embedded systems does not always lend itself to forced-air cooling. Many embedded systems have environmental constraints requiring the system to reside in a sealed housing, which generally prohibits use of a forced-air solution.

- One cooling approach for relatively large embedded systems especially those requiring a sealed enclosure to prevent dust and moisture from penetrating system equipment, is a heat exchanger inside the system enclosure. In this arrangement, two separate fan units (resembling miniature automobile radiators) attach to a wall of the system enclosure. A plenum divides the heat exchanger into interior and exterior parts, each with its own fan. The interior fan circulates air inside the equipment chamber, pulling it through the first radiator, while the exterior fan blows exterior air through the exterior radiator. The heat exchanger's working fluid is contained in many small tubes inside the radiator elements. Interior air motion transfers heat to the working fluid, while exterior air flow rejects heat to the outside air. There is no direct air exchange across the boundary of the equipment enclosure, allowing it to seal out dust and moisture. The heat exchanger removes heat effectively, but no mechanical refrigeration is involved—air temperature inside the enclosure remains elevated relative to outside air temperature. This temperature elevation is often no more than 5–10°C for a properly sized heat exchanger, but the result is a far cooler interior temperature than that which would have resulted with no heat transfer whatsoever. The only energy input to the system involves power for the fans circulating air through the interior volume and the radiator external to the equipment enclosure. Even this power consumption can be minimized by using passive thermostatic switches to control fan operation.

- Another thermal management approach uses compressed air to cool electrical panels and similar equipment in an industrial setting. The operating principle is that of Joule–Thompson cooling, whereby the release of a working gas at above-atmospheric pressure absorbs heat as the pressure drops to atmospheric, cooling the interior of the valve body or expansion chamber. This cooling may be exploited to provide active cooling for a system enclosure. The primary drawbacks are the need for a dependable source of compressed gas—a difficult thing to manage in many embedded systems—and the acoustic noise which accompanies the release of pressurized gas to the atmosphere.

- Finally, an equipment enclosure can be cooled by mechanical refrigeration; in essence, by installing an air conditioner in the equipment enclosure. This option can remove significant heat from an enclosure, but it does so with significant system costs: weight, usually in excess of a hundred pounds (45 kg), electricity to run the fans and compressor, and ductwork needed to insure circulation of cooled air to all parts of the enclosure. Compressor heat must be rejected to external air. This option is typically used inside standard-sized 19″ relay racks containing servers or telephone central office equipment, where the installation is permanent, stationary, and has abundant electrical power available to operate the coolers. This option is not considered practical for small portable or battery-powered units.

## Environmental extremes

Embedded systems have one enormous advantage: silicon is tough. A properly executed design using quality silicon semiconductors and conventional passive electronic components properly operated within their ratings are fully capable of reliably operating in and beyond what is normally considered the "full mil" temperature range of −55°C to +125°C. Not all commercial parts will be useful in demanding environments—for example, certain capacitor types are unable to handle extremes of heat and cold—but components are generally available to span the entire mil-spec temperature range. I can attest that many integrated circuit devices will still function, although they may not meet full specifications and may have shortened operating life, beyond the $+125°C$ rating. As an aside, a petroleum well logging project many years ago demonstrated that simple 4000-series CMOS logic devices finally failed to operate somewhere between $180°C$ and $185°C$. More importantly, they recovered and operated properly again upon return to lower temperatures. Hybrid microcircuits made for petroleum well logging applications have operating ratings approaching $250°C$. Recent developments with silicon carbide semiconductors promise to extend the operating-temperature range above this limit, although at a price.

Unfortunately, other system components have trouble keeping up with the silicon. Capacitors are the biggest limitation, with only a relative few types (primarily those with mica and certain ceramic dielectrics) capable of surviving and functioning well above the mil-spec range. Even then, operating characteristics change significantly because of thermally related parameter changes. As noted in an earlier section, the most temperature-stable capacitors tend to be mica and NPO (C0G) construction, both of which tend to be undesirably large and heavy for many embedded applications. Mylar® and Kapton® dielectric capacitors are possibilities, but such devices are expensive and often must be custom made. At extreme elevated temperatures, even the matter of attaching components to a polyimide or similar high-temperature circuit board (FR4 and G10 boards cannot handle the heat) requires special solders and attachment equipment.

## RFI, EMI, and EMC compliance

Few meaningful general guidelines for RFI, EMI, and EMC compliance issues are available. There are such a great number of standards and jurisdictions (based on product types, environment, and location of intended use) that generalizations are not possible. Even matters that appear to be capable of standardization between jurisdictions—transmitted power, frequencies, and modulation techniques for wireless equipment control, for example—are anything but standardized. Similarly, there are few reliable guidelines to assist circuit developers in initial layout and design of circuit boards. There are any number of references in the literature [6,8–11,18–22], but each project will have its own operating requirements (emissions, susceptibility, transients, and electrostatic shock) and its own physical constraints and characteristics (geometry and enclosure penetrations).

RFI, EMI, and EMC issues can present serious challenges to a project and the project's schedule. Unfortunately, meaningful product testing (i.e., testing capable of demonstrating the product's satisfactory performance relative to the product's RFI/EMI/EMC requirements) can be done only with the test article in its production configuration—that is to say, a production unit or a preproduction unit configured exactly as the production articles will be configured. This means the test article is, in effect, ready for production release except for passing RFI/EMI/EMC testing. It is also an unfortunate truth that many articles do not pass on the first examination, often requiring multiple test sessions before successfully meeting requirements. A test failure usually results in mechanical and electrical design changes, circuit modifications, changes to bills of materials and circuit-board layouts, and—far too often—schedule impacts. *The only real preventative measures to be taken are to allow a generous amount of time for redesign, multiple circuit-board modifications, repeated testing, and adequate testing budget during the project's schedule definition phase*. Subsystems should be tested separately whenever possible to weed out problems at the subsystem level, but this will not guarantee all subsystems will function properly as an interconnected system. You must test the entire system as a single entity. Still, system level testing can be entered with greater confidence if individual subsystems have been demonstrated to be capable of performing adequately. *Test early, test often*.

## Analysis methods

### Worst case

Worst case analysis, as used here, involves analysis of the system at the extremes of specified component manufacturing tolerances, environmental conditions, and operating parameters. Worst case analysis is usually done using published product and component specifications to confirm that the system is capable of satisfying all requirements with any combination of environmental and operating parameters, and with the worst possible combination of component tolerances, with all factors being applied at the same time.

Worst case analysis typically relies on manufacturer's specifications to provide the analytical cases. For example, an AC motor drive application might consider the ability of the motor drive to deliver adequate operating power with minimum input line voltage, minimum power line frequency, maximum ambient operating temperature, and maximum motor load.

### Simulation

Simulation tools are probably the most widely used circuit-design software tools. They provide the means to investigate the performance of different topologies, different components, and a host of design options without time-consuming layout and fabrication issues. Sophisticated commercial circuit-simulation packages are readily available. Many manufacturers provide excellent specialty packages for circuit-board layout, chip design, antenna design, electromagnetics, motor driver design, and similar work. Mechanical design packages are available to treat housing and enclosure design, heat transfer, and linkages to 3D part printers. These tools are essential to contemporary industrial design. Like any other tools, they also have their flaws.

There is a tendency for a designer to put unquestioning faith in the simulation package, but simulation is only as good as the underlying model. Most electrical models are very good, the results are generally reliable and highly valuable, and this often justifies the designer's faith in the simulation. Unfortunately, simulation models are not especially good at depicting the strays, nonidealities, and parasitic effects that often plague systems work, particularly those with demanding packaging requirements or purchased subsystems with overly simplified models. Laboratory results are reality; simulations are guides. This will be especially apparent during RFI/EMI/EMC testing. ***Remember, models are used to simulate reality; they do not constitute reality***.

### Monte Carlo simulations

Of all simulation techniques, Monte Carlo simulations are the most useful. Worst case analysis explores the operational consequences of devices working at the extremes of their specifications—hence the name worst case. Monte Carlo simulations are very useful in examining the consequences of normal production tolerances on a design. Monte Carlo simulations can provide a useful predictor of design robustness. If a production run of sufficient size is anticipated, Monte Carlo simulations can help support strategies for production testing, lot sampling, product tuning and calibration (if necessary), and production yields.

## Testing, qualifications, and conflicts

### Bench tests

Most conventional bench testing is done with a fairly standard set of general instruments (power supplies, voltmeters, oscilloscopes, function generators or arbitrary waveform generators, computer with engineering GUI) and an array of more specialized equipment determined by the nature of the design under evaluation (spectrum analyzer, vector network analyzer, gain/phase analyzer, logic analyzers, test loads, display devices, and so on). A word of caution about the power supplies used for development testing: be sure to transition to a representative power source as soon as possible. This is especially important if the project involves low-level analog circuits. If you spend months testing a prototype using an analog laboratory power supply, then mate the resulting prototype with the small switching power supply you really intend to use, don't be surprised by disappointing results. If your final design will use a switcher, you should use a similar, preferably identical, power supply early in development.

## RFI, EMI, and EMC field tests

Product testing in accordance with EMI/EMC product standards must be done by a properly certified facility. A great deal of development testing can be done, however, before submitting to such a facility. If the first trip to the certified facility is unsuccessful, which is often the case, the design team will have some sort of idea about problem frequencies and the signal levels of interest. This is particularly true of radiated emissions. *Perform EMC testing early and often*.

A spectrum analyzer and reasonably good (and calibrated) antennas will be the primary tools used when tracking down the source of offending emissions. When you know the frequencies of concern, the spectrum analyzer will give relative amplitude readings at the problem frequencies of interest. This will not be sufficient for product acceptance testing purposes, but it will be very helpful for troubleshooting and evaluating circuit and layout changes.

Useful test equipment for RFI/EMI emissions work is not necessarily large and expensive. Relatively inexpensive "sniffer" probes are commercially available, although very useful sniffer probes also can be fabricated quickly for particular problem frequencies [23].

## Environmental tests

The need for environmental testing, and the particular nature and performance limits of the tests to be performed, should be clearly established during the project's concept phase. Certain tests can become key design drivers for the entire project; at the very least, you must know environmental requirements, such as operating and storage temperature ranges for proper electronic component selection. Mechanical shock, vibration, dust, humidity, drop, altitude, salt spray, immersion, fungus, and similar requirements introduce significant complications to what otherwise might be a modest mechanical design effort, and they likewise must be known at the very beginning of the design effort.

In certain cases, unique attributes of the anticipated product deployment environment will give rise to unique test requirements. Testing to unusual, but very real, design requirements can be a substantial design and fabrication burden in its own right. Most environmental test requirements, however, are well known and common to most projects. At a bare minimum, electronic equipment should be tested to verify the ability to meet operating and storage thermal limits. Vibration, mechanical shock, and drop are commonly specified mechanical requirements. Resist the urge to "test by memo." Equipment either will work as specified or it will not. Don't let the customer be the first to find that it will not. In extreme cases, it could cost lives.

Test *limits* will be determined by the design team in accordance with anticipated circumstances of use, storage, and shipment. Most test *procedures*, however, will be found in either a governing test document prepared by the customer or specification-setting group, or by reference to any of several well-known military standards that concern themselves with environmental testing. Probably the best known of these documents in the United States is the latest revision of MIL-STD-810. At the time this chapter is written, the current document is MIL-STD-810F [24]. There are certain to be equivalent IEC and other standards worldwide. The advantage of referencing procedures in MIL-STD-810F or another established standard is the common language and understanding of the procedures in the test community and the availability of suitable test facilities and equipment in test labs around the world.

### Potential conflicts

Dr. Dale Kaufman, a professor of electrical engineering at Kansas State University when I was an undergraduate, said it best: "Engineering is the science of tradeoffs."

Experience teaches us design conflicts are inevitable. Most, if not all, serious conflicts should have been identified and resolved during the product-definition phase if the design team and all other parties did their work well. For the sake of argument, assume one or two issues remain. What might they be? Aside from the more narrow technical issues, many embedded-system efforts (and, indeed, most engineering-development efforts) seem to be vulnerable to at least the following basic conflicts:

### Cost versus performance versus schedule

In many ways, Systems Engineering exists to help resolve the conflicts inherent in these three activities. ***Cost, performance, and schedule seem to be natural antagonists***. Cost is understood to include not only purchased parts and devices needed for the project, but also the human talent and support network needed to work the project. Performance includes the fairly specific objectives of satisfying specifications, but it also has a quality dimension: a robust, reliable design that is easy to build, easy to operate, and capable of operating satisfactorily beyond nominal performance expectations is— by almost any standard—a high-performing design. A design that is complex, fussy, hard to use, and requires a great deal of expensive maintenance just to keep it operating is not considered a good performer. Schedule is the culprit that usually forces questionable decisions and expensive solutions— all of us have experienced the discomfort of doing something we know is an inferior approach only because the schedule doesn't permit the time to pursue a superior option. Schedule often forces hardware engineers to design with parts they can purchase quickly rather than suffer the schedule impact of long lead times needed to obtain superior parts. As the employees of a former employer of mine often reminded their supervisors, "If you want it bad, you get it bad." The classic example of schedule-driven conflict is a 6-month project to prepare the first article of a new product concept for presentation at a trade show. It happens to everyone, but it's a bad idea and a very bad experience.

All too often, product definitions, performance targets, and schedules are made without a clear appreciation of the difficulties that lie ahead. Even if the design team will reengineer an existing product having known performance characteristics and familiar manufacturing processes, problems with part obsolescence, parts shortages, manufacturing capacity, and competition for resources can plague the project. An R&D project, which is a true step into the unknown with a rigid timetable, is a heartache looking for a place to happen.

All conflicts cannot be anticipated and avoided during the planning process. There are always problems and unforeseen circumstances, but a calculated and realistic appraisal of development time, manpower, test effort, and scheduled time for rework and retesting to correct errors or operating deficiencies is essential to minimize schedule conflicts, delays, and project overruns. All too often, planning makes unrealistic assumptions reflecting a sincere (but ultimately misguided) effort to meet the requested schedule. ***Cold-blooded, honest, calculating realism is essential during project planning***.

### Power versus operating life

The conflict between power consumption and operating life frequently arises for embedded systems operating from batteries or secondary batteries in conjunction with certain alternative energy sources (solar cells and wind chargers, for instance). Operating life can be extended considerably by duty-cycling all high-power operations and by holding power consumption to an absolute minimum—these options were discussed earlier in this chapter. The design team often gets into trouble during the product definition phase by calculating operating life using nominal (which, by the way, means "in name only") battery ratings and "typical" power consumption values from semiconductor data sheets. Nominal battery energy ratings depend on the discharge rate. Less obvious is the fact that secondary cell battery capacity is a function of cell history (previous discharge history and number of charge/discharge cycles)—cell energy storage capacity decreases over time. The operating life calculated for a new cell stack is likely to be the maximum operating life. Nominal battery capacity ratings are almost always specified at room temperature; if the application needs to operate in cold environments, the actual usable battery capacity will be significantly reduced. Semiconductor ratings are usually accurate, but some are very temperature sensitive. If power consumption specifications are not guaranteed for the operating range of the intended design, I suggest you calculate power consumption using no less than 125% of typical.

### Size versus function versus cost

Size conflicts commonly arise in consumer products, for increasing functionality in ever-smaller physical devices appears to be the norm and may be expected to continue into the future. Many industrial applications are subject to severe size limitations, as well. I, for example, design equipment used in underground construction and horizontal directional drilling. This operating environment requires downhole equipment residing in a tubular assembly located immediately behind a drill bit. It is not uncommon for the apparatus to be constrained to fit within a cylindrical volume having a diameter of one inch (25 mm) or less.

Whereas consumer products often benefit from economies of scale which make custom chips and custom packages economically viable options, industrial applications often have much smaller production volumes making custom chips and other custom assemblies economically unattractive. When this happens, size, function, and cost become mutually antagonistic and it is often necessary to trade performance for size and cost: for example, many commercial devices for the cell phone market are remarkably small and inexpensive, but they are rarely capable of performing adequately when exposed to the temperatures, mechanical shock, and vibration requirements associated with downhole drilling tools. Devices having the desired technical performance are often too large for the physical volume available. This usually forces the designer to spend engineering effort to circumvent the limitations of a small commercial part when an unacceptably larger part, or an unacceptably costly part, is readily available and better suited for the task.

### Case Study: Design of a DC-GFCI

An industrial R&D project required transmission of approximately 1 kW of DC power to a remote load over hundreds of feet of relatively small-diameter cable. Line resistance was not negligible in these circumstances, which made it necessary to use high-voltage DC (200 VDC) to mitigate the effect of line voltage variations at the remote load. The high-voltage source arrangement satisfied the project's technical demands, but operator safety considerations made a GFCI with low trip current (5 mA) a system imperative. While AC (power line) GFCI devices are readily available from commercial sources, they are not useful in a DC application. DC-GFCI devices are not commonly available, and those few that were identified were too large and costly for the project. Accordingly, it became necessary to develop a small autonomous DC-GFCI device to continuously monitor current through the high-voltage DC conductors and to disable the high-voltage supply and all other processes if the difference current in the HV power conductors exceeded 5 mA. The DC-GFCI device had also had to reside in a sealed HV cabinet where it had to operate with high accuracy and reliability over a wide temperature range. This inaccessible location made device calibration difficult. Furthermore, the device was to be galvanically isolated from other system elements so a failure in the high-voltage circuit would not damage other system elements.

This was a nontrivial problem. Safety considerations made it undesirable to measure current by shunt resistors in direct contact with the high-voltage circuit conductors. An additional complication was the small (5 mA) trip current threshold, which insured the currents in the high-side and low-side high-voltage conductors would be within 5 mA of each other at all times (i.e., a differential current greater than 5 mA must initiate a trip). Because the HV circuit could supply up to 5 Amperes, the trip limit threshold was ±0.1% (1000 ppm) of maximum current. Although not immediately obvious, this requirement means a true 10-bit system would be needed simply to guarantee the ability to resolve the trip current limit if conductor currents were measured directly. Accurate measurement, using the normal engineering 10:1 rule of thumb, would require a true 14-bit or better digitizer.

A solution which did not require direct contact with the high-side and low-side HV conductors was clearly necessary. Magnetoresistive and Hall-effect devices were considered but rejected for lack of adequate sensitivity. Of the few commercially available noncontact DC-responding current measurement devices, the most sensitive was a toroidal magnetic element with associated electronics normally used to measure current on individual conductors in aerospace applications. This sensor produced a nominal output of ±5.00 VDC with a full-scale input current rating of 0 to ±50 mA, for a nominal gain of 100 mV/mA. As used in the DC-GFCI design, both high-side and low-side HV conductors pass through the toroid's annulus (Figure 10.21), along with a much smaller single-turn calibration winding (to be discussed later). In this arrangement, the sensor responds only to the difference current between high-side and low-side conductors. During normal operation, current in high-side and low-side conductors will be identical, which should result in zero net magnetic flux and zero output voltage. The sensor output, however, generally has a DC offset voltage and zero difference current does not produce zero sensor output voltage in a realizable sensor.

Laboratory work over temperature demonstrated sensors of this type were highly linear, even near zero differential current, but the specified output temperature coefficient was significant in the context of the general design problem ( ±0.04% of full scale per degree Celsius, or ±2 mV/°C). The sensor exhibited triangular output ripple at a frequency of several kHz, with output ripple specified to be a maximum of 2% of full scale RMS (100 mV RMS). Measured sensor gains were somewhat less than specified (typically between 90–95 mV/mA rather than 100 mV/mA). Measured DC offsets were generally in the range of ±100 mV. Gain and offset were found to be temperature dependent, a finding which was not unexpected.

The circuit solution used a unity gain, noninverting, third-order Bessel lowpass filter with 8.3 Hz corner frequency to attenuate the triangular output ripple. A precision absolute value circuit and hard limiter converted the sensor signal to a limited unipolar signal suitable for A/D conversion, thereby providing magnitude of the differential current through the sensor (Figure 10.22). A precision Schmitt trigger was used to provide signal polarity information, thereby giving software all information necessary to determine sign and magnitude of the sensor signal. These, and other circuit details, were very successful in conditioning the sensor output for use with a unipolar A/D converter. However, two difficult issues remained—how to treat the sensor's temperature-dependent offset and gain variations. Simply put, the sensor's

**Figure 10.21:**
A photo of the toroidal sensor used in the DC-GFCI. The DC-GFCI uses a commercial DC current sensor appropriate for the current levels involved. Connections to the sensor module are seen at the barrier strip at the left. Three wires are seen emerging from the sensor's central annulus. The heavy wires curving to the reader's left are the two current conductors connected to the high-voltage power supply. The smaller diameter single wire curving to the reader's right is the single-turn calibration winding. *Used with permission of The Charles Machine Works, Inc.—Photo by Mike Gard.*

offset and gain variations contributed unacceptable error to the differential current determination that was essential to the DC-GFCI application.

The solution to the offset and gain variation problems now will be explored in detail. It is an instructive case study, because it contains all elements of a typical embedded system application. It also provides an excellent introduction to the subject of built-in self-test (BIST), which follows.

The sensor's demonstrated linearity is the heart of a direct self-calibration technique which does not require the temperature measurements and correction look-up tables often used to deal with similar problems. Latent in the conventional calibration and look-up table correction is the assumption (often unwarranted) that sensor offset and gain are always the same at a given temperature. Any error in the calibration files (or an unknown shift in the device's operating characteristic produced by a variety of causes) actually makes the situation worse by introducing additional error into a reading through what should be the correction step. The technique discussed in the follow paragraphs treats sensor offset and gain as bounded unknowns at any given temperature. Offset and gain are determined directly each time a measurement is made. This virtually eliminates the risk of corrupted or out-of-date calibration data.

The approach is simple. When the high-voltage supply is off, there is zero current in both of the high-voltage conductors. Any voltage appearing on the sensor's analog channel must be interpreted as DC offset. Offset is assumed to be constant for the duration of the measurements which follow, because they require a few seconds at most. Offset voltage is not read directly at zero current because the signal is very small and is subject to corruption by system noise. Offset is determined another way.

Gain and offset were determined using two known currents through a single-turn calibration winding, which also passes through the sensor annulus. The heart of the calibration circuit is a low dropout voltage regulator known to have very good temperature stability and an

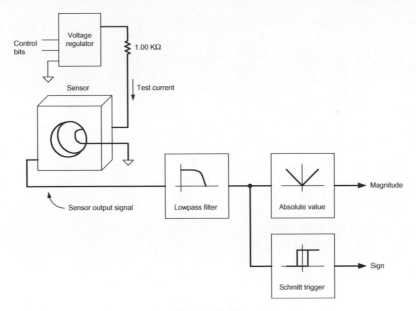

**Figure 10.22:**
Block diagram of the DC-GFCI electronics with self-calibration current loop. This figure illustrates essential functions of the DC-GFCI electronics package. Current conductors from the high-voltage supply are not shown. The test current is injected only during the self-calibration operation described in the text. *Used with permission of The Charles Machine Works, Inc.*

external feedback pin used to determine output voltage. The regulator output is enabled by the first of two control bits only when sensor calibration is desired—otherwise, the regulator output is inhibited and the calibration winding contributes no net flux to the sensor reading. The regulator's enable pin is pulled down to circuit common to insure the regulator's output will be disabled unless it is explicitly enabled.

During the calibration process, the second control bit is used to select one of two possible taps in the resistive divider which establishes the regulator's output feedback loop. Feedback resistors are standard ±1% tolerance devices with lowest available temperature coefficients. The tap voltage applied to the feedback pin of the regulator is determined by the state of the second control bit. The analog switch channel carries no current, so voltage drop across the analog switch is negligible. Resistors in the series feedback string are selected to produce two output voltages in accordance with the tap selection: regulator output will be 5.00 or 10.00 VDC when the regulator output is enabled.

The regulator output is connected to one end of a precision 1.00 K resistor with low temperature coefficient. The other end of the precision 1.00 K resistor is connected to circuit common by the wire passing through the current sensor annulus. Thus, the calibration circuit drives a controlled DC current of either 5.00 or 10.00 mA in the wire through the annulus. Note also that these current values were selected carefully to be equal to the trip current (5 mA) and twice the trip current, which insures the calibration is done at signal levels near the trip current threshold.

The sensor is known to be linear in this current range, so calculations are the same as finding the equation of a straight line $y = kx + b$ with two known points, which is straightforward. The sensor's output voltage, $V_{sensor}$, may be represented by the equation

$$V_{sensor} = kI_{sensor} + V_{offset}$$

where $k$ is the sensitivity, or gain constant, given in mV/mA and the offset voltage, $V_{offset}$, is a static (constant) DC voltage error for the duration of the calibration. Note this offset term includes DC offset voltages from all sources in the analog signal chain.

Let $V_{sensor}(10)$ and $V_{sensor}(5)$ represent the DC-GFCI circuit assembly's output voltages in response to the currents $I_{sensor}(10)$ and $I_{sensor}(5)$ which are 10.00 and 5.00 mA, respectively. Gain is simple to calculate:

$$k = \frac{\Delta V_{sensor}}{\Delta I_{sensor}} = \frac{V_{sensor}(10) - V_{sensor}(5)}{I_{sensor}(10) - I_{sensor}(5)}$$

Offset is likewise simple to calculate:

$$V_{offset} = 2V_{sensor}(5) - V_{sensor}(10)$$

Once these constants are known, the values are stored and the calibration circuit's voltage regulator is disabled (calibration current goes to zero), the high-voltage supply output is enabled and there is electrical current flow through the high-voltage lines. A sensor voltage reading taken with power applied to the load is attributable to the current difference between the high-side and low-side high-voltage conductors and is given by

$$I_{diffsensor} = \frac{V_{sensor} - V_{offset}}{k}$$

While the theory of this case study is straightforward and has been proven reliable in practice, the embedded system designer must address other concerns hidden in the overall measurement problem. These include settling time of the sensor and analog chain, quantization effects of the A/D converter used to make the sensor voltage readings (the mathematics above made no assumptions about quantization), system noise impact on the calibration readings, and the effect of finite precision arithmetic in the calculations. Noise can be controlled by averaging a number of consecutive samples, and it is always wise to establish a qualification range of reasonable values for the gain and offset results. The DC-GFCI is a safety circuit, and the current difference calculation must be done with high confidence in the result.

## Built-in self-test

The case study just completed is a good backdrop for discussing Built-in-Self-Test (BIST)—or, simply, self-test—the final topic of this chapter. Embedded systems often are deployed in remote locations and often are meant for unattended and largely autonomous operation for prolonged periods of time. These circumstances make it highly desirable to provide a means by which the system can execute a self-test procedure to help detect subsystem degradation of any significant subsystems or other issues indicative of pending failure or unacceptable performance. Good self-diagnostics allow the system to flag questionable data or communicate a need for operator or field technician intervention to the central data collection unit.

Embedded systems are good candidates for self-test implementations because they ordinarily have the processing power, memory, data conversion, and computational capability needed for reasonable self-test. The key concept here is "reasonable." It is all too easy for BIST applications to take on a life and meaning of their own, with the resulting complexity becoming a detrimental influence on an otherwise lean and fully functional design. *You cannot test for every possible fault. Keep it simple. Don't overdo it*.

Consider, for example, the calibration step of the DC-GFCI just discussed. The self-calibration circuit's total burden to the system design is a voltage regulator, an analog switch, four essential resistors (six, counting pulldown resistors on the two control lines), a short length of wire, and one or two capacitors. Resistors used to implement the voltage regulator circuit are standard $\pm 1\%$ commercial parts. A careful worst case error analysis and subsequent laboratory measurements suggest the results are good to within a few percent of true values. Experience with this circuit over a period of years has been excellent. The design is simple, straightforward, and good enough.

The reader may reasonably ask, "But how do we know the calibration circuit remains accurate over time?" A truly accurate BIST implementation should be able to determine proper operation without relying on external calibration. At this point, designers will make plans to sample not only the sensor's output signal but also the voltage regulator's 5.00 and 10.00 VDC output signals. The system's A/D converter must have an additional analog channel available to measure the voltage regulator output. It is possible the voltage regulator outputs are correct but the 1.00 K precision resistor used to establish the calibration current could be degraded by aging and heat-related effects, with the resulting resistance shift producing a current error. The A/D converter used to read the voltage regulator outputs can be only as accurate as the converter's reference voltage—how will we determine the quality of the A/D reference voltage? Is the toroidal current sensor response really linear? How do we calibrate the calibrator?

It is at this point in the design cycle the wise designer will remember the admonition "Keep it simple." As mentioned in the discussion of the DC-GFCI, operating software contains a simple test to insure calculated gain and offset values lie within a reasonable range of values—these are determined after learning about typical part performance through operating experience. If calculated gain and offset values fall outside a reasonable range based on analysis and operating experience, that is good reason to recalculate the values or to signal a calibration error—otherwise, the calculated values should be accepted. Resistors rarely fail unless they are overheated or physically damaged. Voltage regulators rarely fail unless they are asked to deliver an abnormally high amount of current and subsequently overheat. An analog switch is a robust part and proper circuit architecture makes the regulator's output voltage almost entirely independent of analog switch channel impedance. If all devices are operated comfortably within their ratings, there is little to fear. Keep it simple. Don't overdo it. "The first philosophers met together and dedicated in the temple of Apollo the first fruits of their wisdom, the famous inscriptions 'Know Thyself', and 'Everything in Moderation'."[25]

## Acknowledgment

Our thanks to Dr. Steven Dyer for reading and reviewing this chapter.

## References

[1]  C.A. Lindquist, Active Network Design, Steward & Sons, Long Beach, CA, 1977, pp. 649–721

[2]  A.B. Williams, Electronic Filter Design Handbook, McGraw-Hill, New York, NY, 1981.

[3]  Basics of Linear Fixed Resistors, Vishay Beyschlag Application Note, Document Number 28771, Revision November 11, 2008, available through <www.vishay.com>.

[4]  P. Fisker, Thick Film Resistors: Failure Mechanisms and Reliability, BB Electronics A/S, August 2011.

[5]  Y. Hernik, T. Troianello, Foil current sense resistors and their applications, IEEE Instrum. Meas. Mag. 15 (6) (2012) 28–30.

[6]  H.W. Ott, Noise Reduction Techniques in Electronics Systems, second ed., John Wiley & Sons, New York, NY, 1988. p. 228.

[7]  www1.eere.energy.gov/manufacturing/tech_assistance/pdfs/10097517.pdf.

[8]  D. Brooks, Signal Integrity Issues and Printed Circuit Board Design, PTR Prentice Hall, Upper Saddle River, NJ, 2003.

[9]  M.I. Montrose, Printed Circuit Board Design Techniques for EMC Compliance: A Handbook for Designers, second ed., IEEE Press, The Institute of Electrical and Electronics Engineers, New York, 2000.

[10]  IEEE Recommended Practice for Powering and Grounding Sensitive Electronic Equipment, IEEE Std 1100–1992, p. 1,256, 1992.

[11]  T. Williams, EMC for Product Designers: Meeting the European Directive, third ed., Newnes, 2001.

[12]  K. Paton, The test, usage and maintenance of power switching subsystems, IEEE Instrum. Meas. Mag. 15 (4) (2012) 8–14.

[13]  K. Fowler, M.F. Gard, Practical issues for installing instrumentation outdoors, part 1, IEEE Instrum. Meas. Mag. 14 (6) (2011) 32–38.

[14]  K. Fowler, M.F. Gard, Practical issues for installing instrumentation outdoors, part 2, IEEE Instrum. Meas. Mag. 15 (1) (2012) 33–41.

[15]  MIL-HDBK-217F, Reliability Prediction of Electronic Equipment, US Department of Defense, Washington, DC, January 2, 1990.

[16]  MTBF Prediction Workbook (ASQ Reliability Review, Vol. 24, No. 1, pp. 18–23, March 2004, with updates including May 14, 2008, update to link MIL-HDBK-217 Update Survey, http://home.comcast.net/~pstlarry/MH217F1.htm.

[17]  Z. Watral, A. Michalski, Selected problems of power sources for wireless sensors networks, IEEE Instrum. Meas. Mag. 16 (1) (2013) 37—43.

[18]  E. Bogatin, Signal Integrity: Simplified, PTR Prentice Hall, Upper Saddle River, NJ, 2004.

[19]  H.W. Johnson, M. Graham, High-Speed Signal Propagation: Advanced Black Magic, PTR Prentice Hall, Englewood Cliffs, NJ, 2003.

[20]  H.W. Johnson, M. Graham, High-Speed Digital Design: A Handbook of Black Magic, PTR Prentice Hall, Englewood Cliffs, NJ, 1993.

[21]  R. Morrison, Grounding and Shielding Techniques, fourth ed., Wiley, New York, NY, 1998.

[22]  MIL-STD-461E, Requirements for the Control of Electromagnetic Interference Characteristics of Subsystems and Equipment, US Department of Defense, Washington, DC, August 20, 1999.

[23]  S. Cole, Minimizing radiated noise in magnetically sensitive applications, IEEE Instrum Meas Mag 16 (3) (2013) 12—16.

[24]  MIL-STD-810F, Test Method Standard for Environmental Engineering Considerations and Laboratory Tests, US Department of Defense, Washington, DC, January 1, 2000.

[25]  Socrates, attributed by various sources including the citation <http://www.stedwards.oxon.sch.uk/Philosophy_introduction.htm>.

## CHAPTER 11

# Software Design and Development

**Geoff Patch**
*CEA Technologies Pty. Ltd.*

## Chapter Outline

Developing and Managing Embedded Systems and Products.
DOI: http://dx.doi.org/10.1016/B978-0-12-405879-8.00011-8

We are surrounded in our daily lives by embedded systems. Modern telephones, automobiles, domestic appliances, and entertainment devices all have one thing in common—large amounts of complex software running on a processor, or processors, embedded in the device. That software implements much of the functionality of the device.

The design and development of software for these systems is a specialized discipline requiring knowledge of constraints and techniques that are not relevant to more conventional areas of software. While the rapid proliferation of embedded systems is being reflected in the gradual introduction of relevant courses into university degrees, in many ways, the efficient production of high-quality embedded software is still an art rather than a science.

This chapter will examine software design and development in the embedded system domain. It explains the processes, tools, and techniques that may be used to overcome the difficulties of development in this environment.

---

**Please remember**

All examples and case studies in this chapter serve only to indicate what you might encounter. Hopefully these examples and case studies will energize your thought processes; your estimates and calculations may vary from what you find here.

---

### Case Study: Loading ammunition

Embedded systems are important. I am a competitive target shooter. To achieve the best possible accuracy during matches I manufacture my own pistol and rifle ammunition. Part of this process involves measuring out the charges of propellant powder that drive the projectiles down the range. These powder charges generate pressures of tens of thousands of pounds per square inch during firing, and propel the projectiles at velocities of thousands of feet per second.

The charges are for historical reasons measured in an old-fashioned unit known as the grain. There are 437.5 grains in an ounce. Typical powder charges are surprisingly small. They must be measured very accurately to achieve consistently high performance from the ammunition. For example, a typical powder charge for a pistol round might consist of 3.5 grains of powder, measured to within one-tenth of a grain.

**Figure 11.1:**
A little less than one-hundredth of an ounce of propellant powder resting on an electronic powder scale. This is a safe amount in the intended bullet, but two-hundredths of an ounce of the same powder would be highly dangerous. *(copyright 2014 by Geoff Patch. All rights reserved. Used with permission.)*

What makes this interesting is that while a charge of 3.5 grains might be perfectly safe, a charge of 4.5 grains might be enough to cause dangerously high pressures during firing that could damage the firearm and perhaps injure me.

To ensure that I measure my powder correctly and safely, I use a high precision digital electronic powder scale specifically designed for this purpose. It features an embedded control system that performs the measurements and drives a user interface consisting of several buttons and a digital display with a variety of readouts. I absolutely rely on the correct operation of this little embedded system, because if it fails me at best I may damage my firearm and at worst I could be seriously injured or killed (Figure 11.1).

***Embedded systems are really important.***

## Distinguishing characteristics

Before contemplating how to approach the development of embedded software, you need to understand what actually distinguishes embedded software from software developed for either the desktop or the server. Understanding the difference between embedded and other domains is extremely important. The distinguishing factors are not trivial; in many cases, they are not obvious. Whether obvious or not, these factors are critical drivers in embedded software development programs; appreciating their impact can differentiate eventual success from failure of a development program.

Not every embedded system has all of these characteristics, of course, and some nonembedded systems will share these characteristics. It would be surprising, however, to find an embedded system that doesn't embody some of the features from the following list.

### Minimal operating system support

A full featured desktop operating system (OS), such as Linux, provides an enormously powerful virtual machine for the desktop software developer to develop against. In most cases, the physical hardware underlying the OS platform is almost entirely abstracted away by large Application Programming Interfaces (APIs) that hide the complexities of dealing with the hardware on which the application is running. Beyond hardware abstraction, desktop OSs provide other abstractions such as virtual memory, multiprocessing, and threading that provide the illusion of multiple independent tasks operating concurrently. In fact these software processes sequentially share a single processor or a small number of cores on a multicore CPU.

The features of a desktop OS provides the desktop developer with enormous power and flexibility, but they come at a cost which includes the amount of storage required to store the OS image and the amount of RAM required to actually run the system. At the time of writing, large desktop OSs may require gigabytes of storage and RAM to operate successfully. As well as their storage requirements, such OSs also consume substantial amounts of processing power to provide their services.

In many cases, it is not feasible to use one of these large-scale OSs in an embedded system because it would make the device under development technically or commercially infeasible. In these circumstances, the developer is faced with two options.

For small systems, the best choice may be to have no OS at all. The embedded software may consist of a program that runs a control loop that polls devices for activity, or responds to flags being set by interrupt handlers. This solution is simple, inexpensive, and is often the best choice for small high-volume applications where minimizing cost is a key driver.

For more complex applications, particularly those where multiple activities must be conducted in parallel, the best choice is a Real Time Operating System (RTOS). An RTOS is a small, efficient, and fast OS specifically targeted at the embedded system domain and designed to meet performance deadlines. RTOSs provide many of the useful capabilities of the desktop OS such as multithreading, while avoiding the plethora of nice-to-have features that consume resources without adding core capabilities to the system.

RTOSs will be discussed in more detail later in this chapter.

## Real-time requirements

A real-time software application must meet all processing deadlines. Ganssle and Barr define real time as "having timeliness requirements, typically in the form of deadlines that can't be missed" [1]. *A system fails if it does not meet either the logical or the timing correctness properties of the system.* While not universal, timing constraints in embedded systems are extremely common.

A common mistake is to believe that a system with real-time requirements must be "fast." A system with real-time requirements simply has to deliver numerically correct computational results while meeting its temporal constraints. A correct numerical result delivered late is a failure.

Determining real-time requirements in an embedded system depends entirely on the nature of the system. In radar systems, timing constraints are frequently specified in units of nanoseconds. In an automobile engine management system, timing constraints may be specified in microseconds or milliseconds. While it may be stretching the spirit of the definition, it's entirely possible that a system may have timing constraints specified in seconds or minutes or even days.

However tight the constraints may be, the addition of timing requirements to a system is frequently a key discriminator between embedded systems and those in other domains. Spreadsheets, word processors, and web browsers work just as fast as they can; the result is delivered when it's available. Users may wish for faster performance from such systems, but a complaint to a spreadsheet vendor that the calculations are performed slowly would most likely fall on deaf ears.

## Real world sensor and actuator interfaces

Real-world interactions drive real-time requirements in an embedded system. This interaction senses some aspect of the external environment around the system; makes decisions based on the information received from the sensors or inputs; then drives change in the external environment through actuators or other control mechanisms.

The pattern of real-world sensing and control, subject to real-time constraints, is extremely common in the world of embedded systems. Moreover, it is one of the most difficult areas to understand fully and implement correctly. The detailed mathematical study of this topic falls under the subject of control systems found in Chapter 8.

## Resource constrained

The computational capacity of most embedded systems is severely constrained. The processor environment is typically much less powerful than the current state of the art in the desktop environment.

This isn't arbitrary. Nothing is free. Heavy duty computational power typically requires large processors in large physical packages, which require large amounts of electrical power. Thus begins a cascade effect on design: large amounts of electrical power require the installation of large power supplies; large amounts of power result in the generation of large amounts of heat, which necessitate heat sinks, fans, and other infrastructure to move the heat away from the processor and peripherals. All these items add complexity, weight, and volume to the embedded system.

As well as limiting the raw computational power of the processor, the physical environment will constrain many other aspects of the software. The amount of nonvolatile memory available to store the executable image often limits the size of the program. Both the amount of RAM available and the speed at which it operates will limit data structures. The availability of interfaces with the appropriate performance constraints will limit the speed of communication with other systems.

Factors that need to be considered in this area are size, weight, power, heat dissipation, and resistance to shock and vibration. (Chapters 8–10 go into the trade-offs for these factors and parameters.) In general, these factors will significantly influence the amount of processing power available to an embedded system software developer.

---

### Case Study: The constrained environment

The constrained environment for hardware provided by most embedded systems *must* be considered during software design and development. The software for embedded systems frequently develops in parallel with the hardware. You first develop and test algorithms in isolation on desktop systems due to the ease of development in a feature-rich environment for software development.

In a number of cases, I have seen systems fail at the integration stage due to algorithms working perfectly in high-performance desktop workstations, but failing to meet their timing requirements when embedded into their target hardware.

The fault in all these cases was not the limitations of the embedded environment. That was a well-known quantity throughout their development. The fault was that the engineers ignored the realities of the resource constraints of their target systems, resulting in the need to perform substantial rework at great cost. **These lessons are both expensive and unnecessary.**

---

### *Single purpose*

Embedded systems often have fixed and single purposes. An engine control module only manages the engine; it has no other function. An industrial process control system will be designed to control a well-defined set of industrial processes. It won't have to download and play MP3 files off the Internet while maintaining the membership database of the operator's basketball club.

Contrast the embedded system's single-mindedness with a desktop system, which is inherently general and multipurpose. Users may install software developed by any third party and at least, in theory, they may run any number of programs simultaneously in any combination.

Educational institutions universally teach the principles involved in developing such general-purpose software as being the correct paradigm to follow for successful application development. This is true outside the embedded systems arena, but following this paradigm can have unexpected and even harmful consequences when used in the development of embedded systems.

For example, dynamic memory allocation techniques allow developers to allocate and deallocate blocks of memory as the load on their application changes. This is advantageous in a general-purpose system performing many different tasks simultaneously because it allows an application to use only the amount of memory it needs, which frees unused memory so that the OS can allocate it efficiently to other tasks. Unfortunately, general-purpose algorithms for memory allocation that work well on the desktop suffer from problems such as temporal nondeterminism and memory fragmentation that can make them totally unsuited for use in an embedded environment. If you are adding rows of cells to the bottom of a spreadsheet, you are unlikely to notice a delay of 10, 20, 30, or even 100 s of milliseconds every now and then when the spreadsheet has to allocate a block of memory to accommodate the new data. On the other hand, if the fuel injection control computer in your vehicle paused for 50 milliseconds to allocate some new memory as you accelerated and failed to supply fuel to the engine for around five engine revolutions, then that is something you would almost certainly notice.

### Long life cycle

Embedded systems, particularly military and large-scale industrial control systems, have life cycles that far exceed those of desktop systems. Some may run as long as 30 or even 50 years! The impact of this long life cycle is that long-term maintainability of embedded system software is extremely important.

If an application such as a game has a life cycle of a year or two, it would be unprofitable to allocate extensive resources to ensuring the long-term maintainability of the software. On the other hand, if an embedded system has a life cycle of 20 years, then it's likely that multiple generations of developers will support and maintain it over that extended time frame.

This extended product life cycle has major implications for the nature of embedded software. From the very start of the project, extensive effort must be put into documentation such as code comments, system interface specifications, and system operational models.

> *If the system is not carefully designed, developed, and documented so that it can be successfully maintained throughout its working life, then it will fail over the long term.*

## Reliability and design correctness

In many cases, the failure of a software system will result in nothing more than some degree of inconvenience to the operator. This is true even with embedded systems such as mobile phones and entertainment devices.

Many embedded systems, however, operate in mission-critical and safety-critical environments. The lives of hundreds of sailors may one day depend on the correct operation of their fire control radar as it guides a missile in the defense of their ship. The continuing heartbeat of a person wearing a heart pacemaker depends entirely on the correct operation of the software and electronics embedded into that device.

Under these circumstances, instead of mere inconvenience, the failure of a system or a design flaw may result in death, injury, economic loss, or environmental damage. These adverse consequences may affect either individuals or entire nations. Therefore, the emphasis in the design and development of such systems must be quality, correctness, and reliability above all else.

## Safety

Operator health and safety in the world of desktop software development is typically limited to consideration of aspects such as choosing screen layouts and character fonts that minimize eyestrain. This is in sharp contrast to the world of embedded system development, where embedded devices frequently control powerful machines and processes at incredible speeds.

Embedded software controlling powerful machinery has the capability to injure and kill people and in notable cases it has done exactly that. Producing software that is reliable is a common design goal in all domains, but producing safe software is critical in the area of embedded systems.

---

### Case Study: Therac-25

When desktop systems go wrong, the usual result is some inconvenience. When embedded systems go wrong, the result may be inconvenience if you're lucky or death you're not.

One of the most well-known and well-studied cases involving loss of life caused by the failure of an embedded system was a device called the Therac-25. The Therac-25 was a radiation therapy machine intended to treat cancer by directing beams of radiation at the tumor site.

The device contained a control system running on a PDP-11 minicomputer with an operator interface connected to a VT-100 terminal.

The embedded control software was poorly implemented and contained a number of defects. As well as the software defects, there were a number of system level defects in the device such as the absence of mechanical safety interlocks in the components responsible for the formation and deployment of the radiation beam.

The combination of software and hardware design flaws and defects meant that under certain rare circumstances, the patient could be subjected to beams of radiation that were around 100 times as strong as the intended dose. Despite the complex nature of the interactions required to trigger the defect, the circumstances occurred on six occasions, on each occasion resulting in the patient receiving a massive overdose of radiation. All of the patients suffered severe injuries involving radiation burns and radiation poisoning, and three of the patients died as a result of the injuries they received.

A detailed study of the Therac-25 case may be found in Ref. [2].

## Standards and certification

Most embedded systems must comply with domain specific standards and achieve certification by relevant authorities. This is particularly true in areas where the incorrect operation of the system may result in harm to people or the environment. Conversely, most desktop software is not developed against particular standards or subject to certification by any regulatory body.

Compliance with external standards has many costs. For example, developers must know the required standards and train to produce software that conforms to the standards. Work products, such as code and documentation, must then be externally audited to ensure compliance.

## Cost

*Embedded systems are expensive to build and maintain!* Implantable medical devices take 5–7 years and between US$12 and 50 million to develop. Industrial and manufacturing test stations may cost millions and support can run into hundreds of thousands of US dollars each year. An automobile recall to simply reprogram an engine control module may only take 15 min per car but at a rate of US$50 per hour for 1 million cars that amounts to US$12.5 million; and that does not include the nonrecurring engineering cost to develop and test the new software!

Companies building embedded systems must employ, train, and then retain highly skilled staff with esoteric expertise. High-quality development tools for embedded software, such as compilers and debuggers, can cost orders of magnitude more than their desktop equivalents. The development of the system may require expensive electronic support tools, such as logic analyzers and spectrum analyzers. The development processes used, and the resulting work products, may be subject to standards compliance audits that increase the time and cost required to perform the development.

### Product volume

Embedded systems usually minimize the cost of the processor environment in large volume manufacturing to save costs and to maximize profits, which helps the product to be commercially competitive. Minimizing the processor environment directly affects the software that operates it. Conversely, the costs of packaging and delivering desktop software, however, are generally small and have no impact on the development of the software.

This cost reduction in embedded systems usually results in the use of less capable devices in the processor environment. You must carefully consider this constraint in the design and implementation of the software.

### Specialized knowledge

Developing software in any environment requires a combination of software development expertise and domain-specific knowledge. If you are going to produce an online shopping web site, you will need web programming and database skills, combined with some knowledge of retail activities, financial transaction processing, tax regulations, and various concerns specific to the business domain.

The same thing applies in the development of embedded systems, except that the domain knowledge required is usually much more specialized and complex than that found in most desktop systems. Esoteric mathematical techniques are commonplace in the world of embedded systems. For example, the fast Fourier transform (FFT) is a digital implementation of a method that extracts frequency information from time-sampled data. It is seldom seen in desktop applications, but may be found in numerous embedded systems, ranging from consumer audio to advanced radar systems.

Similarly, the Kalman filter is a complex algorithm originally devised for tracking satellites. While rare on the desktop, it has become commonplace in embedded systems. It is the basic algorithm for a vast variety of applications such as the Global Positioning System (GPS), Air Traffic Control, and advanced weapon control.

These techniques, along with others such as control system design, require years of advanced study to master. The people who have such advanced skills are in high demand.

## Security

The issue of system security has historically received little attention in the embedded system space. Surprisingly, this was actually a reasonable policy to follow in most cases because embedded systems were typically isolated from potential sources of attack. An engine management system, isolated in the engine bay of a motor vehicle and connected only to the engine, provides few opportunities for compromise by a hostile actor.

Increases in processor power, however, have been accompanied by increases in connectivity throughout the world of computing; now embedded systems have become a part of this trend. As a consequence, previously isolated subsystems tended to be designed so that they could be connected to each other within a larger system. In the automotive world, the widespread adoption of the Controller Area Network (CAN) bus provided a reliable and inexpensive means for linking together the disparate control and monitoring systems within a motor vehicle [3]. This development provided many benefits, but it also meant that the engine management system no longer remained isolated.

The trend toward increasing connectivity has continued. Many embedded systems now have Internet connectivity, which is often implemented to provide remote system diagnostic and upgrade capabilities. Unfortunately, these enhanced capabilities carry with them the risks associated with exposing the components of the system to unauthorized external assaults, i.e., hacking.

The security of embedded systems was once an afterthought, if it was thought of at all. However, embedded systems play a massive part in the operation of our modern world. Many of these devices now have their presence exposed on the Internet or through other means. This means that security must be designed into the systems from the very start if they are to operate robustly in this unpredictable environment. Chapter 12 provides more insight into security issues.

## The framework for developing embedded software

Human beings are inherently orderly creatures. From our earliest years, we are happiest and most prosperous when provided with clear rules and boundaries that are presented in a consistent and coherent manner and that provide tangible benefits as we grow. This desire for order and regularity is reflected in the laws and regulations that we implement to govern our lives. While restricting our actions in many areas, an orderly framework of laws allows a group of disparate individuals to come together as a society and work harmoniously together for the good of all.

The benefits of defining and then working within a commonly understood framework seem obvious, regardless of whether we are referring to an entire society or to a small business unit within an engineering organization. And yet the common experience of software developers in general, whether embedded developers or otherwise, is that their development activities are conducted in an environment that is remarkably unconstrained by rules and regulations.

This lack of structure and constraint is all the more remarkable when software development is contrasted with professional activities in other fields such as law, medicine, and accounting. For example, an accountant working for a corporation will work according to a very clearly defined set of corporate accounting procedures. These procedures will have been derived from the common corporate governance legislation that all companies are subject to within a country, with the result that there is a high degree of standardization within the profession and general agreement on how accounting should be done.

Similar comments could be made about professions such as law and medicine, but the situation with software development is still in a state of flux. Despite activities such as the Software Engineering Body of Knowledge (SWEBOK), there is no common understanding within the profession about how to do good software development [4]. Every organization produces software in a manner that is different to every other organization. Even different departments within the same company may work in completely different ways to produce similar products.

This situation is attributable in large part to the newness of the profession. People have been building bridges for thousands of years. Apart from the occasional mishap such as the Tacoma Narrows collapse, bridge building is a well-understood problem with well-understood and reliable solutions [5]. Large-scale software development, on the other hand, has only been widely conducted over a period of a little more than 40 years, so the common understanding of how to develop good software is still maturing. The growth of this understanding is made more difficult by the incredibly broad variety of applications software and the resistance of many practitioners to the very concepts of regulation and standardization.

In most cases, this resistance springs from a misguided belief that working within such a framework will crush creativity, eliminate elegance, and turn satisfying creative work into soul-destroying production line drudgery. The fact is that nothing could be further from the truth. Written English, for example, is produced within the context of very tight constraints regarding spelling, grammar, and punctuation. Yet within those constraints for written language, you can produce anything from a sonnet to a software user manual. In fact, the widely understood rules of written language do not hinder creativity or elegance, they encourage it.

This section aims to provide guidance for developing and implementing a high-level framework for software within embedded systems. This framework is not a complex or an unproven theoretical construct with burdensome overheads. It is a simple set of ideas that have been refined over many years of engineering practice. It has reliably and repeatedly produced high-quality embedded systems with minimum overhead.

## Processes and standards

Many organizations engaged in the development of embedded systems don't know what they're doing. What I mean by this is that they don't understand the mechanisms that conceive products and then deliver successful products. The result of this lack of understanding is that work is performed on a "cottage industry" basis and difficulties abound. Each project is approached differently. If more than one engineer is involved in a project, they will use different tools and different techniques. Communication between staff may be difficult, components may not integrate well together, and heroic efforts are often required of individuals to achieve deadlines.

Given a fair amount of luck a project that is run like this will succeed, but nobody will be quite sure why it was a success. Without the addition of luck, there will be just as much doubt as to why the effort was a failure.

A number of steps are required to minimize the requirement for large helpings of luck in successful development programs. First, you need to develop an understanding of operations. This involves studying and reflecting upon how development is performed within your organization, determining what works and what doesn't work, and then documenting these observations in a set of operational processes. Second, once these documents have been developed, they should be promulgated within the organization and adopted as the corporate standard to be used by all engineers.

---

### Case Study: Corporate coding style

While it is easy to talk about developing and implementing processes, it's not so easy to do. People are not machines and people who are set in their ways are not as easily reprogrammed as computers are.

Developing, implementing, and enforcing consistent processes within a group requires a high degree of technical skill to distinguish good techniques from bad and to be able to document those techniques successfully. Along with the technical skill, though, is a requirement for substantial management skill involving sensitivity, awareness of human nature, and a willingness to compromise combined with determination to ensure that necessary changes are pushed through.

Most engineers working in unconstrained development environments will have strongly held views about how things should be done and it's highly unlikely that they will all have the same views on every subject. A classic example of this is the placement of parentheses within source code written in the C programming language and its derivatives. There are many different ways to place the parentheses within the code; most engineers quickly adopt a style with which they are comfortable. For many years I wrote C code using the formatting convention described by Kernighan and Ritchie in their classic book "The C Programming Language." When I attempted to introduce a code formatting standard within my team, I assumed that all of my staff would happily adopt that formatting convention. To my surprise, I met passionate resistance to this suggestion. To my further surprise, I met passionate resistance from at least one person to every other formatting style that I suggested.

It didn't take me long to realize that everybody on the team viewed their particular formatting style as the best way to lay out code and that they were very unwilling to adopt any other style. This was a difficult situation, as the only options were to abandon the effort to standardize the code layout or enforce a code layout style that would leave at least some members of the team unhappy.

I was convinced of the benefits of adopting a standard corporate coding style, so I chose the latter course. I consulted extensively with each engineer, so that everybody felt that their voice was being heard. *(This is supported in Chapter 2—Ed.)*, and produced a corporate formatting style that looked good and pleased most staff. As expected, there was some unhappiness within the team, but after a few months everybody came to see the benefits of consistent code layout across projects and the arguments were forgotten.

## One size doesn't fit all

Programs for developing embedded systems come in all shapes and sizes. It's very unusual for one project to closely resemble the next project in dimensions such as scope, complexity, schedule, and staff requirements. For this reason in most cases, it's more important for process documents to describe **what (the ultimate goals and objectives)** should be achieved rather than **how (the exact procedures)** it should be achieved.

For example, the level of system review conducted on a small project that occupies one person for two months will be different to that conducted on a project that occupies six people for three years. Applying a process that is suitable for one of those projects to the other project could be wildly inappropriate. What is appropriate, though, is to have a review process that specifies in part, that regardless of the size of a project, the preliminary design of the system should be documented and externally reviewed prior to the commencement of coding.

Specifying the process in this manner means that the review of the small project could be conducted by a couple of people in a morning, while the large project review might be conducted by ten people over a period of a week. In both cases, the desired business outcome occurs in accordance with the process, with a level of complexity and overhead incurred that is suited to the nature of the project.

While this advice is true in general, there may be cases where you want to be prescriptive, where one size really does have to fit everybody. For example, while some organizations allow projects to use various coding standards, the usual case is for a single standard to be applied at the corporate level. In this case, all code produced within the organization has to comply with the standard, regardless of the nature of the project that it is produced for.

It's also worthwhile considering the nature of the projects that are undertaken by your organization. While there may be substantial differences between individual projects, it's quite likely that most of them will fit into categories such as "small," "medium," and "large," or perhaps "safety-critical" (or mission-critical) versus "nonsafety-critical (or nonmission-critical)." If this is true, then you should generate specific process templates applicable to each of these different categories of project. The result is that at the start of a project, the engineers involved can simply copy the template appropriate to their project, perform some minor tailoring, and have their project process ready to go with minimum effort.

### Process improvement

We live in a rapidly changing world. Change applies to development environments just as much as to any other aspect of life, so you must ensure that corporate processes and their related documentation keep up with evolutionary change. If this does not happen, then either the enforcement of obsolete processes will inhibit necessary and valuable change or changes will occur anyway without the documentation being upgraded so that it eventually becomes useless.

You should examine each recommendation for change to ensure that it is beneficial. In many cases, proposals for process change arise naturally with the introduction of new technology or techniques. *Recognizing and embracing such positive change enhances productivity and enhances staff involvement as they feel that they own the processes.* *(Again, confirmed and supported in Chapter 2—Ed.)*

On the other hand, not all change is good. In particular, you should be aware of the concept of "normalization of deviance," coined by sociologist Diane Vaughan in her book on the Challenger space shuttle disaster. The concept relates to a situation in a process-oriented environment where a small deviation is made from the methods described by a process, with no ill effect [6]. Each time such a process deviation occurs that doesn't lead to a negative result people become more tolerant of that unofficial standard, with the result that actual working methods gradually diverge greatly from the documented processes. It is very likely that at some point, as with the Challenger disaster, the normalization of deviance will produce results that greatly diverge from being harmless to being catastrophic.

Regardless of how it eventuates, work practices will change. You should welcome and incorporate beneficial work practices into the overall process framework while recognizing harmful practices and taking active measures to reject them.

## Process overhead

Complying with corporate processes takes time and effort. In a business environment, this equates to money. Process compliance should cost less than the benefits that are gained from such compliance and if this is not the case then you're not only spinning your wheels, you're also going backward! Therefore, an important aspect of process improvement efforts is studying how processes work and how to modify them, to minimize both the number of processes required for successful operations and the amount of effort required to comply with the processes.

---

### Case Study: Nothing in this life is free

As well as delivering high-quality products, my team wants to ensure the continued quality of our systems as they undergo maintenance and improvement after being fielded. Therefore, my engineering team performs a "last line of defense" review on changes that are applied to software in delivered systems. This review process is known as the Change Control Board (CCB), and its purpose is to ensure that all due diligence has been performed during the design, implementation, and testing of each change. In some cases, we may inspect the actual code itself, but primarily we look for evidence that we have complied with our processes, and that they have been executed professionally. We look for comprehensive evidence of high-quality design, review, and testing activities, with the expectation that these activities will produce high-quality products.

The CCB consists of myself and the four most senior engineers in the group. Depending on the workload, the CCB may meet several times each week. Five staff members meeting for around two hours per week at around $100 per hour for each person means that the CCB is costing the company around $1000 per week.

That's a lot of money, but nothing in this life is free. Is the money well spent? You bet it is! Occasionally the CCB will detect a problem and require more effort to be applied to a work package to fix a defect or improve quality in some way. Each of these problems represents pain that we didn't inflict on our customers, corporate reputation that was not lost, and product recalls or field updates that we didn't have to apply. And ultimately, all of those things represent a lot of money saved for the company.

Ensuring that we maintain a consistently high standard of product quality throughout the life of every delivered product is worth every penny that we spend on it.

---

Unnecessary processes, which don't contribute business value or overly bureaucratic processes with high compliance costs, can actually impede rather than assist progress by wasting staff resources in an unproductive manner.

---

### Case Study: A peer review process needed review

I strongly believe in the benefits of peer review of work products. One of my first process improvement efforts involved the implementation of a process for source code review. The original version of this process involved a great deal of overhead. It described in minute detail how responsibilities were to be allocated during a review, how initial planning meetings should occur, followed by review meetings, followed by washup meetings, how all of these meetings should be scheduled, the nature of the outputs of the review, and much more.

I put a lot of effort into developing this process and I thought it was an excellent piece of work. It was only after I introduced the process and tried to make it operate that I realized it was a complete failure. The process that I had specified was so detailed and onerous that my team would have had to spend far more time reviewing code than writing code to comply with the process. After some initial enthusiasm, the team struggled with the overheads involved and very quickly let me know that we had a problem.

I scrapped the original effort and started from scratch. We obtained a web-based collaboration tool that allows a group of engineers to annotate code listings with comments that are e-mailed to all of the members of the group. The amount of effort required to perform a review using this tool is minimal and the engineers in the team accepted it enthusiastically. Source code reviews using the tool now occur as a matter of course, code quality has improved, and the staff can visibly see the benefits of engaging with the process.

### Process compliance

I have established that developing and implementing processes requires considerable management skill. Maintaining process compliance requires a similar, if not greater, level of management involvement.

No matter how you look at it, process compliance involves extra work and generally it is not nearly as interesting or exciting as writing a cool new piece of code. Furthermore, many of the long-term advantages of working within a process framework are not evident during the early stages of building a process-oriented environment. As a result, people tend to ignore processes, or put process-related work at the "bottom of the pile" so that it never gets done. There is hope! You can take a number of steps to address this problem:

1. Persuade staff members of the benefits of process compliance. This is generally a matter of explaining and demonstrating how a small amount of short-term pain can result in a large amount of long-term gain.
2. Actively involve staff members in process development. Workers will resist irrelevant or burdensome processes that are imposed by management without consultation with the people in the trenches. *(Confirmed in Chapter 2—Ed.)*
3. Encourage and enforce useful and effective processes. You might as well not have the process if it is ignored. If you honestly believe that a process is useful and beneficial for your business, then deviations from that process, no matter how small, must be corrected whenever they occur.

### Case Study: White space

I believe that a consistent "look and feel" is very important when code is produced by multiple developers, so our coding standard is quite prescriptive about where space characters and blank lines are to be inserted in a source code file. There are arguments to be made both for and against this level of detail in a coding standard, but we think it has value so that's how we work.

New staff coming into the group often find it difficult coming to grips with the coding standard for "white space," and they tend to drift from the standard, inserting white space characters as they see fit, rather than in accordance with the standard because they see it as a trivial issue.

In response to this noncompliance we could say "Oh, it's only white space, it doesn't really matter," but we don't. Our standard describes how we work, and we comply with that standard in all regards, both large and small. Our review process catches and corrects these smaller defects, just as it catches the bigger issues.

Automation is also a big help in this area. We now use an open source code formatting tool called "Uncrustify" and a database of formatting rules to mostly eliminate source code formatting problems.

*A word of caution*—always keep in mind that the aim of business is business, not process compliance. Processes are there to support business goals, not as an end in themselves. Do not mistakenly and rigidly follow processes at all times, regardless of the circumstances.

Your processes will be most beneficial for the predictable and routine within the work environment because they provide standard solutions to standard problems. That said, life will always throw you curve balls; unusual situations will arise that aren't covered by your processes, or occasionally following your processes will produce undesirable results. Sometimes, for very good reasons, business demands may dictate that work be performed outside the process framework. Such cases should be atypical, you should violate processes only after due consideration and with full awareness of what you are doing. BUT…when it's necessary don't be afraid to do what has to be done to finish the job.

Under these circumstances, once the emergency has been resolved, it's very tempting to put the unpleasantness behind you and carry on with getting the job done. While this is the easiest thing to do it's also a mistake, because you will lose the opportunity to improve your processes by learning from what happened.

As soon as possible after the event, you should debrief all those involved. Get the players together and talk to them to get their story about what happened. For a variety of reasons, it's likely that everybody will have a different story. Take written notes, work out what actually happened, analyze the event, and learn from it.

Some questions to ask are:

- Was this a one-off event, or is it likely to happen again?
- Could the situation have been handled differently and more effectively?
- Do you (or we) need to adapt your (or our) processes to cater for this happening again?

### ISO 12207 reference process

The International Organization of Standardization (ISO) has produced *ISO 12207—Software Lifecycle Processes* [7]. It is a reference standard that covers the entire software development life cycle: acquisition, development and implementation, support and maintenance, and system retirement.

ISO 12207 details activities that you should conduct during each stage of software development within a project. This standard describes things that should be done during development, but not how to do them. You can best use the standard as a checklist of useful activities that can be tailored to provide a level of process overhead that is suitable for any given project.

### Recommended process documents

The size and scope of your process documentation should depend on the size and nature of your organization. Your organization may only be you, sitting at a computer in your home office, or it may comprise you with a dozen other people working on a ground-breaking piece of new technology in a high-tech start-up.

You should have a documented set of processes that describe the standard approach that you take to standard issues or problems, which lets you concentrate on the unusual or interesting issues that arise. Suppose you started out alone, then one day someone starts working with you; documented procedures will save you time and money if you can say, "Read this. . .it's how I've been doing things." Suppose you are already working in a group, then one day somebody leaves; documented procedures will save you time and money if you can say to the replacement hire, "Read this. . .it's how we do things around here."

In every project, you should tailor the abstract guidance provided by each of these high-level documents into a set of concrete steps appropriate for the project. These procedural documents related to a specific project translate the general, "What do we want to achieve?" into, "This is how we are going to work on this project."

Table 11.1 lists a minimal set of process documents. Developing and implementing this small set of processes will immediately benefit your development activities.

## Table 11.1: A minimal set of suggested process documents

| | |
|---|---|
| **Design** | Describe at a high level what you want your design process to achieve and the artifacts that should result from the design, such as system block diagrams, system use cases, interface specifications, and test procedures. Remember that every project will be different, so don't go into too much detail here. |
| **Implementation** | This is the place to provide an overall description of how your development activities are to operate. Describe the general development life cycle, and the tools such as version control systems (VCSs) that are used to support development. Also provided references to the other supporting standards such as your code formatting standard. |
| **Release** | This is a key activity that often doesn't receive as much attention as it deserves. Describe the process to transfer your newly developed products from your organization to your customers in a controlled and repeatable manner. |
| **Review** | Rigorous peer review of work products throughout the development life cycle is a hallmark of a good software engineering environment. The description of the review process should mandate that these review activities take place. It should also include a requirement that all review activities be documented in some manner both to provide proof that the reviews are taking place and to describe the action items that arise. (See the details of review in Chapter 13.) |
| **Code formatting** | Code formatting standards can be very contentious. This is one of those cases where you should consider being prescriptive, rather than providing general advice. <br><br> No particular formatting standard is significantly better than any other. The important thing is simply to choose something and then stick to it. The long-term maintenance benefits of doing this are dramatic. |
| **Code commenting** | Some people suggest that code should be self-documenting. In an ideal world this would be true, but in the complex and difficult world of embedded system development this is almost never the case. <br><br> Remember that embedded systems typically have long life spans and that large amounts of the cost of the system may be spent on maintenance over the life of the product. Anything that can reduce the effort and hence the cost involved in software maintenance will have substantial long-term benefits. <br><br> As with the code formatting standard, it is best to be prescriptive in this document. |
| **Language subsetting** | The language subset document is specific to the programming language. If your development work involves the use of multiple languages, then each language requires a different standard. The purpose of this document is to codify the best practice in the use of each language. It recommends the use of certain techniques or features of the programming language while forbidding the use of other techniques. |

## Case Study: Embedded systems can last a long, long time

In 1991, I commenced work on the development of the embedded target tracking software for a new radar system. The signal processing hardware environment was very complex, consisting of a custom multi-CPU pipelined parallel processor assisted by a cluster of DSPs which did most of the computational heavy lifting. The software was equally complex, if not more so.

During the next 15 years, this general-purpose tracking system was applied to multiple different problem domains, none of which we anticipated at the time of the original development. Over the course of those years the software was modified, enhanced, and rehosted to different hardware and operating system platforms. I estimate that the maintenance and enhancement of the software over that period of time occupied about 5 times the amount of effort that I spent writing it in the first place.

At the time of writing, that software is still in operational service in a number of locations, and is being supported by people who had only just been born when it was first written.

---

### Case Study: How not to develop code and how to recover (by Kim R. Fowler)

This case study continues from Chapter 1. It is a story with a happy ending but not before we experienced considerable pain getting there. To refresh your memory, my company was the customer and we were buying an instrument to put on an aerospace vehicle. The vendor claimed to have a product that was "Commercial-Off-the-Shelf." My company wanted to modify the design and add a number of sophisticated sensors to the COTS product. (Chapter 1 describes the details, which don't need repeating here.)

For this chapter on software development, you should know that the product had two significant portions to it:

1. An embedded system resided on an aerospace vehicle. The system comprised a processing unit driven by fairly complex sensors. Data streamed from the sensors to the processing unit for data compression and multiplexing. From there the vehicle received the data and transmitted it to the ground in a high-speed serial stream. The embedded system had DSP chips running C code.
2. A ground-based terminal received the high-speed serial stream from the aerospace vehicle. The terminal demultiplexed and then decompressed the data. It filtered and manipulated, displayed, and then stored the data. The terminal was a high-end desktop computer running custom C code.

The first copy of the embedded system was to deliver at the end of August (the terminal was to follow a bit later). In early August, after multiple requests to the vendor, a colleague of mine bluntly asked, "Is it working?!! Demonstrate its current operation to us!" The vendor couldn't.

Several colleagues and I packed up and flew across country to meet with the vendor. We found the following problems:

- Spaghetti code—code that wound through illogical and convoluted paths.
- Nested interrupt routines, in fact the entire program code operated out of these nested routines. *(Never—ever—do this!!!)*
- No common event handlers, each different input had a unique event handler.
- Orphan code segments, there were entire sections of code commented out with no clue as to their original purpose.
- No consistent style; in fact there was no software style guide.
- Legacy code that did not fit our particular project, the vendor was trying to reuse old code and "shoehorn" into the new application.
- The ground-based terminal was a custom program that was in as bad shape as the embedded system.

So what caused all these problems? First the vendor had a senior person who was a single software developer with no accountability. He was an original member of the company with some small ownership in company shares; the company assumed that he knew what he was doing since he was 20 years into his career writing software for aerospace applications. Consequently, the company had no processes, no code reviews, and no planned demonstrations of operations for the product. Actually they had no development plans—design, schedule, or test—of any sort. On top of all this, they were using an unfamiliar DSP processor, with caches, serial ports, and internal memory new to them, and a new development system. It was chaos. The vendor had bitten off more than they could reasonably chew.

We quickly assessed the situation and explained to the vendor how bad it was—for us and for them. We also explained what we expected them to do to deliver a working product to us. We prescribed the following steps and stages to vendor's director of engineering:

1. Establish a schedule and delivery milestones.
2. Prepare staged releases of the software for the embedded system.
3. Hold monthly reviews of the software development with us in person, eyeball-to-eyeball.
4. Institute new software processes that would be defined and followed with code reviews and metrics for production and anomalies.
5. Prepare a complete set of documents.
6. Buy and integrate a commercial software package for the ground-based terminal.

As you know by now, the story turned out well. The vendor, under tutelage of the chastised director of engineering, knuckled under and worked very hard for six months with each employee putting in 70 and 80 hours per week. They instituted clear procedures and styles for developing software. The original developer just never got it. He hung around for several weeks only to resist the changes. Very soon there was a parting of the ways, which was good for the vendor's staff.

We, the customer, rolled up our sleeves and jumped in with the vendor. We helped with the development system and worked to understand the DSP processors with them. We bought and integrated the software for the ground-based terminal. We reviewed their progress and then provided the test facilities and personnel for the environmental testing toward the end of the delivery schedule.

The final product delivered a bit late but it worked as advertised.

So what should the vendor, or any software development organization, do in the first place? Here are some recommended steps:

- Plan, execute, review, report, and iterate each step of the process.
- Prepare and follow development plans.
- Manage the configuration.
- Use a style guide.
- Maintain records of production metrics, bug rates and severity, and status of fixes.
- Perform regular and careful code reviews.
- Perform regular and careful project reviews.
- Deliver incremental releases, if possible, and demonstrate the growing system capability.
- No one—absolutely no one—runs open loop with no accountability.

The embedded system world has displayed considerable interest in language subsetting over the last few years, primarily driven by the widespread use of C and C++ in embedded environments. Two notable efforts in this area come from the Motor Industry Software Reliability Association (MISRA) and Lockheed-Martin.

MISRA is a consortium of motor vehicle manufacturers. In the late 1998, they produced a standard called "Guidelines for the use of the C language in vehicle based software," otherwise known as MISRA-C:1998. This document contains a set of rules for constraining the use of various aspects of the C language in mission-critical systems. While originally targeted at the automotive industry, a wide variety of other industries has adopted this standard. It has undergone a number of revisions, and the current version is MISRA-C:2012. As well as the C language standard, in 2008 MISRA also produced a C++ standard called "Guidelines for the use of the C++ language in critical systems." This document is known as MISRA-C++:2008.

The work by Lockheed-Martin relates to their involvement in the Joint Strike Fighter Program. As part of developing the avionic software for this aircraft, Lockheed-Martin developed a language subset known as the Joint Strike Fighter Air Vehicle C++ Coding Standards, or JSF++ for short.

Both the MISRA and JSF++ standards provide a good basis for study and the development of a good language subset. In particular, the MISRA standards have become so popular that many common compilers and static code analysis tools provide inbuilt standard support that you can easily enable.

### Requirements engineering

If someone were to produce an old-fashioned map showing the embedded systems software development process, then the area entitled "Requirements" would surely also be labeled "Here Be Dragons." There is no part of the process that is more prone to difficulty and no part of the process that is more critical to the development of a system than getting the requirements right. (Chapter 6 goes into great detail on generating and maintaining requirements.)

It's as simple as this. At the start of a project, someone will have a set of ideas in their head about what they want the system to do. Those ideas represent the system requirements. Your job is to construct a coherent list of those requirements and then to build a system that meets the requirements. If you get the requirements wrong at the start, then even if you do **everything** else perfectly you will end up building the wrong thing. You will build something that your customer doesn't want and your project will be a failure. It's simple to describe, but it sure isn't easy to do.

Requirements engineering in embedded system software development has the added difficulty that rather than being the end product, the software is a component that is embedded in a larger system. This means that the system requirements are usually expressed in terms of what the **system** must achieve, instead of what the **software** must achieve, with the result that the detailed software requirements are buried and lost inside the system requirements. It can be very difficult for an engineer to translate a requirement from an application domain, such as radar tracking or engine management, into a meaningful software requirement.

If you are aware of the importance and difficulty of good requirements engineering and if you allocate the necessary resources to perform this part of the process effectively, then you've taken a large step toward mitigating the risks associated with having bad requirements. You also need to keep in mind that even with your best efforts, the inherent difficulty of gathering requirements means that it's unlikely that you'll get it right first time. New requirements will emerge during the development of your product, and you must be prepared to accept these changes and to be flexible enough to incorporate them into your product during the course of the development.

So what do you actually need to perform good requirements engineering? My view is that the critical activities involved in requirements engineering are:

1. **Collection**. This is the process of discussing with all stakeholders and building an initial list of all of the things that the product is supposed to do.
2. **Analysis**. After collecting the initial requirements, you need to study and analyze them for completeness and consistency. Even if only one person specifies the requirements, it's likely that the requirements will contain internal conflicts and inconsistencies that must be resolved at this point.
3. **Specification**. Once you have collected all of the requirements, you need to formalize them into a System Requirements Specification. This document will define each requirement in the most precise language possible and will allocate unique identifiers to each requirement for consistency of reference. The identifiers enable other derived documents, such as test specifications, to remain synchronized with the requirements specification.
4. **Validation**. A final check of the completely defined and identified set of initial requirements contained in the System Requirements Specification to confirm that they meet all of the needs, and the overall intent, of the customer.
5. **Management**. Once you have been specified and validated the initial requirements, you may use them to commence the design activities. From this point, and throughout the project, you will discover that despite your best efforts, your requirements will still contain flaws that need fixing. Some requirements will disappear, new requirements will emerge, and others will change shape. Requirements management involves staying on top of this fluid situation and ensuring at all times that you have a solid specification that precisely defines what you are attempting to build.

**Example: Early models solidify requirements inappropriately**

There are a number of variations on this theme. For example, I have worked with teams who liked to perform graphical system modeling during requirements engineering using the Unified Modeling Language (UML) and other tools. My experience has been that this tends to bring the design process into the requirements phase too early. The effort involved in producing and updating the models can cause people to be unwilling to pursue alternative design options, with the result that a suboptimal design can become fixed too early.

There are a wide variety of software products available to assist in all of these areas of the requirements engineering process. For smaller projects, a spreadsheet and a word processor are probably all that you need. For larger projects with hundreds or thousands of requirements, it can be very difficult to manage the process without the support of dedicated requirements engineering tools.

A good example of such a product is the DOORS requirements tracking tool originally developed by Swedish firm Telelogic, and now supported and marketed by IBM. This consists of a suite of different tools providing various capabilities, but the key functionality supports managing the development of sets of complex requirements for large systems.

### Version control

Automated VCSs are a key tool in any modern software engineering environment. Developing software without the assistance of such a tool is like running a race with a rock in your shoe. Yes, you will reach the finish line eventually, but the race will be slower and more painful than necessary.

VCSs apply to all software development projects and are not specifically related to embedded systems. Regardless, VCSs still deserve some discussion here because their use represents a best practice in modern software development.

The simplest view of VCSs is that they are document management systems specifically targeted at maintaining libraries of text files where the text files typically contain software source code. The systems may be used to store other types of files, but they are less effective when used to manage binary files such as word processor documents.

The conventional method of operation involves storing a master copy of each file in a central repository. A database management engine of some kind typically manages the repository of files. When an engineer needs to make a change to a file, he or she checks the file out of the repository and copies it to a local computer for editing. When the engineer has made the required changes, he or she checks the updated file back into the repository, which simply involves copying the file from the local computer back into the repository.

There are many variations on this basic theme. For example, rather than simply storing entire new copies of changed documents in the repository, some systems encode and store the sequence of changes between one version and the next. When many small changes are made to large files, as is typical in software development environments, this technique can dramatically reduce the storage requirements of the repository and the time taken to access the files. Some VCSs lock the repository copy of a file when it is checked out, which enforces the rule that only one person can be working on a file at any time. While this technique is simple to understand and implement, it can be extremely inconvenient in a team environment as it forces serial access to the file. More advanced systems allow parallel development where multiple users are able to check out files simultaneously for update, with the system assisting to resolve any editing conflicts that might occur at check in time.

Now let's consider the primary benefit of VCS which, as the name implies, is the control and management of different versions of a software product. If the world was a perfect place, VCS would be far less useful. Unfortunately, the world is not a perfect place and Version 1.0 of most software systems is followed by Version 1.1, then Version 1.2...you get the picture.

Management of this in the trivial case where the entire system is built from one source file doesn't present much of a challenge, as there's a one-to-one mapping between each version of the source file and the resulting executable. Unfortunately, few real-world systems are this trivial. I am aware of a major avionics embedded system that consists of over 35,000 source files. Systems of this magnitude are not unusual, and control of the development process in this environment is an extremely complex task that would be virtually impossible without the assistance of a VCS.

The first benefit provided by a VCS is a history of the changes that have been made to each source file, including the date and time of the change, the name of the engineer who made the change, and useful comments about the purpose of the change. This feature is very beneficial in understanding how a system has evolved and who was responsible for the change process, but more importantly, it allows changes to revert when things go wrong, as they inevitably will. This ability to quickly back out of and recover from defective changes can be an absolute lifesaver.

A further complication in a large-scale development environment is that there is very little correlation between the changes that are made to each file because changes are made to different files for different reasons at different times. A new version of the system might require changes to 100 different source files. The next version of the system might require further changes to 35 of those files, with more changes being made to 100 different files, and each succeeding version will require some unpredictable combination of changes to different files. The result is that a snapshot of the state of all the files in the system on one day will most likely be very different to a snapshot taken a week later. Manually reverting to a previous version of the system under these conditions would be an extremely difficult task to perform manually as it would require reverting each individual file to the correct state, but a VCS provides the mechanisms for doing this that make it a trivial process. The usual process is to tag all of the files in a project with the version number or some other unique identifier. Once this tag is in place, it is possible to extract the entire system with a specified tag in a single operation.

The next advantage provided by a VCS involves the coordination of distributed development activities by multiple engineers. As the physical separation between developers increases, the task of coordinating their updates to shared source files increases. The members of a development team these days may be located in different offices within a building, different buildings on a corporate campus, different buildings in different cities, or even in different cities on different continents in different time zones.

Managing the activities of such teams, particularly the very widely distributed teams, is nearly impossible without the use of a VCS. It's still hard even with the use of a VCS, but the difficulties are substantially reduced. Consider two engineers attempting to work on a source file over a period of several weeks when one person is located in Los Angeles and the other is located in Sydney. Thousands of miles and a 7-hour time difference separate these two people. Without a VCS, coordinating their work would require a substantial amount of time and effort to be invested in phone calls at inconvenient times, transfer of files back and forth, and recovery from errors. Automating this process via a VCS will increase productivity by eliminating almost all of that overhead.

Finally, a VCS provides a single convenient target for system backup and recovery processes. Without the use of a VCS, an organization's code base may be spread over multiple folders on a server or over multiple servers. In the worst case, the code may be spread over the individual workstations belonging to the members of the development team. In this sort of environment, the failure of the hard drive in a machine containing the only copies of the source code for an important application can result in a disastrous waste of time and effort.

With the use of a VCS, the source code backup process can be greatly simplified. This simplification means that the process will be inherently more reliable because all that is required is that the source code repository be backed up on a periodic basis to ensure coverage across the entire code base.

The design of a good backup process is beyond the scope of this discussion, but it's worthwhile pointing out that if all of your backup media are stored in the same physical location as the server containing your VCS repository, then you could be heading for trouble. A disaster such as a fire or flood might result in you ending up with no server and no backups either, so always incorporate offsite storage of media into your backup regime.

A final point worth noting is that a VCS is just a tool. Used correctly, it will substantially reduce the costs of developing large-scale software systems. Even when used incorrectly, it will still help, but perhaps not as much as it should. For example, if an engineer checks a source file out of the repository and then works on it intensively for a month without checking it back in to the repository, then many of the benefits of change logging, collaboration, and easy backup are lost. As with most other areas, the development of simple and well-understood VCS management processes are important in ensuring that the benefits of using the tool are fully realized.

### Effort estimation and progress tracking

Software development is an opaque activity. No other form of organized human activity resembles it. There is no other form of organized human activity in which it is so difficult to understand what people involved in a project are doing or achieving.

Consider conventional real-world construction projects that involve lots of metal and concrete. Things built out of concrete can be measured in terms that humans can intuitively understand. The completed length of a freeway, the completed height of a dam, and the number of stories completed on a skyscraper are all things that you can easily express and convey numerically and pictorially. If a freeway construction project is due for completion in three weeks and there are still three miles of unfinished road under construction, then it will be obvious to the project team that the project is in trouble. Very importantly, the problem will also be obvious to outsiders, even those who are completely unfamiliar with any aspect of freeway construction.

Software, as the name implies, is the exact opposite of concrete. The construction material used in software development is human thought, which is an infinitely light, malleable, and flexible material. Software engineers are capable of building large, complex, and incredibly intricate mechanisms out of this material.

Unfortunately, software doesn't come with the intuitions that are associated with conventional construction materials; consequently, the conventional paradigms for human-oriented measurement don't apply to it. You can't pick up a piece of software and heft it in your hand. You can't weigh it or pull out a tape measure to measure its length. And because software doesn't come with the intuitions associated with conventional construction materials, it is very difficult to represent and measure progress within a software development project. A software project may be in a very similar position to the freeway-building project with three miles of road to complete in three weeks without it being obvious to interested outsiders.

There are no magical solutions to this problem, but there are techniques and tools that can mitigate some of the risk associated with tracking the progress of a software project.

The intuitive metric for measuring progress in a software project is the number of lines of code that have been written. If a project requires 10,000 lines of code, and the developers have produced 8000 lines of code, then the project is 80% complete, right? Well, maybe. Or maybe not, because not all lines of code are created equal. The developers may have concentrated on the infrastructure aspects of the project, producing lots of straightforward code to accomplish simple tasks. The remaining 2000 lines of code may represent the core functionality of the application, containing complex algorithms and tricky performance issues that require detailed study, analysis, and testing. In this case it wouldn't be at all unusual to discover that the last 2000 lines of code take as much time to complete as the first 8000. (This is the old rubric that 20% of the work takes 80% of the time.)

In some cases, lines of code may even be a perverse metric that indicates the opposite of what is actually correct. If one engineer can solve a problem in 1000 lines of code, and another can solve the same problem in 100 lines of code that execute 10 times as fast then it would probably be a mistake to consider that the first engineer is more productive than the second. Or perhaps the engineers are equally skilled and productive, but one is working in assembler language while the other is working in a higher level language such as C++, in which case the latter would be able to implement similar functionality in far fewer lines of code.

The keyword here is "functionality." The best technique for approaching this difficult issue lies in assessing progress in terms of the amount of functionality completed, not the lines of code completed. In simple terms, this involves deriving defined points of functionality from the system requirements and allocating a measurement of complexity to each function point. The result is a single metric that represents the total functionality of the system, with greater functional density associated with different function points depending on their complexity.

This technique is known as Function Point Analysis, and it was defined by Allan Albrecht of IBM in the 1970s [8]. Much work has been done on this topic since then, and there are now a number of recognized standards for performing this analysis [9]. Even if you choose to perform your analysis in a less formal way, it is still an invaluable technique for providing insight into the complexity of a project.

Let's assume that you've done a Function Point Analysis on your new project, along with other aspects of your design, and you're ready to set to work. Unfortunately, your problems are not over. In fact, they've only just begun. Setting to work on implementing your function points sounds great, but who is going to do the work? Which engineers will complete which functions? How do you keep track of which function points are completed, which ones are in progress, and which ones haven't been touched yet? Keep in mind also that your iterative testing program will reveal design and implementation defects that will require correction by somebody at some time during the project.

The difficulties of managing this resource allocation and scheduling problem are substantial, as are the risks to your project by getting things wrong in this area. Happily, there are tools available to help you mitigate this risk. They go by a variety of names, including "bug tracking systems," "issue tracking systems," and "ticketing systems." I prefer to use the term "ticketing system" because it's a neutral term that avoids the misconceptions associated with terms such as "bug tracker."

The name "ticketing system" comes from the use of cards known as tickets, which were originally used to allocate work to staff members in a call center and help desk organizations. When a customer called in a problem, the phone operator would fill out a paper ticket with details of the problem and then place it in a queue that was serviced by the responsible engineers and technicians. During the process of working on the task, the engineer would annotate the ticket with details of the job as required. When the task was completed, the ticket would be marked as completed and filed with all of the other completed jobs.

A modern ticketing system is a software implementation of that workflow, with the tickets replaced by database entries, and a software application providing a front-end to the database that allows for queries and updates to the entries. There are a number of ticketing systems that are specifically designed for use in a software development environment. These usually provide specific support for aspects of the software development life cycle, along with features, such as integration into VCSs and report generation systems.

The benefits of using a ticketing system (or whatever you prefer to call it) during a software development project can't be exaggerated. If you use tickets to allocate all of the work at the beginning of a project, then the list of tasks resides with the tickets. If you want to know what you have completed so far, the list of completed tickets is available at the click of a button. The same applies to tickets that are the subject of work in progress, and work that is yet to be started.

Using a high-quality ticketing system to allocate tasks to engineers and to track the progress of those tasks through to completion is a hallmark of good software engineering. If you are developing systems without one of these tools, then you may as well work with one hand tied behind your back.

## Life cycle

A current direction in the life cycles of software development is toward Agile development techniques as espoused in The Agile Manifesto. Agile development techniques are human centered [10]. They recognize the psychology, strengths, and weaknesses of the humans involved as both customers and developers in the software development process.

Agile techniques are particularly well suited to the sort of exploratory software development characteristic of applications that require heavy human interaction, such as web-based transaction processing systems. The software development effort, in such systems, often focuses on the implementation of complex graphical user interfaces (GUIs), which aim to resolve the issues associated with human cognitive strengths and weaknesses. In particular, people generally cannot assess the strengths and weaknesses of a GUI without operating the interface.

One of the particular strengths of this development methodology is that it recognizes that humans are visually oriented creatures. It concentrates, therefore, on prototyping and the rapid experimental development of user facing components of application software. Short cycles of rapid application development combined with intense user feedback and code refactoring can produce very high velocities of development when this technique is used by skilled Agile developers.

Agile development techniques might apply to the development of embedded software. In this domain, Agile development must proceed with great caution since embedded systems fundamentally differ from desktop software, as described in the first part of this chapter.

Agile techniques are diametrically opposed to the "Design Everything First" waterfall methodologies exemplified in the life cycle methodology described in DOD-STD-2167A. This was a large highly prescriptive, documentation-centric methodology promulgated by the US Department of Defense in the 1970s which mandated that development should be conducted as a series of discrete steps, with each step, such as system design, fully completed before beginning the next step [11].

While attractive on the face of it, this methodology fails to recognize the fundamental fact of human fallibility. People make mistakes. Designs will contain errors, and committing to the completion of the entire design before commencing development commits the eventual development process to the implementation of an error prone design. This means that design errors will be propagated through the rest of the development cycle and may not be found until system integration or test time. Fixing these problems at this time will be much more expensive than doing it earlier in the project. Also, being closer to the delivery deadline means that the corrections will be developed under much more stressful circumstances, with it being likely that the quality of the work will be lower due to schedule pressure.

So, the embedded system developer is faced with a conundrum. Highly interactive and highly iterative agile development methodologies with minimal up-front design work well in the development of human centered applications with complex GUIs where user requirements are very uncertain. Unfortunately, most embedded systems don't fit this description, and the application of agile techniques to the development of embedded software can be very problematic.

On the other hand, the heavyweight sequential methodologies that require that everything be designed up-front can also be problematic because the development process chokes on a flurry of design and implementation defects that are only discovered very late in the development cycle.

Unfortunately, there is no magical technique or methodology that provides an answer to this problem because the solution lies in compromise and good judgement developed through experience. You can apply some fundamental ideas though. Don't attempt to design every aspect of your system up-front before commencing development. Conversely, resist the temptation to begin coding before you have thought about your system and produced at least a basic design.

> *I am trying to caution against coding without thought or design and based on inadequate development of requirements. Even now I still hear of teams (not at my company!) who start work on systems by sitting at their computers and writing code without thought for requirement development or design.*

Like Goldilocks and the Three Bears, you need to perform an amount of up-front design that is neither too much, nor too little, but just right. Your experience and good judgment determines what is "just right" for your project and your environment. In some situations, there's no working around it and you will have to do very large amounts of highly detailed design work up-front, while in other cases a more exploratory approach may be entirely suitable.

Whatever the case may be, once the initial up-front design process is completed, commence a series of iterative development cycles. The phrase "design a little, code a little, test a little" best captures the essence of what is involved. The feedback process from the test phases is critical. You must recognize and correct design or implementation errors as early as possible, because such errors take more time and money to correct as the system gradually takes shape. This might be considered a hybrid approach between spiral and Agile development.

Finally, when you complete the development, test your entire system to ensure correct operation; this is validation. Given the testing phases embedded into the development cycles, this testing should actually be a confirmation of correctness as much as a search for errors.

Chapter 1 provides further information on life cycle models, including the waterfall model, the spiral model, and Agile development.

## Tools and techniques

### Real-time operating systems

Fairness is an admirable characteristic in most situations. People undoubtedly wish to be treated fairly in their dealings with other people, and they carry that desire over into their dealings with machines. General-purpose OSs are designed to allocate scarce resources between a variety of user and system level processes. It's unsurprising that one of the primary goals of those systems is achieving a fair and equitable allocation of those resources between all of the competing processes that are running on the computer. Of course, "fairness" may be defined in a variety of ways. For example, an OS may be designed to allocate CPU resources fairly between computationally intensive background processes to achieve high throughput, at the expense of interactive users.

The primary driver for OS design is not a love of fairness on the part of the OS developers, but simply that OSs are general purpose in nature. The desktop operating environment is inherently chaotic and unpredictable. It's highly unlikely that any two general-purpose computers have exactly the same characteristics in terms of applications and data files installed. This in turn means that the computational demands placed on the processors are almost totally unpredictable.

This unpredictability forces the OS designers to make very generic decisions about how their schedulers will allocate system resources between a multitude of competing processes. Fairness is a reasonable starting point when deciding how to slice the pie.

*Embedded systems, however, represent exactly the opposite situation from that encountered on the desktop.* Embedded systems are focused devices dedicated to the solution of a single problem. They highly constrain inputs and outputs and thoroughly define processing. Techniques appropriate to the desktop environment are prone to fail in the embedded world. Fairness as a resource allocation philosophy fails miserably in embedded systems.

Enter the RTOS. The primary distinguishing feature between a general-purpose OS and an RTOS is that an RTOS is not concerned with the fair allocation of system resources. An RTOS allocates each process a priority. At any given time, the process with the highest priority will be executing. As long as the highest priority process requires the use of the CPU, it will get it to the exclusion of other processes. This "highest priority first" scheduling philosophy is an extraordinarily powerful tool for the embedded system software developer. Used well, an RTOS can extract the maximum processing power out of a system with minimal overhead.

There are a large number of commercial and open source RTOS packages available to choose from, with a wide variety of features. The most compelling argument for adopting one of these systems is that they allow your developers to focus on their application domain, rather than on developing infrastructure that, while necessary, doesn't contribute to your business goals.

For example, apart from their key task creation and scheduling functions, RTOSs also provide abstracted software layers and device drivers for processor bootstrap, interprocess communication, onboard timers, external communications ports, and peripherals of all varieties. Getting your expensive engineers to custom build this infrastructure rather than concentrating on your core business applications is almost certainly going to be more expensive and less efficient than acquiring a suitable product from people who are in the business of producing that infrastructure.

## Design by Contract

Design by Contract (DBC) is a powerful software development technique developed by Prof. Bertrand Meyer. He espoused the concept in his book *Object-Oriented Software Construction* [12] and implemented it as a first-class component of the Eiffel programming language. Facilities for using DBC are now available in a wide variety of programming languages, either built into the language or implemented as libraries, preprocessors, and other tools.

DBC uses the concept of a contract from the world of business as a powerful metaphor to extend the notion of assertions that have been a common programming technique for many years. As in a business contract, the central idea of the DBC metaphor involves a binding arrangement between a customer and a supplier about a service that is to be provided.

In terms of software development, a supplier is a software module that provides a service, such as data through a public interface, while a customer is a separate software module that makes calls on that interface to obtain the data. A contract between the two modules represents a mutually agreed upon set of obligations that must be met by both the customer and supplier for the transaction to complete successfully. These obligations represent things such as valid system state upon entry to and exit from the call to the interface, valid input values, valid return types, and aspects of system state that will be maintained by the supplier.

There is a substantial difference between the conventional use of assertions and DBC. The old school use of assertions involves writing code and then adding assertions at various points throughout the code. As the name implies, Design by Contract involves incorporating assertions about the correctness of the program into the design process so that the assertions are written before the code is written.

This is a powerful idea. Creating contracts during system design forces the designer to think very carefully about program correctness right from the beginning. DBC provides convenient tools for guaranteeing that correctness is established and maintained during program operation.

Two standard elements for establishing DBC contracts are the REQUIRES() and ENSURES() clauses. These may be viewed as function calls that take a Boolean expression as an argument. If the Boolean expression evaluates to true, then program execution continues as expected. If the expression evaluates to false, then a contract has been violated and the contract violation mechanism will be invoked. You place one or more REQUIRES () clauses at the start of a function to establish the contract to which the function agrees prior to commencing work. You place one or more ENSURES() clauses at the end of the function to establish the state to which the function guarantees after it has done its work.

The key thing for you to understand is that DBC is intended to locate defects in your code. This might sound like a statement of the obvious, but it's more subtle than it appears at first glance. The contracts that you embed throughout your code should establish the system state required for correct operation at that point of execution. They represent "The Truth" about the correct operation of the system. If a contract is violated, then a catastrophe has occurred and it is not possible for the system to continue to operate successfully. At that point, it is better for the system to signal a failure and terminate, rather than to propagate the catastrophe to some other point of the system where it may fail quietly and mysteriously, resulting in days or weeks of debugging to track backward from the point of failure.

DBC should **_not_** be used for validation of data received over interfaces, either from an operator or from another system. Normal defensive programming techniques should be used at the interface to reject bad data such as operator typographical errors. The difference here is between something bad that might actually happen under unusual circumstances versus something that must never happen at all. You should never invoke a DBC contract on something that could reasonably occur.

One way to look at your contracts is that they are testing for impossible conditions. You shouldn't think, "I won't test for that, because that will **_never_** occur." You should think "I will test for that, because if that does occur, it indicates that there is a defect in the system and I want to trap it right now."

So, the REQUIRES() statement says, "I require the system state to satisfy this condition before I can proceed. If the system is not in this state, then it is corrupt and I'll halt." Similarly the ENSURES() statement says, "I guarantee that inside this function I have successfully transformed the system state from where we were at the beginning, to this state. If this transformation has not been successful, then something has gone very wrong, the system is corrupt, and I'll halt."

You might look at ENSURE at the end of a function and think "Why should I test that? That pointer is never going to be NULL." But the concern is that the pointer will never be NULL if the system is **_working correctly_**. What we're catching with the ENSURE is the case where a maintainer (either you next week or somebody else in 10 years' time) comes along, changes the system erroneously, and introduces a subtle bug that results in the pointer being NULL once every month, which causes the system to mysteriously crash.

To repeat: Ask yourself exactly what the system state should be at the start of a function for subsequent processing to succeed and then REQUIRE that to be so. At the end of the function, ask yourself to what state you've transitioned and ENSURE that you've successfully made that transition.

This process forces you to think more clearly about what you are doing and ensures that you have a much crisper mental model of your system inside your head during design and development. This leads to better design and implementation so that you avoid introducing defects in the first place.

An analogy I like is that DBC statements are the software equivalent of fuses in electrical systems. A blown fuse indicates that something is defective (or failed) while protecting your valuable equipment from an out-of-range value (i.e., too much current). A triggered DBC condition indicates that something is defective while protecting the rest of your system from that corruption.

Some final words of caution:

1.  Nothing comes for free. Using DBC incurs an overhead caused by the computation necessary to evaluate each of the DBC clauses. In systems with limited amounts of spare processing power, it may be the case that the evaluation of the DBC clauses involves a prohibitive amount of overhead. In such cases, it may be necessary to disable DBC evaluation for release after system development and testing has been completed.
2.  The expressions provided in DBC clauses must not have any side effects. If the expressions do have side effects and DBC evaluation is disabled for release, then the released version will operate differently to the debug version at runtime, which is obviously highly undesirable.
3.  It can be very difficult to decide what action to take in an operational system if a DBC clause evaluates to false. When a DBC clause is triggered, it presumably indicates the detection of a defect within the system. The simplest option when this happens is to display or log appropriate errors messages and then terminate execution of the program. While this is a simple solution, it may not be feasible to follow this course of action in a working embedded system. For example, if the system under consideration is the engine management system of a motor vehicle, then terminating the program while the vehicle is traveling down a freeway at high speed is probably not appropriate. On the other hand, doing nothing may be an equally bad choice given the detection of a potentially serious defect within the system.
4.  Realistically, the action to be taken by the DBC processing upon detecting a defect is a choice that can only be made by the developers based on the information they have about the nature of their system.

## Drawings

I want you to visualize a three-dimensional solid object bounded by six square faces and facets or sides with three meeting at each vertex. Take as much time as you need to form an image of that object clearly in your head before you go on.

I'm pretty sure that after reading that sentence (perhaps a couple of times) and thinking about it a little you would realize that it described a cube. I'm also pretty sure that the process of reading that sentence and converting it into a mental image in your head took more than a few seconds. I'm 100% sure that if I'd replaced the sentence with a picture of a cube, you would have looked at the picture, and the concept of "cube" would have entered your thoughts in a fraction of second—much faster than it would take to describe in words. Furthermore, the image of the cube may have had substantial information content that would have been difficult if not impossible to describe in words. For example, the image of the cube may have been a precise shade of pastel blue, and the sides of the cube may have been polished smooth, or textured.

This little exercise conveys a profound insight, which is that the human visual processing system works best at image processing and pattern matching. It's astoundingly good with words, but pictures definitely allow us to communicate some things far more effectively than we can with words. The result of this is that the information content and the speed of delivery of an image may be orders of magnitude greater than a written attempt to communicate the same information.

What's this got to do with software development? Well, in the words of the old joke, a picture is worth 1024 words. When attempting to represent your systems, to model them and to explain them to others, you will undoubtedly need to use lots of words, but you may also use drawings to supplement the words.

This means that effective drawing tools are in important part of every developer's toolbox. For many types of engineering, drawings such as block diagrams (also known as "boxes with lines and arrows"), a general-purpose tool such as Microsoft Visio, is more than adequate. For more complicated drawings, such as UML diagrams and state machines, it is more appropriate to use drawing tools that are dedicated to that particular task.

The state of the art in software system visualization may be found in Model Based Design (MDB) products such as IBM's Rhapsody. This is much more than a drawing tool, as it allows for graphical system modeling using UML, followed by generation of source code in a variety of languages directly from the model.

### Static source code analysis

Software tools called static source code analyzers perform detailed analysis of source code to detect common programming mistakes. The term derives from the fact that the tool analyzes the source files without actually executing the code, in contrast to dynamic analysis that requires code execution. One of the earliest static analyzers was the Unix "lint" program, which was reportedly named because of its role in picking "lint," or potential defects, out of C language source code.

Many of the error detection features of the early analyzers are now implemented as standard components of modern optimizing compilers. A good example of this is the use of an uninitialized variable, which is a defect that is most compilers now detect and flag. Modern high-performance analyzers, however, go far beyond the simple statement-by-statement level analysis performed by Lint and similar tools. These newer tools can analyze across the entire code base of an application and detect errors that are not currently located by compilers alone. As with most things, you get what you pay for. Licences for these tools may cost as little as a few hundred dollars per user, or they may cost hundreds of thousands of dollars.

---

### Example: Tool costs versus engineering time (by Kim R. Fowler)

I, and others in the industry, have seen a strange and illogical reticence by management to buy software tools. The reticence seems myopic because of the simple estimates and calculations can show how much tools can save a project in time and money!

Consider a US$5000 tool that analyzes software within a few seconds. An engineer might spend upward of a week in mind-numbing manual analysis of the same software. If an engineer has a loaded cost of US$100 per hour for salary and benefits, that week of manual analysis just cost US$4000. The breakeven point is less than nine days of engineering time. Furthermore, that is also a delay of a week that the tool could have done in seconds.

---

Like many automated systems, static analyzers have positive and negative aspects. The positive side consists of effectively having a tireless and super-precise language expert on your team who can rapidly and repeatedly examine your code and locate a wide variety of standard defects. Code reviews using these tools may be performed any time, at the click of a mouse button. The downside to the use of these systems is the lack of flexibility caused by removing human intelligence from the analysis process. The result is that the systems may be prone to the detection of "errors" that aren't really errors, otherwise known as false positives (also called false alarms), and the failure to detect actual errors, which are known as false negatives.

False positives make it difficult to separate the wheat from the chaff. If there are too many false positives, the user may be overloaded to the point that actual errors are not discerned among a flood of unnecessary warnings. False negatives, on the other hand, are at least as dangerous as false positives because they can provide the developer with a false sense of security about the correctness of the code that has been analyzed.

Analyzer performance therefore lies somewhere along a spectrum. You usually adjust the sensitivity of the analyzer from one end of the spectrum to the other. One end of the spectrum involves maximum analytical sensitivity, in which all possible errors are detected, along with a potentially very large number of warnings about harmless cases. The other end of the spectrum involves minimum analytical sensitivity, in which few if any false positives are reported, as well as a potentially missing a large number of actual defects that are also not reported.

To reduce the number of false positives in most cases, my experience with static analyzers has been that it's best to reduce the sensitivity and accept the occasional false negative. The reason for this is that an excessive number of false positives (i.e., false alarms) induces the "boy who cried wolf" syndrome in developers, with the result that they may miss the detection of actual errors, if the tool is even used at all. In simple terms, excessive false positives (i.e., false alarms) can result in a tool being used poorly or not at all. A tool that you regularly use, which detects 75% of errors, is better than a tool that detects 80% of errors but is not used because of the workload caused by excessive false positives (i.e., false alarms).

Note that this advice applies to most cases, but not every case. Deciding how to set the analytical sensitivity of a static analyzer must take account of the nature of the system under analysis. For a complex, high value, and high profile system with safety- or mission-critical aspects, you may best use maximum sensitivity; the cost of working through a large number of false positives is likely to be much less than the cost of a system operational failure that results in a high profile mission failure or loss of life. For less critical systems, you may reduce the sensitivity of the analyzer as described to reduce the costs of using the tool, and to ensure that it is actually used.

## Review

Human beings are fallible creatures. Even under the best of circumstances people will make mistakes, and software development is an operating environment that is so far from "the best of circumstances" that it sometimes appears to have been designed to humble us by providing constant reminders of our imperfections.

The problem is caused by the level of abstraction involved in producing a piece of software. People perform at their best when they are doing, rather than thinking. I could toss you a pen, you could catch it without thinking. In fact, thinking about what you need to do to catch the pen would probably cause you to fail to catch it; the same applies to most physical actions that we perform on a daily basis such as walking, talking, and driving a car.

On the other hand, deliberate, extended, careful cognition related to complex concepts is something that humans can do amazingly well, but it isn't natural and it isn't easy. When that intense cognitive effort is devoted to an artificial activity that is almost entirely divorced from real-world physical concepts, then we find ourselves in a position where we are pushing the boundaries of what our minds are capable of achieving. So, we make mistakes. Lots and lots and lots of them. Egregious, colossal blunders, delicate errors of almost gossamer subtlety and complexity, and everything in between.

It seems like an impossible situation. How can we possibly achieve our desired goal of producing reliable software? We are inherently imperfect and prone to errors and we are doing something that is unnatural and difficult.

The first step involves lessons in humility and acceptance. Be humble. Accept that both you and your colleagues will make mistakes. Promote a "mistake friendly" environment, because after all "experience" is just another name for, "I recognize that mistake, because I made it before."

The second step involves, you guessed it, developing a process. In this case, the process required to mitigate the risks associated with the introduction of errors into our systems involves the review of work products by other, qualified people.

> *I strongly believe that the single most effective tool for locating defects during all stages of the software development process is peer review.*

The reason I believe this so strongly is that my experience over many years has been that people are their own worst enemies because they don't like to find their own mistakes...so they often don't. This goes back to the lesson in humility that I mentioned earlier. Most engineers are highly intelligent, highly educated, and capable people. We invest a lot of ourselves, our pride and our ego, into the things we make, so at a deep level it is psychologically very difficult for us to acknowledge that something we have worked hard to produce is imperfect, flawed, or broken in some way. The result of this is that we are frequently blind to our own mistakes because deep in our hearts we aren't humble enough and we really don't want to find those mistakes.

The people sitting next to us, on the other hand, have none of their ego or personal identity involved. Your colleagues will, in a general sense, want you to succeed in your efforts so that your team and your organization as a whole will prosper, but their pride and sense of self-worth comes from their involvement with their work products, not yours. This means that your colleagues are able to view your work dispassionately and without any of the psychological baggage that you bring to it. They will be able to quickly find flaws and suggest improvements that you would struggle to locate, if you could locate them at all.

Ultimately it comes down to a very simple choice. You can get your colleagues to find the flaws in your work or you can get your customers to find them. The first option is simple, effective, and very rewarding. The second option is often disastrous.

In contrast to all this talk of pride in workmanship, there is also another deeply human characteristic that is mitigated by the prospect of peer review. That characteristic is, to put it bluntly, laziness.

Even the most diligent of engineers is subject to this temptation. It would be a very rare individual who hasn't at some time thought, "It's Friday afternoon and I'm tired so I'll just hack something up here. Nobody will ever know." The last sentence in that little trail of thought is the critical one. If it was indeed true that nobody would ever know about the quick and dirty hack, then it would be very easy to submit to this temptation. However, in an environment of high-quality peer review, it makes no sense to do so. The train of thought becomes, "It's Friday afternoon and I'm tired, but if I just hack something up here, the team will pick it up at the review next week and I'll have to do it again, as well being embarrassed." This train of thought has a better outcome.

There's no doubt that reviews need to be performed, but what should be reviewed? In my view, and in an ideal world, all work performed during the development process should be subject to peer review. Requirements, design, implementation, testing, release, and installation. Everything! Unfortunately, few of us live in an ideal world. It generally takes longer to create something than it does to review it, but having every work product independently reviewed in detail by one other person would almost double the size of the team required to do the work. Shrinking budgets and tighter schedules mean that very few organizations are able to devote resources of this magnitude to their review activities.

The best bang for your buck is gained by performing review at all stages of the development cycle, high-quality review of a design is almost pointless if the subsequent implementation is not also subject to review; it could result in a great design that is poorly implemented. Keep in mind also that work doesn't stop on most nontrivial applications once Version 1.0 has been delivered. The processes applied to producing a great product should also be applied to maintaining the quality of the product throughout its life span.

I mention this because of cases I have seen where development organizations have completed projects and then focused all their attention on the next project, to the detriment of the older products. The new work is more interesting than fixing defects in other people's code and the attitude that, "It's only maintenance, so we'll let the intern handle it" has developed, with the result that the quality of subsequent releases diminishes.

***Building a reputation for producing high-quality products takes a long time, but that reputation can disappear in a day.*** You should take care to ensure that this doesn't happen. It may not be glamorous, but applying good processes, including extensive peer review, is just as necessary after the initial delivery of a product as it is during development.

As well as spreading the effort of performing reviews throughout the development cycle, you should allocate the greatest effort first to the areas of greatest risk and complexity within the design. This doesn't mean though, that less critical areas of the code shouldn't be subject to scrutiny if there are resources to do so, as it's not unusual to find a high density of defects in such code. This can be for a variety of reasons such as having less important code being produced by less experienced developers, or simply because it was developed in haste.

## Case Study: Review and tools

There is no doubt that performing effective reviews of other peoples output falls into the category of "hard work." This is particularly the case when reviewing source code, which is renowned as being a difficult and tedious task. There are a variety of techniques for performing code reviews, which historically have been variations on the theme of getting a group of colleagues to read the code and then getting together in a room to work through it line by line on hardcopy listings or on a screen.

Thankfully, software tools have recently been developed by both commercial organizations and open source teams that can increase the productivity of code review teams by an order of magnitude. These *code review collaboration tools* typically provide a means such as a browser interface for a group of developers to share a common view of a set of source modules. Along with the shared view of the code comes the ability to interactively annotate the code in a similarly shared manner. The result is that widely distributed development teams are able to generate rapidly shared threads of comments about aspects of the code in a manner similar to the threads of conversation on an Internet blog.

My experience with these tools has been that they minimize the overheads and tedium of performing reviews while allowing those involved in the reviews to perform their work with great flexibility and efficiency.

An effective peer review process can be a key component for bringing a group of individual workers together and turning them into a cohesive and supportive team. Unfortunately, the converse can also apply. A poorly implemented review process can be divisive and disruptive, and even result in reduced productivity. Unsurprisingly, human nature is once again the source of the problem. *(Again, see Chapter 2 for support and evidence—Ed.)*

Everybody likes being praised. Nobody likes having their failings pointed out. It hurts even more when your foul-ups are diligently noted and dissected in front of your peers, who are some of the people you want to impress most in this world with your expertise and professionalism. Therefore, reviews are inherently a threatening experience at a fairly deep level. Reviews are potentially very uncomfortable for those on the receiving end. From a management perspective, this means that implementing a review process can be very difficult because the people who will benefit most from the process are those who are most likely to resist it. I know of several cases where managers within organizations have attempted to implement review processes and failed due to the resistance put up by the employees. In all of those cases, the employees and the organizations are poorer for the lack of review.

Understanding the extreme sensitivity of this issue is critical to a successful review process implementation. It's unlikely that it will ever be easy sailing, but there are some things that can be done to increase the prospects of success.

The primary reason for introducing a review process is to increase product quality, which translates to a reduced number of defects in the released code. You need to sell the engineers involved in the process on this. The best selling point is that at the end of the day, the credit for producing a high-quality piece of software goes to the developer, not the reviewers. In other words, the reviewers are there to make the developer look good for free. The engineers who embrace this attitude are most likely to be the ones who are smiling at bonus time.

Nobody is ever going to end up smiling though, if they feel like they've been personally attacked in a review. Reviews must be entirely professional and impersonal. Comments have to be about the product, not the person, because there's a world of difference between a comment like, "Reference to uninitialized variable" and "What sort of jerk forgets to initialize memory before using it?"

Senior engineers must embrace the process and participate to set an example. Allowing senior staff to be exempted undermines the process. People still make mistakes no matter how senior they are and junior staff members will be well aware of this. The senior staff members should also be engaged in ensuring the general quality of the process.

Finally, ensuring that reviews are taking place and that the process is effective and productive requires continuous management oversight. Proper management oversight will ensure that reviews are conducted in a spirit of mutual cooperation to maximize their benefits and minimize the possibilities for disruption and ill-feeling between staff members.

Effective peer review of work products really is the silver lining in the cloud of human fallibility that casts its shadow over our attempts to develop high-quality software. I view it as the one great marker of process maturity within a development organization.

---

### Case Study: Ensuring that reviews happen, and that they are effective

When I started introducing more comprehensive reviews into my current team, I had to consider how I could ensure effective management of the process. I wanted to ensure that reviews were actually taking place, I wanted to ensure that they found problems effectively, and I wanted to ensure that they were personally positive rather than destructive for the people involved.

The approach I took was to mandate that I be invited to participate in **every** review, with the understanding that I would actually only be able to participate in a small number of them.

The result of this is that simply by looking at the invitations, I can see what is being reviewed, and by whom. Occasionally I actually participate in a review. This allows me to contribute to the review at a technical level, but more importantly it allows me to evaluate the effectiveness of the ongoing operation of our review process. I can assess whether all staff members involved are participating effectively or not, and I can ensure that the communication with the review is professional, courteous, and supportive.

This simple technique allows me to perform regular reviews of the review process itself, ensuring its ongoing quality and benefit to the organization.

Chapter 13 gives more detail on implementing reviews and review processes.

## Test, verification, and validation

The test regime for your system is the final barrier that stands between your defects and your customers. If you test your products effectively then those memory leaks, intermittent crashes, communication failures, and other embarrassing flaws that weren't picked up by all of your other processes will not enter public consumption. If you don't test your products effectively, then if you are very lucky you will end up with some egg on your face and nothing more. If you are not lucky, which is quite likely, then your organization will surely suffer financial loss and the loss of reputation among your customers. In the worst case, someone may be injured or killed as a result of your poor product testing.

One of the principles of military defensive operations is that defenses should be deep, consisting of multiple, mutually supporting layers. The idea here is that even if an enemy breaks through one layer of defenses, they will be stopped by one of the subsequent layers.

The same principle applies to software system testing processes. You need to have multiple, mutually supporting layers of testing strategically located throughout your development process so that a defect that slips through one layer of testing is caught by a subsequent layer. Failing to do this was the cause of many problems for projects that used waterfall development life cycle models. Detailed testing didn't commence until late in the project when most of the development work was completed. If the tests weren't perfectly comprehensive, and they usually weren't, then quite serious errors could slip through the tests into the released products.

The first line of defense consists of unit tests. As the name implies, unit tests are performed on low-level functional software units, which are typically individual functions within a module. These tests typically ensure the correct operation of a function bypassing in a series of known values to the function, and ensuring that the function correctly transforms the system state and returns the correct value according to the requirements that the function was built against.

There are very useful *unit test frameworks* available for most programming languages that allow for the automatic generation and operation of extensive unit test. These frameworks increase productivity by reducing the workload associated with creating, executing, and evaluating the results of unit tests.

An interesting paradigm that has evolved from the Agile Development Movement is Test Driven Development (TDD) [13]. The essence of this idea is that you should write unit tests before writing the application code, then the aim of writing new application code is to force failing unit tests to pass. Developing code in this manner has a number of benefits, the primary one being that it actually forces developers to produce and maintain their unit tests early. This means that the tests are then always readily available at any time to verify the basic functionality of the system and guard against the introduction of defects into previously working code.

The next line of defense consists of integration tests, which verify the correct interoperation of different independent modules within the system. Integration tests confirm that the modules on either side of an interface specification agree about how the interface should work and then process the interface correctly.

Integration testing frequently produces surprising results with the culprit usually being the imprecision of the English language. It's surprisingly easy for two groups of people to read the same interface specification and come up with completely different interpretations of various areas of the specification. Different specifications mean that various components of the interface don't work together properly on the first try. Because of this, it's extremely important that integration testing thoroughly exercises all areas of the interface.

The third line of defense is system level testing. In the case of most embedded systems, this type of testing isn't specifically software related. The point of system level testing is to forget about how the system functionality is implemented and simply test the system as a whole according to its requirements. At this level of testing, the system could be built from software, programmable logic, discrete components, optics, and mechanics.

System level testing typically revolves around the related concepts of Verification and Validation (V&V). The process of system verification involves ensuring that a system has been constructed according to all necessary regulations, specifications, requirements, and other conditions that may have been imposed on the construction on the product. The process of system validation involves ensuring that a system has been constructed in a manner such that it meets the needs and intent of its customers.

Verification is objective; it quantitatively tests the metrics of the requirements. Validation can be more subjective, "Does the system meet the customer's intent?"

In simple terms, validation asks, "Are we building the right thing?" and verification asks "Are we building the thing right?" It's important to understand the difference between these terms, because to be successful you need to be able to answer yes to both of those questions. If you build something correctly according to all regulations and specifications then it will verify correctly, but if it doesn't pass validation then it simply isn't what your customer wants, and you may well end up back at the drawing board.

An important aspect of the V&V process is that it is most effectively performed by an independent group who have not been involved with the development of the product. This ensures that the results of the V&V process are not subject to the natural human biases of those who have labored to produce the product.

Chapter 14 gives more details on implementing test and integration.

## Conclusion

The military radar systems that my team and I work on are used to shoot supersonic missiles out of the sky with other supersonic missiles, which is not a trivial task. The systems are incredibly complex and the task they achieve is equivalent in difficulty to using a rifle to shoot a bullet out of the sky with another bullet, with the consequence of failure being the deaths of hundreds of people. But we don't fail. The software embedded in our radars works reliably and well, and the reason for this is that the code was developed using the simple and effective techniques described in this chapter.

These techniques have worked for me, and whatever your application domain may be, applying these techniques to your embedded software will help you to build high-quality embedded systems that are delivered on time and on budget.

## References

[1]   J. Ganssle, M. Barr, Embedded Systems Dictionary, CMP Books, San Francisco, CA, 2003, pp. 90–91.
[2]   N.G. Leveson, Safeware: System Safety and Computers, Addison-Wesley, Boston, MA, 1995 (Appendix A).
[3]   ISO 11898 Controller Area Network (CAN).
[4]   IEEE Software Engineering Body of Knowledge (SWEBOK). <www.swebok.org>.
[5]   R. Scott, In the Wake of Tacoma: Suspension Bridges and the Quest for Aerodynamic Stability, ASCE Publications, Reston, VA, 2001.
[6]   D. Vaughan, The Challenger Launch Decision: Risky Technology, Culture and Deviance at NASA, University of Chicago Press, Chicago, IL, 1996.
[7]   ISO/IEC 12207 Systems and Software Engineering—Software Life Cycle Processes.
[8]   A. Albrecht, Measuring Application Development Productivity, Proceedings of IBM Application Development Symposium, IBM Press, 1979, pp. 83–92.
[9]   ISO/IEC 24570, 2005 Software Engineering—NESMA Functional Size Measurement Method Version 2.1—Definition and Counting Guidelines for the Application of Function Point Analysis.
[10]  K. Beck, et al., Manifesto for Agile Software Development. <www.agilemanifesto.org>.
[11]  DOD-STD-2167A, Military Standard: Defense System Software Development, United States Department of Defense, 1988.
[12]  B. Meyer, Object—Oriented Software Construction, Prentice Hall, Upper Saddle River, NJ, 1997.
[13]  K. Beck, Test-Driven Development: By Example, Addison-Wesley Longman Publishing Co., Inc., Boston, MA, 2003.

## CHAPTER 12

# Security

**Eugene Vasserman**
*Kansas State University*

**Chapter Outline**

## *Overview*

Historically, system security has received little attention in embedded systems. Surprisingly, this was actually a reasonable policy to follow in most cases because embedded systems were typically isolated from potential sources of attack. An engine management system, which is isolated in the engine bay of a motor vehicle and connected only to the engine, provides few opportunities to compromise the system by a hostile actor.

Developing and Managing Embedded Systems and Products.
DOI: http://dx.doi.org/10.1016/B978-0-12-405879-8.00012-X

Increases in connectivity, however, have accompanied increases in processor power throughout the world of computing; embedded systems have joined this trend. As a consequence, previously isolated subsystems tended to be designed so that they could be connected to each other within a larger system. In the automotive world, the widespread adoption of the CAN bus [1] provided a reliable and inexpensive means for linking together the disparate control and monitoring systems within a motor vehicle. This development provided many benefits, but it also meant that the engine management system no longer remained isolated.

The trend towards increasing connectivity continues. Many embedded systems now have Internet connectivity, which is often implemented to provide remote system diagnostics and upgrade capabilities. Unfortunately, these enhanced capabilities carry with them the risks associated with exposing the components of the system to unauthorized external assaults, i.e., hacking.

The security of embedded systems was once an afterthought, if it was thought of at all. Now embedded systems play a massive part in the operation of our modern world, and many of these devices expose their presence on the Internet or through other means. This means that you must design security into the systems from the very start if they are to operate robustly in this unpredictable environment.

## Correctness, safety, and security

Security in embedded devices is a daunting task—the challenging problem of security is compounded by resource limits which are far more restrictive in embedded systems than in desktop (or larger) systems. Embedded systems have been designed almost exclusively with the assumption that anyone incorporating them into a larger system will be responsible for security considerations. For this very reason, however, you must not ignore security when designing embedded systems. We have seen time and again how seemingly benign actions on the part of the integrator or system owner have opened the door to potentially catastrophic failures when composing multiple systems or allowing remote control of embedded systems. An operator connecting an industrial control system to a modem may have no way to redesign the system to include security controls, but has a strong incentive to enable remote access, potentially leading to a disaster in terms of safety, simply because security was not considered.

### Security and you (the developer)

Good security design is a necessary part of good safety design, and yet even the most skilled engineers and developers frequently struggle with security, perhaps because of the difficulty in differentiating security from safety. Safety engineering is fascinating and challenging, but successful safety engineering falls short of successful security for one simple reason. Safety generally implies protection from environmental factors and human error. The environment, while unpredictable, is rarely actively out to get you. The environment does not "know" the strengths and weaknesses of our safety systems, and the environment does not consciously plan and seek out problems.

### Other players

Security engineers like to talk about "thinking like an adversary," i.e., an intelligent, conscious, planning entity who knows how to exploit weaknesses, and do so with stealth and style. We can think of security as a game between us (the engineers, designers, and developers) and an adversary (who may, in fact, be several people). If you can place concrete or probabilistic bounds or indefinitely prevent the adversary from interfering with the operation of our system, then you win. Otherwise, the adversary wins.

The safety and security communities appear to have diverged (before my time), and speak different languages and think in different ways. "Security people" consider things (systems, protocols, etc.) to be broken unless proven otherwise. "Safety people" will prove some systems correct, but will design others with as many fail-safes as possible, until they can't think of anything that can go wrong which hasn't been prevented (or which is so rare that preventing it would be too expensive and not worthwhile). Safety proofs of correctness are set in an environment which acts according to some basic assumptions—lightning does not *target* vulnerable components, the probability of a particular environmental event is roughly known, etc. Security researchers will explicitly define adversaries' technological limitations and will assume that adversaries are arbitrarily smart—just because you can't think of a way to break your system, doesn't mean no one can. Therefore, assume a broken system unless proven otherwise. You generally start from either a blank slate or the current state of the world, and add features while ensuring we preserve security. Safety engineers *seem* to work with a given feature set or requirements, and will add mitigation mechanisms for potential safety issues. Security engineers don't necessarily assume that we can enumerate all potential issues, and so safety and security approaches seem to tackle a similar problem from opposite ends.

### Definitions

For our purposes, we can think of security as a set of guarantees that are enforced under a given set of assumptions in the presence of an intelligent adversary. An assumption so common that it's generally not even mentioned is that adversaries are subject to the laws of physics and information theory, i.e., they have bounded communication speeds and computing time at their disposal. In general, it's safest to assume that the adversary controls everything, including the operating environment and communications channels (such as the internal network), and then specifically state assumptions that particular system components, such as the physical hardware of the device, are *not* under the control of the adversary.

## Cryptography

To achieve our desired security properties, we want to approach the problem of security rigorously and scientifically, meaning you will use security engineering and cryptography. They are not to be confused with each other—cryptography is often defined as the process of creating security "codes," while security engineering can be defined more broadly as the science of creating secure systems. Cryptography is a broad set of tools we will use—security engineering is what we will do, cryptography is the bulk of the toolkit we will use to do it. (Security engineering can be an art and a science simultaneously, I'll discuss this in slightly more detail later.)

The reason to use cryptography is twofold: it imposes rigor on your design process, and it lets you use tools created by others. Many people make two critical mistakes when designing security: they rely on their skill alone and don't use known cryptographic concepts (and assume they are smarter than their adversaries), and they reimplement existing cryptographic solutions or tools from the literature. This methodology is problematic. Let's address these points separately.

Designing and building your own security tools is difficult even for seasoned security engineers and architects. Bluntly, *it's hard*! It is difficult to ensure your designs are sound, and difficult to implement those designs correctly. Reading about the problem, although necessary, is not sufficient. (Recommended books—see the recommended reading at the end of the chapter.) Security tools are best built by dedicated teams with years of practice, and should undergo extensive testing and auditing. This is why it is better to use existing tools, whether commercial or free (see Component (COTS) Reuse section). It is always desirable that the code of these tools is publicly available for community audit, although some companies provide code to customers by request, it is difficult (and time prohibitive) to audit external code.

## Careful design and review

"Given enough eyeballs, all bugs are shallow" [2]. This quote brings us to our second point: use cryptography rather than design an *ad hoc* solution. Even if you, as the developer, are very clever (and you must be, given your job requirements), the totality of all adversaries is smarter. Furthermore, once designed, your system is deployed and is difficult to change. You have a finite amount of time to design and build the system, but *adversaries have an infinite amount of time to break it*, after it is deployed. Using known techniques and robust, audited tools saves time, money, and reduces risk.

### Example: Obscurity and review

There is a curious story about a piece of software that provides remote access to workstations. Let's call the manufacturer "Foo," software "Bar," and let's say its current version is 12. Somehow, the source code of Bar version 9 came into the possession of a group of malicious actors, who attempted to extort money from Foo to not release the source code. While these negotiations were ongoing, Foo conducted several code reviews and found critical bugs in Bar version 12. Note that the only barrier between a workstation running Bar and the open Internet is the username/password authentication mechanism used by Bar. We would expect Bar to be highly secure, such that no one without a valid username and password should be able to use the workstation remotely. Yet these are exactly the flaws uncovered by the code review.

Foo continued to perform reviews even after releasing the initial round of patches, and when negotiations were breaking down and it was clear that source code for Bar version 9 would be released, Foo advised all Bar users to stop using Bar. Consider the implications of this: code reviews (done only under threat of exposure of old code) turned up devastating defects in a security-critical piece of software. Given the criticality of Bar's security, reviews of this thoroughness should have been conducted long ago, and continued throughout the software lifecycle. Even worse, Foo was implicitly saying that they had little confidence that Bar version 9 code could not be used to exploit Bar version 12, even after so many iterations of quality assurance testing (versions 9–12). This is a product that must have been vulnerable for more than a decade, and the only thing that led Foo to thoroughly evaluate its security was the threat of source code release. Any malicious person of moderate skill could decompile a binary to extract something resembling source code, therefore vulnerabilities were easy to find if you are looking for them maliciously, but hard to find if you're looking for them for research or evaluation purposes.

## Desired properties

Security is many things to many people and is highly dependent on the specific application. There are many possible security properties, and there is no single, complete definition of "secure." Security is defined according to the application; it includes an adversary with particular resources and capabilities, and security properties which the system will guarantee even in the presence of those adversaries. Here are some basic concepts:

*Confidentiality*: One of the most basic requirements, we want to keep data from being seen by unauthorized parties.

*Integrity*: Data (I will generally refer to data and code interchangeably) has not been altered. This guarantee is necessary whenever devices communicate with each other. Note that the source of the data may be adversarial, we don't know, however, so integrity says nothing about the source of the data. For that we need authenticity.

*Authenticity*: The origin of data is identified. Note that authenticity implies integrity, if integrity is broken, we can't have authenticity and it's unlikely that the data source legitimately produced corrupted data.

*Availability*: The system should remain operational and accessible, even while under attack. This property generally requires quality of service guarantees to be subsumed by the security realm. The system should provide service guarantees even when malicious outsiders or insiders attempt to take the system offline or reduce the resources (e.g., bandwidth) available to the system. This is often a very difficult property to achieve.

*Nonrepudiation*: Another property you may want from authenticity is *nonrepudiability*, which is messages can be traced to their sources using authentication information, and no party other than the source named in the authentication tag could have written the message. This is particularly important for failure analysis.

*Deniability*: Another potentially desirable property, although one rarely used, is deniability, or the ability for a message author to plausibly deny having authored the message. This is a particularly needed requirement in privacy-preserving systems [3].

*Privacy*: An umbrella term for a number of technologies (including deniability), this is not in and of itself a security property, but rather a system goal. All systems should attempt as much privacy as possible unless and until it contradicts other requirements, such as accountability.

*Accountability*: Almost the opposite of deniability, accountability is closely tied to nonrepudiation, but instead of preventing the author from denying writing a message, it guarantees that every message can be traced to a particular source. In some edge cases, you can design systems where you have your cake and eat it too—you are anonymous until such a time as some parties decide that you have misbehaved, in which case your anonymity may be revoked [4].

*Tamper evidence and tamper resistance*: When all else fails, log. However, electronic logs are vulnerable to tampering, so you will want to use tamper-resistant and tamper-evident logging whenever reasonable [5,6].

### Assumptions

Since I have differentiated between "us" and "them," the designers and the opponents, keep in mind that a designer may become the opponent. It is therefore very important to be clear about your assumptions. One of the most important is Shannon's maxim, which is "the enemy knows the system." This is an extension of one of Kerckhoffs' principles, which can be summarized as "the only thing that should be secret about a security system is the secret key itself." If developers "cross to the dark side" then it is clear why you would want to design a system whose architecture can be made public.

There are, however, more subtle reasons to publicize everything even given full confidence in your developers. One is to use the resources of a large community of security researchers who are more than happy to analyze security mechanisms (for free!) and correct any errors that they find, producing an overall better product. More importantly, keeping security mechanisms secret significantly increases the burden for friendly researchers but only slightly increases the burden for unfriendly ones. Remember that adversaries are motivated to break your systems, so they will spend time reverse engineering your security mechanisms.

Friendly researchers, on the other hand, are less inclined to determine how a secret system works. So, exposing the internal architecture (and even code) for your system encourages friendly improvement and makes latter adversarial attack less likely given that a number of well-meaning experts have already examined the system. Any problems that remain are likely to be vulnerable to only the most sophisticated attackers.

Another common assumption is that the adversary controls the network and even the physical environment of the device [7–10]. You make this assumption because you simply don't know how the device will be deployed and used and you can't rule out that your adversary will be your network administrator. Devices that ship with no network connection may later be upgraded to include network interfaces! Consider the difference between control systems in the automotive and aviation worlds: neither was designed with user modification or connectivity in mind, and yet the availability of unobstructed physical access to cars has made attacks against automotive control buses possible and practical [11,12]. One of the few obstacles still standing between aviation control and such attack is the visibility of airline passengers to each other; anyone trying to break into an airliner's avionics is visible to a number of crew and passengers. Note that we can consider this a form of accountability and temper evidence. Visibility makes *all* the difference between avionics and automotive (or other critical protection systems). Because avionics tampering is very visible, the security decisions in avionics are *somewhat* justified. But if you take avionics standards and put them into cars or into medical systems, they break because you've violated an implicit assumption.

### Roots of trust and the Trusted Computing Base (TCB)

A reasonable starting point for designing security is to consider your "root" of trust, meaning the fundamental source of trust in your system. It may be the developer, it may be the owner, it may be a third party, but you must know who it is. Once you do, you can extend that trust to other components of the system. It is crucial to **ensure that the root of trust is trustworthy**, and stays that way, since a malfunctioning or malicious root of trust can destroy trust in the entire system. Note that the root is "trusted by definition," since you must bootstrap trust in the system somehow.

Trusted and trustworthy must align in this situation. Consider the case of trusted hardware, such as smart cards, or the Trusted Platform Module (TPM) [13]. The hardware itself is trusted (although it may not be the root of trust), but it is placed in the hands of the users, who are considered untrusted. This is a disastrous situation from the point of view of security, and we have seen it play out many times in the case of rogue cable and satellite boxes [14,15], smart cards, and gaming consoles (see "chipping" [5]). Giving trusted hardware to untrusted and untrustworthy users makes the hardware itself untrustworthy; how, then, can you expect the hardware to enforce your trust? In fact, the TPM specification explicitly excludes an adversary willing to modify hardware (this is often overlooked or forgotten).

Consider, on the other hand, cryptotokens, commonly used as an additional access control device for corporate networks. They are used to augment users' passwords, as a "second factor" in authentication. They strengthen the usual username/password authentication paradigm rather than replace it. So, just like passwords, a user is incentivized to keep the cryptotoken safe and secure, make sure it's not lost or stolen, and prevent others from tampering with it. In this example, your trusted and trustworthy components align, the user is trustworthy (because you've already allowed them into your networks by giving them a username and password), and they will hold an additional trusted component (the token), which they will want to protect, much like their password.

### Bootstrapping and extending trust

Once you have a root of trust, such as in my last example of a user authenticating with a username, password, and a smart card or cryptotoken, you can extend that trust to create a larger trusted system. TPMs and cryptotokens are meant to provide a hardware root of trust (sometimes in addition to user credentials). An operating system or software may provide a root of trust as well, although this is less reliable, since more components must be nonmalicious for this trust to be warranted. Trusted software running on untrustworthy hardware is fundamentally vulnerable (e.g., directly changing the content of RAM or EPROM using ultraviolet light; also see "chipping" [5]). Most of the time you have no choice—when doing online banking, you assume that your hardware, operating system, and browser are trustworthy, and won't misuse the credentials you type into the bank online portal. In embedded systems, however, you start from the bottom and can ensure a higher level of trustworthiness.

## Security engineering

### Art versus science

Depending on whom you ask (and when), engineering for security is an art, a science, or both [16]. Since you want to be rigorous, since you want some tangible evidence that you designed the appropriate security controls (and that they work correctly), you will use cryptography, one of the major components of security, and a strongly scientific one. Cryptography is generally based on mathematics or at least a very good guess as to the security of a particular algorithm [17,18]. You want to rely on others' assertions regarding the security and soundness of the basic algorithms, so you rely on known-good cryptography implemented as part of operating systems or in external libraries such as OpenSSL, GNUTLS, Mozilla NSS, and libgpg. You must ensure that you are using these libraries correctly [19]. In the highly unlikely case that particular security features you're looking for are not available in commercial off-the-shelf (COTS) tools, published scientific research frequently holds the answers (even if they haven't been implemented).

***It is very dangerous to design cryptographic tools yourself; don't do it!*** If you absolutely must implement some cryptography, make sure you implement vetted, peer-reviewed work and do not build algorithms *ad hoc*. Remember to think like an adversary; just because you can build something you can't break, doesn't mean no one can break it given an arbitrary amount of time and resources.

Good guides for rigorous development exist in the form of standards for building safe and secure software. Frequently, these standards can help you keep track of requirements and trace implementation details to requirements to ensure you're not missing any required security features, nor that you built unneeded security. Unfortunately, standards generally tell you a process to follow; they will not definitively tell you what security properties you need, or when, although they may make suggestions. They will not tell you how to implement the functionality you need or which libraries to use. I will attempt to remedy some of these issues below. Note that this does not make standards useless; standards help you rigorously follow a process to make sure that you don't forget something obvious, but coverage is not complete [20].

Standards are not a replacement for trained security architects or security engineers. Consider that most standards, especially for industrial control and automation, were created under the assumption that the systems being built are isolated. Connecting previously closed control systems to the open Internet, even if you don't advertise their presence, is a recipe for disaster. Previously closed systems become open faster than standards can keep up (also see the "Shodan" search engine for Internet-exposed industrial control systems) [11,21].

### Defining security requirements

***The first step in designating required security and designing security mechanisms for a system is defining the requirements.*** This task should be an exercise for the entire design team, plus at least one experienced security architect. (As discussed before, these individuals are different from safety engineers and cannot be used interchangeably.) You must determine the environment in which your embedded system will be used (e.g., connected to the Internet, connected to a local network, completely disconnected, etc.)

and you will document that as an assumption. You will revisit assumptions periodically to verify that they still hold (otherwise our security design will need to change). (***Editor's Note***: *Defining security in the requirements falls in line with what other authors in this book strongly and emphatically recommend—that important design features need to be defined early in the requirements.*)

It would be helpful to design in a more forward-thinking way, so if the system is initially slated for stand-alone deployment, but Internet or LAN connectivity is a possibility in the future, you should take this into account when defining requirements. You might want to immediately design security for a connected system to allow for future growth. (I have more later on how to justify the "cost" of these future-looking security features.) ***The reason to involve the entire team is so security requirements can be designed alongside functional requirements.***

What are the most important functional requirements of the system other than correctness? For example, if access control is more important than availability, good support for authentication should be built in. If, on the other hand, availability is more important, the system should still include access control, but focus more on logging and accountability of actions, so one may leave access open in case authentication mechanisms somehow fail, but later the logs can be analyzed for what went wrong and what actions were taken. When all else fails, make sure you keep good run-time logs.

Be as pessimistic as possible and define a minimal initial root of trust, extending it as described earlier. Especially distrust cross-component communication and user/network input. Enforce the safety constraints you have identified earlier. Practicing defensive programming? Great! If not, you're asking for trouble. Defensive programming helps you validate assumptions at run-time, protecting vulnerable and fragile code from bad input, helping keep your system in a safe state.

## Reconciling security with functional requirements

Since limited resources are available to implement security, especially in embedded systems, you will want to prioritize your features. For instance, are availability and accountability more important that confidentiality? If so, you will make them your core properties, and you can then leave out confidentiality. However, you should avoid ever eliminating both accountability and authenticity together; at least one of them should be present, otherwise you never know who is interacting with your system.

Remember that money can always be used to express the need for various security features, especially given limited time to design and implement. "How expensive will it be to recover from a failure of property X?" Expense may be reputation rebuilding, developing and issuing software patches, physical infrastructure reconstruction, etc., depending on the security property X. If recovery is infeasible or impractical, a great deal of thought should be put into design of that property—near-infinite costs of failure demand respect and careful thought. (This is called the "**barn door property**" in security systems: once a failure has occurred, it may not be possible to make the disclosed data secret once again, i.e., the horse is out of the barn, and the best you can do is close the door to make sure it never happens again.)

Counterintuitively, you may have more ability to ask for the resources you need to implement security in embedded systems rather than general-purpose (unconstrained, large) systems. The reason is that embedded systems are designed and built for a specific need, so as long as security is not neglected, you can ensure that your system has the resources for good security. Larger systems are built for general purposes, so you are likely out of luck if you need special security considerations to be built in; the system has already been manufactured, and you have no further control over the hardware requirements.

In embedded systems, as long as the arguments for security are sound, there is no reason not to slightly augment hardware requirements to build in strong security. The fractional increased hardware costs are nothing when compared to the disastrous after effects of security breaches [22].

As mentioned earlier, it is important to trace implementation details to requirements (and to trace functional requirements to design requirements) to ensure you do not miss any required security features, nor that you build features unnecessarily.

In terms of availability of security features, it is better to include features you may never need rather than be stuck with an architecture devoid of badly needed components (e.g., after many upgrades). Unused security features rarely impose overhead, but anticipating future need, especially when architecting the overall solution rather than implementing components, can greatly speed latter upgrades, since few or no fundamental changes in architecture would be needed. Reusing your old example of industrial control, it's far better to include authentication and logging and not use it (if there is no external connectivity) than desperately need it later when someone invariably connects the system to the Internet, after repeated warnings not to do so. The earlier that you architect your system to make such changes easy, the lower the cost of the design and implementation. Consider, as always, the cost of supporting a product versus the initial design cost: *up-front design costs, while unpleasant, are exponentially lower than latter costs for support or recovery or upgrade*.

### Planning for inevitable security failures

Even after all that effort, security incidents will happen. Hopefully the causes are easy to fix, such as incorrect configuration, bugs in the implementation, or end users bypassing required security controls to speed up some workflows. Sometimes incidents occur due to more serious issues, such as a violation of an explicit or (worse) implicit design-time, implementation-time, or deployment-time assumption. This is when careful design and contingency plans pay off, and you can put your recovery plan into action.

---

### Example: Thoughts on backups and secondary systems

If you've carefully and diligently performed backups, and your system was compromised in some known time period, you can take the needed time to bring your systems down and recover from a backup predating the incident. You can do this because you've (i) verified your backup plan and (ii) set aside sufficient contingency funds to offset the business losses during the necessary system downtime. An alternate to step 2 would be to keep a known-clean system available (but offline) unless the primary is compromised. You must, however, resist the temptation to put the secondary into use for load balancing purposes (this would make it live, and thus useless for your security purposes).

---

## Building a secure system

The best course of action when designing and implementing any system is to employ a professional security architect as early as possible. There is no replacement for training and expertise; this chapter and all the suggested reading material at the end will neither give you the required expertise nor likely impart the required knowledge without practicing. ***Practicing on products that will go into production is a bad idea.*** Having said this, there are best practice steps you can follow to attempt to raise as little ire as possible from your security specialist(s).

### Security and process standards

A number of standards deal with designing secure systems and with rigorous design, requirement traceability, and developer accountability. We can separate them into "process standards" and "security standards," and they may vary from industry to industry. The go to process standard is ISO 9001 ("Quality Management") and an example for a security standard in the medical space is IEC 80001 ("Application of Risk Management"). Note that your industry may have others, and 80001 is far from the only standard in the medical space [20]. Good examples of security standards are the NIST 800 series for various uses and FIPS 140-2 for the design of software and hardware security products for use in government.

Process standards help you go about your design and implementation in a rigorous and auditable way, while security standards help you chose the security mechanisms to achieve specific goals. As mentioned earlier, standards help you rigorously follow a process to make sure that you don't forget something obvious, but *none of them tell you how to go from design to implementation*. This does not make standards useless; they add rigor to processes which are frequently too hectic, and haste makes it easy to forget even obvious steps [23,24].

In terms of implementation, unfortunately, you're mostly on your own, and this underscores the need to consult trained and experienced professionals and reuse of known-good, audited, off-the-shelf software.

### Component (COTS) reuse

We already discussed reuse of known-good (or at least time-tested) cryptography, so hopefully you have specific algorithms in mind. This does not *necessarily* apply to hardware, as it is significantly harder in practice to examine the internals of hardware implementations than it is to examine software. For a large number of security libraries, source code is readily available for review. ***Cryptographic primitives and algorithms are not by default safe to combine***, so we additionally rely on libraries to give us that functionality. Default settings are generally safe, but always check the documentation to ensure the assumptions made by software developers align with yours. Also, always check documentation to verify what particular functions do, rather than *what you think they do* based on function and variable names [19]. Although libraries have been known to contain mistakes, they are relatively easy to patch. Use hardware implementation when necessary for considering power or CPU or cost, but avoid them when you can use software; you can't easily patch hardware in the field.

### Testing

Although security testing is fundamentally limited—you most likely won't catch bad architecture in tests—it is nonetheless a good idea to put your security mechanisms through the same quality assurance steps as you would for any other hardware and code. Once again, there is no replacement for having a trained security engineer as part of the QA team.

You may catch nonobvious but nonetheless potentially critical flaws by using "fuzz testing," where you expose your products to randomized input of varied size and content. This can point out unforeseen failure scenarios before deployment; bugs of this type are often security vulnerabilities, but would likely be minimized or eliminated when rigorously using defensive programming practices. Testing of this nature is a poor simulation of a smart adversary, who will intelligently probe for bugs rather than randomly.

Sometimes, one can make use of the large "white-hat" (not evil) security researcher community to help you audit your design and implementation. "Break-our-system-for-money" contests are generally not helpful (especially if you keep your design and implementation closed), but openness often attracts helpful feedback. If done before deployment, this can be invaluable.

## Case study: smart meter security

### Introduction to smart metering

Grids supplying electricity and other commodities like gas and water are currently made "smart" by adding information and communication technology (ICT) components. These "brains" may stabilize the grid and optimize the production and consumption of commodities. So-called "smart meters" provide fine-grained, flexible read outs and automatic reporting of the consumed (or produced) quantities.

Smart metering for residential houses comprise a variety of meters and devices, such as metering gateways. These devices and meters contain embedded systems to form nodes in a network. These nodes connect to one or more central head end systems (HES), which collect the metering data, and forward those data to meter data management (MDM) and billing services. Data concentrators (DCs) may gather and forward information communicated between the terminal devices in the homes and the HES. Communication may take place over all kinds of media, including low bandwidth networks like power line communication (PLC), radio signal (GSM), or high-speed lines (DSL). Figure 12.1 depicts a composite scenario.

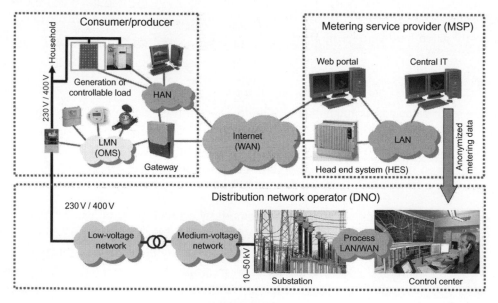

**Figure 12.1:**
Smart metering system overview.

## Security challenges

Smart metering poses functionality, security, and real-time requirements that need to be fulfilled holistically and in technically and economically adequate ways. See Ref. [1] for a more detailed presentation of security requirements and how to fulfill them.

Security threats include the tampering of meter data to manipulate the outcome of billing and the leakage of private information. The biggest concern for private information relates to the lifestyle and monetary situation of consumers. Figure 12.2 shows the local metrological[1] network (LMN), home area network (HAN), and an optional gateway connecting them with a wide area network (WAN). Potential attack points can be grouped into local/physical access and remote access (i.e., over the Internet).

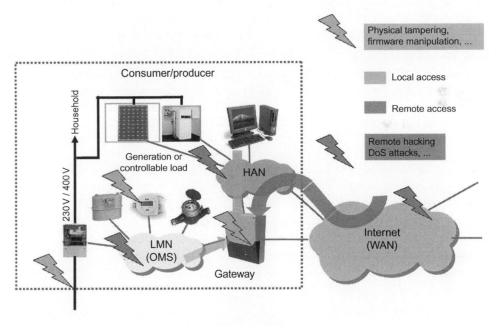

**Figure 12.2:**
Local networks including gateway with attack points.

Countermeasures must protect the overall flow of commands and data among the parties involved. Their effectiveness should be verified by certification, such as that according to the Common Criteria (CC) [2].

Particular challenges arise because of the scale of a smart grid and because its components are widely distributed in the field (and thus expensive to maintain by physical access). For this and other reasons the components need to be very stable and long-lived, which increases the spatial and temporal exposure to potential attackers. Moreover, for the conventional grid (as for many other industrial systems) it has been sufficient to counter security threats by physical means (in particular, simply by locking rooms that house critical

---

[1] Derived from the Greek word for "measurement."

components). Unfortunately, blocking physical access does not hold any more with the substantial inclusion of easily accessible and well-known ICT components.

In several countries, smart meters are already heavily used—Italy currently has the world's largest deployment of smart meters—but with essentially no security in place. For obvious reasons, this has already led to a high volume of fraud producing considerable economic damage (at least as reported for the United States and Puerto Rico) [3]. Local privacy breaches apparently are of minor practical interest, while excessive central accumulation and misuse of personal data could become a real problem [4]. There are currently several national and international groups in research, industry, standardization, and regulatory bodies trying to provide guidance for smart meter security [5,6]. To get a sense just how HUGE these privacy leaks have the potential to be, consider "Multimedia Content Identification Through Smart Meter Power Usage Profiles." This paper claims that you can tell what a person is watching on television just from their smart meter data [7].

### Securing devices in the field

In this case study, I focus on how (not) to secure smart meters and similar embedded devices. For instance, the security regulations for smart metering by the German Federal Office for Information security (BSI) require in their Smart Meter Gateway Protection Profile (SMGW-PP) (as well in the detailed technical implementation guidelines) the use of a so-called security module for secure storage and use of critical key material [8–10]. This has been advocated and advertised by the security chip industry [11]. Yet their claims have not materialized, *"Certified gateways can easily be developed to incorporate smart card controllers. Consequently the implementation of the BSI protection profile should not delay the introduction of a secure smart meter network."*

### Fundamental considerations for use of HSMs

The use of a certified smart card chip reduces the certification effort for the gateway and may improve the personalization process. It does not, however, necessarily improve actual security. Indeed, the use, as specified by the BSI, of a hardware security module (HSM) only superficially increases the security level of the smart meter gateway.

From a security architect's perspective concerning the HSM, there are three types of attackers:

- Remote attackers, which have no access to the gateway hardware. When accessed from remote, a HSM and any alternative software implementation of its cryptographic and storage functionality are interchangeable (except for irrelevant performance differences).
- Local attackers who do not try or do not succeed hacking the gateway (because their motivation or skill level is not high enough). These attackers will not even reach the HSM's functionality, and therefore the hardware HSM is not needed to protect from such attackers.
- Local attackers who do succeed hacking the gateway itself (because they have a sufficiently high attack potential), obtaining control over those hardware and software portions outside the security module. Also if they do not manage to hack into the HSM, they may simply abuse its services and in effect corrupt all uses of the gateway. For instance, even without getting hold of any key stored in the HSM, they may issue signing commands and thus effectively sign any (fake) data in the name of the given gateway.

Consequently, the use of the HSM does not effectively increase assurance.

### Authentication of HSM users

A way to handle the particularly problematic third item would be to authenticate each critical use of the SM. This was not foreseen in earlier versions of the Protection Profile (PP) and Technical Guidance (TR) for the HSM, but has been added to the final versions of the PP andTR-03102-2 [10,12, Part 2].

The HSM can authenticate other parties like the gateway administrator, yet there is no chance for (the rest of) the gateway itself to securely interact with the SM, in particular to authenticate itself using any form of secret stored in the gateway's memory. Despite this problem, the PP contains an assumption that is practically unfulfillable [12].

*Operational Phase*

> *It is assumed that the appropriate technical or organizational measures in the operational phase of the integrated gateway will guarantee confidentiality, integrity, and authenticity of the assets. In particular, this holds for the key and PIN objects stored, generated and processed in the operational phase of the integrated gateway.*

Nevertheless, the use of such a mechanism, called the Password Authenticated Connection Establishment (PACE), is required on the HSM side, and from the HSM perspective, it is assumed that PACE is present on the gateway [10,12, Part 2].

*PACE*

> *The gateway shall securely implement the PACE protocol according to [TR-03110], [TR-03109-3], [TR-03109-2] for component authentication between the GW and the TOE.*

On the other hand, the PACE mechanism is not mentioned in the PP of the gateway [9] and its related TRs. Thus it remains unclear how the PIN needed for PACE is intended to be stored securely. In fact, the fundamental problem of this approach, namely that any attacker who hacks into the gateway also can gain access to any secret like the PIN stored there—and thus misuse its HSM—cannot be overcome.

There is another related design deficiency pertaining to the use of a successful authentication of the gateway administrator. (The same issue was present in a similar form also in the first version of reference SM5 [10, Part 2] available for public commenting.) Man-in-the-middle attacks are possible after the gateway administrator has been authenticated correctly. This is because after successful authentication, the HSM accepts any further command(s) to the SM that arrive via the gateway (on the "trusted" PACE channel between the gateway and the HSM) while there is no direct secure channel between the gateway administrator and the HSM. Instead, the TLS channel with the gateway administrator terminates at the gateway. In other words, after hacking the gateway, as soon as the gateway administrator has successfully authenticated at the HSM, an attacker controlling the gateway may pose as the gateway administrator, intercepting and faking further administration commands. He may even delete and, under certain restrictions, replace any keys stored on the HSM.

There are two other approaches to the problem of local HSM misuse.

### The TPM approach

The TPM approach aims at checking and reporting system integrity [13]. To this end, each node in the system start-up chain, from the CPU to booting the operating system to starting applications, checks the next node in the chain for authenticity and integrity before beginning to execute it. This leads to a hierarchical hash value held on a chip securely bound to the system hardware. In fact, the SM as described by the BSI's PP could be the very same piece of hardware [12].

This approach has been promoted for use in the smart metering domain by several researchers [14,15]. Yet there are some problems with it; the CPU must enforce the use of the TPM chip, any intended changes to the system (e.g., due to firmware updates) must be correctly reflected by an update of the hash value, and any tampering during system run (i.e., after booting) goes undetected. Note that the latter is a fundamental issue limiting the value of the TPM approach because it cannot detect—let alone prevent—temporary manipulations.

### The HSM as security master

The only clean high assurance security architecture that I am aware of for security-critical devices in the field is to allocate all critical operations of the device in an embedded HW security chip. This of course requires more computational resources than a standard HSM, which is a pure crypto slave. Yet with the relatively high performance of state-of-the-art smart card controller (such as the Infineon SLE88, Figure 12.3) and a suitable distribution of tasks between the secure controller and the high-performance main processor, this is feasible. Even the secure off-chip storage of large amounts of data is possible with sufficient efficiency using well-known techniques such as Merkle Trees [16]. This approach has already been used successfully, for instance in the domain of high assurance digital tachographs [17].

*David von Oheimb*
Siemens Corporate Technology, Munich, Germany

**Figure 12.3:**
Example of a TPM.

## Case study references

[1] D. von Oheimb, IT Security architecture approaches for smart metering and smart grid, in: J. Cu'ellar (Ed.), SmartGridSec'12, LNCS, vol. 7823, Springer, Heidelberg, 2013. <http://ddvo.net/papers/SmartGridSec12.html>.

[2] CC, Common Criteria for Information Technology Security Evaluation, ISO/IEC 15408. <http://www.commoncriteriaportal.org/>.

[3] B. Krebs, FBI: smart meter hacks likely to spread, April 2012. <http://krebsonsecurity.com/2012/04/fbi-smart-meter-hacks-likely-to-spread/>.

[4] R. Anderson, S. Fuloria, On the security economics of electricity metering, in: Workshop on the Economics of Information Security, WEIS, June 2010. <http://weis2010.econinfosec.org/papers/session5/weis2010_anderson_r.pdf>.

[5] Open Smart Grid User's Group, Advanced Metering Infrastructure Security. <http://osgug.ucaiug.org/utilisec/amisec/>.

[6] Task Force on Smart Grid Privacy and Security of the Smart Meters Coordination Group, Privacy and Security Approach, Version 0.9, November 2012.

[7] Multimedia Content Identification Through Smart Meter Power Usage Profiles. <http://www.nds.rub.de/media/nds/veroeffentlichungen/2012/07/24/ike2012.pdf>.

[8] BSI, Federal Office for Information Security, Bonn, Germany. <https://www.bsi.bund.de/EN/>.

[9] BSI, Protection Profile for the Gateway of a Smart Metering System, March 2014. <https://www.bsi.bund.de/SharedDocs/Downloads/DE/BSI/Zertifizierung/ReportePP/pp0073b_pdf.pdf?__blob = publicationFile>.

[10] BSI, TR-03109 Smart Energy, <https://www.bsi.bund.de/DE/Themen/SmartMeter/TechnRichtlinie/TR_node.html>, 2012.

[11] M. Klimke, C. Shire, Klimke, Infineon Technologies, Smart Grid Cyber Attacks—Germany Steps Up the Protection, September 2011. <http://silicontrust.wordpress.com/2011/09/23/smart-grid-cyber-attacks-%E2%80%93-germany-steps-up-the-protection/>.

[12] BSI, Protection Profile for the Security Module of a Smart Meter Gateway, October 2013. https://www.bsi.bund.de/SharedDocs/Downloads/DE/BSI/Zertifizierung/ReportePP/pp0077b_pdf.pdf?__blob = publicationFile.

[13] Trusted Computing Group, Trusted Platform Module (TPM). <http://www.trustedcomputinggroup.org/developers/trusted_platform_module>.

[14] R. Petrlic, A privacy-preserving concept for smart grids, in: Sicherheit in vernetzten Systemen: 18. DFN Workshop, pp. B1—B14. Books on Demand GmbH, 2010.

[15] A.J. Paverd, A.P. Martin, Hardware security for device authentication in the Smart Grid, in: J. Cu'ellar (Ed.), SmartGridSec'12, LNCS, vol. 7823, Springer, Heidelberg, 2013.

[16] R.C. Merkle, A digital signature based on a conventional encryption function, in: C. Pomerance (Ed.), CRYPTO'87, LNCS, vol. 293, Springer, Heidelberg, 1988.

[17] Continental Automotive GmbH, Digital Tachograph DTCO 1381, Release 2.0, June 2012. <https://www.bsi.bund.de/SharedDocs/Zertifikate/CC/Digitaler_Tachograph-Vehicle_Unit/0559.html>.

# *Chapter references*

[1] M. Farsi, K. Ratcliff, M. Barbosa, An overview of controller area network, Comput. Control Eng. J. 10(3) (1999) 113—120.

[2] E.S. Raymond, The Cathedral and the Bazaar. <http://www.catb.org/esr/writings/cathedral-bazaar/>, 1999.

[3] N. Borisov, I. Goldberg, E. Brewer, Off-the-record communication, or, why not to use PGP, in: ACM Workshop on Privacy in the Electronic Society (WPES), 2004, pp. 77—84.

[4] J. Camenisch, A. Lysyanskaya, An efficient system for non-transferable anonymous credentials with optional anonymity revocation, in: Advances in Cryptology—EUROCRYPT'01, LNCS, vol. 2045, 2001, pp. 93—118.

[5] R. Anderson, Security Engineering: A Guide to Building Dependable Distributed Systems, second ed., Wiley, New York, NY, 2008.

[6] S.A. Crosby, D.S. Wallach, Efficient data structures for tamper-evident logging, in: USENIX Security Symposium, 2009.

[7] V. Eck, Electromagnetic radiation from video display units: an eavesdropping risk? Comput. Secur. 4(4) (1985) 269—286.

[8] J. Loughry, D.A. Umphress, Information leakage from optical emanations, ACM Trans. Inf. Syst. Secur. (TISSEC) 5(3) (2002) 262–289.

[9] J. Kelsey, B. Schneier, D. Wagner, C. Hall, Side channel cryptanalysis of product ciphers, European Symposium on Research in Computer Security (ESORICS), LNCS vol. 1485 (1998) 97–110.

[10] B. Sunar, W.J. Martin, D.R. Stinson, A provably secure true random number generator with built-in tolerance to active attacks, IEEE Trans. Comput. 56(1) (2007) 109–119.

[11] K. Koscher, A. Czeskis, F. Roesner, S. Patel, T. Kohno, S. Checkoway, et al., Experimental security analysis of a modern automobile, Proceedings of the 2010 IEEE Symposium on SECURITY and PRIVACY, 16–19 May 2010, Berkeley/Oakland, California, pp. 447–462. Published by the IEEE Computer Society, 10662 Los Vaqueros Circle, P.O. Box 3014, Los Alamitos, CA.

[12] S. Checkoway, D. McCoy, B. Kantor, D. Anderson, H. Shacham, S. Savage, Comprehensive experimental analyses of automotive attack surfaces, in: USENIX Security Symposium, 2011.

[13] S.L. Kinney, Trusted Platform Module Basics: Using TPM in Embedded Systems, Newnes, 2006.

[14] USA Today, DirecTV sacks would-be Super Bowl pirates. USA Today. <http://usatoday.com/life/cyber/tech/review/2001-01-26-dtv.htm>, 2001.

[15] K. Poulsen, DirecTV zaps hackers, Security Focus (2001)<http://www.securityfocus.com/news/143>.

[16] <https://www.schneier.com/blog/archives/2013/07/is_cryptography.html>.

[17] A. Bogdanov, D. Khovratovich, C. Rechberger, Biclique Cryptanalysis of the Full AES. Technical report, Microsoft Research. <http://research.microsoft.com/en-us/projects/cryptanalysis/aesbc.pdf>.

[18] National Institute of Standards and Technology, SHA-3 Competition (2007–2012). <http://csrc.nist.gov/groups/ST/hash/sha-3/>, 2012.

[19] M. Georgiev, S. Iyengar, S. Jana, R. Anubhai, D. Boneh, V. Shmatikov, The most dangerous code in the world: validating SSL certificates in non-browser software, in: ACM Conference on Computer and Communications Security (CCS), 2012.

[20] J. King, L. Williams, Cataloging and comparing logging mechanism specifications for electronic health record systems, USENIX HealthTech, 2013.

[21] D. Pauli, Hackers gain 'full control' of critical SCADA systems. SC Magazine. <http://www.itnews.com.au/News/369200,hackers-gain-full-control-of-critical-scada-systems.aspx>, 2014.

[22] A. Greenberg, Lock firm Onity starts to Shell out for Security Fixes To Hotels' hackable locks. Forbes. <http://www.forbes.com/sites/andygreenberg/2012/12/06/lock-firm-onity-starts-to-shell-out-for-security-fixes-to-hotels-hackable-locks/>, 2012.

[23] A.B. Haynes, T.G. Weiser, W.R. Berry, S.R. Lipsitz, A.H.S. Breizat, E.P. Dellinger, et al., A surgical safety checklist to reduce morbidity and mortality in a global population, N. Engl. J. Med. 360(5) (2009).

[24] A. Gawande, The Checklist Manifesto: How to Get Things Right, Picador, 2011.

## Suggested reading

R. Anderson, Security Engineering: A Guide to Building Dependable Distributed Systems, second ed., Wiley, New York, NY, 2008. *The* book to read as a general security primer. It is written in a story-telling style, and very accessible even to those with no prior experience in security. Even experienced security engineers should find compelling.

B. Schneier, Applied Cryptography: Protocols, Algorithms, and Source Code in C, Wiley, 1994. A good primer on implementing security features. Should be complemented with a newer text, such as the one below.

N. Ferguson, B. Schneier, T. Kohno, Cryptography Engineering: Design Principles and Practical Applications, Wiley, 2010. Together with [Schneier 1994], a great way to delve deeper into the challenges and solutions to *implementing* secure systems using modern cryptographic tools.

D. Kleidermacher, M. Kleidermacher, Embedded Systems Security: Practical Methods for Safe and Secure Software and Systems Development, Elsevier, Boston, MA, 2012. A book with good coverage of topics related specifically to security within the embedded systems domain.

## CHAPTER 13

# Review

**Kim R. Fowler**
*IEEE Fellow, Consultant*

## Chapter Outline

Developing and Managing Embedded Systems and Products.
DOI: http://dx.doi.org/10.1016/B978-0-12-405879-8.00013-1

## Introduction to review

Review is a set of feedback paths within system development. The act of review has two primary objectives: (i) to confirm correct design and development and (ii) to expose and identify problems with design, development, or processes.

The US Food and Drug Administration (FDA) defines design review in general terms: "*Design review* means a documented, comprehensive, systematic examination of a design to evaluate the adequacy of the design requirements, to evaluate the capability of the design to meet these requirements, and to identify problems" (21 CFR 820.3(h)) [1]. This definition of design review encompasses a variety of reviews, reviewers, processes, and procedures.

The California Department of Transportation defines technical review along expanded lines: "Technical reviews suggest alternative approaches, communicate status, monitor risk, coordinate activities within multi-disciplinary teams, and identify design defects. Technical reviews can monitor and take action on both technical and project management metrics that define progress" [2].

The Wikipedia definition focuses on using technical peer reviews to find problems: "Technical peer reviews are a well defined review process for finding and fixing defects, conducted by a team of peers with assigned roles. Technical peer reviews are carried out by peers representing areas of life cycle affected by material being reviewed (usually limited to 6 or fewer people). Technical peer reviews are held within development phases, between milestone reviews, on completed products or completed portions of products" [3].

The *NASA Systems Engineering Handbook* describes the mechanisms of technical reviews: "Typical activities performed for technical reviews include (1) identifying, planning, and conducting phase-to-phase technical reviews; (2) establishing each review's purpose, objective, and entry and success criteria; (3) establishing the makeup of the review team; and (4) identifying and resolving action items resulting from the review" [4].

The IEEE 1028−1998 Standard identifies the following five types of reviews:

1. Management reviews, e.g., control gates for decision processes during development
2. Technical reviews
3. Inspections, e.g., identifying errors or deviations from standards and specifications
4. Walkthroughs, e.g., authors of designs present and explain to peers who then examine the requirements or the design
5. Audits, which are part of the configuration management process.

The Project Plan or the Systems Engineering Management Plan (SEMP) should establish the process for conducting reviews. The plan should also describe the content and level of formality for different reviews and how to tailor each review for its purpose and type [2].

### Part of a complete feedback system

Design reviews, coupled with test and integration, are the primary feedback activities to assure adherence to and achievement of the project goals (Figure 13.1). These activities confirm whether your system's design and development are as you planned them to be.

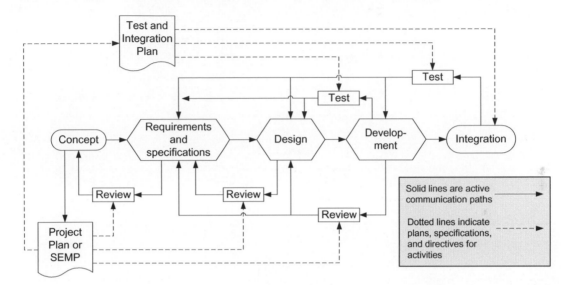

**Figure 13.1:**

Example feedback loops provided by review, test, and integration. © 2013 by Kim R. Fowler. Used with permission. All rights reserved.

### PERRU

Every review should have a set of activities that are both consistent and thorough. Reviews fit into the system engineering structure that can be summarized in the acronym PERRU for plan, execute, review, report, and update. Figure 13.2, which is simplified from the original diagram in Chapter 1, illustrates this structure and data flow. This structure has an interesting self-similarity principle that exists at every level of process, from high-level processes to the low-level processes, finely detailed procedures, within every project or product development.

### Review is necessary

Review is a necessary and critical part of system development. Unfortunately, review receives scant attention in too many situations. ***You should carefully consider and plan review and then perform it regularly***.

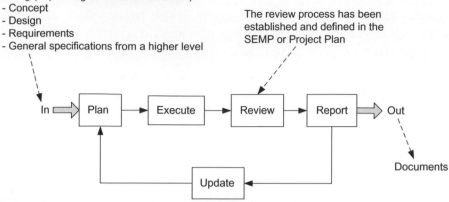

Can be the entire set or a subset of one of the
following (depending on the level of review):
- Concept
- Design
- Requirements
- General specifications from a higher level

The review process has been
established and defined in the
SEMP or Project Plan

In → Plan → Execute → Review → Report → Out

Update

Documents

**Figure 13.2:**
Basic, self-similar unit of operation that suits any level of detail within a project. This diagram is
modified from the diagram in Chapter 1 to focus on review. © *2008–2013 by Kim R. Fowler. Used
with permission. All rights reserved.*

**Please remember**

All examples and case studies in this chapter serve only to indicate situations that you might
encounter. Hopefully these examples and case studies will energize your thought processes;
your estimates and calculations may vary from what you find here.

## General processes and procedures

The sum total of reviews is a multilevel effort within every project. Some reviews are small
groups that meet for a short duration; examples include code walk-throughs and module
reviews. Some reviews are group meetings to assess project progress or design. Some
reviews are formal design reviews and sign-offs. Each has its structure; each has its purpose.

The Defense Acquisition University (DAU) in the United States describes some of the
characteristics and components of a technical review. The web site titles the description
with "Technical Review Best Practices: Easy in principle, difficult in practice." Here are
some concerns [5]:

- Steps, components, and characteristics for best practices in technical reviews include
  the following:
  - Be event-based, e.g., the end of a design phase, such as a preliminary design review
    (PDR).
  - Define objective criteria, up-front, for entry to and exit from each review; the
    *Defense Acquisition Guidebook* provides general suggestions for criteria.

- Involve and engage technical authorities and independent subject-matter experts.
- The Review Chair should be independent of the program team. (*Note*: in small development teams and companies this is seldom practical.)
- Reviews are only as good as those who conduct them.
- Review status of program development from a technical perspective.
  - Involve all affected stakeholders.
  - Involve all technical products, such as specifications, baselines, and risk assessments, that are relevant to the review.
- System-level reviews should occur after the corresponding reviews of the subsystems.

### General outline for review

The California Department of Transportation reinforces these objectives with more thorough details [2]. Figure 13.3 outlines the basic flow and structure of a review. The following description is NOT an exhaustive list of necessary review activities for every occasion; it should serve you well in many situations, though.

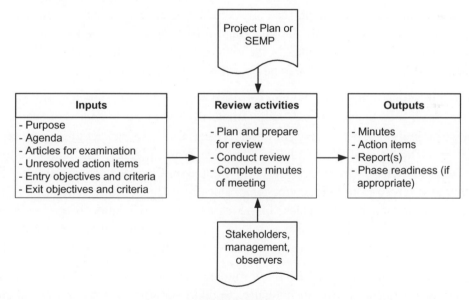

**Figure 13.3:**
Basic diagram for organizing a review. © 2013 by Kim R. Fowler. Used with permission. All rights reserved.

You must clearly establish and publicize the purpose for the meeting and define the expected outcomes. Provide the required review inputs, which are products from the current phase. Also provide the unresolved action items from previous reviews; these action items may be carried over for either continuing discussion or decision. Finally, establish the entry and exit criteria.

The Project Plan (or SEMP) should describe the process to perform technical reviews. Involve the appropriate stakeholders in the review for optimum communication and decision making.

The outputs from a technical review include a number of potential items:

- Action items, which are critical concerns, assigned by the review team to appropriate staff, to be addressed, fixed, modified, eliminated, or reviewed before specified dates.
- Documented review of decisions and action item progress, which includes acceptance, rework with comments, deviations, and waivers to particular developments.
- Feedback to participants and management about the results of the meeting.
- Provide a record of the meeting for their review and comments. Accurate and complete (not necessarily extensive) minutes ensure that decisions, actions, and assignments are accurately documented.

Plan a schedule of reviews. Plan specific activities for each meeting; include who should attend, the technical and formal level of the review, the agenda, the input components, e.g., document drafts, software modules, or hardware, and the output components, e.g. minutes, action items, decisions with criteria, such as 100% consensus agreement or majority agreement or project manager only. In preparing for each review, define the purpose, objectives, and the intended outcomes of the meeting. Identify participants and their roles, invite them, and distribute the agenda and appropriate background material.

Conduct the meeting in the following manner: Start on time. Clearly state the purpose of the review and provide an updated agenda to the attendees. Record the attendees present and their role, e.g., presenter, chairperson, or observer, and if need be, collect up-to-date contact information, such as e-mail, phone number, department, division, organization, or company. Review the ground rules for the review prior to discussion. Conclude one agenda item at a time. Manage discussions to focus on the topic. Document all decisions, actions, and assignments in the minutes. At the close of the review, summarize all decisions, actions, and assignments, review agenda items, and assignments for the next meeting. Confirm date, time, and place of the next meeting. Keep track of the time and then end on time.

After the review, send out a complete set of minutes that include all decisions, action items, and assignments. Follow up with the attendees to make sure the minutes are correct and complete. Periodically check on the progress of critical action items from the review.

### Tailoring your review

Tailor the reviews to the size, type, and formality of the project. For example, if the project is a small COTS integration to control some short-term university experiments, then you may only need monthly reviews. The reviews can be informal and attended by the project engineer and the principal investigator. The meeting minutes may only be a paragraph or two that you e-mail out to the participants. On the other hand, developing a medical device may require weekly status meeting of every engineering team, a monthly status meeting of the entire project, peer review of every completed module, system reviews of modules and integration, phase reviews, and stacks of documents.

## Types of review

*Peer review*: Peer review is a very effective form of review; it involves team members looking over the logic, rationale, and implementation of a module, which can be software, electronic, or mechanical. Peer review generally refers to small, fairly short meetings (less than one and a half hours) where colleagues examine a basic, small unit or module. An example module might be a software routine of less than 60 lines. Another example module might be a circuit board with 50 or fewer components. Even though peer review may be informal, it still should have a stereotyped format including agenda, checklists, action items, and minutes—albeit simple versions of each.

*Code inspections*: Software code walk-throughs are a type of peer review. They are highly effective and an excellent way to encourage proper designs and good development processes. For some types of products, such as medical devices, code inspections are an important part of the formal development. You should still have procedures, which can be simple and straightforward, for recording notes or minutes and then maintain a database of action items.

*Design reviews*: For larger, longer, and more critical programs, regular design reviews monitor progress, status, and conformance to requirements. The end of each development phase almost always has a formal design review; it is a high-level design review that ensures that the project's architecture is balanced and appropriate so that the system's functionality and performance meet the intended need. Design reviews are formal meetings and should have independent reviewers who are not directly associated with the project for more objective assessment.

*Planning review*: Verifies that plans, schedules, and activities are appropriate and timely for the project.

*Concept of operations review*: Ensures that the system operation meets the intended use and addresses the needs of the stakeholders. This review is reserved for large, complex projects where the concept of operations drives many requirements.

*Requirements review:* Ensures that the requirements meet the intent of the project. The requirements review checks that they are complete, appropriate, and address the user needs.

*Component-level detailed design review*: This is a particular form of peer review; it checks the logic, rationale, and implementation of the detailed design for a module. This can be a major review or may be a subset of a larger phase review.

*Test readiness review (TRR)*: A review to confirm that components and subsystems are ready for verification tests. For system integration, the TRR examines the results of the verification tests and addresses concerns for integration.

*Operational review*: Ensures that the system is ready for deployment. This review confirms that training, support, and maintenance for the system are in place.

### Frequency of review

The Project Plan or SEMP should establish a regular schedule for the reviews or at least define when reviews will follow specific events. In some instances, such as with code inspection, a schedule is not possible, but code inspections should occur after the completion of each module. Regardless, the Project Plan or SEMP should state either the timing or the criteria for holding a review.

### Course of action, changes, and updates following review

Following each review, the program manager, system engineer, or appointed staff should monitor the effort to address the action items that derived from the review. An action item should be a clearly defined effort, focused on a single concern. Sometimes an action item can become a job order or a task assignment.

Managing the action items and the results of reviews is critical to a project. No development effort goes forward without some adjustment recommended in review. Here are the primary sources that record issues for adjustment:

- Minutes—peer reviews, design reviews, control board meetings, and failure review board meetings
- Logs—problem reports/corrective actions, engineering change requests, and engineering change notices
- Test results—bench tests, unit tests, system tests, integration, and acceptance tests
- Communications—memos, e-mail messages, notes, letters, and customer audits.

### Roles and responsibilities

Project staff and consultants conduct internal reviews. Customers, audit agencies, and consultants may join project staff for design reviews and external reviews. The Project Plan or SEMP should determine when the reviews and audits may take place; though some audits might be unannounced or developed later in the project development, particularly if the project is mission-critical or safety-critical and certification or approval is required, e.g., FDA approval.

Formal reviews, particularly for project development in mission-critical or safety-critical markets, have a structured set of responsibilities. Less formal development can still benefit from these definitions and this framework.

*Moderator*: Conducts the technical peer review process and collects inspection data. Leads the technical peer review but does not perform the rework [3].

*Inspectors*: Review the design or code or plans and search for defects in work products [3].

*Author*: Responsible for the information about the work product and its design or code or plans; usually the designer of that portion of the project. Responsible for correcting all major defects and any minor defects as cost and schedule permit [3].

*Reader*: Guides the team through the work product during the technical peer review meeting. Reads or paraphrases each work product in detail. Performs duties of an inspector in addition to the reader's role. (This is the one role that is only used in the largest of projects. I have never encountered a situation with a reader.) [3]

*Recorder*: Accurately records each defect found during an inspection meeting; typically each defect becomes an action item. May perform the duties of an inspector [3].

If your company has a Control Board, its function is to confirm implementation of the revisions or updates derived from reviews. For smaller projects (less than 15 or 20 staff members involved), the program manager or system engineer performs this function.

## Components of a review

### Agenda

All reviews should have an agenda, even if it is very brief. The agenda may just be a reference to a company procedure or a short set of e-mailed instructions. Generally, an agenda should have the following:

- Purpose of the review and its goal
- The attendees
- The time
- The date
- The location
- The expected duration.

Figure 13.4 illustrates two examples of agendas for a technical review. The first, in Figure 13.4A, is simple and suffices for many situations. The second, in Figure 13.4A, is formal and detailed; it suits more complex projects.

Most agendas are specific for the topic and the project. Developing templates for the various reviews within a project may be useful for your company or group; this would be a

(A)

```
To: Bob, Joe, Sarah, Jane
From: Rick
Subject: Code Inspection Agenda

Please join me in conducting a code inspection of the DSP
FIR filter module. We will meet in the conference room,
next Monday, July 22, at 9:30 a.m. The review should only
take about 45 minutes.
```

(B)

### (Template) Review Meeting Agenda

Subject of Review/Title: _____

Program/Project: _____

Date: _____ Time: _____ Duration: _____

Location of review: _____

Invited attendees: _____

_____

_____

Type of review: _____

   (Possible types: code inspection, progress and status review, CoDR, PDR, CDR, TRR, control board, failure review)

**Topics and Presentations:**

(Some suggested topics and line items – your needs will vary according to the type of review:

1. Introduction – give purpose and goal(s)
2. Brief overview of meeting
3. First topic – presentation and discussion
4. Second topic – presentation
. . . .

N-2. Discussion and summary of review
N-1. List of action items
N. Wrap up – plan for next meeting (or set of meetings)  )

**Additional comments:**

**Figure 13.4:**
(A) Example of a simple agenda for a review. (B) Example template of an agenda for a formal review in a complex project. *© 2013 by Kim R. Fowler. Used with permission. All rights reserved.*

forward-looking practice to save time on future projects. Unfortunately, not everyone takes the time to do this. If you happen to be reviewing company procedures and processes for quality assurance, the agenda for review or audit should address these questions:

- What procedures and processes are being followed?
- What procedures and processes are not being followed?
- Are the procedures and processes sufficient?
- Do any procedures and processes need updating or revision?
- Should any procedures and processes be eliminated?

## Minutes

Minutes are the archival record of the review. Minutes should identify the primary topics discussed and the main points presented. They should reference any action items generated during the review. Minutes do not need exhaustive detail. Figure 13.5 illustrates one example of minutes. A basic format for the minutes from a review will include:

- The date of the review
- The agenda
- Who attended

```
Subject: Minutes of Code Inspection, DSP FIR Filter Module
Attendees: Bob, Joe, Sarah, Jane
Presenter: Rick
Time: 9:30 to 10:06 a.m.
Date: July 22, 2013
Location: conference room

We conducted a code inspection of the DSP FIR filter module.
The following points were discussed:
   1.  Whether to eliminate the first coefficient in the
discrete filter equation. Bob said that it would shorten the
time to calculate with fewer clock cycles. Sarah pointed out
that doing so would reduce generality and future re-use.
   2.  Jane pointed out some style errors in the coding that
did not follow the company's programming style guide.
   3.  Jane noted that one of the comments had not been
updated to reflect the actual operation of the module.

Three action items were assigned, all to Rick:
ACN6-43: Study eliminating the first coefficient in the
discrete filter equation and present the findings next
Friday, July 26.
ACN6-44: Fix the style errors in the coding to follow the
company's programming style guide.
ACN6-45: Update the comments in the module to reflect actual
operation.
```

**Figure 13.5:**
A simple example of minutes from a review. © 2013 by Kim R. Fowler. Used with permission. All rights reserved.

- Who presented the design
- The lead and independent reviewers
- What major decisions were made
- What action items were generated, their due dates, and who is responsible for each.

### Action items

Reviews should generate action items when identified issues need addressing. All action items should be tracked in a database. Each action item should have the following fields:

- Unique identifier number
- Status (open, closed, in work, in sign-off)
- Date opened
- Brief summary
- Response summary
- Requestor
- Assignee
- Due date.

Figure 13.6 illustrates an example action item.

### Checklist

Here are some items that might go into a checklist to ensure thorough coverage in preparing and conducting reviews [2].

1. Does the Project Plan or SEMP contain a section with a plan for reviews? If so, does the plan contain:
   - The number or frequency of the reviews?
   - The process for carrying out each review?
   - Roles identified for each review?
   - Level of formality identified for each review?
   - All action items will be completed.
   - Any unresolved actions will require program manager sign-off.
2. Here are some items to consider in preparing a larger review—particularly formal reviews:
   - Identify the attendees and their roles as soon as possible.
   - Set the time and location.
   - Check the location of the technical review for size, climate, configuration, equipment, furniture, noise, and lighting.
   - Material preparation:
     - Make sure to identify the purpose and outcomes expected.
     - Develop an agenda and distribute it 10 days to two weeks before the meeting.

---

**Action Item**

ID #: _ACN6-43_  **Title:** _Study eliminating 1st coefficient in DSP FIR  filter_

**Program/Project:** __Motor Control of Stereoscopic Stage_____

**Creation date:** _July 22, 2013_  **Assigner:** _Bob_____

**Completion:** _July 26, 2013_  **Assignee:** _Rick_____

| √ | Area of development | √ | Type of code |
|---|---|---|---|
| √ | Microcontroller and support circuits | | New code |
| | Interface circuits | √ | Modification of existing code |
| | Linear circuits | | Bug fix |
| | Other (specify below): | | Other (specify below): |

**Description of task:**

Study whether to eliminate the first coefficient in the discrete filter equation. Tradeoff shortening the time to calculate with fewer clock cycles with reducing the generality of the module and its future re-use. Prepare a short presentation of the findings.

**Expected output:**

Short PowerPoint presentation.

**Figure 13.6:**

An example of an action item from a review. © 2013 by Kim R. Fowler. Used with permission. All rights reserved.

- Distribute supporting and background material to the attendees 10 days to two weeks before the meeting.
- Bring forward all unresolved assignments, identified in the previous meeting.
- Distribute the ground rules for the meeting beforehand and discuss them before the start of discussion.
- Create an attendance roster for the meeting that requests up-to-date contact information from each attendee.

3.  Following a large meeting, here are some items to debrief:
   - How well did the presenters prepare for the meeting?
   - Were all issues resolved or identified?
   - What were the status and recommended resolutions?

- Did the meeting start on time?
- Did all attendees make introductions, or were they introduced?
- Were the purpose of the meeting and the expected outcomes clearly stated?
- Did you distribute an updated agenda with the priorities assigned for each agenda item?
- Was each agenda item concluded before discussing the next item?
- Did the minutes document all decisions, assignments, and actions?
- Were the meeting and the results summarized at the end of the meeting?
- Did the meeting end on time?
- Were the minutes distributed to the attendees?
- Were all critical assignments and action items followed up between meetings?
- Were the meeting facilities appropriate and useful?
- Were all the food and refreshments appropriately supplied?

4. Some suggested ground rules [2]:
   - Tell it like it is, but respect, honor, and trust one another.
   - Work toward consensus, recognizing that disagreements in the meeting are acceptable and not an indication of failure.
   - Upon agreement, we all support the decision.
   - Hold one conversation at a time.
   - Silence is consent.
   - Focus on issues, not on personalities; actively listen and question to understand.
   - Do not attack the messenger.
   - Start on time and observe the time limits.

## Peer review and inspection

Peer review is extraordinarily useful in catching many of the most difficult problems, such as errors in logic and misinterpreted customer intent. Peer review is primarily a technical discourse; it can address architecture, schematic capture, programming, and code development. Peer review can take several forms—inspection, code walk-throughs, technical presentation, and discussion. Peer review should always collect and archive notes, agenda, and minutes—however brief they may be. These archived items are useful when performing a root cause analysis of problems or preparing for future upgrades.

## Internal review

Internal reviews are those reviews conducted within the project, by company staff, and for the good of the project. While they include peer reviews, they also include meetings to review status and progress. All internal reviews should always collect and archive notes, agenda, and minutes—however brief they may be.

## Formal design review

For larger, longer, and more critical programs, regular design reviews monitor progress, status, and conformance to requirements. Design reviews are formal meetings and should have independent reviewers who are not directly associated with the project for more objective assessment. The review committee for each formal review should consist of at least four members plus a designated chairman; none of whom should be members of the project team. I have to say this rarely happens—it is the ideal but most small companies do not have the people, time, or resources to devote to independent review. Instead, they will put together design reviews with members of the project team presenting and reviewing and invite a customer or the client to attend and critique.

Regardless of the format, you should send a review package to each reviewer about two weeks ahead of the formal design review. The review package should contain a copy of all the slides to be presented at the review along with appropriate background material, the agenda, and ground rules.

### Types of design reviews

A number of design reviews can occur during the development timeline of an embedded system. Each one serves as a gateway to the next phase of the project. Figure 13.7 illustrates a timeline with example design reviews. I will describe some of the activities in these design reviews in the following subsections.

#### Conceptual design review

The Concept of Design Review (CoDR) closes the initial phase of the project. It can also be called the System Design Review (SysDR). The purpose of this phase is to develop a reasonable approach and architecture for the embedded system. The review examines the mission goals, objectives, and constraints; it also examines the first-pass requirements for the project and the approach to meet those requirements.

If the project is a larger, more complex system, such as a vehicle or medical device or spacecraft, you will probably hold another design review called the system requirements review (SRR). Usually the SRR takes place sometime toward the end of the concept phase or the beginning of the preliminary design phase. The US military describes it this way, "The SRR assesses the system requirements as captured in the system specification, and ensures that the system requirements are consistent with the approved materiel solution

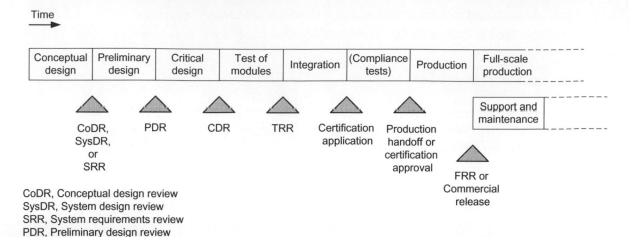

Time

CoDR, Conceptual design review
SysDR, System design review
SRR, System requirements review
PDR, Preliminary design review
CDR, Critical design review
TRR, Test readiness review
FRR, Flight readiness review

**Figure 13.7:**

Timeline of when some design reviews might occur. © 2013 by Kim R. Fowler. Used with permission. *All rights reserved.*

(including its support concept) as well as available technologies resulting from any prototyping efforts" [5].

Here are some example items that a CoDR or SysDR might address:

- Project schedule and budget
- Company or client policies
- Project organizational structure and interfaces
- Project objectives
- Requirements' process and management
- Requirements
  - Mission: environment, available resources, interactions with other systems
  - Performance: technical characteristics and constraints
  - Major system functions and interfaces
  - Safety
  - Manufacturing and logistics
  - Maintenance and support
  - Security
- Research—literature, patent searches
- Development drivers
- Market studies

- Trade studies to identify design constraints and feasibility
- System architecture
  - Concept
  - Hardware components
  - Software components
  - Operations concept
  - Support systems and logistics
- Integration and test plans
- Risk assessment.

*Preliminary design review*

The Preliminary Design Review (PDR) closes the preliminary design phase of the project. A PDR should present the basic system in terms of the software, mechanical, power distribution, thermal management, and electronic designs with initial assessments for loads, stresses, margins, reliability, software requirements and basic structure, computational loading, design language, and development tools to be used in development. Finally, the PDR should present the preliminary estimates of weight, power consumption, and volume.

The US military describes a PDR this way, "The PDR establishes the allocated baseline (hardware, software, human/support systems). A successful PDR is predicated on the determination that the subsystem requirements; subsystem preliminary design; results of peer reviews; and plans for development, testing, and evaluation form a satisfactory basis for proceeding into detailed design and test procedure development" [5].

A PDR presents the design and interfaces through block diagrams, signal flow diagrams, and schematics. Within these topics, a PDR should present first concepts for logic diagrams, interface circuits, packaging plans, configuration layouts, preliminary analyses, modeling, simulation, and early test results. Here are some example items that a PDR might address:

- Technical objectives
- Closure of action items, anomalies, deviations, waivers and their resolution following the CoDR
- Completed research, tradeoffs, and feasibility
- Requirements
  - Functional—what the system does
  - Performance—how well the system does it
  - Connections and interfaces—how it affects other systems
  - Safety
  - Manufacturing and logistics
  - Maintenance and support

- Security
- Analyses
  - Mechanical/structural design, analyses, and life tests
  - Weight, power, data rate, commands, EMI/EMC
  - Electrical, thermal, mechanical, and radiation design and analyses
- Software requirements and design
- Design verification, test flow, and test plans
  - Test and support equipment design
  - Host interfaces and drivers
- Risk and hazard analysis—which might include STPA, Safety Case, ETA, FMECA, and FTA
- Risk management plan
- Contamination requirements and control plan (this tends toward spacecraft development)
- Parts selection and qualification
- Materials and processes.

### Critical design review

The Critical Design Review (CDR) closes the critical design phase of the project. A CDR presents the final designs through completed analyses, simulations, schematics, software code, and test results. It should present the engineering evaluation of the breadboard model of the project. A CDR must be held and signed off before design freeze and before any significant production begins. The design at CDR should be complete and comprehensive.

The US military describes a CDR this way, "The CDR establishes the initial product baseline. A successful CDR is predicated on the determination that the subsystem requirements, subsystem detailed designs, results of peer reviews, and plans for test and evaluation form a satisfactory basis for proceeding into system implementation and integration" [5].

The design should be complete at CDR. The CDR should present all the same basic subjects as the PDR, but in final form. Here are some additional example items, beyond the items in a PDR, that a CDR might address:

- Closure of action items, anomalies, deviations, waivers, and their resolution following the PDR
- Design changes from the PDR
- Final parts list
- Final implementation plans including
  - Engineering models, prototypes, flight units, and spares
  - Software design and process

- Design margins
- Operations plan
- Updated risk management plan
  - Updated risk and hazard analysis
  - Safety requirements
- Test
  - Qualification and environmental test plans
  - Integration and compliance plans
  - Status of procedures and verification plans
  - Test flow and history of the hardware
  - Completed support equipment and test jigs
- Schedule
- Documentation status
- Product assurance
- Identification of residual risk items
- Plans for distribution and support
  - Shipping containers
  - Warehousing and environmental control
  - Transportation.

### Commercial release

Commercial release, or Flight Readiness Review (FRR) (for spacecraft), is the decision gate prior to manufacturing and full-scale production. Sometimes this review is called the Production Readiness Review (PRR). It has multiple purposes: to assure that the design of the item has been validated through the integration and acceptance test program; to assure that all deviations, waivers, and open items have been satisfactorily closed; to assure that the project, along with all the required support equipment, documentation, and operating procedures, is ready for production.

The US military describes a PRR this way, "The PRR examines a program to determine if the design is ready for production and if the prime contractor and major subcontractors have accomplished adequate production planning. The PRR determines if production or production preparations have unacceptable risks that might breach thresholds of schedule, performance, cost, or other established criteria" [5].

Here are some example items that commercial release or a PRR might address:

- Resolution of all anomalies through regression testing or test plan changes:
  - Action items
  - Rework or replacement of hardware
  - Assessment of could-not-duplicate failures and the residual risk
- Compliance certification results

- Measured test margins versus design estimates
- Demonstrate qualification and acceptance of operational and environmental margins
- Demonstrate the failure-free operating time and the results of life longevity tests
- Review of distribution plans
- Post shipment plans.

### Other types of design reviews

Depending on the type of embedded system, you might require a number of different design reviews. The US military has the following definitions [5].

*Alternative systems review (ASR)*—The ASR assesses the preliminary materiel solutions that have been proposed and selects the one or more proposed materiel solution(s) that ultimately have the best potential ultimately to be developed into a cost-effective, affordable, and operationally effective and suitable system at an appropriate level of risk.

*Flight readiness review (FRR)*—The FRR is a subset of the TRR and is applicable only to aviation programs. The FRR assesses the readiness to initiate and conduct flight tests or flight operations.

*Initial technical review (ITR)*—The ITR is a multidisciplined technical review held to ensure that a program's technical baseline is sufficiently rigorous to support a valid cost estimate as well as enable an independent assessment of that estimate.

*In-service review (ISR)*—The ISR is held to ensure that the system under review is operationally employed with well-understood and managed risk. It provides an assessment of risk, readiness, technical status, and trends in a measurable form. These assessments help to substantiate budget priorities for in-service support.

*System functional review (SFR)*—The SFR is held to ensure that the system's functional baseline has a reasonable expectation of satisfying stakeholder requirements within the currently allocated budget and schedule. The SFR assesses whether the system's proposed functional definition is fully decomposed to its lower level, and that preliminary design can begin.

*System verification review (SVR)*—The SVR is held to ensure the system under review can proceed into initial and full-rate production within cost (program budget), schedule (program schedule), risk, and other system constraints. The SVR assesses system functionality and determines if it meets the functional requirements as documented in the functional baseline.

*Test readiness review (TRR)*—The TRR is designed to ensure that the subsystem or system under review is ready to proceed into formal test. The TRR assesses test objectives, test methods and procedures, scope of tests, and safety; and it confirms that required test resources have been properly identified and coordinated to support planned tests [5].

## Change control board

Companies developing mission-critical or safety-critical embedded systems sometimes have a separate function called the change control board (CCB). A CCB comprises company staff from different technical disciplines. It reviews requests for design changes after a design is frozen. The purpose of a CCB is to give technical oversight that manages change efficiently.

NASA has an entire document devoted to problem management that a CCB implements [6]. The US FDA discusses the use of a CCB to review changes to current products, "The Change Board will meet to review the technical content of all proposed changes to released documentation for accuracy and impact on safety, efficacy, reliability, product cost, parts and finished goods inventory, work-in-process, instruction and service manuals, data sheets, test procedures, product specifications, compatibility with existing products, and other factors. . ." [7].

## Failure review board

A failure review board (FRB) is similar to a CCB except that it focuses solely on failures and their causes. Companies developing mission-critical or safety-critical embedded systems sometimes have a separate FRB. Like a CCB, an FRB comprises company staff from different technical disciplines. Its purpose is to review failures in products and pursue resolution.

An FRB often uses analysis tools like failure reporting, analysis and corrective action system (FRACAS). This sort of tool records failures of components, subsystems, and processes; records the results of root cause analysis; identifies corrective actions; provides cognizant people, staff, clients, and customers, with the necessary information to fix the problem [8].

Sematech, Inc. has a document that gives the following description in its executive summary, "Failure Reporting, Analysis, and Corrective Action System (FRACAS) is a closed-loop feedback path in which the user and the supplier work together to collect, record, and analyze failures of both hardware and software data sets. The user captures predetermined types of data about all problems with a particular tool or software component and submits the data to that supplier. A Failure Review Board (FRB) at the supplier site analyzes the failures, considering such factors as time, money, and engineering personnel. The resulting analysis identifies corrective actions that should be implemented and verified to prevent failures from recurring" [9].

## Audits and customer reviews

Audits may be conducted by certification agencies or by customers or by company staff. Their primary purpose is to assure outside parties that your company's processes and procedures are sufficient for the market. The best way for you to minimize the impact of an audit on your project development is to maintain open and clear communications with the agency or customer.

Audits and customer reviews come in many forms from simple online surveys to monthly visits of three and four days at a time. While entertaining a customer takes away time and effort from your development, considering them a partner in your project and using their expertise will help you in the long run. You will develop a better product and will gain a trusted client.

---

**Example: Software Mayhem and customer reviews**

As mentioned in the case study in Chapter 11, I dealt with a vendor who was developing a spacecraft instrument for my company. When the vendor's software was revealed as completely hopeless, we jumped in with regular reviews that required us to travel coast-to-coast once a month. While we could have been seen as an added burden, we offered technical help with specific software programming, technical management, and environmental test facilities. Working together and reviewing progress regularly we finished the task and developed a working instrument.

---

## Static versus dynamic analysis

So far, I have described and discussed static methods to review technical design and progress. Static means that the review is a snapshot of an instance in time of the continuously developing product. The dynamic aspects of review are best handled in test and integration, which Chapter 14 describes in detail.

## Debrief

One of the best things that a team can do is debrief at the end of a project. Ask, "What went right?" and "What went wrong?" Determine how you might do things better in future projects and record your notes and thoughts. This is a long-term view of development and it is an investment in your company's future. Most companies do not take the time to review in retrospect to prepare for the future. This is unfortunate because thorough debriefing will reduce time and effort in future projects.

## Acknowledgments

My thanks to both Geoff Patch and George Slack for their careful and thoughtful reviews, critiques, examples, and suggestions for this chapter.

# References

[1] US FDA design control guidance for medical device manufacturers March 11, 1997, revised as of April 1, 2013. <http://www.accessdata.fda.gov/scripts/cdrh/cfdocs/cfcfr/CFRSearch.cfm?fr = 820.3> (accessed 16.07.2013).

[2] CA DOT. <http://www.fhwa.dot.gov/cadiv/segb/views/document/sections/section3/3_9_10.cfm> (accessed 16.07.2013).

[3] Accessed on July 16, 2013: <http://en.wikipedia.org/wiki/Technical_peer_review>.

[4] NASA Systems Engineering Handbook, NASA/SP-2007-6105 Rev1, December 2007, p. 167.

[5] Accessed on July 16, 2013: <https://acc.dau.mil/CommunityBrowser.aspx?id = 294561>.

[6] Accessed on July 20, 2013: <http://edhs1.gsfc.nasa.gov/waisdata/rel6/pdf/cd61161002s08.pdf>.

[7] Accessed on July 20, 2013: <http://www.fda.gov/MedicalDevices/DeviceRegulationandGuidance/PostmarketRequirements/QualitySystemsRegulations/MedicalDeviceQualitySystemsManual/ucm122605.htm>.

[8] Accessed on July 20, 2013: <http://src.alionscience.com/pdf/RAC-1ST/FRACAS_1ST.pdf>.

[9] Failure reporting, analysis, and corrective action system, Sematech. <http://www.favoweb.com/doc/fracas_sematech.pdf> (accessed 11.11.2013).

## CHAPTER 14

# Test and Integration

**Kim R. Fowler**
*IEEE Fellow, Consultant*

## Chapter Outline

Developing and Managing Embedded Systems and Products.
DOI: http://dx.doi.org/10.1016/B978-0-12-405879-8.00014-3
© 2015 Elsevier Inc. All rights reserved.

## Introduction

Test and integration refer to verifying and validating the design and development effort. While there is some overlap with manufacturing testing, most of the discussion here will focus on design and development.

Happily, Pries and Quigley have recently published a book that addresses test for embedded systems. I refer you to their book for the details and thorough coverage of the subject [1]. For this chapter, I will only hit on specific points that relate to managing the testing of embedded systems.

### The reasons for testing and integration

Since you are already reading this chapter, you may have an inkling that test and integration are important. Let me try to drive it home with an example followed by a quote from Pries and Quigley!

Automobiles are incredibly complex systems. They are loaded with embedded systems and are gaining more each passing model year. Estimates for some current models have 100–150 individual microcontrollers and microprocessor systems within each vehicle. So where are all these embedded systems in a car? Figure 14.1 illustrates some examples of embedded systems found in a car [2]:

- Engine control
- Traction control
- Active suspension

- Electronic steering control
- Tire pressure monitoring
- Instrument cluster
- Environmental controls
- Entertainment systems
- Battery management
- Smart safety
- Locks
- Window controls
- Lighting controls
- Headlamp control
- Indicator controls
- Data bus networking
- GPS or Satellite Navigation.

These are just the high-level systems—engine control may have subsystems with multiple processors handling functions such as throttle control, electronic valve timing, cylinder deactivation, transmission control (or hybrid generation and regenerative braking), and environmental sensing. Even the sensors are migrating to separate, dedicated microcontrollers embedded within the sensor body to supply direct digital data.

Testing is necessary to verify the function of each of these embedded systems. It includes functional tests, parameter variation testing, environmental extreme tests, life tests, and stress tests. That is a lot of testing! But that's not all—testing also includes the integration tests when you put the individual embedded subsystems into the vehicle and verify their interactions with other subsystems. Interoperability is a huge concern for complex systems of embedded modules as found in an automobile. Integration testing is an extensive effort in itself.

Testing can also be for compliance. Extending the example here, you could test for compliance in electromagnetic compatibility or EMC. This would be a significant activity requiring a large effort because so many automobile systems operate at high frequencies or are wireless, which makes interference or susceptibility more likely.

Now that you have seen the automobile example, maybe I can nail shut the case for testing and integration with a quote from Pries and Quigley, "Some enterprises regard testing as the final piece of the development effort rather than as a competitive tool. As a consequence, the test team executes critical test plans as the project is closing-just before launch and well after the time when the design and development teams can perform any kind of rational corrective actions... The situation is aggravated by disengagement between the development and verification groups, as well as a frequent disconnect between project management, development engineering, and test engineering... We owe it to our customers to provide them with... high-quality, reasonably priced, on-schedule, and safe products.

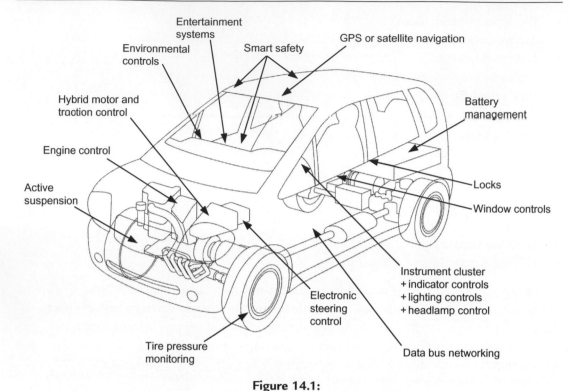

**Figure 14.1:**
Example of an automobile with many embedded systems. © 2013 by Kim R. Fowler. Used with permission. All rights reserved. Illustration from iStockPhoto.

Test engineering is a huge driver for achieving this goal because it is through testing that we reveal the character of our product... Numerous problems in the early product delivered to the field can contaminate the customer's perspective of the product indefinitely" [1].

## Part of a complete feedback system

Test and integration, coupled with design review, are the primary feedback activities to assure the project goals (Figure 14.2). These activities confirm whether your system's design and development are as you planned and hoped that they would be. "...[T]esting should be integrated into the development effort. The reason is that we want corrective actions that are made by developers as well as feedback from the testing and verification group" [1].

A test compares the outcome with the expected value. A proper test and integration plan will specify the review, test, and integration procedures and what parameters will be studied. A closed-loop feedback system that generates a test, or set of tests, for every line item specification is much more likely to provide assurance than *ad hoc* procedures.

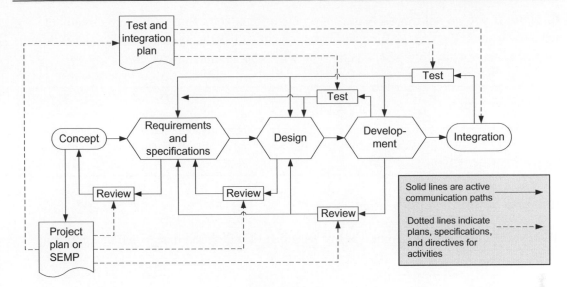

**Figure 14.2:**

Example feedback loops provided by review, test, and integration. © *2013 by Kim R. Fowler.*

Test and integration are integral to every process model, whether it is V-model, spiral, prototyping, or a hybrid approach. Test and integration also make quality assurance (QA) possible. Chapter 1 gives more details on process models and QA and where test and integration fit in.

### The goals for a complete test and integration program

The targets for revelation by test and integration, in order of increasing severity, are as follows [1]:

1. Analyze for readiness
2. Discover and characterize system behaviors; clarify subsystem interactions
3. Discover the product's limits:
   - Verify and validate design limits to customer requirements
   - Characterize failure modes of the product
   - Characterize destructive modes of the product
4. Prevent defects
5. Discover product defects
6. Contain defects to a single release.

The goal is to find problems and noncompliance early in design, when changes and fixes are cheaper and easier. Characterizing a product to its limits gives you an idea of how robust it will be to unexpected and unusual circumstances or abuse.

*Caution*: If you do not test early to find problems then you can expect a progression of unhappy results. First, something will break or fail earlier than you would like. Then it will take time and money to repair. Finally, you will generate unhappy customers.

---

**Please remember**

All examples and case studies in this chapter serve only to indicate what you might encounter. Hopefully, they will energize your thought processes; your estimates and calculations may vary from what you find here.

---

## Overview of test and integration

There are a number of different types of tests that fit within the design and development sequence:

- Bench tests
- Mockup and fit checks
- Unit or module tests
- Fault insertion tests
- Verification
- Validation
- Integration
- Alignment checks
- Calibration
- Field tests
- Compliance
- Environmental tests
- Security tests
- Stress test
- Highly accelerated life test (HALT).

### Bench tests

Bench tests are the initial runs of mechanisms, breadboards, and software code. Bench testing primarily aims to demonstrate feasibility. Hopefully, it reveals any major obstacles in design.

### Mockups and fit checks

Mockups and fit checks are mechanical tests for geometry, interferences, and confirmation that the 3D modeling is correct. These types of checks help verify that the fabricated form factor is according to design. Mockups and fit checks are particularly good for static checks of module location and cable and harness connections and attachments.

*Unit and module tests*

Unit and module tests are verifications of the specifications for mechanisms, electronic circuits, and software functions. These type of tests confirm the behaviors of component subsystems. Unit and module tests help verify that the units and modules meet the design intent.

Unit tests come in two forms: black box and white box. The Systems Engineering Book of Knowledge defines each as follows: "A very common abstraction technique is to model the system as a black box, which only exposes the features of the system that are visible from an external observer, and hide the internal details of the design. This includes stimulus response characteristics and other black box physical characteristics, such as the system mass or weight. A white-box model of a system shows the internal structure and behavior of the system. Black-box and white-box modeling can be applied to the next level of design decomposition in order to create a black-box and white-box model of each system component" [3].

*Fault injection tests*

Fault injection introduces faults or error conditions to test the system behavior to potential errors, faults, and failures. For software testing, fault injection inserts error conditions into code paths; it is useful for exercising code paths for error handling, which are rarely followed, thereby improving test coverage. You can couple fault injection testing with stress testing; this can aid the development of robust software [4].

*Verification and validation*

Verification is the objective comparison of a system or subsystem behavior with its design specification. Every design specification should have at least one paired test to confirm its proper implementation. (See the following example.)

Validation is the (sometimes more subjective) confirmation that the overall system behavior and function meets the original intent of the customer or architect. Validation often includes, as a subset of sign off, the satisfactory completion of all verification tests. (See the following example.)

---

**Example: Verification versus validation**

Several examples of verification might be that the embedded system:

- is within its specification for mass,
- consumes less power than the maximum specified,

> • sufficiently dissipates heat to maintain temperature just below the maximum specified.
>
> An example of validation is that field trials of the embedded system satisfy the customer's intent for operation and the customer then signs off.

## Integration

Integration is the combining of all the subsystems in a considered, pre-planned, stepwise sequence. Integration helps reveal the overall system behavior and confirms the design intent; it is the primary form of validation.

## Calibration and alignment checks

These are adjustments that you perform on components of the embedded system. You perform these adjustments in preparation for other tests or integration. Calibration and alignment checks may also be a standard part of manufacturing.

## Field tests or trials

Field testing puts your embedded system in the hands of potential customers using it in realistic prototype operations. Field testing is not necessarily a completely controlled and rigorous verification. It can, however, reveal unusual or unexpected operations or abuses when potential customers use your embedded system. These forms of tests are sometimes considered the "alpha" phase of acceptance for the final product or system.

## Compliance

Compliance testing usually is a part of commercial certification of a system. You may have to subject your system to safety tests at Underwriter's Laboratory (UL) and get UL approval before commercial release. Or you may have to run your embedded system through EMC tests to demonstrate that it meets a specific set of EM standards and likely will not interfere with other equipment or be susceptible to external disturbances. Compliance often requires independent, third-party testing.

## Environmental tests

Environmental tests are typically qualification tests for your embedded system. They provide a measure of assurance that your system can withstand environmental extremes. Environmental tests may include thermal cycling between extreme hot and cold temperatures or mechanical shock and vibration tests. Environmental tests often require specialized equipment, such as thermal chambers and vibration tables, which can be quite expensive. Often you will use the services of an independent, third party to perform environmental tests.

*Security tests*

As of the writing of this textbook, security design and test is an emerging field. Chapter 12 considers some testing for security.

*Stress*

Stress tests provide indications of margins of robustness within your design. They might be electrical stressing, such as overvoltages or excessive noise on power lines. The environmental tests in the previous paragraph are a form of stress tests. Stress tests can be performed on units or modules before integration in many cases.

*Highly accelerated life test*

HALT and highly accelerated stress screen (HASS) are similar tests. They strive to test a statistically significant population of components, or systems, under environmental extremes and stresses to estimate the expected life of the components or systems. These environmental extremes and stresses serve to speed up the testing and reduce the needed time. Specialized chambers generate the environmental extremes. Often these extremes exceed the specified requirements but are below the physical limits for destruction or failure.

*When to use which tests*

Each of these tests has specific goals and specific applications. Table 14.1 outlines where these tests apply to specific concerns.

*No simulation or manufacturing tests here*

Note that I did not include simulation or manufacturing. Simulation is *NOT* testing; it is modeling based on simplifying assumptions—it can be very helpful in design but it is not testing. Manufacturing test is *NOT* a thorough verification or validation of a system design. Manufacturing test is for quality confirmation and to indicate basic function of subsystems or systems on the assembly line.

## General processes and procedures

Testing and integration need to cover many different dimensions. Just verifying specifications is not sufficient. Pries and Quigley write "Even when an organization tests a product to specification (the minimalist's approach to testing), we generally expect to see subsequent field problems... Testing to specification ... can present the delusion of successful testing" [1].

Besides focusing on different aspects of design and development, the various tests and integration also occur at different times within the phases of a project. Figure 14.3 gives

**Table 14.1: Different tests address different concerns during development**

| | Analyze for Readiness | Verify and Validate Design Limits | Characterize Failure Modes | Characterize Destruction Modes | Prevent Defects | Discover Product Defects | Contain Defects to a Single Release |
|---|---|---|---|---|---|---|---|
| Bench tests | √ | √ | √ | | √ | √ | √ |
| Mockup and fit checks | √ | | | | √ | | |
| Unit or module tests | √ | √ | √ | | √ | √ | √ |
| Verification | √ | √ | | | | | |
| Validation | √ | √ | | | | | |
| Integration | √ | √ | | | | | |
| Alignment checks | √ | | | | | | |
| Calibration | √ | | | | | | |
| Field tests | √ | | | | | √ | |
| Compliance | | √ | | | √ | | |
| Environmental tests | | √ | √ | √ | √ | √ | |
| Stress test | | √ | √ | √ | √ | √ | √ |
| HALT | | √ | √ | √ | √ | √ | √ |

just one example of when the various tests and integration might occur. Different projects and different industries emphasize specific tests and phases; your experience will probably be different from this example, but your projects will probably have elements of each type of test, as shown in Figure 14.3.

## Test plan

The test plan is the important first step following the commitment to perform thorough test and integration. You need to consider questions, such as the following, to prepare a thorough test plan [1]:

- What environments will the embedded system *probably* experience?
- What environments will the embedded system *possibly* experience?
- How do you address these environmental impacts?
- How do you replicate multiple environments?
- How do you detect faults caused by environmental interactions?
- What can bench testing show you about ultimate quality?
- What standards must your system meet and how do you test for them?
- How do you test the user interface?

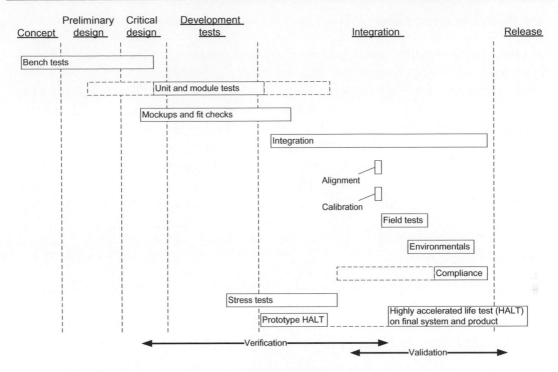

**Figure 14.3:**
One example for when test and integration may take place. Please note that the beginning and ending boundaries for each type of testing are vague. This example also shows a successful development that has no major problems requiring redesign. © 2013 by Kim R. Fowler. Used with permission. All rights reserved.

- What are the most important (or prominent or likely) hazards encountered by your system?
  - Human abuse and injury
  - Environmental assault
  - Mechanical damage
  - Electrical shock, damage, or arcing
- What should be tested to destruction?
- What simulators might you need? How do you verify their operations?
- What types and combinations of tests should you consider?
  - Pass/fail
  - Quantitative and parametric
  - Combinations of factors or environments
  - Statistical.

## Contributors to a test plan

For a big team or company, you might have a test team that develops the test plan. For most of us working on smaller projects, one person is usually responsible for preparing the test plan. Regardless who writes the plan, a number of people should contribute to it. The system engineer or architect may be the author; otherwise, he or she should review it and sign off on it. The program manager, if a separate individual from the systems engineer, should also review and sign off on the test plan. Other contributors include every member of the test team and the designers and developers; these folks should see, review, and be actively involved in the writing of the test plan. Third-party compliance groups can be secondary contributors to the test plan. Government sponsors or customers can be tertiary contributors to the test plan.

## Elements of a test plan

A test plan should describe each necessary test or reference the appropriate procedures if they are company standards. A test plan should cover each specification and indicate the appropriate test or tests. It should also indicate when the tests are conducted and with much more detail than shown in Figure 14.3.

The test plan should call for the right amount of testing at the right times. Early unit and prototype testing often reveal problems sooner and can reduce last minute surprises. It should balance between too little testing, which leaves an embedded system exposed to potential problems when released to the market, and too much testing, which "gold plates" the design and produces only diminishing returns. The balance strives to optimize effort and cost within development of the project.

# Verification

To repeat, *verification* objectively compares the behavior of a component, routine, module, or subsystem with its design specifications. Each specification must have at least one paired test, it may require more tests to completely cover the specification, to confirm its proper implementation. Every verification test must compare the actual outcome with the expected outcome. Verification is necessary to confirming proper implementation of the design, but it is not sufficient.

Verification is easily identified by specification metrics—the quantifiable measures of particular parameters. Each specification should have a clear metric that can be measured such as weight, power, size, or speed. Software unit tests may have a combination of metrics to test; these sometimes can take a long duration for some parameter values to appear. Completed systems can also have verification tests, such as the cornering speed of a sports car at specific accelerations in specific conditions—see the combination of parameters tested at once here?

---

### Example: Mechanism verification

An example verification test of a mechanism is to record its speed, direction, and range of movement following a specific command and compare it to the expected movement. A full suite of verifications for the mechanism would span the range of possible commands and resulting movements.

---

### Example: Electronic circuit board verification

An example verification test of a circuit board is to record its range of power consumption during different operations and compare it to the specified maximum. This is simply recording the instantaneous supplied voltage and current during different operations. Then plot the calculated power and its range for each operation; the plot might be a histogram or it might show maximum and minimum lines. Figure 14.4 gives examples of both a histogram and a set of boundary plots.

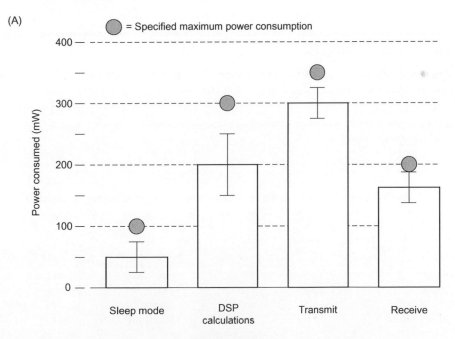

**Figure 14.4:**
(A) Example of a histogram of power consumption for a circuit running in different modes.
(B) Example of a power consumption for a circuit running a sequence of operations. © 2013 by *Kim R. Fowler. Used with permission. All rights reserved.*

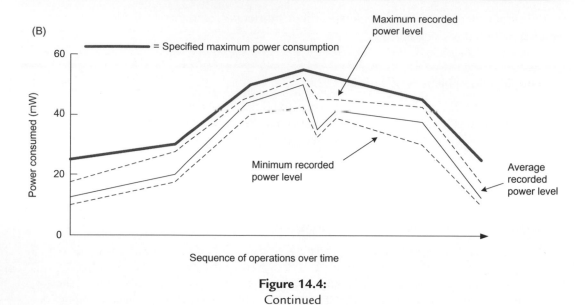

**Figure 14.4:**
Continued

## Example: Software verification

An example verification test of a software routine is to record its data inputs and outputs and compare them to the expected data. A full suite of verifications for the software routine would span the range of possible inputs and resulting outputs. Table 14.2 gives examples of some simple software verification tests. You would also check that it properly rejects out-of-range inputs.

**Table 14.2: Example of several verification tests for a software routine**

| Test ID | Title | Description of Routine's Function and Purpose | Initial Conditions | Procedure | Expected Output | Pass/ Fail |
|---|---|---|---|---|---|---|
| 215 | Generate Signal Waveform—check input parameters, Part 1 | 1. Receive a vector of parameters | — Clear memory cells that contain the parameters<br>— Set underflow threshold | Input values below the underflow threshold | Routine does not accept values, returns an error message, remains in a safe state | |

*(Continued)*

Table 14.2: (Continued)

| Test ID | Title | Description of Routine's Function and Purpose | Initial Conditions | Procedure | Expected Output | Pass/ Fail |
|---------|-------|-----------------------------------------------|--------------------|-----------|-----------------|------------|
| 216 | Generate Signal Waveform— check input parameters, Part 2 | 1. Receive a vector of parameters | — Clear memory cells that contain the parameters<br>— Set overflow threshold | Input values above the overflow threshold | Routine does not accept values, returns an error message, remains in a safe state | |
| 217 | Generate Signal Waveform— check input parameters, Part 3 | 1. Receive a vector of parameters | — Clear memory cells that contain the parameters<br>— Set underflow threshold<br>— Set overflow threshold | Input nominal values between the underflow and overflow threshold | Routine accepts values, moves onto next step | |
| | | 2. Calculate and store data points that represent a complete period of a waveform<br>3. Output the data points in sequence at the frequency specified | | | and calculates the desired data points to represent a waveform<br>and outputs the data points in sequence at specified frequency | |

## Validation

To repeat, *validation* confirms that the system's behavior meets the original intent of the customer or designer. Validation includes the satisfactory completion of all verification tests, field trials, stress tests, HALT, and all compliance tests. Every suite of tests or trials for validation must compare the actual outcomes or behaviors with the expected set of outcomes or behaviors.

Often integration can fulfill the concern for validation. Other times an extended period of operation with customer personnel, much like a field trial but more rigorous and carefully planned, may be necessary for validation:

- Military systems often undergo extended field trials. The first set of trials is called technical evaluations, where the completed and verified prototype is instrumented and monitored by combinations of contractor and military personnel. The second set of trials is called operational evaluations, where the completed and verified production system is operated and monitored by military personnel with occasional consulting from the manufacturer.

- Maritime applications have shakedown cruises to establish the seaworthiness of a craft or ship.
- In the United States, a new model of aircraft must undergo extensive tests, the results of which are carefully reviewed by the Federal Aviation Administration before it is certified for commercial use.

---

### Example: Validation for a satellite

An example validation procedure for a satellite would include many different tests. First, all components and modules must have been verified to all their individual specifications. Second, all components have undergone environmental testing—three days of thermal cycling between temperature extremes ($-40°C$ to $+85°C$) in a vacuum chamber, random frequency vibration tests in all three axes, and selected electronic components have been subjected to heavy ion bombardment to validate radiation tolerance. Third, the satellite, with all subsystems in place, has undergone environmental testing—seven days of thermal cycling between temperature extremes ($-40°C$ to $+85°C$) in a vacuum chamber and then several minutes of random frequency vibration tests in all three axes.

---

### Example: Validation for a medical device

The U.S. Food and Drug Administration (FDA) must approve all medical devices before those devices go on the commercial market in the United States. The FDA requires that all devices demonstrate safety and efficacy. For approval a developer must demonstrate a validation process, which includes proof of concept through university and hospital research; a developer must also show that rigor and discipline are enforced through a QA system, verification tests of all specifications, and extensive clinical tests.

I worked in a small company that developed a medical device. We had 13 years of university research to show safety and efficacy with over 1000 patient records. We hired an experienced QA consultant and instituted a QA system in the company. We used Rational Rose and Clear Case to develop requirements and tests for each requirement. Finally, we prepared and conducted a set of clinical tests of the medical device.

Unfortunately, the clinical tests were poorly conceived and conducted. We only had two medical centers with 20 patients each—the FDA thought that we should have at least three centers or more. The medical expert on our staff closely directed the medical staff in the studies at the two centers, instead of letting the staff operate the devices independently. The company failed to gain FDA approval because of these problems in the clinical studies. In our situation, *the validation was neither sufficient nor complete.*

## Field trial and testing

To repeat, *field trial* and *field testing* operate an embedded system, or its prototype, in a realistic situation and environment with customer staff and personnel. Field testing is neither verification nor validation. Its purpose is to improve the probability of revealing multiple or unusual or unexpected uses, abuses, and operations.

Often inexperienced people will use your embedded system in a field trial. When they do find a problem or concern, they may not describe it completely correctly. They may attempt to interpret the results rather than stating the conditions and facts surrounding the situation. Pries and Quigley say that we should assume that the customer observation is correct but that it may be poorly stated or they have made an erroneous correlation or interpretation [1]. This is where we need to listen very carefully but not necessarily accept all that is said as the whole truth.

### Example: Navy system field test

I designed a safety monitoring system for US Navy ships. The monitoring system had an acoustic alarm for alerting to a problem, should one occur. I designed a built-in-test (BIT) to assure operation of the entire monitoring system and to diagnose quickly any problems; the final part of the BIT was to "beep" the alarm for half of a second. We installed the monitoring system at about shoulder height in a ship for field testing and trained sailors to run the BIT. In a fully functioning system, the BIT took 9.3 seconds to run and finish off the test with an alarm "beep." The sailors quickly figured out how long the test took and how consistent it was at "beeping" the alarm at 9.3 seconds. They would watch for a fellow sailor walking down the passage way towards them, mentally time the arrival of the fellow sailor, press the BIT initiate button, and then hide. As the unsuspecting sailor walked past the safety monitoring system, 9.3 seconds later, the BIT would "beep" the rather loud alarm right in his ear. The other sailors enjoyed the startled response of the unsuspecting sailor.

Ah the things you can find out with field testing! This may seem a trivial example, but it does point to the possibility of unusual or unexpected operations.

### Example: Air force system field trials

I designed a system for uploading data into U.S. Air Force aircraft. The system automated a paper-based catalog of data that Air Force crews would use to tailor operations for missions. The crews originally entered the data manually into the aircraft while flying on the specific mission; it often took 45 minutes to enter all the necessary data. The system, designed by my team, allowed the Air Force crews to prepare the data load in their offices and then support staff uploaded the data to the aircraft; this removed the laborious, error-prone manual entry on the way to the mission.

This was the era just before graphical user interfaces (GUIs) were ubiquitous on computers, though GUIs were available on some computers. Our team developed the data preparation as a command line operation on a desktop computer. It worked correctly and well. The air crews were glad to get it.

Field testing revealed, though, that they did not use the system very often. One reason was that the command line operation was just a bit awkward; a GUI, which we could have programmed, would have been far more acceptable to the Air Force officers using the system. The second reason was that the data upload was very tedious for the enlisted personnel providing the aircraft line support; it took extra time for them to ready the aircraft by uploading the data through a large cable that had somewhat delicate connector pins.

Field testing showed that we had a nice idea—it just was not as practical as we had hoped.

---

### Example: Field testing a process control system

Years ago my dad, also an electrical engineer, designed a control system for equipment used in a process industry. The control system had a panel with industrial indicators and buttons on it to operate the equipment. He had designed the system for the buttons to be pressed in sequence to control operations. His plan was to train plant personnel for the correct operation of the control panel.

As soon as the system was installed for the field trials, the foreman supervisor walked up to the panel, splayed open his hands, and began mashing all the buttons at once. My dad hollered, "Stop it! What are you doing?!? These buttons are to be pressed in sequence!" The foreman simply replied, "That is the first thing the guys on the night shift will do to this panel."

Dad went back to work and redesigned the panel's digital logic with interlocks to prevent incorrect sequencing of the buttons. Only observation of plant personnel and field trials would reveal this sort of unusual or unexpected operations and abuse.

In the same installation, Dad used nylon contact whips on switches to indicate the position of product on the process line. The product had a rough cardboard backing on it, which the nylon contact whips slid across as the product moved down the line. Within the space of a day, the nylon contact whips had worn off because of the abrasion from the rough cardboard backing. He had to go back to the design table and replace the nylon contact whips with stainless steel rollers with ball bearing axles to remove the abrasion problem. Again, only field testing would reveal this problem to an inexperienced designer.

---

## Integration

To repeat, *integration* combines all the subsystems in a considered, pre-planned, stepwise sequence. Integration is the major portion of validation and should reveal the overall system behavior, which should confirm the design intent. Here are some suggested steps for integration (your situation will probably vary at points in this list):

1. Mockups and fit checks might be the first step in integration.
2. Place simulator modules, one at a time, on the mockup and run functional tests of the entire system after each addition. A simulator module is a hardware realization of the actual subsystem and generates a subset of the actual functions; typically it is not fabricated and assembled to the same level as the final design module.
3. If this is a mission-critical system, such as a satellite, you may want to add this step 3— replacing each simulator module, one at a time, on the mockup with an engineering model

and running functional tests of the entire system after each swap-out. An *engineering model* is a complete subsystem built to the same level as the final "flight" hardware but without the final finish, such as conformal coating, encapsulation, or structural enclosure. It also runs the same, final software programs as the "flight" hardware.

4. Replace simulator modules (or engineering models, if you included step 3), one at a time, with production subsystems on the mockup and run functional tests of the entire system after each swap-out.
5. Replace the mockup with a production structure. Add the production, or "flight" wiring and cabling harness to the mockup and run functional and parametric tests on the wiring and cabling.
6. Repeat steps 2 through 4 with a production structure in place of the mockup.
7. Perform calibrations and alignment checks.
8. For extreme environments, you might perform environmental tests of the system, such as thermal cycling or mechanical shock and vibration tests.
9. Field trials may be a part of integration, which you would perform at this point.
10. Commissioning may be a part of integration, or it may be part of market release, which you would perform at this point.

Implicit in all these steps is the feedback through review to specifications and design if problems are found (Figure 14.2). Compliance is not necessarily an integration step, but if it finds a noncompliance and forces a redesign then the feedback concept in Figure 14.2 applies and integration has to be repeated.

Let me be blunt here—*throwing together all the subsystems, in what is called "big-bang integration," and hoping the final system works is stupid!* Big-bang integration is engineering by gambling and luck. I have never known it to work.

## Calibration and alignment checks

Again, *calibration* and *alignment checks* adjust components of the embedded system to tune its performance. Calibration and alignment checks prepare an embedded system for other tests within integration. They typically become a standard part of manufacturing as adjustment steps during final tests for QA and performance.

## Environmental tests

And again, *environmental tests* qualify your embedded system and provide a measure of assurance that your system can withstand environmental extremes. A sampling of types of environmental tests follows:

- Temperature cycling
- Pressure cycling
- Vacuum

- Humidity
- Steam
- Immersion
- Shock
- Vibration
- Dust
- Corrosive chemicals
- Salt spray
- UV exposure
- Deep water submersion
- Algae or microbial infestation.

Environmental testing needs to represent or replicate the actual environment in a controlled manner. "One of the first things is to correlate the real world and the testing needs" [1]. You need to characterize the actual environment, here are some suggestions:

- Field trials might help here, you might instrument a prototype or its future location to record parameters of the actual environment. In an industrial plant you might record the mechanical shocks, vibrations, temperatures, humidity, and concentrations of corrosive agents. By way of example, the corrosive agent may be salt spray in a maritime application.
- These recordings of environmental parameters include the frequency of occurrence, frequency of change, ranges, and durations for each of these parameters.
- Consulting with experienced staff will also help you determine the extremes for these parameters and the unusual events within the environment.

### *Thermal cycling, chamber testing*

Thermal cycling between temperature extremes exercises embedded systems in two ways. First, it causes materials to expand and contract. Second, it causes parameter changes with electronic components. Each of these can generate unexpected consequences if not investigated.

Expansion and contraction can flex materials, particularly metals in enclosures, mechanisms, and wiring. Any latent defects, microfractures, can eventually open up and lengthen resulting in a fracture and a break in the material.

Exacerbating the problem with flexing induced by thermal transients is that every material has its own separate thermal coefficient of expansion or TCE. If two materials with different TCEs are bonded, they will expand or contract to different extension lengths, which produce shears and possibly a fracture. This sort of problem can occur in some circuit boards with different substrates and components, which eventually causes solder joints to fracture and lift leading to failure in circuit function.

Thermal transients also induce parametric changes in the electrical values of electronic components. For analog circuits, which rely on the values of components, operations can change in the signals generated or manipulated.

---

### Example: Laboratory bench test of a circuit

Years ago I designed an embedded system. I placed a prototype system into a thermal chamber to check circuit functions and gain some insight into margins in its operations. Part of the circuit incorporated a set of simple LED drivers; Figure 14.5 illustrates one section of the simple circuit. During low temperatures, starting around $-22°C$, the LEDs would not light.

Apparently the beta, $\beta$, of the transistors decreased with lower temperatures and the transistors would not go into saturation; therefore, they would not pass enough current through LEDs to turn them on. Originally I put in a value of 47 k$\Omega$ for $R$, the base resistor. During low temperature the high value of $R$ prevented the inputs from the processor from driving the transistors into saturation. By changing the value of $R$ to 10 k$\Omega$, the processor could then sufficiently drive the transistors to turn on the LEDs reliably even at low temperatures.

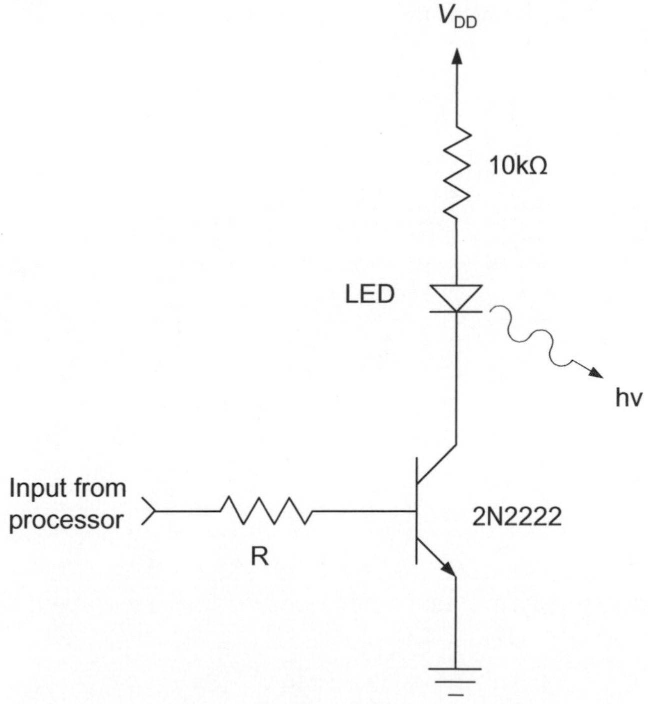

**Figure 14.5:**
LED driver circuit that suffered parametric change at cold temperatures and failed to operate.

The point of this example is that thermal cycling revealed a potential problem with my circuit. Temperature affected the beta, $\beta$, parameter value of the transistors, which I did not recognize at ambient temperatures.

A final reason for thermal cycling tests is to explore margins of operations in your embedded systems, much like I did in the previous example. Often you specify the thermal cycling to exceed the temperature specifications of the system by a small percentage and then run the embedded system at these extremes. Knowing that a system passes these tests adds a degree of confidence in the design. This is done in satellite designs—both with the subsystems and with the final satellite. The case study that follows illustrates a potential thermal cycling test for a satellite subsystem.

### Case Study: Thermal tests

Satellites and their subsystems undergo thermal vacuum tests to provide a measure of assurance in operations. In this case a subsystem is tested in a thermal chamber before integration into the satellite's superstructure (thermal vacuum tests are performed later). The subsystem is specified to operate between $-25°C$ and $+65°C$; the thermal test runs from $-30°C$ to $+70°C$. Here is the sequence of tests, as illustrated in Figure 14.6:

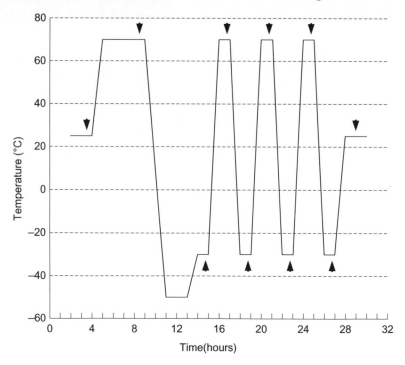

**Figure 14.6:**
Thermal cycling example; the arrow heads point to times when the subsystem turns on and runs a functional test. © 2013 by Kim R. Fowler. Used with permission. All rights reserved.

1. Stabilize the subsystem at an ambient temperature of +25°C.
2. Run a functional test of the subsystem to confirm its operation.
3. Turn off the subsystem and raise the temperature in the chamber.
4. "Soak" the subsystem at the test high temperature (+70°C) for 4 hours.
5. Turn on the subsystem and run a functional electrical test to demonstrate that it operates at the test high temperature (+70°C).
6. Turn off the subsystem and drop the temperature. "Soak" the subsystem again but this time at a very low temperature, −50°C, for 2 hours.
7. Raise the temperature to the test minimum (−30°C) and "soak" the subsystem for 1 hour.
8. After stabilizing the subsystem at the test minimum temperature (−30°C), turn it on and run a functional electrical test to demonstrate that it operates at the low temperature.
9. Turn off the subsystem and raise the temperature to the test high temperature (+70°C).
10. "Soak" the subsystem for 1 hour at high temperature before running its functional test.
11. Cycle the subsystem between the hot and cold temperatures with "soaks" at the extremes before running the functional test at the extreme temperatures.

A functional test should exercise every circuit and mechanism within the embedded subsystem. Assuming the subsystem operates correctly during each functional test, then the integration program continues. If the subsystem does not pass its functional tests, then the development team must investigate the failure with a root cause analysis and fix the problem. (Often this is a mechanical problem, such as an open circuit between a wire and a connector pin.)

Instrumenting and running a thermal test clearly takes time and effort. This case study takes more than a full day to run in three 10-hour shifts. Setting up and taking down the thermocouples and cable connections in the thermal chamber with monitoring instrumentation situated outside the chamber will take another day. A small thermal chamber can be purchased, in used condition, for about US$5000. A larger, new thermal chamber can cost US$30,000 or more. Figures 14.7 and 14.8 are examples of two different types of thermal chambers.

Thermal vacuum tests require a vacuum chamber that can be instrumented with feedthrough wires and tubes for heating, cooling, and monitoring instrumentation. These are expensive assets. Most of us cannot afford to buy a thermal vacuum chamber, so going to a contract test vendor is your only option in most cases. Typically you will need to reserve time in a thermal vacuum chamber months in advance. They will also have personnel who can aid you in developing the appropriate test profiles. Figure 14.9 is an example of a thermal-vacuum chamber.

### Vibration and shock

Vibration and shock testing exercises the mechanical and structural integrity of embedded systems. Vibration and mechanical shock move and flex materials. Movement can cause

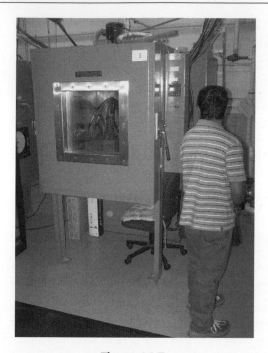

**Figure 14.7:**
An environmental chamber running a test on some products. © *2014 by Kim R. Fowler. Used with permission. All rights reserved.*

**Figure 14.8:**
A satellite being mounted inside a large environmental chamber in preparation for testing. © *2006 The Johns Hopkins University/Applied Physics Laboratory. All rights reserved. For permission to use, modify, or reproduce, contact the Office of Technology Transfer at JHU/APL.*

misalignment and binding of mechanisms. Repeated flexure leads to fractures and breaks in materials. The most susceptible components, in most systems, are the solder joints, wires, and connectors.

**Figure 14.9:**
A thermal-vacuum environmental chamber for testing spacecraft modules. © *2006 The Johns Hopkins University/Applied Physics Laboratory. All rights reserved. For permission to use, modify, or reproduce, contact the Office of Technology Transfer at JHU/APL.*

Interestingly, Matisoff claims that "vibration failures are four times as frequent as mechanical shock failures. Many components and systems can take up to 75 g of shock yet cannot perform under as little as 2 g of sustained vibration" [5].

## Case Study: Vibration table tests

Satellites and their subsystems undergo vibration tests to provide a measure of assurance in operations. In this case, a subsystem is tested on a vibration table before integration into the satellite's superstructure. The subsystem must endure the vibration profile illustrated in Figure 14.10. Here is the sequence of tests:

1. Attach the subsystem to a mounting plate and then attach the assembly to the vibration table.
2. Attach accelerometers with capton tape to various surfaces of the embedded system.
3. Run the vibration profile (often a few minutes of random or swept frequencies) and collect the acceleration data from the accelerometers.
4. While still on the table, connect the embedded system to your test equipment (ground support equipment) and run a functional test to confirm operation. If the test indicates a problem, stop and investigate. It may ultimately lead to a fix or redesign.
5. Assuming the system passes the functional test, continue onto the next axis of vibration. Repeat steps 1 through 4 after mounting in a different axis orientation.
6. If testing the third axis, repeat steps 1 through 4 after mounting in the third axis orientation. This, of course, assumes that the system has passed the previous functional tests.

A functional test should exercise every circuit and mechanism within the embedded subsystem. Assuming the subsystem operates correctly during each functional test, then the integration

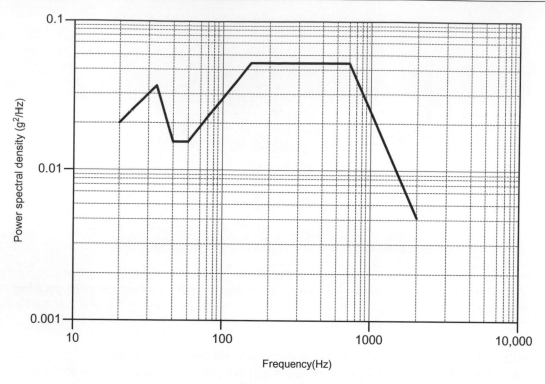

**Figure 14.10:**
Example of a vibration profile. © 2013 by Kim R. Fowler. Used with permission. All rights reserved.

program continues. If the subsystem does not pass its functional tests, then the development team must investigate the failure with a root cause analysis and fix the problem. (Often this is a mechanical problem, such as an open circuit between a wire and a connector pin.)

Instrumenting and running a vibration test takes time and effort, just like a thermal test. This case study could easily take a full day to run. Test personnel must set up and take down the accelerometer connections on the shake table for each axis of vibration. Vibration facilities and shake tables are expensive assets, too. Again, going to a contract test vendor is the only option for most of us. Typically you will need to reserve time in a vibration facility months in advance. They will have personnel who can aid you in developing the appropriate test profiles. Finally, vibration is an interesting concern; your system may be tested to different vibration profiles in each of the three axes of orientation depending on the application. Figures 14.11 and 14.12 show two different examples of vibration tables.

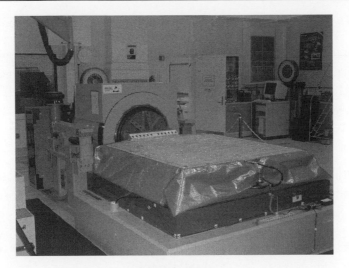

**Figure 14.11:**
An example of a vibration table. © *2006 The Johns Hopkins University/Applied Physics Laboratory. All rights reserved. For permission to use, modify, or reproduce, contact the Office of Technology Transfer at JHU/APL.*

**Figure 14.12:**
A satellite being mounted on a vibration table. © *2006 The Johns Hopkins University/Applied Physics Laboratory. All rights reserved. For permission to use, modify, or reproduce, contact the Office of Technology Transfer at JHU/APL.*

## Humidity, condensation, and salt spray, chamber testing

Humidity and condensation tests evaluate materials, mechanisms, and circuit boards for resistance to moisture incursion and incipient corrosion effects. You conduct these tests in specially designed and constructed chambers that provide moist atmospheres, fogs, and sprays. Table 14.3 lists some of the standards for humidity, condensation, and salt spray tests. If your electronics will be around any humidity or water source, you should use a conformal coat on the circuit boards and exposed pins to seal out the moisture.

Salt spray tests primarily evaluate and compare different coatings and metal finishes for enclosures in marine environments. The goal is to find and demonstrate resistance to salt spray and how well the enclosure protects the electronics and mechanisms inside from

**Table 14.3: Different standards for testing materials for humidity, condensation, and salt spray [6]**

| Standard | Title or Annex | Applications or Comments |
|---|---|---|
| SAE J2334 | Laboratory Cyclic Corrosion Test | Originally developed for automotive cosmetic finishes but applied to a wide variety of materials and finishes |
| ASTM D1735 | Standard Practice for Testing Water Resistance of Coatings Using Water Fog Apparatus | Measures resistance to water penetration to help guage the life expectancy of coating systems |
| ASTM D2247 | Standard Practice for Testing Water Resistance of Coatings in 100% Relative Humdity | Generally a pass/fail test for predicting useful life |
| ASTM B117 | Standard Practice for Operating Salt Spray (Fog) Apparatus | Primarily for testing organic and inorganic coatings on metals, e.g., paint on marine equipment |
| ASTM G85 | Standard Practice for Modified Salt Spray (Fog) Testing | |
| | Annex 1—Acetic-Salt Spray (Fog) Testing | Tests for ferrous and nonferrous metals, organic and inorganic coatings for resistance to more corrosive environments than ASTM B117 |
| | Annex 2—Cyclic Acidified Salt Fog Test | Used for evaluating products in continuously changing environments |
| | Annex 3—Acidified Synthetic Sea Water (Fog) Test | Used for production control of exfoliation-resistant heat treatments of aluminum alloys |
| | Annex 4—Salt/$SO_2$ Spray (Fog) Test | |
| | Annex 5—Dilute Electrolyte Cyclic Fog/Dry Test | Best suited for testing paint on steel |
| ASTM G87 | Conducting Moist $SO_2$ Tests | Produces easily visible corrosion on metals in industrial or marine environments |
| ASTM B368 | Standard Test Method for Copper Accelerated Acetic Acid-Salt Spray (Fog) Test (or CASS Test) | Developed for coated metal products exposed to severe and corrosive environments |

corrosion. MIL-STD-883, Method 1009, calls for 0.5–3% salt solution sprayed on the test article for a defined period that can range from 24 to 240 hours. Then an inspector examines the surface for corrosion under $10\times$ or $20\times$ magnification. ***Please note***—salt spray does not constitute a HASS or HALT. It may be a component of a HASS or HALT but it is in no sense sufficient to be a full HASS or HALT.

The $SO_2$ and copper accelerated acetic acid tests are specific to a small range of applications. These primarily focus on severe and corrosive applications in industrial or marine environments.

## Other concerns

As mentioned above, environmental tests often require specialized equipment, such as thermal chambers, vibration tables, and salt spray chambers, which can be quite expensive. Often you will use the services of an independent, third party to perform environmental tests. Third-party tests are expensive, potentially tens of thousands of dollars per day, and require significant lead time to prepare—upwards of months.

## Stress testing

Stress tests provide indications of robustness within your design. Types of stress tests include electrical, mechanical, and environmental effects. Electrical stressing might be various combinations of undervoltages, overvoltages, spikes, transients, or excessive noise on the power lines. Mechanical stressing might be various combinations of flexing, bending, mechanical shock, or extended vibration. The environmental tests in the previous three paragraphs are forms of stress tests, as well. Stress tests can be performed on units or modules before integration in many cases.

Stress testing indicates that you have removed defects or weaknesses from your design (i.e., *indicates* not *proves*). Modified forms of stress testing can also identify nonconforming product during manufacturing. Table 14.4 gives one vendor's experience at finding defects in design through HASS.

**Table 14.4: Defects and problems found with highly accelerated stress testing [7]**

| Types of Stress | % Defects Uncovered |
| --- | --- |
| 6 DoF vibration and rapid temperature changes | 46 |
| Extreme temperature transitions | 12 |
| High temperature extreme | 12 |
| Low temperature extreme | 30 |

DoF = Degrees of Freedom

For most of us, extended and severe stress testing are best conducted by third-party vendors with the equipment, chambers, and expertise to do so.

## Highly accelerated life test

HALT and HASS are similar tests. They strive to test a statistically significant population of components, or systems, under environmental extremes and stresses to find weaknesses and to estimate the expected life of the components or systems. These environmental extremes and stresses serve to speed up the testing and reduce the needed time.

HALT uses a combination of environmental stresses simultaneously to test an embedded system. HALT may include thermal dwells, rapid temperature changes, vibration, mechanical shock, pressure changes, and even moisture. Table 14.5 gives one vendor's experience at finding defects in design through HALT.

Again for most of us, HALT testing is best conducted by third-party vendors with the equipment, chambers, and expertise to do so. In some cases, HALT testing may only take two to five days, in other cases it can take months.

## Compliance testing

Many different markets require some sort of compliance testing before certification and before approval to sell an embedded system. The following are just examples of what you may encounter in developing an embedded system for a particular market.

### Aerospace

Aerospace subsystems generally must follow processes given in AS9100, while avionics must follow DO-178C for software development and documentation. The SAE provides three DVDs with over 5000 standards, practices, reports, and documents applicable to aerospace [8].

**Table 14.5: Defects and problems found with highly accelerated life testing [7]**

| Types of Stress | % Defects Uncovered |
|---|---|
| Vibration step stress | 45 |
| 6 DoF vibration and rapid temperature changes | 20 |
| Rapid temperature transitions | 4 |
| High temperature step stress | 17 |
| Low temperature step stress | 14 |

DoF = Degrees of Freedom

## FDA

FDA classifies devices into three categories: Class I, II, or III. Class I devices must supply premarket notification, registration, prohibitions against adulteration and misbranding, and rules for good manufacturing practices. Class II devices have the same requirements as Class I plus they must meet performance standards. Class III devices have the same requirements as Class II plus they must gain premarket approval (PMA) from the FDA. Some of these devices may be eligible for 510(k) status, which speeds their approval. 510 (k) status means that the device is substantially equivalent to a device in commercial distribution before May 1976.

The basic regulatory requirements that manufacturers of medical devices distributed in the United States must comply with are as follows [9]:

- Establishment registration
- Medical Device Listing
- Premarket Notification 510(k), unless exempt, or PMA
- Investigational Device Exemption (IDE) for clinical studies
- Quality System (QS) regulation
- Labeling requirements
- Medical Device Reporting (MDR).

### Underwriters Laboratory

UL is a certifying body for safety of products in the United States. UL has more than 1000 Standards for Safety [10]. Most consumer appliances must receive UL approval before sales commence.

### CE marking

CE marking is a compliance mark for products sold in Europe and around the world. The CE mark indicates compliance with the essential requirements out of the European New Approach Directives. The CE marking on products serves as a "passport" into the European market place; the letters "CE" are the abbreviation for "Conformité Européenne." Wikipedia provides a good explanation of CE marking [11]. Another more detailed explanation may be found in Ref. [12].

Compliance for CE marking requires compliance documents, Technical File (TF), users manual, and product labeling. The EC Declaration of Conformity is a document which the manufacturer (or his agent) or importer declares that the product conforms with the essential requirements of the relevant New Approach Directive(s). The declaration contains the following main elements:

- company name and complete address of the manufacturer and, where appropriate, his agent,
- product identification (name, function, model, type, serial number and trade name, and any relevant additional information)
- the directives concerned and where appropriate harmonized standards or other technical standards and specifications,
- appropriate name and address of the person who is responsible to compile the technical file in the community,
- appropriate name, address, and identification number of the notified body,
- identity and signature of the person authorized to act on behalf of the manufacturer or his representative to compile the declaration,
- the date on which the declaration is issued [12].

The TF contains relevant information that must indicate that the essential requirements out of the applicable directive(s) have been met. A TF includes the following:

- a general description of the product,
- a risk assessment (see Chapter 5) to identify health and safety requirements on the applicable product,
- design and fabrication drawings,
- detailed technical information on key aspects of the product,
- a list of standards and essential requirements that are met,
- reporting of calculations and tests,
- certificates and test reports,
- user manual—The information that is given to the user plays an essential role in accident prevention and reducing the risks regarding safety. The preparation and provision of the safety instructions governing the use is fundamental legal safety requirements. The manual should contain all information necessary for the proper and safe use of a product [12].

The cost of complying and obtaining CE marking is hugely variable and specific to each product. Here are some basic questions you need to answer before you can estimate the cost [13]:

- Which CE directive or CE directives apply/applies to the product?
- Which standards apply to the product? And do these standards have the status of European harmonized standards or not?
- Which certification procedure or procedures apply/applies?
- Are you required to involve a third-party certification body, or can you self-certify?
- Is it possible for you to do in-house testing/inspections, or do you need to subcontract this?

- Do you want to combine CE marking with other (private) certifications/markings (e.g., GS, TUV, NEMKO, DEMKO, SEMKO)?
- Can you buy CE approved parts or components that will make your own CE marking effort easier?
- Have you thought about the internal company resources you require for the CE marking? Do these have to be included in your budget? [13].

With this information you can compare quotes from service providers such as test laboratories, certification bodies, and consultants and ensure that will provide the same service [13].

Most companies need help getting the CE marking. Third-party certification to certify products are called "notified bodies," which may be either a private company or a government agency; they are authorized by European countries to serve as independent test labs and perform the steps called out by product directives.

### Military

The US government has moved away from strictly adhering to military standards over the last 20 years. Now military projects use the Statement of Work (SOW) or a Procurement Specification (PS) instead of calling out military standards to provide project specifications.

## Other issues to consider

### Measurement science

Test, integration, and compliance all rely on measurement science. While measurement is historically an old technical subject, it is fundamental to everything that you do. You need to understand, at least implicitly, the definitions for accuracy, precision, and uncertainty of the differences between parameters. At a higher level, you need to define what is a fault, a failure, and an incident; you will also need to set the level of fault detection—will you detect 95% of all faults or 98% or 99.999%? The greater the detection coverage, the higher the cost of test and integration and the longer it takes.

Here are some basic definitions required in test. This is not an exhaustive list.

**Measurand**: The measurand is a specific physical parameter central to the measurement. It can be a property of a material or a condition of a process. A sensor (or a collection of sensors) selectively transduces the desired measurand to produce, in most situations, an electrical signal that enters a system to produce the measurement of the measurand.

**Accuracy**: Describes how much the measured value deviates from the true value (relates to a defined standard) of the measurand. (Please note—accuracy is not the same as resolution! They are related values but not synonymous.)

**Resolution**: Describes the smallest increment of input stimulus that can be sensed, which is not necessarily the accuracy because the transfer function may be nonlinear.

**Linearity**: Describes the form of the transfer function relationship between measurand input and sensor output. A linear relationship means that it is very simple to convert sensor output to the measurement result. This is often the parameter that requires calibration.

**Threshold**: Describes the minimum and maximum input detection levels beyond which the sensor produces no usable output.

**Precision**: Describes how repeatable the measurements are; specifically a measurement repeated with identical input conditions, precision describes how much the results will vary. (Please note—precision is not accuracy. Accuracy describes how close the sensor is to a "static" ideal. Precision describes how results vary dynamically.)

**Uncertainty**: Related to accuracy, resolution, and precision; it describes the variance in repeated samples of the same value or data point.

### Automation versus skilled manual test

Typically a concern with test and integration is whether you will automate the test or will use skilled, expert manual test. The cost of automation, the total extent of the test, and the time required to develop the automated test all factor into the decision between automation or manual test. The level of test, whether mechanical or electrical component or subsystem or system, also plays into the decision as to what is automated and what is manually tested.

### Manufacturing test

Manufacturing test does not validate the system. Manufacturing test confirms QA of systems after manufacture. It often requires additional effort to design and build test jigs—see Chapter 15.

## Acknowledgment

My thanks to George Slack for his careful and thoughtful review, critique, examples, and suggestions for this chapter.

# References

[1]  K.H. Pries, J.M. Quigley, Testing of Complex and Embedded Systems, CRC Press, Boca Raton, FL, 2011, pp. 1–3, 6, 11, 17, 21–23, 38, 39, 187.

[2]  F. Akretch, Automotive electronics systems: trends and impact for test and measurement companies, EDN (2012)<http://www.edn.com/electronics-blogs/scope-guru-on-signal-integrity/4403881/Automotive-electronics-systems-trends-and-impact-for-test-and-measurement-companies-?cid = EDNToday> (accessed 08.01.13).

[3]  The Guide to the Systems Engineering Body of Knowledge (SEBoK), Version 0.5, Released for public review 19 September 2011, p. 170. Copyright 2011 by Stevens Institute of Technology.

[4]  J. Voas, Fault injection for the masses, Computer 30 (1997) 129–130<http://en.wikipedia.org/wiki/Fault_injection> (accessed 06.01.14).

[5]  B.S. Matisoff, Handbook of Electronics Packaging Design and Engineering, third ed., Van Nostrand Reinhold, New York, NY, 1996. p. 163.

[6]  R. Singleton, Accelerated corrosion testing, Met. Finish. (November/December) (2012) 12–19<http://digitaleditions.sheridan.com/display_article.php?id = 1262398> (accessed 28.06.13).

[7]  Envirotronics Company brochure on HALT and HASS chambers and testing. <http://www.envirotronics.com/templates/rt_kinetic_j15/pdf/halthass.pdf> (accessed 28.06.13).

[8]  SAE Aerospace Standards on DVD. <http://store.sae.org/cdstan.htm> (accessed 12.07.13).

[9]  FDA guidance. <http://www.fda.gov/MedicalDevices/DeviceRegulationandGuidance/Overview/default.htm> (accessed 12.07.13).

[10]  UL standards. <http://www.ul.com/global/eng/pages/solutions/standards/> (accessed 12.07.13).

[11]  <http://en.wikipedia.org/wiki/CE_marking> (accessed 28.06.13).

[12]  <http://www.cesolutions.eu/ce-marking.html#10> (accessed 29.05.13).

[13]  <http://www.cemarking.net/what-are-the-costs-of-ce-certification/> (accessed 29.05.13).

# Suggested reading

K.H. Pries, J.M. Quigley, Testing of Complex and Embedded Systems, CRC Press, Boca Raton, FL, 2011.

## CHAPTER 15

# Manufacturing

## Kim R. Fowler
*IEEE Fellow, Consultant*

## Chapter Outline

Developing and Managing Embedded Systems and Products.
DOI: http://dx.doi.org/10.1016/B978-0-12-405879-8.00015-5

## *Overview of manufacturing*

Manufacturing is the transformation from design concept to physical realization. It converts energy, materials, labor, and thought into tangible products that hopefully meet the expectations of customers. Manufacturing is arguably the most obvious and necessary step for ideas to eventually make money. Yet, too often we give manufacturing too little thought when developing a product, which stretches development time and wastes both money and resources. On the other hand, understanding that manufacturing is integral to product development and giving it appropriate forethought will reduce both waste and time.

This chapter aims to make you aware of some aspects of manufacturing that affect embedded systems. It will introduce to you some manufacturing issues in the following areas:

- Circuit boards
- Wiring and cabling
- Enclosures
- Mechanisms.

It will also give examples of how materials, fabrication of components, and assembly of systems interact within manufacturing. These examples should illuminate some of the time and effort involved in producing embedded systems.

---

**Please remember**

All examples and case studies in this chapter serve only to indicate what you might encounter. Hopefully, they will energize your thought processes; your estimates and calculations may vary from what you find here.

---

## Some philosophical issues with manufacturing

The engineering design of an embedded system constrains the manufacturing of the final product and often raises the difficulty of production. Materials, mechanics, packaging, and circuitry all affect manufacturing. Making these different engineering disciplines play together will reduce manufacturing costs and increase quality of the final product.

Let's look at two simple case studies.

---

### Case study: Assembling an enclosure

Assume for this case study the following:

- You have specified a plastic enclosure to house an electronic circuit with a single small cable containing several conductors that exists the box.
- You bought a plastic enclosure that has two halves that fit together.
- The production run is a few thousand units over the space of a year, consequently, assembly will be manual instead of automated or robotic because either would be too costly to set up.
- The assembly costs are $30 per hour for manual labor.

Case 1: Suppose that you bought a plastic enclosure that has four posts for the circuit board and requires four screws to connect the two half shells of the enclosure together. A production line worker places the circuit board on the four posts of one half shell, then places the other half shell on top, and screws the two shells together with an electric tool. Kitting up the screws takes about 4 seconds per unit. Picking and placing the circuit board takes about 5 seconds. Picking and placing the four screws takes about 10 seconds per unit. Driving all four screws takes another 10 seconds. All together, screwing the two shells together consumes 29 seconds per unit; this operation adds US $0.242 per unit in labor costs (Figure 15.1).

Case 2: Suppose that you bought a plastic enclosure that has four posts for the circuit board and snaps together to connect the two half shells of the enclosure. A production line worker places the circuit board on the four posts of one half shell, then places the other half shell on top, and presses the two shells together. Picking and placing the circuit board takes about 5 seconds. Snapping the shells together may take about 4 seconds per unit; this operation adds US$0.075 per unit in labor costs.

Now 24 cents versus 7.5 cents may not seem like a big difference between these two cases. In an inexpensive product, such as a television remote control, 24 cents may be 5–10% of the total cost of the product, whereas 7.5 cents will be about 3%. Other considerations may come into play, of course, and screwing the enclosures together may be preferable, but for the sake of this example where only manufacturing costs are concerned, you see how the difference between types of manufacturing assembly might affect final costs.

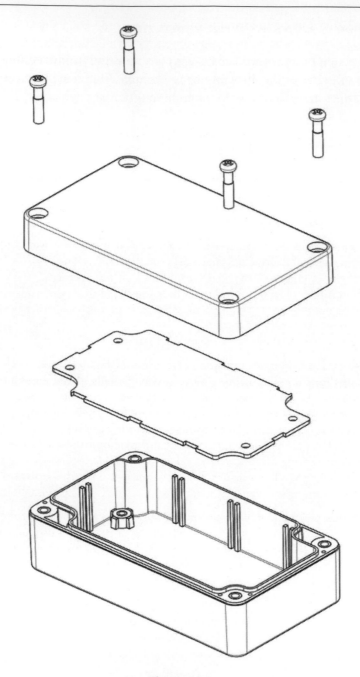

**Figure 15.1:**
Diagram of screwed enclosure shells. © 2013 by Kim R. Fowler. Used with permission. All rights reserved.

The case study illustrates just one example of how labor time equals cost. Making a product simple to assemble reduces labor cost in manufacturing. Reducing the number of steps in fabrication and assembly also reduces costs—see the next case study. Following this line of reasoning, you can see that making a device easy to manufacture is important in selling large quantities of product with low cost for maximum profit.

### Case study: Reducing parts and steps in fabrication

Washington Metal Fabricators, a contract manufacturing company, illustrates how thoughtful consideration can simplify fabrication [1]. One of their online brochures gives a before and after example of metal fabrication for which they provided consulting before fabricating the structure during manufacturing (Table 15.1).

**Table 15.1: Thoughtful design reduces fabrication time**

| Description | Before Consultation | After Consultation |
|---|---|---|
| Number of component parts | 16 | 6 |
| Number of fabrication steps | 41 | 10 |
| Time to weld | 10 minutes | 2 minutes |

This was an 80% reduction in labor time by reducing the number of parts and steps during fabrication. It was the same structure, before and after, with the same function and strength; engineering consultation helped to produce it faster and cheaper. The structure used the same materials but the labor costs were 80% cheaper!

All this means that your team should take time to consider design for manufacturing or DFM. DFM includes plans for fabrication, assembly, and disassembly. One piece castings and press fits rather than multiple pieces screwed together ease assembly. While disassembly means that a piece of equipment can be taken apart for repair or upgrade or disposal, in some situations ease of disassembly can indicate that assembly and manufacture are faster and cheaper, too.

Ultimately, you will have to tradeoff between design time and ease of manufacturing (fabrication or assembly or disassembly or disposal). The design time for a product sold in small volumes may far outweigh the cost of production; moreover, manufacturing for such a product will tend toward more labor-intensive operations because automation is too expensive or impractical for short production runs. Products sold in large volume, however, need consideration to make manufacturing efficient and as inexpensive as possible. If you plan to repair your product in the future, then disassembly becomes important and affects design and manufacturing decisions, too. Conversely, if your product is disposable, design for disassembly will not be important.

These examples may seem simple, even trivial, so why am I spending the time to discuss them? Because ***many companies still do not implement DFM for more efficient production***. Renaissance Services gives several reasons why companies do not implement DFM [2]:

*   Insufficient engineering is performed early in design to assure DFM later. Manufacturing personnel are not involved when they might be most effective in suggesting design changes to improve manufacturability.
*   A disconnect between understanding design, manufacturing processes, and how they fit together.
*   Requirements are too vague, too changeable, or require new technology and materials that do not map clearly to manufacturing.

Renaissance Services says that the cure is to understand the characteristics of the parts and how they fit within the system concept and realization and how those characteristics affect manufacturing processes. Another problem is that companies often bring in the manufacturing engineers too late; they are treated like a "bolt-on" accessory, when they are, in reality, integral to the project. When these manufacturing engineers do find manufacturing problems, then design engineers have to redesign parts and subsystems for the system to be producible—this wastes time and money. The same can be said for the logistics supply chain, bring them in early to leverage their expertise.

## General processes and procedures

Manufacturing divides into fabrication and assembly. Fabrication is the building of individual components and parts. Assembly is the connection of those parts to produce a completed system.

Manufacturing often is a set of parallel efforts to physically realize the final, completed system. Figure 15.2 illustrates the general flow of manufacturing with DFM in mind.

### Electrical and electronic

Manufacturing of electronic circuits is a complex business. For the purposes of this book, I will assume that you will purchase your components and integrated semiconductor circuits. The manufacturing team fabricates printed wiring boards (PWBs) to serve as the foundation for circuit boards before attaching components. Then manufacturing personnel assemble and solder the components on the PWBs. Manufacturing personnel also assemble the wires, cables and connectors, and bundles called wiring harnesses. Finally, manufacturing personnel place, attach, and connect the assembled modules and wiring for the final product.

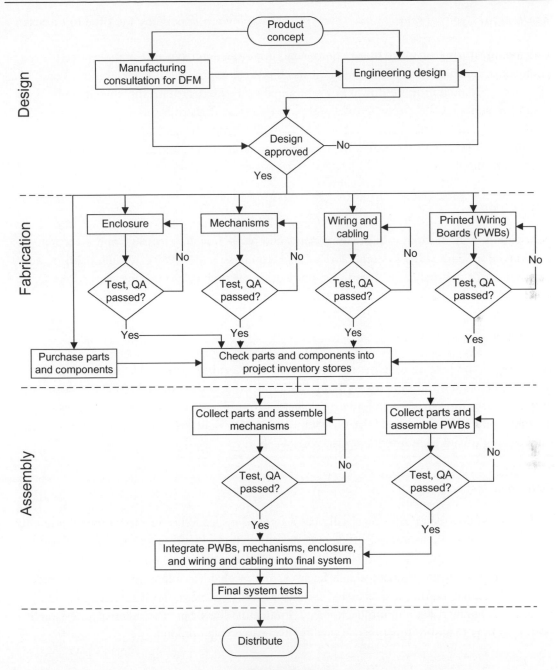

**Figure 15.2:**
General flow of materials and components through manufacturing. © *2013 by Kim R. Fowler.*
*Used with permission. All rights reserved.*

## Mechanical

The manufacturing of mechanical components for embedded systems includes enclosures, mechanisms (e.g., motors, gears, linkages, doors, and latches), circuit board attachment, and cabling tie down. For a minimal embedded system, the mechanical enclosure supports the circuit boards and power supply. If there is any type of human interface, the mechanical enclosure will have to accommodate and support it. Add in any kind of movement generated by a mechanism, and the manufacturing becomes even more involved.

## Fabrication

Fabrication is the building up of components and parts. For the electrical and electronic portions of an embedded system, fabrication encompasses the PWBs, wiring, cabling, and harnesses. For the mechanical portions, fabrication encompasses molding, casting, or machining of the enclosure and component mechanisms.

## Assembly

Assembly is the putting together of components into a subsystem or the final connection of all subsystems into the system—or both. Soldering components onto a PWB is a form of assembly. Building up the enclosure or a chassis from plastic or sheet metal, fasteners, and a backplane PWB is another form of assembly. DFM considerations for assembly focuses on reducing mistakes and reducing the cost of labor.

## Tests and inspections

Each stage within manufacturing usually has a form of inspection or test to monitor for quality. Visual inspection is a useful and efficient tool for finding problems; a technician once told me that he found 60–80% of the problems through visual inspection—incorrect wiring, incorrect component placement, mechanical interferences, and mechanical tolerance problems. Besides inspection, manufacturing tests measure for mechanical tolerances, mechanical alignment, electrical continuity, and circuit functionality. Production lines can have automated tests and power-on tests to assure function and quality of the assembled product.

## Production handoff

Between design and manufacturing should be a review and then an activity called Controlled Release. The purpose of the review and Controlled Release is to ensure that the design of the embedded system "...has been validated through the environmental qualification and the acceptance test program, that all deviations, waivers and open items have been satisfactorily closed and that the project, along with all the required support equipment, documentation, and operating procedures, is ready for production. Here are some example items from a satellite subsystem [3]:

- Rework/replacement of hardware, regression testing, or test plan changes
- Compliance with the test verification matrix
- Measured test margins versus design estimates
- Demonstrate qualification/acceptance temperature margins
- Trend data
- Total failure-free operating time of the item
- Could-not-duplicate failures should be presented along with assessment of the problem and the residual risk that may be inherent in the item
- Project assessment of any residual risk
- Review shipping containers, monitoring/transportation/control plans
- Ground support equipment status
- Post shipment plans
- System Integration Support Plans" [3].

## Specifics of fabrication and assembly

As Figure 15.2 shows, manufacturing should have a number of operations occurring in parallel—enclosure, mechanics, and circuit boards. You can perform fabrication and assembly either in-house or at contract vendors for manufacturing. You may find times where you will do both to handle extra production without increasing the capability within your company. For a specific example, many companies outsource the manufacturing and assembly of their PWBs, which are both labor-intensive and capital-intensive activities. In essence, the product companies become integrators of the final product. For high volume manufacturing, companies might have a dedicated facility or they might send the entire production to low-cost business locations around the world.

### Electronic circuit boards

#### Basic definitions

Let's start with some basic definitions for PWB fabrication and circuit board assembly [4]:

- PWB—printed wiring board.
- Via—hole through all the layers in a PWB, see Figure 15.3.
- Blind via—hole that begins on an external layer of the PWB but ends part way through the PWB.

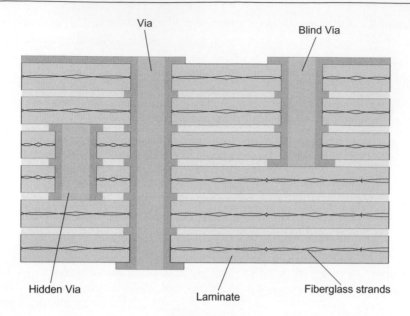

**Figure 15.3:**
Simplified cross section of a PWB. © 2013 by Kim R. Fowler. Used with permission. All rights reserved.

- Hidden via—hole between internal layers of PWB but not through external layers.
- Laminate—layers of glass cloth impregnated with resin such as epoxy or polyimide, which have been pressed together and heated to form an insulating base for circuit boards.
- Prepreg—glass cloth or fabric saturated with epoxy resin and partially cured.
- Copper-clad laminate—laminate with copper foil bonded to one or both sides, sometimes known as core in multilayer circuit boards, also called clad.
- Copper foil—thin layer of copper sheet that bonds to laminate; it is classified according to weight per unit area, commonly in two weights:
  - 1-oz copper = 0.31 kg/m$^2$ (1 oz/ft$^2$) = 0.036 mm (0.0014 in.) thick
  - 2-oz copper = 0.61 kg/m$^2$ (2 oz/ft$^2$) = 0.071 mm (0.0028 in.) thick.
- Copper plating—deposited layer of copper that lines vias and surfaces signal traces; it forms a conducting path between sides of a circuit board.
- Etching—removal of unwanted copper foil from the core by electrochemical reaction and leaves the signal traces and pads.
- Photoresist—a thin photosensitive polymer that supports photographic patterns of the signal traces and pads for etching.
- Solder mask—dark-green coating, called resist, that covers the entire board except the solder pads to prevent solder bridges between traces and to resist moisture and scratches, usually applied with silk screen.

*Fabricating PWBs*

With these basic definitions, I can now provide a general outline for fabricating PWBs [5]:

1. From the electronic circuit schematic, layout the PWB by routing signals using the most appropriate guidelines for your project. Often this uses multiple layers of interconnections through vias and signal traces.

2. The software produces a file suitable for manufacturing a PWB that describes the locations of each part, via, and signal trace on the PWB. This file is often called a Gerber file.

3. Send the Gerber file for the PWB to the manufacturing facility via the internet.

4. Cut the core material, together with prepreg, down to a standard size.

5. Apply photoresist to copper-clad laminate (steps 5 through 11 are for inner layers of the PWB).

6. Place negative-image artwork of the signal traces on the photoresist. The clear areas on the film represent the signal traces or conductors.

7. Expose the artwork and photoresist to ultraviolet light. The light fuses the polymer fibers within the photoresist to form lines that represent the signal traces and pads.

8. Wash the clad to remove the unfused photoresist from areas that are not signal traces or pads. This operation exposes the unwanted copper for later removal. The fused photoresist protects the copper that represents the future signal traces and pads.

9. Etch the exposed copper from the clad in a chemical bath. This operation leaves behind the signal traces and pads.

10. Wash the fused photoresist from the clad.

11. Apply a coating to the remaining copper to prevent oxidation.

12. Repeat steps 5 through 11 for each inner-layer core.

13. Bake the inner-layer cores to remove moisture.

14. Stack the cores with prepreg between each clad. The prepreg provides an insulating dielectric medium.

15. Laminate the cores together to form a PWB with heat, pressure, and vacuum.

16. Sandwich the PWB with aluminum or fiberboard to prevent splintering from drilling.

17. Drill all the holes for vias. (Clearly this is only for full vias, not hidden vias—that is another set of steps that I have chosen to skip in this description.)

18. Plate the internal walls of the holes with copper. (Holes must be greater than a minimum diameter to allow free flow of the plating bath.)

19. Apply photoresist to copper-clad laminate (steps 19 through 27 are for the outer layers of the PWB).

20. Place the positive-image artwork of the signal traces on the photoresist. Now the dark areas on the film represent the conductors.

21. Expose the artwork and photoresist to ultraviolet light. The light fuses the polymer fibers within the photoresist.

22. Wash the clad to remove the unfused photoresist. This operation exposes the copper that represents the future signal traces.
23. Plate the exposed clad with additional copper.
24. Plate the exposed copper with tin. (Tin acts as a resist to the chemicals that wash the photoresist and etch the unwanted copper.)
25. Wash the fused photoresist from the clad.
26. Etch the exposed copper from the clad in a chemical bath.
27. Remove the tin from the copper traces and pads.
28. Apply the solder mask. Leave at least three circular areas, or dots called fiducials, of the bare board uncoated by solder mask, which can be solder coated or gold plated. These fiducials allow automated machines to orient circuits for soldering components.
29. Mark the PWB with an epoxy-based ink to outline and designate components.
30. Electroplate gold onto any contact fingers with an underplating of nickel.
31. Coat the exposed traces and pads with solder.
32. Cut, shear, or "route out" (i.e., mill) the PWB to its final shape.

Just a few notes about PWBs and their fabrication. PWBs can have as many as 24 layers for dense circuit boards with many interconnects and signal traces. Producing PWBs and assembling circuit boards is a capital-intensive business. Assembly of circuit boards (described later) requires pick-and-place machines, reflow solder ovens, and rework stations each can cost US$150,000 or more, which does not account for the etching baths, plating facilities, drilling machines, inspection stations, and facility environmental controls used in PWB fabrication. You could easily spend upwards of US$1,000,000 to set up a PWB fabrication line—and that does not include the support facility!

### Fabricating and assembling commercial circuit boards

After fabricating the PWB (described earlier), the manufacturing process assembles the components, circuit boards, and subsystems. Here is an outline of an example commercial process:

1. Solder components onto the PWB to complete a circuit board. Figure 15.4 illustrates the general configuration of soldered joints between a component and a PWB.
2. Most operations now use a pick-and-place machine to place the components on the PWBs that have flux and solder already deposited on the pads. Figure 15.5 is a photograph of a pick-and-place machine with reels of components ready to place on a PWB.
3. After component placement on the PWB, the conveyor moves the circuit board into a reflow oven to liquify the solder and attach the components firmly to the pads of the PWB. Figure 15.6 is a photograph of the reflow oven that flows the pick-and-place machine. Figure 15.7A is a photograph of a circuit board emerging from the oven.

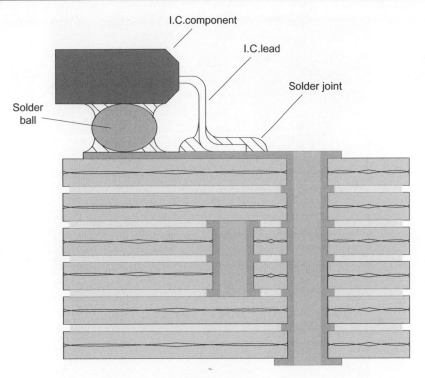

**Figure 15.4:**
Simplified cross section of a PWB with a component that shows two different types of soldered connections. © 2013 by Kim R. Fowler. Used with permission. All rights reserved.

Figure 15.7B is a photograph of several circuit boards showing the blank edges, devoid of circuitry; these edges are for handling the boards during manufacturing and will be sheared or cut off after testing.

4. If the operational environment demands, then the circuit boards are conformally coated to ward off moisture saturating the PWB and components. This is often done for both industrial equipment and consumer appliances used around sources of water and water vapor, e.g., steam irons, mixers, coffeemakers.

5. Once the circuit boards have been completed, a technician connects various components and circuit boards together. Figure 15.8 is a photograph of a technician at a work station assembling components.

6. The last step of manufacturing before placing in inventory for distribution is to test the completed system. Figure 15.9 is a photograph of a technician testing final product.

Besides understanding the actual manufacturing steps, you also need to understand the time involved in manufacturing an embedded system. Tables 15.2–15.4 illustrate examples of the calendar time required to first develop a circuit board and bring it through initial manufacturing and then the calendar time for production once it is running smoothly.

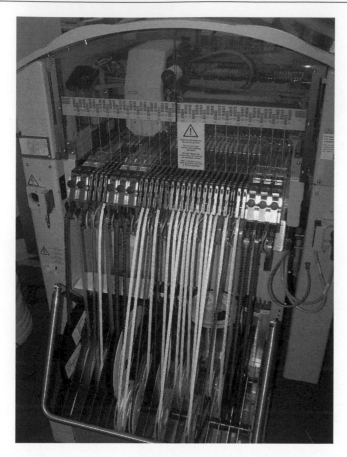

**Figure 15.5:**
A pick-and-place machine with component tape reels. © 2006–2013 by Kim R. Fowler.
*Used with permission. All rights reserved.*

Table 15.2 is an example of the shortest possible calendar time and the average calendar time spent manufacturing the first run of circuit boards. The times in this table are only representative of one example; your projects may require different schedules—and they will probably be longer than these. Please note several caveats:

- These are calendar times, based on business days per month, not staff labor hours. Labor may be greater for some of these activities because several people may be working together.
- Calendar time assumes 21 business days per month.
- The minimum time is rarely, if ever, achieved. That column describes the minimum calendar time for every activity if everything goes well without any problems.

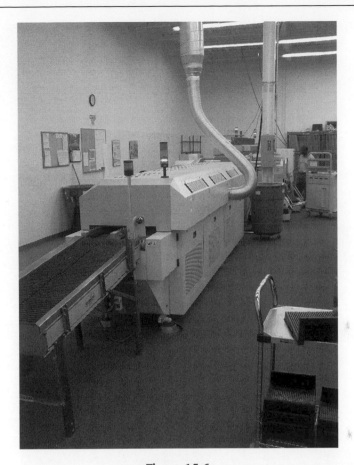

**Figure 15.6:**
A solder reflow oven with the pick-and-place machine behind it. © *2006–2013 by Kim R. Fowler. Used with permission. All rights reserved.*

- The average time is for a fairly reasonable sized circuit board such as the one shown in Figure 15.7. The average time will vary according to the complexity of the circuit and the density of components; more complex and more dense circuits will take longer to design, layout, assemble, and test. More layers can also stretch the time to fabricate a PWB, as well.

For the particular example in Table 15.2, the average time to manufacture and integrate a circuit board is 61 business days; if there is no encapsulation (i.e., a polymer film called conformal coat that covers circuit boards to prevent moisture absorption), this time drops to 58 days; regardless, you are still looking at 3 months of calendar time to complete the first set of circuit boards. The shortest possible time is 13 days; if there is no encapsulation, this time drops to 11 days.

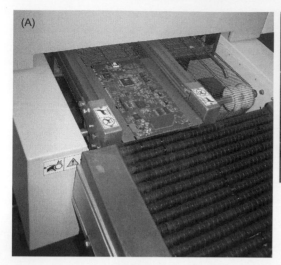

**Figure 15.7:**

(A) A circuit emerging from the solder reflow oven. (B) Circuit boards—note the blank edges on both sides of each board, these are for handling the boards. They will be removed after QA testing. © 2006—2013 by Kim R. Fowler. Used with permission. All rights reserved.

**Figure 15.8:**

A technician assembling electrical components. © 2006—2013 by Kim R. Fowler. Used with permission. All rights reserved.

The average time is due to circuit boards being built, inspected, tested, and assembled in batches. While an individual circuit board spends much less time in fabrication and assembly than the average, it is delayed at each step while other circuit boards are processed. This also explains why calendar time is not equal to labor time per circuit board.

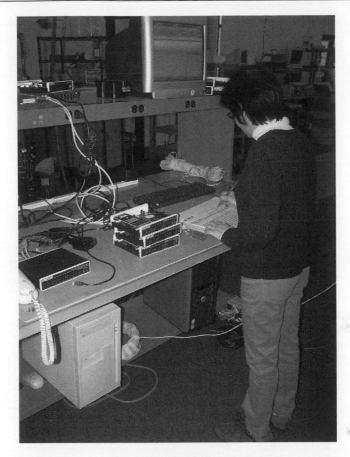

**Figure 15.9:**
A technician testing completed subsystems. © 2006—2013 by Kim R. Fowler. Used with permission.

The labor divides across boards and results in lower actual effort per board. This means that circuit boards spend a lot of time sitting and waiting for the next operation.

Table 15.3 is an example of the shortest possible calendar time and the average calendar time spent manufacturing circuit boards during regular production. The times in this table are only representative of one example; your projects may require different schedules. Please note several caveats:

- These are calendar times, based on business days per month, not staff labor hours. Labor may be greater for some of these activities because several people may be working together.
- Calendar time assumes 21 business days per month.
- The minimum time is rarely, if ever, achieved. That column describes the minimum calendar time for every activity if everything goes well without any problems.

**Table 15.2: Business days and calendar time required to fabricate, assemble, and integrate a commercial circuit board for the first time**

| Category | Description | Minimum Days | Average Days |
|---|---|---|---|
| Layout design | Part placement and board routing | 1 | 5 |
| | Breadboard release | 0.25 | 1 |
| | Engineering Design Review | 2 | 5 |
| | **Design subtotal** | 3.25 | 11 |
| Board fabrication | Request quotes from external vendor(s) | 0.25 | 3 |
| | Place requisition | 0.125 | 3 |
| | Contract released to external vendor | 0.125 | 0.5 |
| | Kitting—request parts from inventory or purchase | 1 | 5 |
| | Fabrication—PWB is not complex | 3 | 10 |
| | Bare board in receiving | 0.25 | 1 |
| | Bare board—inspection | 0.25 | 1 |
| | Prepare task control card | 0.125 | 0.25 |
| | **Board fabrication subtotal** | 5 | 24 |
| Board assembly | Release from kitting | 0.25 | 0.5 |
| | Pick-and-place machine setup | 0.25 | 1 |
| | Assembly of components | 0.25 | 1 |
| | Test and tailor—contingent upon engineer | 0 | 1 |
| | Coating and encapsulation | 0.25 | 2 |
| | Encapsulation curing | 1 | 1 |
| | Final inspection and functional test | 0.125 | 1 |
| | Release to subsystem assembly | 0.25 | 1 |
| | **Assembly subtotal** | 2 | 9 |
| Subsystem assembly | Harness, connector, or miscellaneous assembly | 0.5 | 3 |
| | Module plugin | 1 | 2 |
| | Environmental stress screen | 0 | 10 |
| | Functional tests | 0.125 | 1 |
| | Release to system assembly | 0.25 | 2 |
| | **Subsystem assembly subtotal** | 1.875 | 18 |
| | **Calendar—minimum and average (days)** | 12.6 | 61.3 |
| | **(equivalent time in months)** | 0.60 | 2.9 |

- The average time is for a fairly reasonable sized circuit board such as the one shown in Figure 15.7. The average time will vary according to the complexity of the circuit and the density of components; more complex and more dense circuits will take longer to design, layout, assemble, and test. More layers can also stretch the time to fabricate a PWB, as well.

For the particular example in Table 15.3, the average time to manufacture and integrate a circuit board is 3.6 business days; if there is no encapsulation, this time drops to 2.6 days. The shortest possible time is 1.7 days; if there is no encapsulation, this time drops to about 5.5 hours.

**Table 15.3: Business days and calendar time required to fabricate, assemble, and integrate a commercial circuit board in regular production**

| Category | Description | Minimum Days | Average Days |
|---|---|---|---|
| Board fabrication | Managing contract with external vendor | 0.0625 | 0.25 |
| | Bare board in receiving and inspection | 0.0625 | 0.125 |
| | Prepare task control card | 0.0312 | 0.0625 |
| | Check board into inventory store | 0.0312 | 0.0625 |
| | **Board fabrication subtotal** | 0.19 | 0.5 |
| Board assembly | Load pick-and-place machine | 0.0312 | 0.125 |
| | Assembly of components | 0.0312 | 0.125 |
| | Coating and encapsulation | 0.0312 | 0.125 |
| | Encapsulation curing | 1 | 1 |
| | Final inspection and functional test | 0.0312 | 0.25 |
| | Release to subsystem assembly | 0.0312 | 0.125 |
| | **Assembly subtotal** | 1.16 | 1.8 |
| Subsystem assembly | Harness, connector, or miscellaneous assembly | 0.25 | 1 |
| | Module plugin | 0.0312 | 0.125 |
| | Functional tests | 0.0312 | 0.125 |
| | Release to system assembly | 0.0312 | 0.125 |
| | **Subsystem assembly subtotal** | 0.3 | 1.4 |
| | **Calendar—minimum and average (days)** | 1.69 | 3.63 |
| | **(equivalent time in months)** | 0.08 | 0.17 |

**Table 15.4: A summary comparison of the time required to fabricate, assemble, and integrate a commercial circuit board**

| | Minimum | Average |
|---|---|---|
| Board Development | | |
| Calendar time (days) | 13 | 61 |
| (equivalent time in months) | 0.60 | 2.9 |
| Board in Production (with Encapsulation) | | |
| Calendar time (days) | 1.69 | 3.6 |
| (equivalent time in months) | 0.08 | 0.17 |
| Board in Production (No Loading or Encapsulation) | | |
| Calendar time (days) | 0.66 | 2.59 |
| (equivalent time in months) | 0.03 | 0.12 |

The average time is due to circuit boards being built, inspected, tested, and assembled in batches. While an individual circuit board spends much less time in fabrication and assembly than the average, it is delayed at each step while other circuit boards are processed. This also explains why calendar time is not equal to labor time per circuit board.

The labor divides across boards and results in lower actual effort per board. This means that circuit boards spend a lot of time sitting and waiting for the next operation.

Table 15.4 summarizes and compares the minimum and average times of a first-run circuit board to an identical one in regular production. Some of the steps to verify the design in the first run are removed for regular production. Experience also introduces efficiencies into regular production. In this particular set of examples, the improvement is nearly 20 fold for both minimum and average times!

### Fabricating and assembling space-qualified circuit boards

Circuit boards used in spacecraft have a much different and longer development time. While fabricating the PWB is fairly similar to a commercial-quality PWB, the manufacturing process for assembly is much more involved for space-qualified systems.

Before I describe the production of space-qualified circuit boards, let me provide a short glossary of useful acronyms:

- EDR—Engineering Design Review
- EM—Engineering Model, this is a circuit board or subsystem that is identical to the flight hardware but built to a less rigorous CAD design level and without the final conformal coat or encapsulation. It is used to prove functionality of the final design. The EM can be pressed in to service to replace damaged or failed flight hardware should time be of the essence and the EM judged flight worthy
- FFR—Flight Fabrication Review
- CAD—Computer-Aided Design
- CAM—Computer-Aided Manufacturing
- DRC—Design Rule Check.

Now let's jump into the steps to layout, fabricate, and assemble a space-qualified circuit board. Here is an outline of one company's space-qualified process:

**Design layout:**

1. The CAD department receives the circuit schematic (CAD layout may be a function of the design engineer, but larger companies often have a separate department for circuit layout).
2. CAD places parts on the design layout.
3. CAD routes the signals on the design layout.
4. The project conducts an EDR with the purpose to determine if the design layout is correct. The circuit designer, the CAD designer, and the lead engineer all participate in the EDR. The EDR is called and performed when the design is estimated to be 80% complete (this estimation boundary is set by program management; set too early, such as when the design is estimated to be 65% complete, and you will guarantee redesign

and "churn" in the layout design, thus delaying the project; set too late, such as when the design is estimated to be 98% complete, and the project will move too slowly, again delaying the project).

5. Breadboard release (this occurs in level 1, which is the simplest, least stringent development designation for the project, a breadboard is a good, quick way to verify circuit functionality).

6. EM release (this occurs in level 2a, which is a slightly more rigorous development designation but it still allows redline modifications in the drawings while fabricating circuit boards).

7. The project conducts an FFR with the purpose to determine if the design is manufacturable. Many different people attend the FFR, including CAM, designer, lead engineer, fabrication, EMC engineer, inventory staff, and staff responsible for thermal, packaging, materials, and contamination. The FFR is called and performed when the design is estimated to be 90% complete.

8. The flight design layout then moves to sign-off. Generally, it takes about five working days or more from FFR to sign-off table. The purpose of the sign-off table is to determine if the design is considered complete. Redline corrections are made to the drawings. Many different people sign the flight drawings, including CAM, designer, lead engineer, fabrication, EMC engineer, inventory staff, and staff responsible for thermal, packaging, materials, and contamination. It takes at least two working days (typically five or more) from the sign-off table until all signatures are collected. The biggest problem typically is getting people to review and sign the drawings.

9. After all redlines from sign-off are incorporated, CAD then releases the flight drawings to the product data manager who collects and finalizes the documentation package. This usually takes between 2 hours and 1 day to complete.

**PWB Fabrication:**

10. CAM prepares for fabrication and determines if the PWB will be built in-house or outsourced to a contract manufacturer. CAM performs a DRC, which takes anywhere from 3 to 18 hours. It takes longer if the PWB is complicated with many layers (>8 layers) or many vias ( > 500) or has very tight placement or many signal and bus connections. CAM adds coupons to the circuit board design (a coupon is a test section added to a corner of the PWB panel that provides a measure of quality assurance (QA) for the PWB).

11. If built by a contract manufacturer, then CAM requests quotes for PWB fabrication.
    a. Once CAM receives the quotes, usually within two days, then CAM selects a vendor and places a requisition.
    b. The requisition must be signed, which usually takes several days unless someone walks it around for signature.
    c. CAM then sends the requisition to Purchasing.

12. Kitting—CAM requests parts kits to be assembled, which usually takes two weeks, but it can be done in parallel with the PWB fabrication.

13. Fabrication
    a.  The priorities for fabricating PWBs are set by programs in the queue, otherwise it is first come, first serve.
    b.  Simple boards can usually be built within five days, e.g., those with four layers, not many parts or vias or military processing requirements—primarily for breadboarding experiments.
    c.  Rigid, space quality PWBs typically take 15 working days (3 weeks). An urgent expedite can shorten that to 10 working days (2 weeks) but at double or triple the cost.
    d.  Rigid–flex, space quality PWBs typically take four to six weeks. An urgent expedite can shorten that to three weeks.
    e.  If built in-house, fabrication depends on complexity and the current backlog in queue for fabricating other PWBs. The typical duration is three to five weeks.
    f.  This particular company's in-house quality tends to be extremely good but it is slower than commercial vendors. At times, as many as 80% of the outsourced PWBs may be rejected for manufacturing deficiencies.
14. Bare board complete: Receiving—1 day
15. Bare board test, inspection, coupon test—usually takes about five working days
    a.  Incoming inspection
        i.   Dimensional verification, random check of 10% vias, each board is checked for different vias; after 10 boards, hopefully nearly 100% of all vias and dimensions have been checked across all PWBs.
        ii.  Electrical test for shorts and opens on the netlist.
        iii. Validation to Mil-spec.
        iv.  Workmanship inspection.
    b.  Coupon testing (in parallel with visual inspection)
        i.   Bake coupons to remove moisture, which takes 6 hours, usually this amounts to taking a day.
        ii.  Section (i.e., cut through to expose the edge of) coupons and encapsulate them in epoxy—2 hours minimum.
        iii. Polish encapsulated coupons and microscope inspection—2 hours minimum and depends on the priority fixed for PWB and for other boards already in the queue.
        iv.  Coupon inspection looks for voids between laminates, delaminations between layers, minimum and maximum plating thicknesses for copper, gold, and nickel.
        v.   If a coupon has a single defect of any kind, all boards in that fabrication batch fail and are rejected; the vendor is notified with standard documentation and photographs.
    c.  Prepare a task control card for the PWB, which will become a populated circuit board after assembly.

**Assembly:**

16. Release from kitting—the board is combined with the kitted components, typically takes two days.
17. Assembly—passive components are attached, soldered, and inspected. Depending on the complexity of the board, typically about five working days.
18. Passive component test—released to the design engineer for testing the board with only the passive components installed. Time depends on the engineer's test schedule.
19. Assembly—active components are attached, soldered, and inspected. Depending on the complexity of the board, typically about five working days.
20. Active component test—released to the design engineer for testing the board with all components installed. Engineer then determines the values of the tailor points. Time depends on engineer's test schedule.
21. Tailor process—remaining tailor components are installed and the entire board is inspected. Typically five days to completion.
22. Conformal coating and encapsulation
    a. Components are staked (precoated—typically two days)
    b. Usually the design engineer will test the board between staking and potting the encapsulation
    c. Entire board is encapsulated—typically three to five days
    d. Final inspection.
23. Released to program.

Example fabrication for simple, space-qualified circuit boards

Table 15.5 is one company's record for the shortest possible calendar time and the average calendar time spent manufacturing the first run of simple, space-qualified circuit boards. The times in this table are only representative of one example; your projects may require different schedules. Please note several caveats:

- These are calendar times, based on business days per month, not staff labor hours. Labor may be greater for some of these activities because several people may be working together.
- Calendar time assumes 21 business days per month.
- The minimum time is rarely, if ever, achieved. That column describes the minimum calendar time for every activity if everything goes well without any problems.
- These estimates are for a simple circuit board, which is a board with 8 or fewer layers, less than 500 vias, and not densely packed with components.
- The average time for a simple circuit board will vary according to the complexity of the circuit and the density of components; more layers, more vias, more complexity, and more dense circuits will take longer to design, layout, assemble, and test.

**Table 15.5: Business days and calendar time required to layout, fabricate, assemble, and integrate a simple (<10 layers, <500 vias, not dense), space-qualified circuit board for the first time**

| Category | Description | Minimum Days | Average Days | Backlog Days |
|---|---|---|---|---|
| Design (in CAD) | Received into design | 0.25 | 1 | 5 |
| | Part placement and board routing | 5 | 10 | 10 |
| | Engineering Design Review | 5 | 10 | 0 |
| | Breadboard release | 0.25 | 1 | 0 |
| | Informal Flight Fabrication Review | 5 | 10 | 0 |
| | Engineering Model release | 0.25 | 5 | 0 |
| | Flight Fabrication Review | 5 | 10 | 15 |
| | On-table for sign-off | 5 | 10 | 0 |
| | Flight release to product manager | 0.25 | 1 | 0 |
| | **Design subtotal** | **26** | **58** | **30** |
| Board fabrication | CAM design rule check | 0.375 | 3 | 8 |
| | Request quotes from external vendor | 1 | 3 | 5 |
| | Place requisition and get signatures | 2 | 4 | 0 |
| | Contract released to external vendor | 0.125 | 1 | 2 |
| | Kitting—requests parts kits from inventory | 10 | 10 | 5 |
| | Fabrication—not complex | 5 | 15 | 15 |
| | Bare board in receiving | 1 | 3 | 0 |
| | Bare board—inspection and coupon tests | 3 | 5 | 5 |
| | Prepare task control card | 0.25 | 1 | 2 |
| | **Board fabrication subtotal** | **23** | **45** | **42** |
| Board assembly | Release from kitting | 1 | 2 | 0 |
| | Assembly of passive components | 1 | 5 | 10 |
| | Inspection passive components assembly | 0.5 | 2 | 0 |
| | Test—contingent upon engineer to test | 0 | 5 | 0 |
| | Assembly of active components | 1 | 7 | 10 |
| | Inspection active components assembly | 0.5 | 2 | 0 |
| | Test—contingent upon engineer to test | 0 | 5 | 0 |
| | Tailor process | 1 | 5 | 0 |
| | Inspection | 1 | 3 | 0 |
| | Coating and encapsulation | 2 | 5 | 5 |
| | Inspection after coating | 0.5 | 2 | 0 |
| | Coating touchup | 1 | 3 | 0 |
| | Inspection after coating touchup | 0.5 | 2 | 0 |
| | Final inspection | 0.5 | 2 | 0 |
| | Release to program | 1 | 3 | 0 |
| | **Assembly subtotal** | **12** | **53** | **25** |
| System assembly | Harness assembly | 1 | 5 | 5 |
| | Module plugin | 1 | 2 | 10 |
| | Stake jack screws | 1 | 2 | 0 |
| | Environmental stress screen | 5 | 15 | 0 |
| | Repair of boards | 5 | 10 | 0 |
| | Sign-off | 5 | 10 | 0 |
| | Release to program | 1 | 5 | 0 |
| | **System assembly subtotal** | **19** | **49** | **15** |
| | Calendar—minimum, average, backlog (days) | 79 | 205 | 112 |
| | (equivalent time in months) | 3.8 | 9.8 | 5.3 |

- The minimum time is the ***best*** that can be expected in ***any one*** category. ***DO NOT use minimum time as an estimation tool.*** Use the average time and add extra time for design revision. In essence, the average time total is a bare minimum for scheduling— and that does not account for possible backlog!

For the particular example in Table 15.5, the average time to manufacture and integrate a simple, space-qualified circuit board is 205 business days or 10 months of calendar time to complete the first set of circuit boards. The shortest possible time is 79 days or 4 months of calendar time. An average backlog of PWBs on a project can add up to 5 months of time to produce any one particular circuit board.

Example fabrication for complex, space-qualified circuit boards

Table 15.6 is one company's record for the shortest possible calendar time and the average calendar time spent manufacturing the first run of complex, space-qualified circuit boards. The times in this table are only representative of one example; your projects may require different schedules. Please note several caveats:

- These are calendar times, based on business days per month, not staff labor hours. Labor may be greater for some of these activities because several people may be working together.
- Calendar time assumes 21 business days per month.
- The minimum time is rarely, if ever, achieved. That column describes the minimum calendar time for every activity if everything goes well without any problems.
- These estimates are for a complex circuit board, which is a board with 10 or more layers, more than 500 vias, and densely packed with components.
- The average time for a complex circuit board may vary according to the complexity of the circuit and the density of components; more layers, more vias, more complexity, and more dense circuits will take longer to design, layout, assemble, and test.
- The minimum time is the ***best*** that can be expected in ***any one*** category. ***DO NOT use minimum time as an estimation tool.*** Use the average time and add extra time for design revision. In essence, the average time total is a bare minimum for scheduling— and that does not account for possible backlog!

For the particular example in Table 15.6, the average time to manufacture and integrate a complex, space-qualified circuit board is 210 business days or 10 months of calendar time to complete the first set of circuit boards. The shortest possible time is 84 days or 4 months of calendar time. An average backlog of PWBs on a project can add up to 5 months of time to produce any one particular circuit board.

**Table 15.6: Business days and calendar time required to layout, fabricate, assemble, and integrate a complex ( >8 layers, >500 vias, dense), space-qualified circuit board for the first time**

| Category | Description | Minimum Days | Average Days | Backlog Days |
|---|---|---|---|---|
| Design (in CAD) | Received into design | 0.25 | 1 | 5 |
| | Part placement and board routing (after all revisions) | 5 | 10 | 10 |
| | Engineering Design Review | 5 | 10 | 0 |
| | Breadboard release | 0.25 | 1 | 0 |
| | Informal Flight Fabrication Review | 5 | 10 | 0 |
| | Engineering Model release | 0.25 | 5 | 0 |
| | Flight Fabrication Review | 5 | 10 | 15 |
| | On-table for sign-off | 5 | 10 | 0 |
| | Flight release to PDM | 0.25 | 1 | 0 |
| | **Design subtotal** | **26** | **58** | **30** |
| Board fabrication | Computer aided manufacturing—design rule check | 0.375 | 3 | 8 |
| | Request quotes from external vendor | 1 | 3 | 5 |
| | Place requisition and get signatures | 2 | 4 | 0 |
| | Contract released to external vendor | 0.125 | 1 | 2 |
| | Kitting—requests parts kits from inventory | 10 | 10 | 5 |
| | Fabrication—large, many holes, many layers | 10 | 20 | 0 |
| | Bare board in receiving | 1 | 3 | 0 |
| | Bare board—inspection and coupon tests | 3 | 5 | 5 |
| | Prepare task control card | 0.25 | 1 | 2 |
| | **Board fabrication subtotal** | **28** | **50** | **27** |
| Board assembly | Release from kitting | 1 | 2 | 0 |
| | Assembly of passive components | 1 | 5 | 10 |
| | Inspection passive components assembly | 0.5 | 2 | 0 |
| | Test—contingent upon engineer to test | 0 | 5 | 0 |
| | Assembly of active components | 1 | 7 | 10 |
| | Inspection active components assembly | 0.5 | 2 | 0 |
| | Test—contingent upon engineer to test | 0 | 5 | 0 |
| | Tailor process | 1 | 5 | 0 |
| | Inspection | 1 | 3 | 0 |
| | Coating and encapsulation | 2 | 5 | 5 |
| | Inspection after coating | 0.5 | 2 | 0 |
| | Coating touchup | 1 | 3 | 0 |
| | Inspection after coating touchup | 0.5 | 2 | 0 |
| | Final inspection | 0.5 | 2 | 0 |
| | Release to program | 1 | 3 | 0 |
| | **Assembly subtotal** | **12** | **53** | **25** |
| System assembly | Harness assembly | 1 | 5 | 5 |
| | Module plugin | 1 | 2 | 10 |
| | Stake jack screws | 1 | 2 | 0 |
| | Environmental stress screen | 5 | 15 | 0 |
| | Repair of boards | 5 | 10 | 0 |
| | Sign-off | 5 | 10 | 0 |
| | Release to program | 1 | 5 | 0 |
| | **System assembly subtotal** | **19** | **49** | **15** |
| | **Calendar—minimum, average, backlog (days)** | 84 | 210 | 97 |
| | **(equivalent time in months)** | 4.0 | 10.0 | 4.6 |

Example fabrication for rigid—flex, space-qualified circuit boards

A rigid—flex circuit board has rigid PWBs connected by flex circuitry. Rigid—flex construction provides dense and unusual packaging configurations by allowing circuit boards to fold over on each other or bend various directions. Volume constraints in some spacecraft instruments require rigid—flex circuit boards.

Table 15.7 is one company's record for the shortest possible calendar time and the average calendar time spent manufacturing the first run of rigid—flex, space-qualified circuit boards. The times in this table are only representative of one example; your projects may require different schedules. Please note several caveats:

- These are calendar times, based on business days per month, not staff labor hours. Labor may be greater for some of these activities because several people may be working together.
- Calendar time assumes 21 business days per month.
- The minimum time is rarely, if ever, achieved. That column describes the minimum calendar time for every activity if everything goes well without any problems.
- The average time for a complex circuit board may vary according to the complexity of the circuit and the density of components; more layers, more vias, more complexity, and more dense circuits will take longer to design, layout, assemble, and test.
- The minimum time is the *best* that can be expected in *any one* category. *DO NOT use minimum time as an estimation tool.* Use the average time and add extra time for design revision. In essence, the average time total is a bare minimum for scheduling—and that does not account for possible backlog!

For the particular example in Table 15.7, the average time to manufacture and integrate a rigid—flex, space-qualified circuit board is 230 business days or 11 months of calendar time to complete the first set of circuit boards. The shortest possible time is 104 days or 5 months of calendar time. An average backlog of PWBs on a project can add up to 5 months of time to produce any one particular circuit board.

Summary comparison between different types of space-qualified circuit boards

I have just given you detailed, line-item estimates for manufacturing and assembling space-qualified circuit boards. Here, for your convenience, Table 15.8 compares the bottom line for manufacturing times to be expected for different types of space-qualified circuit boards.

Surprisingly, the times are not very different regardless the complexity. The minimum time to manufacture a board is between 4 and 5 months and the average time is between 10 and 11 months.

Another surprise may be the inversion of backlog times from simple circuit boards to more complex ones. Simple boards have an average backlog of over 5 months. Complex or

**Table 15.7: Business days and calendar time required to layout, fabricate, assemble, and integrate a rigid—flex, space-qualified circuit board for the first time**

| Category | Description | Minimum Days | Average Days | Backlog Days |
|---|---|---|---|---|
| Design (in CAD) | Received into design | 0.25 | 1 | 5 |
| | Part placement and board routing (after all revisions) | 5 | 10 | 10 |
| | Engineering Design Review | 5 | 10 | 0 |
| | Breadboard release | 0.25 | 1 | 0 |
| | Informal flight fabrication review | 5 | 10 | 0 |
| | Engineering Model release | 0.25 | 5 | 0 |
| | Flight fabrication review | 5 | 10 | 15 |
| | On-table for sign-off | 5 | 10 | 0 |
| | Flight release to PDM | 0.25 | 1 | 0 |
| | **Design subtotal** | **26** | **58** | **30** |
| Board fabrication | Computer aided manufacturing—design rule check | 0.375 | 3 | 8 |
| | Request quotes from external vendor | 1 | 3 | 5 |
| | Place requisition and get signatures | 2 | 4 | 0 |
| | Contract released to external vendor | 0.125 | 1 | 2 |
| | Kitting—requests parts kits from inventory | 10 | 10 | 5 |
| | Fabrication—rigid/flex multilayer | 30 | 40 | 0 |
| | Bare board in receiving | 1 | 3 | 0 |
| | Bare board—inspection and coupon tests | 3 | 5 | 5 |
| | Prepare task control card | 0.25 | 1 | 2 |
| | **Board fabrication subtotal** | **48** | **70** | **27** |
| Board assembly | Release from kitting | 1 | 2 | 0 |
| | Assembly of passive components | 1 | 5 | 10 |
| | Inspection passive components assembly | 0.5 | 2 | 0 |
| | Test—contingent upon engineer to test | 0 | 5 | 0 |
| | Assembly of active components | 1 | 7 | 10 |
| | Inspection active components assembly | 0.5 | 2 | 0 |
| | Test—contingent upon engineer to test | 0 | 5 | 0 |
| | Tailor process | 1 | 5 | 0 |
| | Inspection | 1 | 3 | 0 |
| | Coating and encapsulation | 2 | 5 | 5 |
| | Inspection after coating | 0.5 | 2 | 0 |
| | Coating touchup | 1 | 3 | 0 |
| | Inspection after coating touchup | 0.5 | 2 | 0 |
| | Final inspection | 0.5 | 2 | 0 |
| | Release to program | 1 | 3 | 0 |
| | **Assembly subtotal** | **12** | **53** | **25** |
| System assembly | Harness assembly | 1 | 5 | 5 |
| | Module plugin | 1 | 2 | 10 |
| | Stake jack screws | 1 | 2 | 0 |
| | Environmental stress screen | 5 | 15 | 0 |
| | Repair of boards | 5 | 10 | 0 |
| | Sign-off | 5 | 10 | 0 |
| | Release to program | 1 | 5 | 0 |
| | **System assembly subtotal** | **19** | **49** | **15** |
| | **Calendar—minimum, average, backlog (days)** | 104 | 230 | 97 |
| | **(equivalent time in months)** | 5.0 | 11.0 | 4.6 |

**Table 15.8: Comparison of business days and calendar time required to layout, fabricate, assemble, and integrate a space-qualified circuit boards**

|  | Minimum | Average | Backlog |
|---|---|---|---|
| **Simple Board** | | | |
| Calendar time (days) | 79 | 205 | 112 |
| (equivalent time in months) | 3.8 | 9.8 | 5.3 |
| **Complex Board** | | | |
| Calendar time (days) | 84 | 210 | 97 |
| (equivalent time in months) | 4.0 | 10.0 | 4.6 |
| **Rigid—Flex** | | | |
| Calendar time (days) | 104 | 230 | 97 |
| (equivalent time in months) | 5.0 | 11.0 | 4.6 |

rigid—flex boards have an average backlog of about 4.6 months. Upon reflecting about the nature of the boards and the setting of priorities, you may conclude, as I do, that complex circuit boards get higher priority in the queue, which pushes the simple boards back in the queue and thus delays them even further.

The final surprise is that this company and facility, on average with backlog, takes over 15 months to produce any type of high quality circuit board, regardless of complexity. I think this means that the company has pretty well optimized their processes to produce high quality circuit boards for space flight.

Common manufacturing tradeoffs for space-qualified circuit boards

Visual inspection of the PWB's test coupon, through a microscope, provides a verification measure of the quality of fabrication. After encapsulating the coupon in epoxy, slicing it, and polishing the cut surface, an expert staff member examines the surface of the coupon under a microscope. The staff member looks at the distribution of copper plating and solder. Proper PWB manufacture evenly distributes copper plating and solder. Another verification step that you might take is to measure the electrical properties of a coupon, which is a reliable indicator of the characteristic impedance of the circuit board.

Another inspection for PWB quality is to examine the solder joints after assembly. Viewing solder joints with a trained and practiced eye, both with and without a microscope, is very effective for verifying the quality of the solder joints. Three much more expensive forms of inspection are component removal to inspect the pads, X-ray examination of the joints, and laser inspection. Typically, they do not add much advantage to small quantity production runs.

One tradeoff often made in space-qualified production of circuit boards is to decide between manual soldering or using a pick-and-place machine followed by a reflow oven (as in Figures 15.5 and 15.6). For single board quantities, manual soldering by a trained technician is quite effective in achieving high quality results. One company, for which I consulted, uses manual soldering for all their circuit boards; they are confident in the quality of the solder joints; purchasing a pick-and-place machine with a reflow oven is out of its budget; finally, its experience with contract assemblers has not been good for automating the solder process. The company, described in Tables 15.5–15.8, uses manual soldering for single lot quantities but it purchased a pick-and-place machine with a reflow oven to improve throughput of circuit boards, while maintaining quality, when there are five or more identical circuit boards to produce.

### Some basic issues common to manufacturing space-qualified circuit boards

Backlog may be peculiar to the facility described in Tables 15.5–15.8—but I doubt it. The development of spacecraft is episodic; there are times of intense activity, followed by spells of inactivity. The department's staff and assets, such as the CAD area or the PWB etching baths, are undersized purposefully to maintain a viable, expert staff but not price its services out of the market. This, in effect, spreads the work out to maintain the staff and equipment at capacity but it slows the process of producing a satellite. This situation creates backlog, in which PWBs wait in a queue to be manufactured.

The company, described in Tables 15.5–15.8, eases backlog in two ways. First, it outsources the fabrication of PWBs when its own facility is running at capacity. Second, it takes on military projects, which often have similar processing requirements for quality as do space projects. Military projects help maintain the workload within the facility and keep highly expert staff employed.

Outsourcing can be problematic for space-qualified circuit boards. Sometimes, the outsourced PWBs meet the stringent quality requirements and sometimes they do not; this can occur at different times within the same contract manufacturer. The company, described in Tables 15.5–15.8, has found that any one contract manufacturer does not necessarily always maintain a high level of quality; for one project, a specific contract manufacturer may produce PWBs to an acceptable quality; on the next project, the same vendor's quality may suffer. One example is the very clear instruction given to every vendor, on every project, to "route out" (i.e., mill the PWBs) and not to shear off, or cut, the PWBs from the fabrication panels in which they are made. The reason is that routing, which is more expensive an operation than shearing, is much less likely to delaminate the PWBs than shearing. Amazingly, the company has found that all vendors heed the instruction on some projects but ignore it on other projects.

The company, described in Tables 15.5–15.8, has not found a good indicator to gauge whether a specific contract manufacturer is going to produce the needed quality. This situation seems to uphold the contention in the white paper by Sparton Corporation that contract manufacturers have personnel turnover at times, which means quality can occasionally suffer unpredictably [6].

*Fabricating and assembling military or industrial circuit boards*

While military circuit boards endure different environments than space-qualified boards, the manufacturing processes are often very similar. So, even if your company does not manufacture space-qualified circuit boards and focuses on military or industrial-quality products, you can still use these estimates as a guide or sanity check for your manufacturing.

Tables 15.9–15.11 tailor the estimates for industrial or military boards and provide estimates for production times, as well. Note that these estimates do not have all the steps that space-qualified circuit boards might have.

Table 15.9 is an example of the shortest possible calendar time and the average calendar time spent manufacturing the first run of a military or industrial circuit boards. The times in this table are only representative of one example; your projects may require different schedules. Please note several caveats:

- These are calendar times, based on business days per month, not staff labor hours. Labor may be greater for some of these activities because several people may be working together.
- Calendar time assumes 21 business days per month.
- The minimum time is rarely, if ever, achieved. That column describes the minimum calendar time for every activity if everything goes well without any problems.
- The average time for a circuit board may vary according to the complexity of the circuit and the density of components; more layers, more vias, more complexity, and more dense circuits will take longer to design, layout, assemble, and test.
- The minimum time is the *best* that can be expected in *any one* category. *DO NOT use minimum time as an estimation tool.* Use the average time and add extra time for design revision. In essence, the average time total is a bare minimum for scheduling—and that does not account for possible backlog!

For the particular example in Table 15.9, the average time to manufacture a first-run circuit board is 165 business days or nearly 8 months of calendar time to complete the first set of circuit boards. The shortest possible time is 62 days or 3 months of calendar time.

Table 15.10 is an example of the shortest possible calendar time and the average calendar time spent manufacturing military or industrial circuit boards during regular production. The times in this table are only representative of one example; your projects may require different schedules. Please note several caveats:

- These are calendar times, based on business days per month, not staff labor hours. Labor may be greater for some of these activities because several people may be working together.

**Table 15.9: Business days and calendar time required to layout, fabricate, assemble, and integrate a military or an industrial circuit board for the first time**

| Category | Description | Minimum Days | Average Days |
|---|---|---|---|
| Design (in CAD) | Received into design | 0.25 | 1 |
| | Part placement and board routing | 5 | 10 |
| | Engineering Design Review | 5 | 10 |
| | Breadboard release | 0.25 | 1 |
| | Fabrication review | 5 | 10 |
| | On-table for sign-off | 5 | 10 |
| | Release to product manager | 0.25 | 1 |
| | **Design subtotal** | 20.75 | 43 |
| Board fabrication | Request quotes from external vendor | 1 | 3 |
| | Place requisition and get signatures | 2 | 4 |
| | Contract released to external vendor | 0.125 | 1 |
| | Kitting—requests parts kits from inventory, performed in parallel with fabrication | 0 | 0 |
| | PWB fabrication | 5 | 15 |
| | Bare board in receiving | 1 | 3 |
| | Bare board—inspection and coupon tests | 3 | 5 |
| | Prepare task control card | 0.25 | 1 |
| | **Board fabrication subtotal** | 12 | 32 |
| Board assembly | Release from kitting | 1 | 2 |
| | Assembly of all components | 1 | 7 |
| | Inspect components assembly | 0.5 | 2 |
| | Test—contingent upon engineer to test | 0 | 5 |
| | Tailor process | 1 | 5 |
| | Inspection | 1 | 3 |
| | Coating and encapsulation | 2 | 5 |
| | Inspection after coating | 0.5 | 2 |
| | Coating touchup | 1 | 3 |
| | Inspection after coating touchup | 0.5 | 2 |
| | Final inspection | 0.5 | 2 |
| | Release to program | 1 | 3 |
| | **Assembly subtotal** | 10 | 41 |
| System assembly | Harness assembly | 1 | 5 |
| | Module plugin | 1 | 2 |
| | Stake jack screws | 1 | 2 |
| | Environmental stress screen | 5 | 15 |
| | Repair of boards | 5 | 10 |
| | Sign-off | 5 | 10 |
| | Release to program | 1 | 5 |
| | **System assembly subtotal** | 19 | 49 |
| | **Calendar—minimum, average (days)** | 62 | 165 |
| | **(equivalent time in months)** | 3.0 | 7.9 |

**Table 15.10: Business days and calendar time required to fabricate, assemble, and integrate a military or industrial circuit board in regular production**

| Category | Description | Minimum Days | Average Days |
|---|---|---|---|
| Board fabrication | Manage contract with external vendor | 0.0625 | 0.25 |
| | Kitting—requests parts kits from inventory, performed in parallel with fabrication | 0 | 0 |
| | PWB fabrication | 5 | 15 |
| | Bare board in receiving | 0.0625 | 0.125 |
| | Bare board—inspection and coupon tests | 3 | 3 |
| | Prepare task control card | 0.0625 | 0.125 |
| | **Board fabrication subtotal** | 8 | 19 |
| Board assembly | Assembly of all components | 0.0625 | 0.125 |
| | Inspect components assembly | 0.0312 | 0.0625 |
| | Test | 0.0312 | 0.0625 |
| | Tailor process | 0.0312 | 0.0625 |
| | Inspection | 0.0312 | 0.0625 |
| | Coating and encapsulation | 2 | 2 |
| | Inspection after coating | 0.0312 | 0.0625 |
| | Coating touchup | 0.0312 | 0.0625 |
| | Inspection after coating touchup | 0.0312 | 0.0625 |
| | Final inspection | 0.0312 | 0.0625 |
| | Release to program | 0.0312 | 0.0625 |
| | **Assembly subtotal** | 2 | 3 |
| System assembly | Harness assembly | 0.125 | 0.25 |
| | Module plugin | 0.0312 | 0.0625 |
| | Environmental stress screen | 5 | 5 |
| | Repair of boards | 0.125 | 0.25 |
| | Final inspection | 0.0312 | 0.0625 |
| | Release to program | 0.0312 | 0.0625 |
| | **System assembly subtotal** | 5.34 | 5.69 |
| | **Calendar—minimum, average (days)** | 16 | 27 |
| | **(equivalent time in months)** | 0.8 | 1.3 |
| | **Overlap fabrication with assembly, do not stress screen boards** | | |
| | **Calendar—minimum, average (days)** | 2.7 | 3.4 |
| | **(equivalent time in months)** | 0.13 | 0.16 |

- Calendar time assumes 21 business days per month.
- The minimum time is rarely, if ever, achieved. That column describes the minimum calendar time for every activity if everything goes well without any problems.
- The average time will vary according to the complexity of the circuit and the density of components; more complex and more dense circuits will take longer to design, layout, assemble, and test. More layers can also stretch the time to fabricate a PWB, as well.

**Table 15.11: A summary comparison of the time required to fabricate, assemble, and integrate a military or industrial circuit board**

| | Minimum | Average |
|---|---|---|
| Basic Board | | |
| Calendar time (days) | 62 | 165 |
| (equivalent time in months) | 3.0 | 7.9 |
| Board in Production, Sequential Estimation | | |
| Calendar time (days) | 16 | 27 |
| (equivalent time in months) | 0.8 | 1.3 |
| Board Production, Overlapping Fabrication and Assembly, No Stress Screening | | |
| Calendar time (days) | 2.7 | 3.4 |
| (equivalent time in months) | 0.13 | 0.16 |

For the particular example in Table 15.10, the average time to manufacture and integrate a circuit board is 3.6 business days; if there is no encapsulation, this time drops to 2.6 days. The shortest possible time is 1.7 days; if there is no encapsulation, this time drops to about 5.5 hours.

The average time is due to circuit boards being built, inspected, tested, and assembled in batches. While an individual circuit board spends much less time in fabrication and assembly than the average, it is delayed at each step while other circuit boards are processed. This also explains why calendar time is not equal to labor time per circuit board. The labor divides across boards and results in lower actual effort per board. This means that circuit boards spend a lot of time sitting and waiting for the next operation.

Table 15.11 summarizes and compares the minimum and average times of a first-run circuit board to an identical one in regular production. Some of the steps to verify the design in the first run are removed for regular production. Experience also introduces efficiencies into regular production. In this particular set of examples, the improvement can be better than 20-fold for minimum times and nearly 50-fold for average times!

### Electrical wires, cables, and harnesses

Another area of intense labor is the assembly of cables and harnesses. This is often a manual process because automation is so expensive and difficult to change for various configurations in everything except very high volume applications. Humans are good at cutting the wires, routing the wires through a jig or pegboard, crimping connector pins onto wires, and pushing each pin into a connector shell; these operations are simple (though tedious) for humans to perform. Other issues include testing for correctness and quality and limiting the connector mate—demate cycles.

Wiring, cabling, and harnesses should follow these general guidelines [7]:

- Resist breaking from vibration and flexure.
  - Make it neat, sturdy, and short.
  - Space tie-downs 8–10 cm (3–4 in.) apart.
  - Prevent strain on conductors, connectors, and terminals. Use strain reliefs.
  - Allow a reasonable bend radius ( > 10 times outer diameter of cable).
- Avoid damage from cuts and abrasion.
  - Prevent damage to assembled parts and give clearance for moving parts.
  - Use grommets (or rounded edges) on holes and penetrations to prevent abrasion.
  - Don't route over sharp edges, screws, or terminals.
- Another reason for neat, sturdy, and short—permit easy inspection and testing.

Connectors complete conduction paths from male pins through female receptor sockets. The connection is made through a mechanical, gas-tight fit between the pins and the sockets. Mechanical fit also means mechanical wear. Reducing the number of mate–demate cycles between connectors will reduce wear. Spacecraft developers get around this by building "connector savers," which are sacrificial stubs of cable with connectors on each end; connector savers allow developers to test the harness and subsystems with the harness without wearing down the space-qualified connector, rather the connector savers incur the wear. In final integration, these connector savers are removed and the space-qualified connectors are mated; this often means that these connectors have only gone through two mate and one demate cycles.

Cables and harnesses can be expensive business. Twenty-five years ago, I requested a quote for a military-qualified harness. It was small with two 9-pin connectors, 17 conductors, and 8 pigtails (to connect power supplies and filters) for a military data tape unit. A contract manufacturer experienced in military and avionics wiring and cabling gave me a quote of US$8000 for 10 harnesses; I ended having a technician build all 10 in less than a week for less than 20% of that cost.

Table 15.12 is an example of the shortest possible calendar time and the average calendar time spent manufacturing the first run of a wiring harness in-house. Table 15.13 is the same example except carried out by a contract manufacturer. The times in this table are only representative of one example; your projects may require different schedules. Please note several caveats:

- These are calendar times, based on business days per month, not staff labor hours. Labor may be greater for some of these activities because several people may be working together.
- Calendar time assumes 21 business days per month.
- The minimum time is rarely, if ever, achieved. That column describes the minimum calendar time for every activity if everything goes well without any problems.

Table 15.14 is an example of the shortest possible calendar time and the average calendar time spent manufacturing a wiring harness during regular production performed in-house. Table 15.15 is the same example except carried out by a contract manufacturer. The times

**Table 15.12: Business days and calendar time required to prepare and fabricate a wiring harness for the first time in-house**

| Category | Description | Minimum Days | Average Days |
|---|---|---|---|
| Harness and jig design | Harness layout design | 0.5 | 3 |
| | Harness pegboard jig design | 0.5 | 3 |
| | Specific procedures | 0.5 | 1 |
| | Prototype release | 0.25 | 1 |
| | Prepare task control card | 0.125 | 0.25 |
| | Engineering Design Review | 1 | 3 |
| | **Design subtotal** | 2.875 | 11.25 |
| Harness fabrication | Build pegboard jig | 0.25 | 2 |
| | Kitting—request parts from inventory or purchase | 1 | 5 |
| | Fabrication—simple <40 wires and 4 connectors | 0.5 | 2 |
| | Harness in receiving | 0.0625 | 0.5 |
| | Harness—inspection and test | 0.0625 | 0.25 |
| | Final preparation—e.g., encapsulation, insulation | 0.125 | 0.5 |
| | Final inspection and test | 0.125 | 0.5 |
| | **Harness fabrication subtotal** | 2 | 11 |
| | **Calendar—minimum, average (days)** | 5 | 22 |
| | **(equivalent time in months)** | 0.24 | 1.05 |

**Table 15.13: Business days and calendar time required to prepare and fabricate a wiring harness for the first time by a contract manufacturer**

| Category | Description | Minimum Days | Average Days |
|---|---|---|---|
| Design | Harness layout design | 0.5 | 3 |
| | Engineering Design Review | 1 | 3 |
| | **Design subtotal** | 1.5 | 6 |
| Contract fabrication | Request quotes from external vendor(s) | 0.25 | 3 |
| | Place requisition | 0.125 | 3 |
| | Contract released to external vendor | 0.125 | 0.5 |
| | Fabrication at contract vendor | 3 | 5 |
| | Harness in receiving | 0.0625 | 0.5 |
| | Harness—inspection and test | 0.0625 | 0.25 |
| | **Harness fabrication subtotal** | 4 | 12 |
| | **Calendar—minimum, average (days)** | 5 | 18 |
| | **(equivalent time in months)** | 0.24 | 0.87 |

in this table are only representative of one example; your projects may require different schedules. Please note several caveats:

- These are calendar times, based on business days per month, not staff labor hours. Labor may be greater for some of these activities because several people may be working together.
- Calendar time assumes 21 business days per month.
- The minimum time is rarely, if ever, achieved. That column describes the minimum calendar time for every activity if everything goes well without any problems.

Table 15.16 summarizes and compares the minimum and average times of a first-run wiring harness to an identical one in regular production for both in-house and at a contract

**Table 15.14: Business days and calendar time required to fabricate a wiring harness in regular production in-house**

| Description | Minimum Days | Average Days |
|---|---|---|
| Kitting—request parts from inventory or purchase | 0.25 | 0.5 |
| Fabrication—simple <40 wires and 4 connectors | 0.25 | 0.5 |
| Harness in receiving | 0.0625 | 0.25 |
| Harness—inspection and test | 0.0625 | 0.25 |
| Final preparation—e.g., encapsulation, insulation | 0.125 | 0.5 |
| Final inspection and test | 0.0625 | 0.25 |
| **Calendar—minimum, average (days)** | 0.8 | 2.3 |
| **(equivalent time in months)** | 0.04 | 0.11 |

**Table 15.15: Business days and calendar time required to fabricate a wiring harness in regular production at a contract manufacturer**

| Description | Minimum Days | Average Days |
|---|---|---|
| Fabrication at contract vendor | 3 | 5 |
| Harness in receiving | 0.0625 | 0.25 |
| Harness—inspection and test | 0.0625 | 0.25 |
| **Calendar—minimum, average (days)** | 3.1 | 5.5 |
| **(equivalent time in months)** | 0.15 | 0.26 |

**Table 15.16: A summary comparison of the time required to fabricate a wiring harness**

| | In-house | | Contract Manufacturing | |
|---|---|---|---|---|
| | Minimum (days) | Average (days) | Minimum (days) | Average (days) |
| Development | 5 | 22 | 5 | 18 |
| Production | 0.8 | 2.3 | 3.1 | 5.5 |

manufacturer. Some of the steps to verify the design in the first run are removed for regular production. The improvement can be better than 1.5- to 6-fold for minimum times and nearly 3- to 10-fold for average times. Contract manufacturing can take longer simply because of the shipping delay between the manufacturer and your facility.

## Mechanical

### Materials

Many materials are available for producing your embedded systems—plastics, aluminum, and steel—to name just a few, very general categories. Within each type of material are many varieties or alloys, each with specific strengths and weaknesses (see Chapter 9 for more details about materials). Each has unique requirements for use in manufacturing embedded systems.

Selecting materials affects the ease and cost of manufacturing your embedded system. For instance, some may think of plastics as cheap and frail but Figure 15.10 shows plastic enclosures can withstand the thermal cycles, vibration, and corrosion of an automobile engine compartment. Furthermore, paying for custom molds to form plastic enclosures can cost you upwards of US$30,000 for a set of dies. So while plastic per unit costs can be low and can ease assembly, the upfront manufacturing preparation costs can be high; only large manufacturing volumes can justify these costs.

**Figure 15.10:**
A plastic electronics enclosure that endures the extreme environment of an engine compartment.

*Enclosures and circuit board attachment*

Chassis and enclosures are mechanical structures that house embedded systems. Purchased commercial-off-the-shelf (COTS) enclosures eliminate fabrication and shorten assembly drastically—or should. The ideal is for a COTS enclosure or chassis to accept your circuit boards directly; hopefully, the boards just slide into the chassis and are ready to operate. The caveats include what vendors really mean by COTS and does the item really have the correct specifications and tolerances that you require:

- As to the first caveat, you would think COTS means that a vendor can pull a unit from inventory and ship it to you. This is not always the case. Sometimes, the vendor must build the unit before shipping it to you; this is more often the case than you might suspect. Something advertised as COTS in a vendor's catalog may actually take months to deliver. I had that unpleasant surprise years ago when I found a large VME enclosure with specifications that fit my project very well. When communicating with the vendor, I found that it was going to take six weeks to deliver, which was OK because it still fit in my schedule. Somewhere three to four months later, my purchased units finally arrived; by then my project had been delayed. While waiting for delivery, I had to purchase some interim units just to provide hardware functionality so that my software developers were not delayed in developing software for the circuit boards that would ultimately inhabit the VME enclosures.
- For the second caveat, make sure that all the specifications are met. Check out the vendor's reputation for delivering specifications as advertised. (Material on vendor reputation to help ensure satisfaction can be found at the end of this chapter.) Make sure that the enclosure really does meet specifications; this includes mechanical dimensions and tolerances, proper attachment points for the circuit boards, and any associated electronics such as power supplies.

If not purchased as COTS items, then these structures are always custom. Consequently, trying to describe the procedures and estimate the effort and duration of manufacture is difficult to do comprehensively. You can mold plastic enclosures or cast metal structures or mill out an enclosure from raw metal stock. There are many, many options. Before building, check with some suppliers of enclosures, you may find just what you need without having to build it yourself (assuming that you have thoroughly investigated the caveats of the previous paragraph).

For high quality assurance, such as in producing spacecraft, military, or industrial applications, there are steps that are typically recommended. A general outline of how a space-qualified enclosure is fabricated is as follows:

1. Select the raw stock from inventory, usually an aluminum alloy.
2. Set up the CNC milling machine and insert the aluminum work piece.
3. Mill the work piece.

4. QA inspects the milled work piece.
5. Rework the piece to address QA concerns, if needed.
6. Finish to final dimensions.
7. QA inspects the completed work piece.
8. Check out the enclosure and place it in the coating shop for anodizing.
9. QA inspects the completed, anodized enclosure.
10. Rework the piece to address QA concerns, if needed.
11. Check in the completed enclosure for later assembly into the system.

These are sequential steps to allow for QA inspections. Consequently, you can attach an estimated time to each step and add them to determine the overall time for producing an enclosure or work piece.

---

### Example: Basic instrument enclosure

Assume a small rectangular box with a lid and attachment points for a single circuit board. Also assume that the work is done correctly and no rework is required between inspections. This box and lid may take 6 hours to mill, 2 hours to finish to final dimensions, QA inspections take 4 hours after turning in the work piece, and final anodizing takes three days. Then this space-qualified enclosure may take a few hours longer than a work week to complete (Table 15.17).

**Table 15.17: Example of the time required to fabricate a simple, custom enclosure for a spacecraft instrument**

| Step | Description | Time (hours) |
|------|-------------|--------------|
| 1 | Select the raw stock from inventory, usually an aluminum alloy | 0.0312 |
| 2 | Set up the CNC milling machine and insert the aluminum work piece | 0.0625 |
| 3 | Mill the work piece | 6 |
| 4 | QA inspects the milled work piece | 4 |
| 5 | Rework the piece to address QA concerns, if needed | 0 |
| 6 | Finish to final dimensions | 2 |
| 7 | QA inspects the completed work piece | 4 |
| 8 | Place the enclosure in the coating shop for anodizing | 24 |
| 9 | QA inspects the completed, anodized enclosure | 4 |
| 10 | Rework the piece to address QA concerns, if needed | 0 |
| 11 | Check in the completed enclosure for later assembly into the system | 0.0312 |
| | **Total time to build simple, custom enclosure (hours)** | 44.1 |

---

By way of sanity check, Figure 15.11 is a photograph of a system that might take two or three weeks to fabricate and assemble at a contract manufacturer. The time may vary depending on the vendor's workload and availability.

**Figure 15.11:**
An electronics enclosure that might take between two and three weeks to fabricate at a contract manufacturer.

Note that the circuit boards use channel locks to fasten in place. Channel locks use sliding wedges to lock the circuit boards into their channels. Channel locks allow manufacturing personnel to slide circuit boards into a chassis or enclosure and lock in place in fractions of a minute.

### Mechanisms, fluids, and tubing

If you thought that estimating the cost of fabricating enclosures was difficult, mechanical mechanisms are even more so. The variety is astounding—motors, solenoids, "muscles," linkages, gears, hydraulics, and pneumatics. This is one area where purchasing proven components often makes very good sense. On the other hand, your embedded system may just need that unique mechanism; if so, all the caveats and concerns from the previous section on enclosures apply to a greater degree (see Chapter 9 for more details).

Some embedded systems control fluids or pneumatics that comprises systems of reservoirs, actuators, and tubing. Again, just like the mechanisms in the previous paragraph, the variety of such systems and configurations is huge. The same advice applies—purchase the components if you can and recognize the massive uncertainties if you build custom fluid subsystems. This book just does not have the scope or space to cover fluid systems or their fabrication and control.

### Module and subsystem attachment

While each embedded system is different and each configuration of attachment is different, several principles of integration are useful. Those principles include mockups, fit checks with harness and module placed on the mockup, and integration test. Moreover, final integration of an embedded system, just like fabrication and assembly, has two primary stages: development and production.

Development always is more involved and takes more time to ensure the functionality of the final system and to validate its design implementation. A mockup of the final system, which may be constructed from wood and cardboard or sheet metal, provides a good reality check for the correctness in geometry and location. Performing fit checks on the wiring, cabling, and subsystem modules is necessary to correct any shortages, deviations, or interferences. I am always amazed at how difficult the harness for a satellite or vehicle is to design and integrate; the exact configuration often challenges the most expert designers. Admittedly 3D CAD is very good at this now, but a good, physical model sometimes is just hard to beat—it may be faster and cheaper than the time spent generating the 3D model. Besides, 3D modeling tells you what *should be built*, a mockup and fit check tells you what *was built*. Integration and the integration test plan is part of every development—the book devotes Chapter 14 to test and integration in development.

Production does not validate the design but verifies the quality of manufacture to maintain the design intent. Production dispenses with the fit checks and mockups if multiple systems are assembled. Experience with assembling the modules, mechanical components, and harnesses will shorten production times as seen in the case studies above.

### Automated versus robotic versus manual

Only high volumes can justify the expense for an automated or robotic production line. An automated line uses custom machinery to produce one item consistently and very rapidly. Robots on the assembly line allow for quick programming changes to accommodate various products.

Conversely, manual labor is appropriate for meticulous production, such as medical devices, or for custom one-off designs. Humans are good at performing intricate, highly adaptable, and skilled operations. Humans can also spot unusual or unexpected problems very quickly.

## Production test

Production tests assure quality of the manufactured item. Production tests are NOT validation tests of the design—integration tests during development perform extensive validation and verification of the design.

Testing takes time and therefore adds to the cost of manufacturing. Production line tests in semiconductor fabrication, for example, can add as much as 40–60% to the cost of each component. You must balance the level of test coverage with the need for verification.

DFM will ensure that you can easily conduct production tests. DFM will provide for access to the wiring, circuitry, mechanisms, and replaceable substances (e.g., lubricants). DFM will also specify the specialized servicing equipment need to test the product.

Some examples of manufacturing tests include:

- Visual inspection of assembly
  - For correct fit
  - For appropriate finish
- Check of weight and volume
- Check of functionality within limits
  - Voltage levels
  - Electric current consumption
  - Signal levels
  - Data test patterns
  - Memory operations
  - Mechanism operations
  - Human interface—operations of the buttons and display.

### Electronics

#### Circuit boards

The testing of circuit boards occurs in three production stages: bare PWBs, assembled and functioning circuit boards, and assembled systems.

**PWBs**: Tests of PWBs include checking continuity, checking for shorts between wires, checking impedance, and checking for crosstalk between signal conductors. Continuity indicates that a signal can travel from injection at one pad to reception at another pad. The tester can also check for short circuits between wires when none should exist. Testing impedance and for crosstalk help verify that the materials and fabrication are correct.

Testing a PWB requires a fixture that allows probing of various points on the PWB. The test fixture clamps the board, supplies power, and allows access for probes to sample signals from test points on the PWB. This function can be automated with a bed of nails tester.

**Circuit boards**: The tests of circuit boards confirm correct operation. They can take on several different forms: scan, functional tests, and built-in-test (BIT).

In scan testing, a technician or a test appliance stimulates points within a chain of circuitry, records the results, and compares the results with expected responses. The circuitry that implements scan testing simplifies testing at the expense of more complexity and real estate on the circuit board.

Functional tests exercise a subset of all the operations that the circuit may generate. An example of a functional test might be entering a set of stimuli through the buttons on the front panel of the embedded system. Functional tests require less specialized circuitry but may not check all possible modes.

**BIT**: BIT tests the entire system, including the circuit boards. I discuss BIT operation below.

### Cables and wires

Tests for wiring harnesses include checking continuity, checking for shorts between wires, testing the wire insulation with high voltages (called high potential, or hi-pot, tests), and checking for crosstalk between signal conductors. Continuity indicates that you have made the correct connection from one connector pin to another pin on the other end of the wire. Confirming no short circuits between wires means that some mistakes have been avoided. Testing the insulation and for crosstalk help verify that the materials and fabrication are correct.

### Mechanical

The production tests for the mechanical portions of embedded systems are primarily checks for dimensions and tolerances, inspection for fit and finish, and confirmation of mechanism functions.

Manufacturing staff can manually measure dimensions and determine if components are within tolerances. You could also use robotic measurement for dimensions and confirmation of tolerances; these tend to be used more often in large volume production.

Motors, solenoids, and electrical actuators should be measured for power consumption compared to the motion achieved. Belts, linkages, gears, hydraulics, and pneumatics should all be measured and compared to design specifications for range of motion. You can build special test rigs to exercise the various mechanisms to verify their functions.

### ATE versus BIT versus BITE

First some definitions:

- ATE—automatic test equipment, usually found on high volume production lines.
- BIT—built-in-test, circuitry embedded within the product that exercises a circuit or module when activated.
- BITE—built-in-test-equipment, any additional pieces of equipment, such as wrap-around cables that are needed to complete BIT.

ATE is fast and can be very effective in measuring quality. ATE is usually quite expensive and only large volume production justifies its use. You can expect to spend US$100s of thousands, if not millions, on purchasing and operating ATE in production.

When your automobile generates an error code, some sort of BIT was at work to diagnose a problem. BIT usually requires some dedicated circuitry for monitoring and testing the remainder of the circuit board or system. BIT speeds diagnostics at the cost of complexity and lower overall reliability. BITE is BIT with additional equipment required to complete the test; an example would be attaching a cable from an output connector around to an input connector to exercise the outputs and inputs and perform an end-to-end test.

Only complex and expensive embedded systems, produced in volume, can afford ATE, BIT, or BITE.

## Considerations in manufacturing

### Quality systems

A quality system is a defined process for manufacturing that improves the chances of turning out correctly functioning product. If you are in the industrial realm, ISO 9001:2008 is appropriate for many products. For aerospace systems, you probably will need to comply with AS9100:2009. For medical devices and products, you may need to comply with ISO 13485:2003.

### Standards

You will encounter a number of different standards with which your manufacturing may need to comply. You will need to study and then to select the standards that your company will follow for specific applications. Table 15.18 provides just a sampling of some of those possible standards [8].

**Table 15.18: Examples of manufacturing standards that manufacturing may need to follow (not an exhaustive listing!) [8]**

| Manufacturing Area | Industry Standard(s) | Military Standard(s) |
|---|---|---|
| Standard practices for military packaging | | Mil-Std-2073 |
| Gen. spec. insert, screw thread—locked in, key locked | | MIL-I-45914 |
| Hardware torquing and sealing | SAE AS1310 | MIL-S-46163 |
| | | MIL-S-22473 |
| Acceptability of electronic assemblies | IPC-A-610 | |
| Electrostatic discharge (ESD) practices | ANSIESDS2020 1999 | MIL-STD-1686B |
| | | MIL-HDBK-263A |
| Requirements for soldered electrical connections | IPC-A-610 NHB 5300.4 (3A-2) | |
| Soldering | J-STD-004 Fluxes | |
| | J-STD-005 Pastes | |
| | IPC-610 Class. 2, 3 | |
| Rework and repair of printed boards and assemblies | IPC-R-700 IPC-7711 | |
| Requirements for crimping and wire wrap | NHB 5300.4(3H) | MIL-C-22520 |
| Wire cutting, stripping, and tinning | QQ-W-343 | MIL-W-81832 |
| | | MIL-W-22759 |
| | | MIL-W-16878 |
| Wiring, cabling, and harnesses | | MIL-S-83519 |
| | | MIL-C-17 Coax. |
| | | MIL-T-43435 Lacing, Tying Tape |
| Adhesives and bonding | MMM-A-121 | MIL-I-16923 |
| | MMM-A-132 | MIL-A-46146 |
| | MMM-A-134 | RTV MIL-A-8270 |
| | IPC-SM-817 | |
| Conformal coating | IPC-CC-830 | MIL-I-46058 |

## Supply chain

A supply chain is a network of processes and organizations that transfers information, components, and materials [9]. A supply chain should deliver the necessary information, components, and materials in a timely fashion that supports your company's manufacturing of products. The study of supply chains and logistics is a complex and involved endeavor. Managing the supply chain for your product can occupy staff full time. A contract manufacturer may remove that burden from your company if it makes business sense for your company to do so.

## Contract manufacturing

If your company does not have the facilities or if it needs help with a surge in product demand, then contract manufacturing is the answer for you. Even if your company is building small quantities of military, industrial, aerospace, or medical systems, there are specialty contract manufacturers who can do the work to the quality that you specify. Your company can even split the effort between in-house facilities and a contract manufacturer; you may, for instance, outsource the components, modules, or subsystems but do the final assembly and test in-house.

A recent survey suggests that between 50% and 60% of all PWBs, wiring and cabling, and machining is outsourced to contract manufacturers. Contract manufacturers build the enclosure and assemble the mechanical portions occur in over 40% of items produced. Peter Frasso of Segue Manufacturing Services writes "These decisions [for contract manufacturing] are made for many reasons but include the capital equipment investment, the volumetric leverage gained by outsourcing the purchasing and managing of thousands of components, and the knowledge necessary to get consistent repeatable high quality manufacturing" [10].

### Capabilities of contract manufacturing

Contract manufacturers can do piece work, all the way up to turn-key operations. A turn-key service acts as if the vendor is the manufacturing and distribution division of your company. The vendor does everything from sourcing materials for PWBs and enclosures, to stocking components, to manufacturing and testing, to storing in inventory, to drop shipping to your customers. Many contract manufacturers also have design services, should your company not have particular capabilities.

For a small startup firm, a good contract manufacturer can ease the challenge of getting product to market. A good contract manufacturer who knows the applicable standards also takes on the upfront costs of purchasing component and material inventory, which can save the financial bottom line of a small startup; the minor downside to this is that your product may be a bit more expensive than if you did it all yourself—but the tradeoff is getting a product to market and making some money versus possibly not getting to market at all and losing money!

### Concerns for contract manufacturing

Contract manufacturing is not a panacea for solving your production problems. Things can go wrong. Sparton Corporation in its white paper "Contract Manufacturing Pitfalls: What the Wrong Production Process Could *Really* Cost You" lists these areas of concern for contract manufacturing [6]:

1. Employee turnover delays completion times in client projects. When experienced employees leave, the manufacturer must acquire and train new employees and develop in them an understanding of QA. One tip is the accident rate in the facility, which slows production; the higher the rate, the slower the production.

2. Find the vendor's record of timeliness and reputation with other clients. References do make a difference! One time a contract manufacturer gave me the name of a client, who said, when I contacted him, "Well I probably would not use them again" and then proceeded to explain his reasons. What was that vendor thinking? that I wouldn't check references?

3. Understand the vendor's QA system and its effectiveness. Do they inspect and test along the way or at the end when the entire assembly must be scrapped for nonconformance?

4. Demonstration of in-house knowledge and expertise. Can they handle new product technologies?

A friend told me of another outsourcing problem with a foreign vendor. Consistently, the contract manufacturer produced good quality product in the first 100 or 1000 units. Then the tested specifications of sampled product would begin to erode bit-by-bit. The vendor was shaving costs, to improve profit margin, by cutting into the specifications and making a nonconforming product, which was cheaper to manufacture. The vendor knew that its customer would find it difficult to change contract manufacturers in mid-production so was willing to chance contract violations.

*Selecting a contract manufacturer*

So far I have briefly described what a contract manufacturer can do for you and some concerns to beware. Now I will provide some thoughts toward selecting a contract manufacturer. First, here is a case study that illustrates selecting a contract manufacturer requires thorough preparation and follow-up.

---

### Case study: Finding a medical device contract manufacturer

Some years ago my small medical products firm needed to build multiple units of a small drive module and antenna that connected to a pen tablet computer (Figure 15.12). We had the expertise in-house to develop software but we did not have the capability to design the drive module or to manufacture the drive module, antenna, or connecting cables. We had to find a contract manufacturer to design and produce these components. The task for selecting a contract manufacturer fell to me. I developed a multistep process to locate and select a suitable vendor.

First, I searched a manufacturer index for vendors who had the capability to manufacture medical devices. I confirmed the capabilities of each candidate vendor by reviewing the company's website. One metric that I looked for in a vendor was the ability to prepare support documentation to aid in the FDA approval process. Another metric was that they had defined processes for both hardware and software.

Once I had a list of candidate manufacturers who had all the necessary qualifications, I prepared a script of specific questions about their processes, projects that they worked on,

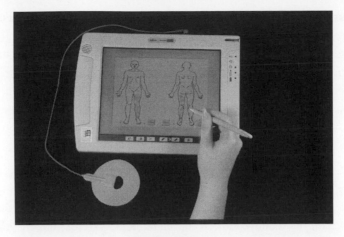

**Figure 15.12:**
A pen tablet computer with an antenna that required contract manufacturing. *From the personal collection of Kim R. Fowler. Used with permission.*

and type of manufacturing they did. Then I called each candidate manufacturer and went through the script. That was enlightening! I remember calling one small company who claimed to develop microcontroller-based medical devices and asked about their software processes—the answer was that they "...have this guy who writes the software... but did they have defined processes? No, this guy was really good at writing software and that was enough." Obviously, they did not understand my concerns or the need for rigor in writing software, so I crossed them off the list.

Through comparing the answers to my questions, I narrowed the candidates to four companies. I hired a consultant to look at the companies and travel with me to visit them. His function was to be an objective, impartial judge of what we saw—he was someone whom I really trusted for both technical expertise and for understanding people. Before meeting face to face with each company, we prepared and sent a short presentation of what we needed and a script of questions that we wanted them to answer. In particular, we asked for a list of former clients who would serve as references for each company.

We set up a three-day trip to visit four companies, one in Florida, two in California, and one in Washington state. The meetings went well; we toured facilities, discussed their capabilities, and got our questions answered. We found that three of the companies were quite suitable. The fourth company gave us a list of client references, but when I called them I found that they had not built medical devices; one reference built exercise machines, while others built various consumer appliances.

In the final analysis, it came down to deciding which company we felt we could work with the best. I think that we could have chosen any one of the three and we would have been satisfied. Regardless, the point of this case study is that thorough preparation and follow-up is very important in selecting a contract manufacturer.

*Selecting a contract manufacturer to partner with you should be an extensive, exhaustive effort*, at least for you. First, determine what it is that your company actually needs. Is it materials sourcing and component fabrication? Or subsystem assembly? Do you need specialized testing and certification, such as UL or CE marking or FDA approval? Do you need design services for particular components or subsystems? Even if you are not sure of the extent of services that you might need, establish a baseline plan for contracted services and be willing to consider a range of services.

The second major step is to do basic research. Look through technical and trade magazines for vendor articles and advertisements. Go to tradeshows and meet vendors. Do the leg work to research the company presence in the market. Review each candidate company's website—do they claim to manufacture components and systems that fit your company's market niche? What industry standards do they meet? You should not only consider recommendations from colleagues but also consider vendors whom you may not have encountered before in your business dealings.

The third major step is to develop a list of candidate vendors. Also prepare a script of questions, *at the same time prepare matching answers that you would hope fulfill your company's need*; these desired answers will provide a basis for determining if the candidate vendor meets the need. Open-ended questions are very good because they tend to reduce "priming" for the desired answer. Vendors can perceive leading questions, which open-ended questions reduce. Here are suggestions for questions [11]:

1. Does the vendor design or manufacture products in my company's market niche?
   * e.g., medical, aerospace, industrial, military
   * What are systems that you have manufactured?
   * What industry standards do you follow?
   * What certifications do the vendor hold to perform the work?
2. What quality system does the vendor follow?
   * e.g., ISO 9001:2008 for industrial products
   * e.g., AS9100:2009 for aerospace systems
   * e.g., ISO 13485:2003 for medical products
3. Will the vendor give you references from clients who are in the same market as you? Can you call those clients and ask questions?
4. Is the vendor technically competent or expert in your particular market? How can they support that claim?
5. Does the vendor have an Enterprise Resource Planning (ERP) system?
   * How well does it work for them?
   * How well do they manage the supply chain?

6. How does the vendor handle the financial aspects of buying components and materials?
     * How do they manage minimum quantity purchases?
     * Are they willing to hold the excess inventory until it is used without charging you upfront?
     * Do they give you the option to amortize the material purchases over the duration of the contract?
7. If they outsource, particularly PWBs, what controls do they have for quality of incoming products?
8. How does the vendor manage configuration?
     * How often do they report to you the progress in manufacturing? weekly? monthly?
     * How do they manage technical changes that you may need or request?
     * How do they manage their vendors?
     * How do they mitigate counterfeit components?
     * How do they manage problems and noncompliances?
9. How clean and orderly are the vendor's operations? (This is a question to reserve for a visit.)
10. How does the vendor interact with you? (These are questions to keep in mind when you visit.)
     * Are they straightforward and honest?
     * Do you sense that they might be trustworthy?
     * Will they "cover your back" and not take advantage of you when you forget something or leave something out or need to change something?

The fourth major step is to call each vendor. You may want to send your basic requirements and your script of questions ahead of the appointed telephone conversation. When you do call each vendor, follow your script of questions that you prepared in the third step! You want to have a consistent basis to compare different vendors.

The fifth major step is to compare the responses from all vendors. Some will immediately fall off the list because they do not meet particular requirements that you have. Try to narrow the list to three or four vendors. In some cases, you may find more than four candidate manufacturers; if so, tighten up your constraints and requirements a bit and compare the vendors again or iterated through all five steps with the list that you currently have. In other cases, you may find that no vendor meets all the criteria that you have set; if so, go back to the first step and reevaluate your goals and scope; you may have to iterate through all four proceeding steps once again to develop possible candidate manufacturers.

The sixth major step is to visit each candidate manufacturer's facility. Give a more detailed presentation of what you are looking to accomplish. Update and refine your script of questions to improve the selection process. When you do visit, make sure to tour their manufacturing facility. You are looking for orderly processes and clean conditions—both indicators of a manufacturer preparing high quality products. (***Please note***: Orderly processes and clean conditions are supporting evidence, not proof, of high quality operations.)

The seventh step is to select the contract manufacturer from your short list. This selection may be obvious or you may need to prepare specific questions and then iterate through phone calls and meetings with several vendors to finalize the selection.

The right contract manufacturing partner can save you a lot of headaches. The wrong one can cause them!

## Design transfer

Design transfer is a phrase of art in medical device manufacturing but it is useful to describe what should be included in moving a design to manufacturing. In medical device parlance, design transfer is a set of "...procedures to ensure that the device design is correctly translated into production specifications" (21 CFR 820.3(h)) [12]. For a medical device, design transfer contains items like the Device Master Record (including the bill of materials (BOM), schematics, layout Gerber files, and instructions), training materials, and manufacturing aids.

If your company has performed the design, then an important part of design transfer is the BOM. You should carefully prepare the BOM according to these suggestions of Peter Frasso of Segue Manufacturing Services [10]:

- Make sure that the BOM is accurate and correct.
- Make sure that the schematics and instructions were clear and properly prepared.
- List the component or materials manufacturer's part number, not the distributors.
  - A contract manufacturer can often negotiate better prices from the component manufacturer than from a distributor.
  - Provide the component manufacturer's contact information—phone number, email, and mailing address.
  - Use a Mil-spec designation over a manufacturer's part number so that the contract manufacturer can often negotiate better prices between material suppliers.
  - Provide alternate part numbers, if possible, to allow the contract manufacturer to get the fastest delivery of components or materials.

## Captive production facility

Having a captive production facility for some companies makes good sense if they have large volume manufacturing or if they want to control quality (e.g., the company showcased earlier with fabricating PWBs). Another reason is if the particular industry needs to protect resource and manufacturing capability for critical functions; as an example, specialized production for the defense industry comes to mind.

The rather large disadvantage is managing vendors, logistics, and the supply chain. This effort usually requires a number of full-time staff, which is a bottom line expense.

## Acknowledgments

I am grateful to Andrzej and Ola Michalski for the lovely, quiet time in their country house to complete this chapter! My thanks to Craig Silver for his insightful comments on this chapter.

## References

[1] Washington Metal Fabricators website, Washington, MO. <http://www.washmetfab.com/product-consulting-engineering> (accessed 28.05.13).

[2] Ensure Affordability with Design for Manufacturing. Renaissance Services white paper, December 17, 2010. <http://www.ren-services.com/white-papers/ensure-affordability-design-manufacturing.html> (accessed 29.05.13).

[3] K. Fowler, What Every Engineer Should Know About Developing Embedded, Real-Time Products, CRC Press, Boca Raton, FL, 2007, pp. 64–65 (Chapter 2).

[4] J.A. DeSantis, An overview of PWB constructions, Printed Circuit Design (1993) 48–53.

[5] B. Davis, A tour through the board manufacturing process, Printed Circuit Design (1993) 11–14.

[6] Contract Manufacturing Pitfalls: What the Wrong Production Process Could *Really* Cost You. White paper by Sparton Corporation, February 2013. <http://sparton.com/resources/wp-cmpit/> (accessed 29.05.13).

[7] B.S. Matisoff, Handbook of Electronics Packaging Design and Engineering, third ed., Van Nostrand Reinhold, New York, NY, 1996.

[8] <http://www.zentech.com/standards-and-capabilities.php> (accessed 02.06.13).

[9] Electronics manufacturing supply chain. Electronics Manufacturing—M528. <http://cnfolio.com/ELMnotes20> (accessed 02.06.13).

[10] The Essential Step-by-Step Guide to Understanding Outsourcing and Establishing a Partnership for Success. An eBook from Segue Manufacturing Services by Peter Frasso, Segue Manufacturing Services, 2012, pp. 4, 36. <http://www.segue-mfg.com/> (accessed 29.05.13).

[11] Top 10 Questions You Should be Asking Your Electronics Contract Manufacturer, White paper by Zentech Manufacturing, July 2012. <http://info.zentech.com/the-top-10-questions-you-should-ask-your-potential-contract-manufacturer> (accessed 27.05.13).

[12] U.S. FDA, Design Control Guidance for Medical Device Manufacturers, March 11, 1997, relates to FDA 21 CFR 820.30 and sub-clause 4.4 of ISO9001. <http://www.fda.gov/cdrh/comp/designgd.pdf>, pp. i, 1, 2, 4, 5, 8, 13, 19, 23, 37, 43.

# Logistics, Distribution, and Support

**Kim R. Fowler**
*IEEE Fellow, Consultant*

## Chapter Outline

Developing and Managing Embedded Systems and Products.
DOI: http://dx.doi.org/10.1016/B978-0-12-405879-8.00016-7

## *Overview of logistics, distribution, and support*

Logistics manages the flow of product and information from manufacturing to customers (Figure 16.1). Logistics integrates packaging, inventory, transportation, warehousing, delivery, and technical sales support. Every one of these arenas folds into the ultimate management of the development of embedded systems; you must consider them early in design because their operations and parameters directly affect the final cost of the system.

Logistics provide goods, raw materials, and commodities that meet requirements for ordering, delivery, quality, and cost. "Logistics is thus a multidimensional value-added activity including production, location, time, and control of elements of the supply chain. It represents the material and organizational support of globalization requiring a complex set of decisions to be made concerning an array of issues such as the location of suppliers, the transport modes to be used, and the timing and sequencing of deliveries" [1].

**Figure 16.1:**
Logistics, distribution, and transportation. *Illustration from iStockPhoto.*

## Business logistics

The Encyclopedia Britannica gives this definition and description of logistics, "...the organized movement of materials and, sometimes, people. The term was first associated with the military but gradually spread to cover business activities. Logistics implies that a number of separate activities are coordinated. In 1991 the Council of Logistics Management, a trade organization based in the United States, defined logistics as: 'the process of planning, implementing, and controlling the efficient, effective flow and storage of goods, services, and related information from point of origin to point of consumption for the purpose of conforming to customer requirements.' The last few words limit the definition to business enterprises. Logistics also can be thought of as transportation after taking into account all the related activities that are considered in making decisions about moving materials" [2].

Basically logistics may be thought of as *"having the right item in the right quantity at the right time at the right place for the right price in the right condition to the right customer"* [3]. The focus for embedded systems in this chapter will be external or outbound logistics from manufacturing to the customer. We will also be considering the optimization of the flow of products and information through shipping and storage.

## Distribution logistics

Distribution logistics looks at the delivery of the finished products to the customer and focuses on order processing, warehousing, and transportation. Distribution is important and a continual challenge because it strives to match the difference in quantity, time, and location of manufacture with the quantity, time, and location of delivery [4].

## Support

Support is the timely distribution of correct information to fix or implement the function of the embedded system. It is also that intangible perception that the supplier knows the situation and is there to help the customer. Support can be as simple as the help desk pointing the customer to the correct page in the installation manual to complete a procedure. Or, support may provide the necessary training for installing and operating the embedded system. Support can also be as complex as designing a fix or workaround when the embedded system does not quite fit the application but its configuration is the nearest component to working acceptably. Support may also be contract employees regularly servicing and maintaining your embedded system. Every embedded system has different requirements for support, but every customer expects the appropriate level of support to use their embedded system.

## Maintenance and repair

All embedded systems need some sort of maintenance or repair. Some systems may not need maintenance; about the only thing to do is repair if it fails, which may be as simple as replacing one box for another and returning the defective unit. Other systems, particularly those with mechanisms or fluidics or pneumatics, may need extensive maintenance; these systems may have bearings to monitor and lubricate or fan filters to replace or optics to realign and recalibrate. Regardless, you, as the designer/developer/manufacturer, have the responsibility to understand and provide for this part of the life cycle of the embedded system.

## Disposal

The last step of the life cycle of an embedded system is disposal. How are customers to remove your embedded system and what will its final disposition be? Recycling is now required by government regulation and control of hazardous materials is mandated.

## Definitions

Table 16.1 gives definitions used for logistics, distribution and support.

**Table 16.1: Definitions used for logistics, distribution, and support**

| | |
|---|---|
| Packaging | The material used to surround and protect the embedded system during shipping, often a cardboard box |
| Inventory | The storage of product until it ships to a distribution point or to a customer |
| Distribution | A superset term to refer to the storage, warehousing, shipping, transporting, and delivery of a product |
| Warehousing | The location where product stores before shipping; it can be an intermediate point between different modes of transportation |
| Transportation | The mode of shipping product from warehouse to customer or distribution point, e.g., via rail or oceangoing cargo ship |
| Replenishment | Filling or replacing a consumable material or component, e.g., lubricants, batteries |
| Repair | Fixing a failed embedded system; can be replacing internal components or replacing the system outright |
| Maintenance | Monitoring system operations, checking tolerances, performing replenishment |
| Condition-based maintenance | A form of maintenance that is performed when conditions demand, it is not a time-based schedule of operation |
| Predictive analytics | A higher level of machine intelligence to interpret conditions and predict maintenance needs and failures |
| Support | Primarily a supply of needed and necessary information for operating the embedded system |
| Technical help desk | One form of support for frequently asked questions and Tier 1 help |
| Training | Another form of support that aids the customer in understanding and operating the system |

*Caveat*

---

**Please remember**

All examples and case studies in this chapter serve only to indicate situations that you might encounter. Hopefully, these examples and case studies will energize your thought processes; your estimates and calculations may vary from what you find here.

---

## Market release

Market release is the "trip point" to introduce a new product to the world and announce its availability. Similar to design transfer, market release conveys a lot of information about how the product will be supplied. It should provide documentation and instructions for inventory, distribution, warehousing, shipping, delivery, sales support, and technical support.

For large systems, such as vehicles, ships, and industry plant control systems, market release of your embedded system should coincide or precede field trials or commissioning of the system.

## Distribution and delivery

Rodrigue and Hesse write "Physical distribution includes all the functions of movement and handling of goods, particularly transportation services (trucking, freight rail, air freight, inland waterways, marine shipping, and pipelines), transshipment and warehousing services (e.g. consignment, storage, inventory management), trade, wholesale and, in principle, retail" [1].

Distribution is the process of making your embedded system available for use by a consumer or a business. Distribution can use either direct or indirect means to put the systems into the customer's hand. Indirect means employ channels, which are separate but interdependent companies that work together to supply the product to the customer. Table 16.2 defines three different types of distribution; you will probably use either selective or exclusive distribution to deliver your embedded systems to customers [5].

**Table 16.2: Types of distribution [5]**

| Distribution Type | Definition |
| --- | --- |
| Intensive | Generally retail where products stock in many outlets, e.g., snack foods or magazines |
| Selective | Manufacturer uses only a few intermediaries to distribute product, e.g., appliances |
| Exclusive | Manufacturer uses only one reseller/distributor or sells directly to customers |

Physical distribution handles inventory through shipping, transportation, and warehousing. Six prominent areas of distribution are as follows [1]:

- **Warehousing**. This is usually the starting point after manufacturing where product resides as inventory until it ships via a transport vehicle.
- **Transportation vehicles**. The general categories of transport vehicles include vans, trucks, trains, airplanes, and ships. These transport modes have been operational thousands of years for oceangoing ships, approaching 200 years for rail traffic, and nearly 100 years for trucks. Changes in the modes of cargo haulage have been dramatic over the past two or three decades, particularly with containers and integrated shipping and with air freight. Maritime shipping has seen an entirely new class of ships and huge economies of scale in container shipping—but these are not directly of concern for us and small, embedded systems, unless, of course, you are delivering large quantities of products.
- **Transportation terminals**. More obvious changes have been seen in recent years with the construction of new terminal facilities moving large volumes of products. New infrastructure and handling equipment speed the movement of products. These terminals are in the distribution channel with the next stop being inland terminals connected by high capacity corridors or distribution centers.
- **Distribution centers**. Distribution centers generally rely on trucking and locate in suburban locations with access to good roads. Rodrigue and Hesse write that these distribution centers service regional markets with a 48-hour service window (lead time) on average [1]. These centers tend to be one floor facilities designed more for throughput than for warehousing. These facilities have specialized loading and unloading bays and sorting equipment.
- **Load units**. Rodrigue and Hesse describe load units as the basic physical management unit in freight distribution and take the form of pallets, swap bodies, semitrailers, and containers.
- **Information technologies**. Distribution and shipping requires product flows linked with information flows. Information technology has improved product information for manufacturers and allowed distributors to adapt quickly to changes in demand.

### Issues with distribution

#### Distribution centers

Distribution centers buffer the stream of manufactured products to batches of shipping. Distribution centers usually serve an area based on the required delivery frequency and response time to order; they often locate near transport routes and terminals [1]. They can also perform light manufacturing activities such as assembly and labeling—this, however, may not be appropriate for embedded systems, which need fairly extensive testing before delivery.

The advent of containers for long distance shipping facilitated expansion of distribution. Containers also ease storage and inventory concerns to some extent because they serve as self-storage units.

### Order and delivery timing

To understand and accommodate the lag between an order and its delivery, you will probably encounter two terms: cycle time and lead time. **Cycle time** is the span of time from the receipt of an order to when the order prepares to ship. Cycle time indicates the level of responsiveness of production. **Lead time**, however, is the time it takes for an order to be fulfilled after production; it includes preparation, packing, and delivery. Lead time indicates the level of responsiveness of distribution [1]. Taken together, cycle and lead time indicate how well your company produces and delivers its embedded systems.

### Lean supply and flow control

Lean supply minimizes inventory levels and keeps most of the inventory in constant circulation. For the electronics sector, lean supply can turn over inventory as many as 10–20 times per year. Lean supply is enabled by flow control and information technology to reduce inventories in manufacturing from several days to as low as several hours [1]. Lean supply chains and flow control generally address issues in much larger volume manufacturing than for most embedded systems.

### Third-party logistics providers

Because logistics are becoming so complex and time sensitive, some producers are turning to contract logistics providers, third-party logistics providers, which are asset-based services (i.e., they have the facilities and equipment to perform the services), and fourth-party logistics providers, which are nonasset-based services (i.e., they only organize the logistics, they do not have any facilities or equipment to perform the services). Outsourcing distribution, and even the entire logistics chain, can relieve smaller manufacturers of the burden of staffing and operating a captive distribution system. Large shipping companies, such as FedEx and UPS, are becoming logistics providers.

### It's all about costs and time

Reducing costs and time in distribution depends on the reliability of distribution, the flexibility of distribution (for quantities and locations and delivery times), and the quality of distribution. The management of distribution can be modeled in terms of flows, nodes, and networks, to which I refer you to the reference by Rodrigue et al. [1]. Other factors include government regulations within and between countries, which can drive levels of inventories and cause shipping delays, and the productivity of regional labor and infrastructure. An unintended consequence of outsourcing to countries with cheap labor is that they might have lower levels of productivity, which affects reliability of lead times, deliveries, and distribution.

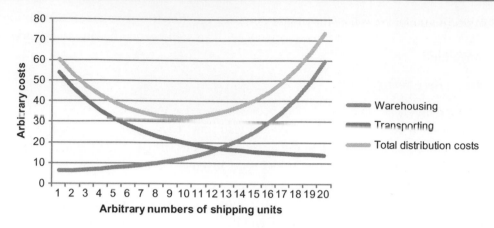

**Figure 16.2:**
One example for the optimum, least-cost region (here it is 9—11) in partitioning number of shipping units in transport versus those in warehousing.

A primary tradeoff that you might make is finding the optimum levels of cost between transporting and warehousing product. Figure 16.2 gives a general overview of the potential differences and how you might find the optimum levels of costs. Figure 16.2 is an example and only for illustration purposes—shipping and warehousing your embedded system may include other considerations, as well.

## Example: Producer of complex computer boards

I once worked for a company that designed and produced complex, multiprocessing boards for specialized computer systems. The company produced from 20 to 40 boards per month, controlled the supply chain, maintained the inventory, warehoused the products, and shipped from their facility after inspection and test. Their production staff not only assembled and tested the circuit boards but also shipped the final product.

This level of shipping and distribution ranged between one and four boards shipped each business day. The complex computer boards sold for prices between $5000 and $30,000 each, depending on the number and type of processors; the boards were standard sizes for an industry standard chassis. The company built into the price of each board the shipping cost of $30—$40 per board. We used FedEx and UPS to ship boxes containing the computer boards.

The example of a small producer of computer boards is not a sustainable model if the product is much cheaper—for instance, a $79 microcontroller board. The shipping and distribution costs are a much larger proportion of the final delivery costs for an inexpensive embedded system, so the customer must pay for the shipping and handling. Furthermore, much greater volume requires separate staff and facility to warehouse, staff, ship, and transport the product. At that point, you may want to consider third-party logistics.

*Some comparisons of costs and delivery times for shipping*

While the example above gave one instance of shipping costs, Tables 16.3–16.5 give examples of the shipping costs alone for various quantities of product.

*Some thoughts on warehouse costs*

Storing product before shipping costs money. Simply put, you have to pay for the space, the equipment, and the furniture. You also have to pay for the labor to perform the packing to prepare for shipping.

**Table 16.3: Example of shipping costs for a 1 lb package anywhere in the United States**

| Time to Delivery | Cost |
|---|---|
| Overnight | $58.00 |
| One day | $26.00 |
| Three days | $20.60 |
| Four days | $9.10 |

**Table 16.4: Example of shipping costs for a 4 lb package anywhere in the United States**

| Time to Delivery | Cost |
|---|---|
| Overnight | $99.00 |
| Two days | $68.00 |
| Two days late evening | $62.00 |

**Table 16.5: Example of shipping costs from factory to distribution warehouse for a load of 4 lb package boxes, each box measuring 15 in. (38 cm) × 9 in. (22.9 cm) × 3 in. (7.6 cm).**

| Number of Package Boxes | Box Weight (lb) | Box Mass (kg) | Description of Transportation | Time to Delivery | Cost of Entire Load ($) | Per Unit Cost ($) |
|---|---|---|---|---|---|---|
| 1000 | 4000 | 1818 | Less than a truck load, trailer is 48 feet (14.6 m) or 53 feet long (16 m) | 3 to 10 days in the United States after pickup | $2650.00 | $2.65 |
| 10,000 | 40,000 | 18,182 | Full truck load, trailer is 48 feet (14.6 m) or 53 feet long (16 m) | 3 to 10 days in the United States after pickup | $10,000.00 | $1.00 |
| 10,000 | 40,000 | 18,182 | Steel shipping container for ocean voyage—40 feet (12.2 m) × 8 feet (2.44 m) × 8 feet (2.44 m) | 2 to 4 weeks from Oklahoma to Rotterdam in the Netherlands | $26,600.00 | $2.66 |

Warehousing costs can be expensive. Space is often leased in a commercial building and most companies begin by leasing. Depending on the location, some commercial spaces can demand upwards of $15 or even $25 per square foot per year ($160–$270 per square meter per year) in the United States.

## Example: Warehouse costs for shipping specialty embedded systems

Assume that your company leases 3000 square feet for $22 per square foot per year (278 square meters at $237 per square meter per year); this results in a yearly cost of $66,000 to lease that space.

Once you have signed on the dotted line for three years of leased space, you have to buy or lease basic equipment and furniture to support your distribution and shipping. Table 16.6 illustrates some of the costs that you might encounter in setting up a very modest distribution and shipping center. These costs do not include the labor to perform the shipping!

**Table 16.6: Example of warehouse costs for a small company to pack specialty embedded systems that ship in 4 lb boxes, each box measuring 15 in. (38 cm) × 9 in. (22.9 cm) × 3 in. (7.6 cm).**

| Item Description | Quantity | Unit Costs ($) | Subtotal Costs ($) |
|---|---|---|---|
| Shelving | 10 | $1500 | $15,000 |
| Desks | 3 | $1000 | $3000 |
| Tables | 4 | $500 | $2000 |
| Desk chairs | 4 | $200 | $800 |
| Laboratory stools | 3 | $900 | $2700 |
| Carpets, antistatic mats | 6 | $400 | $2400 |
| Office equipment (e.g., computer, copier, printer, postage machine) | 1 | $5000 | $5000 |
| Office supplies | 1 | $1500 | $1500 |
| | | **Total assets** | $32,400 |

The labor for packing and preparing to ship small embedded systems is not free. Even a few minutes of time on each package adds to the final costs of an embedded system. Consider the following case study for labor costs to prepare to ship a product.

## Case study: Handling costs for shipping complex computer boards

In the example above of the small firm that built specialty computer boards, we shipped between 20 and 40 boards each month. Production staff packed, labeled, addressed, and arranged shipping of each board.

Assuming that one person spent 15 minutes per board and assuming a loaded labor cost* of $30 per hour ($60,000 per year), then each board had $7.50 in handling costs that included packing a manual, inspection and test certificates, wrapping and sealing the box, and preparing and attaching an address label. These handling costs were in addition to shipping the product.

Over the course of a month, shipping 20 boards requires about 5 hours total time and $150 in handling. Shipping 40 boards would be twice that, or 10 hours time and $300 in handling.

In this example, where the computer boards were priced between $5000 and $30,000 apiece, the additional costs fold into the price of the product without additional charge to the customer. On the other hand, this amount of handling cost for a $79 microcontroller board is not sustainable—it might even wipe out the profit margin. For a $79 microcontroller board, handling needs to reduce to 3 or 4 minutes, so that its costs drops below $2 per board or package. Furthermore, you would probably charge the customer handling and shipping charges to arrive at a final price over and above the advertised $79 price.

*(loaded labor costs = salary + benefits + infrastructure costs)

## Packaging

Many embedded systems are small subsystems that integrate into a larger system, consequently, these sorts of products often ship in small cardboard boxes. Typically, you will protect the product with an antistatic bag and some sort of foam padding within the cardboard box. The antistatic bag prevents electrostatic discharge (ESD) from damaging the electronics of the embedded system during handling. The foam padding protects the product from drops, collisions, and crushing during shipping.

Packaging has costs, too. If you purchase in large quantities and can store materials until you use them for shipping, you can estimate the cost of packing; here are examples:

• cardboard boxes may cost between $0.25 and $0.40 per box,
• antistatic bags may cost between $0.12 and $0.20 per bag,
• foam, without cut outs, may cost between $0.70 and $1.20 per square foot,
• packing tape may cost between $0.001 and $0.01 per strip used.

The packaging cost for one, small embedded module or circuit board can total between $1.07 and $1.81. The handling to pack the box will add several more dollars to the final cost.

Another method for packaging an embedded system is to use and reuse a thermoformed plastic container; this can even be a recycling program. These cases are very sturdy containers that can withstand a lot of physical abuse—dropping, slicing, abrasion, collision, and crushing—while protecting the product. The process is to buy a structurally strong plastic case that is molded in such a way to contain the circuit board or embedded system. Ship the system in the plastic case to a customer with a return label. The customer removes the system, attaches the return label, and mails the case back to you. Though the case is considerably more expensive than a cardboard box, it is stronger and it does not end up in the waste bin. It is used over and over to mail systems to various customers. The next example gives some comparisons for potential cost savings with a reusable plastic case.

---

### Case study: Cost comparisons between a reusable plastic case and cardboard boxes

In the example and case study above of the small firm that built specialty computer boards, we shipped between 20 and 40 boards each month. Production staff packed, labeled, addressed, and arranged shipping of each board. We began buying reusable thermoformed plastic cases to ship the computer boards.

The question for this case study is "How many reuses of the plastic case are required to break even with the alternate method of using disposable cardboard boxes, antistatic bags, and cushioning foam?"

Here are the basic assumptions for this study:

- loaded labor cost is $30 per hour,
- one person takes 4 minutes to pack a board in a plastic case and attach a mailing label,
- one person takes 4 minutes to receive a returned plastic case and store it back in inventory,
- one person takes between 4 and 7 minutes to pack a board in a cardboard box and attach a mailing label,
- a thermoformed, plastic case costs $25.00,
- cardboard boxes cost $0.40 per box,
- antistatic bags cost $0.20 per bag,
- foam costs between $0.70.

Four minutes of handling costs $2.00 while seven minutes costs $3.50. The total material costs for packing your system in a cardboard box is $1.30. The total cost for handling and preparing your system in a cardboard box is between $3.30 and $4.80. The incremental cost for reusing a plastic case is $4.00, which is $2.00 for shipping and $2.00 for receiving the returned case and restocking it.

Figure 16.3A illustrates a breakeven point in cost between reusing the plastic case and using a new, disposable cardboard box each time. If cardboard boxes take 7 minutes to pack, then the breakeven point for costs is after 31 boxes or shipments. If cardboard boxes take 6 minutes to pack, then the breakeven point for costs is out at 84 boxes or shipments. If the cardboard boxes take 5 minutes or less to pack, no crossover, breakeven point exits.

For a company selling small quantities of expensive embedded systems, a reusable plastic case might make economic sense to a small degree. Maybe more important is that you can represent your company as making efforts to be "green."

Now consider the situation where you are selling very small, inexpensive microcontroller boards for $79 apiece. By using the following assumptions, which are

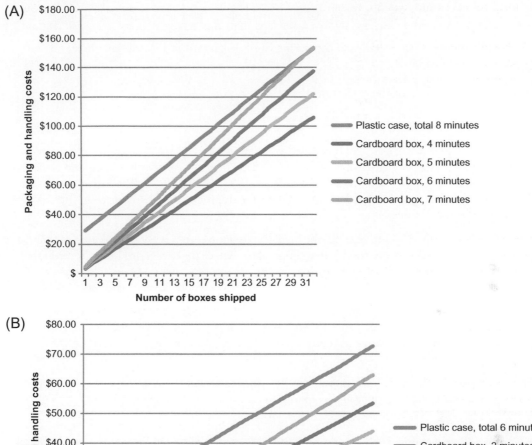

**Figure 16.3:**
(A) Comparison of handling costs between recycled containers and cardboard boxes for an expensive embedded system. (B) Comparison of handling costs between recycled containers and cardboard boxes for a cheap microcontroller board.

more appropriate for this size of embedded system, you may come to a different conclusion:

• loaded labor cost is $25 per hour,
• one person takes 2 minutes to pack a board in a plastic case and attach a mailing label,

- one person takes 4 minutes to receive a returned plastic case and store it back in inventory,
- one person takes between 2 and 5 minutes to pack a board in a cardboard box and attach a mailing label,
- a thermoformed, plastic case costs $15.00,
- cardboard boxes cost $0.25 per box,
- antistatic bags cost $0.10 per bag,
- foam costs between $0.30.

Two minutes of handling costs $0.83 while five minutes costs $2.08. The total material costs for packing your system in a cardboard box is $0.65. The total cost for handling and preparing your system in a cardboard box is between $1.48 and $2.73. The incremental cost for reusing a plastic case is $2.50, which is $0.83 for shipping and $1.67 for receiving the returned case and restocking it.

Figure 16.3B illustrates the approach to one breakeven point between reusing the plastic case and a new, disposable cardboard box for different times taken to pack the cardboard boxes. If cardboard boxes take 5 minutes to pack, then the breakeven point for costs is after 64 boxes or shipments. If the cardboard boxes take 4 or less minutes to pack, no crossover, breakeven point exits.

For a company selling small, inexpensive microcontroller boards, a reusable plastic case does not make economic sense. Furthermore, you may not win any "green points" from your customers because of the hassle to return boxes for such a cheap item—I would guess that they would not even bother.

The conclusion for this case study is that reusable plastic cases only make economic sense when they significantly reduce handling and preparation time and when loaded labor costs are high.

Larger embedded systems, such as a computer or a control rack, need larger shipping containers (obvious, I know). You have several choices for shipping containers: cardboard boxes with styrofoam inserts (popular for desktop computers and peripherals), thermoformed plastic containers, reinforced fiberglass containers with foam interiors, aluminum or steel containers with foam interiors, or custom wooden crates for industrial control panels or satellites.

Most of these containers are specialty items; their prices tend to go up in multiplicative factors as you move to larger, stronger, and more custom configurations. Sturdy, thermoformed plastic container may cost upwards of $200 or $300 in single lot quantities; if you are doing a lot of equipment transport and it needs a moderate degree of protection, these containers can be just right for you. Reinforced fiberglass and metal containers double or triple that price and also afford more protection.

Regular shipping of large equipment really needs more expert advice than I have space for here.

## *Inventory*

For this chapter, inventory is the storage of finished product before shipping. Inventory requires manual labor to operate it (total automation is far outside the scope of both this chapter and most companies providing specialty embedded systems), a warehouse of some sort, and information system with a database to control it. Inventory costs your company because of the space you lease to store and the labor you pay to operate it.

---

### Example: Inventory costs per computer board

In the previous examples and case studies with complex computer boards, the company shipped between 20 and 40 boards each month. Assume the following:

- Loaded labor cost is $30 per hour.
- One person takes 2 minutes to catalog a board in the IT database.
- One person takes 1 minute to place a board in storage on a shelf.
- One person takes 1 minute to retrieve a board from storage on a shelf.
- One person takes 2 minutes to log the removal and shipment of a board in the IT database.
- A board takes up an affective area of 0.05 square foot.
- Lease cost is $20 per square foot per year.
- A board resides in storage for 1 month before shipping.

The total labor cost for handling the board in inventory is $3.00. The total cost for space is $1.00 per year or about $0.09 per month. Without accounting for the IT equipment, the cost of inventory is about $3.09 per board.

If the production flow is steady and a lean supply chain is working really well, then boards only spend an average of a day or less in storage; batch production and shipping may alter this. The IT database for inventory probably resides in a desktop computer but its cost and other uses are nebulous and left out of this example.

---

## Sales support

*Sales are the first line of support. A good sales staff or representative firm solves customer problems by suggesting correct uses of your embedded system.* Sales support helps customers understand the basic, well-known parameters of the product, solves minor problems, and helps customers understand where the embedded system fits in their products or systems.

Sales staff can be captive staff in your company or they can be sales representative firms who get a commission on sales of your products when they are involved. Captive staff is a constant cost, whereas sales representatives live only on commission, which is a contracted fraction of the selling price, and only contribute to cost when a sale occurs.

Sales representative firms make good economic sense for small firms selling specialty products. Once you have large volume sales to fewer customers, sales representative firms begin to cost more than captive staff.

## Technical support

Technical support often "tag teams" with sales support. Technical support provides all the remaining support for problem solving that customers might need—from understanding the details of the principle of operation, to how it integrates in various systems, to identifying and solving customer problems by using your company's embedded system. Technical support problem solving is reactive—it may have to rely on root cause analysis as described in Chapter 7 or on a stable of prefabricated solutions or suggest reference designs using the embedded system. Technical support can be a valuable asset to design engineering for suggesting upgrades, redesigns, fixes, and new product ideas.

### Tier 1, Tier 2, and Tier 3

Many companies provide escalating levels of technical support. These support levels are called Tier 1, Tier 2, and Tier 3.

- **Tier 1**: It provides basic software application or hardware implementation to customers. Typically, the service answers simple "How to" questions or explains fixes for a documented product problem. A call center help desk fulfills this level of support. Examples of Tier 1 support might be how to install software to connect to the embedded system or how to connect several pieces together with commercially available cabling.
- **Tier 2**: It provides more complex subject matter expertise in both application and implementation. Often Tier 2 support escalates after a call to Tier 1 support does not solve the problem. In most cases, Tier 1 and Tier 2 support can be handled by the same support staff (based on their experience and skill); sometimes engineering needs to be involved to solve the problem. Examples of Tier 2 support might be how to install a software fix to overcome an unusual circumstance in the field or simple modification, such as adding a finned heat sink, to resolve overheating in an unexpectedly hot application.
- **Tier 3**: It provides support on complex hardware and software issues. These situations require expert troubleshooting and diagnosis and may require on-site visit to the customer by your engineering staff to fix the embedded system. Tier 3 support can also take much longer time to complete and it depends on the type of problem.

---

**Example: Engineer costs for technical support**

An engineer devoted to customer support can cost your company between $80,000 and $150,000 per year depending on the complexity of the product, experience of the engineer, location, and market. More than that and you are looking at a consultant to provide specific, expert help.

---

### Technical marketing

Many small companies combine technical support with technical marketing, especially if engineers staff those positions. Engineers can be expensive. Furthermore, technical support is neither continuous nor uniformly distributed, which means that support staff cycle between business and downtime. One effective way to balance the load is to have the support staff fill in the lulls in customer service by aiding technical marketing—writing white papers and manuals, attending trade shows, and giving technical presentations about the products.

Technical support staff are the front line of the company and have frequent contact with customers. They can be the source of valuable ideas for new products, upgrades, and fixes. Involve them early in the design of new products—their input can save you money and countless hours of wasted effort.

## Training

Training addresses issues and concerns for the installation, operation, maintenance, and repair of your company's embedded system. You can supply and deliver training via different methods: brochures, documentation (e.g., Users Manual), DVDs, website videos, lecturing, and tutorials at trade shows and conferences. Training is an important component activity that goes with sales and technical support.

Good, basic training instruction can go a long way to making your embedded system more saleable. It can give the customer a good "out of box" experience. This alone can make the difference in a competitive market place when a customer chooses your product.

### Website

The most effective means to train customers is a thoughtfully prepared company website. You can place video tutorials, short customer testimonials, specifications, and even training notes on it. Most potential customers do review a company's website before approaching them for service. A reasonable website can be produced for $5000 to $10,000 and then maintained for several hundred dollars each year.

### Users Manual

A well-written Users Manual is worth the effort and cost. It can explain the installation, operation, and maintenance of your embedded system; this alone can be a selling tool for your product. Once written, you can often reproduce and bind single copies of a 200-page document for less than $25. Even better, you can place the manual or parts of it on the company website for much cheaper distribution; this is the best way for companies selling inexpensive systems like the $79 microcontroller board mentioned previously in several examples and case studies.

### Tutorials

Tutorials are a good way to introduce and showcase an embedded system. Tutorials are also good for instructing basic installation, operations, and maintenance of your company's embedded system. With some care and forethought, you can usually produce tutorials in a simple format for basically your time to write the script, plan it, and film it. Professional production is better for wide distribution of high-ticket systems; then you might expect to spend thousands of dollars to produce the tutorial.

## Maintenance and replenishment

Maintenance is the service of components that might wear out, such as motors, bearings, batteries, and elements that flex or rub. Maintenance also addresses components that need regular calibration, such as sensors in process industries, or alignment, such as optics, mechanical linkages, lids, and hinges. Replenishment is the service to disposable materials and elements such as filters (for fans or oil or fluids), lubricants, fluids, fuels, and batteries. Table 16.7 illustrates one example of a checklist for maintenance and replenishment.

Often the customer for your embedded system can perform the maintenance and replenishment tasks. More complicated procedures may need expert personnel within the industry or help from your company; should your company supply maintenance services these may be under contract to generate revenue for your company. Sometimes, all that is needed is the expert eye of a consultant to do a routine audit and check of the system to monitor for proper function.

**Table 16.7: Example checklist for maintenance and replenishment**

| Activity | Check and Inspect | Monitor and Test | Replenish | Fix and Adjust |
|---|---|---|---|---|
| Cleanliness, dust, particulates | √ | | | |
| Friction and wear | √ | | | |
| Alignment | √ | √ | | √ |
| Vibration | √ | √ | | |
| Acoustics | √ | √ | | |
| Fluid levels and leaks | √ | | √ | |
| Pneumatic pressure and leaks | √ | √ | | √ |
| Lubrication | √ | | √ | |
| Battery life | | √ | | √ |
| Sensor calibration | | √ | | √ |

This is not an exhaustive list, it is only for illustration.

The type of maintenance described thus far and in Table 16.7 is *preventative maintenance*. It is performed on a regular, scheduled basis. Service personnel should keep logs of the maintenance performed. These might be computer files or even recorded directly on tablet computers for transmission to a server.

There are two other types of maintenance paradigms that are emerging: *condition-based monitoring and maintenance* and *predictive analytics*. These forms of maintenance and service are data based and require IT services and support.

### Condition-based monitoring and maintenance

A newer type of maintenance that is gaining acceptance, particularly with process industries, is condition-based monitoring and maintenance. This type of maintenance requires instrumented equipment to monitor its operation. An example would be permanently installed acoustic sensors to "listen" to bearings on an electric motor; should the bearings begin to wear excessively, the acoustics change, which indicates wear and potential, imminent failure. The advantage of condition-based monitoring and maintenance over regularly scheduled maintenance is that condition-based monitoring and maintenance reduces cost; it replaces excessively repetitive procedures for maintenance with less frequent, but more targeted maintenance for components that need attention.

### Predictive analytics

The most recent development for maintenance is predictive analytics, which uses machine intelligence to interpret monitored conditions. "...predictive analytics encompasses a variety of techniques from statistics, modeling, machine learning, and data mining that analyze current and historical data to predict future events" [6]. While predictive analytics are not maintenance procedures, *per se*, they indicate when maintenance might circumvent impending problems. They provide tools for higher order diagnostics.

Predictive analytics might also cover the loss of experienced personnel. "...reductions to headcount and the anticipated loss of expertise due to the industry's aging workforce undermine the industry's ability to maintain effective and efficient production levels... Gone will be that valuable combination of knowledge and intuition that understood the quirks of the plant environment—the visible and audible clues that are not listed in an operating manual and that only come with experience... condition-based monitoring... are incapable of applying past experience to current operations" [6].

The advantage of predictive analytics over condition-based monitoring and maintenance is that predictive analytics can use a constellation of parameters to estimate trends and faulty operations. Predictive analytics easily handle multivariable problems, while condition-based monitoring and maintenance does not. Predictive analytics can forecast problems far in advance of actual failure, while condition-based monitoring and maintenance generally have a much shorter time frame to prediction of failure. Predictive analytics are more than just adjusting alarm limits, which are "...not a solution to this type of multivariable problem. Tightening alarm limits on vibration would have resulted in too many nuisance alarms—another problem widely faced by manufacturers" [6].

Condition-based monitoring and maintenance use "...existing model-based approaches... [to apply] a mathematical approach based on equations" [6]. Predictive analytics apply "...statistics and machine learning in a purely data-driven approach..." Predictive analytics still need the embedded sensors and detection methods of condition-based monitoring to supply the empirical data upon which the machine intelligence operates to make predictions. Consequently, these two techniques, condition-based monitoring and predictive analytics, can "...be highly complementary tools for maintaining asset reliability" [6].

### Some further issues in maintenance

Design issues for maintenance often are similar to those in manufacturing, assembly, and disassembly. Usually whatever makes assembly or disassembly easier will make maintenance easier. Maintenance requires access to the internals: circuitry, mechanisms, and replaceable substances (e.g., lubricants). Maintenance may also require specialized equipment to monitor or to test the product or to replenish its consumables.

Specialized service equipment and access bring up another issue—supportability. The next section in this chapter addresses the concern for supportability.

## Diagnosis and repair

Diagnosis and repair are critical concerns for customers of embedded systems. Repair means that the customer is losing money from lost operations. An average of 5% of worldwide plant production is lost annually to downtime; the best manufacturers experience less than 2% downtime, whereas the worst average nearly 15% downtime each year [6]. Downtime of 5% for the case study below would be $18.25 million lost each year.

---

### Case study: Two different scenarios for the cost of downtime for repair

Suppose that you are running an industrial complex that generates $1,000,000 per day in revenue, which is $365 million per year. Now suppose that you have the possibility of buying an embedded control module that controls a critical choke point in the process from one of two different manufacturers. Let's look at the cost to you if that embedded control module fails.

Embedded control module #1: When it fails, it takes an average of 8 minutes to diagnose the problem and then another 12 minutes to fix or replace it. The total time from failure to fix and running again is 20 minutes, which equals 0.0139 day; with revenue at $1,000,000 per day that is a loss in revenue of $13,900. This assumes that there are no start up problems or extra material or product to scrap to contribute even more lost income.

Embedded control module #2: When it fails, it takes an average of 4 hours to diagnose the problem and then another 12 minutes to fix or replace it. The total time from failure to fix and running again is 4.33 hours, which equals 0.175 day; with revenue at $1,000,000 per day that is a loss of $175,000. This assumes that there are no start up problems or extra material or product to scrap to contribute even more lost income.

Time is money and lost time is lost money!

This case study, while clearly contrived, still indicates that capable and correct diagnosis can be an important part of maintenance and repair.

Components, circuit boards, wiring, cabling, and mechanisms failure or breakdown necessitate diagnosis and repair. Often, the customer for your embedded system can perform the diagnosis and repair. More complicated fixes may need expert personnel within the industry, such as a consultant, or help from your company; should your company supply repair services these may be under contract to generate revenue for your company.

Design issues for diagnosis and repair are similar with those in maintenance, manufacturing, assembly, and disassembly. Usually whatever makes assembly or disassembly easier, will make diagnosis and repair easier. Diagnosis and repair requires access to the internals: circuitry, mechanisms, and replaceable modules. Maintenance may also require specialized equipment to test the product for diagnostics.

## Supportability

Supportability is important for easing diagnosis, repair, maintenance, and replenishment of an embedded system. The US Department of Defense (DoD) has the following definition for supportability:

> **Supportability**: *The inherent quality of a system—including design, technical support data, and maintenance procedures—to facilitate detection, isolation, and timely repair/replacement of system anomalies. This includes factors such as diagnostics, prognostics, real-time maintenance data collection, 'design for support' and 'support the design' aspects, corrosion protection and mitigation, reduced logistics footprint, and other factors that contribute to optimum environment for developing and sustaining a stable, operational system [7].*

The US DoD works toward supportability by "…addressing issues pertaining to:

a)   Commonality (physical, functional, and operational);
b)   Modularity (physical and functional);
c)   Standardization (system elements and parts, test and support equipment);
d)   Diminishing Manufacturing Sources and Material Shortages (DMS and MS); and

e)  Technology maturity and refreshment, Commercial Off The Shelf (COTS) technology maturity, open system standards, proprietary issues, single source items [7]."

Supportability is not just about ease of access but also about understanding the system and recognizing its component interactions. Good training and documentation are aspects of supportability.

The military perspective for supportability contains the following elements [8]:

- System training
- System documentation
- Supply support and spares
- Sustainment planning
- Test and support equipment
- Facilities
- Packaging
- Handling
- Transportation
- Manpower.

## Recalls, patches, and updates

From time to time, embedded systems do need updates or fixes. Often these can be software patches or updates that can be posted to your company's website; customers then download them and install them in their systems.

Patches and updates can be performed either by your company's technical support or by the customer. These changes need careful development, testing, verification, and validation before release. Unintended or unforeseen consequences are not welcome by anyone—obviously.

The cost in time and effort is usually more than we expect. A simple recall on an automobile engine control module can cost your company millions of dollars, as seen in this automobile recall example originally printed in Chapter 7.

---

### Example: Automobile recall

An automobile recall to simply reprogram an engine control module may only take 15 minutes per car but at a rate of US$50 per garage service hour for one million cars that amounts to US$12.5 million, and that does not include the nonrecurring engineering cost to develop the new software!

---

More serious, not only in terms of cost but in reputation and legal liability too, is a consumer recall required by the government. A government recall of your company's embedded system is expensive and time-consuming for you; it involves regulatory compliance issues and notifications.

## Reverse and green logistics and disposal

Some new terms and processes are entering the logistics chains; they are *disposal logistics, reverse logistics*, and *green logistics*. All three are interrelated. The jury is still out as to which one or two will prevail and incorporate the other terms.

Disposal logistics is the reduction in logistics costs and enhancement of services during the disposal of waste during production. Reverse logistics has to do with reuse of products and their remanufacture or refurbishment [9]. Green logistics is the minimization of ecological impact by logistics [10].

Each of these forms of logistics strives to reduce environmental impact through efficiency and optimization. As an example, one area of impact is reducing $CO_2$ emissions when shipping product. They can optimize the load and route with sophisticated software planning tools and IT support to gain efficiency and reduced $CO_2$ emission. Load optimization might be further aided by efficient packing of product [10].

A number of people and stakeholders are involved in these new forms of logistics [10]:

- Government officials enforcing national and international regulations
- Customers wanting to do business with an environmentally and socially responsible company
- Employees working for an environmentally and socially responsible company
- Lenders, investors, insurers, and investors.

### WEEE Directive

The European Community Directive 2002/96/EC set targets for recycling of electrical goods; it is called the **Waste Electrical and Electronic Equipment Directive** or WEEE Directive. The WEEE Directive has 10 separate reporting categories, including large and small household appliances, IT equipment, medical devices, and control instruments [11]. Many embedded systems would fall into one of these categories.

### RoHS

The Reduction of Hazardous Substances or RoHS is a directive that took effect in July 2006. It is coupled with the WEEE Directive [12]. RoHS restricts the use of six hazardous materials in the manufacture of electronic equipment; these substances include [13]:

- Lead (Pb)
- Mercury (Hg)
- Cadmium (Cd)
- Hexavalent chromium ($Cr^{6+}$)—used in chrome plating and chromate coatings

- Polybrominated biphenyls (PBB)—flame retardants in some plastics
- Polybrominated diphenyl ether (PBDE)—flame retardants in some plastics.

These substances have been found in cables, solders, PWBs, display screens, and batteries.

*Please note*: Aerospace and military applications usually exempt products from RoHS compliance because tin whiskers can cause shorts. Lead prevents or greatly reduces the growth of tin whiskers in solder joints. These applications are a miniscule proportion of all electronic products and can easily be exempted from the requirements of RoHS.

## Acknowledgment

My thanks to Craig Silver for his insightful comments on this chapter.

## References

[1] J.-P. Rodrigue, C. Comtois, B. Slack, The Geography of Transport Systems, Hofstra University, Department of Global Studies & Geography, 2012 <http://people.hofstra.edu/geotrans> (accessed 06.06.13).

[2] <http://www.britannica.com/EBchecked/topic/346422/logistics#ref528537> (accessed 07.06.13).

[3] S. Mallik, in: first ed., H. Bidgoil (Ed.), The Handbook of Technology Management: Supply Chain Management, Marketing and Advertising, and Global Management, vol. 2, John Wiley & Sons, Inc., Hoboken, NJ, 2010, p. 104.

[4] <http://en.wikipedia.org/wiki/Logistics> (accessed 06.06.13).

[5] <http://en.wikipedia.org/wiki/Distribution_(business)> (accessed 06.06.13).

[6] R. Rice, R. Bontatibus, Predictive Maintenance Embraces Analytics. <http://www.isa.org/InTechTemplate. cfm?template = /ContentManagement/ContentDisplay.cfm&ContentID = 92287&utm_medium = kimf%40ieee. org&utm_source = Eloqua&utm_campaign = PLMA%20Email_Existing-17APR13-AW>, January/February 2013 (accessed 17.04.13).

[7] Designing and Assessing Supportability in DOD Weapon Systems: A Guide to Increased Reliability and Reduced Logistics Footprint, Prepared by the Office of Secretary of Defense, October 24, 2003, pp. 12, 14. <https://acc.dau.mil/CommunityBrowser.aspx?id = 32566> (accessed 08.01.13).

[8] D. Verma, B. Gallois, Graduate Program in System Design and Operational Effectiveness (SDOE): Interface Between Developers/Providers and Users/Consumers, International Conference on Engineering Design (ICED 2001), Glasgow, UK, August 2001.

[9] <http://en.wikipedia.org/wiki/Reverse_logistics> (accessed 13.06.13).

[10] <http://en.wikipedia.org/wiki/Green_Logistics> (accessed 13.06.13).

[11] <http://www.environment-agency.gov.uk/business/topics/waste/139283.aspx> (accessed 13.06.13).

[12] <http://www.dtsc.ca.gov/hazardouswaste/rohs.cfm> (accessed 13.06.13).

[13] <http://en.wikipedia.org/wiki/Restriction_of_Hazardous_Substances_Directive> (accessed 13.06.13).

## Suggested reading

J.-P. Rodrigue, C. Comtois, B. Slack, The Geography of Transport Systems, third ed., Routledge, New York, NY, 2013 A very readable and interesting textbook with good sections on distribution.

M. Levinson, The Box, How the Shipping Container Made the World Smaller and the World Economy Bigger, Princeton University Press, Princeton, NJ, 2006. A delightful and entertaining read on the vast changes in worldwide shipping over the past 50 years.

## CHAPTER 17

# Agreements, Contracts, and Negotiations

Craig L. Silver

*Director—Strategic Initiatives/General Counsel, Amches, Inc.*

**Chapter Outline**

## Interpretation of contracts generally

*You interpret what you perform.* You're a nice guy or gal. Most people think that they are nice. So you are nice, too. So, after the contract for your embedded project has been signed, put in a drawer, never to be seen again, the parties get cranking on the project. About four months into the project, the specification has changed in that the customer says in an e-mail that they need more memory and the response time on outputs needs to be faster. You, being a nice guy engineer, e-mail, "OK, we can do that." A discussion ensues about the ability to map the FPGA differently so that the system will work better. Later, the customer's project manager e-mails and asks if you are sure that you "can do that" as you told his engineer. You, being a nice guy, say "yes," as you really do think you can do it and besides, you're a nice guy. So, off you go. Later, it comes to your attention that the specification documents that came with the FPGA had an error in it and it looks like the new mapping idea may not work after all. In fact, it is not working now. You, being a nice guy, sweat it out for a few weeks before you go to your boss and say that the new mapping is not going to work and "we are late" and the spec sheet was wrong. Your boss, who is not a nice guy, e-mails the customer and says that the spec sheet was wrong and there are problems. The customer,

Developing and Managing Embedded Systems and Products.
DOI: http://dx.doi.org/10.1016/B978-0-12-405879-8.00017-9
© 2015 Elsevier Inc. All rights reserved.

no longer a nice guy himself, says that they are really going to be in a hard place and that you can expect their legal team to sort it out and, by the way, they are not nice guys.

At this point, the first thing the attorney for either party is going to do is to check the contract. You know, the one that was put in the drawer never to be seen again?[1] The contract, as it turns out, provides for a way to make changes that requires the representatives of the respective companies to approve in writing any material changes to the contract or the contract deliverables. "Whew," you say, "I am glad of that, because we never did that." You think you are out of the woods. However, your attorney then begins to ask some tough questions, like "are there any documents, e-mails, etc. that exist that demonstrate that you agreed to change the project? Were there any phone calls that might constitute an oral promise or give rise to an 'estopple'?"[2] You explain that "Yes, we sent them all kinds of e-mails saying that we could do it as we thought we could, even though it was going to cost us more money and it was going to impact the schedule." "What is it with you guys? Are you just nice?" he asks.

The issue here is that it is important to realize that how a contract is performed is just as important as the negotiation aspects. In fact, if the parties drift too far away from the negotiated purposes and statements of the contract, a court can find that the contract was ignored by both parties or, in the end that an oral contract was in force. The e-mail trail here may act as written modifications to the contract or confirmation of the existence of an oral contract. I have seen this happen when an engineering contract had a term of six months and was to expire, with no automatic extension, at the end of the six months. However, the parties kept working beyond the stated term. The court found that the parties were just performing an oral contract as the written one had expired.

Additionally if a *product* is being delivered and not only *engineering services*, the *Custom and Usage* aspects of the Uniform Commercial Code[3] may come into play. This possibility

---

[1] It is often found that parties to contracts cannot even locate a signed version of the contract and by "signed" I mean a version with all party's signatures on the contract, not just yours but the other side as well. Or, to the contrary, you might have the opposing side's signatures, but not a version of your own signatures. Did the contract go in the drawer too soon or with an assumption that all had signed?

[2] The elements of estopple by acts or representations are: reliance by a person entitled to rely on the acts or representations, the misleading of such person, and, in consequence, a change of position to his detriment, so that the person responsible for the misleading will not be permitted to deny the truth of his own statements or actions. This brings up another issue that is often ignored and that is, if you are relying upon statements, e-mails, phone calls, etc., that have contract implications, then the question needs to be asked if the statements are coming from an authorized person. So, in the example above, the customer may not have had the right to rely upon the engineer who first said that it could be done if the contract clearly states who can bind the parties or if the engineer is not an officer or one with apparent authority to bind the corporation. Note how often a company is referred to as "they" in your experience. "They" said such and such. Or "they agree." The prudent manager needs to ask "Who is they? and "Says who?"

[3] All states have enacted the Uniform Commercial Code (UCC) which pertains to transactions for goods, not services, between merchants. It is designed to fill in terms and conditions that are not otherwise stated between the parties.

allows the opposing side to testify that it is customary in the industry to make changes to specifications as the parties move along in the project and to compensate for "additional costs," but that the engineers would be expected to perform as promised.

These types of problems ensue from not referring to the contract during contract performance and getting legal advice when problems arise. As you might appreciate, answering the question as to "who is right?" in the matter can quickly become problematic and not susceptible to an easy answer. Attention needs to be paid during the negotiation stage and throughout the performance stage. It is also important to close out the contract in a right way. Suppose everything goes right (when was the last time that happened?) and the product is performing well and your customer says "It is working great and we have appreciated working with you folks. You sure are nice guys!" Now is the time to get what is known in the law as a "waiver certificate"[4] or, at least, some kind of acknowledgement that you have fully performed the contract and that the product meets the specification.

Now, the customer may balk at that. This can be rectified by having clear acceptance criteria and time limits in the contract. If you draft the contract to say that "the product will be deemed accepted unless, within 30 days after delivery, a written exception to performance is delivered to the offices of the undersigned specifying each and every deficiency claimed." Courts, and thus, attorneys know that strict time lines are enforceable and courts especially like strict time lines as it is an easy way to rule on a mess and be right without too much concern if the matter is appealed. To be sure, it is advisable to put in language that says, "Time is of the essence to all time periods required in this Agreement." "Time is of the essence" is a legal "word of art" that courts will enforce which means, in the most practical and legal sense, that even if performance is one day late, then there has been a breach of the agreement. This phrase is often found in engineering contracts for just this purpose, but is often overlooked as mere "legal gobbledygook" put in by not-nice attorneys.

How beautiful it is to report that the clouds of litigation are parting and the sun is shining, as your opposing force did not comply with a strict time limit that is supported by the appropriate legal language.[5]

Since tomorrow's litigation is today's screw up, and as has been demonstrated above, it would seem that much more attention should be given, and more time allocated, to the contract negotiation phase than is commonly done. The specification should be given many looks from many different disciplines, to make sure it is complete, without surprises. Engineers, of course, should review. So should the product managers, marketing managers,

---

[4] A waiver is an intentional relinquishment of a known right.

[5] Note that "time is of the essence" language, if not associated with one item, cuts both ways. If such language is put in a contract generally and you are the one who is supposed to deliver the product, just being one day late for one deliverable would constitute a breach of the agreement.

executives, and the legal office. If it is a fixed price contract, greater specificity is needed. In fact, I have found it to be desirable, if the customer will allow to provide for a *first phase* where the customer pays to work with the engineering team to produce the specification. This way the spec tends to get the appropriate attention and detail it needs in order to avoid problems later. It is also important to define terms. This is especially true if the project goes across international boundaries where language and cultural differences exist.

## The signing of agreements

You might think that such a topic would not be the subject of much concern. You might be surprised by how much litigation has involved the simple act of putting pen to paper and signing a name. What are the common ways to mess this up?

1.  Well, if you have a corporation or LLC, it is important to have the name of the corporation above the signature. Even if the very top of the agreement states that the agreement is between ABC Corporation and DEF LLC, it is imperative that the signature page contains the name of the company right above the name of the party signing. Why? Because sometimes individuals sign too. Individuals might guarantee the obligation of the company and be asked to so sign. In such a case, it is customary to have the person sign with the word "individually" set out next to their name. When signing individually, the person is personally agreeing to all of the obligations of the company and agreeing to be held liable for such.

    An example of signatures with corporate and individual signatures is shown in Figure 17.1.

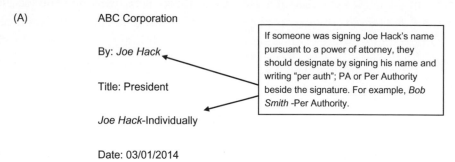

**Figure 17.1:**
(A) Signing a contract and (B) signing to distinguish authority and capacity.

2. Another issue concerns who is actually supposed to sign the document. (See the section on Agency.) The person who is signing must have statutory, actual, or apparent authority to bind the company. If you ever plan on enforcing an agreement, you want to make sure that the person signing for the other side is authorized to bind the institution. So often, especially in the case of Nondisclosure Agreements (NDAs), the engineer in charge of the project signs, and the engineer may not be binding the company to anything. Examples of "safe" signatories are the President, Secretary, anybody called "partner" (in the case of partnerships or joint ventures), executive Vice Presidents, or Directors. However, be advised that many companies have lots of Vice Presidents or Directors and they are not statutory officers, just title holders. Unless you have the organizational chart of the company you may not know how high up this person goes or they have authority to sign contracts at all. Additionally, you may want to watch what corporate entity is signing. Perhaps you have been negotiating with ALL ABOUT US TECHNOLOGY. Later, when the final contract is submitted for signature you note that the contract is between you and ALL ABOUT THEM TECHNOLOGY. When questioned, your counterparts just want to shrug it off and say, we always use our offshore Bahamian entity for "this type of deal." Clearly, if the offshore entity has no assets, is not available for service of process, or has no active business presence, then you are not going to get any relief should the matter have to be litigated. This issue is especially important in China where it is hard to know who the real legal entity is. In fact, because China is a communist nation, to make an agreement in China truly enforceable, it must be approved by the government official who is authorized to bind the whole country.

3. Titles, like in England, are important. If you want only your company to be bound and not you personally, you want to make sure that you put your title down. This demonstrates that you are signing in a corporate capacity and not personally. Among the states, there are various cases which have resulted in a person being found to be personally liable on a contract, simply because they did not list their title and sign in close proximity to a corporate designation.

4. What about seals? You probably have lost your signet ring and your wax impression kit. Therefore, signing your name next to the word "seal" is no big deal, right? Well, you may want to know that in most states, signing your name next to the word "seal" creates what the law calls a "specialty." This means that a longer statute of limitations is going to be applied to the transaction represented by the document. Instead of three years, for example, to which a normal contract might be subject, depending upon the state, a full 12- or 15-year statute of limitations is going to apply. To avoid this, you can cross out the word "seal," but make sure you do not circle the word, as this indicates acknowledgement of the seal.

5. What about dates? (Not the kind that are hard to get). I am talking about what date you put down next to your name. You put down the same date as the day that you are signing the document. ***Never back date a document***. There is never a legitimate reason to back date a document. If the parties have been operating without a written agreement, then it is acceptable to make the agreement "effective as of" a certain date by just putting in such a provision in the document. But back dating is usually designed to make an action appear to have occurred at a time when it did not. Learn from the events of 1994 where the *Secretary of State of Missouri* was impeached and convicted for back dating a document. History shows that President Nixon back dated some documents which hurt him plenty. Unfortunately, some of the most eager advocates of back dating documents are lawyers and accountants. This is because there are often large monetary incentives to back date, such as getting capital gains treatment of a stock option or making it appear that a certain document was filed with an agency when it was not. And, even if you are not a lawyer, you will have a hard time explaining why you back dated a document. This is why it is a good idea to never sign your name to a document and leave the date blank. Now, that is a date that you can get!

## The ubiquitous NDA

For parties who might be contemplating a business transaction, it is a common practice throughout high technology business to first sign an NDA. This agreement is intended to keep the parties from acting opportunistically, i.e., stealing business ideas or technology during the negotiations pending a deeper business arrangement. I wonder who the first guy was who wrote an NDA because, at first blush, they seem to be cut from the same cloth. However, after reviewing hundreds of these documents, I have come to believe that there are some important issues that can trap the unwary. For instance:

• Is it really just an NDA? Or, are there other provisions in the document that might implicate your company beyond what one might expect in the agreement? For example, sometimes a party will sneak a *noncircumvention* clause into the document while delivering the agreement with the words, "Here is our standard NDA." The significance here is that the noncircumvention clause might contain language that prohibits you from pursuing certain customers that they consider to be "theirs," or it may prohibit you from doing business in certain geographic areas or even prohibit your pursuit of a particular market or technology segment. The NDA might also contain a nonsolicitation clause prohibiting the offer of employment to key employees. Remember, the law generally does not give much authority to the ***title*** of a document. The law looks into what is ***contained*** within the "four corners" of the agreement!

- Another concern is to ask exactly what is prohibited! So many of the form NDA's floating around say that one is not to "disclose" to third parties, but fails to state that the protected technology is not to be "used" by the company with whom you are directly contracting! It is advisable that any NDA clearly states that the protected technology will not be used internally for any purpose not agreed between the parties.

- Forever... Forever... Like that movie with the big dog,[6] you should be concerned about the duration of the agreement. Some NDAs say nothing about the duration or what causes the agreement to terminate. The general rule should be that the more onerous the terms, the shorter the time you want to be subject to its teeth. Additionally, many engineering and high technology companies come and go along with key personnel, a long-term agreement makes it more likely that somebody who has been exposed to the information may be working for another company that may not be in "privity" with the original agreement.

- Closely associated with the "term of the entire agreement issue" as stated above is the concern about the keeping of confidential information itself. Suppose the agreement is for three years. What happens if confidential or proprietary information is conveyed in the last month of the agreement? Therefore, it is important that the agreement contains a separate clause that states that any confidential information disclosed be kept confidential (and not used) for a certain duration, measured in years—regardless of the duration of the agreement itself. By the way to be sure of enforcement, a clause like this also has to state that it will remain in effect beyond the termination of the agreement.

- How is confidential information going to be designated? Does oral information count? I think it is a bad practice to be on the receiving end of an agreement which allows oral information to be subject to the NDA. Anybody could claim later that "we told them about that and that is our proprietary design!" The better practice is that if oral information is going to be considered confidential, then the party furnishing the information must send something in writing describing what was said and what was confidential.

- Is it really mutual? Some NDAs report at the top of the document that it is a "mutual" NDA, supposedly implying in such a statement that what is good for the goose is good for the gander. In other words, every clause that applies to one party applies to the other. However, many NDAs say "mutual" at the top and then, upon closer inspection, it is revealed that there are inequalities or additional provisions applying to one party and not the other. Some "mutual" NDAs contain no mutual provisions whatsoever.

- Beware Residuals! Some NDAs reserve the right to *not* consider any information confidential that is considered "residuals." By that they mean any information that someone may be able to remember without notes or being documented, recorded, or

---

[6] The Sandlot, released 1993 by 20th Century Fox.

preserved. Of course, this guts the entire point of having an NDA. I would suggest that you strike out any provision that contains a "residuals" clause.

An interesting case has come out of the First Circuit.[7] The case is captioned *Contour Design Inc. v Chance Mold Steel Co. Ltd.* (September 2, 2012). It may be a situation where the "exception proves the rule" in that this dispute principally involved an NDA that had a 20-year term. It also contained a provision that the supplier shall not "develop any other product derived from or based on" the company's product. The issue in the case was whether certain derived products were covered by the NDA. The court ruled that they were covered using the reasoning that since the NDA had a 20-year term, then the parties must have contemplated that other products would be added and consequently, subject to the NDA. So, here is a case that demonstrates that NDAs often contain other types of clauses and where the length of the term was used to interpret the intent of the parties.

## MOU means IOU

The questionable legal animal Memorandum of Understanding (MOU) has entered the high technology world in force. The concept appears to be that the parties don't want to get lawyers involved, but they want to memorialize enough of a deal to keep everyone honest. It is common to hear phrases like "We don't have a contract, but we have an MOU." We are not married, just engaged. Many sites on the internet will advise that an MOU is some kind of legal document that is less binding than a contract. The problem is that *the law does not recognize this lesser type of engagement*. The issue is further complicated in many Universities, and State and Municipal governments use the phrase MOU. However, their MOUs are meant to be binding in all respects and are really just contracts. Maybe the reason why Universities like to use the term MOU is because some are state institutions and they have a type of immunity. Often, it is hard to successfully sue a state institution. The reason for this is that a lot of state courts have held that the state institution cannot pay monies that have not been allocated by the legislature. So your contract with the State, may very well be just an MOU if you cannot get paid.

The law will recognize an oral contract. The law will recognize a written contract. The law will sometime recognize an oral contract that has some kind of written confirmation.[8] And,

---

[7] There are 13 federal circuit courts called the U.S. Courts of Appeals Circuit Courts. Twelve have territorial jurisdiction over cases being appealed from the lower U.S. District Courts within their regions. The 13th, the U.S. Court of Appeals for the Federal Circuit, has nationwide subject-matter jurisdiction over certain classes of cases, such as patent disputes and international trade issues. Disputes between engineering companies in different states often end up in the federal court system.

[8] In North Carolina, "the parties' failure to execute a written contract does not preclude the creation of an enforceable agreement" *Walker v. Goodson Farms, Inc.*, 369 S.E.2d 122, 127 (N.C. App. 1988). Acceptance of an offer by a course of conduct is valid and effective even in the absence of a signed agreement. *Snyder v. Freeman*, 266 S.E.2d 593, 602 (N.C. 1980).

sometimes the law will recognize an MOU as a contract—even if the parties thought that they were just getting engaged.

To appreciate the issue with MOUs, let's look at how courts interpret contracts. The courts will start with the idea that the purpose of the document is to ascertain the intention of the parties. However, the courts also want to provide certainty in written contracts. Thus, the courts will tend to look at the document as a whole—at least at first blush. For MOUs, things can go downhill from there. For instance, most state courts will not give too much consideration to how the document is titled. So, if the title is "MOU between X and Y With the Intent to Not be Bound Herein" that will be given just a little weight in the analysis. Likewise, an MOU may have a bunch of "Whereas" clauses.

As in the title of the document, the "Whereas" clauses are given little weight. In some states, the "Whereas" clauses are given no effect, unless in the body of the document, there is some kind of recitation that incorporates the "Whereas" clauses into the document. So, you might have a document that so far looks like this:

**MOU between X and Y with the Intent to Not be Bound Herein**

> **Whereas**, the parties just want to be engaged but not married and
>
> **Whereas**, the parties just want to reduce to writing the good feelings that they have had by e-mail and
>
> **Whereas**, the parties will enter into an agreement someday if they so choose...
>
> Then, what can happen next gets interesting. The parties may state further in the body of the document that X agrees to support the project by doing all of the firmware and the software that is to be ported onto the "XY47 processor." Y agrees to develop the hardware and to do all the "bring up" on the hardware. The parties may agree that this project is going to have to be in sync and completed by May 2014. The parties may further agree to share royalties on a 50/50 basis.
>
> Later, when the project gets into trouble because Y decided that another hardware project was more important, X may go to a lawyer who may bring suit. Then the court may have no trouble saying "Well, the title to the document does not control. The 'Whereas' clauses are of no effect. However, all the requisite elements of a contract are present in the body of the document. There is an agreement to be bound, mutual promises and detriments to that effect, and thus 'consideration' has been given and therefore, a binding contract ensues."

Letters of Intent fall into the same category as MOUs. Just because someone says that their *Letter of Intent* is nonbinding, does not mean that the same pitfalls as demonstrated in MOUs cannot take place. So what can be done to prevent intentional nonbinding documents from becoming binding?

To begin with, it is important to put any recitation of the nonbinding language in the body of the agreement itself, not only in the title but also in the "whereas" clauses. Next, it is advisable to put in language that clearly states that if, at any time, the parties desire to enter into a binding arrangement, they will enter into a definitive agreement to that end. Furthermore, it is a good idea to set out a list of the items that would find their way into the definitive agreement. If you list a bunch of items that would have to be negotiated, it is difficult for a court to agree with a party that there was a meeting of the minds that covered the subject matter—hence no enforceable contract. Additionally, you can add a time limit for discussions, stating that a definitive agreement shall be entered into by a date certain or else, all negotiations and discussions shall be terminated. This acts as a boundary so as to limit any improper future use of the nonbinding document.

## A word on negotiations of contracts

*Quite a few engineers or managers of technology companies fancy themselves as "good negotiators" or even "tough negotiators." For some reason, people think that if they have the ability to shout no, whine, or get huffy, that constitutes good negotiating on the part of the enterprise. I beg to differ. The best negotiators are the ones that accomplish the most for their company on the most important matters. The best negotiators spend more effort accomplishing the goal and less effort being "Self-aware" and wondering how they are perceived by the other party. Good negotiators bring a professional veneer to the table that does not reveal all that may concern them. Good negotiators remember why they are there in the first place and they stuff their ego accordingly. What should be the goal of the negotiations?*

1.  For starters, unless you think you are going to get your own TV show, it is best to leave the other party with some dignity and a sense that they have also "won" the negotiations. There are several prudent reasons for this. One is that no matter how thick the resulting contract may be, in almost every transaction there is some item that nobody thought of or that arises a few weeks later. If you have applied slash and burn techniques throughout the negotiations and left the other party in their underwear, they are going to say "kiss my (now-exposed) posterior" and not give a single inch. Another reason is that if the contract ever goes to litigation, the other party is going to tell their attorneys that "you can't deal with them, so just sue them for all they are worth." Additionally, most people realize when their dignity has been snared and they will make sure that this is the last deal that they ever do with you.
2.  The important issues should be the ones that bring out the knives. A sure sign of a neophyte in negotiations is someone who negotiates each and every issue as if it was designed to bring peace to the Middle East.

3. "Bet the Company" issues should be brought up at the earliest opportunity. If you are dealing with a large enterprise many times your size, they are going to have all kinds of "forms" or "company policies" that they are going to try to ram down your throat. Contained in the forms will be provisions that have the effect that, if anything were to go wrong in the deal, it will bring down your entire operation. Indemnification clauses belong to this category. Even limitations on damages, which sound good at first blush, can be problematic. For instance, many contracts will contain limitations on damages with words like "neither party will be responsible for any indirect, punitive, liquidated, or exemplary damages of any kind." The problem is that when dealing with a large project, even the direct damages can dwarf the size of a smaller company's annual revenue. You may need to have a cap even on direct damages.

   Therefore, it is best to flag very early in the negotiations what are going to be deal-killer issues. You may want to ask if there are form contracts that you would be asked to sign and get a copy of these early in the negotiations.

---

The typical practice is that one technical person from company A will have lunch with a technical person from company B and say "Hey, if we port our software to your hardware we could have a killer product. Let's tell marketing!" The marketing managers sign off on the deal and decide to have some kind of exclusivity involved and then divide up market segments. After extensive correspondence and e-mails, they submit the matter to the lawyers as some kind of afterthought and necessary evil. When your lawyer gets it he indicates that if anything goes wrong you could lose the company and besides, the way the markets are being divided up, it looks like there could be antitrust implications. "I hope you haven't been e-mailing each other about this!"

---

   So, the best practice would be to have the attorney involved from the very beginning. The attorney could make, as part of an initial term sheet, demands for indemnification or limitations on direct damages so that the ultimate viability of the enterprise is not threatened. Additionally, he could ask for an opinion letter from the large company's legal counsel that the endeavor would not violate antitrust rules.

   Another reason for bringing up the most concerning issues early is that human nature accomplishes the most in meetings when people are fresh and not tired. Also, parties are less likely to walk away from the negotiations early in the negotiation process. That tends to happen later when they feel that they have given in on a host of other issues and feel that some kind of "last straw" is being placed upon their backs.

4. Use the Moral High Ground. Even though the world seems to be in a bit of a mess, people like to think that they are moral and fair. Criminals, when caught in an act that is going to land them in some place confined, frequently attempt to justify their acts.

Even mass murderers will justify their acts based upon their "morality" and will seek ways to explain their actions. Remember Virginia Tech or the "Unibomber?" Chances are that whoever you are negotiating with has at least as much morality as a mass murderer! The goal here is to appeal to their sense of morality and fair play. For instance, if the company web site claims that their vendor relationships are of prime importance to them, you might weave this in during an appeal for better trade terms. If you are demanding a certain arbitration term and they are saying "we don't like to litigate anything anyway" (but you have done some research and know that the company is currently engaged in litigation with a vendor), you have an opportunity to politely indicate the contrary. Since they have already attempted to distance themselves from the concept of litigation, you can take advantage of their "moral view" and save them from being hypocritical by adopting your arbitration clause.

Likewise, suppose you are negotiating an employment contract for the position of vice president. The company may say that they do not want to have any contract whatsoever. If you know that the last three Vice Presidents have all departed after about 14 months, pointing this out challenges their sense of fair play. The company does not want to think that there might be something undesirable about their company, and thus by appealing to their moral high ground, they will want to agree to a contract.

You can see from the above that it is imperative to have done research into the object of your negotiations! You should know from the start: the company history, who the key players are, any past or pending lawsuits, the company's financial health, the rate of growth, and the key suppliers and partners with whom they are associated.

5. Keep Negotiations Just as Negotiations. With e-mails, instant messages, faxes, video conferencing, voice mail, web pages, etc., we are all leaving lots of electronic footprints. Oral contracts are enforceable as long as they do not violate the statute of frauds. Oral contracts can be confirmed by any kind of writing to "take it out of the statute of frauds."[9] So one goal of the negotiations is to make sure that whatever is finally agreed upon is the real deal and not subject to further interpretation, because 500 e-mails say something differently. Practical ways of minimizing this problem are:

    a. Limit the number of departments or personnel that are authorized to communicate during the negotiations.

    b. Make sure that the final document has what is known as an "integration clause" stating that all communications that have been exchanged before are merged or integrated into the final document.

    c. Be clear at the onset that the negotiations contemplate that a final written agreement is to be agreed upon.

---

[9] States have "statute of frauds" laws that are referred to as such and are designed to keep someone from saying they are owed money when there is no written evidence to support such. Therefore, oral contracts are limited to certain types of activities, of limited duration and are limited to how much money can be involved.

    d.   Communicating personnel should be cautioned about making statements that indicate that they can bind the company if they are not authorized to do so. (See the sections about MOUs and Agency.)

6.   Be willing to walk away. A sure way to get behind in contract negotiations is to be in a position, or to telegraph a position, that no matter what, you need an agreement to result. It is a sure sign of a neophyte who just gushes to the opposing party how wonderful it will be to do business with them and how necessary the program is going to be to his future growth and that his boss told him to "make it happen."

    Even if you do walk away that does not mean that you can't come back.

> I can recall a situation involving a large telecom provider who needed a project done. Their contract manager put out an RFP which set out the basic parameters. A tight deadline was given for not only responding to the RFP but for doing the project. After reviewing the RFP, it was noted that the telecom giant stated that no NRE[10] would be paid and that the provider would have to allocate development costs across the production run. Significantly, the giant would not agree to any kind of binding forecast or minimum order quantity. We decided to no-bid the matter. In doing so, we asked for a teleconference. We asked first if we had understood the criteria as stated—we did. We asked if the requirements were cast in stone—they were. So we politely told them that we would no-bid the deal, because nobody in the business would agree to such terms and that our bosses would shoot us in the head if we ever agreed to such a deal. A month later we heard that all vendors had no-bid the deal and that the contracts manager had been reassigned. Also, we were chosen to do the deal upon terms that were reasonable and in accordance with normal business practices.

    Sometimes, the opposing side needs to see that you have the guts to walk away. However, if you do walk away, make sure that they know exactly why you did so and what the pressure points were.

## Humble negotiations with the Big Guy (reprinted with permission from the September 2001 IEEE Instrumentation and Measurement Magazine, by Craig Silver)

What happens when you must negotiate with a powerful entity? Whether that entity is a large corporation, an institution, government, or your boss, it can pose as an intimidating bulwark against the object of your desire. If you are not careful, the very fact that you dare negotiate could be considered "bad form" as our British friends might say, or "stupid," as your boss might say. To avoid this, consider following this pattern.

---

[10]  Nonrecurring engineering costs. These are the development costs that are of a one-time nature and not recurring, like those that would be associated with ongoing production.

## A lop-sided negotiation

You may have valid reasons to suggest that your company change its direction, yet you may not be in the position to suggest those changes. You have a lop-sided negotiation with two elements: an opposing party (the company) and a disadvantaged party (you). Consider the most lop-sided negotiation in history where God was the opposing party! (Genesis 18:16–33) Abraham, the—eh—disadvantaged party, is going to initiate and sustain negotiations with an All-Knowing, All-Powerful, and All-Present Creator. How does Abe approach this situation? How would you? Appeal to what is a right, i.e., internal corporate values. Very powerful parties are cognizant of the fact that they hold tremendous power. Many also claim that they want to do what is right and not take advantage of the weak. Consequently, Abraham wisely makes an appeal. "Would you also destroy the righteous with the wicked?" This appeal is to a central character of God—His justice. "Suppose. . ." Follow your appeal with a suggestion beginning with "Suppose. . ." You are attempting to continue a negotiation this way. Many fruitful negotiations can ensue after obtaining a commitment from a supposition. Instead of asking for the moon with the intention of settling for less, as is common among the less initiated, ask for less with a view of getting more. Consider as an example, a situation where you are soliciting venture capital. You ask the "suppose" question such as "suppose I can obtain a co-investor, would you match an investment by Highway Robbery Ventures?" Abraham's supposition is classic. "Suppose there were fifty righteous within the city. Would You also destroy the place und not spare it for the fifty righteous that were in it?" Abraham makes no attempt to get God to compromise on one of his central attributes—Righteous Judge. A large corporation will rarely compromise a central tenant. Microsoft is not going to trade software development for pencil manufacturing; Hewlett-Packard won't begin producing sporting equipment; London is not going to move the London Bridge to Arizona. . . well, I said, rarely.

## Commitment is gold

When dealing with a large entity, a firm commitment is like gold. God said to Abraham, "If I find in Sodom fifty righteous within the city, then I will spare all the place for their sakes." Fantastic! Abraham caused negotiations to take place! Note that he got a firm commitment where he just started out with a supposition. Do not move to equality! Never telegraph to a powerful party that, by virtue of the exchange, you are now seated next to the throne of power! Remember that the powerful allowed you access in the first place. You are not ready to go golfing with the Chairman of the Board!

Abraham humbly approaches the Other Party. "Indeed now, I who am but dust and ashes have taken it upon myself to speak to the Lord: Suppose there were five less than the fifty righteous: Would you destroy all of the city for the lack of five?" Here he realizes that to continue to move God is rather presumptuous. By humbling himself, he demonstrates that he is not equal to God and that he has not forgotten who holds the real power.

### Stick to what you say

Last minute changes are frustrating, like haggling with a car dealer over price and then finding numerous hidden miscellaneous charges added to the bottom line. Most negotiators find the act of adding terms to be highly offensive. If you must add a term, then index the change to some other term that has been negotiated during the proceedings. For example, Abraham humbly got more and more commitments from God as he bargained down to ten. At one point, he states that he will "speak but once more." Whatever you tell the powerful, stick to it! To change position, ask for more, waffle, or just confuse the situation has the effect of diminishing the value of the whole matter. Clarity begets charity.

Be humble and negotiate in good faith.

## CHAPTER 18

# Dealing with the Government

## Craig L. Silver

*Director—Strategic Initiatives/General Counsel, Amches, Inc.*

**Chapter Outline**

## Considerations in US federal government contracts

Because embedded engineers are smart, there is a tendency to assume that their smartness is naturally extendable to other disciplines. There is nothing wrong with applying oneself to the task at hand. However, you should be warned that government deals have several significant hazards and some nonintuitive ones at that. For instance, there are many government contract clauses that can be read into a contract, even if they have been inadvertently omitted by the government, or even their prime contractor. This is the *Christian Doctrine* as stated in *G.L. Christian Associates v. US*.[1] Therefore, a project manager who may refer to a subcontract may find himself in the unenviable position of having to consider clauses that are not set out on paper. So, no matter how smart you are, it is hard to analyze contract provisions that are not even there.

## The government's right to change

If only it would. Anyway, regarding contracts, one frequent hazard to be navigated is the government's right to make changes. If an item is being made to government specifications, the government may order changes to the drawings, design, specifications, method, shipment, or place of delivery. A contractor is entitled to an equitable adjustment to the contract price if

---

[1] 160 Cl. Ct.1 (1963).

Developing and Managing Embedded Systems and Products.
DOI: http://dx.doi.org/10.1016/B978-0-12-405879-8.00018-0

ordered changes cause the contractor to incur increased costs. A more likely scenario occurs when actions of the government cause extra work, but is not set out as formal changes. These "constructive changes" have been recognized by courts and administrative appeals boards and may take the form of defective specifications; informal extra work directives; failure of the government to cooperate during contract performance; and simply wrong contract interpretations taken by the government. Like many other areas of the law, it does not pay to "sleep on one's rights," as a contractor should assert the right to an equitable adjustment within 30 days after being advised of a contracting officer's requested change, or in the case of a constructive change, before final payment. Unlike most commercial dealings, one does not have the right to stop work if there is a dispute about changes.

## The government's right to terminate

Don't temp me. Anyway, another hazard to watch for is that the government can terminate the contract at any time. One should be aware of this possibility—especially as a subcontractor. FAR[2] clause 52.249-2 gives the right to terminate if the contracting officer determines that such an event is in the government's interest and, as long as the determination is not an abuse of discretion or in bad faith, it will be allowed. The prime contractor who receives such a notice is obligated to also terminate all the subcontractors on the project. The prime will have a year to petition for contract termination fees which can include performance to the date of the termination, the costs of idle facilities, and a reasonable profit. A prime contractor often "flows down" to the subcontractor all FAR clauses that it is subject to fulfill. It is important for the sub to, in a sense, "flow up" the same benefits that the prime may receive in the event of termination, so that it is fairly compensated for any termination. It is a good idea to try to flow up as much as possible so that the interests of the prime and the sub are aligned.

## Ethical issues in government contracts

There are ethical areas of concern which markedly depart from what is acceptable practice in commercial transactions. In commercial practice, a company may publicize its gift policy and ask that vendors comply. Unless the policy finds its way into a supply contract, the vendor is not subject to a very high risk if it were to violate the policy. For government arrangements, there are criminal and civil regulations that severely restrict certain practices. These restrictions apply to virtually anything of monetary value such as meals, travel, lodging, gifts, and discounts. There are exceptions, such as it is acceptable to offer food and drinks if they are not part of a meal, and discounts if they are available to the public or government at large. Likewise, the law restricts commissions and contingent fees to be

---

[2]  Federal Acquisition Regulations that have the effect of agency law upon the relationship between the US government and its contractors. The defense department has similar regulations known as the DFARs.

paid to anyone—even if they are not a government employee and even if that person is not "in the business" generally.[3] In other words, if there is a company that helps to advise companies on obtaining government business, it would be acceptable for them to receive a commission on contract awards obtained by their clients. However, it would not be acceptable for a commission sales person to receive a SPIF,[4] where the SPIF was paid by an employer that was a prime on government contracts or to pay with government funds.

> *I recall how amazed I was, shortly after the events of 9/11, when one would have thought that the FBI would have been very busy doing other work, to have two FBI agents investigate a client who received a trip to a resort as a sales incentive for landing business with another commercial prime which performed government contracts. Note, nothing of monetary value was given to any government employee!*

## Some criminal statutes relevant to government contracting

Title 18 of the US Code, where many criminal statutes reside, contains section 1001. This statute makes it a criminal offense to make any false, fictitious, or fraudulent statement to a government agency. It has even been used to convict persons of lying to private entities who are subject to government regulation! For instance, if in connection with obtaining a loan from a bank, a person overstates their income or understates their obligations; they can, and have been, successfully prosecuted. It is a favorite arrow in the quiver of prosecutors who might otherwise have a difficult case to prove, as there is no requirement that intent to defraud be proven, just that a false statement was made and that it was known to be false. Like we have all been saying since Watergate,[5] it is the cover-up that gets people in trouble, more often than the underlying offense.

Furthermore, the False Claims Acts[6] provide for sanctions for persons or entities who make false statements in relation to a government contracts. For instance, if one bills the government and the bills are false in a material way, and the government pays the bills, there can be civil or criminal liability. It is important to understand that the civil actions allow for anyone who obtains knowledge of the false event to file a civil action known as a *Qui Tam* action, i.e., to act as a "private attorney general" and essentially act on behalf of the government to obtain relief for the government. They also get relief for themselves, as the statutes allow for a percentage of the recovery, as well as the payment of their attorney fees. This is a lucrative area of the law for legal practitioners, and there are many

---

[3] Meaning here a "bona fide employee or agency." Bona fide agency or employee means a commercial or selling agency or person that secures business where it does not exert "improper influence" to obtain government contracts or holds itself out as being able to obtain government contracts through improper influence.

[4] Sales Promotion Incentive Fund.

[5] The scandal that started with a "third-rate burglary" and ultimately resulted in the impeachment and resignation of President Richard Nixon.

[6] 31 U.S.C. sections 3729–3733 and 18 U.S.C. section 287.

law firms that specialize in this type of action. Managers who like to lie and tick off employees in the process should be especially wary of the False Claims Acts and the possibility of *Qui Tam* actions.

In case you skipped kindergarten, it is statutorily prohibited to offer a kickback to a government employee for the purposes of obtaining a contract award or any favorable treatment in conjunction with the award.[7] Likewise, just as it is wrong to somehow get a copy of the exam before the test date, it is wrong to obtain opposing bid information or agency source selection information from the government. In short, *anything that assaults the integrity of the bid and proposal effort is suspect and most likely prohibited by statute.*

## The government contractor defense

State tort[8] claims are preempted where the plaintiff's injury is caused by the allegedly defective design of military equipment manufactured by the defendant pursuant to a contract with the federal government. See *Boyle v United Technologies Corp.*, 487 U.S. 500 (1988). This case established the government contractor defense for tort liability. The Court reasoned that although state tort law may allow products liability claims against military manufacturers, this is an area of uniquely federal concern, regardless of the lack of federal legislation specifically claiming the immunity. In an effort to determine the scope of this defense, the Court stated that this preemption is limited to areas of "significant conflict" between federal policy and state law. For guidance on the extent of "significant conflict," the court applied the discretionary function exception of the Federal Tort Claims Act (FTCA) which states that state law will be preempted wherever it threatens a discretionary function of the federal government. In the *Boyle* case, the design of the allegedly defective product was a military discretionary decision.

In sum, *Boyle* established that state law which imposes liability on a military manufacturer is preempted when (i) the United States approved reasonable and precise product specifications, (ii) the equipment conformed to those specifications, and (iii) the supplier warned the United States of the known dangers of using the equipment.

An important issue is whether the defense applies only to contracts with the military, or whether it can be used by other government contractors. The Supreme Court has employed language hinting that it may apply to all contractors but has never ruled directly on the issue. See *Hercules, Inc. v. U.S.*, 516 U.S. 417, 421 (1996). ("The Government contractor defense... shields contractors from tort liability for products manufactured for the Government in accordance with Government specifications, if the contractor warned the United States about any hazards known to the contractor but not to the Government.")

---

[7] 41 U.S.C. §§51–58 and implemented in FAR 3.502.

[8] A tort is any civil wrong, such as negligence, fraud, and misappropriation of trade secrets distinguished from civil breach of contract actions and, of course, criminal matters.

## CHAPTER 19

# Agency and Getting Paid

### Craig L. Silver
*Director—Strategic Initiatives/General Counsel, Amches, Inc.*

**Chapter Outline**

## Agency

Everyone has heard of agents, e.g., real-estate agents,[1] talent agents, secret agents, but there is less apprehension of the law of agency. It may surprise you that you are most likely an agent for somebody. A wife can be the agent for her husband. A neighbor could be an agent for a next door neighbor. If you work for a company as an employee, and not as an independent contractor, you are the company's agent. In fact, even as an independent contractor you could be an agent, depending upon how you act or what is said to third parties.

## Why are agency relations so important?

So, what is so important about agency? An agent legally binds the principal. The law of agency is based on the Latin maxim "Qui facit per alium, facit per se," which means "he who acts through another is deemed in law to do it himself." Also, for contract purposes, if you are acting as an agent for a principal, then you are not personally liable on the contract.

---

[1] It is surprising to many that a real-estate agent is an agent for the seller. Even if you have approached one to show you offices around town, they represent the seller. This confusion has caused the creation of "buyer's agents" and laws that require clear disclosure as to who is an agent for whom.

Developing and Managing Embedded Systems and Products.
DOI: http://dx.doi.org/10.1016/B978-0-12-405879-8.00019-2

*693*

Agency can be created by contract or through the operation of law. For instance, suppose you and a friend decide to start an embedded engineering company. You form a limited liability company (LLC). If you are the Vice President, you are an agent for the company and by statute, common law or judicial decision; you are authorized to bind the company and sign contracts. This assumes that you are an Executive Vice President and not Vice President of the water closet. Suppose further that in the very beginning phases of flight and before you formed the LLC, you and a friend just start a business and decide to split profits 50/50. You are partners in the endeavor. The law says each partner can bind the other. So, if while you are at lunch your partner buys an expensive copier, you are on the hook to pay for the copier, as much as your partner. If your partner backs out of the business or dies, you are on the hook to pay for the copier by yourself.

> *Your friend may have a nephew who needs to earn a few bucks. He joins your operation for the summer. The aforementioned copier is out of ink because it is new and the new copiers are always just half full of ink. Your friend says that he will send his nephew to the store to get ink. Unfortunately, on the way to the store, the nephew strikes a pedestrian with his car. The law of agency kicks into high gear. The law will say that the nephew was doing official partnership business at the time of the accident, was further acting as the agent of the partners and thus, the partnership, as well as the individual partners are liable for the damages. This scenario can also be problematic from an insurance point of view. The nephew's carrier is going to say that the accident happened while employed in a commercial enterprise and the car was not insured for commercial purposes.[2]*

Another scenario that arises frequently in the high-tech world involves the issue of "apparent authority" or "holding out" as an agent. You appoint a sales representative firm to help sell your latest board. The sales agent may make promises about the performance of your board, prices, or delivery schedules which would bind your company. You can protect yourself, to a degree, by putting language in the representative contract that they are not authorized to act as an agent and are not authorized to hold themselves out as an agent who is able to bind the principal, i.e., your company on such matters. However, suppose you learn that, contrary to the agreement, indeed the sales representatives are going around making all kinds of promises about performance or delivery schedules. The doctrine of "apparent authority" could decide the matter by declaring that the "agent" had the *apparent* authority to bind your company and thus innocent third parties would have the right to rely upon the "agent's" assertions. Therefore, the law puts the burden upon the principal to rein in his "agents" when they are exceeding their authority should he have reasonable knowledge of their improper actions. How does he rein them in? By notifying the third parties who may

---

[2]  In fairness, the carrier may have to pay if this was a "one off" occurrence. But if the nephew was running daily errands, they may be able to escape payment.

have been misled by the agents that the agents were not authorized to make, e.g., performance promises.

The same type of problem can arise with the use of subcontractors who have customer contact or are placed on a customer site. Often, the principal may not want the customer to know that so and so is really a 1099 subcontractor[3] and not an employee of the company. Just be aware that if you present the subcontractor in such a way so as to not make any distinction between them and your real employees, the customer will have the legal right to assume that they are your employee and an agent of the company.

The converse is also true. You may not have the right to rely upon statements of an agent, even one with overt authority, if it is not reasonable to have so relied. If a sales agent tells you that a major DSP chip supplier is coming out with an eight core 10 GHz model next month, you could not reasonably rely upon such a statement without getting the information from a higher source within the company.

Choose your partners carefully because partners are agents of each other too.

### Scope of agency

Just because someone may be an agent, the legal analysis does not stop there. A more granular questioning is in play. The scope of the agency needs to be ascertained. The scope criteria ask the question "He is an agent to do what?" In the previous examples, the Vice President would have the authority to bind the company in all matters except extraordinary matters which would require a corporate resolution. Even more so with the partner in a partnership because resolutions are not really normative to partnerships, essentially anything done by one partner is going to bind other partners as well as the partnership itself. In the case of the sales representative, the scope of the alleged agency could extend to issues such as delivery, but it would not extend to representations of company policy regarding the willingness to sell a company division. The nephew in his car would be acting within the scope of his agency to pick up ink, but not to go visit his girlfriend. The subcontractor may have authority to state that the processor only draws 7 W, but not have authority to state what the list price is going to be when it is finally sold.

As a practical matter, it is important to make sure that employees have business cards with titles that reflect the actual extent of their authority and *be trained to not exceed authority*. There should be clear company statements as to who is authorized to sign for the company on contracts. Contracts with third parties should clearly state whether or not any form of agency is contemplated and designate the scope of the agency.

---

[3] Refers to IRS form 1099 which reports payments to independent contractors.

## Getting paid

Getting paid on engineering contracts is getting more and more complicated. With the so-called "global economy" and the internet allowing everyone to have a presence next to yours, customers are becoming more cagey about paying. Here are some considerations to keep in mind for getting paid on domestic and international contracts.

1. *Delineate the currency.* If you have a foreign customer, you should not assume that they plan on paying in dollars or euros. With wild swings in currency exchange rates taking place, you may need to account for these swings if the project is going to go on for a long period of time. You can index the payment to an established bank rate so that you are protected if the currency in play tanks.

2. *Watch the terms.* Many companies pay outside of agreed terms. If you agree to 60 days, you may not see your money until 80 or 90 days. You should be diligent to have an interest rate applied to any balance that is outside of agreed terms, e.g., 1% per month. Add that the customer will pay any attorney fees should the matter go to collection. At least you can threaten the interest charges when you pursue them for payment. You may want to give a discount for prompt payment. In the case of the US government, they are required to take the discount and pay within those terms.[4] However, many commercial entities will try to take the discount even when they are paying outside the requisite terms. For government or commercial entities, the invoice should contain:
   - Name of the contractor
   - Invoice number and date
   - Contract number or purchase order number and delivery order number or task order (if applicable)
   - Description, unit prices, and extended (exactly as written on the purchase order)
   - Quantities, shipping terms, and payment terms (followed as written on the purchase order and clearly stated on the invoice)
   - Line items on the invoice that agree with line items on the procurement document
   - Complete remittance address
   - Other information as required in the contract, i.e., statement of work , standard of performance, or contract data requirements.

3. *How do you know you have delivered?* You should have very tight understanding as to what you are delivering and what the acceptance criteria will be for the deliverables. If not, you can end up debugging the project till Taylor Swift settles down with one guy. A good idea is to try to structure the deal with only a small milestone payment due at

---

[4] 5 CFR 1315 The Prompt Payment Final Rule (formerly OMB Circular A-125, "Prompt Payment") requires Executive departments and agencies to pay commercial obligations within certain time periods and to pay interest penalties when payments are late.

the end of the project. That way your customer will not have you over a barrel if the project is in trouble. The project may have problems that are not related to you but they will want you to be their partner in the project and not pay until you help them out.

4. *Watch their system.* A common trick by customers is to say that your invoice is not in their system and thus will not be on that week's check run. Of course, you may not have known about that, eh, issue until day 75 of your 60-day agreed term. The way to head this off is to start making inquiry of your customer on day 45 and ask if they have all the invoices and wire transfer information that they need in order to pay on day 60. Ask further if day 60 is in phase with the next check run. It can help to email, fax, and mail your invoice. When they lie to you, they will know that they are lying to a savvy company and may feel that they should pay eventually.

5. *Collections.* If you do need to file a collections case, it is desirable to be able to do so in your local court. That is why you want, if at all possible, to have the underlying contract to be construed under the laws of your state and to have your customer agree to submit to the jurisdiction of your local court. These types of provisions are considered by many to be nuisance boilerplate that nobody cares about, but for the savvy, they are opportunities to greatly enhance the efficacy of collection efforts.

6. *Getting paid from overseas.* In some ways, getting paid on overseas deals is easier than domestic sales, because overseas, nobody trusts any person and they don't expect credit, at least at first. Therefore, the way to get paid is to either ask for the money up front, or arrange for a letters of credit (LOCs). LOCs follow the forms and procedures as contained in Uniform Customs and Practice (UCP 600) by the International Chamber of Commerce. A bank issues the LOC and it is the bank that pays you. The arrangement is that as long as the documentation is complete and the bank can get confirmation that conforming goods have been delivered, then the bank is supposed to pay the LOC. If you do not trust the customer's local bank, you want the issuing bank to have the LOC confirmed by a stateside bank. This is called a confirmed LOC. Many banks in the United States have "correspondent" arrangements with foreign banks and can facilitate a confirmed LOC. Unlike the United States, some countries still allow private banks. Private banks can go bankrupt just like anybody else. Therefore, be wary of any transaction that can only have an LOC issued from a private bank.

## Documentary collection

Another way to get paid is to use a procedure called "documents against payment" or D/P for short. Basically, it is a procedure where you don't hand over the documents necessary for the goods to be transferred to your customer until the customer has paid through his local bank. The steps are outlined here:

1.  You, as the exporter, fill out a sight draft[5] and documents and send to your freight forwarder or shipper.
2.  Freight forwarder ships goods and obtains bill of lading signed by the actual carrier doing the shipment.
3.  Freight forwarder sends documents back to you, and you forward these documents to your bank.
4.  Your bank sends the documents to the importer's bank.
5.  Importer accepts the documents and approves payment. The importer is given the documents by his bank, and this is proof the goods belong to him so the carrier will release the goods.
6.  Importer's bank wire transfers money to exporter's bank. Your bank puts the payment into your account.

There is another procedure called "documents against acceptance"(D/A) which is similar to the above, but it allows for additional time to be given to the customer to pay the bank. The advantage of the D/P or D/A method of transaction for the customer is, unlike the LOC, the customer funds are not required to be on account or paid in advance. The bank fees for these types of transactions are less as well.

## Bankruptcy—what does his problem have to do with me?

If you are anywhere on the planet, eventually a company you are associated with, or dealing with, will file for bankruptcy reorganization (Chapter 11) or seek an outright discharge (Chapter 7) of the bankruptcy code. Bankruptcy law is one area of the law that many fair-minded folks have a hard time with, as the laws are designed to achieve a certain result that can work hardships on seemingly innocent parties.

Suppose you have been selling your product to the Sinking Ship Company and they have been paying net 45 days terms on credit. They come to you and say, "Hey, we could use that delivery real soon. If you can get it out by next week we would appreciate it." Now you happen to have heard from the parts vendors, and all parts vendors gossip, that Sinking Ship just might be ready to hit the shoals and you are concerned about their continued viability. Proud of your inside knowledge, you say, "eh, ahem, we are going to have to insist upon cash or at most net 10-day terms." Sinking Ship needs your stuff so they say OK, net 10. Because you are sharp, you make sure that there is some kind of writing agreeing to the new credit terms. What has just happened here? If Sinking Ship files for bankruptcy protection and a trustee gets involved, the trustee who works for the US Justice Department will start sniffing around and is statutorily obligated to look for *preferences* and

---

[5] A "sight draft" is the equivalent of a check payable upon demand. *Mt. Vernon Nat. Bank v. Canby State Bank*, 276 P2d 262.

to sue to set aside any transactions that can be qualified as a "fraudulent conveyance." What will be looked at is the pattern of activity for a minimum of 90 days prior to the filing of bankruptcy. If the pattern indicates that for years you were extending 45-day terms and all of a sudden terms go to net 10-day terms, the trustee may succeed in having the transaction set aside because the transaction does not appear to have occurred in the "ordinary course," but as some kind of deal in "contemplation of bankruptcy." So the trustee can get a judgment against your company and demand the return of all consideration, e.g., money, that you received from the last deal. In my experience, some of the most bitter reactions are observed under this scenario. It sounds so unfair as the engineering company complains, "All we did was act like prudent businessmen and sell some stuff. And now we have committed 'fraud?' And we have to pay back the money and we don't get our product back?" It is best to duck when the answer is, "That's right."

So what might be the way to handle the situation from the beginning? One way might be to keep the terms the same and hope that they do not file before you get paid. Another idea is to make them buy from somebody else selling your equipment, e.g., "You can buy from our distributor in Indiana." Asking for an LOC from a bank or personal guarantees from the principals of the company would also be viable solutions. Still another option is to structure the deal in such a way as to provide "new value." The new value defense is a statutory animal that basically recognizes that if some other consideration is provided, in addition to what has been typically been the pattern between the parties, then it may very well be that the transaction was not done in contemplation of bankruptcy. An example of new value might be agreeing to update the firmware of a hardware device at the request of the soon-to-be bankrupt company.

## CHAPTER 20

# Intellectual Property, Licensing, and Patents

Craig L. Silver

*Director—Strategic Initiatives/General Counsel, Amches, Inc.*

## Chapter Outline

## *Software licensing, source code, and somebody going broke*

Software engineers are well advised to understand the following, because in many small companies the role of the software engineer tends to morph over to the licensing aspects of business deals. Additionally, the software engineer may be on the "front line" of vendor/licensee relationships and have opportunity to do some action which may have severe consequences or benefits. So, let's analyze the situation from a real-world perspective.

You work for a small company, "Developments-R-Us" (DRU), as a software developer/manager. You are tasked with sourcing a small RTOS kernel to be embedded on a CPCI board. After endless debates within your circle of peers, you settle on the RTOS you focused on two months ago, a new product from "Zippy Source RTOS" (ZSR). The ZSR product seems really hot and is getting a lot of notoriety in the trade press. However, they

Developing and Managing Embedded Systems and Products.
DOI: http://dx.doi.org/10.1016/B978-0-12-405879-8.00020-9

have only been in business for two and a half years and you are concerned about their viability as you would not be getting a source code license.

Because you read books, you are aware of the following issue. ZSR might go into bankruptcy some day and if that happened, DRU would be in trouble if it could not rely upon ZSR to perform the contract with its associated licensing provisions. Additionally, DRU may have signed agreements with its customers that obligates it to certain supply terms for its CPCI boards. You lean back in your quiet cube and contemplate that in 1988, the Intellectual Property Bankruptcy Protection Act[1] was enacted which added to §365 of the US Bankruptcy Code. You recall that bankruptcy law has as its goal the maximizing of the "estate" (the assets of the debtor in bankruptcy) for the benefit of the secured and general creditors while providing relief for the debtor from the creditors. It is the balancing of these opposing interests that will provide complexity and consequences to ZSR and DRU.

When a company goes into bankruptcy, the company (or the trustee on its behalf) has the right to accept or reject a contract that is "executory." An executory contract contains clauses that have material and important provisions that both parties must perform in the future or on an ongoing basis. A software license is *executory* because there are future obligations to pay royalties or fees, provide updates, perform maintenance, indemnify for Intellectual property infringement, etc.[2]

You started to doze off in you chair, someone dropped a book, you startle and continue your thoughts. You understand that if the trustee decides that if it would be too burdensome upon ZSR to perform the contract and that the benefit to the estate would be inconsequential, he may reject the contract. It is upon this rejection that the provisions of the Intellectual Property Bankruptcy Protection Act kick in. Under §365(n), DRU can consider two options:

1.  Treat the rejection of the contract by the trustee as a material breach of the contract, claim any damages as a general creditor of ZSR, and terminate the contract.
2.  Retain the license rights under the contract with significant modifications. The law states that DRU would have all its rights to the software as of the date that ZSR filed for bankruptcy minus any obligation on the part of ZSR to provide maintenance, support, indemnification, etc.[3]

---

[1] Enacted by Congress in response to the *Lubrizol Enterprises, Inc. v. Richmond Metal Finishers* case 756 F.2d 1043 (4th Cir. 1985).

[2] If you just bought the software outright with no obligation to provide updates by the vendor, then the contract would not be executory as all provisions would have been fully performed.

[3] The reader may note that it is the executory functions that are modified here. The full extent of what intellectual property rights that are available is somewhat in flux as of the time this is being written as there is a conflict among the federal circuits in light of the 7th Circuit's ruling in *Sunbeam Products, Inc. v. Chicago American Manufacturing* (Case No. 11-3920 July 9, 2012) which involved trademarks which are not specifically mentioned in §365(n).

Oh, by the way, you recall that DRU would have to continue to pay royalties as per the contract. This gives rise to an important contract drafting consideration. It is desirable to separate out the cost of maintenance obligations from the license fee. This way DSU will not end up paying the full cost of the past royalty arrangement while getting less in return.

As you sit in your cube ignoring the ultimate consequences of a sedentary lifestyle, you formulate an idea. You think, "Wait a minute, a week won't go by when I won't talk to somebody from ZSR. I would have plenty of advance notice if they were getting into trouble. Also, we buy a lot of parts from the parts vendors and they always have the latest gossip on who is in trouble. DRU doesn't need the source code now and I can avoid those discussions until the time comes." The problem with this thought process is that the bankruptcy code has this type of thinking targeted. The law would say that any arrangement that is atypical or in contemplation of bankruptcy can also be set aside by the trustee. The law uses nefarious sounding phrases like *fraudulent conveyance*.

So, if DRU had been paying a monthly royalty of $5000 to ZSR for an object code license and later, the contract was changed to include a source code license, this too could be set aside by the trustee as not being in the "ordinary course" of business. An exception to this is if DRU were to provide "new value," e.g., an additional $50,000 for a source code license. At any rate, it would be important for DRU to have rights to the source code as of the date that ZSR filed for bankruptcy. Since under this scenario, DRU can expect nothing from ZSR, the source code license should include rights to modify, copy, adapt, and to make derivatives.

Putting the source code into escrow is a common step that is designed to thwart attempts by subsequent bankruptcy trustees to interfere with the source code. A source code escrow arrangement typically involves a third party who agrees to receive the code and keep it in a safe place. In this case, ZSR, DRU, and the escrow company would sign a three-way agreement which would release the software from escrow upon certain conditions, such as ZSR going bankrupt. Of course, the time to set up a source code escrow is early in the relationship, so that the source code arrangement itself does not become the object of concern by the trustee during the "look back" analysis. Another reason to have a separate escrow arrangement is that at least you know where the source code is. After a trustee takes over, he may know how to identify a desk, but if you wanted to buy the source code for a project, he may not know where it is and may not care.

### Software licensing in general

It is important to note what kind of rights you are obtaining by licensing software in the first instance and what kind of rights you can extend to third parties. Consider the case where you are buying a Software Tool Kit (STK) or a Software Developer Kit (SDK). It most likely will come with some kind of license terms. If you are designing for an embedded system to be deployed and manufactured for the use of third parties, you need to make sure that you can extend the host software that you developed from or upon the Kit to your end customer. In other words, you need the right to distribute the software that came in the Kit or that you developed from the Kit. Many manufacturers distinguish between their SDK, and its associated license, from their *embedded* license which you buy for an additional fee, while others can be silent about the issue altogether.

In some cases, it can be hard to find appropriate language that can clearly state whether you have the right to embed the software into a product for general distribution. The right to distribute the resultant code, either as a derivative or upon sublicense terms, is important, because later it can come up in your customer's terms and conditions of purchase of your product. You may be asked to warrant that any software supplied is either yours to license, or, if from a third party, is being distributed in accordance with the originator's license terms. Sophisticated buyers may ask to see the licenses and review the terms themselves and not just rely upon warranty language. Typical language that might be found in a software license that allows for the resultant code to be distributed may be:

> **Grant of License**—*Subject to the terms and conditions of this Agreement, Licensor hereby grants to Licensee, a worldwide, non-exclusive, non-transferable license:*
>
> (i)   *to integrate the Licensed Technology into Integrated Products;*
> (ii)  *to reproduce the Licensed Technology as so integrated into integrated Products; and*
> (iii) *to distribute the Licensed Technology as integrated into Integrated Products solely to End-Users who are subject to an End-User License Agreement. Licensee shall make no use of any copies of the Licensed Technology except as provided in this Section. Licensee may sublicense the distribution rights granted under this Section solely as described in this Agreement. All rights not specifically granted herein shall be retained by Licensor.*

Note here that this *license grant* specifically contemplates that the licensee will make sure that the end users of the products will be subject to another license, called here the *End-User License.* Sometimes, the Licensor will dictate the terms of this End-User License and even supply the copy and sometimes they just want approval of such a license and sometimes they just mention that your end users should be subject to such a license.

A typical End-User License may contain these limitations which may not be in the Kit license:

## End User Restrictions

> *You may not, and you may not permit others to:*
>
> (a)   *reverse engineer, decompile, decode, decrypt, disassemble, or in any way derive source code from, the Software;*
> (b)   *modify, translate, adapt, alter, or create derivative works from the Software;*

   *(c)   copy (other than one back-up copy), distribute, publicly display, transmit, sell, rent, lease or otherwise exploit the Software; or*

   *(d)   distribute, sublicense, rent, lease, loan [or grant any third party access to or use of] the Software to any third party.*

Further contents of a license may indicate whether or not it is to be exclusive or nonexclusive. The license can be further limited to a list of products or particular hardware processors for instance. The license grant language may restrict the license to a particular geographical area or to a time period. It is common to put in language as to whether the license is irrevocable or not, and if any further royalty is required. Typically, the royalty will be discussed elsewhere in the contract between the parties in which case the license grant language will call the license "paid up." Thus, you will need to make sure that royalty payments are clearly stated. Vagueness in royalty payment language is the source of much pain in the high-tech industry. If you put in a clause that says that you will pay a dollar per port, be sure to define what a port is. If you are to pay a 1% royalty, be sure to ask, "1% of what?" Gross sales? Sales net of shipping and taxes? If the royalty payment is to be applied to a certain product line, make sure that new products that are added to the product line are covered or that, if you think that they should not be, why not? If you are buying chips and the price includes a royalty payment for a certain codec that is licensed, you may want to include language that reduces the price of the chip if you end up using the chip in a product that will not use the codec. Experience teaches that you cannot get too wordy when defining terms and imagining changing circumstances where royalties are concerned.

## Protection of intellectual property

What is meant by intellectual property? In the integrated circuit world they mean the basic layout or topography of an *integrated circuit*. However, in this chapter we are talking about patents, copyrights, trademarks, and trade secrets, exclusive of integrated circuits "mask works" that are of concern to the embedded engineer at the board level.

## Copyrights and the embedded engineer

As a general rule, computer programming languages are not subject to copyright. By this it is meant that copyright law does not give the copyright holder control over all the ways to implement a function or specification. The Copyright Act confers copyright ownership over the specific way in which the author wrote out his version. Others are free to write their own implementation to accomplish the identical function. Ideas, concept, and functions cannot be monopolized by copyright. This was summed up by Judge Alsup in the ongoing *Google v. Oracle* lawsuit that concerns the attempts by Oracle to get Google to pay for Android.

## *Protection of trade secrets*

People, and that includes engineers, like to think that everything they work on is unique to some extent and thus, subject to protection in the marketplace. Companies who employ people are apt to consider all work-product as secret. So, how does the law address trade secrets in most jurisdictions? Most states have adopted, either outright or a variation, of the model Uniform Trade Secrets Act (UTSA). Accordingly, trade secrets are defined as formula, pattern, compilation, program, device, method, technique, or process. Additionally, the trade secret should, in and of itself, derive actual or potential economic value to the party claiming its secrecy. Furthermore, in order to be considered a secret, it should be, well, a secret. The law defines "secret" as something that is not generally known and has been the subject of reasonable efforts to be maintained as a secret. Lastly, the "secret" should *not* be readily ascertainable by proper means, e.g., read about in trade press by competitors who can obtain economic value from the secret's disclosure or use thereof. This is why it is important to coordinate the activities of the marketing department with the product managers. Often it is to the consternation of the company to attempt to assert a trade secret claim against a competitor only to find out that previous detailed press releases or manuals have disclosed the "secret" to the public.

In my experience, many companies decide that something was a secret after they see their competitor using some aspect of their company work-product. To counter this practice, some engineers and company systems put "proprietary" or "company confidential" on every e-mail and document that they produce. This practice might be better than nothing, but is subject to attack in litigation when some attorney waives an e-mail showing that the announcement for the company picnic was also labeled "proprietary."[4]

A concern is presented when a company *comes into possession* of a trade secret. Generally, for violations to be claimed, the trade secret has to come by way of "improper means." An event that is becoming very common concerns those e-mails that contain a company secret that gets sent outside the company, because somebody hit "reply all." In such a case, it might be successfully argued that, as a listed recipient, it was not acquired by improper means. What did your mother say you should do if you found a wallet on the street? Like your mother, the UTSA states that improper acquisition of a trade secret can be obtained through "accident or mistake." So, one might fairly deduce that an improper acquisition occurred through "accidental" means when your competitor sent their pricing proposal to

---

[4] Absolute secrecy is not required. A limited disclosure for a restricted purpose or disclosures under a nondisclosure agreement will not waive protection. See e.g., *Trandes Corp. v. Guy F. Atkinson Co.*, 996 F.2d 655 (4th Cir. 1993).

you by hitting "reply all." Therefore, your mother would have you return the acquisition and not make use of the contents.

Another frequent event occurs when a company hires a former employee of a competitor. This employee may come with knowledge of secret formulas, customer lists, programs, etc. The fact that the former employee is eager to dispense with his common law duty of loyalty and is now freely dispensing a competitor's trade secrets on a "silver platter," does not entitle his new employer to the secret according to the UTSA.

Since any litigation involving trade secrets will inevitably focus on what the allegedly aggrieved party did to *maintain the secrecy*, it is interesting to see what others have claimed in actual cases where the matter went to court. A statistical study was undertaken by attorneys writing in the Gonzaga Law Review. They reported their findings as to what was claimed by various litigants in State and Federal trade secret cases. As reported by the authors:

> *As detailed in the federal study and elsewhere, a trade secret owner is not entitled to protection unless the owner took reasonable measures to protect its trade secrets. There is no bright-line rule for the number or type of measures necessary to support a finding that such measures are reasonable. For example, in adopting the Economic Espionage Act, Congress stated that "what constitutes reasonable measures in one particular field of knowledge or industry may vary significantly from what is reasonable in another field or industry." (Citations Omitted) Both the state and federal studies seek to provide objective evidence of the measures courts cite most often, and of the measures that are associated with a finding that the trade secret owner took reasonable measures. Specifically, for those cases in which the court decided whether the trade secret owner engaged in efforts that were reasonable to maintain the secrecy of an alleged trade secret, we coded for the types of measures the plaintiff undertook.*

> *State Study Table presents the data from the state study, and table 18, the Federal Study.*

State Study Table—**types of measures used by trade secret owner**
**(1995–2009 Data)**

| | |
|---|---|
| Confidentiality agreements with employees | 11% (39) |
| Confidentiality agreements with third parties | 3% (11) |
| Computer-based protections | 6% (22) |
| Physical-based protections | 8% (28) |
| Education of employees about secrecy | 2% (6) |
| Label confidential documents | 2% (7) |
| Record Keeping | 1% (2) |
| Interviews | 3% (1) |
| Surveillance | 3% (1) |
| Written Policies | 2% (6) |

Federal Study Table—**types of measures used by trade secret owner.**

|  | 1950–2007 | 2008 |
|---|---|---|
| Confidentiality agreements with employees | 9% (24) | 17% (20) |
| Confidentiality agreements with third parties | 6% (17) | 11% (13) |
| Computer-based protections | 4% (12) | 13% (16) |
| Physical-based protections | 7% (18) | 3% (4) |
| Education of employees about secrecy | 2% (5) | 2% (2) |
| Label confidential documents | 2% (6) | 4% (5) |
| Record Keeping | 0% (1) | 0% (0) |
| Interviews | 0% (1) | 0% (0) |
| Surveillance | 0% (1) | 0% (1) |
| Written Policies | 1% (2) | 4% (3) |

*Confidentiality agreements with employees are the reasonable measure that courts cite most often in both federal and state cases. Specifically, courts cited such agreements in both 11% of state court cases and 11% of cases in federal courts.*[5]

Therefore, if you are the company owner or one charged to protect your company's secrets, you can employ any or all of the methods cited in the tables above. Having employees sign confidentiality agreements that specify the company's secrets is the most recognized way to protect secrets in both federal and state courts.

Most courts will follow the type of reasoning that this particular Virginia court did in *MicroStrategy, Inc. v. Li*, 601 S.E.2d 580, 589 (Va. 2004). The appellate court first notes that

> *[T]he determination whether a trade secret exists ordinarily presents a question of fact to be determined by the fact finder from the greater weight of the evidence.*

What this means, and it may come as a surprise to the embedded software engineer, is that although there are categories of items that *can* be considered a trade secret, that does not mean that for any given item, they are automatically to be considered as such. In other words, source code[6] *could* be a trade secret, but only if there has been compliance with the other provisions of the statute.

The court in *MicroStrategy* went on to state:

> *Under the definition of "trade secret" set forth in Code §59.1-336, MicroStrategy was required to prove that the software components at issue: (1) had independent economic*

---

[5] D.S. Almeling, D.W. Snyder, M. Sapoznikow, W.E. McCollum, J. Weader, Trade secret litigation statistics: a statistical analysis of trade secret litigation in state courts, Gonzaga Law Rev. 46 (2010) 57 et seq.

[6] This is how one court defines source code. "Source code" is a document written in computer language, which contains a set of instructions designed to be used in a computer to bring about a certain result. See *Trandes Corp. v. Guy F. Atkinson Co.*, 996 F.2d 655, 662–63 (4th Cir. 1993).

*value from not being generally known and readily ascertainable by proper means by persons who could obtain economic value from their disclosure; and (2) were the subject of reasonable efforts to maintain their secrecy. To prove a "misappropriation" of alleged trade secrets under the Act, based on its theory of the case, MicroStrategy was required to establish two factors: (1) that the defendants disclosed or used trade secrets developed by MicroStrategy without its express or implied consent; and (2) that the defendants knew or had reason to know that their knowledge of the trade secrets was either acquired under circumstances giving rise to a duty to maintain their secrecy, or derived from or through a person who owed such a duty to MicroStrategy.*

Another case which is of interest, though it be unpublished, is from the 4th Circuit Court of Appeals. The case is captioned No. 09-2300 *Decision Insights, Inc. v. Sentia Group, Incorporated*. Let's look at part of the court findings and see what can be learned.

The facts:

*Decision Insights, Inc. (DII) filed a complaint in June 2006 against Sentia Group, Inc. (Sentia) and the four individuals that founded Sentia (collectively, the defendants). DII alleged in its complaint that Sentia's development of a competing software application was based on materials obtained from the defendants' misappropriation of DII's trade secrets. DII also alleged that several of the individual defendants breached contractual and fiduciary obligations owed to DII by disclosing DII's confidential and proprietary information, including the "source code" for DII's software, reports containing marketing and research material, information contained in the user manual for the DII software at issue, and certain information pertaining to DII's clients.*

Already you can start to see how the essential elements of a trade secret case may be established. From the courts recitation of facts, not only is the defendant in possession of the Plaintiff's software, but we can see how the software may have come into the defendant's possession establishing the necessary element of "improper means" as we saw in the *MicroStrategy* case above.

Further in the court's opinion:

*In June 2007, the district court granted the defendants' motion for summary judgment on all DII's claims. In an unpublished opinion issued in February 2009, this Court affirmed the judgment of the district court in part and reversed in part. Decision Insights, Inc. v. Sentia Group, Inc., 311 F. App'x 586 (4th Cir. 2009) (per curiam). We remanded the case with instructions to the district court to consider, among other issues, whether DII's software application, as a total compilation, could qualify as a trade secret under Virginia law. Id. at 593-94. On remand, the district court again granted the defendants' motion for summary judgment, holding that DII failed to develop facts during discovery that would establish that DII's software, as a compilation, is not generally known or ascertainable, and that all DII's claims must be dismissed in light of that holding. Upon review of the district court's judgment on remand, we vacate the district court's judgment, and we remand the case for further proceedings consistent with this opinion.*

Here we can note that the court is focusing on a "compilation" which you will recall is one of the categories of subjects that can be considered a trade secret. This can be important to the embedded engineer. Various files or software modules may be available from any number of sources, including public sources. They may be compiled by someone into a single program. This means that someone's particular *compilation* of the modules can be protected as a trade secret!

The court went on to say:

> *Sentia initially sought to obtain a software license from DII for use of the EU Model. However, negotiations between the parties did not result in an agreement. Sentia later hired Carol Alsharabati, a former consultant for DII who worked extensively on the source code for DII's EU Model, to develop software for a Sentia product that would compete directly with DII's software. Alsharabati completed this task in about six weeks, which, according to DII, could only have been accomplished by using DII's source code to create the Sentia software.*

> *DII filed a complaint in the district court against the defendants, alleging that Abdollahian, Efird, Kugler, and Alsharabati disclosed DII's trade secrets to Sentia in violation of the Virginia Trade Secret Misappropriations Act, Virginia Code §§59.1-336 through -343 (the Act).[4] According to DII, the software developed by Alsharabati for Sentia is almost identical to DII's EU Model, both in terms of method and in the results obtained when the respective programs are executed. DII asserted that Sentia's software could not achieve results equal to DII's software unless all the parameters, variables, and sequencing associated with the programs are equal. Although the EU Model uses certain mathematical formulas that are in the public domain, DII asserted that the combination and implementation of these formulas in DII's source code for the software constitutes a trade secret.*

Well, the above answers the question that might be raised in the lunch room, "How are they ever going to prove that we got their software if a lot of it is online?"

More from the court on compilations:

> *As stated in our prior opinion in this case, we held in* Trandes *that a "plaintiff's alleged software compilation trade secret is to be analyzed separate and apart from other software trade secret claims, and that production of source code is an acceptable method of identifying an alleged compilation trade secret." 311 F. App'x at 594 (citing* Trandes, *966 F.2d at 661-63.).*

So a compilation of code can be its own trade secret separate and apart from other software claims!

Since this is the second time the case had been up on appeal, the court states what happened on the first remand regarding the software compilation and elements of proof:

> *On remand, the district court first held that DII met its burden under <u>Trandes</u> to demonstrate that DII's source code was unique. However, the district court concluded that DII failed to satisfy its burden to show that DII's software, as a compilation, was not generally known or readily ascertainable by proper means. In reaching this conclusion, the district court found that DII "failed to distinguish which aspects of its software, as a compilation, are publicly available or readily ascertainable and which are not." The district court did not address the other relevant criteria under the Act, including whether the compilation had independent economic value, and whether the compilation was subject to reasonable efforts by DII to maintain the secrecy of the information.*

The court is going to take issue with the lower court which found against the Plaintiff and states in part:

> *We have recognized that a trade secret may be composed of publicly-available information if the method by which that information is compiled is not generally known. For instance, in <u>Servo Corp. of America v. General Electric Co.</u>, <u>393 F.2d 551</u>, 554 (4th Cir. 1968), we held that a trade secret "might consist of several discrete elements, any one of which could have been discovered by study of material available to the public."*

The court then relates, in part here, the type of proof that had been supplied at trial, some through experts:

> *DII also presented the report and deposition testimony of Gary Slack, a current DII employee and co-author of DII's software program. Mr. Slack stated in his report and during his deposition testimony that many aspects of the source code, and hence the compilation of the source code as a whole, were not public knowledge or readily ascertainable by proper means. In his report, Mr. Slack identified and described 13 proprietary processes in the source code, and stated that "[t]he collection of these processes as a whole and the sequence of these processes also serve as a proprietary aspect of [DII's EU Model software]." Mr. Slack's expert report also included a "flow chart," which set forth the sequencing of DII's source code, including its organization and structure. Mr. Slack testified that portions of this process, as well as the entire sequencing of the process, were not known to the public.*

> *During his deposition testimony, Mr. Slack also identified numerous variables that are part of DII's software code for the EU Model, noting that the set of these variables had "never been disclosed to anyone else." Mr. Slack further testified that while a few of these individual variables were in the public domain, numerous other of these variables were not in the public literature or known outside of DII. Additionally, Mr. Slack testified that "[n]one of the code has ever been shared with anybody that has not signed a confidential[ity] agreement."*

You can see why it is so important to have nondisclosure agreements (NDAs) with third parties! At any rate, the court found that the lower court had erred:

> *In light of the foregoing evidence offered by Dr. Bueno de Mesquita and Mr. Slack concerning the nonpublic and proprietary nature of DII's software code, we conclude that the district court erred in holding that DII failed to satisfy its evidentiary burden to show that DII's software compilation was not generally known or readily ascertainable by proper means. Therefore, we conclude that DII adduced sufficient evidence during discovery to render this issue appropriate for decision at a trial.*

Importantly, the court further noted:

> *We note that separate consideration of DII's various breach of contract claims will be necessary regardless of the outcome on remand of DII's trade secret claims concerning its software compilation. Those breach of contract claims do not require a finding that the materials at issue qualify as a trade secret, because the respective nondisclosure clauses apply to "any confidential or proprietary information... owned or used by" DII.[7] This contractual language is broader than the definition of a "trade secret" under the Act and, thus, the nondisclosure language may apply to the software code and other proprietary materials at issue even if those materials are not covered by the Act. Therefore, we direct the district court to consider the language of the respective nondisclosure clauses when the court analyzes DII's breach of contract claims.*

This last statement demonstrates that even if one may not be able to maintain a suit for trade secret misappropriation under a state statute, you can still claim that the subject matter, as listed in your NDAs, is a secret and recover under a simple breach of contract claim!

## Trademarks

A trademark is a phrase or word that is not merely descriptive, through use or intent to use in interstate commerce, has become associated with a particular product or company. Suppose you design a fast DSP analog-to-digital converter board and you decide to call it "DSP digitizer." You then try to register it with the United States Patent and Trademark Office (USPTO). Such a scenario is going to pose several problems. For starters, the Trademark Office is going to look at the product and most likely reject the requested registration, because they are going to see it as "merely descriptive." A better word might be to call it "The Chopper." Since chopping may not normally be associated with an A/D DSP boards and it has a marketable ring, it most likely would be approved. Conversely, if you were selling axes, it would be subject to rejection by the Office. By the way, *initial* registration does not mean that the US government says that you have exclusive rights to that mark or that it is conclusively determined that you have a superior right to the mark. It does give notice to the world when you started to use the mark and give you a right to sue.

The Lanham Act,[7] §33, provides that federal trademark registration is *prima facie* evidence of the validity of the registered mark, the registrant's ownership of the mark, and the registrant's exclusive right to use the registered mark in connection with the goods and services specified in the registration. The foregoing applies to initial registration. The Lanham Act further provides that a trademark registration can become "incontestable" five years after the date of registration. A registration that is incontestable is immune to challenge on certain grounds, including that the registered mark is merely descriptive.[8]

You also will have to show the intent to use the mark in commerce. You will need marketing brochures or data sheets to demonstrate this.

The process for filing a trademark with the USPTO consists of the following:

1. The filing of an application in the USPTO.
2. The examination of the application by the USPTO and a search of their records to determine if there are any conflicts with preexisting marks. There may be correspondence and arguments made by the filer or their attorney with the USPTO regarding the "likelihood of confusion."
3. If the application is accepted and no conflicts are found, the application for the mark is published in the USPTO's Official Gazette. This commences a 30-day period during which third parties can file opposition proceedings against the application.
4. After 30 days, if no opposition is filed, if the mark is in use, the USPTO will issue the trademark registration certificate.

The symbols ®, ™, and ^SM provide notice to the world that you are claiming trademark rights in any mark using these symbols. You may use the ™ on marks identifying goods, and the ^SM on marks identifying services. You do not have to have a federal or state registration to use the ™ or ^SM symbols. However, the ® symbol, that provides "statutory notice" can only be used if your mark is federally registered by the USPTO.

### The use and misuse of trademarks

The law generally looks at two main aspects of trademarks where issues of enforcement and infringement are involved. The first is to ask if there is a "likelihood of confusion" in the marketplace. Two companies can have similar trademarks and if one is selling fish and the

---

[7] The Lanham Act, 15 U.S.C. §1051, is the federal statute that provides for civil suit remedies for trademark infringement, false advertising and trademark dilution.

[8] There are state trademark registration procedures too. The federal system has preeminence over the state system. However, if you have a local competitor who you are particularly concerned with, e.g., your ex partner, you may want to register in your local state, especially if you need to file quickly to establish an earlier filing date for the mark.

other HD television, the court is going to be inclined to say that there is no way someone is going to confuse the two.

The second aspect concerns a trademark holder's duty and right to make sure that his trademark is not being used in such a way as to "hypothecate" the mark. Hypothecation is the watering down of the mark, perhaps by allowing licensees of the mark to use the mark without the appropriate indication of trademark status or attribution in favor of the trademark holder.

It is important to protect marks from finding their way into the public domain. This can happen in several ways. One is to fail to alert the world to your use of the mark, such as neglecting to place the ™ symbol next to the word or phrase. Another is to allow the use of the word in some kind of generic fashion. At the time of this writing, Google is wincing and challenging the use of its trademark as a verb. "Why don't you Google that?" is the type of scenario and phrasing that they want to avoid.

Presently, one fertile ground of trademark litigation concerns the way that Internet search engines are being used. The search engine companies have been earning large sums of money by selling "key words" to companies. These key words can be purchased in such a way that the word will either come up higher in the search return ranking or bring up a paid advertisement that is prominently displayed among the fist returns of the search. Likewise, many companies have been burying their competitor's trademarks or product identifiers, e.g., model numbers among the Metatags of the HTML code on their company web pages. The result of all this is that if consumer A goes looking for the "Model X7" from company B, the internet search will tend to steer him to company C if company C has purchased the words "Model X7" or has put "Model X7" as a Metatag in the C company website. The question for the courts is to determine if such actions and activities constitute trademark infringement or an unfair trade practice under the Lanham Act or other statutes.

Even before one applies the "likelihood of confusion" test, the threshold question is if the alleged offender has "used in commerce" the trademark. This was the issue in the *Rescuecom v. Google* case.[9] Here, the Second Circuit Court of Appeals decided that Google's AdWords and the pay-per-click scheme *did* constitute a "use in commerce." This is a significant decision, because the court noted that Google was an intermediary and yet, had met "display," "offer," and "sale" aspects that would satisfy the use of another's trademark requirement, and would implicate the relief provided by the Lanham Act.

Another important and recent decision is the *Rosetta Stone v. Google* case.[10] This case overruled a lower federal court decision and revived the case. Unlike the Rescuecom case, the court assumed that "use in commerce" requirement had been satisfied and first focused

---

[9] *Rescuecom Corporation v. Google, Inc.*, 562 F.3d 123 (2nd Cir. 2009).

[10] *Rosetta Stone, Ltd. v. Google, Inc.* 676 F.3d. 144 (4th Cir. 2012).

on Google's intent by finding that the AdWords program was very likely to result in market confusion. Rosetta Stone further had produced surveys and testimony that actual market confusion had taken place. The court went on to rule that just because someone can navigate the internet, that does not conclusively mean that the consumers are of such sophistication that they cannot be confused by the AdWords practice.

This area of the law is still in flux and does not indicate that one should sell their stock in Google anytime soon. The courts are still analyzing the AdWords practice on a case-by-case basis. Although I think it is safe to say that the majority of jurisdictions have come to the point where they acknowledge that third-party implementers, like a website, can be held liable and that their particular practice can constitute a "use in commerce." The courts are also coming to the conclusion that just because the internet is in use, this does not mean that consumers are so sophisticated as to not be misled by the various practices. A company should obtain legal counsel prior to engaging in an online Metatag or key word purchase.

An example of a trademark battle that follows traditional lines of dispute would be East Carolina University's suit against Cisco.[11]

In January 2013, East Carolina University filed a suit against Cisco for trademark infringement regarding the use of the phrase "Tomorrow Starts Here." Here is a one relevant paragraph of the suit:

> *This is an action for: (i) trademark infringement and false designation of origin under the Trademark Act of 1946, also known as the Lanham Act, codified at 15 U.S.C. §1051 et seq.; (ii) common law trademark infringement and unfair competition; and (iii) unfair or deceptive trade practices pursuant to N.C. Gen. Stat. §75-1.1 et seq.*

Note how the suit includes claims for statutory rights being violated pursuant to the Lanham Act, Common Law trademark infringement and unfair competition, and a claim for unfair or deceptive trade practices under a North Carolina statute. The case typifies the type of claims that one could expect in a trademark case no matter where you reside.

It is also interesting to note that the drafters of the suit seem to have anticipated an argument that Cisco will no doubt raise. As stated previously, the salient issue in a trademark case is whether there is the likelihood of confusion in the marketplace. So, Cisco can be expected to argue that their product line, regarding routers and gateways, has nothing to do with educational services and that nobody is going to be confused in dealing with Cisco thinking that they are dealing with East Carolina University. Therefore, in apparent anticipation of this argument, the suit contains this paragraph:

> *Plaintiff has also acquired extensive common law trademark rights in the Mark. By way of example, Plaintiff has utilized "TOMORROW STARTS HERE" in association with*

---

[11] Case # 4:13-cv-0003-FL, US District Court for the Eastern District, North Carolina.

*Plaintiff's College of Technology and Computer Science, research, software development, intellectual property and licensing, in addition to commercialization of technology and software long prior to Defendant's adoption of the identical mark for overlapping goods and services.*

Therefore, East Carolina has tried to link the trademarked term with specific technology, such as computers, that would tend to define a market in common with Cisco, so as to negate Cisco's defense on the "confusion" issue.

## Patents

Abraham Lincoln stated in his Second Lecture on Discoveries and Inventions:

*Next came the Patent Laws. These began in England in 1624; and, in this country, with the adoption of our constitution. Before then, any man might instantly use what another had invented; so that the inventor had no special advantage from his own invention. The patent system changed this; secured to the inventor, for a limited time, the exclusive use of his invention; and thereby added the fuel of interest to the fire of genius, in the discovery and production of new and useful things.*

Some like to philosophize that patents are evil and, with the way things are working now, restrict commerce. But since a patent system is required in Article I section 8 of the US Constitution, it is fair to say that patents will be with us for some time, even if congress or judicial rulings temper the situation in the future.

Patent attorney Gene Quinn in his blog *at IP Watchdog* sets out the reason for obtaining a patent on software and not relying upon copyright alone:

*That there are an infinite number of ways to translate a desired set of functionalities into software is why you want to obtain a patent on the software, rather than rely on copyright protection. With a patent you will be able to lock in exclusive rights with respect to any possible way the software is coded to accomplish what is recited in you patent claims. This make rights obtained via a patent very different and far broader and stronger than the rights one can obtain in software from a copyright. With a copyright you only get protection of the specific code, so to receive the same protection via copyright alone as is offered in a patent you would have to write the software in those infinite number of way (sic) and copyright all of them. For this reason patent claims should focus on the core process steps and not discuss the software code.*

### What is patentable and patent litigation

To begin with, in order to patent anything, an invention must:

1. be statutory,
2. be new,

3. be useful, and
4. be nonobvious.

- Be statutory means that data structures or programs per se (these are considered "functional descriptive material," that impart functionality when employed as a computer component, but are mere descriptive material when claims standing alone). These items may be patentable when claimed in a different form to include computer-readable medium. Electromagnetic signals, which are considered forms of energy and as such are nonstatutory natural phenomena (which is why computer programs are patentable when embodied in something physical, such as a computer-readable medium, but computer programs are not patentable when embodied in a computer-readable signal stream). These items are considered indistinguishable from abstract ideas and laws of nature, and therefore are unpatentable.
- Be new, or novel, means that the invention has not been known or used by others in the United States, or patented or described in a printed publication in the United States or a foreign country, before the current applicant filed for his patent.
- Be useful means that the invention serves a useful purpose. Being useful also includes operational effectiveness, meaning that an invention must operate or perform its intended purpose.
- Be nonobvious means that the invention should not have been obvious to a person having ordinary skill in the area of technology related to the invention.

At the time of this writing, a key case that is in progress is *CLS Bank v. Alice*, that has at issue, the following language of the subject patent:

> *A method of exchanging obligations as between parties, each party holding a credit record and a debit record with an exchange institution, the credit records and debit records for exchange of predetermined obligations, the method comprising the steps of:*
>
> *(a)  creating a shadow credit record and a shadow debit record for each stakeholder party to be held independently by a supervisory institution from the exchange institutions;*
>
> *(b)  obtaining from each exchange institution a start-of-day balance for each shadow credit record and shadow debit record;*
>
> *(c)  for every transaction resulting in an exchange obligation, the supervisory institution adjusting each respective party's shadow credit record or shadow debit record, allowing only these transactions that do not result in the value of the shadow debit record being less than the value of the shadow credit record at any time, each said adjustment taking place in chronological order; and*
>
> *(d)  at the end-of-day, the supervisory institution instructing ones of the exchange institutions to exchange credits or debits to the credit record and debit record of the respective parties in accordance with the adjustments of the said permitted transactions, the credits and debits being irrevocable, time invariant obligations placed on the exchange institutions.*

The issue in the case is whether, according to section 101 of the Patent Act, such an "invention" contains any core inventive concept. Further at issue, is what test or procedures should a court use in determining patentability of a software-type patent. At any rate, and unlike Europe, software patents are still alive and well in the United States. Of course it is important to note that patents are not labeled as a "software patent," either in their application or issuance. So, as a developer or employee of an entity, it is desirable to have a general understanding of what is patentable and how litigation can ensue.

The law has been rather tumultuous in recent years regarding what can be patented. You may find it interesting how the US Patent Office goes about determining if an invention meets the statutory guidelines. In doing so, it has issued to its examiners its own guidelines. Having a familiarity with the Patent Office's guidelines can assist the engineer in drafting preliminary patent documents which can save money at the attorney's office. Figure 20.1 outlines the guideline steps for the eligibility test for process claims.

As of 2009, the Patent Office was using these guidelines for determining patentability:

*AUGUST 2009: Interim Instructions*

*INTERIM EXAMINATION INSTRUCTIONS FOR*

*EVALUATING SUBJECT MATTER ELIGIBILITY UNDER 35 U.S.C. §101*

*I. OVERVIEW*

*The state of the law with respect to subject matter eligibility is in flux. The following interim instructions are for examination guidance pending a final decision from the Supreme Court in Bilski v. Kappos. These examination instructions do not constitute substantive rulemaking and hence do not have the force and effect of law. Rejections will be based upon the substantive law, and it is these rejections that are appealable. Consequently, any perceived failure by Office personnel to follow these instructions is neither appealable nor petitionable.*

*35 U.S.C. §101 establishes the threshold for patentability by setting requirements for subject matter that is eligible for patenting. To pass the threshold eligibility inquiries of §101 for patent protection, a claimed invention must be directed to statutory subject matter and must be useful.*

*§101 also provides the basis for the prohibition against double patenting.) Thus, under §101 two separate patent eligibility considerations are raised: (1) subject matter and (2) utility. To evaluate utility or the real world use of an invention, follow the detailed "utility" guidelines in MPEP 2107. Following the utility guidelines, the claims and supporting disclosure must be reviewed to evaluate whether the claimed invention has an asserted or well-established utility that is specific, substantial and credible. The usefulness of the invention must be commensurate with the broadest reasonable interpretation of the claimed invention in light of the specification as it would be interpreted by one of ordinary skill in the art.*

(A)

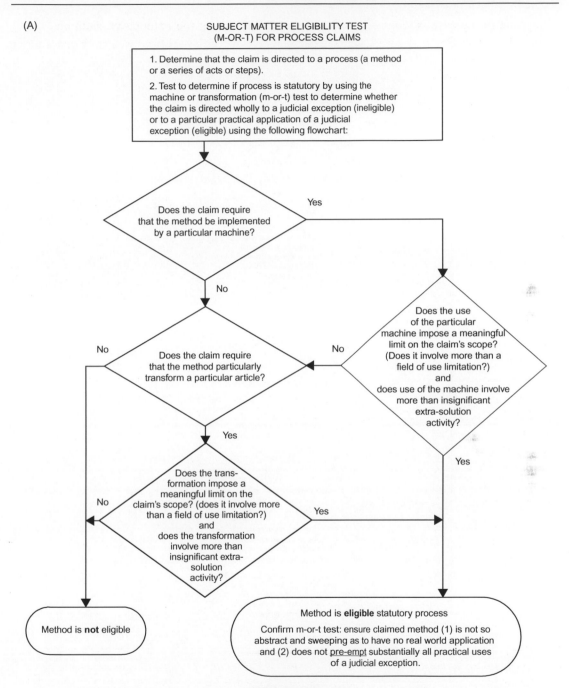

**Figure 20.1:**
(A) Flowchart for process-related claims. (B) Flowchart for non-process-related claims.

(B)                              SUBJECT MATTER ELIGIBILITY TEST

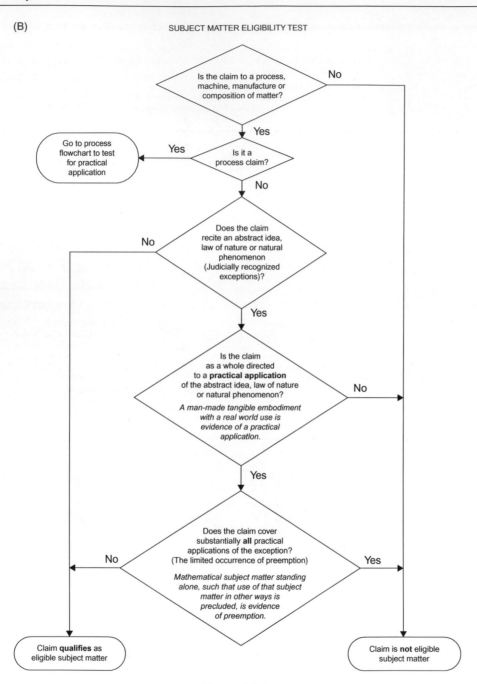

**Figure 20.1:**
*(Continued)*

*This document includes instructions for evaluating subject matter eligibility and, for the interim, should be used for examining claims under 35 U.S.C. §101 for subject matter eligibility. Since the subject matter eligibility guidelines set forth in MPEP 2106 were prepared, case law has developed that has necessitated revision to the guidelines. The following instructions supersede previous guidance on subject matter eligibility that conflicts with the Instructions, including MPEP 2106(IV), 2106.01 and 2106.02.*

*Subject Matter Eligibility: There are two criteria for determining subject matter eligibility and both must be satisfied. The claimed invention (1) must be directed to one of the four statutory categories, and (2) must not be wholly directed to subject matter encompassing a judicially recognized exception, as defined below. The following two step analysis is used to evaluate these criteria...*

As you may note, these guidelines were interim because they were issued in contemplation that a key Supreme Court case was going to address patentability issues associated with "software patents," i.e., *Bilski v. Kappos*, 561 U.S. ___ (2010). The issue in the *Bilski* case was whether a computer program that helped users hedge risks in energy markets produced anything that was useful, or concrete, or tangible within the meaning of the patent laws. A lower court had said no and affirmed the original patent examiner's rejection who had said that the claimed invention was not implemented on a specific apparatus, merely manipulated an abstract idea, and solved a purely mathematical problem. Before arriving at the Supreme Court, another court had ruled that a claimed *process* is patent eligible if: (i) it is tied to a particular machine or apparatus, or (ii) it transforms a particular article into a different state or thing. Concluding that this "machine-or-transformation test" is the *sole* test for determining patent eligibility of a "process" under the patent laws, that lower court applied the test, and held that the application was not patent eligible.

The Supreme Court upheld the lower court. However, the Supreme Court rejected that this "machine-or-transformation-test" was to be the *sole* test for patentability. Regarding the use of the word "process," the court stated that this word should be given its "ordinary meaning." The Court further stated, "that it is unaware of any ordinary, contemporary common meaning of *process* that would require it to be tied to a machine or the transformation of an article."

Thus, the Patent Office issued new guidelines that sought to address the "too rigid" interpretation that was present in the above-mentioned 2009 guidelines. In Figure 20.2 the new guidelines state, in part:

Note that the new guidelines still makes mention of the 2009 guidelines and also reflects the Supreme Court guidance of the *Bilski* decision with phrases, such as "but not limited" or "as a whole."

A hundred and fifty years later, Lincoln's "fire of genius" stoked the fires of litigation. Commentators estimate that about 50% of all patent litigation is concerning software

## 101 Method Eligibility Quick Reference Sheet

The factors below should be considered when analyzing the claim **as a whole** to evaluate whether a method claim is directed to an abstract idea. However, not every factor will be relevant to every claim and, as such, need not be considered in every analysis. When it is determined that the claim is patent-eligible, the analysis may be concluded. In those instances where patent-eligibility cannot easily be identified, every relevant factor should be carefully weighed before making a conclusion. Additionally, no factor is conclusive by itself, and the weight accorded each factor will vary based upon the facts of the application. These factors are not intended to be exclusive or exhaustive as there may be more pertinent factors depending on the particular technology of the claim. For assistance in applying these factors, please consult the accompanying "Interim Guidance" memo and TC management.

### Factors Weighing Toward Eligibility:
- Recitation of a machine or transformation (either express or inherent).
  - Machine or transformation is particular.
  - Machine or transformation meaningfully limits the execution of the steps.
  - Machine implements the claimed steps.
  - The article being transformed is particular.
  - The article undergoes a change in state or thing (e.g., objectively different function or use).
  - The article being transformed is an object or substance.
- The claim is directed toward applying a law of nature.
  - Law of nature is practically applied.
  - The application of the law of nature meaningfully limits the execution of the steps.
- The claim is more than a mere statement of a concept.
  - The claim describes a particular solution to a problem to be solved.
  - The claim implements a concept in some tangible way.
  - The performance of the steps is observable and verifiable.

### Factors Weighing Against Eligibility:
- **No recitation of a machine or transformation (either express or inherent).**
- Insufficient recitation of a machine or transformation.
  - Involvement of machine, or transformation, with the steps is merely nominally, insignificantly, or tangentially related to the performance of the steps, e.g., data gathering, or merely recites a field in which the method is intended to be applied.
  - Machine is generically recited such that it covers any machine capable of performing the claimed step(s).
  - Machine is merely an object on which the method operates.
  - Transformation involves only a change in position or location of article.
  - "Article" is merely a general concept (see notes below).
- The claim is not directed to an application of a law of nature.
  - The claim would monopolize a natural force or patent a scientific fact; e.g., by claiming every mode of producing an effect of that law of nature.
  - Law of nature is applied in a merely subjective determination.
  - Law of nature is merely nominally, insignificantly, or tangentially related to the performance of the steps.
- The claim is a mere statement of a general concept (see notes below for examples).
  - Use of the concept, as expressed in the method, would effectively grant a monopoly over the concept.
  - Both known and unknown uses of the concept are covered, and can be performed through any existing or future-devised machinery, or even without any apparatus.
  - The claim only states a problem to be solved.
  - The general concept is disembodied.
  - The mechanism(s) by which the steps are implemented is subjective or imperceptible.

**Figure 20.2:**
US Patent Office guidelines for method eligibility, the quick reference sheet.

<u>NOTES:</u>

**1) Examples of general concepts include, <u>but are not limited,</u> to:**
   - Basic economic practices or theories (e.g., hedging, insurance, financial transactions, marketing);
   - Basic legal theories (e.g., contracts, dispute resolution, rules of law);
   - Mathematical concepts (e.g., algorithms, spatial relationships, geometry);
   - Mental activity (e.g., forming a judgment, observation, evaluation, or opinion);
   - Interpersonal interactions or relationships (e.g., conversing, dating);
   - Teaching concepts (e.g., memorization, repetition);
   - Human behavior (e.g., exercising, wearing clothing, following rules or instructions);
   - Instructing "how business should be conducted."

**2)      For a detailed explanation of the terms machine, transformation, article, particular, extrasolution activity, and field-of-use, please refer to the Interim Patent Subject Matter Eligibility Examination Instructions of August 24, 2009.**

**3)**      When making a subject matter eligibility determination, the relevant factors should be weighed with respect to the claim **as a whole** to evaluate whether the claim is patent-eligible or whether the abstract idea exception renders the claim ineligible. When it is determined that the claim is patent-eligible, the analysis may be concluded. In those instances where patent-eligibility cannot be easily identified, every relevant factor should be carefully weighed before making a conclusion. Not every factor will be relevant to every claim. While no factor is conclusive by itself, the weight accorded each factor will vary based upon the facts of the application. These factors are not intended to be exclusive or exhaustive as there may be more pertinent factors depending on the particular technology of the claim.

**4)   Sample Form Paragraphs:**

**a.**      Based upon consideration of all of the relevant factors with respect to the claim <u>as a whole</u>, claim(s) [1] held to claim an abstract idea, and is therefore rejected as ineligible subject matter under 35 U.S.C. § 101. The rationale for this finding is explained below: [2]

>    1. In bracket 2, identify the decisive factors weighing against patent-eligibility, and explain the manner in which these factors support a conclusion of ineligibility. The explanation needs to be sufficient to establish a *prima facie* case of ineligibility under 35 U.S.C. § 101.

**b.**      Dependent claim(s) [1] when analyzed as a whole are held to be ineligible subject matter and are rejected under 35 U.S.C. § 101 because the additional recited limitation(s) fail(s) to establish that the claim is not directed to an abstract idea, as detailed below: [2]

>    1. In bracket 2, provide an explanation as to why the claim is directed to an abstract idea; for instance, that the additional limitations are no more than a field of use or merely involve insignificant extrasolution activity; e.g., data gathering. The explanation needs to be sufficient to establish a *prima facie* case of ineligibility under 35 U.S.C. § 101.

**Figure 20.2:**
*(Continued)*

patents. There has been exponential growth in the issuance of "software patents" and associated litigation. See the Chart from James Bessen in Figure 20.3.

## Beware the troll

It is important to know thy enemy. Two classes of enemies are likely. The first is the nonpracticing entity (NPE), the so-called "Patent Troll." These outfits are becoming more and more sophisticated and, if large enough, will operate as follows. They will look for

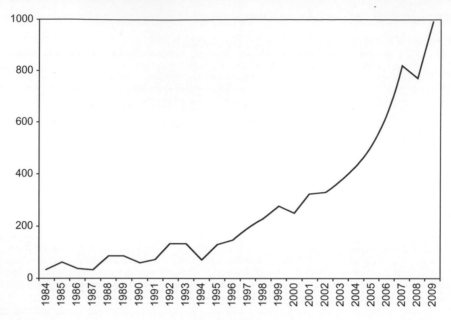

**Figure 20.3:**
Number of patent lawsuit filings involving software patents. *J. Bessen, Software Patents: A Generation of Software Patents, Boston University School of Law, Working Paper No. 11-31, June 21, 2011, p. 19.*

elements of any portion of a market that appears to have been successful. They will acquire patents from individuals or companies that perhaps have never even been associated with the targeted market. Then the "troll" will form, for instance, a Delaware corporation which will essentially have no assets except for the requisite patent. Next, they will file a patent infringement suit in some of the perceived plaintiff-friendly jurisdictions, like Delaware or the Eastern District of Texas. The troll might sue 10 or 20 of the industry leaders in the targeted market with the goal of settling with some of the smaller companies.[12] Then those funds are used to pay the lion's share of the ongoing litigation with a much larger industry leader. It is difficult to defend against the NPE, because they have no skin in the game except their costs. However, here are a few suggestions:

A.  Your company can purchase expensive patent infringement insurance which is a special rider to the General Commercial Liability Policy.
B.  You can try to avoid overstating your market presence in advertising. When you claim to be the "market leader," it should be worth the revenue that such a claim brings when compared to the sharks such blood in the water may attract.

---

[12]  However, there are new limits going into effect regarding this practice under the American Invents Act. Now, everybody being sued should be selling or making the same product.

C.   You can petition the US Patent Office for reexamination of the patent.[13] Some courts will allow a stay of the proceedings while such an examination is taking place. The disadvantage of such a procedure is that should the Patent Office uphold the patent, then you are denied the right to further contest the validity of the subject patent in the infringement proceedings.[14] This presents a possible opportunity to get some possible relief from codefendants. A smaller entity may threaten to reexamine and this strategy may conflict with what the larger defendants may desire, especially when a likely result may be a settlement with a license to the patent-in-suit. Therefore, the large entity may agree to indemnify or at least pay the litigation costs of the smaller entity in exchange for not derailing a litigation strategy.

D.   A smaller company can appeal to the fact that the damage-base of the suit is minimal.[15] If the patent-in-suit is only a US patent, and if your company has mostly foreign sales, then the damage-base may not be very large, even for a "market leader."

## Beware the non-troll

The second enemy is a real practicing entity that might be a competitor. In such a case, it helps to have your own patent portfolio in your back pocket so that you can counter-sue. These are called "defensive patents" that are great to have when a competitor comes threatening. This is why wise companies are zealous to have their engineers patent even seemingly trivial inventions. Suppose your back pocket is empty? You can go buy a patent somewhere and use it to counterclaim against the competitor. It is important for any company that acquires a patent that may have application to its product line to make sure to mark the products with notice of an issued or pending patent. This is to avoid the allegation that any patent that may have been applied to the product has been waived or relinquished.

Turning our attention to patents generally, there have been some very important changes to patent law.

## The America Invents Act

The America Invents Act has many significant provisions which had gone into effect in March 2013. One of the most significant changes is that the United States will join Europe and others and become a "first-to-file" jurisdiction. For centuries, the United States had a "first-to-invent" scheme. No surprise here to assume that lawyers will benefit from this change as there will now be a race to the patent office to file. This may further result in

---

[13]   Note the new procedures and processes, *supra*, as they are contained in the America Invents Act.

[14]   It is important to understand that just because a patent has issued, that does not mean that it cannot be attacked in the infringement suit on several grounds, including obviousness and the existence of prior art.

[15]   Generally, a plaintiff can go back six years in calculating the damage-base that is the amount of gross sales of infringing product multiplied by a percentage royalty, e.g., 2%, multiplied by 6.

piece-meal approaches, where multiple applications are filed, even though more work needs to be done on the invention. If a small company cannot afford to pursue this type of approach, then keeping inventions secret longer and deeming the invention a trade secret will be the order of the day.

A very important change in US patent law as contained in the America Invents Act is the "prior use" defense clause as codified in 35 U.S.C.273. The defense is only available for patents that have issued on or after September 16, 2011.

This is so significant; it is set out in its entirety here:

(a)   **In General.**— *A person shall be entitled to a defense under section 282 (b) with respect to subject matter consisting of a process, or consisting of a machine, manufacture, or composition of matter used in a manufacturing or other commercial process, that would otherwise infringe a claimed invention being asserted against the person if—*

   (1)   *such person, acting in good faith, commercially used the subject matter in the United States, either in connection with an internal commercial use or an actual arm's length sale or other arm's length commercial transfer of a useful end result of such commercial use; and*

   (2)   *such commercial use occurred at least 1 year before the earlier of either—*

      (A)   *the effective filing date of the claimed invention; or*

      (B)   *the date on which the claimed invention was disclosed to the public in a manner that qualified for the exception from prior art under section 102 (b).*

(b)   **Burden of Proof.**— *A person asserting a defense under this section shall have the burden of establishing the defense by clear and convincing evidence.*

   *What this means is that if you were using the claimed invention in business for a year prior to the patent filing, you will have a defense. This provision grants "prior user rights" as a personal defense to patent infringement. It is a personal and non-transferable defense.*

(c)   **Additional Commercial Uses.**—

   (1)   **Premarketing regulatory review.**— *Subject matter for which commercial marketing or use is subject to a premarketing regulatory review period during which the safety or efficacy of the subject matter is established, including any period specified in section 156 (g), shall be deemed to be commercially used for purposes of subsection (a)(1) during such regulatory review period.*

   (2)   **Nonprofit laboratory use.**— *A use of subject matter by a nonprofit research laboratory or other nonprofit entity, such as a university or hospital, for which the public is the intended beneficiary, shall be deemed to be a commercial use for purposes of subsection (a)(1), except that a defense under this section may be asserted pursuant to this paragraph only for continued and noncommercial use by and in the laboratory or other nonprofit entity.*

*This means that Universities do not get to use this defense if they are in business.*

(d) **Exhaustion of Rights.**— *Notwithstanding subsection (e)(1), the sale or other disposition of a useful end result by a person entitled to assert a defense under this section in connection with a patent with respect to that useful end result shall exhaust the patent owner's rights under the patent to the extent that such rights would have been exhausted had such sale or other disposition been made by the patent owner.*

(e) **Limitations and Exceptions.**—

    (1) **Personal defense.**—

        (A) **In general.**— *A defense under this section may be asserted only by the person who performed or directed the performance of the commercial use described in subsection (a), or by an entity that controls, is controlled by, or is under common control with such person.*

        (B) **Transfer of right.**— *Except for any transfer to the patent owner, the right to assert a defense under this section shall not be licensed or assigned or transferred to another person except as an ancillary and subordinate part of a good-faith assignment or transfer for other reasons of the entire enterprise or line of business to which the defense relates.*

        *This will prevent someone from going out and buying a business that has been using the invention as a way of gaining the defense*

        (C) **Restriction on sites.**— *A defense under this section, when acquired by a person as part of an assignment or transfer described in subparagraph (B), may only be asserted for uses at sites where the subject matter that would otherwise infringe a claimed invention is in use before the later of the effective filing date of the claimed invention or the date of the assignment or transfer of such enterprise or line of business.*

    (2) **Derivation.**— *A person may not assert a defense under this section if the subject matter on which the defense is based was derived from the patentee or persons in privity with the patentee.*

    *This is designed to prevent someone from stealing the invention, putting it to use and later claiming the use of the defense.*

    (3) **Not a general license.**— *The defense asserted by a person under this section is not a general license under all claims of the patent at issue, but extends only to the specific subject matter for which it has been established that a commercial use that qualifies under this section occurred, except that the defense shall also extend to variations in the quantity or volume of use of the claimed subject matter, and to improvements in the claimed subject matter that do not infringe additional specifically claimed subject matter of the patent.*

    (4) **Abandonment of use.**— *A person who has abandoned commercial use (that qualifies under this section) of subject matter may not rely on activities*

> *performed before the date of such abandonment in establishing a defense under this section with respect to actions taken on or after the date of such abandonment.*
>
> (5)  *University exception.—*
>
>    (A)  **In general.—** *A person commercially using subject matter to which subsection (a) applies may not assert a defense under this section if the claimed invention with respect to which the defense is asserted was, at the time the invention was made, owned or subject to an obligation of assignment to either an institution of higher education (as defined in section 101(a) of the Higher Education Act of 1965 (20 U.S.C.1001 (a)), [1] or a technology transfer organization whose primary purpose is to facilitate the commercialization of technologies developed by one or more such institutions of higher education.*
>
>    (B)  **Exception.—** *Subparagraph (A) shall not apply if any of the activities required to reduce to practice the subject matter of the claimed invention could not have been undertaken using funds provided by the Federal Government.*
>
> (f)  **Unreasonable Assertion of Defense.—** *If the defense under this section is pleaded by a person who is found to infringe the patent and who subsequently fails to demonstrate a reasonable basis for asserting the defense, the court shall find the case exceptional for the purpose of awarding attorney fees under section 285.*
>
> (g)  **Invalidity.—** *A patent shall not be deemed to be invalid under section 102 or 103 solely because a defense is raised or established under this section.*
>
> *The Act also expands the scope of the defense to all patents as previous law only allowed prior user rights to patents covering computer business methods. However, unlike previous law, the use of the claimed invention will not be allowed to invalidate the patent generally.*

There is a challenge here in complying with the "prior use" defense of section 273. When trying to establish the commercial use of the technology as a type of trade secret, you risk establishing the patent holder's infringement case. One's evidence may not be clear and convincing for the defense, and yet, it may satisfy the preponderance-of-evidence standard for infringement.

The practical effect of this is that a company would be well advised to carefully document its use of technology. As we have mentioned earlier, the America Invents Act brought the United States in line with the practice of most foreign countries patent systems by initiating a "first-to-file" system. To some laypersons and public commentators, there are visions of a race to the Patent Office as petitioners file for patents on inventions that they heard about at a trade show or that they stole at lunch. To be fair, this could happen, but not without some perjury involved. Even under the "first-to-file" system, a petitioner still has to be the actual inventor and so state under oath. The system could more properly be called a "first inventor to file" system.

But, what if the layperson and public commentators are right? The new Act provides for a new procedure called a "derivation" proceeding that allows for an aggrieved party, i.e., the guy who thinks his invention was ripped off, to challenge an issued patent. To do so, the actual inventor would file a petition within one year of the first publication of the invention by the USPTO or the World Intellectual Property Organization. The petition must follow the statutory guidelines that include:

A.   A request for relief.
B.   A demonstration that the application or patent was derived from the inventor who is named in the petition. Thus, there is a requirement that the petitioner state how the offending party came into possession of the inventor's information.
C.   A statement that the petitioner did not authorize the filing of the application or patent.
D.   A presentation why the claimed patent or application is the same or substantially the same as the subject invention.

If these criteria are met, the Director may institute a derivation proceeding. This determination is non-appealable.

As a practical matter, you can see why the use of NDAs, document tracking, electronic access controls, dating of documents, designation of authorized persons, etc. becomes more essential under the new first-to-file system.

Somewhat akin to the derivative proceeding is the new *inter partes* and *post grant* procedures. The *post grant* procedure only applies to patents that were filed after March 16, 2013. The *inter partes* procedure can apply to any issued patent, subject to statutory time frames.

At any rate, for those of you who are constantly amazed at what gets patented and are saying, "everyone already knew that," or "we did the same thing eight years ago and published an article on it," here is your chance to put a stop to an issued patent. To do so, a petition would be filed that would be based upon the acceptable statutory grounds, such as obviousness, the existence of prior art in the form of other patents, or printed publications. The legal standard to be applied to the petition is the "preponderance of the evidence" test. This test is less onerous for the petitioner than the standard that would be applied in a civil action that sought to overturn a patent—the "clear and convincing evidence" test. The patent owner will have the right to respond to any *inter partes* petitions. *Inter partes* review may then be authorized, if the Director makes a threshold determination that there is a reasonable likelihood that the petitioner would succeed on at least one of the challenged claims. Unless otherwise necessary in the interest of justice, discovery will be limited to depositions of witnesses who have submitted affidavits or declarations. The patent owner will have one opportunity to file an amendment to cancel challenged patent claims and to replace them with a reasonable number of substitute claims. It is further required that the challenger be provided with at least one opportunity to file written comments. Each party has the right to an oral hearing. The petition may not be brought earlier than nine months after the patent of concern has issued. Oh, by the way, the petition fee for the patent office is $27,200, if the number of claims in question is limited to twenty.

The *post grant* procedure is similar to the *inter partes* process and is summarized below:

A. Proceedings may be instituted only within a period of nine months from the grant of a patent or its reissue.

B. Post grant review may be sought on any ground of invalidity that can be raised as a defense in an infringement action.

C. It is explicit that the challenger bears the burden of proving its case by a preponderance of the evidence.

D. Petitions for post grant review will be subject to an initial review by the Director to determine whether "the information presented. . . if not rebutted would demonstrate that it is more likely than not that at least one of the claims challenged in the petition is unpatentable." Review may also proceed if the initial determination shows that the petition raises a novel or unsettled legal question that is important to other patents or patent applications. There is to be no review of determinations made on this issue.

E. Discovery is to be permitted, subject to rules made by the Director, but limited to "evidence directly related to factual assertions advanced by either party."

F. Amendments will be permitted during post grant review proceedings only to cancel a challenged claim or replace a challenged claim with a reasonable number of claims.

G. Final decisions of the PTAB (Patent Trial and Appeals Board) are to be made within one year of the date of institution of the proceeding, subject to a possible six month extension "for good cause."

H. At any time before the PTAB's decision, the parties may settle. However, they must file full details of the settlement that will be available to "government agencies on written request or any person showing good cause," but otherwise can be kept from public view.

I. Decisions of the PTAB are to be appealable to the Court of Appeals for the Federal Circuit by either party.

J. Final decisions in favor of the patent owner have an estoppel effect in other proceedings for "any ground that the petitioner raised, or reasonably could have raised" during the post grant review.

Oh, by the way, the fees for this are $35,800.

US inventors have traditionally been the beneficiaries of a one-year grace period that allowed an inventor to publicize the invention without waiving the right to file for a patent. Under the new America Invents Act, the grace period of one year remains, but puts limitations on the types of public disclosures that an inventor may make. Under the new law, an inventor's early filing date is important, because any disclosure by a third party prior to the inventor's filing date will be seen as an expression of prior art that negates the patentability of the subject matter. However, there are two major exceptions to this general rule. First, a third party's disclosure within one year of the applicant's filing date does not constitute prior art, if either the inventor had already disclosed the invention prior to the third-party disclosure, or second, the third-party disclosure was somehow derived from the inventor. Many practitioners now argue that these new requirements provide incentives to publically disclose inventions as a means of defeating a prior art allegation. The downside to this course of action is that a public disclosure can foreclose the right to pursue a non-US (foreign) patent, because most countries do not allow for disclosure prior to filing.

## CHAPTER 21

# Open-Source Software

Craig L. Silver

*Director—Strategic Initiatives/General Counsel, Amches, Inc.*

## Chapter Outline

## Best read in a Volkswagen minibus

The "Open-Source Initiative" is really an interesting animal. Open-source software has revolutionized the embedded software industry. Most of the licenses that are of interest to the embedded engineer can be boiled down into a few items of inquiry. They are as follows:

1.  What copyright (or copyleft[1]) notice must be maintained?
2.  What access to source code must be given and how is it conveyed?
3.  What modifications to source code can be made and how are they identified?
4.  What licenses and associated language must be maintained and propagated?
5.  What considerations must be made if the open-source software is linked with a proprietary product?

To understand this further, and in keeping with the 1960s feel of the "initiative," imagine some hippies sharing a certain type of cigarette. The first guy, say his name is Mort (known for a good stash), starts up a cigarette and passes it around. Mort doesn't mind passing it around as long as he gets the credit for the hit. So, as it passes, the other users have to say "This is from Mort." That is the copyright or "copyleft" notice. Now suppose Herb in the group adds his special herbal additive. It is a different color and the group can clearly see

---

[1] Appears to be an anti-establishment term meant to mirror "copyright." In other words, obtain the benefits of legal society without acknowledgment?

Developing and Managing Embedded Systems and Products.
DOI: http://dx.doi.org/10.1016/B978-0-12-405879-8.00021-0
© 2015 Elsevier Inc. All rights reserved.

Herb's stuff mixed in with Mort. That is OK with everyone as long as it is publicized along the lines of "This is Mort's stash, but Herb has added his #4 herbal additive." That is a modification to the source code and notice. If others like the #4 additive, then Herb is obligated to tell what is in his #4 contribution, if asked, and even give the recipe. No one in the group should ask for money for the cigarette. Now, someone in the group can charge for the *act* of passing the hit along or for the paper wrapper. That is cool. And you don't have to share the "bread" with the group. However, suppose someone decides to mix Mort's cigarette with his own cigarette to the point that it is no longer clear where Mort's stash stops and the other's begins. That's uncool.

As you can see, over time, people in the group may want to make sure that they are only getting Mort's stash and they may want to make sure that all of the known additives have been documented. Someone in the group discovers that even though you can get the stuff free, there is a market for making sure that it is from Mort. Furthermore, in that market all the recipes of Mort are correct and if there is a problem, then there is someone to call and ferret out problems and troubleshoot issues. Hence, the creation of companies sells open-source software services.

Historically, the creation of open-source software is traced to the GNU[2] project which eventually resolved into the Free Software Foundation. Any embedded engineer should be aware that there are other types of "free software" and projects that are not affiliated with the GNU initiative, for instance, The Debian Project software. Not all open-source license terms are the same. At any rate, the GNU project advocated four fundamental "freedoms":

1. The freedom to run the program for any purpose.
2. The freedom to study how the program works and change it to make it do what you wish.
3. The freedom to redistribute copies so you can help your neighbor.
4. The freedom to improve the program and release your improvements (and modified versions in general) to the public so that the whole community benefits.

It is important to understand that these "freedoms" under GPL (General Public License) have nothing to do with issues regarding your desire, or not, to charge for *services* associated with the software. You certainly can charge. You must, however, offer the source code and derivatives of the software for free.[3]

The GNU GPL permits the licensee to copy, modify, i.e., create derivative works, and distribute the licensed software subject to certain conditions. These are the same rights that

---

[2] GNU means "G'noo not Unix."

[3] One individual sued IBM, Red Hat, and Novell for antitrust violations by alleging a conspiracy to keep out competitors who would like to offer to the market a competing OS for a price! He alleged that free Linux kept out businesses who would like to make money. The court found the claim to be without merit. *Wallace v. IBM* et al. 467 F.3d 1104 (7th Cir. 2007).

the Copyright Act of 1976 protects in 17 U.S.C. §106. Since the GPL is a license, the Licensor has the right to set conditions of the license that may not be contained in any law. For instance, the GPL requires that any software distribution be accompanied by access to the source code. Additionally, the GPL allows for modifications to the software and the subsequent distribution of modifications, as long as the modifications are also licensed under the terms of the GPL. Thus, there are two conditions where the GPL must be applied:

Condition 1: A derivative work is created.

Condition 2: A derivative work is distributed.

One never-ending issue in open-source software development is trying to determine where the open-source software starts and where it ends. In other words, where and how can proprietary software interface with the open-source code and not be subject to the license requirements. This issue has its origin, the legal interpretation of derivative works in traditional copyright law. A derivative work belongs to the original copyright owner and is defined as a work that is based upon, is similar to, or contains a substantial amount of the pre-existing work.[4] It is a difficult and complex question in law and license interpretation to determine if the code is a derivative of the licensed code (and thus subject to the GPL) or not subject to the GPL when proprietary links are established. Proprietary links may be code that calls to various modules or libraries.

You can see a simple representation depicting where the open-source software is delineated and how the proprietary software interfaces in Figure 21.1.

Is the proprietary code (the code that you may not want to be subject to the GPL) so intertwined with the existing GPL code, so that it must be considered part of the licensed source code or is it sufficiently separated? In other words, is it a derivative work or not? To

**Figure 21.1:**
A simple representation delineates the open-source software—GPL—and how the proprietary software interfaces to it.

---

[4] Under the Copyright Act, a derivative work is defined as "a work based upon one or more pre-existing works, such as a translation, musical arrangement, dramatization, fictionalization, . . ., or any other form in which a work may be recast, transformed, or adapted. A work consisting of editorial revisions, annotations, elaborations, or other modifications which, as a whole, represent an original work of authorship, is a 'derivative work'" (17 U.S.C. §101).

try to solve this question, many in the industry have tried to determine how the proprietary code interfaces with the GPL code and thus make this the determining factor. To this end, they employ a static or dynamic linking analysis. Under this analysis, static linking is considered to be a derivative and dynamic linking is not. This analysis is not settled among practitioners of trademark law and the courts. Nor does it take into account, in my opinion, the fact that section 102(b) of the Copyright Act does not provide copyright protection for processes and methods.

At any rate, consider link editors as being defined as linkers. The compiler automatically invokes the linker as the last step in compiling a program. The linker inserts code (or maps in shared libraries) to resolve program library references and/or combines object modules into an executable image suitable for loading into memory.

Static linking is the result of the linker copying all library routines used in the program into the executable image. The linker is given the user's compiled code with unresolved references to library routines. Static linking does not require the presence of the library on the system where it is run. A static link outputs a combined file from the compiled code and the libraries.

Dynamic linking is accomplished by placing the *name* of a sharable library in the executable image. When a linker builds a dynamically linked application, it resolves all the references to library routines without copying the code into the executable image. Actual linking with the library routines does not occur until the image is run, when both the executable image and the library are placed in memory. Multiple programs can share a single copy of the library.[5]

This concept of distinguishing static linking from dynamic linking has some compelling logic as, after all, the goal is to determine the degree that the software program makes use of the copyrighted original work.

The courts tend to begin with a broader analysis. They will look to see how the licensed work is incorporated into the alleged derivative work. Next, they look for substantial similarity. Third, the courts will analyze the market impact of the alleged derivative. This embodies the idea that if demand for the original work has not been impacted, then it is unlikely that the alleged derivative has made much use of the original work. Lastly, the courts like to construe the definition of a derivative *narrowly* so as to not chill the market from new innovations. Some similarities relating to the basic functioning of computer systems (e.g., subroutine entry and exit code, external interfaces) can occur by coincidence or intentionally, because "that's the way computers have to work." In other cases, program functions are coded in a particular way, because that is the only (or most effective or the

---

[5]  Indiana University Knowledge Base—developed in part pursuant to National Science Foundation (NSF) grant OCI-1053575 (http://kb.iu.edu/data/akqn.html).

industry standard) way to implement that specific function on that particular computer architecture. Such source code must be excluded from the comparison because it is not entitled to copyright protection; instead, it is an idea that has merged into expression and is thereby rendered uncopyrightable.[6]

There can be an interesting intersection between the Digital Millennium Copyright Act or DMCA (see more regarding the DMCA under laws that can nail the embedded engineer below) and open-source licensing schemes, in that at least one court has ruled copyright information that was removed in violation of *open-source license terms* was a removal of *copyright management information* under the DMCA! This demonstrates how many legal aspects can be implicated in a simple event such as failing to maintain a copyright notice.

## Top 20 most commonly used licenses in open-source projects

Table 21.1 illustrates the top 20 licenses that are used in open-source projects.

**Table 21.1: Top 20 licenses that are used in open-source projects according to the knowledgeBase**

| Rank | License | % |
|------|---------|---|
| 1 | GNU General Public License (GPL) 2.0 | 32.65 |
| 2 | Apache License 2.0 | 12.84 |
| 3 | GNU General Public License (GPL) 3.0 | 11.62 |
| 4 | MIT License | 11.28 |
| 5 | BSD License 2.0 | 6.83 |
| 6 | Artistic License (Perl) | 6.27 |
| 7 | GNU Lesser General Public License (LGPL) 2.1 | 6.19 |
| 8 | GNU Lesser General Public License (LGPL) 3.0 | 2.62 |
| 9 | Eclipse Public License (EPL) | 1.61 |
| 10 | Code Project Open 1.02 License | 1.33 |
| 11 | Microsoft Public License | 1.32 |
| 12 | Mozilla Public License (MPL) 1.1 | 1.08 |
| 13 | Common Development and Distribution License (CDDL) | 0.31 |
| 14 | BSD 2-clause "Simplified" or "FreeBSD" License | 0.30 |
| 15 | Common Public License (CPL) | 0.26 |
| 16 | zlib/libpng License | 0.23 |
| 17 | Academic Free License | 0.20 |
| 18 | GNU Affero GPLv3 | 0.16 |
| 19 | Microsoft Reciprocal License (Ms-RL) | 0.14 |
| 20 | Open Software License (OSL) | 0.14 |

*http://osrc.blackducksoftware.com/data/licenses/, accessed February 22, 2013*

---

[6] L. Rosen, Open Source Software, 2004, p. 285 (Chapter 12).

**Table 21.2: Top 10 Most recent projects to be converted to GPLv3, LGPLv3, or AGPLv3 (as of February 22, 2013)**

| Project Name | Project Version | Project License |
|---|---|---|
| a-space-game | trunk-20121010-svn | GNU GPLv3 |
| 315-snow-bros-evolution | trunk-20121009-svn | GNU GPLv3 |
| aetna | trunk-20121008-svn | GNU GPLv3 |
| 515-planarity-testing | trunk-20121007-svn | GNU GPLv3 |
| 3rd-party-services-mybb | trunk-20121006-svn | GNU GPLv3 |
| abbdd | trunk-20121005-svn | GNU GPLv3 |
| 15-puzzle-jc209673 | trunk-20121004-svn | GNU GPLv3 |
| abc70400 | trunk-20121003-svn | GNU GPLv3 |
| 3bf615be75f76d7667854639a5d86f17 | trunk-20121002-svn | GNU Lesser GPL |
| 2dbe | trunk-20121001-svn | GNU Lesser GPL |

## Most recent projects to convert to GPLv3, LGPLv3, or AGPLv3

Table 21.2 highlights the 10 most recent projects to be converted to GPLv3, LGPLv3, or AGPLv3. This list updates daily.

## Public domain and shareware

The internet has ratcheted up the issue as to what exists in the public domain in regard to copyrights. The way the law exists now is that almost anything that is written down, i.e., "fixed in a medium" is automatically given copyright protection without claiming the right or even by giving copyright notice. It is often asked "What is in the public domain?" Or, is artfully stated, "It is shareware so, it is public domain." Because of the automatic copyright attachment, there is really only one way to be sure that the shareware has really been dedicated to public use. That is to have the originator actually file with the copyright office! 37 CFR 201.26 which implements 17 USC 201.4 provides for recordation: any document clearly designated as a "Document Pertaining to Computer Shareware" and which governs the legal relationship between owners of computer shareware and persons associated with the dissemination or other use of computer shareware may be recorded in the Computer Shareware Registry.

## Litigation and an open-source license

A very interesting case which demonstrates the type of modern litigation that engineers can find themselves embroiled in, and that demonstrates that even open-source licenses can be problematic, would be exemplified by *Jacobsen v. Katzer*.[7] The case settled for a $100K

---

[7]   535 F.3d 1373 (Fed. Cir. 2008).

payment by the defendant, but not before more than 400 filings were made in federal district court proceedings, that included two appeals to appellate courts. There were allegations of patent infringement, as well as cyber squatting proceedings, before a WIPO[8] tribunal. But what should interest us here are the open-source license issues. First, the facts are described.

Jacobsen managed an open-source software group called the Java Model Railroad Interface (JMRI). The group created an application called DecoderPro which allowed model railroad enthusiasts to use their computers to program the decoder chips that control model trains. This application was available for download under the Artistic License[9] which contained, among others, two salient provisions: (i) Jacobsen's copyright notice was to be maintained and (ii) any changes to the files must contain a description of how the files or code had been changed. Defendant Katzer, it was alleged, when offering his competing Decoder Commander program used the open-source files and did not include (i) the author's names, (ii) JMRI copyright notices, (iii) references to the COPYING file, (iv) an identification of JMRI as the original source of the definition files, and (v) a description of how the files or code had been changed from the original source. All of these actions would have constituted a violation of the Artistic License.

The court solidified the concept that even though the copyright holders of the software at issue offered the software for free that did not mean that valuable benefits, including economic benefits, would not ensue. The court stated that the originator may enhance its reputation or increase market share by offering certain components for free. Additionally, the court recognized that improvements and bug fixes may be obtained faster and from experts in the community of programmers.[10]

In ruling upon the matter, one of the threshold questions for the court was to determine whether the terms of the Artistic License are *conditions* of the license or *covenants* contained in a contract. The court noted that a copyright owner who grants a nonexclusive license to use his copyrighted material waives his right to sue the licensee for copyright infringement and can sue only for breach of contract. The court further stated that if a copyright *license* is limited in scope, and the licensee acts outside the scope, the licensor can bring an action for copyright infringement. The distinction regarding a condition or covenant is important, because the remedies that are available to a Plaintiff will hinge on the type of claim that can be brought. For instance, a breach of contract action could be

---

[8] According to them, "The World Intellectual Property Organization (WIPO) is the United Nations agency dedicated to the use of intellectual property (patents, copyright, trademarks, designs, etc.) as a means of stimulating innovation and creativity." There are international treaties that give enforcement right to the Organization.

[9] More on this license can be found at http://opensource.org/licenses/.

[10] The court cited *Planetary Motion, Inc. v. Techsposion, Inc.*, 261 F.3d 1188, 1200 (11th cir. 2001) in this regard.

claimed, where damages would be hard to prove by one who offered his software for free, or a claim could be filed for infringement where, at least, injunctive relief can be obtained, i.e., to seek a court order to stop the infringing activity of Katzer.

The court had no trouble finding that the terms of the Artistic License held various *conditions* of use which, if breached, would be outside the scope of the granted license and thus, a case was made for copyright infringement rather than just breach of contract.

Significantly, the court stated further that

> Copyright holders who engage in open source licensing have the right to control the modification and distribution of copyrighted material. As the Second Circuit explained in Gilliam v. ABC, 538 F.2d 14, 21 (2d Cir. 1976), the unauthorized editing of the underlying work, if proven, would constitute an infringement of the copyright in that work similar to any other use of a work that exceeded the license granted by the proprietor of the copyright. Copyright licenses are designed to support the right to exclude; money damages alone do not support or enforce that right. The choice to exact consideration in the form of compliance with the open source requirements of disclosure and explanation of changes, rather than as a dollar-denominated fee, is entitled to no less legal recognition. Indeed, because a calculation of damages is inherently speculative, these types of license restrictions might well be rendered meaningless absent the ability to enforce through injunctive relief. In this case, a user who downloads the JMRI copyrighted materials is authorized to make modifications and to distribute the materials provided that the user follows the restrictive terms of the Artistic License. A copyright holder can grant the right to make certain modifications, yet retain his right to prevent other modifications. Indeed, such a goal is exactly the purpose of adding conditions to a license grant.

This case should stand as a beacon of understanding to all software developers who use open-source licenses that there are licensors that will jealously guard the terms of the open-source license and make that free software very expensive.

## CHAPTER 22

# Laws That Can Nail Embedded Engineers

Craig L. Silver

*Director—Strategic Initiatives/General Counsel, Amches, Inc.*

## The Digital Millennium Copyright Act

In 1998, Congress passed the Digital Millennium Copyright Act (DMCA) which contains criminal provisions for production and dissemination of technology, devices, or services intended to circumvent digital rights management or that control access to copyrighted works. It also criminalizes the act of circumventing access control, whether or not there is actual infringement of the copyrighted work itself. Curiously, the law allows the Librarian of Congress the right to grant exemptions to the law.[1] These exemptions are of interest to

---

[1] The Library summarizes the law as follows: The Digital Millennium Copyright Act (DMCA) was enacted to implement certain provisions of the WIPO Copyright Treaty and WIPO Performances and Phonograms Treaty. It established a wide range of rules for the digital marketplace that govern not only copyright owners, but also consumers, manufacturers, distributors, libraries, educators, and online service providers. Chapter 12 of Title 17 of the United States Code prohibits the circumvention of certain technological measures employed by or on behalf of copyright owners to protect their works ("technological measures" or "access controls"). Specifically, Section 1201(a)(1)(A) provides, in part, that "[n]o person shall circumvent a technological measure that effectively controls access to a work protected" by the Copyright Act. In order to ensure that the public will have the continued ability to engage in non infringing uses of copyrighted works, however, subparagraph (B) limits this prohibition. It provides that the prohibition shall not apply to persons who are users of a copyrighted work in a particular class of works...

Developing and Managing Embedded Systems and Products.
DOI: http://dx.doi.org/10.1016/B978-0-12-405879-8.00022-2

embedded engineers who like to "Jailbreak" devices. The current exemptions are in force until October of 2015.[2] The first exemption applies to disabled persons, such as the blind, which allows for the circumvention of controls on e-books, assuming that the copyrighted work has been legitimately purchased.

Another exemption allows for the circumvention of "computer programs that enable wireless telephone handsets to execute lawfully obtained software applications" where it is understood that the purpose of the circumvention is to accomplish interoperability of programs with the telephone handset. The circumvention of tablets is not allowed, as it was ruled that a sufficient definition of "tablet" was not available.

A major change from previous exemptions has occurred regarding cell phone unlocking. Previously, phones could be unlocked so that they could be used on a different carrier's network. While phone unlocking is allowed to be covered under the latest exemptions, the phone must have been acquired before January 2013.

There is no recognized exemption for "jailbreaking" a video game console. There is not a recognized exemption for defeating controls in order to install another legitimately obtained OS on a personal computer.

There is an exemption to allow for the defeating of controls on a DVD in order to make use of short portions of the motion pictures for the purpose of criticism or comment in certain instances, such as noncommercial video or documentary films. However, the practice of "space shifting" where a user defeats controls in order to watch a film on a device that lacks a DVD drive is not a recognized exception. The astute reader may conclude that the types of exemptions that the Librarian is allowing are the type of actions that may have constituted "fair use" under copyright laws. This may mean that the fair use defense to certain actions has been narrowed to include *only* those that the Librarian of Congress allows. If you are not defeating DMCA technology controls, decompiling copyrighted object code is "fair use," but only pursuant to delineated conditions: "where disassembly is the only way to gain access to the ideas and functional elements embodied in a copyrighted computer program and where there is legitimate reason for seeking such access, disassembly is a fair use of the copyrighted work, as a matter of law."[3]

> *Laws in the EU also permit decompilation under specific circumstances: (a) the person or entity must own a license to the software; (b) the decompilation must be necessary to achieve interoperability with other application; and (c) the decompilation must be limited, if possible, to the source code sections needed to achieve interoperability.*
>
> **EU 2009 Software Directive.**

---

[2] The final rule can be found at http://www.copyright.gov/fedreg/2012/77fr65260.pdf.
[3] *Saga v. Accolade* (977 F.2d 1510).

## Stored Communications Act

Every engineer is eventually presented with the opportunity or obligation to share technical information with other engineers. Often an "FTP site"[4] is set up with the understanding that the information will have limited access. Alternatively, an engineer may be the host of a blog or message board where certain information is not intended to be generally available to the public. At first glance, comfort may be found in the Stored Communication Act (18 USC § 2511 et seq.) that provides for protection of facilities from unauthorized access. The Act prohibits access to e-mail inboxes, wikis, weblogs, networking sites, and the like by those who have not received authorization or who have exceeded an authorization. Apparently, the Act has been interpreted by federal courts as requiring more than just a routine registration. For instance, in *Snow v. DirecTV*, lawyers of DirecTV accessed a site which, by its terms, specifically barred access by employees or agents of DirecTV. Nevertheless, the court found that there must be some further allegation that the site provided some "screening" of intended users beyond the mere registration in order to be subject to the Act. The Act *does* specify that it is not intended to replace state laws, that in some state jurisdictions have found that merely exceeding the authorized access or stated purpose of the site, will give rise to liability.

## The Computer Fraud and Abuse Act 18 USC § 1030

This statute has been used by prosecutors to charge defendants with a crime who violate a term of service on a computer network. This is an example of the type of federal statute that comes into play, when engineers or hackers sit down to play. Most states also have laws that mirror, or are more severe than, the federal statutes. By "more severe" I am not only talking about penalties, but the type and number of elements that a prosecutor would need to prove in order to establish guilt. A "more severe" statute would be one where the prosecution would *not* have to prove a lot of elements. For instance, the federal statute requires that the hackers damage[5] the system, have fraudulent intent, be exposed to confidential data, or convey passwords. It also requires that the computer be used in interstate commerce. However, many state statutes make criminal the simple act of intentionally accessing a computer without authorization. Note further that §1030 is the federal statute that is specifically aimed at computers. There are other statutes, such as the wire-fraud statute, that can be used to charge a person with a crime when any scheme with

---

[4] **File Transfer Protocol (FTP)** is a standard *network protocol* used to transfer files from one host or to another host over a TCP-based network, such as the Internet.

[5] The term "damage" means any impairment to the integrity or availability of data, a program, a system, or information.

nefarious intent uses interstate data communication networks. Be advised that attempts to commit a federal crime or conspiracy to commit a crime are, in and of themselves, separate offenses that are often charged. Also, one who counsels, commands, aids or abets, or otherwise acts as an accessory before the fact is liable as a principal for the underlying substantive offense to the same extent as the individual who actually commits the offense.

> *§1030a6 – knowingly and with an intent to defraud trafficking in, i.e. to transfer, or otherwise dispose of, to another, or obtain control of with intent to transfer or dispose of a computer password or similar computer key either of a federal computer or in a manner that affects interstate or foreign commerce. This section is designed to pursue those individuals who think it is cool to hack the password of a computer and then post the password on the internet or otherwise allow others access to the hacked password.*

> *The Penalties: Imprisonment for not more than 1 year (not more than 10 years for repeat offenders) and/or a fine under title 18 (the higher of $100,000 for misdemeanors/ $250,000 or felonies or twice the amount of the loss or gain associated with the offense, 18 U.S.C. 3571).*

## Torts and the engineer

### Negligence

People tend to associate lawsuits with negligence. Everybody has seen a sitcom or two making fun of hurt necks and rumored acts of negligence. Most engineers tend also to think that unless they are a professional civil engineer, they are far removed from allegations of negligence. And of course there is that famous phrase like, "I am incorporated so that I cannot be sued." Understand that being incorporated does not insulate you from your acts of negligence. The corporation or LLC only insulates you from *contractual* liability where you have signed in a corporate capacity. But the law in every state says that an individual actor is responsible for their own acts of negligence. Let's look at the elements of negligence and apply them to the engineer.

1. *Duty.* The law says that in order to be liable for negligence a person must first be under a duty to act in a reasonable or non-negligent way. For instance, if you are designing a power supply, and it is understood that the output is going to carry 400 amps, a designer would have a duty to make sure that the power supply is not going to catch fire. So the duty here would be to design power supplies that are reasonably designed to not catch fire when under design loads and installed according to specifications.
2. *Breach.* Now if the supply does catch fire and the reason the supply caught fire is that 20 gauge wire was used in the output, it might be said that there has been a breach of the duty, because a reasonable man would not use 20-gage wire in the output of a power supply carrying 400 amps.

3. *Foreseeable harm.* Yes, the 'ole "no harm, no foul" exists in the law. In most applications, the law does not apply the law of negligence to potential harm or even the apprehension of potential harm. Negligence law concerns itself with *actual* harm and damages whether they be for personal injuries or economic loss.

4. *Proximate cause.* The Breach of the Duty must be the Proximate Cause of the Harm. This means the act of negligence would be a foreseeable result in harm that is closely related to the breach of the duty. Sometimes the acts of negligence are too remote in effect, action, or time. In such cases the negligence is said to have been attenuated. If the 20-gage wire causes a fire, and the fire department's water damages a car, and the owner of the car hurts himself repairing the damage, the law would say that the owner's injuries are too remote from the breach so as to not constitute proximate cause.

Regarding the Duty as mentioned above, what standard is applied to the computer programmer? There are cases that are becoming more and more commonplace that find that a programmer is a professional, that is one "who holds themselves out to the world as possessing skill and qualifications in their respective trades or professions impliedly represent they possess the skill and will exhibit the diligence ordinarily possessed by well-informed members of the trade or profession." The closer that computer programmers come to being considered professionals, means that they will be held to a higher standard of care. Thus, the chances increase to find that that there was a duty to act professionally, and that the duty was breached by the way the software performed.

### Limiting exposure

Embedded engineers can take some steps to minimize their risks in regard to negligence actions. Whether you are a small company or an independent contractor, there are some steps that you can take. For instance, you can avail yourself of the almighty indemnification clause. This is also known as a "hold harmless" provision. What it is designed to do is to shift the costs, not the liability, to a third party that has deeper financial pockets. A typical provision may read:

> *The undersigned,* Deep Pockets *agrees to indemnify, hold harmless and defend* Joe Embedded *from any and all judgments, costs, fees, including attorney fees, causes of action, threatened causes of action, and the like which arises out of, or concerns the negligence of,* Joe Embedded, *where such claim or action or the like is brought by a third party.*

This is a very simple example. Note that in the example above, that this covers an issue that might be brought by a third party. Nothing prohibits an indemnification provision from being established in a contract directly between the two contracting parties, in this case *Deep Pockets* and *Joe Embedded*. In fact, you can expect to find indemnification provisions in almost any contract that you sign. It is a common practice to have a large company ask for indemnification by saying, "It is our boilerplate," and make you feel like you are being unreasonable, as *Joe Embedded*, to ask that perhaps, they should be indemnifying you! After all, if anything goes wrong they have a host of attorneys and resources to cope with the matter. However, you may actually be put out of business if a claim is litigated, especially for a long time.

Another consideration is to be aware who gets indemnified. It is advisable to add, as a designated party, "and his agents, subcontractors and assigns." Of course, you would want your corporate entity to be named, along with its officers and employees. Also, you can see in the example above, this indemnification is limited to acts of negligence. However, there are other forms of liability that can arise. For instance, patent infringement. If you are the indemnitee, you want the broadest possible language that would include all forms of loss, including intellectual property infringement. Note further, that the example language above includes *threatened* matters. Many indemnification clauses only kick in to play if there is some kind of award or judgment rendered. In such a case, you may have to foot the bill for the entire litigation, and maybe a negotiated settlement, if the clause is limited accordingly.

There are other aspects of an indemnification clause which are desirable, if you are the indemnitee. For instance, it is appropriate to set out the procedures for notifying the indemnitor of a claim and to set out time limits for them to act on the matter. Thus, it is also desirable to have the right to defend the claim, and bill the indemnitor ultimately, if they should fail to act in accordance with the contract. Still another consideration in this type of clause is to be aware of how such an indemnification interacts with other provisions of the contract. If the contract has a limitation on damages, and only allows for, e.g. direct damages, such a provision could render the indemnification clause a nullity. Therefore, it is important to "carve out" an exception to the limitation on damages.

Are there limits on the type of indemnification in your jurisdiction? How does such a clause work with the indemnitor's insurance? Would property damages be covered if the provision just says "negligence"? These are all valid questions and direct the sharp engineer to the attorney's office.

There are other provisions that the software developer can use to limit their legal exposure. One is to limit the monetary damages that are available to the aggrieved party. If you clearly state in the contract that the customer's remedies for any breach of the agreement is limited to repair or replacement of the defective software, then they may not be able to recover other damages. Likewise, you can put limits on the amount of recovery. A clause can state:

> *Any damages claimed for breach of this agreement shall be limited to the amount of monies paid for the software in the year preceding the event which gave rise to the claim of breach.*

This type of limit can get interesting, because it may be that only *maintenance fees* were paid in the preceding year, and there can be all kinds of discussions as to what "event" gave rise to the claim.

You can limit damages, preferably in all capital letters, generally by stating:

*IN NO EVENT WILL VENDOR BE LIABLE FOR ANY DAMAGES, INCLUDING LOSS OF DATA, LOST PROFITS, COST OF COVER OR OTHER SPECIAL, INCIDENTAL, CONSEQUENTIAL OR INDIRECT DAMAGES ARISING IN ANY WAY OUT OF THIS CONTRACT, OR FROM THE USE OF THE SOFTWARE OR ACCOMPANYING DOCUMENTATION, HOWEVER CAUSED AND ON ANY THEORY OF LIABILITY. THIS LIMITATION WILL APPLY EVEN IF VENDOR HAS BEEN ADVISED OF THE POSSIBILITY OF SUCH DAMAGE AND NOTWITHSTANDING THE FAILURE OF ANY LMIITED REMEDY PROVIDED HEREIN. CUSTOMER ACKNOWLEDGES THAT THE LICENSE FEE REFLECTS THIS ALLOCATION OF RISK.*

The law is not settled regarding the application of the Uniform Commercial Code (UCC) to software. The UCC generally applies to products and not services. If the software is a product, you can, with the blessing of the UCC, limit the time frame that a customer could bring suit. You can shorten the time to 1 year by contract, even though the statute of limitations in your jurisdiction may be 3 years.

Another way to limit exposure is to put up gates that the customer has to go through in order to not be in breach of the contract themselves. You can add provisions that say, for instance:

*Vendor shall be notified within 10 days after the software is put into operation of the discovery of any bug or defect that materially interferes with the customer's use of the software. Notification shall be made by registered mail to the head engineer listed below.*

Since everyone seems to forget to read the contract, you can imagine how often a customer would fail to comply with such a clause.

Warranty disclaimers are another important tool in the developer's kit that help to limit legal exposure. There are important *words of art* that should be employed. The type should be large. Example:

*VENDOR MAKES NO WARRANTIES, EXPRESS, IMPLIED, STATUTORY OR IN ANY COMMUNICATION WITH YOU, AND VENDOR SPECIFICALLY DISCLAIMS ANY IMPLIED WARRANTY OF MERCHANTABILITY OR FITNESS FOR A PARTICUALR PURPOSE. VENDOR DOES NOT WARRANT THAT THE SOFTWARE WILL OPERATE UNINTERRUPTED OR ERROR FREE.*

You are advised to pay particular attention to disclaim the warranty of "fitness for a particular purpose." This *phrase of art*, if not disclaimed, can give rise to any and all claims that the software was defective for the customer's application and for uses that you may not have anticipated. Note that the limitations that are being advocated here are for the commercial realm. These limitations would not be allowed for consumer applications.[6]

## Products liability

So many embedded systems are used in safety critical applications that some understanding of products liability is healthy, especially since the seeds of future consequences happens during the design and execution phase. We like to say, "today's mess-up is next year's litigation."

Products liability is a close cousin of negligence law and uses a rather nefarious concept of "strict liability." What makes products liability such a headache for engineers, who might be defendants, is that the law has a fairly low standard for plaintiffs to meet in pursuing a products liability claim. Most states have adopted, either by statute, or by judicial decree, the language of Section 402 of the *Restatement of Torts*, that states that anyone who places a product into the "stream of commerce" that, due to a defect in design or manufacture, that is unreasonably dangerous, and is unchanged when it injures the ultimate consumer is *strictly liable* to the injured person.[7]

## Public policy

The courts and legislatures have advanced public policy reasoning regarding the concept of strict liability. Some are listed below:

1. The consumer finds it too difficult to prove negligence against the manufacturer.
2. Strict liability provides a valuable incentive to manufacturers to make their products as safe as possible.
3. Reputable manufacturers do in fact stand behind their products, replacing or repairing those which prove to be defective, and many of them issue express agreements to do so. Therefore, all should be responsible when an injury results from a normal use of the product.
4. The manufacturer is in a better position to protect against harm, by insuring against liability for it, and by adding the cost of the insurance to the price of his product, and to pass the loss onto the general public.

---

[6] These limitations may not be applied by a court if the software project was a complete failure so as to be "so total and fundamental" that limitations on damages would not be recognized. *RRX Industries, Inc. v. Lab-con, Inc.*, 772 F.2d 543 (1985).

[7] Since products liability concerns personal injury, as a matter of public policy, this type of liability cannot be disclaimed by the manufacturer.

5. Strict liability already can be accomplished by a series of actions. The consumer first recovers from the retailer on warranty. Liability on warranties is then carried back through the intermediate dealers to the manufacturer. The process is time consuming, expensive, and wasteful. There should be a shortcut.

6. By placing the product in the market, the seller represents to the public that it is fit, and he intends and expects that it will be purchased and consumed in reliance upon that representation. The distributors and retailers are no more than a conduit through which the thing sold reaches the consumer.

7. The cost of accidents should be placed on the party best able to determine whether there are means to prevent the accident. When those means are less expensive than the cost of such accidents, responsibility for implementing them should be placed on the party best able to do so.

## Elements of products liability

In the first instance, products liability is concerned with personal injury. Purely economic damages are not considered to be within the purview of products liability; that is the concern of product warranties.

In order to claim that a product is "unreasonably dangerous" under products liability theories of strict liability it is still necessary to claim:

1. That the product was defective and unreasonably dangerous to the consumer at the time of production and at the time of sale,[8]
2. That the product was expected to and did reach the ultimate consumer without substantial change in the condition from the time it was sold,
3. That the defective condition of the product proximately caused the plaintiffs injury.

How do the courts determine if the product was defective or unreasonably dangerous? They seek to determine if the manufacturer departed from proper standards of care. Questions can be asked, such as, did the manufacturer know of the defect? If it did, would it be reasonable for it to introduce the product into the stream of commerce? These are the type of questions that have been litigated in the famous cigarette cancer cases. Sometimes the situation can call for an application of the Risk/Utility Test. Perhaps, there was a foreseeable danger in the product that would not be obvious to the consumer. In such cases, the manufacturer is obligated to warn of the "defect" or the danger. What is obvious can be in dispute and this

---

[8] However, the manufacturer is under no duty "to make a product that will last forever or will withstand abuse or lack of maintenance" or that is "foolproof," *Tri-State Insurance Company v. Fidelity & Casualty Insurance Company*, 364 So.2d 657, 660 (La. App. 2d Cir.).

is why manuals and products may have a lot of warning stickers. Under a Risk/Utility Test the courts have required the Plaintiff to prove variations or combinations of the following:

1. That the state of the art, at the time the product was produced was such that a safer design was available,
2. That the safer design is described as...,
3. That the safer design was cost feasible and was known in the industry.

Sometimes the courts will apply a Consumer Viewpoint Test that attempts to ascertain if the average, reasonable consumer would have been aware of the defect, and the risk inherent in the product. If the answer is yes, the product is not considered defective. Note here that contributory negligence is not a defense in products liability cases.

An interesting case which concerns the misuse of a product and the "failure to warn" issue is found in the Maryland case of *Moran v. Faberge, Inc.*[9] The action involved a 17-year-old girl and a friend who decided to pour some Faberge cologne on a candle to make it scented. The vapors from the cologne flashed and severely burned the plaintiff. This was largely a "failure to warn" case. As would be expected, Faberge argued in effect, "how are we expected to anticipate every crazy thing someone may do with our product." However, it was proven at trial that the flash point of this particular concoction was only 73°F, essentially room temperature. Therefore, the highest court in Maryland said:

> ...*we hold that it was unnecessary, to support a verdict in favor of the petitioner, that there be produced evidence, as demanded by the Court of Special Appeals, "which would tend to show or support a rational inference that Faberge foresaw or should have foreseen that its [Tigress] cologne would be used in the manner [(pouring the cologne on the lower portion of a lit candle in an attempt to scent it)] which caused the injuries to Nancy Moran...rather, it was only necessary that the evidence be sufficient to support the conclusion that Faberge, knowing or deemed to know that its Tigress cologne was a potentially dangerous flammable product, could reasonably foresee that in the environment of its use, such as the home of the Grigsbys, this cologne might come close enough to a flame to cause an explosion of sufficient intensity to burn property or injure bystanders, such as Nancy.*[10]

What is important to note here is that the "failure to warn" may involve an inherent aspect of the product itself, and not the most obvious improper use scenarios that the product development team may be able to anticipate. Additionally, some of the most legally effective warnings are those that not only warn, but suggest a different course of action by

---

[9] 273 Md 538, 332 A. 2d 11 (1975).

[10] It is interesting to note that almost 20 years later, this time a 14-year-old girl, is cutting into a can of a product by Faberge and suffering injuries from a resultant explosion from the home environment and the company is being found liable under failure to warn and design defects. *Nowak v. Faberge*, 32 F.3d 755 (1994).

the consumer. A warning concerning a motion controller which says, "Shut down apparatus immediately and consult manual if smoke, vibration or unusual sounds are heard," may be more effective than a general warning that says, "Verify proper operation at the start of each shift."

For the embedded software engineer, we have to ask if software is a product. Pure software has traditionally not been considered a product, but a service that is licensed. There are numerous writers and legal pundits who are advocating that software be considered a product and thus, apply products liability law to software. This issue may be of little concern to the **embedded software** engineer. After all, "embedded" usually means embedded onto some hardware processor and it is the operation of the software in conjunction with the hardware where the human—machine interface takes place, and also where the injury will be traced. The fact that software engineers and hardware engineers blame each other over the product's performance highlights the fact that the blame can be difficult to ascertain. As far as I know, no defendant in a products liability case has been able to escape strict liability by claiming that the cause of a personal injury was due *solely* to software. The most famous case where "software" in isolation, but as data, was treated like a product, and resulted in a successful products liability suit was the Jeppesen case.

The Jeppesen case is noteworthy not only because it concerns data, but the source of the data was the US Government itself. This involved the September 8th, 1973, crash of a DC-8 in Cold Bay, Alaska. The following is an excerpt from NTSB accident report #1-0018:

> *According to CVR - 10 - data, the only indication that the crew noticed an irregularity was the first officer's remark that his DME was "not good". That remark coincided with the aircraft's descent through 4,500 feet and may have been prompted by intermittent operation of the first officer's DME such as that described by the crew involved in the approach incident near Cold Bay. The crew of Flight 802 were apparently unaware of the terrain-related restrictions of the navigation signals and made no attempt to climb higher associated with loss of the DME, the captain apparently kept relying on for better signal reception. Despite his monologue about the problems of 37, 35, 29, 24 and 20 miles were all accurate. The probability that the flight's right turn to a westerly heading, about 20 seconds before impact, was based on erroneous azimuth information receiver... which was tuned to the Cold Bay VORTAC.*

All of this is to say that when the pilot descended too low he was not getting reliable navigation signals—whether they be the DME (distance measuring equipment) or the azimuth signal supplied by the VORTAC.

Naturally, and as is typical, the NTSB found the pilot at fault for not complying with the published approached procedure or being familiar with the published minimum sector altitudes. However, it was later determined by a federal jury that the cause of the accident was an error in the chart.[11] The Jeppesen case is a wonderful introduction to products liability, because it can spotlight some interesting aspects of products liability law.

A.   Jeppesen may not have thought that they were selling a "product."
B.   Jeppesen was relying upon government-supplied data.
C.   Jeppesen would have found it very difficult to correct any alleged defect in the chart.

However, as the court found, Jeppesen maintained an office in Washington, DC, and worked closely with the FAA, and had access to the people who could make changes. Therefore, Jeppesen was found liable for its product, even though the data that was used was government supplied. This type of case has helped trigger a lot of Geographical Information Systems litigation as a result.

### Minimizing risk in embedded system product development

There are several steps that can be taken to minimize risk from a products liability perspective:

1. There should be a quality system in place that documents that the development is being performed pursuant to such a system.
2. Employees should be trained on the system; the training should be verified that it has taken place; that the employees are actually using the system. It is not uncommon for organizations that have complicated systems to find that employees spend a fair amount of creativity in performing runs around the system. Of course, employees should be qualified, and the standard as to who is qualified, should be determined prior to hire, so as to avoid the temptation to hire Uncle Fred's nephew who is 14 years old, but an embedded genius.
3. Warnings should be part of the GUI in some cases and not just hidden in the manual.
4. Final testing should be done by an independent group, or at least a different department. The test plan and final verification should be part of the overall product approach, rather than at the end of product development.
5. Reviews should be held to look at the product specifications and synch those with the product design, especially if the specification has changed over time.
6. Documentation and control of engineering changes should be in force. This is important to have so that defects can be quickly fixed, if they become known after general release. It stands to reason that some critical defects need to be addressed as soon as they become known, as it would be litigation "gold" if it were to be revealed that the company knew of the defect, but waited until the next annual release to fix the problem. To this end, the company needs to have a plan in effect to "push" critical fixes to users. Does the company even know who the users are?

---

[11] A federal statute now provides for indemnification of chart makers because of this case-49 U.S.C. § 44721. One of the alleged defects was the absence of a transition arc which would have given guidance from the initial to the intermediate phases of flight on an instrument approach.

7. There needs to be an effective plan for receiving complaints from end users and for escalating complaints to the highest level in the company for those that may indicate a safety issue. There should be a procedure in place for product recall.

8. Documents should be preserved pursuant to an established document retention system. Documents should not be destroyed, except pursuant to policy. One of the first events that occurs when litigation is threatened or imminent is to receive a "preservation letter" demanding that all relevant documents, etc. be preserved. Of course, you don't have to obey. The risk is that if any documents or code are missing, the adversary can claim that there has been a "spoliation of evidence." It is possible, that if a court agrees, then the case could essentially be over and a judgment entered, at least as it concerns liability. It would be asserted that the claimant could not prove their case, due to the actions of the company by destroying key evidence.

9. Warnings regarding the use of the product should be intense enough to convey the gravity of the perceived harm, and should further be evaluated for the following issues:

    a. Does one part of the warning dilute another part?

    b. Has the warning been communicated in a manner designed to reach the person to whom it is directed? Have important warnings been permanently affixed to the product?

    c. Have warnings been undercut by advertising or promotional materials?

    d. Is the warning clear, conspicuous, and unambiguous?

    e. Does the warning tell the user what to do in the event of an incident?

    f. Does the warning describe with adequate specificity the nature of the risk so that the end user knows what consequences to expect? In other words, will the user appreciate the danger so as to want to heed the warning?[12]

    g. Does the warning overload the senses with far-fetched risks?

    h. Does the warning indicate how to avoid the hazard?

    i. Will the product go through middlemen or other handlers who may not convey the warning?

    j. Are the warnings repeated in the instruction manual?

    k. How do the warnings compare to competitive products?

    l. How are postsales warnings pushed out to the customers, if certain defects are discovered later?

10. If the product has been developed as part of a recognized government approval process, make sure that the design is fixed as per the government submissions, and that the product has been manufactured as per the design. This way, you can benefit from the government review process.[13]

---

[12] Do you remember the movie Ghost Busters? One warns about "crossing the streams". Another asks, "What happens if you cross the streams?" Cataclysmic events, that's what. It is summed up, "OK, Important safety tip."

[13] Such as the PMA or premarket approval as required by the FDA for medical devices. Under the Medical Device Act the Supreme Court has ruled that such approvals preempt state tort law. *Riegel v. Medtronic.*

## CHAPTER 23

# Corporate Operations

## Craig L. Silver
### *Director—Strategic Initiatives/General Counsel, Amches, Inc.*

**Chapter Outline**

## *The charter*

A corporation is granted a charter from the state in which it is incorporated. Therefore, it is important to keep the charter intact. Most states are pretty liberal in letting certain matters slip. However, there are ways to jeopardize one's corporate status. For example, if the corporation mixes personal funds with the corporate accounts, or never acts as if the corporation exists as a separate entity, such as never having an annual meeting, does not keep separate books or records, does not file its annual return, does not use the corporate designation in correspondence, etc., there is the risk that at the time you need it most, that

Developing and Managing Embedded Systems and Products.
DOI: http://dx.doi.org/10.1016/B978-0-12-405879-8.00023-4

someone may find a way to "pierce the corporate veil" and attach personal liability to the owner of the company.[1]

## Shares and stocks

Another fertile ground of litigation concerns the issuance of shares. Most states recognize that the corporate stock ledger is the official record of who owns what shares. Most engineers who may have shares issued to them, never see or have heard of the official stock ledger. Sometimes, they don't even get the actual stock certificate. Every state recognizes a modicum of minority shareholder rights and one of these is to inspect the corporate books. Asking to see the corporate books is a good way to find out if anyone even knows where the shares are, or if there have been appropriate entries on the stock ledger.

Speaking of shares, before you go home and tell your wife that you are a shareholder and what a wonderful thing this is, let's consider what factors might take the shine off that crystal chandelier. For starters, shares of stock carry income tax implications. If the company is an ongoing concern and is actually making money, and you are issued shares outright, you will have to pay ordinary income tax on the pro rata value of the ongoing enterprise. That is why stock warrants are so popular. A warrant is basically a subscription for a certain number of shares to be granted at a later time and at a predetermined strike price. This avoids the immediate realization of ordinary income. This is also why some are tempted to illegally backdate the options, as one can get capital gains type treatment for options that are exercised and held for a period of time.

Another factor to think about is what kind of stock the company is generously giving the engineer. There might be more than one kind of stock. If the company has a Venture Capitalist funding the venture, then it is a common practice to allow the venture company *preferred stock* which means that they would get dividends payable to them first, before the *common stock* is paid out as a dividend to the engineer. Sometimes there is participation with the common stock along with the preferred stock on some kind of ratio basis.

Still another issue concerns the company's ability to "water down" your stock to a very small percentage. Most states allow stock to be watered down for business purposes and as a means to raise capital. Therefore, minority shareholders are allowed a *prescriptive* right to maintain their percentage share, but that requires the minority shareholder to ante up some cash and pay for the stock on the same basis as the new offering.

Another issue that can arise is that once you are a stockholder, you become an owner. You now run the risk that if you own too much, then you could become personally liable for certain statutory obligations that owners of companies are obligated to perform.

---

[1]  Many states require some fraudulent act in order to pierce the veil.

For instance, the payment of IRS Form 941 withholding taxes. So you can see how, what at first looked like an occasion to go to dinner with only your wife, can turn into an occasion to bring along your accountant, lawyer, financial consultant, spiritual advisor, and marriage counselor.

Limited liability companies (LLCs) are designed to provide for contractual limited liability of a corporation, along with the freedom that a partnership provides. Just like a partnership, the LLC allows for the participants or members, by way of the operating agreement to allocate and designate:

- the members' percentage interests in the LLC,
- the members' rights and responsibilities,
- the members' voting powers,
- how profits and losses will be allocated,
- how the LLC will be managed,
- rules for holding meetings and taking votes,
- buy/sell provisions, that determine what happens when a member wants to sell his interest, dies, or becomes disabled.

There is great freedom in what an LLC's operating agreement can say. Therefore, it is advisable to put some effort into customizing the agreement for your particular operation, and your particular concerns. In the absence of an operating agreement, your state default rules would be controlling. If your friend asks you to join an LLC and makes promises to you that you will have 20% of the company, you want to make sure that you actually see that the operating agreement has been modified to reflect that important assertion. You will want to know what tax elections have been filed with the IRS. An LLC will, in the first instance, be taxed like a partnership, with income being allocated to the members, unless the LLC has elected to be treated like a corporation for tax purposes. In other words, there is no such thing as retained earnings in a partnership without such an election. Income is offset by expenses and any profit goes straight to the individual's return, whether or not you actually received the cold, hard cash. If the LLC elects to be treated as a corporation, then retained earnings can be left off of the individual's returns.

## Hiring or contracting with foreigners

A common practice, even among small entities, is to hire an employee who will work from home in a "foreign" state, or to contract with a distributor in another state. These activities appear benign at first. Wise businessmen will proceed with caution and with accountant and attorney advice. One issue of concern is whether having that employee in the foreign state will require you to apportion income to that state, and pay income tax to that state. This can, depending upon the location, increase direct costs of that employee, as well as the accounting costs. Similarly, by having that employee in that state means that, under most analysis, you now have a legal "presence" in the state for the purpose of getting sued for contract and for tort issues. If you are in Texas and you hire an employee in California, you might be surprised to learn that your normal non-compete employee contract would not be enforceable in California, as that state considers non-competes against public policy, embodied in a statute. Speaking of California, it has a very expansive attitude toward what constitutes a presence. So, if you just have a contract with a sales representative firm in that state, or you have a "1099" contractor in the state, they will try to assess sales taxes based upon those sales that can be attributed to activities within the state.

## So you want to export

Many firms find that exporting presents many advantages and opportunities, but there are special considerations worthy of notice.

### Bribery

What if somebody wants a bribe? In the United States, only Congress is allowed to be bribed. But for overseas opportunities, bribes are illegal under the Foreign Corrupt Practices Act (FCPA)[2] and it is enforced by the US Department of Justice. This law generally divides payments into two types of categories: one is legal and the other is illegal. The legal bribes that are allowed are the kind demanded by low-level clerks doing ministerial jobs. For instance, maybe there is a customs dude who puts a routine stamp on a document. He generally does not do his job unless he gets an extra $100. In fact, the host country knows he charges this extra fee and they pay him accordingly. In such cases, the law indicates that this type of payment is allowed under the rationale that all that you are paying for is to get someone to expedite something that was supposed to happen anyway.

The illegal bribe is the kind that is paid to a government official to make something happen that, if not for the payment, would not occur. Paying an official to make sure that his brother in the Ministry of Defense awards a contract to you fits this category. Yes, these laws are actually enforced. Studies show that one of the chief reasons for so many poor third world countries being unable to advance economically is the rampant corruption that invades every level of their society. So, how might you get caught? Usually by your competitor reporting you after they did not get the job, because you offered the official more money than they did.

---

[2] 15 U.S.C. §§78dd-1, et seq. Note further that if you are part of a public company, the SEC (Securities and Exchange Commission) has the ability to levy fines under the books and records provision of the law. In other words, unless the public reporting by your company has an entry for "bribes for foreign officials" then the books of the company have not been accurately reported. The DOJ (Department of Justice) usually has hundreds of active investigations ongoing regarding the FCPA.

The general elements of the FCPA antibribery provisions are as follows:

- Giving, offering, or promising to give anything of value
- "corruptly"
- to an officer, employee, or agent of a foreign government or international organization, or instrumentality of that government...
- while knowing that the gift, offer or promise to give is
- for the purpose of influencing or inducing an act, or decision, securing any improper advantage, or inducing such foreign official to use his influence
- in order to assist in obtaining or retaining business for, or with, or directing business to any person...

## Export restrictions

US Department of State has export control over defense-related items. The US Department of Commerce through the Bureau of Industry and Security (BIS) has authority over most commercial exports, including so-called "dual use" items, e.g., items that can also have a military application, yet are intended for the commercial market. There are other government agencies that also control exports, depending upon what it is that you are exporting. For instance, controlled drugs, the kind that you should have avoided in college, are under the province of the DEA, but regular drugs, the kind you take after college, are FDA. A full list of regulatory agencies is available as Supplement 3 of Part 730 of the BIS Export Administration Regulations (EAR). In this age of terrorism, you should also be aware that the US Treasury Department's Office of Foreign Asset Control also has lists of debarred countries, targeted individuals, and sanctions programs that require compliance. In order to comply with the various statutes, at a minimum you should:

- Properly classify your products with the correct Schedule B or Harmonized Tariff Codes, as well as determine whether or not your product has an export control classification number.
- Screen your products against the US EAR to determine if they require an export license.
- Check for debarred persons against the BIS Restricted Parties lists to make sure you are not shipping to someone you shouldn't be.

## Cryptography issues

Wikipedia actually gives good direction on this point:

> *Cryptography exports from the U.S. are now controlled by the Department of Commerce's Bureau of Industry and Security. Some restrictions still exist, even for mass market products, particularly with regard to export to "rogue states" and terrorist organizations. Militarized encryption equipment, TEMPEST-approved electronics, custom*

*cryptographic software, and even cryptographic consulting services still require an export license. Many items must still undergo a one-time review by or notification to BIS prior to export to most countries. The regulations, though relaxed from pre-1996 standards, are still complex, and often require expert legal and cryptographic consultation. Other countries, notably those participating in the Wassenaar Arrangement, have similar restrictions.* [Accessed on October 9, 2013 at http://www.wassenaar.org/]

You generally will not have any problem getting a license for AES and DES3 encryption as it is believed that the real reason for requiring a license is so that the US government can identify the exported devices and the associated encryption scheme embedded on the device.

## ITAR issues

ITAR refers to the International Traffic in Arms legislation. This is worth mentioning because it can be misleading to some in the market. You may think that you don't traffic in arms so you can skip on. However, the law can apply to any item that has been modified for a military or security purpose. A ruggedized computer board that has been modified to meet military specifications can be subject to the law. Products that are used by police for internal security can be subject to this law's reach. Importantly, you are supposed to register if you are making any kind of "arm," even if you have not exported. The law applies not only to devices but also to the data associated with the device. Furthermore, be advised that the law is greatly concerned with who is exposed to the device or the data in that foreign nationals, even if working in the United States, are not allowed access to the foregoing. You should seek legal advice if you are not sure if your project is controlled or not so as to not see an announcement like this one from 2008 by the Department of Justice:

KNOXVILLE, TN—A federal grand jury in the Eastern District of Tennessee returned an 18-count indictment today charging J. Reece Roth, a Professor Emeritus at the University of Tennessee, and Atmospheric Glow Technologies Inc., a Knoxville-based technology company, with conspiring to defraud the US Air Force and disclose restricted US military data about unmanned aerial vehicles, or "drones," to foreign nationals without first obtaining the required US government license or approval.

## Export of high-performance computers (Section 732.3 of the EAR)

The US government regulates the export and re-export of high-performance computers (HPCs) based upon the adjusted peak performance (APP) of the computer.[3] Computers with

---

[3] You might find it interesting to know that the whole export scheme is allowed to be regulated pursuant to the International Emergency Economic Powers Act. This act allows the President to declare an emergency and control commerce in the process. The *Export Administration Act of 1979*, with all of the regulations like the ones mentioned here, expired in 1994. So, congress passed a law, it expired, and then the President in 1994 declared an emergency and the law is back on the books!

APP exceeding 0.75 weighted teraflops are called HPCs and are subject to stricter US export controls. According to the government:

> *APP is simple, can usually be calculated with publicly available vendor literature, does not require actual benchmarks, and provides a reasonable degree of accuracy in ranking HPCs. Like CTP, it produces a peak number which can be thought of as a "not to exceed" value, independent of memory and I/O considerations. The only thing that matters is the computer's ability to produce 64-bit or larger floating-point arithmetic results per unit time.*

- *APP is measured in Weighted Teraflops or Tera Floating point operations per second. Determine how many 64 bit (or better) floating point operations every processor in the system can perform per clock cycle (best case). FPO(i).*
- *Determine the clock frequency of every processor. F(i).*
- *Choose the weighting factor for each processor: 0.9 for vector processors and 0.3 for non-vector processors. W(i).*
- *Calculate the APP for the system as follows: APP = FPO(1) \* F(1) \* W(1) + . . . + FPO(n) \* F(n) \* W(n).*

As a practical matter, scalability can become an issue where an individual computer may have a weighted teraflop rating way below the 0.75 threshold, but the ability to put a bunch of computers in a chassis, such as in blade server configurations, may present the argument that a restriction should apply, even if you are not planning to ship a total system. At least this is how certain officials have acted in the past even though the calculations above and the exact letter of the law argues a contrary result.

### Controls on HPC exports

For HPC exports and re-exports, the US government has divided the world into three tier groups based on US policy and security interests. Each country's tier group has a maximum APP limit. These are the three tiers:

| | |
|---|---|
| Tier 1 | There is no maximum limit |
| Tier 2 | Less than 0.75 weighted teraflops |
| Tier 3 | Restricted-export license required |

If you find yourself in the export business of an HPC system, it is highly advisable to be sure to know your customer and your customer's customer so as to determine the tier where your customer is located so as to not run afoul of the export regulations. Additionally, you should not be cavalier in your attitude in assuming that a seemingly benign item, to a seemingly benign country, would not be export controlled. Consider that in March of 2005, Metric Equipment Sales pled guilty in the Northern District of California to one felony count of exporting digital oscilloscopes "controlled for nuclear nonproliferation reasons" to

Israel without a BIS license. The item of concern? Oscilloscopes with sampling rates exceeding 1 GHz. The penalty? A criminal fine of $50,000, an administrative penalty of $150,000, and a five-year suspension of export privileges.

Another example of, "Gee, I didn't know that that would be wrong" is exemplified by Cabela's Inc., the outdoor equipment outfitter in Nebraska. They were fined $680,000 for selling riflescopes to Argentina, Brazil, Canada, Chile, Finland, India, Ireland, Malaysia, Pakistan, etc.

If you or your company accidently export in violation of the law, you should seek legal counsel. He may advise that you voluntarily report the violation, as voluntary disclosures with promises of stepping up compliance, go a long way with the investigators.

## Antiboycott considerations (ignoring, "I told you not to play with her!")

The Arab League has engaged in a boycott of Israel. In doing so, Arab states often require trading partners to participate in the same boycott. The United States, as a friend of Israel, has enacted provisions in the Export Administration Act and amendments to the Tax Reform Act of 1976, that prohibit US persons from participating in any such boycott. You, as an exporter, are required to keep records of boycott requests and to report same to the BIS.

An example: According to BIS, "during the period of 2001 to 2005, Colorcon Ltd, of the United Kingdom, a wholly owned subsidiary of Colorcon, Inc of West Point, Pennsylvania, furnished to persons in Syria ten items of prohibited information about another person's business relationships with boycotted countries. Also, Colorcon knowingly agreed to refuse to do business with another person and failed to report its receipt of a boycott request." They ended up paying a $39,000 civil penalty. Note that the US company is not immune from the law, just because certain events were transacted through a subsidiary in a foreign country. It is also interesting to speculate how commercial interests may have played into the actions that were of concern here.

If you are dealing with the Arab states, you should note carefully the terms and conditions on innocuous documents, such as Terms of Purchase. Often, buried in the fine print, will be some kind of acknowledgment by seller that they have not shipped a similar product to Israel.

## Arbitration clauses under international contracts

There are some good reasons to give careful consideration to how a dispute may be arbitrated in dealing with international agreements.[4] What good is it to sign an agreement and even litigate the matter to a successful end, if the resultant judgment is not going to be

---

[4]  It can be a good idea to put arbitration obligations into domestic contracts as the Supreme Court has ruled that the Federal Arbitration Act is the supreme law of the land and supersedes state court jurisdiction.

able to be collected? Or, conversely, you may not want to be subject to trial before a tribunal that is parochial in its intent, that is often found in foreign courts where Americans are concerned. Fortunately, about 150 countries are signatories to the Convention on the Recognition and Enforcement of Foreign Arbitral Awards which is commonly known as the New York Convention of 1958. You and your trading partner can agree to private arbitration or arbitration before some formally constituted commissions or associations such as:

Court of Arbitration of the International Chamber of Commerce (Paris based)
Inter-America Commercial Arbitration Commission (comprised of national sections for the Americas)
American Arbitration Association through its international arm, the International Centre for Dispute Resolution (based in New York City).

Therefore, if you are engaged in a commercial transaction with a foreign entity, ask your attorney that reviews the agreement if it contains an arbitration clause. Ask further how the arbitration would work, what law would be applied, where would the arbitration take place, and how would a mandatory clause seeking arbitration be enforced.

## Insurance

Most of the jokes about insurance are true. It is quite a racket. There is a reason some of the tallest buildings in the world have insurance company names on them, and it is not because they pay all their money to claims. Unfortunately, business operations today generally will require you to have insurance. Even if you have just started up a small engineering firm, the landlord will most likely require that you carry a certain amount of liability insurance. Even your larger customers will require that you have a commercial liability policy. For smaller businesses that may not be in a position to approach Lloyds of London, you are going to be offered a commercial policy following the standard form commercial general liability. All the insurance companies follow the basic form policy with various addenda designed to comply with individual state laws. A cynical, but not far-from-the-truth view, is to understand that the policy does not cover much. Without special riders and the payment of additional premiums, most of the real threats to your business are not covered. For instance, the frequently litigated sexual harassment claims are not covered. Patent infringement, trademark infringement, and other intellectual property infringement matters are not covered. Products liability is not covered. Even if you buy the rider for products liability, there are additional exceptions to coverage, such as aircraft products or medical products claims. Most forms of professional liability, such as the obligation to adhere to a professional creed, or code required of accountants, lawyers, and professional engineers, and violations of the professional standard to which they are subject are not covered. So what is covered you ask? Personal injuries for inland marine accidents for boats of less than 25 feet in length; fire damage (as long as not terrorism related) and personal injuries to third parties due to your negligence on your premises, such as slips and falls. Your employees are not covered for their injuries because that is covered under worker's compensation. There are a few other commercial sounding coverages such as "advertising injury" if you happen to accidently slander somebody on your web site.

So, how can a company or the individual engineer protect themselves? Since insurance is said to be "risk spreading," what you want to do is spread the risk around with the firms or entities with whom you are dealing, sort of like your own private insurance pool. As an employee, if you are fortunate to be in a position to negotiate an employment contract, you want to ask for indemnification by the employer for any acts you commit within the scope of your employment. The company wants to negotiate an indemnification agreement with its suppliers for items, such as intellectual property infringement. The company may want to limit its liability to the amount of consideration it receives on any particular deal. An engineering firm may want to make sure that contracts contain provisions that require the other party to pay its attorney fees in the case that a matter goes to court or arbitration. An interesting approach is to put a requirement in contracts or purchase orders that say that any claims must be brought within one year, and the company must be notified within one week of any alleged defect or problem. The idea is to spread risk, off-load risk, minimize damages, and make litigation less attractive.

## Compliance—or why won't you comply?

There is a host of electrical-type compliance standards worldwide. It is often asked by embedded engineers, "What compliances do we have to have?" This is a complex issue, because the answer depends upon the type of equipment, where it is going to be installed, what country it is going to be sold into, and what type of risks are being sought to avoid. For instance, there is no US nationwide general standard for safety of electrical products. There are many local government standards for certain electrical products. Accordingly, to minimize products liability, it is a good practice to apply some standard to electrical products. How would it look to a jury to have a company admit to the fact that it designed and manufactured a product to *no* standard, but the product passed all its tests! Although it does not amount to a safe harbor, it is advantageous to be able to say that the product passed some tests that were administered to a standard, and that the tests were administered by a third party.

### Typical compliance certifications done by a representative US datacom manufacturer for new telecom products:

**EMC:**
 Emissions: EN55022 Class A (commercial) or B (residential)
 Emissions: FCC Part 15 class A or B (for USA)
 Immunity: EN55024

**Safety:**
> International: EN60950-1 CB-Scheme
> USA: NRTL testing according to CSA/UL60950-1

**Telecom** (most general):
> USA: FCC Part 68
> Europe: TBR21

Note that this company seeks to pass the EN60950 or the UL 60950[5] that are very similar safety standards, designed principally to avoid fire and shock hazards. In the list above, note further that this company will seek NRTL certification meaning a Nationally Recognized Test Lab will be used to meet the CSA/UL standard. A resultant issued CSA mark with the indicator "US" or "NRTL" means that the product is certified for the US market to the applicable US standards.[6]

Underwriters Laboratories (UL) makes a distinction between products that are "Listed" and components that are "Recognized." A UL Listing Mark contains the following elements:

- UL in a circle mark,
- the word "LISTED" in capital letters,
- an alphanumeric control number,
- the product name.

To be *listed* also implies that the product is subject to the UL Follow-Up Service (FUS). The FUS is part of an agreement you might have with UL that allows for audits by UL to determine that the product is being manufactured, and the components are being used as per the original listing. To not run afoul of the FUS agreement, and, if you need to have certain components second-sourced, You should have these components approved, or have the agreement acknowledge the possibility of certain substitutions.

A UL Recognized Component Listing is commonly found on individual components, such as plastics, wire, and printed wiring boards, that may be used in specific products or in a wide range of applications. Note that having a product comprised of a bunch of Recognized Components does not imply that the entire product is approved, or that the resultant conglomeration is necessarily safe.

Note further that international certification will be sought pursuant to the "CB Scheme." The CB Scheme is a global approach to proving compliance for product safety. Instead of an "NRTL" for the United States, an approving laboratory for international

---

[5] UL stands for Underwriters Laboratories. They develop standards and testing systems to, in the first instance, assist the insurance industry in managing products liability risk.

[6] Canadian Standards Association. CSA International is an example of an accredited NRTL by the Occupational Safety and Health Administration (OSHA).

certification is called the Competent Body Test Lab (CBTL). A CBTL laboratory could also be your local NRTL and issue CB Test Reports. There is no mark or sticker to apply to your product for CB Scheme approvals. However, once your product is evaluated, a CBTL issues you a report called the "CB Scheme" that can be presented to any CB Member Country in the world, whereby you receive their marks, logos or approvals accordingly.

The approval process with a CB Scheme report varies by country. Some, like India, require nothing more than a certificate proving compliance. Others (like the UL in the United States or CSA in Canada) require a fee, complete a review of the report, and reserve the right to look at a sample. The CB Scheme is an alternative method to receiving an NRTL mark.[7]

EN stands for Euro Norm. It means that the standard has been adopted as a European standard. Variously, standards are sometimes IEC standards, EN standards, and country specific standards, such as British Standards, so a product may end up with multiple prefixes such as BS EN 60950. This is a British Standard and is also a Euro Norm.

### CE mark

The CE (Conformite Europeene) marking is mandatory for certain product groups in the European economic area, consisting of the 27 Member States of the EU and EFTA countries Iceland, Norway and Liechtenstein.

Sometimes a customer will ask for a Declaration of Conformity. A declaration of conformity is a declaration that the product will comply with the EU tests from a given standard when installed as instructed. The declaration can then be used by the purchaser to apply the CE mark themselves without further tests being performed. The usual practice is to have such a declaration signed by an individual. If you are the person being asked to sign such a declaration, you want to be sure that the assertions are true or else get someone else to sign.

It is a general practice to use third party representatives in administering the CE mark. Thus, a US manufacturer may use a NRTL to test a product, issue the documentation and render a Declaration of Conformity to the third party representative who adds the CE mark. In the first instance, the third party representative is taking on the conformance liability.

---

[7] Further information on the CB Scheme can be found at http://www.iecee.org/cbscheme/cbfunct.pdf.

Regarding the datacom company mentioned above, for the CE test, they will need to show compliance to the following standards:

- EN55022, EN55204 (EMC Directive 2004/108/EC),
- EN60950-1 (Low-Voltage 2006/95/EC),[8]
- EN50581 (RoHS Directive 2011/65/EU),
- ECO Directive 2009/125/EC if power supply furnished by third party.

They will test through a NRTL for the first two items. They will have internal documentation set up for RoHS compliance from their parts vendors, and will have datasheets/reports for power supplies to support the ECO directive.

Certain international standards are available from the IEC http://www.iec.ch/ or IHS http://www.global.ihs.com/.

---

[8] LVD stands for the low-voltage directive. This directive sets the way in which low-voltage products are to be harmonized to allow their free trade across the borders of member states of the EC. The full name for the LVD is Directive 73/23/EEC Council Directive of 19 February 1973 on the harmonization of the laws of Member States relating to Electrical Equipment designed for use within certain voltage limits.

# Case Studies

## Chapter Outline

## Introduction

Here are five very different projects. They fit particular situations but each may give some insights toward designing and developing systems that you may encounter.

## Two case studies from the Oak Ridge National Laboratory: development of real-time instrumentation systems

Authors:
**Kenneth W. Tobin, Ph.D., Dwight A. Clayton, Bogdan Vacaliuc**
*Measurement Science and Systems Engineering Division, Oak Ridge National Laboratory, Oak Ridge, TN*

Developing and Managing Embedded Systems and Products.
DOI: http://dx.doi.org/10.1016/B978-0-12-405879-8.00033-7

## Introduction

The Oak Ridge National Laboratory (ORNL) is the US Department of Energy's (DOE) largest science and energy laboratory, applying signature strengths in neutron science and technology, materials science and engineering, computational science and engineering, and nuclear science and technology in conducting research and development for DOE, other federal customers, and US industry. The Measurement Science and Systems Engineering (MSSE) Division at ORNL performs applied research and development in nationally important areas of energy and security, tying together ORNL's strengths through the development of novel measurement instrumentation and systems. Through its applied research programs, MSSE provides pathways for the translation of basic science to engineering applications. This is accomplished through the creation and realization of foundational capabilities and technologies in electronics, sensors, signal processing, and integrated systems. The two case studies that follow represent examples of the creation of unique instruments systems to address challenging measurement problems. The first describes a robust, portable chemical and biological mass spectrometer (CBMS) for use on the battlefield. The second case study describes a real-time radar simulation environment designed to test and validate radar systems for the US Army. Both of these systems integrate commercial off-the-shelf technologies with unique components, software, and control strategies to produce one-of-a-kind solutions to challenging measurement problems.

## ORNL case study 1—development of the CBMS

### Statement of the situation

The US Army's Soldier and Biological Chemical Command tasked ORNL with the development of the next generation of deployable mass spectrometers (CBMS) capable of simultaneously detecting and identifying chemical and biological warfare agents on the battlefield. Since this new mass spectrometer would be installed on a variety of military vehicles, the environmental storage and operating parameters were quite rigorous—the CBMS system has to work anywhere the US Army might deploy. In addition to temperature extremes and vibrations, the CBMS system was required to operate after being exposed to nuclear radiation.

While specially educated chemists typically operate mass spectrometers, the CBMS operators would be soldiers, not scientists, so the CBMS needed to be relatively easy to use. In addition, false negatives (i.e., not detecting a warfare agent when one or more were present) and false positives (i.e., detecting a chemical or a biological warfare agent when none were present) had to be minimized since lives were at risk either through exposure by failing to detect and identify an agent present or possible military response by alarming when an agent was not present.

As is often the case in the development of complex systems, this development required a multidisciplinary team. The CBMS project needed experts in mass spectrometry, microbiology, aerosol sampling, computer science, and electrical engineering. Experts in the mass production of complex systems also helped meet the goal of deploying the CBMS systems at an affordable cost.

### Issues

The extreme storage and operating environmental conditions for the CBMS limited the selection of some key components. Many of these components are not routinely screened for these extreme conditions: extremely cold storage temperatures, off-road use of a turbomolecular vacuum pump and radiation tolerance of electronic components. While mass spectrometry analysis typically takes place in a laboratory setting using fixed and calibrated equipment, the CBMS might operate in vehicles moving on the battlefield. The battlefield environment includes many sources of interference, such as diesel exhaust, smoke, and dust, that are not normally present in a laboratory.

Good communications between team members are always essential to the successful completion of a complex R&D project. Five different scientific and technical disciplines residing in different divisions at ORNL and a commercial partner who manufactured the first production run outside of ORNL communicated during the CBMS development. Eventually this commercial partner would manufacture the CBMS systems without assistance from the experts who designed and developed the system.

After the terrorists' attacks of September 11, 2001, the sense of urgency for the CBMS system increased dramatically applying pressure to the development and deployment schedule. Although the original development schedule included many parallel activities, we reexamined the schedules to identify additional overlap to shorten the time before the CBMS systems could be deployed.

### Solution

Early during the development process, we performed environmental tests on several key components to assist in the selection. For example, we tested turbomolecular vacuum pumps from two different vendors to determine how well they would perform while operating in an off-road vehicle. In addition, the end of the cold war reduced demand for dedicated radiation-hardened chip fabrication facilities, which became scarce and expensive. Finally, the temperature requirement for extremely cold storage limited the selection of some components—for example, common LCD displays freeze at low temperatures. Once we selected the key components and fabricated a working prototype, we placed the prototype system in an environmental chamber to determine if the prototype would withstand the storage and operating environmental requirements.

While the final system only needed to detect and identify chemical and biological warfare agents, the development and testing phases for the prototype required the additional detailed data of an expert mode. To address the need for additional data, we made accommodations for a separate keyboard and monitor.

We developed and maintained Interface Control Documents (ICDs), which allowed the various subteams of scientific and technical experts to proceed somewhat independently from the other teams. Weekly team meetings discussed issues that affected the interface to other subsystems. Often the solutions to these issues required multiple subteams to make adjustments.

Since producing a relatively large number of systems was the end goal, a commercial partner was competitively selected early in the program with the understanding that the selected commercial partner would build the first production units. The commercial partner had an employee work alongside ORNL staff members during the early months to facilitate information and knowledge exchange.

### Evaluation of effectiveness

We identified and eliminated nonconforming components early in the design phase by testing subsystems for various environmental compliances. Although this approach required additional resources early in the project, the final system-level testing resulted in no major surprises. This approach minimized schedules.

The largest barrier experienced was interpersonal; some project team members were reluctant to discuss difficult issues early—delaying the resolution of the issue. It is essential that all team members feel free to disclose technical issues as early as possible; often an issue can be solved by making minor modifications to another subsystem.

The early involvement of a commercial partner was essential to the successful completion of the CBMS project. While ORNL has long experience in developing unique complex systems, ORNL does not mass produce these systems. Active participation by the commercial partner early in the project kept the manufacturing team informed of the reasons behind design decisions on a continual basis. Likewise, manufacturing considerations, expressed by the commercial partner, influenced many design decisions throughout the project. The transition from the R&D phase to the manufacturing phase is often difficult for a complex system; the CBMS project benefited from the collaborative nature of the partnership between a DOE national laboratory and a private commercial company.

Table 24.1 summarizes some of the best practices used by ORNL in developing the CBMS project. Figure 24.1 is a photograph of the CBMS system.

**Table 24.1: Summary of best practices at ORNL for the CBMS project**

| Issue | Solution | Effectiveness |
|---|---|---|
| Extreme environmental requirements | Verify key components and subsystems met required environmental conditions prior to having a working prototype of the entire system | While testing key components and subsystems does not eliminate the need to test the entire system, the selection of key components can be greatly facilitated |
| Complex system to be operated by nominally trained soldiers | Developed sophisticated detection and identification algorithms with a simple user interface | Worked well and verified with a thorough system test program plan |
| Keeping development activities for a large multidisciplinary team on schedule and budget | Extensive use of Interface Control Documents (ICDs) with weekly team meetings | ICDs worked well by identifying interface boundaries. Team meetings were ideal events to identify any needed changes to the ICDs |
| Production units manufactured by a commercial partner | A commercial partner was selected via a competitive bid process early during the development phase | Early involvement of the commercial partner helped to smooth the transition from ORNL to the commercial partner |
| Minimize the delay between the completion of the ORNL prototypes and the manufacturing of the commercial partner's production units | Have the commercial partner participate in the weekly team meetings via videoconferencing | Involvement of the commercial partner in the weekly team meetings allowed timely information exchange |

### ORNL case study 2—development of the Common Radar Environment Simulator

#### Statement of the situation

The US Army's Project Manager—Radars (PM-RADARS) tasked ORNL with the development of a new-generation test equipment that would address the limitations of the current state-of-practice. When a radar system is developed, a design and development team codeveloped special hardware called the radar test environment simulator (RTES), along with the radar signal processor. The RTES interacts with the radar electronics and ensures that the radar performs correctly under various conditions and scenarios. Each RTES is purpose built for the radar it supports.

The common radar environment simulator (C-RES) was conceived to allow the US Army's test and evaluation mission to work seamlessly with the radar system architects and scientists as well as mission planners to take advantage of the full capability offered by the advances in computing and radar science. The C-RES enables the US Army to design missions incorporating scenarios with new target and behaviors observed on the

**Figure 24.1:**
Early production CBMS unit.

battlefield and locate the simulation *on the battlefield itself*. In this way, the behavior of the radar, when deployed, can be predicted and increase the likelihood of mission success [1].

As is often the case in the development of state-of-the-art computer systems, this development required the integration of new software and new hardware from many different vendors that were not explicitly designed to work together. The development of C-RES required tools for modeling and simulation, code generation, real-time operating systems, vendor-supplied source code compilers, and interprocessor communication libraries. ORNL assembled a team of modeling and simulation experts, radar signal processing experts, embedded systems experts, and systems integration specialists.

*Issues*

The C-RES project required functionality unavailable in the industry: a multiprocessor, heterogeneous computer system simulator that reacted to stimulus with a very short response time while simultaneously maintaining fidelity to actual behavior. To develop the desired functionality, the design team had the goal to integrate a variety of software tools and hardware from several vendors into a seamless process. The team faced many obstacles in achieving this functionality.

The functionality for generating code for general purpose and digital signal processors (DSP) from high-level simulation models was mature. However, the generation of code for field programmable gate arrays (FPGA) had just been introduced. The C-RES design relied on combining the operations of DSP and FPGA codes. Although largely functional, the new FPGA code generators exhibited one of three types of problems: (i) unsupported functions, (ii) incorrect code generation, and (iii) failures due to interdependent limitations in several software tools used to generate the logic.

The architecture analysis showed that C-RES required a DSP/FPGA hybrid, but few hardware vendors had developed such products. The C-RES team also faced immaturity in the hardware design and support, primarily in the inability to communicate reliably between the different processors. Simple test cases would succeed; however, long-term stress tests of communications exposed bit errors and communication halts due to faulty logic and thermal or board design problems, such as crosstalk.

Even after development, the system did not achieve the reactive response within the time required. Although the use of digital communication signaling offers some amount of tolerance in the timing of the response, every radar signal processor has predefined limits on when a response must be received. The C-RES must react to the waveforms it receives digitally from the radar signal processor to generate a return, so it must guarantee a response within that time window. This is the reason why most RTES systems are developed using either fixed hardware logic or FPGA technology. It is also the reason for limits in both functionality and flexibility.

Finally, the C-RES team faced two sociological problems in the latter days of the development: the demonstration mindset[1] and the purist syndrome [2]. The demonstration mindset is the drive for early capability because key program decisions rest with minimum capability demonstrations. This leads to an urgency of development and pressure to move on despite failures in the components. This was exemplified in the C-RES program by a reluctance to move to newer versions of the software tools because of uncertainty involved in the effort required to move to those tools. The purist syndrome is the goal to write perfectly structured code in every situation; in the C-RES program, this was manifest in the insistence on using the immature *vendor-neutral* FPGA generation tool when another alternative, but *vendor-specific* tool could have been used.

---

[1] "Report of the Panel on Reducing Risk In Ballistic Missile Defense Flight Test Programs," Defense Technical Information Center, p. 10, February 1998, available at http://www.dtic.mil/cgi-bin/GetTRDoc? AD=ADA354163.

*Solution*

The C-RES team was able to work around unsupported functions in the code generators by recasting the needed computation using simpler, supported building blocks and mathematical theory. Incorrect code generation was harder to discover because it only manifested when the application was run on the hardware and not in the simulation of the system in the modeling environment. The solution for this was to use classic model-driven engineering practice to develop and use unit tests that compared hardware execution with system simulation. Any discrepancies could be analyzed, identified, and eliminated in a simplified system configuration. Failures due to interdependent limitations were perhaps the most vexing issue faced by the team, since there was no error *per se* in the expressed code. The problems surfaced downstream when the software tools ran to produce the firmware. The FPGA code generation tool, for example, always generated synchronous reset for a lookup-table component. The FPGA vendor's synthesis tool was unable to place a lookup table into efficient block random access memory if the lookup-table code included a synchronous reset, and used distributed memory instead. The model had a lot of lookup tables, and so the synthesis would fail after spending several hours trying to fit into the device. The work-around was to manually modify the generated code prior to synthesis. Subsequent versions of both the code generation tool and the vendor's synthesis tool eliminated the issue.

The solution to the hardware communications failure was to commission the design of custom communication firmware that incorporated a calibration feature to adjust for the variability in the cabling and board design problems. Performing a detailed system thermal simulation and ensuring proper airflow over the required environmental operating conditions managed the thermal issues.

*Evaluation of effectiveness*

Solving the performance challenge is perhaps the best showcase for the power of the model-driven engineering process. The initial computational model of the environment simulator's reactive processing did not meet the performance target. To address this, the C-RES team instrumented the multiprocessor implementation to obtain timing and event information and ran it at speed. The timing information had to be normalized so that the behavior of all the processors including the FPGA could be plotted on the same time axis. Observing the behavior of the model running at speed allowed the team to visualize the progress of the model and to gain understanding of the bottlenecks producing the performance limitations. The solution reorganized the computational model over a series of iterations as individual bottlenecks were systematically eliminated.

Table 24.2 summarizes some of the best practices used by ORNL in developing the C-RES project. Figure 24.2 is a diagram of the C-RES system.

**Table 24.2: Summary of best practices at ORNL for the C-RES project**

| Issue | Solution | Effectiveness |
|---|---|---|
| Immature software tools | Carefully gauge the maturity of the software tools in use and use that knowledge to decide where to focus problem analysis efforts | Minimized the time spent searching for a problem in the wrong place |
| Immature hardware communication interfaces | Insist on confidence test programs from the vendor or develop and use them to rule out hardware problems | Minimized the time spent in searching for software problems that were due to hardware issues |
| Performance | Obtain or develop profiling and system analysis capability early in the integration process | Using traces captured during system execution, bottlenecks were identified and the computational model was modified to meet performance requirements |
| Demonstration mentality | Resist the temptation to disqualify solutions because of time constraints | The correct solution is always the one that leads to long-term benefit. If needed, spin off a separate team to develop the "quick-fix" implementation |
| Purist syndrome | Resist the temptation to "remain pure" to the ideal of the project when a less technically elegant solution could be effective | In the face of serial and systemic adversity to achieving success, it is prudent to honestly consider a less flexible solution that achieves the requirement |

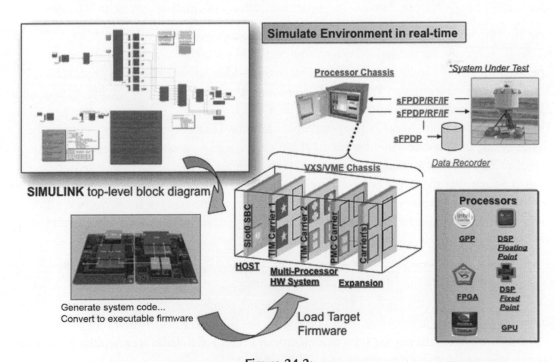

**Figure 24.2:**
Model-driven engineering and automated firmware generation for C-RES.

## Case study 3: design of a parallel computer-based, streaming digital video instrument

**Authors:**
**Lee Barford, Hong-Liang Xu, Chun-Hong Zhang**
*Measurement Research Laboratory, Agilent Technologies*

Analog television transmission and reception equipment (including cable television and common antenna systems with amplifiers and other active components) could be tested with standard instruments. Power meters verified signal strength. Oscilloscopes identified timing problems. Spectrum analyzers helped technicians spot and gain information about interfering signals. Television receivers provided subjective verification of received video and audio quality.

Today, manufacturing test of digital television equipment and commissioning and maintenance of digital television systems requires instrumentation specialized for each transmission standard. This is because in addition to radio frequency (RF) measurements (e.g., power spectral density (PSD), signal-to-noise ratio), measurements of the quality of the specific digital modulation such as error vector magnitude, bit error rate, and constellation diagram display are needed. Making these digital measurements requires the instrument to perform tasks such as identifying and synchronizing to data frames and extracting symbols. These tasks require detailed information concerning the signal format, specifically how symbols, frames, and synchronization information appear in the signal.

One instrument should provide all these RF and digital measurements and display the received video in one package. To catch intermittent problems, the instrument should operate in a streaming mode, that is, the incoming video signal should be processed without any temporal gaps. To achieve gapless measurement, the average processing rate cannot fall below the rate at which the video signal is sampled; it must achieve a minimum throughput. Another temporal consideration is latency, the delay between the signal being sampled and the display of the corresponding measurement results. To be useful to technicians debugging on a bench top or in the field, the latency must be small enough that the delay between making a physical action such as connecting a probe and seeing the measurement results change must not be annoyingly long. So, a digital video instrument has a hard requirement on throughput but a subjective requirement on latency.

This case study describes the construction of such an instrument. Figure 24.3 shows its hardware architecture, which comprises a preamplifier and downconverter, a high speed analog to digital sampler, and a PC used to do signal processing, compute the measurements, decode the video, and display the measurement results and received video (Figure 24.4). We selected the RF components and digitizer to have bandwidths sufficient for the highest bandwidth measurement, in this case PSD.

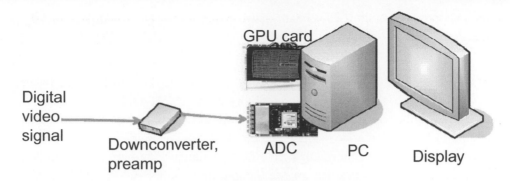

**Figure 24.3:**
Hardware architecture of the digital video instrument.

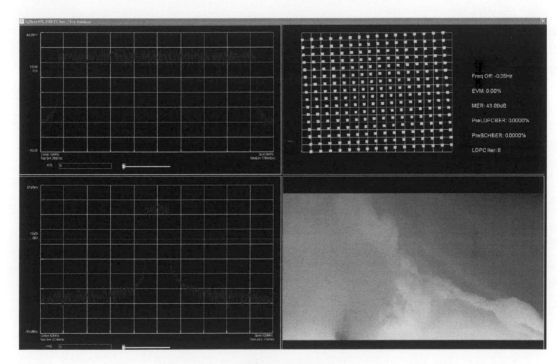

**Figure 24.4:**
Measurement and video display of the instrument: narrow bandwidth PSD centered on the video signal (upper left), modulation measurements including a constellation diagram (upper right), real-time display of the video program (lower right), and a wider bandwidth PSD (lower left).

A particular constraint of this design is that all computational processing was to be done using standard components from the information technology industry. To support the high computation load of signal processing, measurements, and video decoding and display, the most powerful desktop computing components readily available were chosen: a quad-core $\times$ 86 CPU and a graphics processor unit (GPU) board optimized as a massively parallel, floating point processor. We installed a second video card so as not to burden that GPU with driving the display.

It is necessary to collect samples into buffers and then process the buffers to amortize software overheads like function calls and memory management operations over the processing of a number of samples. When using traditional serial processing, the main consideration in buffer size selection is that there be at least as many samples in a buffer as that needed by the measurement requiring the highest number of consecutive samples. Had we used serial processing in this instrument, the minimum buffer size would have been determined by the number of samples needed to compute a PSD with the desired resolution bandwidth.

Using parallel processing adds overhead: coordination and synchronization among the CPU cores, among the GPU cores, and between the CPU and the GPU. Using rather large buffers, even of millions of samples, results in the highest throughput (samples processed per second) [3,4]. However, raising buffer size increases latency because measurement results depending on the first sample in the buffer are not obtained until the last sample in the buffer is acquired and the entire buffer has been processed. So when using parallel processing in measurement, there is a roughly linear trade-off between throughput and latency.

This instrument required a minimum throughput, as was discussed above. Modifying the value of a constant in the code and recompiling the instrument software changes the buffer size. So, once we completed a prototype of the instrument software, we iteratively increased the buffer size until we attained the required average throughput. We checked the latency by comparing the video display of the instrument with the display of a TV being fed the same signal. The TV display could barely be seen to be ahead of that of the instrument. This amount of latency was acceptable.

The GPU performed the parallel signal and measurement processing on following the principles discussed in Ref. [5]. Each signal processing or measurement computation was divided into a sequence of steps, each one of which was straightforward to do in parallel. References [3] and [6] give detailed examples of this process.

In addition to using parallelism in computation, the instrument also required parallelism in data handling to achieve gapless, streaming measurement. At any given moment, it filled one buffer with samples from the data acquisition card, while it used a second buffer to perform signal processing, measurement computation, and video decoding and decompression. It updated the measurement display from the prior computed measurement results. Finally, it updated the video display at the video frame rate. This parallel data handling was obtained using multiple threads running on the multicore CPU.

This work demonstrates that parallel computation can advantageously be used in instrumentation. The primary design consideration that is introduced is the need to make trade-offs between parallel compute performance and real-time performance as measured by throughput and latency.

## Case study 4: troubleshooting a boiler points out the need for good, comprehensive design and development

**Author:**
**Jake Brodsky**
*PE, Control Systems Engineering*

We have a Munchkin 140M condensing boiler configured for natural gas that, with routine maintenance, ran reliably for 3 years. One morning, it stopped heating, and the controller display showed a code of F13.

I design control systems for industrial processes. I would have jumped in to this thing right away except that at that time I was traveling. My wife called upon three different, reputable service firms to fix this boiler. The first one simply reset the unit. The second gave us the equivalent of an intelligent shrug, saying that "we don't service these things any more" and the third told me that he thought I had a bad blower, but just to be sure he wanted to replace the control board as well for nearly US$1000. Even then, he wouldn't guarantee his work and he really thought I should replace the whole unit to the tune of nearly $11k.

That is when I got involved. I looked up the code on the display in the service manual. Supposedly it meant that the blower in this boiler wasn't spinning fast enough to safely open the gas valve and begin ignition. The problem was that, according to the diagnostics display on the controller, it actually was spinning at least 200 rpm faster than the lowest speed for ignition. If I reset the controller, it would begin ignition, heat the water, satisfy the demand, and then shut down with that mysterious error code again. Sometimes, it would inexplicably shut down before demand was met.

I was suspicious about the controller board. There was little that could explain what it was balking at. I decided that I needed to make sure all the sensors were working as they should. One by one, I tested them to make sure the controller responded appropriately. The controller reacted pretty much as I would have expected. While doing this, I noticed that many key parts weren't made by Munchkin. The blower was made by EBM Pabst of Germany; the gas solenoid was made by Dungs of Denmark; the controller was made by S.I.T. Controls of Italy; the rest was apparently assembled in the United States. This thing reminded me of the Tower of Babel story in the Bible.

Though there wasn't much data about the controller, EBM Pabst has a web site with lots of data about their blowers. I had presumed that the blower was operating with some degree of control because when I unplugged the control lines while powered up, it went to full speed. From EBM Pabst, I found out that if I grounded the pulse width modulated speed control line, it should go to a standby mode. I tried that, and the blower didn't shut down. Now I had reason to spend hundreds on a new blower. After replacing the blower and checking for leaks, I put it to the test. Everything worked properly again.

Looking back, I have to wonder whether S.I.T. Controls considered this failure mode. The F13 code appeared to be triggered because the blower speed was not modulating properly, not because it was under-speed, as the error code documentation suggested. The controller also wasn't consistent about why or how it tripped, leaving everyone wondering what the controller was balking at.

I give EBM Pabst low marks for quality control. S.I.T. Controls gets failing marks for erratic firmware behavior and poor Internet presence. However, Munchkin completely flunked because of its manual.

Despite detailed performance data, the boiler manuals did not have a ***Control Narrative*** section. The Control Narrative explains in gory detail exactly what the automation does in response to all stimuli in every automation state that it could have. It is a key deliverable to clients. As a registered controls engineer myself, I wouldn't dream of designing a potentially dangerous device such as a gas boiler without explicit documentation of this sort.

Without that narrative, even experienced, reputable service firms have little idea what the unit is supposed to be doing, why an error code might be there, or what tests one might use to diagnose it. It is not enough to fail-safe on devices like this. ***Customers or their designated repair firms deserve a reasonable explanation of what kept a device from working.***

These firms were not populated by fools. They were all knowledgeable, reputable, and capable people who wanted to do the right thing. What they lacked is some way to reason through this problem. There may not have been a way that Munchkin themselves could have known that the F13 code documentation was wrong. And it turns out that Munchkin is not alone in this. Many firms build their products out of the pieces and parts from other companies.

There were other hints that the control board from S.I.T. Controls had been adapted from other projects that didn't really resemble this one. There were mysterious configuration registers with instructions not to change them. There were hints of an external MODBUS control system, but no indications of how it should be used.

This is the hazard of modern controls: unless you define exactly what a system is supposed to do at every step of the way and what stimuli it expects, there is no way to know that anything you have is in specification or that it is even broken.

In contrast, one of the projects I worked on where I'm employed is a filter backwash system. We have backwash control narratives with over a dozen steps. We carefully document each of these steps for the operators. If a step gets stuck there will be an alarm and the operators can know exactly how to respond to get things moving again.

The lack of any such documentation in this application shows that this company was too sloppy to care, and they're hardly alone in this. In the last several years, I had problems with a dryer, an oven, and a cook-top. The common thread in all of these products is a missing set of documents to explain how the thing is supposed to do its job. It is no wonder modern "technicians" are reduced to board swapping—it's all they can do.

The only reason I was able to fix this boiler was because I reasoned my way through the entire process narrative. In effect I had to reverse engineer this thing. Expecting a technician to spend time doing that is, well, optimistic. Were it not for the motive of saving over $10k, I don't think I would have bothered either.

I really wanted to do a postmortem on the blower, but it was assembled in a way that made removal of the control board very difficult for someone with just hand tools. My best guess is that perhaps one or more of the power devices in an H-Bridge shorted out. This might cause the motor to be less controllable, without actually failing.

As an aside, note that the world's market for electronic devices has always been fraught with knock-off imitations that are not as robust. From what I've read in various electronics magazines, these bogus devices continue to find their way into many distribution channels.

This goes to show just how difficult it can be to build a reliable product. And where profit margins are very narrow, most people simply do not want to be bothered with this level of diligence. Unless it is mandated by regulation it probably won't happen.

## *Case study 5: debugging of electromagnetic compatibility issues*

**Authors:**
**Daryl Beetner, Natalia Bondarenko, Shao Peng, and Tom Van Doren**
*Missouri University of Science and Technology*

Despite the design engineer's best efforts, many embedded systems will have electromagnetic compatibility (EMC) problems that must be solved before they hit the market. Debugging these issues can be challenging.

One of our partners recently asked us to help with such a problem on a display module used in transportation systems. While the emissions from the module were well below Federal Communications Commission (FCC) limits, the module had to meet much lower limits in some bands for the particular application. Of particular concern was the 1.2 GHz GPS band. Even very low emissions from the display module could prevent a GPS receiver in the vehicle from receiving a strong signal. The following study shows how we found and eliminated the primary source of 1.2 GHz emissions.

A diagram of the printed circuit board (PCB) used in the display module is given in Figure 24.5. All communications and power are routed through two connectors. The board

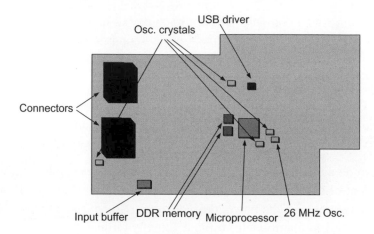

**Figure 24.5:**
Diagram of the PCB used in the display module.

sits in a plastic enclosure. Several components work at frequencies that could cause emissions, including a microprocessor (clock frequencies of 33.3, 100, and 400 MHz), a USB chip (60 MHz), DDR memory (133.3 MHz), and several oscillator crystals (32, 1, 8, and 26 MHz). Each of these clocks has harmonics around 1.2 GHz, so there are many possible sources of emissions.

EMC issues can be solved by reducing the strength of the source of the emissions energy, reducing the effectiveness of the coupling path between the source and the victim, or by making the victim more robust against the emissions. Since the victim cannot be changed, here we must first find the source or the coupling path that generates the strongest emissions and then mitigate the emissions from this source. Changing clock frequencies so that they do not have harmonics at 1.2 GHz is sometimes a possibility, but was not an option for this board.

For many systems, the first step in reducing radiated emissions is to find parts of the system that might serve as good, unintentional antennas. Good antenna parts are conductors (or slots in conductors) with dimension on the order of a quarter wavelength (e.g., the length of a power wire connected to a PCB, the PCB return plane, or a slot in an enclosure). At low frequencies (e.g., 100 MHz), only a few good antenna parts are generally present so they are easy to identify. At 1.2 GHz, however, a quarter wavelength is only 6.25 cm. There are many possible antennas on this board.

Lacking obvious antennas, another good first step is to determine what components may be possible sources of the emissions. Peaks in the radiated emissions are one clue. Figure 24.6

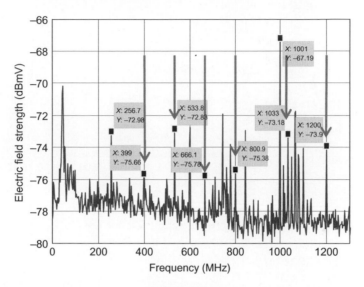

**Figure 24.6:**
Radiated emissions from the display module.

shows the radiated emissions from the display board as measured in a 3 m semi-anechoic chamber. Large peaks occur at harmonics of 33.3, 100, 133.3, and 400 MHz, suggesting the source is associated with the main processor or the DDR memory.

To further identify the source of emissions, the crystal oscillator frequencies were modified slightly while observing the emissions at 1.2 GHz. The oscillator frequency was modified by warming the crystal with a finger. Figure 24.7 shows an example of the radiated emissions measured by a spectrum analyzer in max. hold mode, while warming the crystal oscillator driving the microprocessor, DDR memory, and USB ports. The radiated emissions normally showed only a single peak around 1.2 GHz. When a finger was placed on the part, however, the peak would move over time as the crystal got warmer. The "broadband" pattern in max. hold mode was created by the sweeping oscillator frequency. Since this change in 1.2 GHz emissions was not observed when warming any other crystal, the emissions source must be associated with the microprocessor, DDR memory, or USB port.

The emissions source can sometimes be found using a near-field probe. A simple near-field probe can be constructed relatively easily using a coaxial cable as shown in Figure 24.8,

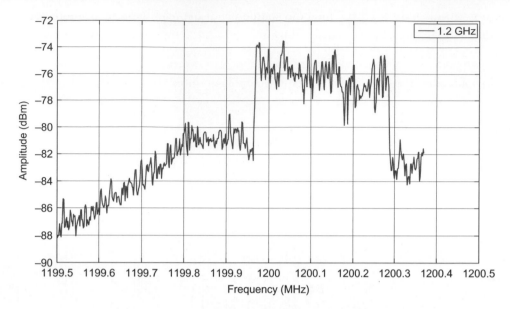

**Figure 24.7:**
Radiated emissions while warming the crystal responsible for the clock to the microcontroller, DDR memory, and USB driver.

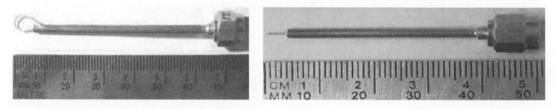

**Figure 24.8:**
Homemade near-magnetic field probe (left) and near-electric field probe (right).

where the center conductor is connected to the shield to form a magnetic loop probe, or where the center conductor is left open circuited and protrudes a short distance out of the shield to form an electric field antenna. Such a probe was used to investigate where strong magnetic or electric fields might appear around the PCB, by connecting the probe to a spectrum analyzer and scanning it near parts around the board.

Strong 1.2 GHz fields were found around a single pin of the input buffer shown in Figure 24.5. This buffer feeds a signal to the microprocessor.

Strong near-field emissions do not mean that the part generated strong radiated emissions, but do mean it could be a potential source of these emissions. One method to eliminate the pin as a source is to connect it to an efficient antenna. If the radiated emissions do not increase when the wire is added, then the source is not the main culprit. If the radiated emissions increase, however, it only means that the source may warrant additional investigation.

A wire was attached to the suspected pin as illustrated in Figure 24.9. The wire was 6 cm long—roughly one quarter wavelength at 1.2 GHz. Radiated emissions increased by 20 dB

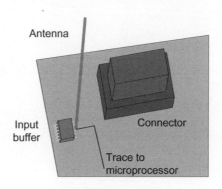

**Figure 24.9:**
The potential of the buffer pin to be associated with the primary source of emissions was tested by connecting an antenna to the pin and checking for an increase in radiated emissions.

at 1.2 GHz after attaching the wire, showing the pin could be a source. The trace between the microprocessor and buffer was cut and the test repeated with the "antenna" first connected to the buffer and then to the microprocessor side of the trace. The emissions only increased when the wire was connected to the microprocessor side of the trace.

The input buffer was connected to an input pin of the microprocessor. While an input pin can be a source of emissions, it is unusual to be so strong. The 1.2 GHz energy was probably coming from somewhere else. Traces near to the trace between the buffer and microprocessor were good possibilities.

Nearby traces were identified and the noise voltage on these traces was measured with a direct contact probe, shown in Figure 24.10. When the probe is placed on a trace, current flows through the probe tip, through the measurement instrument, and back through the cable shield, where it returns to its source through the capacitive connection between the shield and the board return plane. This configuration allows one to quickly make measurements of the *relative* magnitude of the voltage among the traces without significantly loading the trace drivers.

An example result is shown in Figure 24.11. Several traces near to the input buffer trace had similar voltage spectra, with strong emissions at harmonics of 100 MHz. These traces

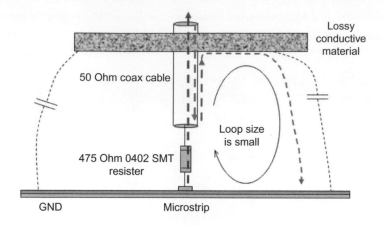

**Figure 24.10:**
Direct contact probe used to capacitively measure the voltage on each pin.

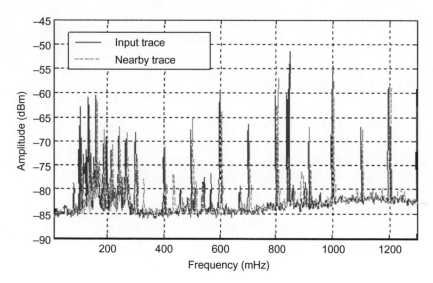

**Figure 24.11:**
Measured voltage spectrum on the input buffer trace and another nearby trace.

were all connected to input pins or low-frequency output pins of the microprocessor, suggesting that a single source was coupling energy to all of these traces.

To demonstrate whether any of these measured traces might itself be the source—or at least be part of the path between the source and the antenna—the traces were shorted at high frequency by connecting a capacitor between the trace and the return plane. If the trace was responsible for capacitively coupling energy to another nearby trace or antenna, the emissions should decrease. If it was responsible for inductively coupling energy, the emissions should increase. No change in emissions would suggest the trace was not a critical part of the emissions mechanism. While shorting these traces with a capacitor significantly reduced the voltage on the trace, no change was observed in the 1.2 GHz radiated emissions from the board. We still had not found the primary source of emissions.

Since the microprocessor operates at 100 MHz, it was a good potential source for the observed noise. An automated near-magnetic field scan of the microprocessor was performed to indicate which pins this energy might be coming from. The advantage of an automated scan is that it can be used to generate a map of the strength of the near field around the IC, as shown in Figure 24.12. The area in red (the shaded area underneath the " + " fiducial) shows the location of the strongest 1.2 GHz magnetic fields. Interestingly, many of the traces investigated earlier originated from this area.

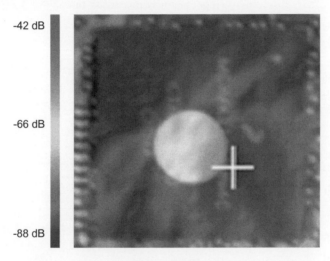

**Figure 24.12:**
Near-magnetic field over the microprocessor.

The direct contact probe was used again to investigate the microprocessor pins where there were strong 1.2 GHz magnetic fields. One pin, in particular, had strong emissions at harmonics of 100 MHz, as shown in Figure 24.13. This pin was terminated with a low impedance to see if this had an impact on the radiated emissions. When "shorted," the voltage on this trace and all the traces studied earlier dropped by more than 16 dB. The far field emissions reduced by 6 dB (Figure 24.14). We had found the source.

The pin responsible for emissions turned out to be a test pin that was not used after the board was shipped. Turning off this pin after testing resulted in an 8 dB reduction in emissions at 1.2 GHz, a simple fix with a big impact. If the pin could not have been turned

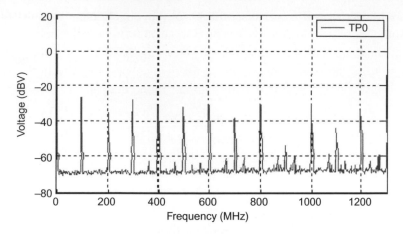

**Figure 24.13:**
Measured voltage on microprocessor pin TP0.

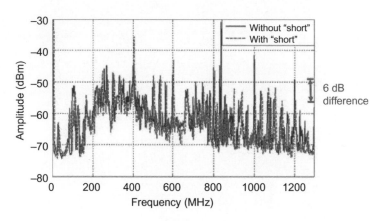

**Figure 24.14:**
Far field emissions from the module with and without the TP0 traced "shorted."

off, there were several other options that could have been investigated. Increasing the rise or fall time of the 100 MHz signal on this pin could significantly reduce the energy at 1.2 GHz. This change can sometimes be implemented in the microprocessor through the pin configuration. It could also be done by adding a filter to the pin—possibly by simply adding a ferrite or resistor in series with the pin. If the rise or fall time could not be increased, then one must prevent the 1.2 GHz energy on this pin from coupling to an efficient antenna. Some options for preventing this coupling include keeping a large separation between this trace and other traces or running this trace close to and between the power and return planes. It is critical that this trace be given a low-impedance return path.

A typical method for ensuring a low-impedance return path is to run the trace close to an unbroken return plane and that the trace is kept as short as possible.

Turning off the test pin significantly reduced emissions at 1.2 GHz but did not eliminate all emissions. Another source now dominates. Unfortunately, that is often the case. One rarely eliminates all emissions at once. When the strongest emissions source is eliminated, it is time to find and eliminate the next strongest source.

## *References*

[1]  M.A. Buckner, et al., From requirements capture to silicon: a model-driven systems engineering approach to rapid design, prototyping and development in the Oak Ridge National Laboratory's Cognitive Radio Program, Proceedings of the SDR '08 Technical Conference, Washington, DC, November 2008.

[2]  F. Baker, Organizing for structured programming, in: C. Hackl (Ed.), Lecture Notes in Computer Science, Springer, pp. 64, ISBN 978-3-540-07131-0, available at http://dx.doi.org/10.1007/3-540-07131-8_22.

[3]  V. Khambadkar, L. Barford, F.C. Harris, Massively parallel localization of pulsed signal transitions using a GPU, in: Proc. 2012 IEEE International Instrumentation and Measurement Technology Conference, Graz, Austria, pp. 2173−2177.

[4]  L. Barford, Parallelizing small finite state machines, with application to pulsed signal analysis, in: Proc. 2012 IEEE International Instrumentation and Measurement Technology Conference, Graz, Austria, pp. 1957−1962.

[5]  L. Barford, Multicore programming for measurement: a tutorial, IEEE Instrum. Meas. Mag. 14(2) (2011) 34−40.

[6]  T. Loken, L. Barford, F.C. Harris, Massively parallel jitter measurement from deep memory digital waveforms, in: Proc. 2013 IEEE International Instrumentation and Measurement Technology Conference Minneapolis, MN, (under review).

# Dependability Calculations

**Kim R. Fowler**

*IEEE Fellow, Consultant*

## Brief overview

Dependability describes how well and how long a system operates. It has analytical foundations in measurement science and stochastic processes. ***This short appendix only serves to whet your appetite for much more in-depth studies found in the recommended reading.***

Dependability has the following components:

- Reliability
- Testability
- Safety
- Availability
- Maintainability
- Performability

***Reliability*** is defined as the probability that the system will fully and correctly operate for a specified continuous time duration under specified conditions. If anything fails or degrades, then the system is considered to have failed. Reliability is an "all-or-none" statistic but it is useful and it supplies important information to several other components.

***Testability*** defines the ease of test for certain attributes within a system. [1] Testability also describes the proportion of potential faults covered and usually addresses the ease of diagnosing a potential problem or fault. Testable architectures generally are not automatic and do not provide continuous monitoring—either an operator initiates testing or a trigger periodically initiates testing.

"... ***[Safety]*** is the probability that a system will either perform its functions correctly or will discontinue the functions in a manner that causes no harm." [1] This is a fairly narrow definition of safety. Leveson argues for a much broader context to describe safety. [2] Safety designs often use redundant, dissimilar operations or subsystems to check operations.

***Availability*** is defined as the time a system is in service compared to the total time; the total time is the time in service plus the time out of service. Reliability is an important factor in the availability metric. Availability describes the percentage of "up time" that a system is operating. A system may be considered highly available even though it experiences frequent periods of nonoperation, which are extremely short in duration. "In other words, the availability of a system depends not only on how frequently it becomes inoperable but also, how quickly it can be repaired." [1] Availability allows for a less reliable system, but one with easy and quick maintenance to be put back into service, to have the same value as a more reliable, but more difficult to service, system. An example of a highly available system is a data server farm; individual disk drives are not terribly reliable but each one is easily and quickly swapped out for a very short repair time.

*Maintainability* describes how easily a system is maintained; it defines the ease of system maintenance, as well as the repair for a failed system. Quantitatively, it is the probability that a failed system or one down for maintenance will restore to operation within a set period of time. [1] Testability, BIT, diagnostics, and repairability all are intertwined components within maintainability.

*Performability* describes the level of performance or degradation for fault tolerant operation. Fault tolerance (an aspect of performability) means that a system continues operating in some form when a component fails or an operational fault occurs; the performance may decrease but the system continues operating. The difference between reliability and fault tolerance (performability) is that reliability defines the likelihood that *all* of the functions perform correctly, while fault tolerance (performability) defines the likelihood that a *subset* of the functions performs correctly. Degradation is an aspect of fault tolerance whereby the system reduces its performance during a fault. [1]

## Observed failure rates

Different systems have different rates of observed failures. Figure A.1 illustrates two examples of failure rates; one for mechanical systems and the other for electronic systems. Note that early in development or manufacturing of these components or systems they tend to have higher failure rates until the "infant mortalities" shake out. [3]

## First approximation: simplified failure rates

Many people use a first approximation for reliability calculations with simplified failure rates. The modeling assumptions are:

- Semiconductor failure rates are constant (between regions of infant mortality and the back of the bathtub in Figure A.1).
- Mechanical wear and failure rates have a mean expected life, $\mu$, and a standard deviation, $\sigma$, after debugging (i.e., after infant mortality).

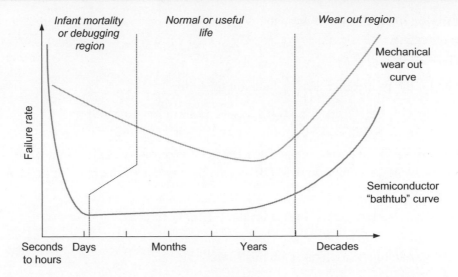

**Figure A.1:**

Examples of failure rate curves. © 2013–2014 by Kim R. Fowler. Used with permission. All rights reserved.

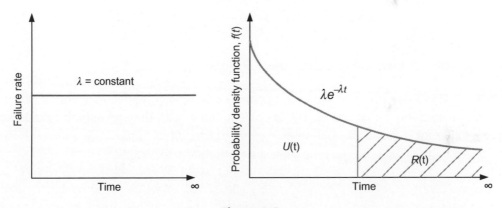

**Figure A.2:**

Assuming a constant failure rate for electronic systems gives an exponential probability density function. © 2013–2014 by Kim R. Fowler. Used with permission. All rights reserved.

The implication of the assumption of a constant failure rate for electronic systems is that if a component has not failed yet, it is still as good as new. This also means that the probability that the component remains working decreases exponentially with time, see Figure A.2. Reliability, $R(t)$, is a Poisson probability function:

$$R(t) = e^{-\lambda t}$$

where $\lambda$ = failure rate per unit time.

The probability of failure is unreliability, $U(t)$, which is:

$$U(t) = 1 - R(t) = 1 - e^{-\lambda t}$$

Consequently, the probability density function for failure:

$$dU(t)/dt = \lambda e^{-\lambda t}$$

A useful metric is the expected Mean Time Between Failures or MTBF. In a large population of identical components MTBF equals the inverse of the failure rate (assuming a constant failure rate):

$$\text{MTBF} = 1/\lambda$$

Consequently, reliability, $R(t)$, takes on this form:

$$R(t) = e^{-\lambda t} = e^{-t/\text{MTBF}}$$

Probability of a component not working at a time $t$ has this form:

$$U(t) = 1 - R(t) = 1 - e^{-\lambda t} = 1 - e^{-t/\text{MTBF}}$$

### Reliability with multiple components: simplex system

So far the analysis has been for single components. Multiple components within a system, each with its own reliability and MTBF can be combined with the appropriate calculations to give a value for the system. A simplex system, illustrated in Figure A.3, with independent components has the following system failure rate, $\lambda_{\text{sys}}$:

$$\lambda_{\text{sys}} = \lambda_1 + \lambda_2 + \lambda_3 + \cdots + \lambda_n$$

The critical failure rate is a fraction of the system failure rate. The critical failure rate describes the rate of those important failures that actually make the system nonoperational. The equation that describes the critical failure rate, $\lambda'$, is as follows:

$$\lambda' = f \cdot \lambda_{\text{sys}},$$

**Figure A.3:**
An example of a simplex system.

The value, $f$, is the fraction of failures that make system inoperable. Then the reliability of the simplex system is described by this equation (remember, the failure rate, $\lambda$, is assumed constant):

$$R(t) = e^{-\lambda't} = \exp(-f\lambda_{\text{sys}}t)$$

The unreliability of the simplex system becomes:

$$U(t) = 1 - \exp(-\lambda't) = 1 - \exp(-f\lambda_{\text{sys}}t)$$

### Reliability with multiple components: identical parallel units in the system

A system with $n$ identical parallel units, illustrated in Figure A.4, with independent components has the following system unreliability number, $U(t)$, and reliability figure, $R(t)$:

$$U(t) = U_1(t) \cdot U_2(t) \cdots U_n(t) = [U_1(t)]^n$$
$$R(t) = 1 - U(t) = 1 - [U_1(t)]^n$$

Should the parallel system have $m$ active units, with $d$ dormant units held in spare, the MTBF becomes:

$$\text{MTBF} = [(d+1)/m] \cdot \text{MTBF}_i$$

(Remember, the failure rate, $\lambda$, is assumed constant in these calculations.)

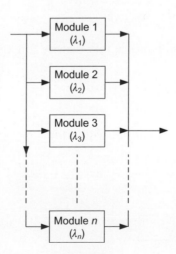

**Figure A.4:**
An example of a system with $n$ identical parallel units.

## Maintainability

Maintainability uses the metric of Mean Time to Repair or MTTR. This is the expected time between repairs.

$$MTTR = (\Sigma\, T_{repair})/n$$

where $n$ is the number of repairs and $T_{repair}$ is the repair time for each repair. The repair rate, $\mu'$, has the following equation:

$$\mu' = 1/MTTR$$

## Availability

Availability compares the time that the system is out of service in repair to the total time of system use. The Mean Time to Failure (MTTF) is:

$$MTTF = 1/\lambda'$$

Then calculate the time that the system is unavailable as:

$$Unavailability\ (U) = MTTR/(MTTF + MTTR) = \lambda'/(\lambda' + \mu')$$

Once you have unavailability, $U$, you can calculate the availability as follows:

$$Availability = 1 - U = 1 - [\lambda'/(\lambda' + \mu')]$$

## Defining reliability: mechanical wear out

Mechanical wear has the following probability density function for failure:

$$f(t) = \frac{1}{\sigma\sqrt{2\pi}} \exp\left\{ -\frac{1}{2}\left(\frac{t-\mu}{\sigma}\right)^2 \right\}$$

Unreliability is the probability of wear out within time, $t$:

$$U(t) = \frac{1}{\sigma\sqrt{2\pi}} \int_0^t \exp\left\{ -\frac{1}{2}\left(\frac{t-\mu}{\sigma}\right)^2 \right\} dt$$

Reliability is the probability of no failure over time, $t$:

$$R(t) = 1 - U(t) = 1 - \frac{1}{\sigma\sqrt{2\pi}} \int_0^t \exp\left\{ -\frac{1}{2}\left(\frac{t-\mu}{\sigma}\right)^2 \right\} dt$$

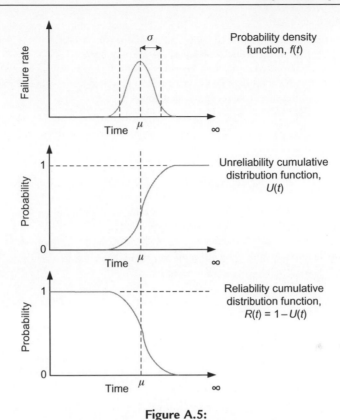

**Figure A.5:**

Examples of curves that could model mechanical wear out. *© 2013–2014 by Kim R. Fowler.*
*Used with permission. All rights reserved.*

Figure A.5 illustrates these set of curves for mechanical wear out.

## Experimental analysis

Clearly not all systems have constant failure rates. Goble in Appendix D of his textbook has
a nice example of how to calculate reliability from experimental data [3]. Basically you run
components or systems to failure and record the failures per unit time. From these data you
can easily calculate the reliability, unreliability, and failure rates.

## Recommended Reading

W.R. Dunn, Practical Design of Safety-Critical Computer Systems, Reliability Press, Solvang, CA, 2002.
Chapter 6, pp. 229–290. This book is a good, basic introduction to designing safety-critical systems.
Chapter 6 covers reliability calculations, to a degree it is like Goble but not to the depth of Goble. I still
like the book for its ease of reading.

W.M. Goble, Control Systems Safety Evaluation and Reliability, third ed., International Society of Automation, Research Triangle Park, NC, 2010. This is a good, all around textbook on reliability and dependability. Goble writes clearly and well. Have this book handy on your book shelf.

P.D.T. O'Connor, Practical Reliability Engineering, fourth ed., John Wiley & Sons, Ltd., West Sussex, England, 2001. This book is thorough and in-depth. O'Connor covers practical test issues and reliability management in much more detail than the other two recommended books.

## *References*

[1]  D.K. Pradhan, Fault-Tolerant Computer System Design, Prentice Hall PTR, Upper Saddle River, NJ, 1996, pp. 4—6.

[2]  N.G. Leveson, Engineering a Safer World: Systems Thinking Applied to Safety, The MIT Press, Cambridge, MA, 2011, pp. 14—15, 31, 33, 36, 47, 50, 51—53, 56.

[3]  W.M. Goble, Control Systems Safety Evaluation and Reliability, third ed., International Society of Automation, Research Triangle Park, NC, 2010, pp. 435—440.

# *Index*

Note: Page numbers followed by "*b*", "*f*" and "*t*" refer to boxes, figures and tables, respectively.